ERGEBNISSE DER MIKROBIOLOGIE IMMUNITÄTSFORSCHUNG UND EXPERIMENTELLEN THERAPIE

FORTSETZUNG DER ERGEBNISSE DER HYGIENE
BAKTERIOLOGIE · IMMUNITÄTSFORSCHUNG UND EXPERIMENTELLEN THERAPIE
BEGRÜNDET VON WOLFGANG WEICHARDT

HERAUSGEGEBEN VON

W. KIKUTH
DÜSSELDORF

K. F. MEYER
SAN FRANCISCO

E. G. NAUCK
HAMBURG

A. M. PAPPENHEIMER JR.
CAMBRIDGE/MASS.

J. TOMCSIK
BASEL

ZWEIUNDDREISSIGSTER BAND

MIT 26 ABBILDUNGEN

SPRINGER-VERLAG BERLIN HEIDELBERG GMBH 1959

ISBN 978-3-540-02385-2   ISBN 978-3-662-42618-0 (eBook)
DOI 10.1007/978-3-662-42618-0

Inhaltsverzeichnis

Corrigenda

zu

Die Rückfallfieber in: Ergebnisse der Mikrobiologie, Immunitätsforschung und Experimentellen Therapie **31**, 184—228 (1958)

S. 189, 2. Zeile von oben, sollte es heißen: „. . . in neugeborenen *Kaninchen* gehaltene Stämme . . .“ anstatt „. . . in neugeborenen Ratten gehaltene Stämme . . .“

S. 194, 1. Abschnitt, letzte Zeile, sollte es heißen: „. . . in Persien aus B. microti und *B. persica,* usw.“ statt „. . . in Persien aus B. microti und B. tholozani, usw.“

S. 200, 4. Zeile von oben, sollte es heißen: „. . . in das Lumen einer Coxa *infizierter* Zecken ein.“ anstatt „. . . in das Lumen einer Coxa gesunder Zecken ein.“

S. 202, 4. unterste Zeile, sollte es heißen: „. . . *mit B. persica infizierte O. tholozani* . . .“ statt „. . . mit O. tholozani infizierte Zecken . . .“

S. 203, 2. Abschnitt von unten, 1. Satz, sollte heißen: „Die eigentliche Entdeckung der lokalen Spezifizität verdankt man DAVIS u. HOOGSTRAAL *(1949) sowie* BALTAZARD et al. *(1954).*“ anstatt „. . . verdankt man BALTAZARD et al. (1950c).“

S. 206, 4. Abschnitt, 1. Zeile, sollte es heißen: „. . . *B. hispanica* . . .“ statt „. . . O. hispanica . . .“

S. 211, 1. Abschnitt, 20. Zeile, sollte es heißen: „. . . wenn man mit BALTAZARD et al. *(1950c)* annimmt, . . .“ anstatt „. . . wenn man mit BALTAZARD et al. (1954) annimmt, . . .“

S. 211, 2. Abschnitt, 2. Zeile, sollte es heißen: „. . . die kleine Rasse von *O. erraticus* . . .“ statt „. . . die kleine Rasse von O. moubata . . .“

S. 216, 1. Abschnitt, 4. unterste Zeile, sollte es heißen: „. . . in dieser Gegend von *Uganda* fielen . . .“ anstatt „. . . in dieser Gegend von Ulanga fielen . . .“

On the Mechanism Underlying Initiation of Influenza Virus Infection

By

ALFRED GOTTSCHALK[1]

Table of Contents

I. The life cycle of virulent viruses

In a delightful essay LWOFF (1957), scrutinizing the various types of infection, concludes "that the essence of infection is not the disease, but the introduction into an organism of a foreign entity able to multiply, to produce a disease and to reproduce infectious entities." In the case of infection by viruses the foreign entity entering the host cell is essentially the genetic material of the virus, i.e. nucleic acid, either naked or very nearly so as with bacteriophages, or bound to protein as with tobacco mosaic virus and polio virus, or associated with protein and surrounded by a coat of mucoprotein and lipid as with fowl plague and influenza viruses.

The essence of virus multiplication is the replication within the host cell of the genetic material carrying the markers of the virus specific components. It is the virus genetic material which channels the activities of the pre-existing enzymes of the host cell into the formation of virus specific nucleic acid and virus specific protein from host cell building blocks. This redirection may be visualized as the result of competition between virus polynucleotide and host cell polynucleotide for the apoenzymes involved in the synthesis of nucleic acid and protein; both types of polynucleotide are assumed to act as coenzymes in the complete enzyme systems and to determine by their template the sequence and spatial arrangement of the bases and amino acids respectively (GOTTSCHALK 1957d). The vegetative

---

[1] The Walter and Eliza Hall Institute of Medical Research, Melbourne, Australia.

phase comprising the synthesis of the virus components ensures the reproduction of the virus true to type. It is followed by the assembly of the component parts to the mature virus particle and by the liberation of the mature virus from the host cell.

Preceding the vegetative phase is the attachment of the virus particle to the surface of the host cell and the penetration of the virus into the host cell. In the case of bacteriophages the bulk of the nucleic acid (DNA) of the phage particle passes into the bacterial cell, nearly all of the protein remaining attached to the surface of the cell. For fowl plague and influenza viruses it seems very probable that during or subsequent to penetration the virus particle is fragmented enzymatically into subunits.

These various phases — adsorption, penetration, fragmentation, replication, maturation, liberation — are common to the life cycle of all virulent viruses. The background information for the individual phases derives from biophysical, biochemical and biological, especially genetic, studies of bacterial, plant and animal viruses. Whereas for bacteriophages the penetration mechanism and the vegetative replication of the germinal substance are best understood, it would appear that for the influenza virus the co-ordinated approach by biological, biochemical and biophysical technique has shed much light on the initial phases of infection. It is the aim of this review to present the relevant data and to derive from them a coherent concept for the mechanism underlying the initiation of infection by influenza virus.

II. The influenza virus particle

According to ANDREWES (1954) the agents of influenza, mumps, Newcastle disease and fowl plague which share the property of having a specific affinity for certain mucins are grouped together as "myxovirus." Of the members of the group the influenza virus is the one most extensively studied at all levels. Four immunologically different types are known, A, B, C, D showing no serological crossreactions. Within the type A influenza and type B influenza several strains are discernible. These strains though serologically closely related to the prototypic virus show some antigenic differences (JENSEN 1957). Most probably each individual antigenic character residing in some specific protein structure of the influenza virus surface is genetically controlled. The great tendency of the influenza virus to transform its character has been related to the facts (1) that man only is its natural host and (2) that the rate of transmission is high. Exhaustion of susceptible hosts will give a mutant not neutralizable by the host's antibodies, produced in a previous influenza virus infection, a better chance of survival than the strain with which the host had already past experience.

The elementary particles of influenza virus appear as small spheres in the electronmicrograph, their diameter being about 85 mμ. Roughly they are made of an outer coat containing protein, mucoprotein and lipid possibly in a mosaic-like pattern. Inside this glyco-lipo-protein coat is situated a ribonucleoprotein. It is only after acid and trypsin treatment of the virus particle that ribonuclease will act on the viral ribonucleic acid (VALENTINE and ISAACS 1957). Purified influenza A virus contains about 60% protein, 32% lipid, 4.5% non-aminosugar, 2.5% aminosugar and 1% ribonucleic acid (ADA and PERRY 1954; FRISCH-NIGGEMEYER and HOYLE 1956; ADA and GOTTSCHALK 1956).

III. On the anatomy and physiology of the respiratory tract

The influenza virus infection spreads by droplets passing from individual to individual. The virus causes a disease mainly of the upper respiratory tract involving the nasopharynx, trachea and bronchi from which parts infective virus has been isolated. The respiratory tract is lined by stratified columnar epithelial cells which are ciliated. Scattered among these ciliated cells are mucus producing goblet cells ("Becherzellen"). Whether the goblet cells are really a class of their own is an unsettled question. HERS (1955) is of the opinion that mucus production is a potential property of the ciliated cells and that they are able to assume the goblet (wine glass) form. Embryonically the ciliated epithelial cells and the goblet cells are derived from the same layer. Embedded in the deeper layers of the mucosa (submucosa) are mucus secreting glands, their ducts leading from the acini to the surface of the mucosa. The combined activity of the various mucus producing cells keeps the surface of the lining cells covered by a thin layer of mucus. According to observations by FLOREY (1954) the cilia of the nasal passages and of the trachea move a thin layer of mucin in a well defined direction at a very appreciable rate. Such streams of mucus were found able to remove carbon particles dropped experimentally onto the mucosa. Inflammatory stimuli increase the secretion of mucin. From observations on both the intestinal and respiratory tract it seems rather sure that the mucus covering the ciliated epithelial cells functions as a protective coat. The mucus is able to trap foreign particles including bacteria, to wrap them up and to hand the parcel over to the garbage removing machinery. With inanimate material and bacteria this protection is afforded predominantly, if not exclusively, by the physical properties of the viscous, sticky mucus; inhibition of bacterial growth by mucus has not been demonstrated. As pointed out by FLOREY (1954), the efficiency of the protection provided by the mucus is well illustrated by the fact that the bronchi are usually sterile whereas the nasal cavity is not.

From the preceding remarks it is evident that influenza virus containing droplets inhaled by a person or by an experimental animal are lodged primarily on the mucus layer lining the upper respiratory tract. Obviously the virus particle in order to make contact with its host cell, the ciliated columnated epithelial cell, has to break through the mucus barrier. We will, therefore, first describe what is known about the interaction between influenza virus and mucus.

IV. Interaction between influenza virus and mucoproteins

As suggested by MORGAN (1953), the term "mucin" is best used in a physiological sense to denote a viscous secretion. These secretions represent a salt solution of a variety of proteins and carbohydrate-containing proteins. The latter are conveniently designated mucoproteins, if they exhibit predominantly protein character, or mucopolysaccharides, if they are predominantly of polysaccharide nature. Mucoproteins are conjugated proteins with a polysaccharide as prosthetic group. Several mucoproteins have been isolated in an electrophoretically homogeneous state and it is with some of these pure mucoproteins that their relationship to the influenza virus has been investigated.

1. Biological data

Two sets of experimental results at the biological level have provided the basis for a biochemical approach to the problem of influenza virus-mucoprotein relationship:

1. a) When influenza virus particles are added to fowl or human red blood cells at 4^0 C, adsorption of the virus to the surface of the red cells and agglutination of the red cells take place. This phenomenon may be observed for more than 15 hours at 4^0 C (HIRST 1942a).

b) When the temperature is raised to 37^0 C, spontaneous elution of the adsorbed virus soon sets in and is completed in about two hours. The eluted virus is functually intact, whereas the red cells are rendered inagglutinable by renewed addition of virus (HIRST 1942b).

c) An exo-enzyme obtained from the culture filtrate of Vibrio cholerae also renders the erythrocytes inagglutinable (BURNET and STONE 1947).

d) Influenza virus, when heated for 30 min at 55^0 C in an appropriate medium, still adsorbs to and agglutinates red blood cells, but fails to elute spontaneously (BRIODY 1948, STONE 1949a).

2. The same series of events has been demonstrated when certain mucoproteins of human or animal origin, eventually obtained in pure state, are allowed to interact with influenza virus. The results may be summarized as follows:

a) When heat-inactivated virus is added to these mucoproteins, the latter attract and bind the inactive virus, as is seen from the inhibition of haemagglutination upon addition of red blood cells to the system (for review see BURNET 1951).

b) Interaction between living influenza virus and haemagglutinin inhibitory mucoproteins at 35^0 results in the loss of their biological activity (BURNET 1951).

c) Interaction between the vibrio cholerae enzyme (RDE) and the inhibitory mucoproteins at 37^0 has the same result as under 2b (BURNET 1951, GOTTSCHALK 1954b).

The inference drawn from these findings is that the influenza virus particle attaches itself to a receptor substance at the surface of the red cell and to a segment of the mucoprotein respectively. By virtue of an enzyme embedded in the virus coat the receptor substance of the red cell and the segment of the mucoprotein respectively are altered in such a manner as to be no longer able to attract and bind the virus. The vibrio cholerae enzyme imitates this effect. Heating the virus at 55^0 for 30 min destroys the activity of the viral enzyme without seriously impairing the structural feature instrumental in the attachment of the virus to the red cell and mucoprotein respectively. For this reason heat inactivated virus (indicator virus) is a very convenient preparation to demonstrate haemagglutination and its inhibition by certain mucoproteins (ANDERSON et al. 1948).

The mucus of the respiratory tract belongs to the group of influenza virus haemagglutinin inhibitory mucins. Human bronchial secretions contain two inhibitory mucoproteins; one of them representing 25-35% of the dry weight of the secretion has been obtained electrophoretically pure. RDE treatment of the inhibitor reduces its antihaemagglutinin titres for Mel, Lee and WSE indicator viruses from 9000, 4000 and 8000 respectively to 10. Concomitant with the loss of inhibitory titre is a reduction of the electrophoretic mobility in phosphate buffer (p_H 7.0) from 6.9×10^{-5} cm^2 sec^{-1} volt^{-1} to 5×10^{-5} cm^2 sec^{-1} volt^{-1}. The second inhibitor has not been prepared free from contaminants. After RDE treatment the haemagglutinin inhibitory activity of this fraction is lost and the mobility of a considerable proportion of the fraction is reduced (MARMION, CURTAIN and PYE 1953).

It is of great interest that the influenza virus haemagglutinin inhibitor present in bronchial washings of normal mice disappears in the early stages of infection of the animal with influenza virus. The disappearance of the inhibitor shows close correlation with the presence of active virus in the respiratory tract. Thus it was found that the experimental infection of mice with influenza virus can be recognised as readily by this criterion as either by the isolation of the virus from the lung tissue or, retrospectively, by the demonstration in serum of specific antibodies during convalescence (FAZEKAS DE ST. GROTH 1950a, 1950b).

These observations stimulated similar investigations on humans. As was established by STONE (1949b) for several mucoids, the pattern of inhibitory titres obtained when tested against a series of indicator viruses varies from mucoid to mucoid and is a characteristic of the mucoid. This indicator spectrum is independent of the absolute titres and may, therefore, be expressed as the ratio of titres. Contact with living virus results in differential reduction of titres against the various indicator viruses, thus distorting the normal pattern. For normal human nasal secretions it was found that the average ratio of titres against

WSE: Mel: Lee is 10:24:9 and that the "inhibitor index" $10 \times \dfrac{\text{Mel-titre} + \text{Lee-titre}}{\text{WSE-titre}}$,

an abbreviated version of the indicator spectrum, is 33 with a standard deviation of ± 8. Since the inhibitory capacity of nasal mucus against WSE indicator virus is the one most readily destroyed, the inhibitory index rises when the mucus is acted upon by living virus (FAZEKAS DE ST. GROTH 1952a, 1952b). During an epidemic of influenza A-prime in Melbourne in 1950 good agreement, though not complete agreement, was observed between rise in index values and the results of the accepted laboratory tests for influenza virus (isolation of virus from throat washings, antibody titration and estimation of complement-fixing antibody). A variety of respiratory diseases other than epidemic influenza do not cause similar changes (FAZEKAS DE ST. GROTH 1951).

2. Biochemical data

The chemical reaction underlying the "destruction" of the haemagglutinin inhibitory capacity of mucoproteins by the living influenza virus has been elucidated over the last decade. In 1949 GOTTSCHALK and LIND presented evidence that the loss of biological activity of inhibitory mucoproteins upon treatment with living influenza virus at 37⁰ is coincided by the release of a low molecular weight compound. This compound, referred to as split product, was found to have reducing power, to contain nitrogen and to form in the Morgan-Elson reaction (indirect Ehrlich reaction) a purple-coloured substance similar to that formed from N-acetyl-D-glucosamine under same conditions. However, in contrast to N-acetylglucosamine the split product gave also a direct Ehrlich reaction, i.e. without alkali pretreatment, on heating and a positive BIAL's orcinol test. The outstanding feature of the split product was its rapid decomposition with humin formation on heating with dilute mineral acid (GOTTSCHALK and LIND 1949a, 1949b; ODIN 1952). The same product was obtained whether semi-purified ovomucin or electrophoretically homogeneous human urine mucoprotein was used as substrate for the living virus. Appropriate controls with heat-inactivated virus were negative (GOTTSCHALK 1951). These data taken together with the finding that RDE released from the above mentioned mucoproteins a compound indistinguishable from the split product left little doubt on the enzymic nature of the process. It was also established that the influenza virus enzyme is an intrinsic part of the influenza virus particle and that a similar enzyme is not present in the host cell of the influenza virus (GOTTSCHALK and PERRY 1951; GOTTSCHALK 1954b).

Two compounds with properties similar to those of the split product had been isolated previously from quite different sources by chemical procedures. BLIX (1936) had described the preparation in crystalline form of an acid from bovine salivary mucin by mild acid treatment and KLENK (1941) had reported the isolation of the methoxy derivative of an acid, termed neuraminic acid, from brain gangliosides upon treatment with water-free methanolic hydrochloric acid. BLIX' compound, later on named sialic acid (BLIX, SVENNERHOLM and WERNER 1952), resembled the split product in all its properties and was shown to have the composition $C_{13}H_{21}O_{10}N$ (BLIX, LINDBERG, ODIN and WERNER 1955). It contains an N-acetyl group, an O-acetyl group, very easily split off, a reducing group and five hydroxyl groups (after loss of the O-acetyl), one of which is an α-hydroxy group, another one a primary alcohol group (BLIX et al. 1956). KLENK'S methoxyneuraminic acid is similar to sialic acid and the split product in all those reactions carried out in acid medium and at 100⁰ (direct Ehrlich reaction, orcinol reaction, decomposition with humin formation). In alkali, however, the compound is stable; it does not reduce FEHLING's solution and does not produce a colour with Ehrlich reagent in the cold after heating with alkali. It is free of acetyl groups and it gives a ninhydrin reaction. On elementary analysis KLENK assigned the formula $C_{11}H_{21}O_9N$ to methoxyneuraminic acid. The same compound was obtained from bovine salivary mucoprotein (BSM) and from human urinary mucoprotein (UM) when heated with 5% methanolic hydrochloric acid (KLENK and LAUENSTEIN 1952). The close relationship between the split product, sialic and neuraminic acid became more evident when KLENK and FAILLARD (1954) isolated N-acetylneuraminic acid (NANA) from bovine salivary mucoprotein. NANA displays all properties described above for the split product; it differs from sialic acid apparently only by the possession of one acetyl group (N-acetyl) instead of two. The formula given to NANA was $C_{12}H_{21}O_{10}N$; on BLIX' formula for sialic acid it should have been $C_{13}H_{21}O_{10}N$ minus $C_2H_2O = C_{11}H_{19}O_9N$ (for detailed discussion see GOTTSCHALK 1956b). Finally, KLENK, FAILLARD and LEMPFRID (1955) confirmed GOTTSCHALK's (1951) experimental results on the digestion of urine mucoprotein with active influenza virus and identified the split product as NANA.

The first indication of the structure of NANA was afforded by the isolation of pyrrole-2-carboxylic acid from alkali treated BSM (GOTTSCHALK 1953, 1955a). This finding was followed by the alkali degradation of the split product to pyrrole-2-carboxylic acid (GOTTSCHALK 1954a) and by the alkali degradation of NANA to

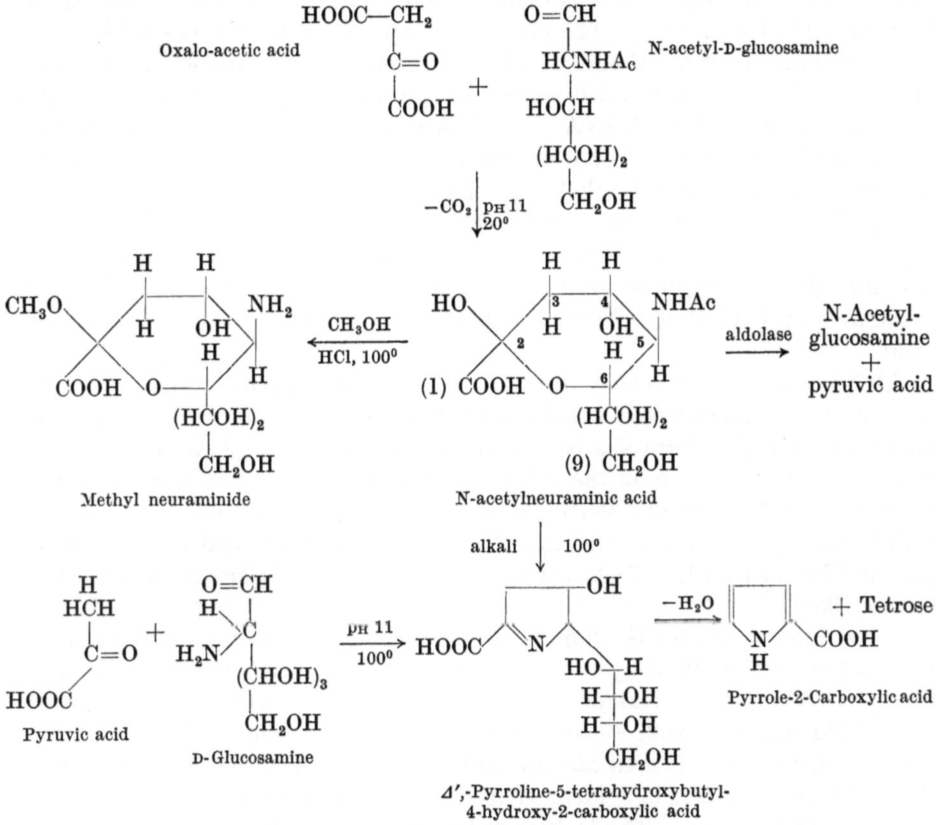

Diagram 1. N-Acetylneuraminic acid: its structure, synthesis and transformation products

the same pyrrole derivative (KLENK and FAILLARD 1954). Concerning the mechanism of the degradation of NANA to pyrrole-2-carboxylic acid it was of relevance that the same pyrrole was obtained in good yield by condensation of D-glucosamine and pyruvate in dilute alkali (GOTTSCHALK 1955b, 1957a). On the basis of these data the following structures and reaction mechanisms were proposed for the compounds and reactions discussed at the time without indication of the stereochemistry (GOTTSCHALK 1955b, 1956b, 1956c, 1957d).

As may be seen from diagram 1 NANA is visualized as the aldol condensation product of N-acetylhexosamine and pyruvic acid; on deacetylation it will yield neuraminic acid. On alkali treatment this acid having a primary amino group in γ-position to a keto group will form an internal SCHIFF's base, the 5-substituted Δ^1-pyrroline-4-hydroxy-2-carboxylic acid, also produced by a KNORR type condensation of D-glucosamine and pyruvic acid. The unstable pyrroline derivative undergoes spontaneous rearrangement to pyrrole-2-carboxylic acid concomitant with reverse aldolization. Methoxyneuraminic acid is the alkali stable methylglycoside of neuraminic acid.

Further work has substantiated and qualified the structures proposed. NANA has been fragmented to N-acetyl-D-glucosamine and pyruvic acid both chemically (KUHN and BROSSMER 1956b; ZILLIKEN and GLICK 1956) and enzymatically (HEIMER and MEYER 1956). The chemical synthesis of NANA was achieved by aldol condensation at 20^0 C and p_H 11.0 of N-acetyl-D-glucosamine[1] and carboxylated pyruvic acid (oxaloacetic acid), the crystalline synthetic product being identical with crystalline NANA of biological origin (ovine NANA) in melting point, specific rotation, infrared spectrum and X-ray diffraction pattern. On elementary analysis the synthetic product was shown to have the formula $C_{11}H_{19}O_9N$ (CORNFORTH, FIRTH and GOTTSCHALK 1958). KUHN and BROSSMER (1957) elucidated the steric configuration at C_4 of NANA as shown in diagram I and also the configuration at C_2 of crystalline NANA. In bovine diacetylneuraminic acid the O-acetyl group is most probably attached to C_7 (BLIX et al. 1956).

Acetylated neuraminic acids are components of all influenza virus haemagglutinin inhibitory mucoproteins tested so far (for review see GOTTSCHALK 1958). NANA in crystalline form has been obtained by the action of the viral enzyme and/or the vibrio enzyme on the urine mucoprotein, human serum mucoprotein, meconium mucoprotein and BSM, the latter containing apparently a mixture of NANA, diacetylneuraminic acid and N-glycolylneuraminic acid (KLENK, FAILLARD and LEMPFRID 1955; FAILLARD 1957; BÖHM et al. 1957; ZILLIKEN et al. 1957; KLENK 1958).

It has been shown for the human urine mucoprotein and for BSM that sialic acid, in the new terminology the group name for all acylated neuraminic acids (BLIX, GOTTSCHALK and KLENK 1957), is localized in the carbohydrate-prosthetic group (GOTTSCHALK 1952, 1957c). Since the viral enzyme, RDE and gentle treatment with mild acid split off only sialic acid from mucoproteins (GOTTSCHALK 1956a, KLENK 1956), it is evident that sialic acid occupies a terminal position in the prosthetic group. The lack of reducing power of BSM containing about 17% sialic acid indicates masking of the reducing group by engagement in a ketosidic linkage (GOTTSCHALK 1956a; HEIMER and MEYER 1956). These observations characterize the viral enzyme and the isodynamic enzyme of Vibrio cholerae as "neuraminidase" cleaving the ketosidic linkage between the reducing carbon atom of neuraminic acid and the appropriate grouping of the adjacent

---

[1] *Added in proof:* Recently D. G. COMB and S. ROSEMAN [J. Am. Chem. Soc. 80, 497 (1958)] presented evidence that aldolase splits NANA at p_H 7 to pyruvic acid and N-acetyl-D-mannosamine, epimerization of the latter to N-acetyl-D-glucosamine and of N-acetyl-D-glucosamine to N-acetyl-D-mannosamine under alkaline conditions being responsible for the previons results.

sugar unit (GOTTSCHALK 1957b, 1957d). In BSM the partner in the ketosidic linkage is N-acetylgalactosamine (GOTTSCHALK 1957c) and recent investigations have shown that sialic acid is linked glycosidically to C_6 of the aminosugar. The disaccharide sialyl $(2 \rightarrow 6)$ N-acetyl-galactosamine (diagram 2) is the main prosthetic group of the mucoprotein (GOTTSCHALK and GRAHAM 1958).

It is interesting that in sialyl-lactose, a trisaccharide isolated from rat mammary glands (TRUCCO and CAPUTTO 1954) and from cow colostrum (KUHN and BROSSMER 1956a) sialic acid is joined to the galactose moiety of lactose (GOTTSCHALK 1957b). In this case the linkage is to C_3 of galactose involving the α-form of sialic acid (KUHN and BROSSMER 1958). Sialyl-lactose is split by the viral

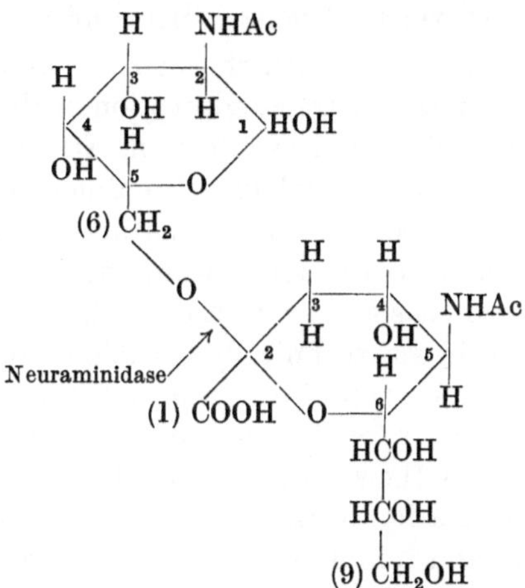

Diagram 2. 6-(N-Acetyl-α-D-neuraminyl)-N-acetylgalactosamine

and bacterial neuraminidases into sialic acid and lactose (KUHN and BROSSMER 1956a; GOTTSCHALK 1957b). These data qualify the influenza virus and Vibrio cholerae neuraminidases as α-glycosidases and the linkages susceptible to them as α-ketosidic ones.

Table 1. *Composition of human nasal and bronchial mucus**

Material	Contents in per cent of dry weight				
	Nitrogen	Hexosamine	Galactose	Fucose	Sialic acid
Nasal secretion	10.3	11.2	7.8	3.0	4.4
Bronchial secretion	12.9	8.3	3.6	3.0	3.8

* Compiled from the experimental data of WERNER (1953).

As may be seen from Table 1 compiled from a paper by WERNER (1953), human nasal mucous secretions and human tracheobronchial mucin contain sialic acid; they also contain hexosamines, in the case of sputum shown to be a mixture of glucosamine and galactosamine. These epithelial mucins are virus haemagglutinin inhibitors susceptible to the influenza virus enzyme and/or RDE (s. Section IV, 1). Both the purified influenza virus particles and purified

RDE preparations are not known to exhibit any other enzymic functions than neuraminidase activity. It seems, therefore, justified to conclude that the influenza virus particle, after landing on the mucus layer of the respiratory tract and threatened to be wrapped in a coat of respiratory mucus and to be disposed of, breaks through the host's first line of defence by means of the neuraminidase units embedded in its surface. By splitting off sialic acid from a number of mucoprotein molecules adsorbed to and separating the virus particle from its host cells, the products of hydrolysis will fall off from the virus particle as they fall off from the enzyme surface in any other hydrolase action. This brings the virus face to face with its host cells, the lining epithelium of the respiratory tract.

V. Interaction between influenza virus and red blood cells

When the influenza virus has pierced through the mucus barrier protecting the epithelial cells of the respiratory tract, adsorption of the particle to the host's cell outer surface is the next step in virus infection. It is now generally accepted that specific receptors are instrumental in attracting and adsorbing the virus to the cell surface. Our knowledge about the nature of this receptor is mainly derived from investigations into the relationship between red blood cells and influenza virus. Though red cells are not host cells, the data obtained seem to be highly relevant. Whenever it was technically feasible to substitute host cells for red cells in a given set of experimental conditions, similar results were obtained. In fact red cells are ideal objects for the study of the mechanism of virus adsorption, just because adsorption of the virus is not complicated by its penetration into the cell. Because of this isolation of the adsorptive phase the close relationship between cellular receptors and inhibitory mucoproteins was first recognized and then elaborated with the system mucoprotein-influenza virus-red cell.

1. Biological data

In most of the work with red cells chicken or human erythrocytes have been used though they are not the only ones subject to agglutination by influenza virus.

The primary attachment of the virus to the red cell is an adsorption phenomenon with three characteristic features:

A. The adsorption of the virus onto red cells is temperature independent (HIRST 1942b). For the system PR8 influenza virus (indicator state) and guinea-pig red blood cells it was shown (BATEMAN et al. 1955) that the adsorption of virus to the cells is a diffusion limited reaction, the apparent bimolecular velocity constant at 25^0 being 1.72×10^{-8} cm^3 min^{-1}/red cell, in good agreement with the value of 1.3×10^{-8} cm^3 min^{-1}/red call for the systems swine influenza virus (indicator state)-fowl red blood cells and active Lee virus-fowl red cells (LANNI and LANNI 1952). Approximately 7500 receptor sites are present at the surface of the guinea-pig red cell (BATEMAN et al. 1955) and about 5500 at the surface of the chicken red cell (DAWSON and ELFORD 1949).

B. The adsorption of the virus to the red cell proceeds only in a suitable ionic environment. In isotonic glucose solution, free of electrolytes, influenza virus is not adsorbed to red cells. With the system influenza virus-red cells-sodium chloride solution, it was shown that adsorption is roughly proportional to a function of the electrolyte concentration. Below 0.003 M concentration no adsorption takes place, at 0.15 M maximum adsorption occurs. The dependency of virus adsorption to red cells on electrolyte concentration was demonstrated with influenza A and B strains (in the active and indicator state), for human and chicken red cells, at 4⁰ and 37⁰ C (LOWELL and BUCKINGHAM 1948; FLICK, SANFORD and MUDD 1949; DAVENPORT and HORSFALL 1948).

Whereas the nature of the anion is indifferent for adsorption, lower molar concentrations of Ca and Mg salts are required than of Na and K salts (BURNET and EDNEY 1952). For influenza A (FM1 strain) and influenza B (Lee strain) evidence has been presented that at p_H 4,7 adsorption to chicken red cells is nearly completely inhibited (DAVENPORT and HORSFALL 1948). The same holds for NDV (LEVINE and SAGIK 1956).

C. The adsorption of the virus to the red cell is reversible. This is proven by the following results:

a) Indicator virus adsorbed to red cells at 4⁰ C or at room temperature may be displaced by RDE (BURNET 1952).

b) Indicator virus (influenza A) adsorbed to fowl red cells can be eluted partially by the addition of homologous antiserum (ISAACS 1948).

c) Indicator virus of some but not all strains may be removed from red cells by addition of inhibitory mucoproteins (BURNET 1952).

d) Active influenza A virus (PR8 strain) adsorbed to human red cells at 4⁰ C is readily eluted from the cells by salt-free water at 4⁰ C without receptor destruction (FLICK, SANFORD and MUDD 1949).

e) RDE displaces active haemagglutinating virus at the red cell surface at a rate incompatible with spontaneous elution of the virus.

f) Addition of a cation exchange resin to a suspension of active PR8 virus adsorbed to red cells at 3⁰ C effects transfer of the virus from cell to resin (PUCK and SAGIK 1953).

g) Active Lee virus adsorbed to cat erythrocytes at 5⁰ C desorbs spontaneously without receptor destruction upon raising the temperature to 24⁰ C in a suitable ionic medium (TAMM 1954).

2. Biochemical data

The common denominator of all types of union between influenza virus (in the active phase or indicator state) and red cell receptors or soluble mucoprotein receptors is that by treatment with RDE the atomic groupings instrumental in attracting and binding the virus are removed. The enzymic action of purified RDE has been shown to be that of a neuraminidase splitting off the terminal N-acylated neuraminic acid from an oligo- or polysaccharide. It was therefore expected and it was substantiated that the red cell receptor is chemically similar to the inhibitory mucoproteins, that it contains sialic acid and that by RDE action this acid is released. McCREA (1953) has isolated from human erythrocyte stroma by treatment with pentane a receptor substance chemically closely related to UM and BSM (for analysis of UM and BSM s. GOTTSCHALK 1952 and GOTTSCHALK and ADA 1956). The substance contains 10% nitrogen, total reducing sugar 14.7% (calculated as glucose after hydrolysis with 0.5 N HCl for 24 hr at 100° C) and 9.5 % hexosamine. The component sugars were chromatographically identified as galactose, fucose, galactosamine and glucosamine. McCREA'S preparation gives a strongly positive direct Ehrlich reaction indicative of sialic acid (GOTTSCHALK, unpublished). The receptor substance, in 1% solution, inhibits haemagglutination by Lee indicator virus to a titre of 10,000, a property lost upon incubation with active virus or RDE. Recently a potent water-soluble inhibitor substance containing 7—11% sialic acid has been obtained from suspensions of human erythrocyte stromata. Up to 50% of the sialic acid is rendered dialysable by treatment with infective influenza virus, as determined colorimetrically (HOWE, ROSE and SCHNEIDER 1957). KLENK and LEMPFRID (1957) succeeded in splitting off NANA from human erythrocytes by RDE and in crystallizing the compound. These contributions prove beyond doubt the close analogy between red cell receptors and soluble haemagglutinin inhibitory mucoproteins. This analogy is also reflected in the reduction of the net negative surface charge of both human erythrocytes and inhibitory mucoproteins upon treatment with active influenza virus and/or RDE. The action of RDE on human red cells reduces their electrophoretic mobility from the normal value of -1.30×10^{-6} cm² sec⁻¹ volt⁻¹ to a value of -0.17×10^{-6} cm² sec⁻¹ volt⁻¹, i.e. by 87%, indicating that the net negative surface charge of human erythrocytes is mainly due to the dissociated carboxyl group of sialic acid ($pK^1 = 2.60 \pm 0.05$ at 20° C and 0.05 N). Treatment of UM with active virus (Lee and Mel strains) or with RDE results in mobility reduction from -8.4 to about -6.5×10^{-5} cm² sec⁻¹ volt⁻¹ (HANIG 1948; ADA and STONE 1950; TAMM, PERLMANN and HORSFALL 1952; PYE 1955).

From the data presented it would appear that for the primary union between influenza virus and red cell surface the sialic acid present at the receptor site provides the main anchoring groups; removal of the acid renders adsorption of the virus to the red cell impossible. Since dissociation of the carboxyl group of sialic acid is a prerequisite for virus adsorption (s. p. 11), electrostatic attraction between complementary configurations at the virus and red cell surfaces seems most likely, the monovalent or divalent cations of the medium required for adsorption reducing the energy barrier resulting from the repulsive electrostatic potentials of the two interacting particles (PUCK and SAGIK 1953). The participation of hydroxyl (polar) groups of sialic acid in binding the virus to the receptor is suggested by experiments with periodate, a chemical known to oxidize within ten minutes at 20° C adjacent cis-OH groups in sugars. Thus it was shown that amounts of periodate in the range of 0.2 to 2.0 mg/gm packed fowl red cells modify the cellular receptors in a characteristic manner: they still adsorb active viruses of the influenza group; however, the adsorbed virus neither elutes spontaneously nor is displaced by RDE (FAZEKAS DE ST. GROTH 1949). Larger amounts of periodate (10 mg/gm packed red cells) destroy the ability of red cells to adsorb virus (HIRST 1948). After treatment of the red cells with RDE no effect of periodate at various concentrations is demonstrable; the cells fail to adsorb virus (FAZEKAS DE ST. GROTH 1949), indicating that sialic acid is the main anchorage for the virus also in the case of suitably periodated red cells. These results are in accordance with the observation by BURNET (1948, 1949) that there is a critical concentration of KIO_4 at which ovarian cyst mucoid is converted into an inhibitor of haemagglutination by active virus (WSE strain). Doubling the critical concentration causes loss of the capacity of the mucoid to act as inhibitor of virus haemagglutination. NANA bound in the prosthetic group of BSM consumes 2 moles of IO_4^- (GOTTSCHALK and GRAHAM 1958). It is tempting to conclude from these experiments that the presence of unaltered OH-groups attached to C_7, C_8 and C_9 of NANA is essential for viral neuraminidase action and that an intact OH-group at C_7 but not necessarily at C_8 and C_9 is a prerequisite for bindung the virus to the receptor or to the inhibitor. Such an interpretation would account for the fact that gently prepared BSM whose prosthetic group contains NANA with an additional O-acetyl group at C_7 (diacetylneuraminic acid) inhibits only haemagglutination by PR8 indicator virus, but not the haemagglutinins of other influenza virus strains. It would also explain the findings that active PR8 and Lee viruses do not reduce the electrophoretic mobility of genuine BSM and that RDE, a simple protein molecule, does so only after prolonged action (CURTAIN and PYE 1955).

In essence, then, the reversible virus-red cell adsorption closely resembles the reversible formation of the enzyme-substrate complex in an ordinary enzyme reaction: $E + S \rightleftharpoons ES$. This step is followed by the activation of the enzyme-substrate complex resulting in the formation of products and free enzyme, in our case sialic acid, stabilized red cell and free virus.

The virus neuraminidase is associated with the virus haemagglutinin. With fowl plague virus, a typical haemagglutinating virus of the myxo group, SCHÄFER and ZILLIG (1954) were able to fragment the particle by ether treatment. In this way a spherical haemagglutinin, 30 mμ in diameter, and a nucleoprotein were obtained. The haemagglutinin, composed of protein and carbohydrate and reacting with homologous antiserum, was found to liberate NANA from UM (SCHÄFER and MOHR 1958).

Most probably several enzymically active patches are present at the virus surface, thus enabling one virus particle to combine with more than one red cell. This process, when carried out with appropriate quantities of virus and erythrocytes, results in agglutination, a phenomenon not observed with RDE. The multiplicity of enzymic patches distributed over the virus surface may also account for the observation that an influenza virus particle adsorbed to a red cell and having acted upon one receptor molecule is not immediately released but acts on an adjacent one and continues to act until all receptors accessible to the respective virus have lost their sialic acid (ANDERSON et al. 1948). The concept of a single virus particle "browsing" over the red cell surface requires a considerable density of receptor molecules. According to BATEMAN et al. (1955) the average distance between receptor centers at the surface of guinea-pig red cells is about 130 mμ. Obviously the boundaries of adjacent receptors must be even closer together than 130 mμ. Since the diameter of the influenza virus particles ranges from 80 to 100 mμ, any collision between virus and red cell will bring the former within a distance of 30 mμ or less from a receptor, i.e. close enough for electrostatic forces to become effective.

Influenza viruses in the active phase and in the indicator state differ in two ways: 1. In the active phase influenza virus adsorbs to and elutes spontaneously from the red cell surface; adsorption to red cells results in haemagglutination, elution effects reduction of the overall net negative charge of the cells (ADA and STONE 1950). Indicator virus only adsorbs to and agglutinates red cells. 2. Certain mucoproteins inhibit haemagglutination by indicator virus, but in general fail to inhibit the agglutinin of active virus, even if the experiment is carried out at 0^0 C to suppress enzymic action (ANDERSON at al. 1948). The first difference is easily understood. Since heating at 56^0 C for 30 min will almost certainly denature the protein of the virus neuraminidase, activation of the enzyme-substrate complex does not take place. The denaturation, however, does not proceed far enough to impair the electrostatic attraction between enzyme and substrate. The second difference is not as easily explained. SMITH and WESTWOOD (1950) have shown that living virus and heat-inactivated virus, agglutinating dose for agglutinating dose, have the same combining capacity for rabbit serum inhibitor and combine with this inhibitor at identical rates. Even in the presence of agglutinable red cells equal haemagglutinating doses of living and of heat-inactivated virus exhibit equal avidity for the inhibitor. The authors conclude that a given amount of inhibitor molecules will blanket the combining groups of the heat-distorted virus surface to an extent sufficient for haemagglutinin neutralization, whereas the equivalent amount of inhibitor is insufficient to blanket living virus to the same extent. In other words, the exact apposition of the virus combining groups to receptor sites embedded in the mosaic of the red cell surface is much more sensitive to distortion (by heat) of the geometry of the respective virus groups than is the union between virus and the isolated mucoprotein molecules present in an inhibitor solution.

VI. Interaction between influenza virus and host cells

The receptor substance of influenza virus susceptible cells has not been isolated and no chemical work on the interaction between host cell receptors and active virus has yet been carried out. However, the close structural relationship between the host cell receptor and the red cell receptor is evident from their similar behaviour under a variety of conditions. The relevant experimental results may be summarized as follows:

1. Adsorption of influenza virus to and spontaneous elution from susceptible respiratory cells was demonstrated by HIRST (1943) using the excised ferret lung.

2. The excised mouse lung adsorbs active Lee virus in 15 min nearly completely; at 20^0 C spontaneous elution begins 50 min after administration and at the end of 3 hr 75% of the inoculum is recovered. Heat-inactivated virus adsorbs to the lungs but fails to elute (FAZEKAS DE ST. GROTH 1948 a).

3. Administration of RDE to the excised mouse lung prior to virus installation renders the respiratory surface incapable of adsorbing the virus (FAZEKAS DE ST. GROTH 1948 a).

4. RDE displaces active Lee virus adsorbed to the receptors of the respiratory cells of the excised mouse lung (FAZEKAS DE ST. GROTH 1948 a).

5. When the allantoic cavity of the chick embryo is washed with formalin in order to kill the embryo, active virus adsorbs to and spontaneously elutes from the allantoic endothelium of the chorioallantois. RDE displaces both active and heat-inactivated virus adsorbed to the endothelium. Pretreatment of the

membrane with RDE prevents adsorption of active or indicator virus (STONE 1948a, FAZEKAS DE ST. GROTH 1948c).

6. Pretreatment of the allantoic cavity with suitable amounts of periodate results in the loss of the ability of adsorbed active virus to elute spontaneously and of the capacity of RDE to displace the virus adsorbed to the modified receptor (FAZEKAS DE ST. GROTH 1948c, FAZEKAS DE ST. GROTH and GRAHAM 1949).

These findings correspond in detail to the results obtained with red cells and influenza virus and they fully justify the use of red cells as models for the study of host cell receptors. However, though the respiratory cells and the endothelial cells of the chorioallantoic membrane are host cells for the influenza virus, under the conditions of the test they are not viable. It was, therefore, of importance when evidence became available showing the same close similarity of the receptors of viable host cells and red cells. Thus it is possible to provide a limited protection against influenza virus infection of the intact animal by pretreatment of the susceptible cells with RDE. This has been shown for mice inoculated intranasally and for chick embryos infected in the allantoic cavity with various strains of influenza A and B (STONE 1948a, 1948b, FAZEKAS DE ST. GROTH 1948b). The dgree of protection depends on the virus strain used for infection, on the infective dose and in each case is of short duration due to rapid regeneration of the degraded cellular receptors. CAIRNS (1951) was able to protect mice by a small local injection of RDE in one hemisphere against a subsequent intracerebral challenge with the neurotropic strain NWS.

The recorded data on the influenza virus — host cell combination are mainly at the biological level. However, the experimental results with both viable and non-viable host cells imitate so closely those with red cells that there seems scarcely an alternative to the interpretation that host cell receptors are sialo-mucoproteins. Though RDE has not yet been obtained crystalline or even electrophoretically homogenous, there is no indication that highly purified RDE has any other effect than to release a terminal sialic acid from an oligosaccharide or polysaccharide. It is, therefore, most probable that sialic acid is the main anchorage for the influenza virus particle at the host cell surface as it is at the red cell surface and at the inhibitory mucoproteins.

The main difference between viable host cells on the one hand and non-viable host cells and red blood cells on the other hand is that the latter do not support virus growth. Therefore, with the latter cells spontaneous elution of the adsorbed virus involving enzymic action has not to compete with any other reaction engaging the virus. With viable host cells the situation is different. Here the penetration of the virus into the interior of the cell does compete with the elution of the virus from the host cell surface. In this case spontaneous elution of the virus is difficult to demonstrate unambiguously since there is no direct way of distinguishing between residual virus not previously adsorbed and virus eluted subsequent to adsorption. Some data pertinent to this question may be quoted. ISAACS and EDNEY (1950a) have shown that even a large inoculum of heat-inactivated Lee virus (4800 agglutinating doses) is completely taken up by the allantoic cells within 18 hr at 35° C. In contrast, with non-infectious but enzymically active virus (irradiated) a constant ratio of non-adsorbed to adsorbed virus is observed over a period of 24 hr (HENLE and HENLE 1944; LIU and HENLE

1951), and with infectious virus this ratio (about 0.4) remains constant until newly formed virus is liberated from the infected cells whether the inoculum is 10^2 or 10^9 ID_{50} (HENLE, HENLE and ROSENBERG 1947). As pointed out by HENLE (1953) this difference between enzymically active and enzymically inactive virus may suggest spontaneous elution of some portion of the active virus adsorbed to viable cells of the chorioallantois. This interpretation is supported by the ability of RDE to displace some of the active virus adsorbed to the living chorioallantois (FAZEKAS DE ST. GROTH 1948c; ISAACS and EDNEY 1950b).

Of the large proportion of the infectious influenza virus absorbed by the living chorio-allantoic membrane of the chick embryo one hour after inoculation a minute fraction only (1—3%) can be recovered from ground suspensions of the membrane (HOYLE 1948, HENLE 1949). Displacement by RDE of virus still adsorbed to the surface of the endothelial cells prior to grinding reduces the recovery to less than 1% of the amount taken up by the membrane (ISAACS and EDNEY 1950b). The inference from these findings is that the bulk of the virus has entered into close association with the host cell and is no longer accessible to RDE.

Adsorption of the virus onto the host cell is followed by penetration into the cell. FAZEKAS DE ST. GROTH (1948c) has advanced the concept that influenza virus adsorbed to susceptible and viable cells enters the cytoplasma by a process analogous to the phenomenon of colloidopexis, the uptake by cells of submicroscopical particles, and he termed this particular event "viropexis". As soon as the virus is taken into the cell, it is fragmented thus losing its identity. This is evidenced by recent investigations with ^{32}P labelled fowl plague virus (WECKER and SCHÄFER 1957) and with ^{35}S labelled influenza A virus (HOYLE and FINTER 1957).

VII. The role of influenza virus neuraminidase in the initiation of infection

In contrast to HIRST's (1942b) original suggestion enzymic "destruction" of host cell receptors does not seem to be necessary for virus penetration into the cell. When the viable host cells of the respiratory tract (mouse) or of the allantoic cavity (chick embryo) are pretreated with periodate so as to make them insusceptible to neuraminidase action without impairing their adsorbing capacity (s. Section V 2), these modified cells are still infected by active virus. There is identity of infectivity end-points in periodated and normal mice and chick embryos (FAZEKAS and GRAHAM 1949). Conversely, indicator virus, i.e. haemagglutinating virus devoid of enzymic activity, is adsorbed to and then taken into the normal lining cells of the allantoic cavity (FAZEKAS 1948c; ISAACS and EDNEY 1950b). From a biochemical point of view it would be difficult to think of an advantage the virus may derive from enzymic release of sialic acid prior to penetration. There is no indication that removal of sialic acid, present most probably as a terminal unit or side chain in the prosthetic group of the cellular receptor, effects a discontinuity in the cell membrane such as the enzyme at the tail of the coliphage produces in the cell walls of Escherichia coli (WEIDEL 1951; BARRINGTON and KOZLOFF 1956).

Though viral enzyme action is apparently not a prerequisite for infection, the enzyme protein, whether catalytically active or not, provides the complementary structure instrumental in making intimate contact with the sialic acid containing segment of the cellular receptor resulting in adsorption. This concept is based on the following facts:

1. RDE, a bacterial neuraminidase, splits off a terminal sialic acid residue from inhibitory mucoproteins and from red cell receptors.

2. After loss of sialic acid neither mucoproteins nor red cells combine with influenza virus.

3. The influenza virus has embedded in its surface several neuraminidase units acting on mucoproteins and red cells in the same manner as the "free" bacterial neuraminidase does.

4. Influenza virus particles adsorbed to red cells, to the respiratory cells of the excised lung or to the endothelium of the chorioallantois (viable or not viable), are displaced by RDE.

5. When active influenza A virus (Melbourne strain) and RDE are together inoculated intranasally in mice, specific lung lesions are reduced by more than thousandfold as compared with the lesions in the control mice inoculated with the same virus dose without RDE (STONE 1948b).

There can be no doubt that the bacterial enzyme forms an enzyme-substrate complex with the sialic acid containing segment of the cellular receptor. Since the influenza virus has an isodynamic enzyme combining with and acting upon the same receptor, as shown conclusively with red cells, and since RDE displaces the virus from the common substrate, the conclusion seems logical that the adsorption of the virus particle to the cellular receptor is effected by the enzyme protein patch of the virus surface. One may elaborate more complicated mechanisms; however, they will not explain more and they will not explain better. The high collision efficiency of the system influenza virus and red cells (SAGIK, PUCK and LEVINE 1954) is adequately accounted for by the two complementary electrostatic configurations, by the large number of receptor sites uniformly distributed over the red cell surface and by the multiplicity of enzyme patches at the virus surface.

Two vital functions of the viral neuraminidase in the initial phases of infection are thus evident: firstly, to disengage the virus particle from the coat of mucus which will enfold it soon after its landing on the respiratory tract; secondly, to attach the virus particle to the surface of the host cell.

VIII. Summary

In this review on the initial stages of influenza virus infection an attempt has been made to interpret the information available at the biological level in terms of the underlying chemical structures and catalytic reactions. Since a limited amount of precise chemical and biochemical data had to cover a large area of biological observations, occasionally an assumption had to be invoked to fill a gap in the chemical information. The most serious gap is the still missing isolation of the receptor substance present at the host cell surface. However, it is felt that the circumstantial evidence is so overwhelmingly in favour of its close similarity to the red cell receptor that its eventual isolation will only prove what seemed almost certain before.

The key reaction in the invasion of the natural host by the influenza virus
is the formation of an enzyme-substrate complex, the enzyme possessed by the
virus, the substrates present a) in the protective mucus layer of the respiratory
tract and b) in the outer membrane of the host cell. In the encounter with the
protective mucus the enzyme-substrate complex is undergoing activation yielding
products and free virus particle. In the following step the formation of the enzyme-
substrate complex is the mechanism underlying adsorption of the virus particle
to the surface of the host cell. In this case a minor proportion of the complexes
undergoes activation as above, resulting in release of the virus into the medium.
The larger proportion of the virus fixed by its enzyme to the receptor substrate
of the host cell enters the cell, thus evading elution. This step of penetration is
least understood and its attention by biologists and/or biochemists seems very
desirable. On entering the cell the virus is fragmented into sub-units.

References

ADA, G. L., and A. GOTTSCHALK: The component sugars of the influenza virus particle.
Biochem. J. **62**, 686 (1956).

—, and B. T. PERRY: The nucleic acid content in influenza virus. Aust. J. exp. Biol. med. Sci.
32, 453 (1954).

—, and J. D. STONE: Electrophoretic studies of virus-red cell interaction: Mobility gradient
of cells treated with viruses of the influenza group and the receptor-destroying enzyme of
V. cholerae. Brit. J. exp. Path. **31**, 263 (1950).

ANDERSON, S. G., F. M. BURNET, S. FAZEKAS DE ST. GROTH, J. F. McCREA and J. D. STONE:
Mucins and mucoids in relation to influenza virus action. VI. General discussion. Aust.
J. exp. Biol. med. Sci. **26**, 403 (1948).

ANDREWES, C. H.: Nomenclature of viruses. Nature (Lond.) **173**, 620 (1954).

BARRINGTON, L. F., and L. M. KOZLOFF: Action of bacteriophage on isolated host cell walls.
J. biol. Chem. **223**, 615 (1956).

BATEMAN, J. B., M. S. DAVIS and P. A. McCAFFREY: A note on the reaction of heated influenza virus protein with red blood cells. Amer. J. Hyg. **62**, 349 (1955).

BLIX, F. G., A. GOTTSCHALK and E. KLENK: Proposed nomenclature in the field of neuraminic
and sialic acids. Nature (Lond.) **179**, 1088 (1957).

— G.: Die Kohlenhydratgruppen des Submaxillarismucins. Hoppe-Seylers Z. physiol.
Chem. **240**, 43 (1936).

— E. LINDBERG, L. ODIN and I. WERNER: Sialic acids. Nature (Lond.) **175**, 340 (1955).

— — — — Studies on sialic acids. Acta Soc. Med. upsalien. **61**, 1 (1956).

— L. SVENNERHOLM and I. WERNER: The isolation of chondrosamine from gangliosides and
from submaxillary mucin. Acta chem. scand. **6**, 358 (1952).

BÖHM, P., J. ROSS u. L. BAUMEISTER: Über die Abspaltung von N-Acetylneuraminsäure aus
Serum durch das „Receptor-Destroying-Enzyme" aus Vibrio cholerae. Hoppe-Seylers Z.
physiol. Chem. **307**, 284 (1957).

BRIODY, B. A.: Characterization of the enzymic action of influenza viruses on human red cells.
J. Immunol. **59**, 115 (1948).

BURNET, F. M.: Mucins and mucoids in relation to influenza virus action. IV. Inhibition by
purified mucoid of infection and haemagglutination with the virus strain WSE. Aust. J.
exp. Biol. med. Sci. **26**, 381 (1948).

— The effect of periodate on the virus inhibitory qualities of mucoids in solution. Aust. J.
exp. Biol. med. Sci. **27**, 361 (1949).

— Mucoproteins in relation to virus action. Physiol. Rev. **31**, 131 (1951).

— Haemagglutination in relation to host cell-virus interaction. Ann. Rev. Microbiol. **6**, 229
(1952).

BURNET, F. M., and M. EDNEY: Influence of ions on the interaction of influenza virus and cellular receptors or soluble inhibitors of haemagglutination. Aust. J. exper. Biol. med. Sci. **30**, 105 (1952).

—, and J. D. STONE: The receptor destroying enzyme of V. cholerae. Aust. J. exp. Biol. med. Sci. **25**, 227 (1947).

CAIRNS, H. J. F.: Protection by receptor destroying enzyme against infection with a neurotropic variant of influenza virus. Nature (Lond.) **168**, 335 (1951).

CORNFORTH, J. W., M. E. FIRTH and A. GOTTSCHALK: The synthesis of N-acetylneuraminic acid. Biochem. J. **68**, 57 (1958).

CURTAIN, C. C., and J. PYE: A mucoprotein from bovine submaxillary glands with restricted inhibitory action against influenza virus haemagglutination. Aust. J. exp. Biol. med. Sci. **33**, 315 (1955).

DAVENPORT, F. M., and F. L. HORSFALL: The associative reactions of pneumonia virus of mice and influenza viruses: the effects of pH and electrolytes upon virus-host cell combinations. J. exp. Med. **88**, 621 (1948).

DAWSON, I. M., and W. J. ELFORD: The investigation of influenza and related viruses in the electron microscope by a new technique. J. gen. Microbiol. **3**, 298 (1949).

FAILLARD, H.: Über die Abspaltung von N-Acetylneuraminsäure aus Mucinen durch das „Receptor-Destroying-Enzyme" aus Vibrio cholerae. Hoppe-Seylers Z. physiol. Chem. **307**, 62 (1957).

FAZEKAS DE ST. GROTH, S.: Destruction of influenza virus receptors in the mouse lung by an enzyme from V. cholerae. Aust. J. exp. Biol. med. Sci. **26**, 29 (1948a).

— Regeneration of virus receptors in mouse lungs after artificial destruction. Aust. J. exp. Biol. med. Sci. **26**, 271 (1948b).

— Viropexis, the mechanism of influenza virus infection. Nature (Lond.) **162**, 294 (1948c).

— Modification of virus receptors by metaperiodate. I. The properties of IO_4-treated red cells. Aust. J. exp. Biol. med. Sci. **27**, 65 (1949).

— Studies in experimental immunology of influenza. I. The state of virus receptors and inhibitors in the respiratory tract. Aust. J. exper. Biol. med. Sci. **28**, 15 (1950a).

— Influenza: a study in mice. Lancet **1950** (b) I, 1101.

— Quick test for the early diagnosis of influenza. Nature (Lond.) **167**, 43 (1951).

— Nasal mucus and influenza viruses. I. The haemagglutinin inhibitor in nasal secretions. J. Hyg. (Lond.) **50**, 471 (1952a).

— Nasal mucus and influenza viruses. II. A new test for the presumptive diagnosis of influenza infection. J. Hyg. (Lond.) **50**, 492 (1952b).

—, and D. M. GRAHAM: Modification of virus receptors by metaperiodate. II. Infection through modified receptors. Aust. J. exp. Biol. med. Sci. **27**, 83 (1949).

FLICK, J. A., B. SANFORD and S. MUDD: The effect of salt concentration on the interaction of influenza A virus and erythrocytes. J. Immunol. **61**, 65 (1949).

FLOREY, H.: Lectures on general pathology. Melbourne: University Press 1954.

FRISCH-NIGGEMEYER, W., and L. HOYLE: The nucleic acid and carbohydrate content of influenza virus and of virus fractions by ether disintegration. J. Hyg. (Lond.) **54**, 201 (1956).

GOTTSCHALK, A.: N-Substituted isoglucosamine released from mucoproteins by the influenza virus enzyme. Nature (Lond.) **167**, 845 (1951).

— Carbohydrate residue of a urine mucoprotein inhibiting influenza virus haemagglutination. Nature (Lond.) **170**, 662 (1952).

— Presence of 2-carboxypyrrole in mucoprotein and its relation to the viral enzyme. Nature (Lond.) **172**, 808 (1953).

— The precursor of 2-carboxypyrrole in mucoproteins. Nature (Lond.) **174**, 652 (1954a).

— The influenza virus enzyme and its mucoprotein substrate. Yale J. Biol. Med. **26**, 352 (1954b).

— 2-Carboxypyrrole: its preparation from and its precursor in mucoproteins. Biochem. J. **61**, 298 (1955a).

— The structural relationship between sialic acid, neuraminic acid and 2-carboxypyrrole. Nature (Lond.) **176**, 881 (1955b).

— The linkage of sialic acid in mucoprotein. Biochim. biophys. Acta **20**, 560 (1956a).

Gottschalk, A.: Neuraminic acid: the functional group of some biologically active muco-proteins. Yale J. Biol. Med. 28, 525 (1956b).
— The chemistry of neuraminic acid and its derivatives. Aust. J. Sci. 18, 178 (1956c).
— The synthesis of 2-carboxypyrrole from D-glucosamine and pyruvic acid: its bearing on the structure of neuraminic acid. Arch. Biochem. 59, 37 (1957a).
— Neuraminidase: the specific enzyme of influenza virus and Vibrio cholerae. Biochim. biophys. Acta 23, 645 (1957b).
— The structure of the prosthetic group of bovine submaxillary gland mucoprotein. Biochim. biophys. Acta 24, 649 (1957c).
— Virus enzymes and virus templates. Physiol. Rev. 37, 66 (1957d).
— Neuraminidase: its substrate and mode of action. Advanc. Enzymol. 20, 135 (1958).
—, and G. L. Ada: The separation and quantitative determination of the component sugars in mucoproteins. Biochem. J. 62, 681 (1956).
—, and E. R. B. Graham: The structure of the disaccharide of bovine salivary mucoprotein. Z. Naturforsch. 13b (1958).
—, and P. E. Lind: Product of interaction between influenza virus enzyme and ovomucin. Nature (Lond.) 164, 232 (1949a).
— — Ovomucin, a substrate for the enzyme of influenza virus. I. Ovomucin as an inhibitor of haemagglutination by heated Lee virus. Brit. J. exp. Path. 30, 85 (1949b).
—, and B. T. Perry: Ovomucin, a substrate for the enzyme of the influenza virus. III. Enzymic activity as an integral function of the influenza virus particle. Brit. J. exp. Path. 32, 408 (1951).
Hanig, M.: Electrokinetic change in human erythrocytes during adsorption and elution of PR 8 influenza virus. Proc. Soc. exp. Biol. (N.Y.) 68, 385 (1948).
Heimer, R., and K. Meyer: Studies on sialic acid of submaxillary mucoid. Proc. nat. Acad. Sci. (Wash.) 42, 728 (1956).
Henle, W.: Studies on host-virus interactions in the chick embryo-influenza virus system. I. Adsorption and recovery of seed virus. J. exp. Med. 90, 1 (1949).
— Multiplication of influenza virus in the entodermal cells of the allantois of the chick embryo. Advanc. Virus Res. 1, 141 (1953).
—, and G. Henle: Interference between inactive and active viruses of influenza. II. Factors influencing the phenomenon. Amer. J. med. Sci. 207, 717 (1944).
— — and E. B. Rosenberg: The demonstration of one-step growth curves of influenza viruses through the blocking effect of irradiated virus on further infection. J. exp. Med. 86, 423 (1947).
Hers, J. F. Ph.: The histopathology of the respiratory tract in human influenza. Leiden: H. E. Stenfert Kroese N. V. 1955.
Hirst, G. K.: The quantitative determination of influenza virus and antibodies by means of red cell agglutination. J. exp. Med. 75, 49 (1942a).
— Adsorption of influenza haemagglutinins and virus by red blood cells. J. exp. Med. 76, 195 (1942b).
— Adsorption of influenza virus on cells of the respiratory tract. J. exp. Med. 78, 99 (1943).
— The nature of the virus receptors of red cells. I. Evidence on the chemical nature of the virus receptors of red cells and of the existence of a closely analogous substance in normal serum. J. exp. Med. 87, 301 (1948).
Howe, C., H. M. Rose and L. Schneider: Enzymic action of influenza virus on human erythrocyte stroma components. Proc. Soc. exp. Biol. (N. Y.) 96, 88 (1957).
Hoyle, L.: The growth cycle of influenza virus A. A study of the relations between virus, soluble antigen and host cell in fertile eggs inoculated with influenza virus. Brit. J. exp. Path. 29, 390 (1948).
—, and N. B. Finter: The use of influenza virus labelled with radio-sulphur in studies of the early stages of the interaction of virus with the host cell. J. Hyg. (Lond.) 55, 290 (1957).
Isaacs, A.: Reactivation of neutral mixtures of influenza virus and serum by virus inactivated by heat. Brit. J. exp. Path. 29, 529 (1948).
—, and M. Edney: Interference between inactive and active influenza viruses in the chick embryo. II. The site of interference. Aust. J. exp. Biol. med. Sci. 28, 231 (1950a).

ISAACS, A. and M. EDNEY: Interference between inactive and active influenza viruses in the chick embryo. IV. The early stages of virus multiplication and interference. Aust. J. exp. Biol. med. Sci. 28, 635 (1950b).

JENSEN, K. E.: The nature of serological relationships among influenza viruses. Advanc. Virus Res. 4, 279 (1957).

KLENK, E.: Neuraminsäure, das Spaltungsprodukt eines neuen Gehirnlipoids. Hoppe-Seylers Z. physiol. Chem. 268, 50 (1941).

— Chemie und Biochemie der Neuraminsäure. Angew. Chem. 68, 349 (1956).

— Neuraminic acid. Ciba Foundation Symposium on the Chemistry and Biology of Mucopolysaccharides. London: J. &. A. Churchill 1958.

—, u. H. FAILLARD: Zur Kenntnis der Kohlenhydratgruppen der Mucoproteide. Hoppe-Seylers Z. physiol. Chem. 298, 230 (1954).

— — u. H. LEMPFRID: Über die enzymatische Wirkung von Influenzavirus. Hoppe-Seylers Z. physiol. Chem. 301, 235 (1955).

—, u. K. LAUENSTEIN: Zur Kenntnis der Kohlenhydratgruppen des Submaxillarismucins und Harnmucoproteids. Die Isolierung von Neuraminsäure als Spaltprodukt. Hoppe-Seylers Z. physiol. Chem. 291, 147 (1952).

—, u. H. LEMPFRID: Über die Natur der Zellreceptoren für das Influenza Virus. Hoppe-Seylers Z. physiol. Chem. 307, 278 (1957).

KUHN, R., u. R. BROSSMER: Über O-Acetyl-lactaminsäure-lactose aus Kuh-Colostrum und ihre Spaltbarkeit durch Influenza-Virus. Chem. Ber. 89, 2013 (1956a).

— — Abbau der Lactaminsäure zu N-Acetyl-D-glucosamin. Chem. Ber. 89, 2471 (1956b).

— — Die Konfiguration der Lactaminsäure. Angew. Chem. 69, 534 (1957).

— — Die Konstitution der Lactaminsäurelactose; α-Ketosidase-Wirkung von Viren der Influenza-Gruppe. Angew. Chem. 70, 25 (1958).

LANNI, F., and Y. T. LANNI: A quantitative theory of influenza virus haemagglutination-inhibition. J. Bact. 64, 865 (1952).

LEVINE, S., and B. P. SAGIK: The interactions of Newcastle disease virus (NDV) with chick embryo tissue culture cells: attachment and growth. Virology 2, 57 (1956).

LIU, O. C., and W. HENLE: Studies on host-virus interactions in the chick embryo-influenza virus system. IV. The role of inhibitors of haemagglutination in the evaluation of viral multiplication. J. exp. Med. 94, 269 (1951).

LOWELL, F. C., and M. BUCKINGHAM: A comparison of the effect of various salt concentrations on the agglutination of red cells by influenza A virus and antibody. J. Immunol. 58, 229 (1948).

LWOFF, A.: The concept of virus. J. gen. Microbiol. 17, 239 (1957).

MARMION, B. P., C. C. CURTAIN and J. PYE: The effect of human bronchial secretions (sputum) on the haemagglutinin and infectivity of influenza virus. Aust. J. exp. Biol. med. Sci. 31, 505 (1953).

McCREA, J. F.: Studies on influenza virus receptor substance and receptor substance analogues. II. Isolation and purfication of a mucoprotein receptor substance from human erythrocyte stroma treated with pentane. Yale J. Biol. Med. 26, 191 (1953).

MORGAN, W. T. J.: Some aspects of the chemistry of mucins. Proc. roy. Soc. Med. 46, 783 bis 785 (1953) (Section of Comparative Medicine, pp. 29—31).

ODIN, L.: Carbohydrate residue of a urine mucoprotein inhibiting influenza virus haemagglutination. Nature (Lond.) 170, 663 (1952).

PUCK, T., and B. SAGIK: Virus and cell interaction with ion exchangers. J. exp. Med. 97, 807 (1953).

PYE, J.: Assay of inhibitors of influenza virus haemagglutination by electrophoresis. Aust. J. exp. Biol. med. Sci. 33, 323 (1955).

SAGIK, B., T. PUCK and S. LEVINE: Quantitative aspects of the spontaneous elution of influenza virus from red cells. J. exp. Med. 99, 251 (1954).

SCHÄFER, W., and E. MOHR: Personal communication, 1958.

—, u. W. ZILLIG: Über den Aufbau des Virus Elementarteilchens der klassischen Geflügelpest. I. Mitteilung. Gewinnung, physikalisch-chemische und biologische Eigenschaften einiger Spaltprodukte. Z. Naturforsch. 9b, 779 (1954).

SMITH, W., and J. C. N. WESTWOOD: Influenza virus haemagglutination. The mechanism of the Francis phenomenon. Brit. J. exp. Path. **31**, 725 (1950).

STONE, J. D.: Prevention of virus infection with enzyme of V. cholerae. I. Studies with viruses of mumps-influenza group in chick embryos. Aust. J. exp. Biol. med. Sci. **26**, 49 (1948a).

— Prevention of virus infection with enzyme of V. cholerae. II. Studies with influenza virus in mice. Aust. J. exp. Biol. med. Sci. **26**, 287 (1948b).

— Inhibition of influenza virus haemagglutination by mucoids. I. Conversion of virus to indicator for inhibitor. Aust. J. exp. Biol. med. Sci. **27**, 337 (1949a).

— Inhibition of influenza virus haemagglutination by mucoids. II. Differential behaviour of mucoid inhibitors with indicator viruses. Aust. J. exp. Biol. med. Sci. **27**, 557 (1949b).

TAMM, I.: Influenza virus — erythrocyte interaction. I. Reversible reaction between Lee virus and cat erythrocytes. J. Immunol. **73**, 180 (1954).

— G. E. PERLMANN and F. L. HORSFALL: An electrophoretic examination of a urinary mucoprotein which reacts with various viruses. J. exp. Med. **95**, 99 (1952).

TRUCCO, R. E., and R. CAPUTTO: Neuramin-lactose, a new compound isolated from the mammary gland of rats. J. biol. Chem. **206**, 901 (1954).

VALENTINE, R. C., and A. ISAACS: The structure of viruses of the Newcastle disease-mumps-influenza (myxovirus) group. J. gen. Microbiol. **16**, 680 (1957).

WECKER, E., u. W. SCHÄFER: Studies mit ^{32}P-markiertem Virus der Klassischen Geflügelpest. I. Mitt. Untersuchungen über das Verhalten des Virus beim Eindringen in die Wirtszelle. Z. Naturforsch. **12b**, 483 (1957).

WEIDEL, W.: Über die Zellmembran von Escherichia Coli B. I. Präparierung der Membranen. Analytische Daten. Morphologie. Verhalten der Membranen gegenüber den Bakteriophagen der T-Serie. Z. Naturforsch. **6b**, 251 (1951).

WERNER, I.: Studies on glycoproteins from mucous epithelium and epithelial secretion. Acta Soc. Med. upsalien. **58**, 1 (1953).

ZILLIKEN, F., and M. C. GLICK: Alkalischer Abbau von Gynaminsäure zu Brenztraubensäure und N-Acetyl-D-Glucosamin. Naturwissenschaften **43**, 536 (1956).

—, G. H. WERNER, R. K. SILVER and P. GYÖRGY: Studies on the enzymatic properties of influenza viruses. I. The action of influenza B virus and RDE on the hemagglutinin inhibitor of human meconium. Virology **3**, 464 (1957).

Das kulturell-biochemische und serologische Verhalten der Cryptococcus-Gruppe [1,2,3]

Von

H. P. R. SEELIGER

Mit 8 Abbildungen

Inhalt

Einleitung

Seit Entdeckung des Erregers der Cryptococcose durch BUSSE (*24, 25, 26*) und BUSCHKE (*23*) vor rund 65 Jahren (*7*) ist diese weltweit verbreitete, durch einen Sproßpilz verursachte Krankheit annähernd 500mal beim Menschen nachgewiesen worden [vgl. (*37, 42, 66, 111* u. a.)], sporadisch und epizootisch ferner beim Milchvieh sowie vereinzelt beim Affen, Schwein, Pferd, Hund, Katze, Meerschweinchen und bei anderen Tierarten [Zusammenfassung s. (*61, 111*)].

Während des vergangenen Jahrzehnts wurde diese gefährliche und zweifellos oft übersehene Mykose auch in West- und Mitteleuropa mit zunehmender Häufigkeit festgestellt (*92, 161, 162* u. a.), wahrscheinlich als Folge der sich rasch verbessernden Diagnostik und des erhöhten Interesses für Pilzinfektionen. Eine Reihe einschlägiger Fälle wurde allerdings erst auf dem Sektionstisch und durch die histologische Untersuchung von Organschnitten erkannt. Unzulängliche Züchtungsmethoden sowie die noch vielfach vorhandene Unkenntnis über das morphologische, kulturelle, biochemische und serologische Verhalten des Erregers sind wohl die Hauptursachen für seine in unseren Breiten bisher relativ selten geglückte Isolierung bzw. Differenzierung. Umständliche und zeitraubende Testverfahren erschweren den Gang der Untersuchung ebenso wie mannigfache Verwechslungsmöglichkeiten mit verwandten oder ähnlich erscheinenden Sproßpilzarten. Hinzu kommen noch gewisse Unklarheiten in der Nomenklatur und

---

[1] Aus dem Hygiene-Institut der Rheinischen Friedrich Wilhelms-Universität Bonn/Rh. (Direktor: Prof. Dr. Dr. H. EYER).

[2] Herrn Prof. Dr. E. RODENWALDT zum 80. Geburtstag gewidmet.

[3] Ein Teil der experimentellen Untersuchungen wurde mit Mitteln der Deutschen Forschungsgemeinschaft durchgeführt.

systematischen Einordnung, wodurch sich nicht unbedeutende Schwierigkeiten für eine vergleichende Beurteilung des umfangreichen vorhandenen Schrifttums ergeben.

Durch Anwendung verschiedener neuartiger Untersuchungsmethoden, die sich zum Teil bereits auf anderen Gebieten der Mikrobiologie bewährt haben, ist nun innerhalb weniger Jahre die Biochemie und Serologie der *Cryptococcus*-Gruppe ganz wesentlich bereichert worden. Diese neuen Wege der *Cryptococcus*-Diagnostik eröffnen manche Möglichkeiten für eine bessere Kenntnis phylogenetischer Zusammenhänge, die bisher mehr ein Gegenstand der Vermutung als des exakten Wissens waren. Ein Großteil der jüngsten Forschungsergebnisse ist in den zusammenfassenden Berichten und Monographien von BENHAM (*12, 13*), CONANT et al. (*37*), COX und TOLHURST (*42*), LITTMAN und ZIMMERMAN (*111*) sowie LODDER und KREGER-VAN RIJ (*117*) noch nicht oder nur zum Teil berücksichtigt.

Nach Abschluß mehrjähriger eigener Untersuchungen, die teils in Bonn und teils in den USA durchgeführt wurden[1] und sich auf Ergebnisse an über 100 frisch isolierten und Sammlungsstämmen der *Cryptococcus*-Gruppe sowie verwandter Gattungen stützen, soll im folgenden der heutige Stand unserer Kenntnisse über die Morphologie, das kulturelle und biochemische Verhalten sowie die Serologie der *Cryptococcus*-Gruppe erörtert werden.

In Anbetracht des Erscheinens zahlreicher ausführlicher Darstellungen (*37, 42, 66, 92, 111, 133, 137, 164, 169, 177, 178* u. a. m.) wird die Klinik der Cryptococcose in diesem Rahmen nicht besprochen; desgleichen wurden die Fragen der Epidemiologie und Epizootologie, der Pathogenese und Pathologie sowie der experimentellen Tier- und Therapieversuche ausgeklammert.

Da sich der anschließende Bericht in erster Linie an den medizinisch ausgerichteten Mikrobiologen wendet, stehen die *Krankheitserreger* der *Cryptococcus*-Gruppe, insbesondere *C. neoformans*, im Vordergrund. Die verwandten Arten werden nur insoweit besprochen werden, als es zum Verständnis der phylogenetischen Zusammenhänge und der für die Praxis wichtigen Differentialdiagnose nötig ist.

Geschichte und Nomenklatur

Die Entdeckung und Reinzüchtung des Erregers der Cryptococcose erfolgte 1894 durch BUSSE (*24, 25, 26*) und BUSCHKE (*23*) anläßlich der Untersuchung einer subperiostalen, gumma- bzw. sarkomähnlichen Tibiageschwulst. Etwa zur gleichen Zeit isolierte SANFELICE (*147, 148, 149*) eine ähnliche Keimart von Pfirsichen und aus vergorenem Fruchtsaft.

Während BUSCHKE die beobachteten Zelleinschlüsse für Coccidien („Krebscoccidien") hielt, wurden sie von BUSSE (*24*) als „Hefen" angesehen, wobei BUSSE einem Hinweise LOEFFLERs folgte, dem einige Präparate vorgelegt worden waren. Später übernahm COSTANTIN den schon vorher (*25*) benutzten Namen *Saccharomyces hominis*. Die erste taxonomisch gültige Bezeichnung lautet jedoch auf

[1] Für gebotene Arbeitsmöglichkeiten, Überlassung zahlreicher Pilzstämme sowie für wertvolle Hinweise und wohlmeinende Kritik danke ich Herrn Prof. Dr. KLUYVER †, Delft, Frl. Dr. SLOOFF, Delft, Prof. Dr. L. AJELLO und Dr. L. GEORG, Communicable Disease Center, Mycology Division, Chamblee, Georgia, sowie Prof. Dr. TH. PAINE und Prof. Dr. E. E. EVANS, Department of Microbiology, University of Alabama, Birmingham, Alabama.

Grund einer eingehenden Beschreibung *Saccharomyces neoformans* [SANFELICE *(147)*]. Der italienische Autor hielt diese Art für identisch mit BUSSEs Pilz, eine Ansicht, der sich seither fast alle Nachuntersucher angeschlossen haben.

Da die Originalkulturen verschiedentlich mit gleichlautendem Ergebnis überprüft worden sind, ist an der Identität kaum zu zweifeln. Auffallend ist nur, daß BUSSE (1894) selbst schreibt, er habe bei den von LOEFFLER angeratenen Kulturversuchen innerhalb weniger Tage reichliche Kohlensäurebildung und Fermentation beobachten können, ein Verhalten, das unseren Kenntnissen über die Biochemie der zu besprechenden Sproßpilze widerspricht.

Der Name *Cryptococcus neoformans* wurde erstmals von VUILLEMIN (1901) *(187)* für die von BUSSE und SANFELICE beschriebene Hefeart gebraucht, nachdem der französische Forscher das 1833 von KÜTZING [zit. *(117)*] geschaffene, ziemlich heterogene Genus *Cryptococcus (Globuli gonomici, minutissimi, solidi, mucosi, in stratum indefinitum aggregati)* auf die nichtsporenbildenden parasitären Hefen eingeengt hatte.

In der Folgezeit wurde jedoch der Gattungsbegriff *Cryptococcus* für so viele verschiedene, meist parasitäre, teils fermentierende, teils aber auch nichtfermentierende Sproßpilze gebraucht, daß er von LODDER *(115)* 1938 als *nomen dubium et confusum* abgelehnt wurde. Dem entsprach auch die von LODDER (1934) *(114)* bzw. DIDDENS und LODDER (1942) *(46)* geübte Benennungsweise. Hierbei wurde der 1934 von LODDER *(114)* der Vergessenheit entrissene Gattungsname *Torulopsis* (BERLESE 1894) für *Cryptococcus neoformans* angewandt. Damit kamen die Ergebnisse früherer Studien von CIFERRI (1925) *(35)* und ein Vorschlag von REDAELLI (1931) *(138)* zum Tragen. Auf letzteren geht die auch heute noch vielfach gebrauchte Bezeichnung *Torulopsis neoformans* zurück.

Durch diesen Schritt wurde der bereits 1902 von WEISS *(190)* geprägte Name *Torula neoformans*, ein Synonym der 1916 von STODDARD und CUTLER *(177)* beschriebenen *Torula histolytica*, hinfällig. Allerdings hat sich auch diese Bezeichnung ungewöhnlich lange gehalten, insbesondere im klinischen Schrifttum, wo die Krankheit heute noch vielfach unter dem Begriff Torulose läuft.

Nur am Rande sei vermerkt, daß der gleiche Pilz auch zu den Gattungen *Blastomyces* [SASAKAWA 1922 *(150)*, ARZT 1924 *(4)* — wahrscheinlich auf Vorschlag von BENEDEK (zit. *117*)] und *Atelosaccharomyces* [DE BEURMANN und GOUGEROT 1909 *(45)*, VERDUN 1912 *(186)*] gerechnet wurde.

Ausgehend von TODDs und HERRMANNs *(182)* sowie eigenen Befunden (vgl. S. 30) überführte REDAELLI c.s. *(139)* den Pilz 1937 zur Gattung *Debaryomyces* unter dem Artnahmen *Debaryomyces neoformans*, ohne sich hiermit jedoch durchsetzen zu können.

In eingehenden Darstellungen begründeten BENHAM *(11)* und vor allem SKINNER *(173)* die Vorzüge und Priorität des u. a. auch von DODGE *(47)* anstelle von *Torulopsis* gebrauchten Gattungsnamens *Cryptococcus*. Es erwies sich als nötig, die Genera *Torulopsis* und *Cryptococcus* neu zu definieren. Dies geschah 1952 unter Berücksichtigung zusätzlicher biochemischer Kriterien [LODDER und KREGER-VAN RIJ *(117)*]. Ein Großteil der nichtfermentierenden asporogenen Hefearten wurde im Sinne VUILLEMINs zur *Cryptococcus*-Gruppe zusammengefaßt. Diese wurde durch die Kriterien der Kapselbildung und Stärkesynthese auf eine neue biologische Basis gestellt. Der heute klar umrissene *Cryptococcus*-Begriff hat inzwischen weitgehende internationale Anerkennung gefunden.

Allerdings ist zu beachten, daß eine von Rivolta (1873) veröffentlichte Definition der Gattung *Cryptococcus* in den meisten Betrachtungen übersehen wurde [Ciferri 1951 (*36*)]. So gehört nach Rivolta bzw. Ciferri beispielsweise das in der Veterinärmedizin so wichtige *Histoplasma farciminosum* zur Gattung *Cryptococcus*. Da die Bezeichnung *Cryptococcus* in der Tiermedizin sowohl für *Cryptococcus farciminosus* als auch für *Cryptococcus neoformans* angewandt wird und die beiden ätiologisch völlig verschiedenen Krankheiten unter dem Begriff Cryptococcose laufen, sind mannigfaltige Verwechslungsmöglichkeiten gegeben.

Zahlreiche kulturelle und morphologische Varianten von *Cryptococcus neoformans* wurden mit mehr oder weniger Grund neu benannt. Aus der umfangreichen Liste der angeblichen Synonyma, die 1939 von Giordano (*76*) publiziert wurde, haben Lodder und Kreger-van Rij (*117*) nach kritischer Überarbeitung und Sichtung die Namen zusammengestellt, die mit ziemlicher Sicherheit identisch mit *C. neoformans* sind (Tabelle 1).

Tabelle 1. *Synonyma von Cryptococcus neoformans nach* Lodder *und* Kreger-van Rij (1952)

Saccharomyces neoformans Sanfelice (1895)
Saccharomyces lithogenes Sanfelice (1895)
Saccharomyces hominis Costantin (1901)
Saccharomyces plimmeri Costantin (1901)
Saccharomyces blanchardi Guiart (1910) ?
Saccharomyces breweri (Verdun) Neveu-Lemaire (1921)

Torula neoformans (Sanfelice) Weiss (1902)
Torula plimmeri (Costantin) Weiss (1902)
Torula klein Weiss (1902)
Torula histolytica Stoddard et Cutler (1916)
Torula nasalis Harrison (1928)

Torulopsis neoformans (Sanfelice) Redaelli (1931)
Torulopsis lithogenes (Sanfelice) de Almeida (1933)
Torulopsis hominis (Vuillemin) Redaelli (1931)
Torulopsis costantini (Froilano de Mello et Gonzaga Fernandes) de Almeida (1933)
Torulopsis plimmeri (Costantin) de Almeida (1933)
Torulopsis breweri (Verdun) de Almeïda (1933)
Torulopsis histolytica (Stoddard et Cutler) Castellani et Jacono (1933)
Torulopsis hominis (Vuillemin) Red. *var. honduriana* Castellani (1933), Castellani et Jacono (1933)
Torulopsis neoformans (Sanfelice) Red. *var. sheppei* Giordano (1935)

Blastomyces neoformans (Sanfelice) Arzt (1924)
Blastomyces lithogenes (Sanfelice) Sasakawa (1922)

Debaryomyces neoformans (Sanfelice) Redaelli, Ciferri et Giordano (1937)
Debaryomyces hominis (Vuillemin) Todd et Herrmann (1936)

Atelosaccharomyces hominis (Vuillemin) Verdun (1912)
Atelosaccharomyces busse-buschki de Beurmann et Gougerot (1909)
Atelosaccharomyces breweri Verdun (1912)

Cryptococcus lithogenes (Sanfelice) Vuillemin (1901)
Cryptococcus hominis Vuillemin (1901)
Cryptococcus costantini Froilano de Mello et Gonzaga Fernandes (1918)
Cryptococcus plimmeri (Costantin) Neveu-Lemaire (1912)
Cryptococcus kleini (Weiss) Cohn (1904) apud Guéguen)
Cryptococcus breweri (Verdun) Castellani et Chalmers (1913)
Cryptococcus cerebriloculosus Freeman et Weidman (1923)
Cryptococcus nasalis (Harrison) C. W. Dodge (1935)
Cryptococcus hondurianus Castellani 1933, Castellani et Jacono (1933)
Cryptococcus hominis Vuill. *var. hondurianus* Castellani 1933, Castellani et Jacono (1933)
Cryptococcus psicrophilicus Niño (1934)
Cryptococcus meningitidis C. W. Dodge (1935)

Nach dem derzeitigen Stand der Kenntnisse ergibt sich folgende *Klassifizierung* der *Cryptococcus*-Gruppe im engeren Sinne:

Classis (?):	*Fungi imperfecti (Deuteromycetes)*
Ordo:	*Cryptococcales*
Familia:	*Cryptococcaceae*
Subfamilia:	*Cryptococcoideae*
Genus:	*Cryptococcus*

Die *Gattung Cryptococcus* ist charakterisiert „durch vorwiegend runde oder ovale Sproßzellen ohne oder mit rudimentärem Pseudomycel. Echtes Mycel fehlt stets. Die Vermehrung erfolgt durch multilaterale Sprossung. Die Zellen sind von einer mehr oder weniger dicken Kapsel umgeben, in der unter besonderen Bedingungen ein stärkeähnliches Polysaccharid gebildet wird. Die Kulturen haben ein muköses Aussehen. Die Atmung ist strikt oxydativ."

Nach LODDER und KREGER-VAN RIJ (*117*), denen auch die vorstehende Definition zu danken ist, gehören zur Gattung *Cryptococcus* 5 Species und 3 Varietäten. Die Synonymie der nicht zu *C. neoformans* gehörenden Arten ist in Tabelle 2 wiedergegeben.

Tabelle 2. *Synonyma der von* LODDER *und* KREGER-VAN RIJ (1952) *aufgeführten Cryptococcus-Arten außer C. neoformans, zusammengestellt nach (111, 117)*

Species	Synonym mit
Cryptococcus laurentii (KUFFERATH) SKINNER	*Torula laurenti* KUFFERATH (1920) *Torula heveanensis* GROENEWEGE (1921) *Torula aurea* SAITO (1922) *Torulopsis laurentii* (KUFFERATH) LODDER (1934) *Torulopsis heveanensis* (GROENEWEGE) MAGER and ASCHNER (1947) *Torulopsis carnescens* VERONA et LUCHETTI (1936) *Candida heveanensis* (GROENEWEGE) DIDDENS and LODDER (1942) *Chromotorula aurea* (SAITO) HARRISON (1928) *Rhodotorula aurea* (SAITO) LODDER (1934)
C. laurentii var. flavescens	*Torula flavescens* SAITO (1922) *Torulopsis flavescens* (SAITO) LODDER (1934) *Cryptococcus flavescens* (SAITO) SKINNER (1947)
Cryptococcus albidus (SAITO) SKINNER	*Torula albida* SAITO (1922) *Torula gelatinosa* SAITO (1922) ? *Torula alpina* nomen nudum *Torulopsis albida* (SAITO) LODDER (1934) *Torulopsis albida* (SAITO) LODDER *var. japonica* LODDER (1934) *Torulopsis liquefaciens* SAITO et OTA (1934) *Torulopsis acris var. granulosa* MARCILLA et FEDUCHY
Cryptococcus luteolus (SAITO) SKINNER	*Torula luteola* SAITO (1922) *Chromotorula luteola* (SAITO) HARRISON (1928) *Torulopsis luteola* (SAITO) LODDER (1934)
Cryptococcus diffluens (ZACH) nov. comb.	*Torulopsis diffluens* (ZACH) WOLFRAM et ZACH (1934) *Torulopsis nadaensis* SAITO et OTA (1934) *Cryptococcus neoformans var. innocuus* BENHAM (1935)

Zum Verständnis der mykologischen Klassifizierungssysteme ist wichtig, daß morphologische Gesichtspunkte stets im Vordergrund stehen, bei den Hefen außerdem die biochemische Leistung. Diese, für die praktische Diagnostik wertvolle Betrachtungsweise wird allerdings den tatsächlichen, phylogenetisch fundierten Verwandtschaftsverhältnissen gerade im Bereich der Sproßpilze nur unvollkommen gerecht [Seeliger (*161*)].

Das Formgenus *Cryptococcus* unterscheidet sich aber von zahlreichen anderen Gattungen — erinnert sei nur an das Formgenus *Candida* — durch eine beachtliche Einheitlichkeit der wesentlichen Kriterien. Das schließt nicht aus, daß eine ganze Anzahl enger Verwandter der *Cryptococcus*-Gruppe aus taxonomischen und praktisch-diagnostischen Erwägungen bei anderen Gattungen, insbesondere bei *Candida und Trichosporon*, untergebracht wurde. Hierauf wird weiter unten noch Bezug genommen werden (s. S. 62).

Außer den von Lodder und Kreger-van Rij aufgeführten *Cryptococcus*-Arten (vgl. Tabelle 2) wurden von Skinner (*173*) noch weitere Species genannt, die nach Ansicht der holländischen Verfasser (*117*) teils zur Gattung *Torulopsis* und teils zum askosporogenen Genus *Lipomyces* gehören. Auch die von Benham (*13*) als *Cryptococcus glabratus* bezeichnete Art wird heute generell zur *Torulopsis*-Gruppe (*Torulopsis glabrata*) gerechnet. Sie ist wahrscheinlich eng verwandt mit *Torulopsis inconspicua* (*117*). Beide letztgenannten Arten werden im deutschen Schrifttum als Vaginalcryptococcen diskutiert (*134*).

Demgegenüber handelt es sich bei Benhams *Cryptococcus neoformans var. innocuus* bzw. *C. innocuus* um eine echte *Cryptococcus*-Species, die vom Verf. (*154*) als identisch mit *Cryptococcus diffluens* erachtet wird [vgl. auch (*12, 111, 117, 161*)]. Schließlich sei der *Cryptococcus mucorugosus* Benham (*12, 13*) erwähnt, über dessen Synonymie zur Zeit noch nichts Näheres bekannt ist. Vielleicht handelt es sich hierbei um eine Intermediärform zwischen *Cryptococcus sensu stricto* und Verwandten der *Trichosporon*-Gruppe.

Inzwischen wurde die *Cryptococcus*-Gruppe um eine weitere Art *Cryptococcus terreus* [Di Menna (*46a*)] vermehrt (vgl. Nachtrag).

Cytologie

Charakteristisch für das Zellbild der meisten *Cryptococcus*-Arten sind runde bis ovale Sproßpilze, die von einer mehr oder weniger deutlich ausgeprägten Kapsel eingehüllt sind. Die *Größenverhältnisse* der einzelnen Pilzzellen sind recht unterschiedlich. Sie wechseln mit dem Alter und der Zustandsform der Kultur und werden auch durch die Zusammensetzung des Nährbodens sowie durch die jeweiligen Wachstumsbedingungen beeinflußt. Sie variieren bereits bei verschiedenen Abimpfungen des gleichen Stammes. Dementsprechend zeigen auch verschiedene Stämme derselben Art keine absolute morphologische Übereinstimmung. Noch ausgesprochener gilt dies für Angehörige der als *C. luteolus* und *C. laurentii* abgegrenzten Arten.

Der mittlere Durchmesser schwankt — in Abhängigkeit von den eben geschilderten Umständen — bei der am gründlichsten untersuchten Species *C. neoformans* nach Angaben von Benham (*10, 12, 13*), Drouhet und Couteau (*53*), Negroni und Briz de Negroni (*126*), Lodder und Kreger-vanRij (*117*) sowie Littman und Zimmerman (*111*) zwischen 2,5 und 10 μ. Bei einzelnen Stämmen überwiegen Zellen mit einem Durchmesser von 5—8 μ, während andere durchschnittlich nur 3,5 μ messen. Glatt (S)-Formen weisen häufig einen relativ konstanten Durchmesser von 4—6 μ auf (*53, 111*). Ausnahmsweise wurden im Gewebe Zelldurchmesser bis 15 μ (Kapsel nicht eingerechnet) festgestellt (*111*).

Tabelle 3 vermittelt einen Überblick über die Größenverhältnisse, wie sie bei den verschiedenen *Cryptococcus*-Arten nach dreitägigem Wachstum bei 25⁰ C in Malzextrakt gefunden wurden (*117*).

Der bei anderen pathogenen Pilzarten häufig anzutreffende *Dimorphismus* wird bei der *Cryptococcus*-Gruppe stets vermißt [SCHERR und WEAVER (*151*)].

Obwohl die *Zellform* innerhalb der einzelnen Arten zahlreiche Übereinstimmungen aufweist, ist sie nicht ganz konstant. Runde und ovale Zellen sind ein gemeinsames Charakteristikum von *C. neoformans*, *C. albidus* und *C. diffluens*. Das Cytoplasma ist mit doppeltbrechender Granula durchsetzt, die bei *C. neoformans* grobkörniger sind als bei den beiden anderen Arten. Nach FREEMAN (*75*) bestehen die Granula zum Teil aus Lipoidkörperchen. Bei der sog. var. *uniguttulatus* [WOLFRAM und ZACH (*198*)] sind im Zellkörper meist nur ein bzw. wenige Öltröpfchen vorhanden. Es ist jedoch mehr als fraglich, ob diese Eigentümlichkeit zusammen mit dem Fehlen einer deutlichen Kapsel eine taxonomische Sonderstellung rechtfertigt. *C. laurentii* zeigt in ausgewachsenem Zustand 2 polar angeordnete Tröpfchen [BENHAM (*12*)], während junge bzw. noch in der Sprossung befindliche Zellen meist fein granuliert erscheinen.

Pseudomycel fehlt fast immer. Nur beim Sproßvorgang kommt es gelegentlich zur Entwicklung langgestreckter Keimschläuche. Manchmal hängen langgestreckte Zellen in kurzen Ketten zusammen [LITTMAN und ZIMMERMAN (*111*)], so daß der Eindruck eines rudimentären Pseudomycels entstehen kann. Echtes *Mycel* wird nie gebildet.

Tabelle 3. *Zellgröße und Morphologie der Cryptococcus-Arten nach 5tägiger Bebrütung in Malzextrakt bei 25⁰ C, zusammengestellt nach* LODDER *und* KREGER-VAN RIJ (*117*), *ergänzt nach* BENHAM (*13*) *sowie* LITTMAN *und* ZIMMERMAN (*111*) *und* DI MENNA (*46a*)

Art bzw. Variation	Vorherrschende Form	Zelldurchmesser in μ
C. neoformans	rund, selten rudimentäres Pseudomycel	4—6 Minimum 2,5 Maximum 15
C. albidus	rund, kurzoval	4,5—7 × 4,5—8 Minimum 3—6 Maximum 5,5—10
C. diffluens	rund, oval	3,5—5,5 × 4,2—6,5
C. luteolus	oval, länglich rudimentäres Pseudomycel	3,5—6 × 6—11
C. laurentii	klein, rund, oval bis länglich	2,5—5,5 × 4—6,5 Maximum 9—12
C. laurentii var. *flavescens*	oval, gelegentlich gebogen und polymorph	3—5 × 6—12
C. laurentii var. *magnus*	rund, kurzoval	4—7,2 × 6—9
C. terreus	rund, suboval	7,5—5,0 × 7,0—3,0

Die Tochterzellen entwickeln sich teils uni-, nicht selten aber auch multilokulär (*10, 27, 56, 111, 117* u. a.). NEGRONI u. Mitarb. (*126*) haben diesen Vorgang bei *C. neoformans* unter Verwendung eines einfachen Mineral-Mediums mit einem Zusatz von Thiamin genau untersucht. Danach erfolgt die Sprossung an einer „offenen" Wachstumsstelle — meristematischer Punkt genannt —, die mit einem kurzen Hals und einem Kragen versehen ist. Hier wird der „meristematische" Sproß gebildet, der seine vegetative Entwicklung fortsetzt und dementsprechend nicht als echte Conidie (conidium verum) anzusehen ist. Mehrere Sproßzellen werden im Laufe mehrerer Tage am selben meristematischen Punkt aus der Mutterzelle entlassen, und jedesmal bildet sich ein neuer schalenförmiger Kragen innerhalb des vorhergehenden. Nach etwa 5 Tagen verlangsamt sich bei ausgewachsenen Zellen die Reproduktionsfähigkeit.

Neben Polysacchariden und Lipoidtröpfchen enthält das Zellinnere *Eiweiß*, dessen Gehalt an Aminosäuren von Uzman et al. (*184*) bei je einem stark bekapselten und einem wenig bekapselten Stamm von *C. neoformans* näher untersucht wurde (Tabelle 4). Dabei zeigten beide Kulturen einen hohen Gehalt an Aminosäuren mit zwei Carboxylgruppen und das Fehlen von Methionin.

Wiederholt wurden bei *C. neoformans*-Stämmen *Innenkörper* beobachtet, deren Natur bisher allerdings umstritten geblieben ist. Todd und Herrmann (*182*) berichteten 1936 über den Nachweis eigenartiger, dreieckig-ovaler Zellen mit einer dicken Wand und einem, meist nahe der Spitze gelegenen rundlichen Innenkörper, den sie als *Askospore* deuteten. Auf Grund dieses Befundes wurde gefolgert, die Gattung *Cryptococcus* sei das imperfekte Stadium des askosporogenen Genus *Debaryomyces* — daher der Name *D. hominis* (vgl. S. 25 und Tabelle 1). In Bestätigung der vorstehenden Ergebnisse glauben auch Redaelli, Ciferri und Giordano (*76, 139*), bei *C. neoformans* eine echte Sexualität festgestellt zu haben. Sie beschrieben einen als Befruchtung erachteten Vorgang und die daraus resultierende Bildung eines der Askospore von Todd und Herrmann in jeder Hinsicht gleichenden Sphäroidkörpers. Langeron (*108*), Lodder und De Minjer (*116*), Emmons (*56*), Lodder und Kreger-van Rij (*117*) konnten sich auf Grund sorgfältiger eigener Untersuchungen diesen Deutungsversuchen nicht anschließen. Emmons (*56*) sieht die frag-

Tabelle 4. *Aminosäure-Gehalt des Körpereiweißes bei 2 C. neoformans-Stämmen, bestimmt mittels der quantitativen Ionenaustausch-Chromatographie* [nach Uzman, Rosen und Foley (*184*)]

Aminosäure	Stamm C—JM (kleine Kapseln)		Stamm C—94 (große Kapseln)	
	Micromol pro 10 ml Hydrolysat	% des Gesamtaminosäuregewichts	Micromol pro 10 ml Hydrolysat	% des Gesamtaminosäuregewichts
Asparaginsäure	2,60	10,0	2,58	9,3
Threonin . . .	1,66	5,7	1,65	5,5
Serin	2,06	6,0	1,81	5,2
Glutaminsäure .	2,60	11,0	2,72	10,9
Prolin	1,58	5,2	1,15	3,6
Glycin	2,02	4,3	2,08	4,3
Alanin	2,44	6,3	2,64	6,3
Valin	1,90	6,3	2,08	6,5
Cystin	<0,1[1]		<0,1[1]	
Methionin . .	0	0	0	0
Isoleucin . . .	1,75	6,7	2,24	7,9
Leucin	2,98	11,2	3,16	11,2
Tyrosin	0,39	2,0	0,49	2,5
Phenylalanin .	1,01	4,9	1,26	5,7
Histidin. . . .	0,42	2,0	0,61	2,5
Lysin	1,86	7,8	2,00	7,9
Arginin	1,20	6,0	1,07	5,2
Ammoniak . .	8,65[2]	4,3	11,05[2]	5,2

[1] Nur mittels Papierchromatographie nachweisbar
[2] Zum Teil möglicherweise aus Tryptophan stammend.

liche Askospore als zusammengelaufene Lipoidkörperchen an, läßt aber auch die Möglichkeit offen, daß es sich um gespeicherte Nahrung bzw. um Kapselsubstanz handelt. Analog dazu fanden auch Negroni und Briz de Negroni (*126*) sowie Vanbreuseghem (*185*) keine Anhaltspunkte für das Vorhandensein von Askosporen. Immerhin sah Vanbreuseghem (*185*) nach 25—30tägiger Bebrütung im Tusche- und Ziehl-Präparat einzelne Zellen, deren besonders dichte Kapseln sich an einem Zellpol verdichtet hatten. Aus diesen Strukturen ragte die schwach säurefeste Pilzzelle wurstförmig verlängert heraus. Der belgische Autor spricht von Expulsionsformen, während Negroni u. Mitarb. (*126*) einen ähnlichen Vorgang als „moulting phenomenon" beschreiben. Danach bleibt die alte Zellmembran an einem Ende der regenerierten Pilzzelle als leere Scheide hängen. Diese dürfte den „Ghost-cells" entsprechen, die nach Benham (*12, 13, 14*) wie leere, zusammengefallene Säcke (Asken) mit dünnen, durchsichtigen Wänden aussehen. Sie sind teilweise an normale Zellen angeheftet, teilweise flottieren sie auch frei im Gesichtsfeld herum. Entgegen den Vorstellungen von Negroni u. Mitarb. (*126*) sowie Vanbreuseghem (*185*) sollen sie nach Benham (*14*) aus den normalen Zellen durch Sprossung entstanden sein. Nach 2 Wochen bis 2 Monate langer Bebrütung einer Mischung von insgesamt 4 *Cryptococcus*-Stämmen auf Würze- oder Noyes-Agar zeigten sich in diesen „Ghost cells" 2, 4 oder 8 Rundkörper, die als Sporen bzw. Askosporen erachtet wurden. BENHAM (*13, 14*) wies auf die große Ähnlichkeit dieses Vorgangs mit der Askosporenbildung bei *Lipomyces starkeyi* bzw. *Lipomyces lipoferus* hin, 2 Arten, die nach CONNELL et al. (*40*) wie LODDER u. Mitarb. (*117*) viele andere Übereinstimmungen mit *Cryptococcus*-Stämmen aufweisen.

Diese wichtigen Beobachtungen erfordern weitere cytologische Untersuchungen, vor allem an *C. neoformans*-Kulturen vom Menschen, zumal sich BENHAMS Befunde zunächst nur auf einen Stamm stützen, der aus einem Hundehirn isoliert worden ist.

Wir selbst haben — in Bestätigung anderer (*111, 117*) — bei der Überprüfung zahlreicher *Lipomyces*-Stämme sowohl im Nativpräparat wie auch mittels der Sporenfärbung (*169a*) auf einfachen wie zur Sporenbildung empfohlenen Nährböden (*117*) stets reichlich typische Askosporen gefunden, im Gegensatz zu mehr als 20 pathogenen und apathogenen *Cryptococcus*-Stämmen verschiedenster Herkunft. Bei einigen der letzgenannten waren nach monatelanger Züchtung Sphäroidkörper zu sehen; aber diese Strukturen färbten sich nicht in gleicher Weise selektiv wie die *Lipomyces*-Sporen.

In Anbetracht dieser Resultate und anderweitiger biochemischer wie serologischer Unterschiede, die weiter unten besprochen werden (s. S. 50 u. 60), ist die Frage einer möglichen Identität zwischen *Cryptococcus* und *Lipomyces* zur Zeit noch ebenso ungeklärt wie die seit Jahren umstrittene Askosporenbildung bei *C. neoformans*.

Die Zellen der *Cryptococcus*-Gruppe sind *einkernig* (*126*) und stimmen im cytologischen Aufbau mit vielen anderen Sproßpilzen überein.

Sie unterscheiden sich jedoch durch ihre mehr oder minder vorhandene Fähigkeit zur *Bildung von Kapseln*, die im Gewebe gelegentlich einen Durchmesser bis 50 μ erreichen. Allerdings ist die Kapseldicke besonders großen Schwankungen unterworfen. Sie ist u. a. abhängig vom Stamm und den Ernährungsbedingungen. Manche Kulturen bilden auch auf einfachen Medien wie SABOURAUD- oder GRÜTZ-Agar regelmäßig mächtige Kapseln, bei anderen ist die Kapselbildung auf künstlichen Substraten kümmerlich, gleichgültig, ob es sich um pathogene oder apathogene Arten handelt.

Nach DROUHET und COUTEAU (*53*) bestehen enge Beziehungen zwischen dem Ausmaß der Kapselbildung und der *kulturellen Zustandsform:* S-Formen (s. S. 28) besitzen meist nur eine relativ kleine Kapsel, während die R-Form kapsellos erscheint. Die mukösen M-Sektoren bzw. M-Kolonien bestehen dagegen fast ausschließlich aus stark bekapselten Zellen. Die Anwesenheit von Maltose scheint die Kapselentwicklung zu fördern, während sich Dextrose nachteilig auswirken soll. Darüber hinaus fanden NEGRONI und LANATA (*127*), daß Mannose, Mannitol, Trehalose und Dextrose in Gegenwart von 0,5% Ammoniumsulfat die Kapselbildung begünstigen. Einen ähnlichen Effekt scheint auch Pepton in Anwesenheit von 1% Dextrose auszuüben. Demgegenüber blieb die Bekapselung auf einem Histidin-Medium kümmerlich.

Mit Sicherheit gelingt der Kapselnachweis bei pathogenen *Cryptococcus*-Stämmen nur im *Gewebe* bzw. in *Körperflüssigkeiten*. Das wahre Ausmaß der Bekapselung wird oft erst nach Verimpfung auf die weiße Maus und nachfolgende Untersuchung des Peritonealexsudats erkennbar.

An den Stellen der Bildung von Tochterzellen ist die Kapsel zunächst meist recht dünn. Sie verdickt sich erst mit zunehmendem Alter der neugebildeten

Zelle. Nicht selten findet man aber ganze Sproßzellverbände umgeben von einer mächtigen Lage Kapselmaterial.

Selbst sehr große *Kapseln* sind im *Nativpräparat* mit Kochsalzlösung bei durchfallendem Licht *unsichtbar*. Nach Segretain et al. (*167*) lassen sie sich — in Übereinstimmung mit den Angaben von Tomcsik und Guex-Holzer (*183*) — auch mittels des Phasenkontrastmikroskops nicht darstellen. Tomcsik und Guex-Holzer (*183*) erreichten die Sichtbarmachung der Kapseln im Lichtmikroskop mit einem einfachen Verfahren: Nach Zusatz von Eiweiß — z. B. in Form von Normalserum, Casein und Hämoglobin, aber nicht von Gelatine, Eialbumin, Pepsin oder Trypsin — wird die Kapsel in einem schmalen p_H-Bereich auf der sauren Seite des isoelektrischen Punktes sichtbar. Zusatz einer kleinen Menge 2%iger Essigsäure bewirkt eine salzähnliche Verbindung der Kapselpolysaccharide mit dem Serumeiweiß und damit an der Kapseloberfläche eine Veränderung der Lichtdurchlässigkeit, die sich in einer typischen *Kapselreaktion* äußert (vgl. Abb. 1). Mikroskopisch ähnelt dieser Vorgang der Kapselreaktion nach Zugabe von Immunserum täuschend (s. S. 55).

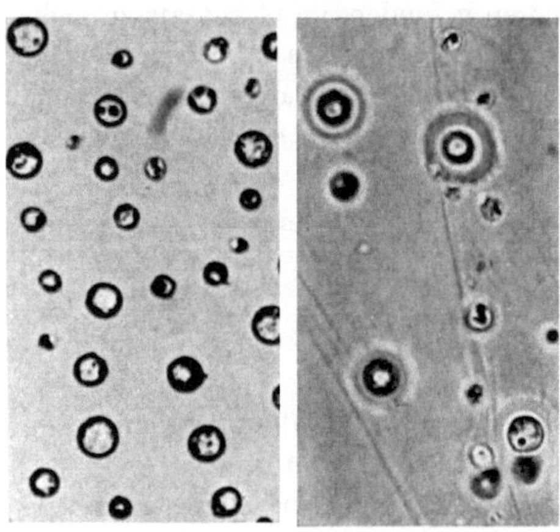

Abb. 1. *Cryptococcus neoformans* im Nativpräparat. Links: Aufschwemmung in physiologischer Kochsalzlösung; rechts: Kapselnachweis in Serum nach Zusatz von Essigsäure

Die *chemische Zusammensetzung* des Kapselmaterials war Gegenstand eingehender Untersuchungen. Aschner u. a. (*5*) sowie Mager und Aschner (*119, 120*) berichteten 1945—1947 über das Vorhandensein einer sich mit Jod blau färbenden, extracellulären Stärke

Tabelle 5. *Chemische Zusammensetzung der löslichen Polysaccharide von Cryptococcus neoformans und Cryptococcus diffluens (innocuus)*

Pilzart	Stickstoffgehalt in %	Reduzierender Zucker nach Hydrolyse in %	Zuckerart	Autoren
C. neoformans Kapselsubstanz	0,3	60	Glucuronsäure, Mannose, Xylose	Drouhet und Segretain
C. neoformans- Typ A, B und C Kapselsubstanz	weniger als 0,01	75	Glucuronsäure, Mannose, Xylose, Spuren Galaktose	Evans und Mehl
C. neoformans Typ B, gereinigte Kapselsubstanz	weniger als 0,01	75	Glucuronsäure, Mannose, Xylose	Evans und Theriault
C. diffluens (C. innocuus)	0,3	47,5	6,7 % Glucuronsäure, 18,1 % Dextrose, 31 % Pentose	Einbinder, Benham und Nelson
C. diffluens	0,3	63,0	Glucuronsäure, Mannose, Xylose, Dextrose	Burcik und Beutman

in der Kapselsubstanz, die außerdem ein d-Xylose-haltiges Polysaccharid enthält. KLIGMAN (*98*) isolierte aus Kapseln ebenfalls ein Polysaccharid, das nicht antigen wirkte (s. S. 53). HEHRE et al. (*87*) extrahierten das sich mit Jod blau färbende Kapselmaterial und identifizierten das kristalline Produkt als Polysaccharid der Amylose-Klasse. Außer dieser Fraktion A erhielten sie noch eine Fraktion B, wahrscheinlich ein mehrere Pentosen enthaltendes Polysaccharid, das auch die serologische Spezifität bedingt (s. S. 56). Möglicherweise ist in der Kapselsubstanz auch Chitin enthalten [MOLNAR (*124*)].

Über die Chemie des nicht zur Amylose gehörenden Polysaccharids liegen Berichte von DROUHET, SEGRETAIN und AUBERT (*50, 52*), EVANS u. Mitarb. (*64, 67*), EINBINDER et al. (*55*) sowie von BURCIK und BEUTMANN (*199, 200*) vor. Die zum Teil übereinstimmenden Ergebnisse sind in Tabelle 5 zusammengefaßt.

Bei der serologisch aktiven Kapselsubstanz handelt es sich demnach um ein *Polyosid*, das der Hyaluronsäure zwar nahe steht, sich aber durch das Fehlen von Glucosamin und einige andere Eigenschaften deutlich unterscheidet (DROUHET, briefliche Mitteilung). Dieses Polyosid ist quantitativ für die *Virulenz* der Stämme verantwortlich (*52*); es übt auf Leukocyten eine phagocytosehemmende Wirkung aus [DROUHET und SEGRETAIN (*54*)] und hemmt bei Virustitration auf *Nicotina glutinosa* das Tabakmosaikvirus [HIRTH und DROUHET (*89*)].

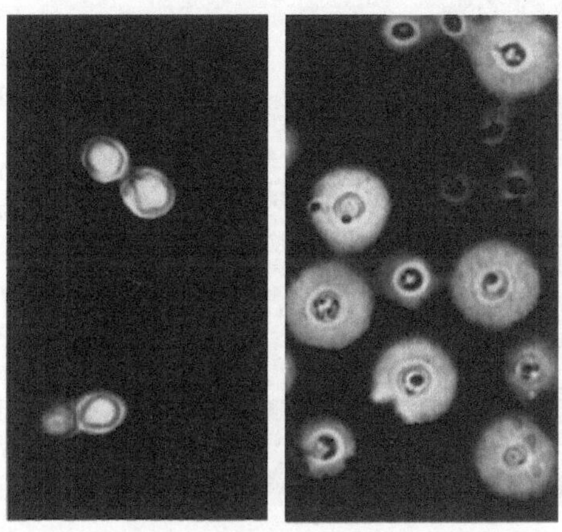

Abb. 2. *Cryptococcus neoformans* im Tuschepräparat. Links: geringe Kapselbildung; rechts deutliche Kapselbildung

Viel diskutiert wurden Befunde von DROUHET und SEGRETAIN (1040) (*50, 51*), wonach sich der Kapseldurchmesser durch Behandlung mit Hyaluronidase geringfügig vermindere und die Zellen besser agglutinabel würden. Ein enzymatischer Abbau wie bei Streptokokken ließ sich jedoch nicht nachweisen [NEGRONI und LANATA (*127*), EVANS und THERIAULT (*67*)]. In Übereinstimmung mit EINBINDER et al. (*55*) fanden auch FOLEY und UZMAN (*74*) durch Behandlung von *Cryptococcus*-Polysaccharid-Lösungen mit Hyaluronidase keine meßbare Viscositätsänderung. Dies wurde durch LITTMAN [zit. (*111*)] erhärtet, zumal eine Dekapsulation lebender *Cryptococcus*-Stämme weder bei 20° noch bei 37° C im sauren (pH 4) oder neutralen Milieu durch Einwirkung von 150 trübungsvermindernden Hyaluronidase-Einheiten erzielt werden konnte. Auch durch Einwirkung von 10 n NaOH oder 0,5n HCl ist *keine* Verminderung der Kapselgröße zu erreichen [NEGRONI und LANATA (*127*)]. Der von DROUHET u. Mitarb. beschriebene Hyaluronidase-Effekt ist demnach *unspezifisch*. Er beruht auf einem physikalisch-chemischen Phänomen: der Kombination von Eiweiß mit dem Polyosid (DROUHET, briefliche Mitteilung 1956).

Zur *Darstellung* der Kapseln im flüssigen Milieu (Kulturaufschwemmungen, Untersuchungsmaterial aus Körperflüssigkeiten) ist das Tuschepräparat am besten geeignet (Abb. 2).

Die Zellen färben sich auch nach der Methode von GRAM und mit GIEMSA-Lösung. Da sich dabei die Kapseln nicht oder nur ausnahmsweise darstellen, erscheinen die Pilzzellen oft recht atypisch und können leicht mit Blutbestandteilen verwechselt werden. Die reifen Zellen sind nicht säurefest.

Zum *histochemischen Nachweis* der Cryptococcen, insbesondere des *C. neoformans*, eignen sich mehrere Färbemethoden. Die Kapseln und Zellwände lassen sich nach LITTMAN und ZIMMERMAN (*111*) sowie IVERSON (persönliche Mitteilung 1957) besonders gut mit *Mucicarmin* und der RHINEHART-ABDUL-HAJ-Technik (*143*) für saure Mucopolysaccharide nachweisen. Ausgezeichnete Dienste leistet auch die Färbung nach BAUER (*9*). Vorzüglich bewährt haben sich ferner die *Perjodsäure-Schiff-Färbung* [HOTCHKISS (*93*), KLIGMAN et al. (*101*), McMANUS (*121*) u. a. m.] sowie die von GRIDLEY (*82*) ausgearbeitete Modifikation und die GOMORI-Färbung [GROCOTT (*83*)]. LITTMAN und ZIMMERMAN (*111*) belegten dies durch zahlreiche hervorragend gelungene Abbildungen. Im übrigen gelang auch durch Verwendung von Toluidinblau der Nachweis von Hyaluronsäure in den Kapseln nicht.

Nach MOLNAR (*124, 125*) zeigt die *Cryptococcus*-Kapsel im Polarisationsmikroskop eine positive Doppelbrechung von sehr geringer Intensität. Im Wassermedium beträgt der Gangunterschied bei $5\,\mu$ dicken paraffinierten Schnitten, gemessen mit der Senarmontschen Kompensation, 20—22 mμ und manifestiert sich in Form eines negativen Polarisationskreuzes. In Medien mit einem höheren Brechungsquotienten als Wasser, z. B. Canadabalsam, verschwindet dieser Effekt vollständig. Dies veranlaßte den ungarischen Autor, die von SCHMIDT (*153*) schon 1932 beschriebene *dichroïtische* Färbung zu versuchen, mit dem Ergebnis, daß sich die Kapseln mit 12 Farbstoffen anfärben ließen. Am besten bewährten sich Bismarckbraun und Akridinorange. Nach Behandlung mit Acridinorange, im Verhältnis von 1:1000 in Wasser gelöst, leuchten die Kapseln im Polarisationsmikroskop goldgelb und im Fluorescensmikroskop rot auf (*124*). Mittels der von KLATZO und GEISLER (*203*) beschriebenen Färbemethode läßt sich *C. neoformans* im histologischen Präparat besonders gut bei Beleuchtung mit polarisiertem Licht darstellen: dabei leuchtet das Cytoplasma rosa, umgeben von doppeltbrechendem radiär angeordnetem Kapselmaterial.

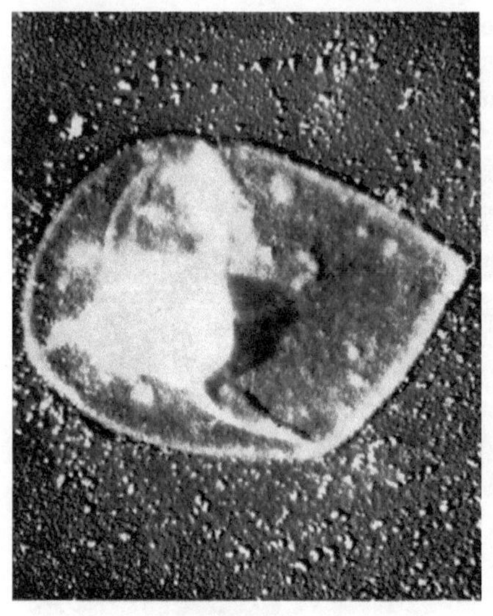

Abb. 3. *Cryptococcus neoformans*. Elektronen-optische Darstellung nach Behandlung mit Äthyläther und nach Waschen mit Wasser. Annähernde Vergrößerung 23000fach (mit freundlicher Genehmigung von Dr. E. RIBI, Hamilton, Montana). [Aus RIBI, SALVIN u. BROWN: Exp. Cell Res. **10**, 401 (1956)]

Weitere Aufschlüsse erbrachten elektronenoptische Untersuchungen von RIBI und SALVIN (*142*). Bekapselte *C. neoformans*-Zellen zeigten nach Bedampfung mit Osmiumsäure amorphes Kapselmaterial, dessen Dicke nach außen zu asymptotisch abnahm und keine deutliche Grenze erkennen ließ. Wenig oder gar kein Kapselmaterial war im elektronenoptischen Bild erkennbar, wenn die Zellen vor der Präparation gründlich mit Wasser gewaschen wurden. Ähnliches ergab sich nach Schütteln von Zellsuspensionen mit 2 Volumina Äthyläther und anschließendem scharfem Zentrifugieren. Die Oberfläche der Zellwand zeigte nach dieser Behandlung deutlich feinkörnige Granula und praktisch kein Kapselmaterial mehr (Abb. 3).

Kulturelles Verhalten

Alle *Cryptococcus*-Arten wachsen auf den einfachen, in der mykologischen Praxis üblichen Nährböden wie Sabouraud-Agar, Grütz-Agar, Malzextrakt-Agar, Würze-Agar usw.

Koloniegröße und -morphologie, Konsistenz, Pigmentierung und Wachstumsgeschwindigkeit werden vor allem vom p_H des Mediums (s. S. 38), der Bebrütungstemperatur (s. S. 37) und von der chemischen Zusammensetzung des Substrats nachhaltig beeinflußt. Für den geübten Untersucher ergeben sich bei Verwendung einheitlicher Nährböden und Methoden aus dem Aussehen der Kolonien, insbesondere der längere Zeit bebrüteten Riesenkolonien, wertvolle Rückschlüsse auf die Artzugehörigkeit der gezüchteten *Cryptococcus*-Stämme.

Auf synthetischen Nährböden, wie sie von Negroni u. Mitarb. (*126*) sowie Reid (*140*) benutzt wurden, läßt sich das Wachstum durch Zusatz von 1—2% Dextrose, nach Drouhet und Segretain (*49*) auch durch Maltose fördern. Während die französischen Autoren (*49*) durch Zugabe von Laktoflavin keine Wachstumssteigerung, eher eine durch das Lösungsmittel bewirkte Hemmung beobachteten, konnte Reid (*140*) eindeutig einen wachstumsstimulierenden Effekt durch Vitamin B_1 (Thiamin) nachweisen. Dem trugen auch Hehre c.s. (*87*) in ihrem synthetischen Substrat Rechnung.

Aus der Fülle der synthetischen Medien, die das Wachstum von *Cryptococcus*-Arten gestatten, sei die von Schmidt et al. (*152*) verwandte Grundformel mitgeteilt. Auf 1 Liter Aqua dest. kommen folgende Zusätze:

Dextrose	20,0 g	$MgSO_4 \cdot 2H_2O$ . . . . 1 mg
$(NH_4)_2SO_4$	0,17 g	$FeCl_3 \cdot 6H_2O$ . . . . 1 mg
$(NH_4)_2HPO_4$	1,0 g	$CuSO_4 \cdot 5H_2O$ . . . . 0,2 mg
KH_2PO_4	1,5 g	$Na_2B_4O_7 \cdot 10H_2O$ . . . . . . 0,2 mg
NaCl	1,0 g	$ZnSO_4 \cdot 7H_2O$ . . . . 0,14 mg
$MgSO_4 \cdot 7H_2O$	0,3 g	KJ . . . . 0,1 mg
$CaCl_2$	0,2 g	Molybdänsäurepulver (85%) . 0,02 mg

Optimales Wachstum erfolgt aber erst nach Zusatz von Thiamin (0,1—1 γ je 5 ml). Purine und Pyrimidin sind für Cryptococcen keine essentiellen Wuchsstoffe. Aminosäuren üben nur eine leicht stimulierende Wirkung aus.

Cryptococcus neoformans ist zwar zur Pyrimidinsynthese imstande, aber nicht zur Thiazolsynthese. Er ist auch nicht in der Lage, aus beiden Substanzen das Thiamin-Molekül aufzubauen (Schmidt et al.). Andererseits bewirken die Thiamin-Hemmstoffe Pyrithiamin und Neopyrithiamin keine Hemmung des Thiamin-Verbrauchs.

Die Beschreibungen von Benham (*10, 12, 13*), Lodder und Kreger-van Rij (*117*), Cox und Tolhurst (*42*) sowie die mit zahlreichen instruktiven Abbildungen versehene Darstellung von Littman und Zimmerman (*111*) vermitteln ein anschauliches Bild der *Cryptococcus-Koloniemorphologie* auf *Oberflächenkulturen.*

Nach verlängerter Bebrütung und entsprechender Beimpfung eines 2—3 cm messenden Areals der Nährbodenoberfläche bilden sich bei den meisten *C. neoformans*-Stämmen charakteristische *Riesenkolonien* mit *typischen Sektoren.* Letztere lassen nach Drouhet und Couteau (*53*) drei vorherrschende Hauptformen erkennen:

a) Die *Glatt (S)-Form*. Sie besteht aus cremefarbenem, undurchsichtigem, hefeähnlichem Wachstum. Abimpfungen entwickeln sich zu glattrandigen, etwas unregelmäßig konturierten Kolonien mit glatter, erhabener, glänzender Oberfläche. Mikroskopisch bestehen sie aus Sproßzellen mit meist kleinen Kapseln (s. S. 31). Nach 10tägigem Wachstum nehmen diese Kolonien mehr und mehr durchsichtiges Aussehen an. Sie werden gleichzeitig runder und erhabener, meist auch schleimiger. Auf Schrägagar-Kulturen entsteht aus den Einzelkolonien ein konfluierendes Wachstum. Die Zellen sind dann meist stark bekapselt. Gelegentlich, vor allem bei Abwesenheit von Dextrose oder Maltose im Nährmedium, bleibt aber der reine S-Typ erhalten, so daß die Kolonien anderen Hefearten, z. B. *Candida albicans*, täuschend ähnlich sehen.

b) Die *Rauh (R)-Form*. Sie ähnelt zunächst weitgehend der S-Form. Die Zellen zeigen aber mit zunehmendem Alter weder Anzeichen von Kapselbildung noch muköses Wachstum. Vielmehr wird die Oberfläche trocken und unregelmäßig gefaltet. Mikroskopisch findet sich nicht selten ein mehr oder weniger deutliches, meist rudimentäres Pseudomycel.

c) Die *muköse (M)-Form*. Sie zeigt die schon vorstehend bei der S-Form beschriebenen Veränderungen: runde, glasig-schleimige, durchsichtige, erhabene Kolonien mit glattem Rand, die leicht zusammenfließen und aus Zellen mit mächtigen Kapseln bestehen. Sie sind von schleimig-zäher Konsistenz und schwer zu verreiben im Gegensatz zum pastösen Wachstum der S-Form. Bei manchen Stämmen ist diese Wuchsform von Anbeginn vorhanden und bleibt auch nach jahrelangen Passagen auf künstlichen Substraten unverändert erhalten.

Die *Pigmentierung* der Kolonien ist unterschiedlich. Man findet alle Übergänge von weißcreme über gelblichorange bis zu beige und bräunlichen Farbtönen, deren Intensität sowohl vom Nährmedium wie von der Bebrütungsdauer als auch von stammspezifischen Eigentümlichkeiten bestimmt wird.

In Übereinstimmung mit Littman und Zimmerman (*111*) fanden auch wir (*165*) mehrfach *Cryptococcus neoformans*-Stämme menschlicher Herkunft (Fisteleiter, Liquor), bei denen sich auf künstlichen Nährböden stets nur reine S-Kolonien entwickelten. Erst nach Tierpassagen und verlängerter Bebrütung in Anwesenheit von Thiamin stellte sich auch muköses Wachstum ein. Bei einer Kultur wuchsen aber trotz starker Kapselbildung im Versuchstier Subkulturen wieder nur wenig bekapselt in Form der S-Kolonien.

Betreffs der Wuchseigentümlichkeiten auf Spezialnährböden wie Littmans Rindergalle-Agar, Hirn-Herz-Infusions-Agar (meist nur kümmerliches Wachstum) und Littmans Leber-Milz-Dextrose-Agar mit Zusatz von 20 OE Penicillin und 50 E Streptomycin/ml Substrat sei auf die Darstellung der amerikanischen Autoren (*111*) verwiesen.

Mit *C. neoformans* leicht zu verwechseln sind die Kolonien der eng verwandten, aber apathogenen Arten *C. diffluens (C. innocuus)* und *C. albidus*. Obwohl bei *C. diffluens* pastöses, cremefarbenes Hefewachstum meist vorherrscht — wir beobachteten es ausnahmslos bei 6 frisch isolierten Kulturen —, finden sich Sammlungsstämme mit typisch mukösem Wachstum, deren Kolonien von *C. neoformans*-M-Kolonien makroskopisch und mikroskopisch nicht unterscheidbar sind. Ähnliches gilt für *C. albidus* [vgl. auch (*12, 111, 117, 165*)].

Benham (*12, 13*) weist allerdings darauf hin, daß das Wachstum von *C. neoformans* meist stärker ist als das seiner apathogenen Verwandten, insbesondere bei der als *C. innocuus* bezeichneten Art, die wir als identisch mit *C. diffluens* ansehen. Nach unseren Erfahrungen gilt dies nicht für das Wachstum bei Zimmertemperatur (22—24⁰ C), wohl aber für Temperaturen über 30⁰ C. Deutlichere Unterschiede kommen beim Vergleich der Stämme auf WICKERHAMs Morphologie-Agar für Hefen zum Ausdruck. Auf diesem Medium nehmen *C. neoformans*-Riesenkolonien nach dreiwöchiger Bebrütung eine dunkelorange bis bräunliche Färbung an. Gleichzeitig entwickeln sich radiäre Auswüchse, so daß die ganze Kolonie ein sternförmiges Aussehen erhält. Im Zentralteil zeigen sich überdies muköse Erhebungen. Demgegenüber bleiben *C. innocuus*- (*C. diffluens*-) Kulturen meist cremefarben oder gelblich und flachrandig. Gelegentlich entsteht aber ebenfalls die typische Sternzeichnung.

Deutlich abgrenzbar ist hingegen *C. laurentii*, der in glänzend mukösen gelblich-orangefarbenen Kolonien mit gelegentlicher Brauntönung wächst und im Gegensatz zu dem glasigen Aussehen der *C. neoformans*-M-Kolonien undurchsichtig-trüb erscheint.

Demgegenüber zeichnet sich *C. luteolus* durch ein feuchtes, konfluierendes, gelbliches Wachstum aus, das dem von *C. neoformans* sehr ähnelt (*12, 13, 117*).

Die von BENHAM kreierte Art *C. mucorugosus* läßt sich durch ihre trockenen, faltigen Kolonien mit gelblich-brauner Farbtönung leicht von den Vorgenannten unterscheiden (*12*).

In *flüssigen Nährböden*, wie Malzextrakt oder 2%iger Dextrose-Bouillon, wachsen die verschiedenen *Cryptococcus*-Arten ohne allzu wesentliche Unterschiede (*117*). Junge Kulturen zeigen regelmäßig eine homogene Trübung des Substrats. Erst bei verlängerter Bebrütung in Zimmertemperatur bildet sich bei den meisten Arten ein mehr oder minder kräftiges Sediment unter gleichzeitiger Ringbildung an der Oberfläche. Gelegentlich entwickeln sich an der Oberfläche flottierende Inselchen, die zu einem Häutchen zusammenwachsen können, das dann meist von schleimiger Konsistenz ist, z. B. bei *C. laurentii*. Ausgesprochene Häutchenbildung findet sich hingegen, meist schon nach kurzfristiger Bebrütung, bei einigen mit der *Cryptococcus*-Gruppe eng verwandten Arten, die aus morphologisch-taxonomischen Erwägungen zur Zeit noch zu anderen Genera gerechnet werden, so z. B. bei *Candida curvata*, *Candida humicola* und *Trichosporon cutaneum*. Daß es sich hierbei um echte Verwandte der Cryptococcen handelt, wird weiter unten noch näher begründet werden (s. S. 62).

Stämme mit kräftiger Schleimbildung wandeln flüssige Substrate, die assimilierbare Zucker enthalten, insbesondere Malzextrakt, nach mehrwöchiger Bebrütung in eine schleimige Masse um. Im Gegensatz zu den übrigen *Cryptococcus*-Arten zeichnet sich *C. albidus* durch vorherrschendes Oberflächenwachstum bei nur mäßigem, teilweise sogar fehlendem Bodenwachstum aus.

Recht charakteristisch ist das Verhalten der einzelnen *Cryptococcus*-Arten gegen *Temperatureinflüsse*. Die optimale Wachstumstemperatur liegt zwischen 25 und 29⁰ C. Höhere Temperaturen werden praktisch nur von *C. neoformans* vertragen, der im Unterschied zu den übrigen *Cryptococcus*-Arten als einziger in der Lage ist, bei 37⁰ C ungehemmt zu wachsen. Diese Eigenschaft ist eines der wichtigsten differentialdiagnostischen Merkmale dieses mit Sicherheit pathogenen Vertreters der *Cryptococcus*-Gruppe (*12, 13, 111, 117, 160* u.a.m.). Allerdings gilt diese Aussage nach eigenen Erfahrungen nicht uneingeschränkt

(*165*). Es zeigt sich nämlich, daß gelegentlich auch apathogene *Cryptococcus*-Stämme, die zu *C. diffluens* gehören und mit Benhams *C. neoformans*, var. *innocuus* identisch sind, in der Erstzüchtung bei 37° C anwachsen, vor allem dann, wenn sie aus dem Körperinneren isoliert werden. Aber schon nach 1—2 Passagen bei 37° C geht diese Fähigkeit verloren, und die Kulturen sterben ab, wenn sie nicht bei niederen Temperaturen weitergezüchtet werden.

Von erheblichem Interesse — sowohl im Hinblick auf die vielfach diskutierte Fiebertherapie der disseminierten bzw. zentralnervösen Cryptococcose als auch auf die wirksame Pasteurisierung infizierter Milch — ist die *Hitzetoleranz* von *C. neoformans*. Die übrigen *Cryptococcus*-Arten zeichnen sich, wie aus Vorstehendem zu ersehen, durch eine nur geringe Hitzetoleranz aus. Kuhn (*105, 106*) berichtet auf Grund eingehender Beobachtungen an infizierten Tieren und Kulturen, daß *C. neoformans* bei 39,4° C langsam abstirbt und bei 40,6° C die meisten Keime innerhalb von 24 Std zugrundegehen. In Bestätigung dessen fanden auch Kligman und Weidman (*99*) nach 6tägiger Bebrütung bei 40° C keine lebenden

Tabelle 6. *Hitzeempfindlichkeit von Cryptococcus neoformans*
(nach Littman und Zimmerman, ergänzt)

Temperatur in Celsius	Einwirkungsdauer	Ergebnis	Autoren
40,6	24 Std	meiste Zellen abgestorben	Kuhn (*105, 106*)
40,0 alkal.pH		kein Wachstum	Mosberg et al. (*125a*)
40,0	6 Tage	keine lebenden Zellen mehr	Kligman und Weidman (*99*)
42,0	30—50 Std	Abtötung	Benham (*11*)
50,0	42 min	Abtötung	Crone c.s. (*42a*)
55,0	5 min	Abtötung	Emmons (*59*)
60,0	5 min	Abtötung	Crone c.s. (*42a*)
60,0	5 min	Abtötung	Cox und Tolhurst (*42*)
62,8	30 min	Abtötung	Emmons (*59*)
73,9	1 min	Abtötung	Pounden et al.(*135*)

Zellen mehr. Mosberg et al. (*125a*) zufolge wird dieser Temperatureffekt durch ein alkalisches pH noch gesteigert. *Alkali* allein ist in Konzentrationen, wie sie bei den Anreicherungsverfahren für Tuberkelbakterien verwandt werden, nicht gegen *C. neoformans* wirksam [Ajello u. Mitarb. (*1*)]. Dagegen wird der Pilz nach Simon (*171*) durch n-NaOH innerhalb von 15 min abgetötet. In Tabelle 6 sind die Ergebnisse über die Hitzeempfindlichkeit von *C. neoformans*-Stämmen, die vom Menschen, Tier und aus Bodenproben isoliert wurden (*58, 59, 135* u. a.), zusammenfassend wiedergegeben. Versuche an Hühnerembryonen erwiesen, daß die Überlebensdauer mit *C. neoformans* infizierter Embryonen beträchtlich vermehrt ist, wenn diese für 8 Tage einer Temperatur von 40° C ausgesetzt werden [Kligman u. Mitarb. (*102*)].

Gegen ein *saures* pH in festen und flüssigen Nährböden sind Cryptococcen in Übereinstimmung mit vielen anderen Hefearten relativ unempfindlich. Sie wachsen ungehemmt in einem pH-Bereich zwischen 5,0 und 7,5, bevorzugen aber offensichtlich Säurewerte um 6,0—5,6. Auch eine zunehmende Säuerung im flüssigen Milieu, wie sie bei der Zuckerassimilation eintritt, übt keine nachweisbare Wachstumshemmung aus.

Die bisher meist fehlgeschlagenen Versuche einer gezielten Behandlung der *C. neoformans*-Infektion mit chemischen und antibiotischen Substanzen gaben zu zahllosen *in vitro*-Untersuchungen Anlaß, in denen die *Hemmbarkeit von*

C. neoformans durch Derivate von *chemischen Verbindungen* geprüft wurde, bei denen antibakterielle oder antimykotische Eigenschaften bekannt oder zu erwarten waren.

Hierbei haben sich neben (*16, 19, 27, 28, 32, 34, 44, 72, 78, 79, 81, 84, 94, 97, 103, 104, 109, 110, 111, 113, 118, 122, 123, 125b, 152, 163, 165, 170, 171, 174, 178, 179, 180, 197*) vor allem KLIGMAN und WEIDMAN (*99*), SCHMIDT u. Mitarb. (*152*) sowie BOCOBO et al. (*17, 43*) durch die Testung einer besonders großen Zahl von Substanzen ausgezeichnet.

Das Vorhaben, die *in vitro*-Befunde der vorstehend aufgeführten Autoren tabellarisch zusammenzufassen, wurde aufgegeben, da ein Vergleich der Untersuchungsergebnisse in den meisten Fällen nicht möglich ist.

Die *Untersuchungsmethoden*, auf die sich die mitgeteilten Hemmwerte stützen, waren ebenso uneinheitlich wie die Bewertungsmaßstäbe: Das Ausgangsmaterial —Nährböden wie Teststämme — war bei praktisch jedem Untersucher ein anderes. Neben der Testung im flüssigen Milieu kamen vielfach auch Gußplattenverfahren zur Anwendung. Die Wirksamkeit der einzelnen Substanzen wurde im Reihenverdünnungstest und im Agar-Diffusionstest festgestellt. Je nach der Löslichkeit wurden die verschiedensten Lösungsmittel benutzt. Die Bestimmung der kleinsten Hemmdosis erfolgte nach einer von 2—10 Tagen schwankenden Beobachtungsdauer bei Temperaturen zwischen 20 und 37° C teilweise durch Kontrolle der Wachstumshemmung, teilweise aber auch durch Feststellung der Wachstumsverminderung gegenüber dem Kontrollwachstum mittels der Trübungsmessung. Die jeweils benutzten Keimeinsaaten variierten beträchtlich. Während in den meisten Berichten die komplette Wachstumshemmung (Fungistase) als Grundlage der Wertbestimmung eines gegebenen Mittels dient, beziehen sich die Hemmwerte von LARSH et al. (*109, 110*) lediglich auf eine 50%ige Wachstumsverminderung, gemessen am Wachstum in *Gewebekulturen*. Es ist verständlich, daß sich die Hemmwerte einzelner Substanzen bei so verschiedenartigen Untersuchungsmethoden um oft eine bis mehrere Größenordnungen unterscheiden.

Soweit es sich beispielsweise um die zur Hemmung von *C. neoformans* nötigen Sulfonamidkonzentrationen handelt, blieben bei den meisten Untersuchungen die in den Nährböden in Form von Pepton enthaltenen *Sulfonamidantagonisten* unberücksichtigt.

Wie eigene Untersuchungen (*163*) in Bestätigung früherer Arbeiten zeigten, sind auch günstige *in vitro*-Hemmwerte — komplette Wachstumshemmung nach 10tägiger Bebrütung bei 37° in Anwesenheit von 1—20 γ/ml Medium — meistens irrelevant, da ein Großteil der *Cryptococcus*-hemmenden Substanzen einen beträchtlichen *Eiweißfehler* aufweist.

Schon ein Zusatz von 1% Albumin, 10% Vollblut, 10—30% Serum oder Ascitesflüssigkeit führt vielfach zu einer erheblichen Wirkungseinbuße [BOCOBO et al. (*17, 43*), KEENEY et al. (*97*), KLIGMAN und WEIDMAN (*99*), LITTMAN und ZIMMERMAN (*111*), SCHMIDT et al. (*152*), SEELIGER (*163*) u. a.].

Umgekehrt müssen aber auch relativ hohe Konzentrationen, die zur Hemmung *in vitro* erforderlich sind, nicht unbedingt eine klinische Wirksamkeit ausschließen. Manche Mittel wie die aromatischen Diamidine scheinen sich im Entzündungsgewebe anzureichern, so daß dort die zur Pilzhemmung *in vivo* nötigen Konzentrationen noch erzielt werden. Bisher ist eine solche gegen *C. neoformans in vivo sicher* wirksame Substanz allerdings noch nicht gefunden worden.

Eine Anzahl von Verbindungen, die nach dem günstigen Ausgang von *in vitro*-Versuchen tierexperimentell geprüft wurde, hat leider die in sie gesetzten Erwartungen in keinem Falle erfüllt [KLIGMAN und WEIDMAN (*99*), LITTMAN und ZIMMERMAN (*111*), KÖNIGSBAUER (*104*) u. v. a.].

Die zur *Abtötung* von *C. neoformans* nötigen Konzentrationen sind in der Regel meist erheblich höher als die zur Fungistase erforderlichen.

Soweit nach den vorliegenden Ergebnissen zu urteilen ist, wirken die gängigen *Desinfektionsmittel*, insbesondere die Phenolverbindungen, in den zur Bakterienabtötung üblichen Konzentrationen und Einwirkungszeiten auch auf

C. neoformans fungizid (*165*). Allerdings scheinen weitere Untersuchungen in dieser Richtung angezeigt. 30% H_2O_2 und 70% Äthylalkohol töten *C. neoformans* erst nach einer Einwirkungsdauer von 15 min sicher ab [SIMON (*171*)]. Ähnlich wirkt 10%iges Natriumhyposulfit [TAGER et al. (*179*)].

Hinsichtlich der Unwirksamkeit alkalischer Substanzen, die zur Anreicherung von Tuberkelbakterien Verwendung finden, sei auf die bereits zitierten Angaben von AJELLO (*1*) verwiesen. Auch Schwefelsäure ist nur bedingt wirksam. Nach STAB (*204*) wurden drei *C. neoformans*-Stämme durch 8 bzw. 12 Vol.-% H_2SO_4 abgetötet, wogegen eine Kultur von *C. albidus* noch nach 30 min lebensfähig war.

Die bisher praktisch erfolglose Suche nach *C. neoformans*-Hemmstoffen, die auch *in vivo* wirksam sind, spiegelt sich in zahlreichen weiteren Versuchen wider, die mit *antibiotischen Substanzen* angestellt wurden.

Übereinstimmend wird festgestellt, daß die nachfolgend aufgeführten *Antibiotica keine* oder nur *unzureichende Hemmwirkungen* auf *C. neoformans* entfalten:

Tabelle 7. *In vitro-Hemmung von C. neoformans-Stämmen durch Antibiotica und Antimycotica*

Mittel	Hemmdosis in γ/ml	Autoren
Acti-dione	0,2	WAKSMAN (*188*)
(Cycloheximid)	0,2	CARTON und LIEBIG (*31*)
	0,24	WHIFFEN c.s. (*191*)
	3,13	BROWN und HAZEN (*20*)
	<10,0	KLIGMAN und WEIDMAN (*99*)
	50,0	WILSON und DURYEA (*196*)
Actinomycin	0,6	REILLY c. s. (*141*)
Allicin	<10,0	KLIGMAN und WEIDMAN (*99*)
Amphotericin A	0,31	LARSH et al. (*110*)
	3,1	GOLD et al. (*77*)
Amphotericin B	0,1—0,4	LARSH at al. (*110*)
(Fungizone)	0,6	GOLD et al. (*77*)
Anisomycin	0,06—0,1	LARSH et al. (*110*)
Antimycoin	1,4	WAKSMAN (*188*)
Ascosin	1,2 Einh.	HICKEY et al. (*88*)
Biformin	20,0	KLIGMAN und WEIDMAN (*99*)
Candicidin	<0,1	LARSH et al. (*110*)
	0,5—1,0	KLIGMAN und LEWIS (*102a*)
Circulin	8,0—15,0	THAYER c. s. (*181*)
Endomycin	10,0	LARSH et al. (*110*)
Eulicin	0,25	LARSH et al. (*110*)
Filipin	0,95	GOTTLIEB (*80*)
Fluvomycin	13,0—20,0	CARVAJAL (*30*)
Fradicin	3,0	WAKSMAN (*188*)
Fungistasin	14,0—21,0	HOBBY et al. (*90*)
Mycostatin	1,0	WAKSMAN (*188*)
(Nystatin,	1,56	HAZEN und BROWN (*20, 86*)
Fungicidin)		
Mycosubtilin	5,0	WALTON und WOODRUFF (*189*)
Mycoticin	6,0—8,0	BURKE et al. (*22*)
Neomycin	25,0	SIMON (*171*)
Nepera 1968	1,0	LARSH et al. (*110*)
Pleurotin	<10,0	KLIGMAN und WEIDMAN (*99*)
Polymyxin	3,0	FLORESTANO und BAHLER (*73*)
	25,0	SIMON (*171*), RATCLIFFE und COOK (*137*)
Rimocidin	5,0—10,0	CARTON und LIEBIG (*32*)
Thiolutin	5,0—25,0	CARTON und LIEBIG (*32*), SENECA et al. (*168*)
Ustilagosäure	10,0—15,0	HASKINS und THORN (*85*)

Actinorubrin (*99*), Amebicid-Lilly 16726 (*116*), Aspergillinsäure (*99*), Bacitracin (*99*), Bacillomycin (*99, 107*), Chloramphenicol (*99, 111, 165, 178*), Citrinin (*99*), Clavicin (*141*), Crepin (*99*), Eumycin (*95*), Gladiolensäure (*99*), Gliotoxin (*99, 118, 141*), Gramicidin (*99*), Kojisäure (*99*), Lavendulin (*99*), Penicillin und Derivate (*99, 111, 152, 165, 166, 178* u. a.), Protoanemonin (*99*), Streptomycin (*99, 111, 152, 165, 166, 178* u. a.), Streptothricin (*99, 141*), Subtilin (*99*), Tetracyclin und Derivate (*99, 111, 152, 165, 178* u. a.), Tyrocidin (*99*), Tyrothricin (*99*), Viridin (*18, 99*).

In Tabelle 7 wurden die *Antibiotica* bzw. *Antimycotica* zusammengestellt, deren *in vitro*-Hemmwirkung auf *C. neoformans* in Mengen unter $25\,\gamma/\mathrm{ml}$ beschrieben wurde. Hinsichtlich der teilweise beträchtlichen Streuung der Hemmwerte sei auf obige Ausführungen über die mangelnde Einheitlichkeit der Untersuchungsmethodik verwiesen.

Wie Tabelle 7 erkennen läßt, handelt es sich bei der überwiegenden Mehrzahl der gegen *C. neoformans* aktiven Substanzen um Antimycotica, die aus Streptomyceten bzw. Bodenaktinomyceten gewonnen wurden. Bei *in vivo*-Versuchen [KLIGMAN und WEIDMAN (*99*), Zusammenfassung siehe bei LITTMAN und ZIMMERMAN (*111*)] erwies sich leider ein beträchtlicher Teil der vorstehenden Stoffe als toxisch oder besaß bei ausreichender Verträglichkeit keine Schutz- oder Heilwirkung gegen die *C. neoformans*-Infektion. —

Die *Stammhaltung und Aufbewahrung* von *Cryptococcus*-Kulturen bereitet keine besonderen Schwierigkeiten. Auf Würze-, SABOURAUD- oder Mycophil-Agar (BBL) lassen sich die Stämme als Schrägkulturen gut konservieren (*111, 165* u.a.). Je nach dem Zustand der Nährböden müssen sie in Abständen von 3—6 Monaten überimpft werden. Die Aufbewahrung soll bei Zimmertemperatur oder besser im Kühlschrank bei $+4^{0}\,\mathrm{C}$ erfolgen. Temperaturen über $30^{0}\,\mathrm{C}$ können die nicht zu *C. neoformans* gehörigen Kulturen leicht irreversibel schädigen. Nach 5jähriger Fortzüchtung gingen von über 50 *Cryptococcus*-Kulturen im Laboratorium lediglich 2 *C. albidus*-Stämme zugrunde, möglicherweise infolge der vorübergehenden Verwendung einer Nährbodencharge, die keine ausreichende Menge von Thiamin enthielt. Die von AJELLO et al. (*3a*) sowie von BUELL und WESTON (*21*) empfohlene Konservierung von Pilzstämmen unter sterilem Mineralöl (Paraffinum liquidum puriss.) hat sich im Laufe einer 3jährigen Anwendung auch bei *Cryptococcus*-Kulturen vorzüglich bewährt (*165*). Die wahrscheinlich beste und wirtschaftlichste Methode stellt jedoch die Gefriertrocknung dar, die sich auch bei der Konservierung von Pilzstämmen zunehmend durchsetzt [RAPER und ALEXANDER (*136*), WICKERHAM und FLICKINGER (*194*) u. a.].

Biochemisches Verhalten

Die Sproßpilze der *Cryptococcus*-Gruppe können nach den heute vorliegenden Kenntnissen als eine *ökologische Einheit* angesehen werden.

Nachdem durch die Untersuchungen von EMMONS (*57, 58, 60, 61*), AJELLO (*2, 3*) sowie ASCHNER und GODINGER (*8*) mittels einer Modifikation der von STEWARD und MEYER (*176*) angewandten Technik zur Isolierung pathogener Pilze aus Erdproben erwiesen ist, daß auch die menschen- und tierpathogene Art *C. neoformans* in Bodenproben verschiedenster Provenienz wechselnd häufig vorkommt, wird man in allen bisher bekannten *Cryptococcus*-Arten *primäre Bodenbewohner* vermuten dürfen. Das Vorkommen pathogener und apathogener *Cryptococcus*-Stämme im menschlichen und tierischen Körper wie auf seiner Oberfläche stellt wahrscheinlich nur einen Sonderfall dar. In dieser Hinsicht zeigen die *Cryptococcus*-Arten eine gewisse ökologische Verwandtschaft zu der auch in manchen anderen Punkten ähnlichen, aber askosporogenen Gattung *Lipomyces*, deren Vertreter ebenfalls auf der menschlichen Haut nachgewiesen werden konnten [CONNELL u. Mitarb. (*39, 40*)].

Die biochemische Einheitlichkeit der *C. neoformans*-Stämme als medizinisch wichtigsten Vertretern der *Cryptococcus*-Gruppe wurde durch vergleichende Untersuchungen saprophytärer, menschlicher und tierischer Kulturen hinreichend gesichert [Emmons (*58, 59*), Carter und Young (*29*), Benham (*10, 12, 13*), Pounden et al. (*135*), Littman und Zimmerman (*111*), Ajello (*3*), Verf. (*154*) u. a.], so daß sich eine getrennte Betrachtung der biochemischen Verhaltensweise nach der Herkunft der Stämme erübrigt.

Die von Benham (*10*), Diddens und Lodder (*46*), Lodder (*14*) u. a. in den dreißiger Jahren erarbeiteten Grundmerkmale, die zur biochemischen Abgrenzung der Cryptococcen von anderen Sproßpilzen dienten, wurden inzwischen nach ausgedehnten Untersuchungen von einer schärfer umrissenen biochemischen Definition der Gattung *Cryptococcus* abgelöst, über die in den Darstellungen von Lodder und Kreger-van Rij (*117*), Benham (*12, 13*) sowie Littman und Zimmerman (*111*) keine Meinungsverschiedenheit mehr herrscht. Die differentialdiagnostischen Möglichkeiten einer Abgrenzung der *Cryptococcus*-Gruppe von anderen Hefen mit biochemischen Mitteln wurden in jüngster Zeit durch einige Verfahren bereichert, die zusätzlich auch interessante Rückschlüsse auf mögliche Verwandtschaftsbeziehungen zu physiologisch sich ähnlich verhaltenden, taxonomisch aber anders eingeordneten Pilzarten eröffnen.

Tabelle 8. *Assimilationsvermögen von 27 Cryptococcus-Stämmen, geprüft nach* Wickerhams *Methode*

Art	Zahl der Stämme	Dextrose	Maltose	Saccharose	Lactose	Galaktose
C. neoformans	12	+	+	+	−	+
C. diffluens .	10	+	+	+	−	+
C. albidus . .	3	+	+	+	±	+
C. luteolus .	1	+	+	+	−	+
C. laurentii .	1	+	+	+	+	+
C. terreus . .	1	+	+	−	+	+

Zuckerspaltung: Übereinstimmend wurde festgestellt, daß die Angehörigen des Genus *Cryptococcus* nicht zur Gasbildung aus Zuckern befähigt sind. Im mykologischen Sprachgebrauch wird deshalb die *Cryptococcus*-Gruppe als *nichtfermentativ* bezeichnet (*12, 13, 38, 111, 117, 172, 195*), wobei hervorzuheben ist, daß unter dem Begriff Fermentation in der Mykologie — im Gegensatz zur medizinischen Bakteriologie — stets die Zuckerspaltung unter Bildung von Gas zu verstehen ist.

Demgegenüber sind die *Cryptococcus*-Arten in unterschiedlicher Weise zur *Zuckerassimilation* imstande, in deren Verlauf durch oxydative Kohlenhydratspaltung nachweisbare Säuremengen freigesetzt werden.

Allerdings wird bei Prüfung der Fähigkeit zur Kohlenhydratassimilation der Säurebildung meist nur geringe Beachtung geschenkt. Im Vordergrund steht die von Beijerinck (*15*) schon 1889 inaugurierte Methode des Agar-Auxanogramms, die neuerdings zunehmend durch das von Wickerham c. s. (*193, 195*) ausgearbeitete Prüfungsverfahren im flüssigen Substrat verdrängt wird. In beiden Verfahren erfolgt die Beurteilung des Assimilationsvermögens durch Beobachtung des Pilzwachstums in Anwesenheit bestimmter Zucker in einem sonst kohlenhydratfreien Medium. Bei richtiger Untersuchungstechnik führten beide Methoden bei der Testung zahlreicher *Cryptococcus*-Stämme zu weitgehend gleichartigen Ergebnissen.

Das Zuckerassimilationsvermögen ist eine für jede *Cryptococcus*-Art charakteristische Eigenschaft.

In weitgehender Übereinstimmung mit den Angaben von Lodder und Kreger-van Rij (*117*) wurden bei der Prüfung von 27 *Cryptococcus*-Stämmen, die

Angehörige aller anerkannten Arten einschließen, die in Tabelle 8 niedergelegten Befunde erhoben.

Zu praktisch gleichen Ergebnissen gelangten auch EMMONS (*61*) und BENHAM (*12, 13*). Letztere vermißte jedoch die für *C. albidus* charakteristische Fähigkeit zur Lactose-Assimilation, die auch in den eigenen Versuchen nur geringfügig eintrat.

Wie aus Tabelle 8 ersichtlich, ist mittels der Zuckerassimilation eine Trennung pathogener und apathogener *Cryptococcus*-Arten nur ausnahmsweise, z. B. bei *C. laurentii*, gewährleistet. Wie wichtig andererseits die Feststellung des Assimilationsvermögens ist, wird durch eine Mitteilung AJELLOS (*3*) belegt, wonach die Artdiagnose *C. laurentii* bei einem aus einer Erdprobe in Nigeria isolierten, stark bekapselten Stamm, der auch bei 37° C noch schwach wuchs und deshalb zunächst für *C. neoformans* gehalten wurde, erst auf Grund einer deutlichen Lactose-Assimilation gestellt werden konnte.

C. laurentii zeigt in seinem Zuckerassimilationsvermögen völlige Übereinstimmung mit einigen weiteren gaslosen Hefe-Arten, die zur Zeit bei anderen Gattungen eingeordnet werden: *Torulopsis aeria* (1 Stamm)[1], *Candida humicola* (2 Stämme)[2], *Candida curvata* (3 Stämme), *Trichosporon cutaneum* (4 Stämme) und BENHAMS *Cryptococcus mucorugosus* (*12, 13*). Hierbei scheint es sich nicht um einen Zufall zu handeln; denn — wie im weiteren ausgeführt werden wird — lassen auch andere biochemische und serologische Reaktionen eine auffällige Verwandtschaft erkennen. Mit den lactosenegativen *Cryptococcus*-Arten stimmt hingegen *Lipomyces starkeyi* und — wie Untersuchungen von CONNELL (*40*) und Verf. (*165*) erwiesen — auch *L. lipoferus* assimilatorisch überein [vgl. aber hierzu (*144*)].

Äthanol-Verwertung: LODDER und KREGER-VAN RIJ (*117*) benutzten als weiteres biochemisches Klassifizierungsmerkmal die Fähigkeit zur Verwertung von Äthylalkohol als einziger Kohlenhydratquelle im Medium. Soweit es sich hierbei um die Arten-Trennung der Cryptococcen handelt, sind die Ergebnisse bei den Species *C. diffluens*, *C. albidus* und *C. laurentii* variabel, vorwiegend allerdings negativ, während *C. neoformans* und *C. luteolus* auf dem Äthanolmedium üppig gedeihen.

Citratassimilation: Die Prüfung von *27 Cryptococcus*-Stämmen, in Anwesenheit von Natriumcitrat als einziger C-Quelle auf SIMMONS Citratagar mit einem Zusatz von Thiamin zu wachsen, ergab bei einer 10tägigen Beobachtungsdauer stets negative Befunde (*165*).

Arbutin-Spaltung: Die in Delft geübte Methode des Nachweises der Arbutin-Spaltung zeigte im Genus *Cryptococcus* ziemlich variables Verhalten (*117*).

Stärke-Synthese: Einen wesentlichen methodischen Fortschritt in der Hefe-Differenzierung stellt die von MAGER und ASCHNER (*119, 120*) sowie ASCHNER u. Mitarb. (*5, 6*) beschriebene Fähigkeit *aller Cryptococcus*-Arten dar, auf einem synthetischen Dextrose-Thiamin-Agar, p_H 5, Stärke zu synthetisieren. Die taxonomische und diagnostische Bedeutung dieser Eigenschaft wurde von allen Nachuntersuchern bestätigt (*12, 13, 111, 117, 165* u. a. m.).

Die Methodik ist einfach: 10 Tage alte, bei Zimmertemperatur auf ASCHNERs Medium gewachsene Pilzkulturen werden auf der Schrägfläche mit einigen Tropfen Jod-Jodkali-

---

[1] Zahl der vom Verfasser geprüften Stämme.
[2] Synonym mit *Candida suaveolens* (6).

Lösung versetzt. Bei vorhandener Stärkesynthese tritt rasch eine Blaufärbung des Pilzrasens ein. Bei negativem Ergebnis färbt sich die Kultur mit der Lugolschen Lösung bräunlichgelb.

Der besondere Wert dieser Reaktion beruht darauf, daß sie nur noch bei wenigen anderen Sproßpilzarten positiv ausfällt, so bei *Candida curvata*, *Candida humicola* und *Trichosporon cutaneum* bzw. *Trichosporon infestans* (6, 165). Dies steht im Einklang mit dem Ergebnis eigener Untersuchungen an den auf S. 43 bereits genannten Pilzstämmen.

Nach Skinner und Connell (173) ist auch *Torulopsis aeria* wegen der nachweisbaren Fähigkeit zur Stärkeproduktion als *Cryptococcus*-Art anzusehen, zumal Arten der *Torulopsis*-Gruppe nicht zur Stärkebildung in der Lage sein sollen (117). Bei der Prüfung eines aus Delft erhaltenen *T. aeria*-Stammes zeigte sich nach 10tägiger Bebrütung nur eine schwache Jodstärke-Reaktion (165).

Darüber hinaus stellten Connell, Skinner und Hurd (40) auch die Synthese einer stärkeähnlichen Verbindung bei den von ihnen geprüften *Lipomyces*-Kulturen fest, allerdings erst nach einer Beobachtungsdauer von 4 Wochen und bei Verwendung des Wickerhamschen Nährbodens. In Übereinstimmung mit Lodder und Kreger-van Rij (117) beobachtete Verf. bei 10 *Lipomyces*-Stämmen niemals Stärkesynthese nach 10—14tägiger Bebrütung auf Aschners Medium. — Bei Anwendung der von Aschner c. s. beschriebenen Methodik lassen sich somit *Cryptococcus*- und *Lipomyces*-Stämme im Hinblick auf ihr Stärkebildungsvermögen sicher unterscheiden.

Das schließt eine phylogenetische Verwandtschaft nicht aus, die möglicherweise die Gattungen *Cryptococcus* mit *Lipomyces* einerseits und *Lipomyces* mit *Taphrinia* andererseits verbindet [Wickerham (195)].

Wichtig für die Klärung taxonomischer Fragen erscheint auf jeden Fall die Feststellung, daß sich die Fähigkeit zur Stärke-Synthese übereinstimmend bei den Sproßpilzarten findet, die auch im Hinblick auf ihr Kohlenhydrat-Assimilationsvermögen und die weiter unten noch zu besprechende Harnstoffhydrolyse viele Gemeinsamkeiten zeigen. Offenbar ist es so, daß innerhalb dieser großen Gruppe einzelne Arten konstant und rasch innerhalb weniger Tage Stärke bzw. stärkeähnliche Substanzen bilden, während andere Arten (z. B. *Lipomyces*) hierzu erheblich längere Zeit benötigen.

Nitratassimilation und Nitritbildung: Analog zur Kohlenhydrat-Assimilation wird schon seit Jahrzehnten bei der Differenzierung von Hefearten stets geprüft, ob sie in Anwesenheit von Kaliumnitrat als einziger Stickstoffquelle wachsen können.

Beijerincks Gußplatten-Methode in einem Stickstoff-freien, aber zuckerhaltigen Substrat (mit Pepton zur Wachstums-Kontrolle) diente lange Zeit als Standard-Prüfverfahren.

Gleich gut geeignet und manchmal sogar überlegen ist die Prüfung der Nitratassimilation in Wickerhams Spezialmedium (192), dem als Stickstoffquelle 0,078% KNO_3 zugefügt wird. Zur Kontrolle wird das gleiche Substrat mit 1% Ammoniumsulfat oder 1% Pepton benutzt. Bei den submers gelegentlich relativ schlecht wachsenden Cryptococcen ist dieses Verfahren der Gußplatten-Methode vorzuziehen.

Besonders einfach ist ein von Benham (12, 13) inauguriertes Platten-Ausstrichverfahren. Als Nährboden wird Wickerhams Hefe-Kohlenhydrat-Basalmedium[1] mit einem Zusatz von KNO_3 und von 2% gereinigtem Agar verwandt. Die Reinigung des Agars dient der Entfernung N-haltiger Bestandteile.

---

[1] Als Trockenpräparat ebenso wie Wickerhams Hefe-Stickstoff-Basalmedium (Difco) im Handel erhältlich.

Zu eigenen Versuchen diente ein ähnliches, in Anlehnung an die Rezeptur WICKERHAMs hergestelltes, synthetisches Substrat, in dem die Vitaminlösung durch „Protovit" (Hoffmann-La Roche) ersetzt wurde.

Grundsubstrat

Dextrose	10,0 g	Inositol	0,3 g
KNO$_3$	0,78 g	Para-Aminobenzoesäure	0,05 g
KH$_2$PO$_4$	1,0 g	Vitaminlösung „Protovit"	2,5 ml
MgSO$_4$ · 7 H$_2$O	0,5 g	Agar	15,0 g
CaCl$_2$ · 2 H$_2$O	0,1 g	Aqua dest.	1000,0 ml
NaCl	0,1 g		

pH 7,0

Herstellung. Die Bestandteile werden in Aqua dest. gelöst, ausgenommen die Vitaminlösung. Nach Zugabe des Agars wird das Substrat 15 min auf 110° C erhitzt. Die Vitaminlösung wird dem kochendheißen Medium zugesetzt.

Das fertige Substrat wird in Petri-Schalen ausgegossen und nach dem Erstarren 18 Std bei 37° C zur Prüfung der Sterilität bebrütet.

Nicht benötigte Mengen werden in Flaschen abgefüllt und vor Gebrauch verflüssigt.

Verwendung. Eine Normalöse einer 24 Std alten Agarschrägkultur wird in 1 ml Aqua dest. aufgeschwemmt. Eine Öse dieser Aufschwemmung wird dann auf einen Sektor der Agaroberfläche in üblicher Weise ausgestrichen. Bebrütung erfolgt bei 20—30° C.

Beurteilung. Die Fähigkeit zur Nitratassimilation wird durch *kräftiges* Wachstum angezeigt. Es empfiehlt sich, stets eine Kontrollplatte mit einem bekannt positiven und negativen Stamm mitlaufen zu lassen.

In Übereinstimmung mit BENHAM (*12, 13*) sowie den Angaben von LITTMAN und ZIMMERMAN (*117*) zeigten alle nitratassimilierenden *Cryptococcus-Stämme* auf obigem Medium üppiges Wachstum (vgl. Abb. 4 und 5).

Kümmerliches Wachstum ist allerdings gelegentlich auch bei Arten zu beobachten, die in den Standard-Prüfverfahren nicht zur Nitratassimilation befähigt sind (*12, 13*). Dies beruht teils auf geringfügigen N-Beimengungen im Agar, möglicherweise auch in der Vitaminlösung, vor allem aber in der Überführung von Nährstoffen aus den zur Anzüchtung bzw. Vorkultur benutzten Nährböden bei Beimpfung der Plattenoberfläche.

Die Plattenausstrichmethode erlaubt innerhalb von 3 bis höchstens 5 Tagen bei Bebrütung von 20—30° C eine einwandfreie Beurteilung des Nitratassimilationsvermögens.

Nach NICKERSON (*132*) ist die Assimilation von KNO$_3$ bei Hefen stets mit der Fähigkeit zur *Nitritbildung* gekoppelt, so daß der Assimilationstest durch den Nitritnachweis ergänzt oder ersetzt werden kann. LODDER und KREGER-VAN RIJ (*117*) haben dieses Verfahren nicht routinemäßig angewandt, da die Ergebnisse nicht konstant genug gewesen seien (persönliche Mitteilung). — Betropft man die Oberfläche des vorstehend beschriebenen Nitratmediums mit Griesschem Reagens, so kommt es bei kräftigem Wachstum stets zu intensiver Rotfärbung durch das freigesetzte Nitrit [BENHAM (*13*), Verf. (*165*)]. Die Nitratreduktion läßt sich bei Sproßpilzen in gleicher Weise wie bei Bakterien auch auf 1%igem Nitrat-Pepton-Agar und in Nitratbouillon durchführen. Wie aus Tabelle 9 hervorgeht, laufen die Ergebnisse denen der klassischen Assimilationsproben weitgehend parallel.

Aus bisher noch unbekannten Gründen besteht aber keine völlige Übereinstimmung. Bei einzelnen nitratassimilierenden *C. diffluens*-Stämmen sowie bei *C. albidus* war die Nitritbildung relativ schwach, obwohl die Untersuchung über 10 Tage ausgedehnt und täglich kontrolliert wurde. Daß dies nicht durch den Verbrauch allen Nitrats im Medium bewirkt war, wurde durch entsprechende Kontrollen mit Zinkstaub sicher ausgeschlossen.

Abb. 4. Nitratassimilation von *Cryptococcus*-Stämmen bei 22⁰ C (Ausstrich-Verfahren)

Abb. 5. Nitratassimilation von *Cryptococcus*-Stämmen bei 30⁰ C (Ausstrich-Verfahren)

Ergänzend zu Vorstehendem sei noch erwähnt, daß die Arten *Candida curvata, Candida humicola, Trichosporon cutaneum, Lipomyces lipoferus* und *Lipomyces starkeyi* — in Übereinstimmung mit (*117*) — auch bei der Nitritprobe stets negativ reagierten. Nach Benham (*13*) ist *C. mucorugosus* zur nitritpositiven Gruppe zu rechnen. Das gleiche gilt auch für *Torulopsis aeria* (s. Tabelle 9).

Tabelle 9. *Vergleichende Prüfung der Nitratassimilation und Nitritbildung bei 28 Cryptococcus-Stämmen*

Art	Zahl der Stämme	Assimilationsverfahren			Nitritbildung	
		Auxano-graphie	flüssiges Substrat (WICKERHAM)	Platten-ausstrich verfahren	flüssiges Medium	fester Nährboden
C. neoformans	12	—	—	—	—	—
C. diffluens .	10	++	++	+++	++	++/+++
C. albidus . .	3	++	++	+++	++	+/+++
C. terreus . .	1	++	++	+++	+	++
C. luteolus . .	1	—	—	—	—	—
C. laurentii. .	1	—	—	—	—	—
Torulopsis aeria	1	++	++	++	+	+

Die vorliegenden Ergebnisse berechtigen zu folgenden Schlüssen: Innerhalb des Genus *Cryptococcus* und einiger, mit ihm möglicherweise verwandter Arten erlaubt die Fähigkeit zur Nitratassimilation bzw. zur Nitritbildung die Aufgliederung in 2 Untergruppen (vgl. Tabelle 9) (*13, 111*):

Untergruppe A. Die zu dieser Untergruppe gehörenden Arten *C. neoformans*, *C. luteolus* und *C. laurentii* sind weder zur Nitratassimilation noch zur Nitritbildung befähigt.

Untergruppe B. Sie schließt die nitratassimilierenden bzw. nitritbildenden Species *C. diffluens*, *C. albidus*, *C. terreus* und *C. mucorugosus* ein. Offenbar gehört auch *T. aeria* zur gleichen Untergruppe.

Da die Stämme der *Untergruppe B* apathogen sind, stellt somit der Nachweis der Nitratassimilation bzw. der Nitritbildung ein wichtiges Merkmal zur raschen Erkennung pathogenetisch bedeutungsloser Stämme dar. Der Wert gerade dieser Proben hat sich beim Nachweis von 5 *C. diffluens*-Kulturen aus Liquores im Laboratorium des Verf. eindrucksvoll bestätigt.

Hinsichtlich der optimalen Untersuchungsmethodik sei jedoch bemerkt, daß die Nitritprobe trotz aller Vorteile nur bei positivem Ausfall diagnostische Bedeutung besitzt. Es empfiehlt sich, negative Befunde durch die herkömmlichen Assimilationsteste zu erhärten.

Die auffallende Beobachtung, daß einige andere Hefearten, die als nicht nitratassimilierend beschrieben wurden, beim Nitritnachweis positiv reagierten, mahnt zur Vorsicht und ist Gegenstand weiterer Untersuchungen.

Gelatineverflüssigung: Nach Cox und TOLHURST (*42*) verflüssigt *C. neoformans* innerhalb von 6—9 Wochen Gelatine geringfügig. Weitere Untersuchungen, insbesondere unter Einschluß hochempfindlicher Testverfahren, stehen noch aus und versprechen interessante Aufschlüsse.

Harnstoffhydrolyse: Die Harnstoff-Assimilation wurde von zahlreichen Untersuchern [Zusammenfassung s. WICKERHAM (*192*)] auf ihre Verwendbarkeit bei der Klassifizierung von Hefen geprüft, ohne daß dabei die unterschiedliche Fähigkeit der Sproßpilze zur aktiven Harnstoffhydrolyse differentialdiagnostisch ausgewertet worden wäre.

Bei der Prüfung der Brauchbarkeit bakteriologischer Untersuchungsmethoden für die Sproßpilzdiagnostik beobachtete SEELIGER (1953), daß eine Anzahl nichtfermentierender Hefen — darunter alle vorhandenen *Cryptococcus*- und

Rhodotorula-Stämme — im Gegensatz zu zahlreichen fermentierenden Hefearten auf Christensens Harnstoffagar (*33*) zur raschen Harnstoffhydrolyse befähigt sind. Nach Diskussion der Ergebnisse mit amerikanischen Mykologen wurden die Untersuchungen im Herbst 1955 am Communicable Disease Center, Chamblee, Georgia, durch Prüfung von mehr als 30 weiteren *Cryptococcus neoformans*-Stämmen ergänzt. Es handelte sich dabei um Kulturen, die in den USA und anderen Ländern aus menschlichem Untersuchungsmaterial, von Tieren und aus Bodenproben isoliert worden waren. Einzelheiten über diese Stämme finden sich in einem Bericht von Ajello (*3*). Auf Grund der eindeutigen Befunde an 214 Hefekulturen aus 39 Arten berichtete 1956 Seeliger (*159, 160*) über die differentialdiagnostische Bedeutung der Harnstoffprobe bei der Erkennung von *Cryptococcus*-Stämmen. Übereinstimmend erwähnen im gleichen Jahre auch Littman und Zimmerman (*111*) — allerdings ohne Angabe der Untersuchungsmethodik —, daß 9 *Cryptococcus*-Stämme die Eigenschaft der Harnstoffhydrolyse gezeigt hätten.

Tabelle 10. *Häufigkeit der Urease-Bildung bei Hefen und hefeähnlichen Pilzen auf* Christensens *Harnstoff-Agar*

Artzugehörigkeit	Anzahl der geprüften Arten	Kräftige Urease-Bildung in 4 Tagen	Keine oder minimale Urease-Bildung in 4 Tagen
Askosporogene Hefe-Arten[1] . .	29	1[2]	28[3]
Asporogene Hefe-Arten[1] . . .	54	23	31
Ballistosporogene Hefe-Arten[1].	12	12	—
Geotrichum-Gruppe.	4	—	4
Pullularia-Gruppe	3	3	—
Sporotrichon-Gruppe	2	2	—

[1] Bestimmt nach Lodder und Kreger-van Rij (*117*).

[2] *Schizosaccharomyces pombe.*

[3] Einschließlich *Lipomyces lipoferus* und *Lipomyces starkeyi* (vgl. hierzu Text S. 50).

Tabelle 11. *Häufigkeit der Urease-Bildung bei den asporogenen Arten der Familie Cryptococcaceae auf* Christensens *Harnstoff-Agar*

Unterfamilie bzw. Genus	Anzahl der geprüften Arten	Kräftige Urease-Bildung in 4 Tagen	Keine oder minimale Urease-Bildung in 4 Tagen
Cryptococcoideae:			
Cryptococcus	6 +2 Var.	8	—
Torulopsis	9	1	8[1]
Brettanomyces	1	—	1
Candida	21	2	19[2]
Kloeckera	1	—	1
Trichosporoideae	7	3	4
Rhodotoruloideae	7	7	—

[1] Schwach positiv *T. famata.*

[2] Schwach und unregelmäßig positiv *C. lipolytica* und *C. krusei.*

In Erweiterung des Vorstehenden wurden vom Verf. im September 1955 im Laboratorium von Dr. E. E. Evans, Birmingham, Alabama, zusätzlich 72 *C. neoformans*-Stämme und im Bonner Institut bis Ende 1957 weitere 12 Kulturen der *Cryptococcus*-Gruppe sowie zahlreiche andere Pilzarten auf das Vorhandensein einer Urease überprüft. Die Ergebnisse sind zusammenfassend in Tabelle 10 und Tabelle 11 wiedergegeben und stehen weitgehend im Einklang mit den 1957 von Littman (*112*) publizierten Befunden.

Bevor auf die Ergebnisse und einige scheinbare Divergenzen in den vorstehend zitierten Arbeiten näher eingegangen wird, seien die *Untersuchungsmethoden* kurz erläutert.

a) Harnstoffagar nach CHRISTENSEN (*33*) (2% Urea, 0,1% Pepton, 0,1% Dextrose, 0,5% NaCl, 0,2% KH_2PO_4 mit Phenolrot als Indicator, pH 6,8) wird auf der Schrägfläche mit reichlich Kulturmaterial beimpft und bei 30 und 37⁰ C bebrütet. Nach 2, 4, 6, 24, 72 und 96 Std Bebrütung wird die Harnstoffhydrolyse am Maß der Alkalisierung der Schrägfläche und des übrigen Mediums beurteilt. Aktive Urease-Bildung ist gekennzeichnet durch intensive Rotfärbung, oft schon nach 2—4 Std, spätestens nach 4 Tagen. — Zur Kontrolle dient das gleiche Medium ohne Harnstoff.

b) Harnstoffagar nach LITTMAN (*112*) (1% Urea, 0,1% Thioton, 0,5% Dextrose, 0,5% NaCl, 0,3% KH_2PO_4, 0,3% Na_2HPO_4, 1,5% Agar, Phenolrot 0,012 g/l, pH 7,0) wird in gleicher Weise wie CHRISTENSENs Medium beimpft und 60 Tage bei 20⁰ C bebrütet. Ablesung täglich wie unter a). Als Kontrolle dient das Grundsubstrat ohne Harnstoff.

Der wesentliche Unterschied zwischen den beiden Medien liegt in der höheren Zuckerkonzentration in LITTMANs Urea-Agar, die nach Angaben des Autors für das Harnstoffspaltungsvermögen einzelner Pilzarten vorteilhaft sein soll. Dem gleichen Zweck soll die Verwendung von Thioton anstelle von Pepton dienen. Soweit es sich um die Prüfung von Sproßpilzen handelt, haben wir in vergleichenden Untersuchungen (s. Tabelle 12) keine wesentlichen Unterschiede zwischen den beiden Substraten nachweisen können, so daß auch für mykologische Zwecke kein Anlaß besteht, auf das bewährte Medium von CHRISTENSEN zu verzichten.

c) Urease-Schnelltest. Er wird mit einer 5%igen Lösung von Harnstoff in Aqua dest. durchgeführt, die in Mengen von 3 ml pro Röhrchen abgefüllt wird. Das pH soll 6,8 betragen; als Indicator dient Phenolrot. Als Kontrolle läuft stets ein Phenolrot-Wasserröhrchen mit.

Die Testflüssigkeiten werden unmittelbar vor Gebrauch hergestellt, *nicht* erhitzt und nach dem Abfüllen mit einigen Ösen einer aktiv wachsenden, 24—48 Std alten Kultur beimpft. Das Material ist gut zu verreiben, damit eine möglichst homogene Suspension entsteht. Die Bebrütung erfolgt — je nach der Art — bei 30 oder 37⁰ C. Die Ablesung kann stündlich vorgenommen werden, da bei aktiven Urease-Bildnern stets innerhalb von 2—6 Std kräftige Harnstoffhydrolyse einsetzt.

Diese wird durch kräftige Rötung der Harnstofflösung bei unveränderter Gelbfärbung des Kontrollröhrchens, das ebenfalls mit dem zu prüfenden Stamm beimpft wurde, angezeigt.

Das freigesetzte Ammoniak wird außerdem mit NESSLERs Reagenz in Filterpapierstreifen nachgewiesen, die im Röhrchen über dem flüssigen Substrat angebracht werden. Bei positivem Ausfall tritt eine deutliche Gelbbraunfärbung auf.

Die Ergebnisse vergleichender Untersuchungen an repräsentativen Stämmen der *Cryptococcus*- und *Lipomyces*-Gruppe sind bei Anwendung der 3 vorstehend geschilderten Methoden in Tabelle 12 gegenübergestellt.

Tabelle 12. *Vergleich verschiedener Untersuchungsmethoden bei der Urease-Bildung von Cryptococcus- und Lipomyces-Stämmen*

Art	CHRISTENSENs Harnstoff-Agar	LITTMANs Harnstoff-Agar	Urease-Schnelltest		
			Alkalibildung	NESSLERs Reagenz	Leer-kon-trolle
C. neoformans (6)[1]	+++ 1–2 Tg	+++ 1–2 Tg	+++ 2 Std	+++ 2–4 Std	—
C. diffluens (4)	+++ 2–4 (3)	+++ 2–4	+++ 2 Std	+++ 2 Std	—
	++ 4–5 (1)	++ 4–5	++ 6 Std	++ 6 Std	—
C. albidus (3)	+++ 2–4	+++ 2–4	+++ 2 Std	+++ 2 Std	—
C. laurentii (1)	++ 4	++ 4	++ 2 Std	++ 2 Std	—
C. luteolus (1)	++ 4	++ 4	++ 2 Std	++ 2 Std	—
C. terreus (1)	+++ 4	+++ 4	+++ 2 Std	++ 2 Std	—
L. lipoferus (2)	+ 5	+ 5	— 2 Tg	— 2 Tg	—
L. starkeyi (10)	+ 5	++ 5	+ 2 Tg (4) / — 2 Tg (6)	— 2 Tg	—

[1] In Klammern Zahl der geprüften bzw. reagierenden Stämme.

Die Brauchbarkeit der Harnstoffprobe als diagnostisches Hilfsmittel bei der Bestimmung von Sproßpilzen ist nunmehr durch die Kontrolle mehrerer hundert einwandfrei klassifizierter Hefen und hefeähnlicher Kulturen hinreichend gesichert. Das Merkmal der Urease-Bildung ist sicherlich art-, vielleicht sogar gattungsspezifisch. Anhand von 120 *C. neoformans*-Kulturen aus allen Teilen der Erde und verschiedenster Provenienz sowie Vertretern aller übrigen *Cryptococcus*-Arten ist erwiesen, daß die Harnstoffhydrolyse zu den *charakteristischen Eigenschaften* eines jeden *Cryptococcus*-Stammes gehört. Sie ist auf CHRISTENSENs wie LITTMANs Medium innerhalb von 1 bis maximal 4 Tagen deutlich nachweisbar und im Schnelltest oft schon nach 2 Std zu erkennen.

Lediglich eine als *Torulopsis rotundata* (REDAELLI) bezeichnete Kultur, die von holländischer Seite als zu *C. diffluens* gehörig angesehen wird (*117*), gab nur eine schwach positive Reaktion und bildete im Schnelltest erst nach 6 Std Ammoniak.

Demgegenüber sind die nichtfermentierenden Arten der Genera *Debaryomyces* und *Torulopsis* — mit Ausnahme der wahrscheinlich zur *Cryptococcus*-Gruppe gehörenden *T. aeria*, die positive Urease-Reaktion zeigte — bei viertägiger Beobachtung harnstoffnegativ.

Von besonderem Interesse sind die Befunde im Hinblick auf eine mögliche Abgrenzung der sich in mancher Hinsicht ähnelnden Genera *Cryptococcus* und *Lipomyces*.

SEELIGER (*159*) hatte bei der Prüfung eines Stammes von *L. lipoferus* bei viertägiger Beobachtungsdauer keine Anzeichen von Urease-Bildung festgestellt. LITTMAN (*112*) fand hingegen bei einem *L. lipoferus*- und 4 *L. starkeyi*-Stämmen nach 5 Tagen eine schwache bis deutliche Harnstoffhydrolyse, angezeigt durch Rötung des Mediums. Dem entsprechen auch die Ergebnisse der in Tabelle 12 niedergelegten Untersuchungen an insgesamt 12 *Lipomyces*-Kulturen. Diese reagierten innerhalb von 4 Tagen auf den beiden Agar-Medien einwandfrei negativ und zeigten erst nach 5 Tagen eine beginnende Rötung, die jedoch niemals die Intensität und das Ausmaß der Alkalisierung bei *Cryptococcus*-Stämmen erreichte. Im Schnelltest trat bei den *Lipomyces*-Kulturen nur zum Teil eine schwache Alkalibildung — oft erst nach einigen Tagen — auf, während der Ammoniak-Nachweis mittels der Filterpapiermethode — im Gegensatz zu den *Cryptococcus*-Kulturen und anderen starken Urease-Bildnern — stets mißlang.

Es hat demnach den Anschein, daß bei Einhaltung der von SEELIGER (*159, 160*) benutzten Untersuchungsmethodik eine Trennung der beiden in Rede stehenden Arten auch mittels der Harnstoffhydrolyse möglich ist. Andererseits unterliegt es keinem Zweifel, daß die *Lipomyces*-Stämme unter bestimmten Bedingungen Harnstoff hydrolysieren, was mit der wiederholt diskutierten Möglichkeit einer stammesgeschichtlichen Verwandtschaft vereinbar wäre. Offenbar ist diese Eigenschaft bei den *Lipomyces*-Arten an zusätzliche Wachstumsfaktoren gekoppelt, die von Cryptococcen nicht benötigt werden.

Nur wenige weitere, nichtfermentierende, weiß wachsende Hefearten wurden Urease-positiv gefunden, darunter die schon mehrfach diskutierten Arten *Candida humicola*, *Candida curvata*, *Trichosporon cutaneum* und *Trichosporon infestans*, deren Verwandtschaft zur *Cryptococcus*-Gruppe hierdurch eine weitere Stütze erfährt; ferner aber auch *Trichosporon pullulans* und *Bullera alba*. Schwächer positiv reagierte *Candida lipolytica*. Geringfügige Alkalisierung zeigten auch noch einzelne nicht oder nur schwach fermentierende Kahmhefen (*165*).

Demgegenüber sind alle bisher kontrollierten roten Hefen der Gattungen *Rhodotorula* und *Sporobolomyces* starke Urease-Bildner, desgleichen ein Großteil der *Dematium*-Arten sowie die menschen- und tierpathogenen dimorphen Pilze *Histoplasma capsulatum, Blastomyces dermatitidis, Blastomyces brasiliensis* und *Sporotrichon schenckii (112, 165)*.

Möglicherweise läßt die Fähigkeit zur Harnstoffhydrolyse Rückschlüsse auf den *normalen Standort* der ureasebildenden Pilze zu. Denn stark positive Reaktionen finden sich vornehmlich bei den Arten, die saprophytisch im *Boden*, auf *Pflanzen* und auf *Abfällen* vegetieren. Nachdem bestimmte pathogene Pilze — wie *Cryptococcus neoformans* und *Histoplasma capsulatum* — mit zunehmender Häufigkeit in Bodenproben nachgewiesen wurden, die mit Tierexkrementen angereichert sind *(2, 58, 60, 61* u. a.), ist anzunehmen, daß der Harnstoffreichtum solcher Böden — z. B. in Hühnerhöfen, unter Taubenschlägen usw. — das saprophytäre Wachstum dieser Pilze begünstigt, ebenso wie hohe Zuckerkonzentrationen Voraussetzung für das Gedeihen anderer Hefearten sind.

Da zahlreiche andere Pflanzen- und Bodenbewohner aus dem Reich der Pilze, z. B. die *Cladosporium-*, *Penicillium-*, *Aspergillus-*, *Nocardia-* und manche *Streptomyces-*Arten ebenfalls aktiv Harnstoff hydrolysieren, ist es denkbar, daß es sich hierbei um ein *kennzeichnendes Merkmal der primären Pflanzen- und Bodenbewohner* handelt.

Die diagnostische Bedeutung der Harnstoffhydrolyse bei der Sproßpilz-Differenzierung wird durch die nicht ganz seltenen schwach positiven Reaktionen nicht geschmälert. Sie ist von entscheidendem Wert, wenn nur stark positive Reaktionen innerhalb von 4 Tagen gewertet werden. Das gleiche Prinzip gilt auch in anderen Bereichen der Mikrobiologie, wo sich mit erhöhter Empfindlichkeit der biochemischen Untersuchungsmethoden deren diagnostischer Wert im gleichen Maße vermindert.

Die auffallende Einheitlichkeit des Verhaltens der Sproßpilze gegenüber Harnstoff sollte Anlaß sein, der ausgeprägten Fähigkeit der Harnstoffhydrolyse bei künftigen Klassifizierungsversuchen besondere Beachtung zu schenken.

Serologisches Verhalten

Obwohl sich zahlreiche Untersucher mit der Serologie der *Cryptococcus-*Gruppe beschäftigt haben, sind erst während der letzten 15 Jahre Ergebnisse erzielt worden, die einen tieferen Einblick in die Antigenstruktur dieser Pilzgattung gestatten. Noch 1956 schreibt BENHAM *(13)* trotz aller erzielten Fortschritte auf diesem Teilgebiet der Pilzserologie: "More extensive studies of the types described using a larger group of cryptococci are needed for a more clearcut picture and a truer understanding of this important genus."

Die ersten serologischen Untersuchungen von SANFELICE *(147)* und später SHEPPE *(165)* erbrachten keine Anhaltspunkte für eine Antigenität der geprüften *Cryptococcus-*Stämme, eine Erscheinung, die sich auch in vielen nachfolgenden Berichten über die geringen Titer von *Cryptococcus-*Seren widerspiegelt [ASCHNER u. a. *(5)*, MAGER und ASCHNER *(120)*, DRAKE *(48)*, KLIGMAN *(98)*, NEGRONI und LANATA *(127)*], gleichgültig, welche Untersuchungsmethode zur Anwendung gelangte. Lediglich RAPPAPORT und KAPLAN *(136a)* erhielten Antiseren mit niedrigen Titern (1:80), desgleichen COX und TOLHURST *(42)* (1:128).

BENHAM *(10)* gebührt das Verdienst, als erste im Agglutinationsversuch mit spezifischen Antiseren den Beweis für eine serologische Sonderstellung der Cryptococcen innerhalb der Sproßpilze geführt zu haben (Tabelle 13).

4*

Tabelle 13. *Gruppierung der Cryptococcen nach* Benham *(10, 13)*

Gruppe	Koloniebild	Zellbild	Gasbildung in Zuckern	Agglutination in Serum der Gruppe			
				I	II	III	IV
1	Glatt oder wenig gefaltet, pastös, weiß oder cremefarben	Rund bis oval, 1,5—4,5 × 2,5—4,5 μ. Sprossung einzeln, in Ketten oder Gruppen. Keine Kapsel	+ oder 0	++++	0	0	0
2	Gefaltet, cremefarben bis bräunlich	Rund oder oval, 3—6,5 × 4,5—6,5 μ. Dicke Wand, geringe Kapsel	0	0	++++	+	0
3	Glatt, mucoid, cremefarben bis gelb oder bräunlich	Rund, 4,5—6 μ. Dicke Wand, mächtige Kapsel	0	0	+ oder 0	++++	0
4	Glatt oder gefaltet, rosa bis rot	Oval, 2,5—4 × 3—5μ. Dicke Wand, geringe Kapsel	+ oder 0	0	0	0	++++ oder 0

Dabei ist jedoch zu beachten, daß Benhams Gruppe I heute zu *Torulopsis* und verwandten Arten und die nichtfermentierenden Stämme der Gruppe IV zur Gattung *Rhodotorula* und nicht zu *Cryptococcus* gerechnet werden (*13, 161*).

Die Stämme der Gruppe II schließen saprophytische *Cryptococcus*-Kulturen ein, während Gruppe III aus apathogenen (*C. innocuus*) und pathogenen Stämmen (*C. neoformans, C. hominis, T. histolytica* und *Saccharomyces tumefaciens*) besteht.

Einwandfrei geht aus diesen Befunden die serologische Verwandtschaft apathogener und pathogener *Cryptococcus*-Stämme hervor.

Benham (*10*) erzielte agglutinierende Seren mit einem Maximaltiter von 1:160 durch 10—20 wöchentliche Injektionen von Zellen, die mit Salzsäure behandelt worden waren. Sie schrieb das Ergebnis in erster Linie der chemischen Behandlung zu, was sich allerdings bei wiederholten Nachuntersuchungen nicht bestätigen ließ.

Nicht weniger erfolgreich waren Hoff (*91*) und vor allem W. Kaufmann (*96*), dessen 1944 erschienener Beitrag im gesamten einschlägigen Schrifttum bis 1954 leider unberücksichtigt blieb (*154, 161*).

Kaufmann (*96*) bewies an einer ausreichenden Zahl von Sammlungsstämmen im Agglutinations-, Präcipitations- und Absättigungsversuch die weitgehende Gleichheit der Körperantigene von *C. neoformans, C. albidus* und *C. albidus var. japonica*, die sich serologisch von Redaellis *T. rotundata* abgrenzen ließen. Wesentlichen Schwierigkeiten scheint der Autor bei der Serumherstellung nicht begegnet zu sein. Seine Schlußfolgerungen sind wichtig für die gesamte Hefeserologie, zumal die Ergebnisse darauf hindeuten, daß die botanische Klassifizierung von *Cryptococcus*-Stämmen durch serologische Methoden zu vereinfachen und vielleicht auch zu revidieren ist.

Auch Salvin (*145*) berichtet wenig später über *Cryptococcus*-Seren mit einem Agglutinationstiter zwischen 1:32 und 1:256 und über positive Komplementablenkungen in Anwesenheit homologen Antigens in Serumverdünnungen bis 1:4095. Im Vergleich mit anderen menschenpathogenen Pilzen zeigte *C. neoformans* eine ausgeprägte serologische Eigenstellung.

Die unterschiedlichen Befunde bei der Immunserumherstellung sind, wie man durch die Untersuchungen von Neill c. s. (*128, 130*), Evans u. a. (*62, 63*) und Seeliger (*154, 161*) weiß, zum Teil durch eine unterschiedliche Antigenität der benutzten *Cryptococcus*-Kulturen und durch das wechselnde Verhalten der Versuchstiere bedingt. Von größter Bedeutung für die Gewinnung guter Tierseren erwies sich die jeweilige Bekapselung der Immunisierungsstämme. Mit stark bekapselten Kulturen gelingt im allgemeinen die Immunserumherstellung nicht. Voraussetzung ist die Verwendung schwach bekapselter Kulturen oder Varianten in der S-Form.

Neill u. Mitarb. zufolge (*128, 130*) erhält man hochwertige Antiseren durch 13 intravenöse Injektionen von je 250 Millionen schwach bekapselter Cryptococcen in täglichen Abständen. Zur Vermeidung von anaphylaktischen Reaktionen erfolgt die Entblutung der Tiere am 20. Tag.

Obwohl *C. neoformans* über ein *Endotoxin* verfügt (*146*), vertrugen in eigenen Versuchen Kaninchen bei monatelanger Immunisierung — ganz im Gegensatz zu vielen anderen pathogenen Pilzen — heroische Keimmengen. Dies steht in Übereinstimmung mit den Erfahrungen von Stanley (*175*), wonach das die serologische Spezifität bedingende Polysaccharid keine erhebliche Endotoxicität aufweist. Die endotoxische Wirkung läßt sich jedoch durch Zusatz von getrockneten Tuberkelbakterien steigern (*146*).

Die Fehlschläge zahlreicher Untersucher sind wahrscheinlich meist methodisch bedingt. Da das reine *Cryptococcus*-Polysaccharid als *Hapten* fungiert (*154, 175*), ist es z. B. verständlich, daß Kligman (*98*) bei Injektion dieser Substanz — auch nach Verwendung von Schleppern — keine Antikörperbildung erreichte.

Dieses *Polysaccharid*, das u. a. von Stanley (*175*) und Hehre c. s. (*87*) ebenso wie von Kligman (*98*), Evans c. s. (*65*), Seeliger (*154*), Burcik und Beutmann (*199, 200*) u. a. extrahiert werden konnte, ist bei Verwendung hochwertiger Antiseren serologisch aktiv, im Gegensatz zur nichtantigenen und inaktiven *Cryptococcus*-Amylose (*87, 175*).

Stanley (*175*) stellte dieses Polysaccharid sowohl mittels physikalischer als auch chemischer Methoden dar und erhielt mehrere Fraktionen (P, S, N und F benannt), die sich chemisch und serologisch trotz Verschiedenheit der Herstellungsverfahren nicht wesentlich unterschieden. Noch in hohen Verdünnungen reagiert das *Cryptococcus*-Polysaccharid im Präcipitationsversuch mit *C. neoformans*-Serum (*65, 154, 161, 175*) und anderen *Cryptococcus*-Seren unter rascher Bildung von Präcipitinringen und flockigen Präcipitaten.

Demgegenüber verhielt sich die *C. neoformans*-Ribonucleinsäure im Präcipitintest einwandfrei negativ. Es gelang jedoch Stanley (*175*), die Ribonucleinsäure an die Oberfläche von Schafsblutkörperchen zu adsorbieren und mit diesen sensibilisierten Erythrocyten eindeutige *Hämagglutinationsreaktionen* durch homologes Antiserum auszulösen.

Interessanterweise war es nicht möglich, Schafserythrocyten mit dem *C. neoformans*-Polysaccharid zu sensibilisieren (*175*). Nach Evans und Haines (*69*) lagert sich die S-Substanz, das Kapselpolysaccharid, auch nur schlecht an die Oberfläche von Kollodiumteilchen an. Ihre negative Ladung bewirkt jedoch, daß sie leicht von den kationischen Gruppen eines Anionenaustauschers angezogen wird (*69*).

Solche Anionenaustauscher, das ,,Amberlite IR-4B" und das ,,Amberlite IRA-400", wurden in pulverisiertem Zustand und nach spezieller Vorbehandlung zu einschlägigen

Versuchen herangezogen. Gleiche Mengen dieser Partikel und von SB-Polysaccharid werden unter 30 min langem Schütteln gut vermischt und dann zweimal mit Aqua dest. sowie einmal mit 0,2% NaCl-Lösung gewaschen. Das Sediment wird mit 0,2%iger NaCl-Lösung auf das Ausgangsvolumen aufgefüllt und dient als Antigen. Dieses wird im Reagenzglas zu gleichen Teilen mit der jeweiligen Serumverdünnung gemischt und nach 2 Std Bebrütung bei 22 bis 28° C auf Agglutination kontrolliert. — Die Titer lagen nach Evans und Haines (69) höher als bei Verwendung des Vollantigens.

Allem Anschein nach haben aber die allgemein niedrigen bzw. oft unzureichenden *Cryptococcus*-Antikörpertiter noch eine weitere Ursache. Fehlende oder nur geringfügige Antikörperbildung ist nämlich nicht auf die kapselbildenden *Cryptococcus neoformans*-Stämme beschränkt, sondern findet sich auch bei zahlreichen anderen kapselbildenden bzw. schleimig wachsenden Sproßpilzen. Seeliger (*154, 161*) beobachtete diese Erscheinung z. B. bei Stämmen der Gattung *Lipomyces*, aber auch bei *Rhodotorula glutinis* und *Bullera alba*, wobei jeweils 2 bis 3 verschiedene Kulturen geprüft wurden. Sofern überhaupt Antikörper nachgewiesen werden konnten (nach 3—5 Monaten Immunisierung), waren diese nur in geringer Menge vorhanden. Möglicherweise beruht dies auf einem der „immunologischen Paralyse" (Stark) entsprechenden Vorgang. Danach wird die Antikörperbildung zwar nicht gehemmt, aber die Persistenz der Polysaccharide führt im Organismus zu einer laufenden Bindung der Antikörper und deren anschließender Zerstörung.

Daß tatsächlich große Antigenmengen im Organismus kreisen können, wird durch den Ausgang tierexperimenteller Versuche von Neill c.s. (*129*) sowie durch Beobachtungen am cryptococcosekranken Menschen von Neill u. a. (*131*) sowie Seeliger und Christ (*162*) belegt. Bei einwandfrei negativem serologischem Verhalten der Infizierten gelang der Nachweis löslicher *Cryptococcus*-Antigene im Liquor zweier Patienten mittels des Präcipitintests unter Verwendung von *C. neoformans*-Antiserum. Diese spezifische Neill-Reaktion wird als Hilfsmittel der Cryptococcose-Diagnostik empfohlen (*162*). Die übrigen serologischen Verfahren versprechen keinen Erfolg (*111, 157, 158, 161*).

Über die Menge der im Körper des immunisierten Versuchstieres kreisenden bzw. verbleibenden Antigene bzw. Polysaccharide liegen noch keine sicheren Angaben vor. Vielleicht gelingt es, durch Verwendung fluorochromierter Antikörper in Verbindung mit spezifischen Färbemethoden, die bereits manche Aufschlüsse über den Verbleib der *Cryptococcus*-Antigene vermittelten (*111, 131*), diese Frage zu lösen.

Für eine Blockierung der Antikörper im Versuchstier spricht auch die wiederholte Beobachtung des Verf. (*154, 161*), daß nach kurzfristiger Immunisierung mit lebenden, formolisierten, gekochten oder autoklavierten Aufschwemmungen von *C. neoformans, C. diffluens* und *C. albidus* (stark bekapselte Stämme) innerhalb von 21 Tagen Serumantikörper mit Titern von 1:10 bis 1:80 auftraten, bei weiterer Immunisierung aber anstelle des erhofften Antikörperanstiegs stets ein Titersturz bis zum völligen Verschwinden erfolgte.

Eine direkte *Abhängigkeit der Serumtiter von der Kapselgröße* der Immunisierungsstämme ist zweifellos *nicht immer nachweisbar*. Seeliger (*161, 165*) konnte in Übereinstimmung mit Kröger (briefliche Mitteilung) und Holtz (*202*) mit kapselarmen *C. neoformans*- und *C. diffluens*-Stämmen nur Seren mit einem geringen Antikörperspiegel gewinnen, während umgekehrt die Immunisierung mit bekapselten Kulturen von *C. laurentii* und *C. luteolus* innerhalb von 21—28 Tagen zur Gewinnung hochwertiger Antiseren führte.

Die Ursachen für die unterschiedlichen Befunde bei der *Cryptococcus*-Serumherstellung sind somit sicher komplex. Weitere Untersuchungen erscheinen geboten, bevor eine Aussage über die *optimale Methodik* möglich ist. Auch bei der Antigenbereitung sind noch manche Verbesserungen zu erwarten. So ließ sich die Reaktionsfähigkeit und Spezifität von Pilzantigenen durch *Extraktion mit Pyridin* steigern (*71*). Derart vorbehandelte *C. neoformans*-Antigene reagierten im Modellversuch bei der Komplementablenkung spezifisch und nicht antikomplementär.

Außer den bereits erwähnten serologischen Methoden hat die *Kapselreaktion* nach NEUFELD beträchtliches Interesse in der *Cryptococcus*-Diagnostik erlangt.

EVANS (*62, 63, 65*) sowie NEILL et al. (*128, 130*) haben 1949 und in den darauffolgenden Jahren unabhängig voneinander über Kapselreaktionen an *C. neoformans*-Stämmen berichtet, EVANS (*62, 63, 65*) unter Verwendung von homologen Immunseren, hergestellt mit Formolantigenen, und NEILL c.s. (*128, 130*) zunächst unter Benutzung eines *Oospora*-Serums (Nr. 1936) und später von selbst hergestellten *Cryptococcus*-Seren. Regelmäßig fanden sich in den Seren der genannten Autoren beim Vorhandensein kapselreaktiver Substanzen agglutinierende und präzipitierende Antikörper. Durch Absättigung derartiger Seren mit den löslichen Polysacchariden ließ sich auch die Fähigkeit zur Kapselreaktion entfernen, woraus gefolgert wurde, daß die an den Kapselreaktionen beteiligten Antigenbestandteile mit dem Präcipitinogen (S-Substanz von EVANS) identisch sind.

Die von NEILL u. a. (*128*) beobachtete Kapselreaktion ist ein Beispiel einer heterologen Kreuzreaktion von *Cryptococcus*-Antigenen mit dem Antiserum eines, leider verlorengegangenen, saprophytären *Oospora*-Stammes. Sie war mit Antiseren anderer *Oospora*-Kulturen nicht nachweisbar und gelang auch nicht in eigenen Versuchen an Hand von drei *Geotrichum* (*Oospora*)-Seren und 28 *Cryptococcus*-Kulturen (*154, 161*).

Neuere Befunde des Verf. (*161*) haben jedoch gezeigt (s. S. 60), daß mit *C. neoformans* antigenverwandte Stämme sowohl innerhalb der *Candida*- wie der *Trichosporon*-Gruppe zu finden sind. Da die Gattung *Trichosporon* nach WICKERHAM (briefliche Mitteilung) zwei phylogenetische Entwicklungsrichtungen — nach dem Genus *Cryptococcus* und dem Genus *Geotrichum* (*Oospora*) — erkennen läßt, wird man vermuten dürfen, daß es sich bei NEILLs *Oospora* 1936 vielleicht um ein *Trichosporon* gehandelt hat.

Die *Methodik der Kapselreaktion* darf als bekannt vorausgesetzt werden, so daß sich ein näheres Eingehen auf sie hier erübrigt [Einzelheiten s. (*62, 63, 128, 130, 161*)]. Voraussetzung für das Gelingen des Versuchs ist, daß die *Antigene maximal bekapselt* sind. Dies wird bei *C. neoformans* am leichtesten durch intraperitoneale Impfung von Mäusen erreicht. Das Antigen wird aus dem Peritonealexsudat gewonnen und vor Durchführung der Probe 1- bis 2mal in Wasser gewaschen. Formolisiert halten sich Aufschwemmungen stark bekapselter Stämme mehrere Monate lang bei Aufbewahrung im Kühlschrank. Das Waschen der Stämme bzw. Antigene ist erforderlich, um Präcipitationen der löslichen Kapselsubstanz mit dem Antiserum zu vermeiden. Bei geringem Antikörpergehalt kann durch Präcipitation gelöster Antigene die Kapselreaktion abgeschwächt oder ganz blockiert werden.

Die Kapselreaktion selbst ist nicht durch eine „Quellung" und Vergrößerung des Kapseldurchmessers charakterisiert, sondern lediglich durch eine *Präcipitation der Antikörper an der Kapseloberfläche.*

Zur Klärung dieser Frage unternommene Versuche von Evans und Seeliger (70) erwiesen, daß nach Kontakt zwischen gewaschenen *C. neoformans*-Zellen und homologem Antiserum keine Zunahme der Kapselgröße erfolgt, gleichgültig, ob die Vergleichsmessungen mikroskopisch (Tabelle 14) oder volumetrisch vorgenommen werden. Bei sehr kapselreichen Antigenen kommt es nach der Antigen-Antikörper-Reaktion sogar zu einer scheinbaren Verminderung des Kapseldurchmessers. Werden die Antigene vor dem Versuch nicht gewaschen, kommt es zu einer Copräcipitation der gelösten Kapselantigene durch das Antiserum, wodurch bei volumetrischen Bestimmungen eine Vergrößerung des Kapseldurchmessers vorgetäuscht werden kann.

Unter Verwendung der aufgeführten Testmethoden erzielte Evans (63, 65) mit absorbierten Seren eine Unterteilung von 19 *C. neoformans*-Stämmen in 3 *Kapseltypen A*, *B* und *C* (s. Tabelle 15).

Tabelle 14. *Kapseldurchmesser in C. neoformans-Präparaten mit Tusche (T) und Antiserum (AS) nach*
Evans, Seeliger, Kornfeld und Garcia (70)

Präparation	Mittlerer Durchmesser[1] in μ	Standardabweichung	Standardabweichung vom Mittelwert	Wahrscheinlichkeit einer Fehlberechnung[2]
T	36,72	2,18	0,40	0,001
AS	34,62	2,00	0,37	
im hängenden Tropfen				
T	38,43	3,22	0,59	0,001
AS	34,90	2,66	0,49	
T	38,00	3,43	0,63	0,001
AS	34,76	2,49	0,46	
in Zählkammer				
T	43,86	2,28	0,42	0,001
AS	41,16	2,21	0,40	
auf Objektträger mit Deckglas				

[1] 30 Kapselmessungen mit Okularmikrometer.
[2] T-Methode von Student.

In den unabgesättigten Seren sind die Kreuzreaktionen zwischen den Kapseltypen teils unilateral, teils reziprok. Manche Vollseren des Typs A reagieren praktisch typenspezifisch, andere greifen — in Abweichung von Tabelle 15 — auch auf die anderen Kapseltypen über (161). Zur Absättigung sind im letzteren Falle B-Zellen nötig. B-Seren erfassen unabsorbiert auch A-Antigen. C-Seren reagieren ebenfalls übergreifend, behalten nach der Absättigung aber noch typspezifische Antikörper.

Bei der Herstellung von Kapselseren zeigte sich in den Versuchen von Evans (63) wie Seeliger (154), daß die Kaninchenimmunseren trotz weitgehend gleichartiger Herstellungsbedingungen teilweise recht beträchtliche qualitative und quantitative Unterschiede im Antikörpergehalt aufweisen können.

Tabelle 15. *Kapselreaktionen in C. neoformans-Seren vor und nach Absättigung* [nach Evans (63) vereinfacht]

Antiserum	abgesättigt mit	Kapselreaktion mit		
		Typ A	Typ B	Typ C
Typ A	—	++++	—	—
	Typ B	+++	—	—
	Typ A	—	—	—
Typ B	—	++++	++++	+
	Typ A	—	—	—
	Typ B	—	—	—
Typ C	—	++	++	++++
	Typ A	—	—	+++
	Typ C	—	—	—

So fand sich beispielsweise in einem polyvalenten *Cryptococcus*-Serum und einem spezifischen Typ A-Serum, die Verf. von Dr. Evans erhielt, eine kräftige Kapselreaktion mit einem Sammlungsstamm von *C. luteolus,* obwohl der gleiche Stamm durch ein in Bonn hergestelltes *C. neoformans*-Typ A-Serum nicht erfaßt wurde und sein Antiserum alle geprüften *C. neoformans*-Stämme unbeeinflußt ließ.

In der Regel kommt es bei positiven Kapselreaktionen stets auch zur nachfolgenden Agglutination. Bei schwach bekapselten Stämmen steht die Agglutination im Vordergrund.

Papierchromatographisch zeigen die typenspezifischen Kapselpolysaccharide nach EVANS und MEHL (*64*) keine auffallenden Unterschiede.

Soweit aus den bisher vorliegenden Berichten ersichtlich, gehört die Mehrzahl der serologisch überprüften *C. neoformans*-Kulturen zum Kapseltyp A (*62, 63, 161, 165*). Dabei handelt es sich um Stämme, die vom Menschen, aus den Eutern mastitiskranker Kühe und aus Bodenproben isoliert worden waren. Eine Trennung der Stämme nach ihrer Herkunft scheint somit serologisch nicht möglich und ist bei der wahrscheinlichen Infektkette Boden → Mensch bzw. Tier auch nicht zu erwarten.

SEELIGER (*161*) erweiterte die vorstehenden Untersuchungen — ab Herbst 1955 in Zusammenarbeit mit EVANS, der seine Kapselseren zur Verfügung stellte — durch serologische Überprüfung von Standardstämmen und frisch isolierten Kulturen aller übrigen *Cryptococcus*-Arten. Wie schon ausgeführt, wurden bei der Herstellung von Immunseren gegen *C. diffluens* und *C. albidus* ähnliche Schwierigkeiten wie bei *C. neoformans* beobachtet, und es gelang nicht, gegen die bekapselten Kulturen dieser Arten brauchbare Antiseren zu erzeugen. Zusätzliche Untersuchungen erstreckten sich auch auf die Wirkung von Antiseren der Arten *Candida curvata*, *Candida humicola* und *Trichosporon cutaneum*, deren in vorstehenden Abschnitten geschildertes biochemisches Verhalten mit der Annahme einer Verwandtschaft mit der *Cryptococcus*-Gruppe vereinbar ist. In die

Tabelle 16. *Kapselreaktion* (K) *und Agglutination* (A) [1] *von Cryptococcus- und verwandten Arten in homologen und heterologen Seren*

	Antiserum												
	C. neoformans												
	Poly-valent	Evans-Typ											
Antigen		A	B	C	*C. diffluens*	*C. luteolus*	*C. laurentii*	*C. terreus*	*Candida curvata*	*Candida albicans*	*Trichosporon cutaneum*	*Geotrichum candidum*	*Lipomyces lipoferus* / *Lipomyces starkeyi*
C. neoformans A (7) [2]	KA	KA	KA	—	—	—	—	—	A	—	—	—	—
C. neoformans B (1)	KA	—	KA	—	—	—	—	—	a	—	—	—	—
C. neoformans C (1)	KA	—	—	KA	—	—	—	—	a	—	—	—	—
C. diffluens (1)	KA	KA	—	—	—	—	—	—	—	—	—	—	—
C. diffluens (2)	A	A	a	—	A	—	a	—	KA	—	—	—	—
C. diffluens (5)	A	a	—	—	a	—	a	—	a	—	—	—	—
C. diffluens (1)	a	—	—	—	—	—	—	—	—	—	—	—	—
C. albidus (2)	KA	KA	KA	—	—	—	a	—	KA	—	—	—	—
C. albidus (1)	A	A	a	—	—	—	a	—	A	—	—	—	—
C. albidus (1)	—	—	—	—	—	—	—	—	—	—	—	—	—
C. laurentii (1)	a	—	—	—	—	—	A	—	A	—	—	—	—
C. luteolus (1)	KA	KA	—	—	—	K [3]	—	—	KA	—	—	—	—
C. terreus (1)	—	—	—	—	—	—	—	A	(a)	—	—	—	—
T. aeria (1)	A	—	—	A	—	—	—	—	—	—	—	—	—
L. lipoferus/*L. starkeyi* (10)	—	—	—	—	—	—	—	—	—	—	—	—	KA/A

[1] A = starke Agglutination, a = schwache Agglutination im 1:10 verdünnten Serum.

[2] Zahl der untersuchten Stämme.

[3] Keine Agglutination bei mehrfach wiederholtem Ansatz im Röhrchen und auf dem Objektglas.

Untersuchungen wurde ferner ein Standardstamm von *Torulopsis aeria* einbezogen.

Zur Gegenkontrolle dienten Antiseren von mehr als 20 Sproßpilz- und Hyphomyzeten-Arten [vgl. (*154, 161*)], aus denen für die vorliegende Darstellung lediglich *Candida albicans* und *Geotrichum candidum* herausgegriffen wurden.

Die an anderer Stelle (*154, 161*) und vervollständigt in Tabelle 16 niedergelegten Ergebnisse beweisen eindeutig, daß *C. neoformans* und verwandte Arten innerhalb des Pilzreichs eine weitgehende *serologische Eigenstellung* einnehmen, untereinander aber durch übergreifende Gruppenantigene verbunden sind, die sich teils in Kapsel-, teils in Agglutinationsreaktionen und oft durch beide Reaktionstypen zu erkennen geben.

Aus dem Rahmen fallen lediglich ein *C. albidus*-Stamm, der aus Flachs isoliert worden war, und die zu *Cryptococcus* gehörend erachtete *T. rotundata*. Beide Kulturen wurden von keinem *Cryptococcus*-Serum erfaßt.

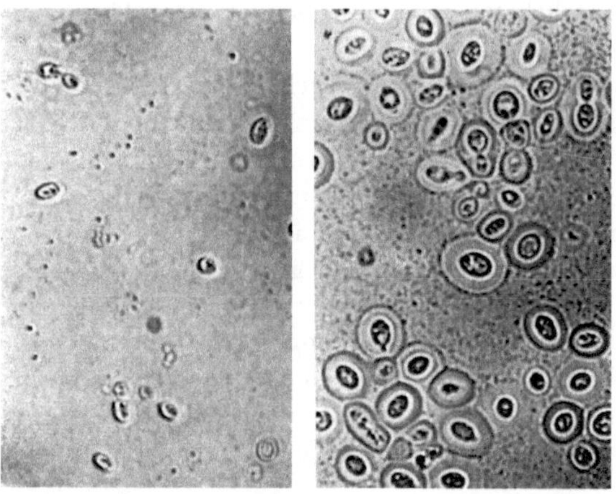

Abb. 6. Kapselreaktion mit *C. diffluens*-Antigen. Links: Antiserum gegen *L. lipoferus*; rechts: Antiserum gegen *C. neoformans*

C. terreus reagierte nur mit seinem Eigenserum (Agglutinintiter 1 : 160), das seinerseits auf keinen der übrigen Stämme übergriff und nur im empfindlichen Präcipitintest (s. weiter unten) geringfügige Gruppenantigengemeinschaften zeigte. Die biochemische Sonderstellung dieser neu beschriebenen Art (*46a*) wird somit auch serologisch bestätigt.

Zwei kapseltragende *C. albidus*-Stämme und eine stark bekapselte *C. diffluens*-Kultur reagierten ebenso wie eine *C. luteolus*-Kultur in der Kapselreaktion und in der Agglutination kräftig mit einem polyvalenten und einem Typ A-Serum von *C. neoformans*, die beiden *C. albidus*-Stämme außerdem mit dem Typ B-Serum. Die übrigen *C. diffluens*-, *C. albidus*- und *C. laurentii*-Stämme wurden von *C. neoformans*-

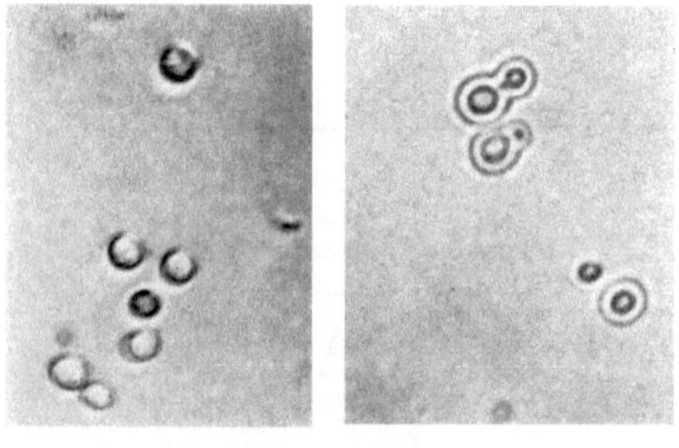

Abb. 7. Kapselreaktion mit *C. albidus*-Antigen. Links: Antiserum gegen *L. lipoferus*; rechts: Antiserum gegen *C. neoformans*

Serum mehr oder minder kräftig agglutiniert. Die Kapselreaktionen je eines *C. diffluens*- und *C. albidus*-Stammes mit *C. neoformans*-Serum sind in den Abb. 6 und 7 festgehalten.

Alle *C. neoformans*-Stämme konnten durch die Kapsel- oder Agglutinationsreaktion mit einem der 3 Evans-Typen identifiziert werden.

Nach diesen Befunden scheint es beim derzeitigen Stand der Kenntnisse noch nicht möglich, pathogene und apathogene *Cryptococcus*-Arten sicher mit serolo-

gischen Mitteln voneinander zu trennen, da sich einzelne bekapselte *C. diffluens*-
und *C. albidus*-Stämme serologisch wie der Kapseltyp A von *C. neoformans* verhalten. Wie die Erfahrung zeigt, fällt die Kapselreaktion bei *C. luteolus* nicht
mit allen *C. neoformans*-Seren positiv aus (s. S. 56).

Weitere Aufschlüsse sind von *Absättigungsversuchen* zu erwarten. Die Absorption von Antiseren bekapselter *Cryptococcus neoformans*-Stämme mit wenig
bekapselten *C. neoformans*- und *C. diffluens*-Antigenen führte in eigenen Versuchen
meist nicht zur Erschöpfung der K-Antikörper. KRÖGER (briefliche Mitteilung)
gelangte zu ähnlichen Ergebnissen. Die Kreuzreaktionen beruhen somit wahrscheinlich auf der *Komplexität der Antikörper*, die sich teils aus K-, teils aus
O-Antikörpern zusammenzusetzen scheinen. Eine sichere Trennung dieser
Komponenten ist Voraussetzung für die Aufstellung eines versuchsweisen Antigenschemas. Eine Klärung der noch offenen Fragen dürfte sich durch Behandlung
bzw. Markierung der Antikörper mit Flurochromen und einschlägige serologische
Versuche ermöglichen lassen. Erste erfolgversprechende Ergebnisse wurden bereits von EVELAND et al. (*201*) bekanntgegeben.

Von Interesse ist weiter das Verhalten von *Torulopsis aeria*, deren Kultur-
wachstum eine kräftige Agglutinationsreaktion mit EVANS Typ C- und poly-
valentem *Cryptococcus*-Serum ergab (*165*). In Übereinstimmung mit den biochemischen Testergebnissen besteht kein Zweifel, daß es sich bei diesem Stamm um
eine *Cryptococcus*-Art handelt, die serologisch dem Kapseltyp C von *C. neoformans*
nahesteht. Acht andere Torulopsis-Arten wurden von *Cryptococcus*-Seren nicht
beeinflußt, die Mehrzahl dagegen von *Candida albicans*- und *Saccharomyces
cerevisiae*-Antiseren.

Das einzige *C. diffluens*-Serum, dessen Herstellung gelang, agglutinierte lediglich 7 von 9 *C. diffluens*-Stämmen, meist nur in niedrigen Titern (bis 1:40).
Da keine der bekapselten Kulturen mit diesem Serum reagierte, wird vermutet, daß dieses vorzugsweise O-Antikörper enthielt, die beim Vorhandensein
starker Kapseln serologisch nicht in Erscheinung treten können.

Wieder anders liegen die Dinge beim *C. laurentii*-Antiserum, das bei einem
homologen Agglutinintiter von 1:640 in Verdünnungen von 1:20 bis 1:80
8 wenig bekapselte, überraschenderweise aber auch 2 bekapselte *Cryptococcus*-
Stämme agglutinierte. Diese Kreuzagglutination betrifft ausschließlich apathogene Stämme und wurde durch Absättigung mit *C. diffluens* (S 9) zum Verschwinden gebracht. Das Restserum reagierte spezifisch nur noch mit *C. laurentii*,
dessen serologische Eigenstellung damit ebenfalls erwiesen ist.

Die gleiche Folgerung gilt auch für *C. luteolus*. Obwohl der Teststamm mit
einzelnen *C. neoformans*-Seren eine kräftige Kapselreaktion gibt, verhält sich das
C. luteolus-Serum mit einem homologen K-Titer streng spezifisch. Auf dem Objektträger wie im Reagenzglasversuch kam es trotz spezifischen K-Titers nie zur Agglutination, ein Befund, der anzeigt, daß ein „negatives" Agglutinationsresultat bei
bekapselten Sproßpilzstämmen stets mikroskopisch auf eine Kapselreaktion
kontrolliert werden muß. Dies ist bei allen oben geschilderten Untersuchungen
auch geschehen.

Die vermutete Verwandtschaft von *Candida curvata* mit der *Cryptococcus*-
Gruppe wurde durch den eindeutigen Ausfall der antigen- und seroanalytischen
Untersuchungen erhärtet. Bei einem Eigentiter von 1:1280 (die Ablesung wurde

allerdings durch eine geringfügige Spontanagglutination des Teststammes erschwert) erfaßte das *C. curvata*-Serum in der Agglutination ein Großteil der geprüften *Cryptococcus*-Stämme einschließlich *T. aeria* und führte bei 2 *C. diffluens*- und 2 *C. albidus*-Stämmen sowie bei der *C. luteolus*-Kultur zu einer deutlichen Kapselreaktion.

Die bisher aufgetretenen Schwierigkeiten bei der Herstellung von Antiseren gegen *Candida humicola* und *Trichosporon cutaneum* erlauben noch keine endgültige Aussage darüber, ob die Antigenbeziehungen dieser beiden Arten zur *Cryptococcus*-Gruppe unilateral oder reziprok sind. Die bisher nur mäßigen Titer eines geprüften *Trichosporon cutaneum*-Serums (Tabelle 17) sprechen zwar für eine echte Antigengemeinschaft, erfordern aber weitere klärende Untersuchungen. Da es nicht gelang, im *Candida humicola*-Serum nach dreimonatiger Immunisierung homologe Antikörper nachzuweisen, ist die Beweiskette der echten Partialantigengemeinschaft dieser Art mit der *Cryptococcus*-Gruppe noch nicht endgültig geschlossen.

Daß es sich hierbei nicht um Zufallsergebnisse handelt, läßt sich nicht nur aus weitgehenden biochemischen Ähnlichkeiten folgern, die die besagten Stämme miteinander verbinden. Es entspricht auch dem Verhalten zahlreicher Antiseren gegen andere Sproßpilze, die biochemisch von der *Cryptococcus*-Gruppe verschieden sind [vgl. (*154, 161*)]. Als Beispiele für viele andere sind in Tabelle 16 lediglich die negativen Befunde mit *Candida albicans*- und *Geotrichum candidum*-Seren aufgeführt, die auch mit den Ergebnissen von Beutmann (199) übereinstimmen.

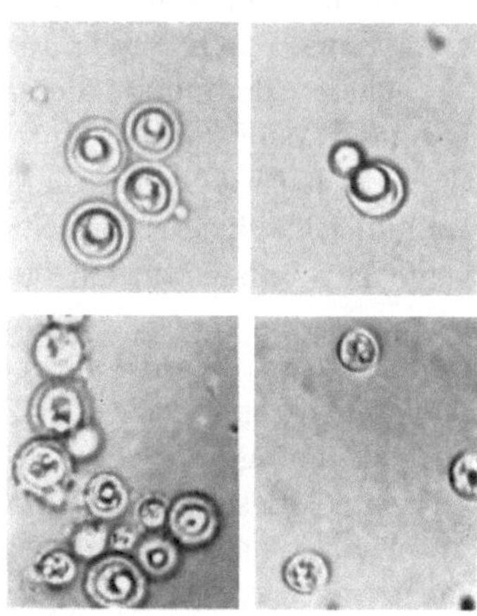

Abb. 8. Serologische Unterschiede in der Kapselreaktion bei Prüfung von *Lipomyces*-Stämmen in *Cryptococcus*- und *Lipomyces*-Seren. Obere Abbildung: *L. lipoferus*-Antigen; untere Abbildung: *L. starkeyi*-Antigen; linke Abbildung: positive Kapselreaktionen in Gegenwart von *L. lipoferus*-Antiserum; rechte Abbildung: negative Kapselreaktionen in Gegenwart von *C. neoformans*-Antiserum

Die vorstehend wiederholt angeschnittene Frage einer Verwandtschaft oder möglichen Identität zwischen dem anascosporogenen *C. neoformans* und ascosporogenen Stämmen der *Lipomyces*-Gruppe (*14, 40, 112, 159*) gab Anlaß, zur Beantwortung auch serologische Methoden heranzuziehen.

Trotz beträchtlicher Herstellungsschwierigkeiten gelang es schließlich, von *L. lipoferus* und *L. starkeyi* je ein Antiserum mit einem mäßigen Agglutinintiter (1:320 bzw. 1:160) zu gewinnen. Ein Teil der *Lipomyces*-Stämme ergab — im Gegensatz zu allen geprüften *C. neoformans*-Kulturen — neben mehr oder minder deutlichen Agglutinations- und Präcipitationsreaktionen mit homologem Serum der *Lipomyces*-Arten eindeutige Kapselreaktionen (Abb. 8) (*154, 161*), die durch keines der in Tabelle 16 aufgeführten *Cryptococcus*-Seren ausgelöst werden konnten. Überraschend war hingegen eine übergreifende, unilaterale Reaktion eines *L. lipoferus*-Serums auf *Geotrichum candidum* und antigenverwandte Arten (*154, 161*).

Trotz gewisser morphologischer und mancher kulturell-biochemischer Ähnlichkeiten liefern die serologischen Untersuchungsergebnisse somit keine Anhaltspunkte, die sich im Sinne einer nahen Verwandtschaft oder gar Identität zwischen *Lipomyces-* und *Cryptococcus-*Stämmen interpretieren lassen.

Die serologischen Untersuchungen auf breiter Basis wurden schließlich durch *Präcipitationsversuche mit Polysaccharid-Extrakten* [Herstellung s. (*65, 98, 161, 165* u. a.)] abgerundet. Im Gegensatz zu anderen Pilzarten konnten hierbei durch Verwendung der Präcipitation im Agar-Gel bei der O-Antigenanalyse (*155*) keine methodischen Verbesserungen erzielt werden. Die in Tabelle 17 zusammengestellten Ergebnisse gründen sich auf den klassischen Ringtest (Ablesung nach 30 min) und nachfolgende Beobachtung der Präcipitatbildung der in Capillaren gemischten Antiseren und Extrakte nach 24 Std im Kühlschrank.

Tabelle 17. *Präcipitationsversuche mit Polysaccharidextrakten*[1] *von Cryptococcus- und anderen Arten in homologen und heterologen Immunseren*

Antigen von (Zahl der geprüften Stämme in Klammern)	C. neoformans	C. luteolus	C. laurentii	C. terreus	C. curvata	T. cutaneum	C. albicans	G. candidum	L. lipoferus	L. starkeyi
C. neoformans (9)	+++	—	+++	+	+++	++	—	—	—	—
C. diffluens (9)	+++	—	+++	±	+++	++	—	—	—	—
C. albidus (4)	+++ oder —	—	++ oder —	±	+++	++	—	—	—	—
C. luteolus (1)	— oder +[3]	+++	—	—	+++	+	—	—	—	—
C. laurentii (1)	++	—	+++	—	++	±	—	—	—	—
C. terreus (1)	—	—	—	++	++	—	—	—	—	—
T. aeria (1)	—	—	—	—	++	—	—	—	—	—
C. curvata (3)	—	—	++	—	+++	+ oder —	—	—	—	—
C. humicola (1)	—	—	++	—	++	—	—	—	—	—
C. albicans (10)	—	—	—	—	—	—	+++	—	—	—
T. cutaneum (3)	+	+	+++	—	++	+++	—	—	—	—
T. infestans (1)	+	+	++	—	+	++	—	—	—	—
L. lipoferus (2)	—	—	—	—	—	—	—	—	+++	+
L. starkeyi (8)	—	—	—	—	—	—	—	—	—, + oder ++	+++
G. candidum (3)	—	—	—	—	—	—	--	+++	+++	n.d.[2]

[1] FULLER- und Alkohol-Waschwasser-Extrakte.
[2] n. d. = nicht durchgeführt.
[3] 2 Stämme.

Sie bestätigen im wesentlichen die in der Kapsel- bzw. Agglutinationsreaktion erhobenen Befunde, weichen allerdings in manchen Einzelheiten aber nicht unerheblich von den in Tabelle 16 aufgeführten Resultaten ab. Dies gilt z. B. für den Stamm von *T. aeria,* dessen Polysaccharidextrakte trotz wiederholter Versuche nicht von *C. neoformans-*Antiserum präcipitiert wurden.

Ein Großteil der Unterschiede ist darauf zurückzuführen, daß die Präcipitationsreaktion sowohl durch die polysaccharidhaltigen Kapselsubstanzen (*65, 128, 161*) wie auch durch die löslichen O-Antigene, die ebenfalls Polysaccharidcharakter besitzen, zustande kommt. Eine Trennung dieser beiden Anteile ist derzeit noch nicht sicher erfolgt. Es ist aber beim Fehlen von sichtbaren Kapseln höchstwahrscheinlich, daß sich die Antiseren von *C. curvata, C. humicola, T. cutaneum* und vor allem von *C. laurentii* hauptsächlich gegen die O-Antigene der betreffenden Stämme richten. Wie aus Tabelle 17 ersichtlich, erfassen einzelne dieser Seren in der Präcipitation kräftig die überwiegende Mehrzahl der *Cryptococcus*-Antigene. Diese *Gruppenreaktion* ist im wesentlichen auf die *Cryptococcus*-Gruppe und die mit ihr verwandten Arten beschränkt und wird durch ein gemeinsames *Gruppenantigen* bedingt. Auch im Präcipitationsversuch kommt es nicht zu einer Verwandtschaftsreaktion zwischen der *Cryptococcus*- und *Lipomyces*-Gruppe.

Aus den vorstehenden Befunden (Tabellen 15—17) ergeben sich für die Antigenstruktur des Genus *Cryptococcus* folgende Schlüsse: Die Antigenstruktur ist komplex, sowohl im Hinblick auf die Kapselsubstanzen als auch auf die somatischen Antigene, die beide den Charakter von Polysacchariden aufweisen. Die O-Antigene sind vornehmlich für die gruppen(gattungs)-spezifischen Reaktionen verantwortlich, die besonders deutlich in den Antiseren von *C. laurentii* und *C. curvata* zutage treten. Der gleiche Kapseltyp kommt bei verschiedenen Spezies im Sinne der botanischen Systematik vor, eine Erscheinung, die auch aus der Bakteriologie geläufig ist. Bis zur serologischen Identität reichende Verwandtschaftsreaktionen bestehen zwischen *C. neoformans* einschließlich der Variante *uniguttulatus* und einzelnen *C. diffluens*- bzw. *C. albidus*-Stämmen, während andere Kulturen der letztgenannten Arten abweichend reagieren. *C. laurentii, C. luteolus* und *C. terreus* sind durch Partialantigene mit den meisten *Cryptococcus*-Arten verbunden, verfügen aber über artspezifische Antigene, die die Herstellung spezifischer Seren ermöglichen.

Demnach decken sich die Serotypen (Kapseltypen) der *Cryptococcus*-Gruppe nur zum Teil mit den Biotypen der Systematik. Die pathogenen *Cryptococcus*-Stämme können mit hochwertigen Immunseren zwar sicher erfaßt, aber nicht ausreichend von apathogenen *Cryptococcus*-Arten abgegrenzt werden. Deshalb hat die Serodiagnose nur einen bedingten Wert.

Entsprechend dem biochemischen Verhalten lassen einzelne, von *Cryptococcus* morphologisch erheblich abweichende Sproßpilzarten aus den Gattungen *Candida* und *Trichosporon* echte reziproke Antigengemeinschaften mit der *Cryptococcus*-Gruppe erkennen, als deren stammesgeschichtliche Verwandte *Candida curvata, Candida humicola, Trichosporon cutaneum* und *Trichosporon infestans* aufzufassen sind.

Demgegenüber lassen sich von der Serologie her keine Anhaltspunkte für eine Verwandtschaft oder gar Identität von *Lipomyces*-Arten und *C. neoformans* nachweisen.

Wie auch in anderen Bereichen der mikrobiologischen Serodiagnostik gibt es auch *in der Pilzserologie keine absolute serologische Spezifität (154, 161)*. So beobachteten Evans et al. *(68)* in einzelnen *Cryptococcus*-Seren die Mitagglutination von *Candida*- und *Saccharomyces*-Stämmen. Das *Candida curvata*-Serum des Verf. *(165)* reagierte im Präcipitationsversuch übergreifend auf eine größere Zahl verschiedener Hefearten. Auch waren manche *Cryptococcus*-Seren imstande, Trichophytin, *Pneumococcus*-Typ 2-Polysaccharid und Tragantgummi zu präcipitieren bei negativen Reaktionen mit Histoplasmin, Blastomycin, Coccidioidin, Blutgruppe A-Substanz und Kapselmaterial von Klebsiellen, Meningokokken, *Hämophilus-* und anderen Bakterien-Arten (*68*).

Zusammenfassende Schlußbetrachtung

1. Die Nomenklatur der *Cryptococcus*-Gruppe, vertreten durch den tier- und menschenpathogenen Sproßpilz *C. neoformans* und mehrere verwandte apathogene Arten, hat sich durch die Ergebnisse taxonomischer Untersuchungen stabilisiert. Sie findet in zunehmendem Maße internationale Anerkennung.

2. Das kulturell-biochemische und serologische Verhalten ermöglicht die Aufgliederung der Gattung *Cryptococcus* in mehrere Arten.

3. Durch kombinierte kulturell-biochemische, morphologische und serologische Untersuchungen ist eine sichere Abgrenzung der *Cryptococcus*-Stämme von anderen Hefearten in 24—96 Std gewährleistet.

4. Die Differentialdiagnose des menschenpathogenen *C. neoformans* von verwandten *Cryptococcus*-Arten ist durch kulturell-biochemische und serologische Verfahren nur bedingt möglich (vgl. Tabelle 18) und erfordert bei der Erstzüchtung stets den *beweisenden Tierversuch*.

Tabelle 18. *Schnelldiagnose von C. neoformans und Verwandten (156, 160)*

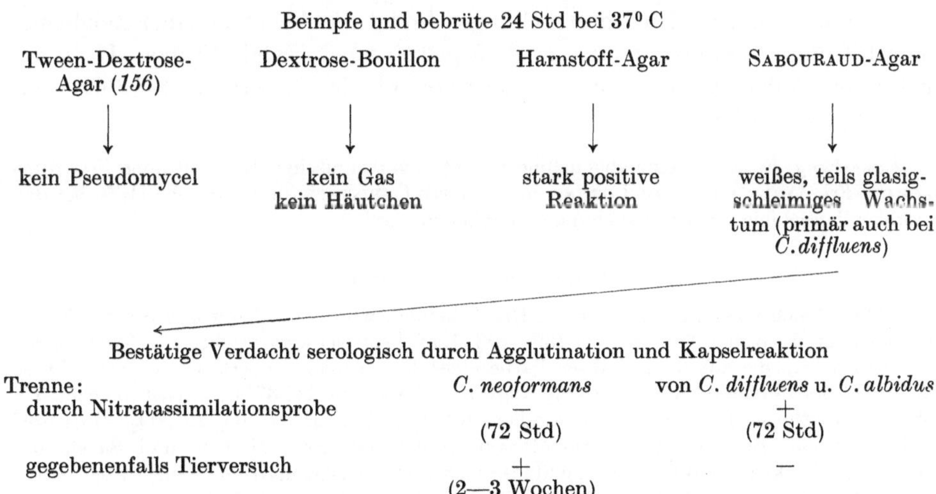

5. Die Art *Torulopsis aeria* zeigt morphologische, kulturelle, biochemische und serologische Eigenschaften, die ihre Überführung in die Gattung *Cryptococcus* rechtfertigen.

6. Die Arten *Candida curvata, Candida humicola, Trichosporon cutaneum* und *Trichosporon infestans* weisen zahlreiche biochemische und serologische Beziehungen zum Genus *Cryptococcus* auf. Sie sind — ungeachtet der bei der Klassifizierung wertvollen morphologischen Unterscheidungsmerkmale — als phylogenetische Verwandte der Gattung *Cryptococcus* anzusehen.

7. Trotz gemeinsamer morphologischer und biochemischer Merkmale (Kapselbildung, Zuckerassimilation, Stärkesynthese usw.) lassen sich die Gattungen
Cryptococcus und *Lipomyces* mit Sicherheit voneinander abgrenzen (Tabelle 19).

Tabelle 19. *Unterscheidungsmerkmale für die Diagnose von C. neoformans, C. diffluens, C. albidus und Lipomyces spp.*

Merkmal	*C. neo-formans* (9)[1]	*C. diffluens-, C. albidus-* Gruppe (14)	*L. lipoferus-, L. starkeyi-* Gruppe (10)
Wachstum bei 37° C	+++	— oder +	—
Nitratassimilation, Nitritbildung	—	+++	—
Starke Harnstoffhydrolyse in 24—96 Std	+++	+++	—
Stärkesynthese	+++	+++	— oder +
Serologisches Reaktionsvermögen mit *C. neoformans*-Antiserum	+++	+++ oder vereinzelt —	—
Serologisches Reaktionsvermögen mit *Lipomyces*-Antiserum	—	—	meist ++ bis +++
Tierpathogenität	+++	—	—

[1] Zahl der Stämme in Klammern.

8. In den vorliegenden Untersuchungen findet die Annahme einer möglichen
Identität von *C. neoformans* und *L. starkeyi* keine beweisende Stütze. Es liegen
auch keine Anhaltspunkte vor, *C. neoformans* als das imperfekte Stadium von
L. starkeyi anzusehen.

Anmerkung. Es ist mir eine angenehme Pflicht, den technischen Assistentinnen Frl. Sulz-
bacher, Frl. Jager und Frau Lorenz sowie Herrn Christeit für technische Hilfe bei den
vorstehend geschilderten Untersuchungen herzlich zu danken.

Nachtrag bei der Korrektur

Während sich vorstehende Arbeit im Druck befand, wurde von Rich und Stern [Myco-
pathologia et Mycologia Applicata, 9, 189—193 (1958)] eine schwarz wachsende *Cryptococcus*-
Art beschrieben, die aus abgeschabter Farbe eines Benzintanks isoliert worden war. Diese
Art erhielt den Namen *C. nigricans.* Sie zeigt kein Wachstum bei 37° C, wächst aber gut bei
28° C, ist apathogen für die Maus, assimiliert KNO_3 und reduziert NO_3 zu NO_2. Harnstoff
wird gespalten. Bei fehlender Fermentation werden Dextrose, Maltose und Saccharose
assimiliert, Galaktose, Laktose und Melibiose nicht. Die anaskosporogene Kultur zeigt multilaterale Sprossung und deutliche Kapselbildung. Die Kolonien sind tiefschwarz, anfangs
schleimig, glattglänzend, später gefaltet und sehr zäh. In flüssigen Kulturen werden ein
schleimiges Häutchen und Polysaccharide gebildet. — Über die Jodstärke-Reaktion finden
sich keine Angaben. Es wird künftigen Untersuchungen vorbehalten bleiben, zu klären, inwieweit diese neue Art Verwandtschaften zu anderen schwarzen „Hefen" (z. B. *Pullularia,
Cladosporium, Margarinomyces heteromorphus*) aufweist.

Literatur

1. Ajello, L., V. Q. Grant and M. A. Gutske: The effect of tubercle bacillus concentration
procedures on fungi causing pulmonary mycosis. J. Lab. clin. Med. **38**, 486—491 (1951).
2. — Soil as a natural reservoir for human pathogenic fungi. Science **123**, 876—879 (1956).

3. AJELLO, L., Occurrence of Cryptococcus neoformans in soils. Amer. J. Hyg. 67, 72—77 (1958).

3a. — V. Q. GRANT and M. A. GUTSKE: Use of mineral oil in the maintenance of cultures of fungi pathogenic for humans. Arch. Derm. Syph. (Chicago) 63, 747—749 (1951).

4. ARZT, L.: Zur Klinik und Pathologie der Sproßpilzerkrankungen. Arch. Derm. Syph. (Berl.) 145, 311—312 (1924).

5. ASCHNER, M., J. MAGER and J. LEIBOWITZ: Production of extracellular starch in cultures of capsulated yeasts. Nature (Lond.) 156, 295 (1945).

6. —, and A. CURY: Starch production in the genus Trichosporon. J. Bact. 62, 350—352 (1952).

7. ASHBEL, R., and A. DOLB: 60 years discovery of C. neoformans (SANFELICE) VUILLEMIN (Torula histolytica). Harefuah, 49, 108—113 (1955).

8. ASHNER, M., and D. GODINGER: An ecological study of yeasts of the genus Cryptococcus. Harefuah 49, 107—108 (1955).

9. BAUER, H.: Mikroskopisch-chemischer Nachweis von Glykogen und einigen anderen Polysacchariden. Z. mikr.-anat. Forsch. 33, 143—160 (1933).

10. BENHAM, RH.: Cryptococci — their identification by morphology and serology. J. infect. Dis. 57, 255—274 (1937).

11. — Cryptococcosis and blastomycosis. Ann. N.Y. Acad. Sci. 50, 1199—1214 (1955).

12. — The genus Cryptococcus: the present status and criteria for the identification of species. Trans. N.Y. Acad. Sci., Ser. II, 17, 418—429 (1955).

13. — The genus Cryptococcus. Bact. Rev. 20, 189—196 (1956).

14. — Cryptococcus neoformans: An ascomycete. Proc. Soc. exp. Biol. (N.Y.) 89, 243—245 (1955).

15. BEIJERINCK, M. W.: L'auxanographie, ou la méthode de l'hydrodiffusion dans la gelatine appliquée aux recherches microbiologiques. Arch. néerl. Sci. 23, 367—372 (1889).

16. BIEBERDORF, F. W.: A study of the growth response of some fungi to various concentrations of isonicotinic acid hydrazide. Antibiot. and Chemother. (Basel) 3, 513—520 (1953).

17. BOCOBO, F. C., A. C. CURTIS, W. D. BLOCK, E. R. HARRELL, E. E. EVANS and R. F. HAINES: Evaluation of nitrostyrenes as antifungal agents. Antibiot. and Chemother. (Basel) 6, 385—390 (1956).

18. BRIAN, P. W., and J. C. MCGOWAN: Viridin: a highly fungistatic substance produced by Trichoderma viride. Nature (Lond.) 156, 144—145 (1945).

19. BROCKMANN, D. D.: The in vitro effect of atabrine on Cryptococcus neoformans. Amer. J. trop. Med. 28, 295—297 (1948).

20. BROWN, R., and E. HAZEN: Nystatin and actidione: Two antifungal agents produced by Streptomyces noursei. In STERNBERG and NEWCOMER, Therapy of fungus diseases, p. 164—167. Boston and Toronto: Little, Brown & Comp. 1955.

21. BUELL, C. B., and W. H. WESTON: Application of the mineral oil conservation method to maintaining collection of fungus cultures. Amer. J. Bot. 34, 555—561 (1947).

22. BURKE, R. D., J. H. SWARTZ, S. S. CHAPMAN and W. Y. HUANG: Mycoticin, a new antifungal antibiotic. J. invest. Derm. 23, 163—168 (1957).

23. BUSCHKE, A.: Über eine durch Coccidien hervorgerufene Krankheit des Menschen. Dtsch. med. Wschr. 1895, 14.

24. BUSSE, O.: Über parasitäre Zelleinschlüsse und ihre Züchtung. Zbl. Bakt., I. Abt. Orig. 16, 175—180 (1894).

25. — Über Saccharomycosis hominis. Virchows Arch. path. Anat. 140, 23—46 (1895).

26. — Experimentelle Untersuchungen über Saccharomycosis. Virchows Arch. path. Anat. 144, 360—372 (1896).

27. CALDWELL, D. C., and S. S. RAPHAEL: A case of cryptococcal meningitis. J. clin. Path. 8, 32—37 (1955).

28. CARSON, L. E., and C. C. CAMPBELL: The inhibition effect of three antihistaminic compounds on the growth of fungi pathogenic for man. Science 111, 689—691 (1950).

29. CARTER, A. S., and J. L. YOUNG: Note on the isolation of Cryptococcus neoformans from a sample of milk. J. Path. Bact. 62, 271 (1950).

30. CARVAJAL, F.: Fluvomycin: an antibiotic effective against pathogenic bacteria and fungi. Antibiot. and Chemother. (Basel) 3, 765—772 (1953).

31. Carton, C. A.: Treatment of central nervous system cryptococcosis; a review and report of four cases treated with actidione. Ann. intern. Med. 37, 123—154 (1952).
32. —, and Ch. S. Liebig: Treatment of central nervous system cryptococcosis; laboratory studies. Arch. intern. Med. 91, 773—783 (1953).
33. Christensen, W. B.: Urea decomposition as a means of differentiating Proteus and Paracolon cultures from each other and from Salmonella and Shigella types. J. Bact. 52, 461—466 (1946).
34. Christison, J. G., N. F. Conant and C. K. Bradsher: The sensitivity of Histoplasma capsulatum to rhodanine and related compounds. In Sternberg and Newcomer, Therapy of fungus diseases, p. 268—278. Boston and Toronto: Little, Brown & Comp. 1955.
35. Ciferri, R.: Studi sulli torulopsidaceae. Atti Ist. Bot. Univ. Pisa, Ser. III 2, 147—301 (1925).
36. — A neglected definition of the genus Cryptococcus made by Rivolta (1873). Mycopathologia (Den Haag) 6, 135—136 (1951).
37. Conant, N. F., D. T. Smith, R. D. Baker, J. L. Callaway and D. St. Martin: Manual of clinical mycology, 2. edit. Philadelphia and London: W. B. Saunders Company 1954.
38. Connell, G. H., and C. E. Skinner: The origin of some species of the genus Candida in non-pseudomycelium-producing anaskosporogenous yeasts. Mycopathologia (Den Haag) 6, 65—71 (1951).
39. — — The external surface of the human body as a habitat for nonfermenting non-pigmented yeasts. J. Bact. 66, 627—633 (1953).
40. — — and R. C. Hurd: Lipomyces starkeyi on the skin surface of the human body. Mycologia (N. Y.) 46, 12—15 (1954).
41. Costantin, M. J.: Sur les levures des animaux. Soc. Mycol. France 17, 145—148 (1901).
42. Cox, L. B., and J. C. Tolhurst: Human Torulosis, p. 149. Melbourne: University Press 1946.
42a. Crone, J. T., A. F. de Groat and J. G. Wahlin: Torula infection. Amer. J. Path. 13, 863—879 (1937).
43. Curtis, A. C., F. C. Bocobo, E. R. Harrell and W. D. Block: Further studies on antifungal activity on some cinnamic acid derivatives and nitrostyrenes. Arch. Derm. Syph. (Chicago) 70, 786—798 (1954).
44. Danowski, T. S., and M. Tager: Thiourea and the inhibition of growth by fungi. J. infect. Dis. 82, 119—125 (1948).
45. Beurmann, L. de, et R. Gougerot: Les exascoses, endomycoses et parendomycoses (Muguet), saccharomycoses (mycose de Busse-Buschke) et parasaccharomycose-zymonématoses (mycose Gilchrist). Révision et démembrement de l'ancien groupe des blastomycoses. Trib. méd. Paris 42, 501—505, 517—524 (1909).
46. Diddens, H. A., u. J. Lodder: Die anaskosporogenen Hefen, 2. Hälfte. Amsterdam: Noord-Hollandsche Uitgevers Mantschappij 1942.
46a. Di Menna, M. E.: Cryptococcus terreus n. sp., from soil in New Zealand. J. gen. Microbiol. 11, 195—197 (1954).
47. Dodge, C. W.: Medical mycology. St. Louis: C. V. Mosby Comp. 1935.
48. Drake, C. H.: Antigenicity of Cryptococcus neoformans. Soc. Amer. Bact., Proc. 1948, p. 57.
49. Drouhet, E., et G. Segretain: Sur l'action de la lactoflavine à l'égard de Torulopsis histolytica (T. neoformans) et d'autres champignons pathogènes. C. R. Soc. Biol. (Paris) 142, 316—318 (1948).
50. — — Action de hyaluronidase sur la capsule de Torula histolytica. C. R. Acad. Sci. (Paris) 228, 424—425 (1949).
51. — — Biologie et pouvoir pathogène de torulopsis neoformans (Torula histolytica). Rev. Path. comp. 50, 37—43 (1950).
52. — — et J. P. Aubert: Polyoside capsulaire d'un champignon pathogène Torulopsis neoformans. Relation avec la virulence. Ann. Inst. Pasteur 79, 891—900 (1950).
53. —, et M. Couteau: Sur les variations sectorielles des colonies de Torulopsis neoformans. Ann. Inst. Pasteur 80, 456—457 (1951).

54. DROUHET, E., et G. SEGRETAIN: Actions réciproques entre phagocytes et Torulopsis neoformans. Bull. sci. roum. Sect. Sci. méd. 1, 40—44 (1952).

55. EINBINDER, J. M., RH. W. BENHAM and C. T. NELSON: Chemical analysis of the capsular substance of Cryptococcus neoformans. J. invest. Derm. 22, 279—283 (1954).

56. EMMONS, CH.: In HENRICI, Molds, yeasts and actinomycetes, 2. edit. New York and London: John Wiley & Sons 1947.

57. — The natural occurrence in animals and soil of fungi which cause disease in man. Proc. 7. Int. Bot. Congr., Stockholm 1950, p. 416—421.

58. — Isolation of Cryptococcus neoformans from soil. Bact. Proc. 2, 113 (1951). — J. Bact. 62, 685—690 (1951).

59. — Cryptococcus neoformans strains from severe outbreak of bovine mastitis. Mycopathologia (Den Haag) 6, 231—234 (1952).

60. — The significance of saprophytism in the epidemiology of the mycoses. Trans. N. Y. Acad. Sci., Ser. II 17, 157—166 (1954).

61. — Saprophytic sources of Cryptococcus neoformans associated with the pidgeon (Columbia livia). Amer. J. Hyg. 62, 227—232 (1955).

62. EVANS, E. E.: An immunologic comparison of twelve strains of Cryptococcus neoformans (Torula histolytica). Proc. Soc. exp. Biol. (N. Y.) 71, 644—646 (1949).

63. — The antigenic composition of Cryptococcus neoformans. I. A serologic classification by means of the capsular and agglutination reactions. J. Immunol. 64, 423—430 (1950).

64. —, and J. W. MEHL: A qualitative analysis of capsular polysaccharides from Cryptococcus neoformans by filter paper chromatography. Science 114, 10—11 (1951).

65. —, and J. F. KESSEL: The antigenic composition of Cryptococcus neoformans. II. Serologic studies with the capsular polysaccharides. J. Immunol. 67, 109—114 (1951).

66. —, and E. R. HARRELL jr.: Cryptococcosis (Torulosis): A review of recent cases. Univ. Mich. med. Bull. 18, 43—63 (1952).

67. —, and R. J. THERIAULT: The antigenic composition of Cryptococcus neoformans. IV. The use of paper chromatography for following purification of the capsular polysaccharide. J. Bact. 65, 571—577 (1953).

68. — L. J. SORENSEN and K. W. WALLS: The antigenic composition of Cryptococcus neoformans. V. A survey of cross-reactions among strains of Cryptococcus and other antigens. J. Bact. 66, 287—293 (1953).

69. —, and R. F. HAINES: The agglutination of ion exchange resin particles coated with polysaccharide. J. Bact. 68, 130—131 (1954).

70. — H. P. R. SEELIGER, L. KORNFELD and O. GARCIA: Failure to demonstrate capsular swelling in Cryptococcus neoformans. Proc. Soc. exp. Biol. (N. Y.) 93, 257—260 (1956).

71. FISCHER, J. B., and N. A. LABZOFFSKY: Preparation of complement fixing antigen from C. neoformans. Canad. J. Microbiol. 1, 451—454 (1955).

72. FISHER, A. M.: The clinical picture associated with infections due to Cryptococcus neoformans (Torula histolytica). Report of three cases with some experimental studies. Bull. John Hopk. Hosp. 86, 383—414 (1950).

73. FLORESTANO, H. J., and M. E. BAHLER: Antifungal properties of the polymyxins. Proc. Soc. exp. Biol. (N. Y.) 79, 141—143 (1952).

74. FOLEY, G. E., and L. L. UZMAN: Studies on the biology of Cryptococcus. IV. Isolation of a highly polymerized polysaccharide from encapsulated strains. J. infect. Dis. 90, 38—43 (1952).

75. FREEMAN, W.: Torula meningo-encephalitis; comparative histopathology in seventeen cases. Trans. Amer. neurol. Ass. 203—217 (1930).

76. GIORDANO, A.: Studia micologica de Debaryomyces neoformans (SANFELICE) REDAELLI, CIFERRI et GIORDANO e significata della specie nella patologia animale. Mycopathologia (Den Haag) 1, 274—304 (1939).

77. GOLD, W., H. A. STOUT, J. F. PAGANO and R. DONOVICK: Amphothericins A and B; antifungal antibiotics produced by a streptomycete. Antibiot. Ann. 1955/56, 576—586.

78. GONZALEZ-OCHOA, A., L. F. BOJALIL-JABER y R. SOTO PACHECO: Acciòn estimulante de las sulfonamidas — cobre y de la penicillina sobre el desarrollo de Cryptococcus neoformans y Sporotrichum schenckii, respectivamente. Rev. Soc. Mex. Hist. Nat. 11, 35—41 (1950).

79. Gonzalez-Ochoa, A., L. F. Bojalil-Jaber y R. Soto Pacheco: Actividades „in vitro" de complejos sulfa-cotre sobre algunos hongos pathogénos. Rev. Inst. Salubr. Enferm. trop. (Méx.) 11, 79—91 (1950).

80. Gottlieb, D.: Filipin, an antibiotic inhibiting fungi. In Sternberg and Newcomer, Therapy of fungus diseases, p. 142—146. Boston and Toronto: Little, Brown & Comp. 1955.

81. Granits-Thurner, J., u. H. Schirmer: Über die Wirkung von Zimtsäurederivaten auf hautpathogene Pilze. Mykosen 1, 60—63 (1957).

82. Gridley, R. G.: A stain for fungi in tissue sections. Amer. J. clin. Path. 28, 303—307 (1953).

83. Grocott, R. G.: A stain for fungi in tissue sections and smears using Gomori's methenamin-silver nitrate technic. Amer. J. clin. Path. 25, 975—979 (1955).

84. Grunberg, E., and R. J. Schnitzer: A chemotherapeutic study of antimonials in the experimental infection of mice with Cryptococcus neoformans. Yale J. Biol. Med. 26, 132—138 (1953).

85. Haskins, R. H., and J. A. Thorn: Biochemistry of the ustilaginales. VII. Antibiotic activity of ustilagic acid. Canad. J. Bot. 29, 585—592 (1951).

86. Hazen, E. L., and R. Brown: Fungicidin, an antibiotic produced by a soil actinomycete. Proc. Soc. exp. Biol. (N.Y.) 76, 93—97 (1951).

87. Hehre, E. J., A. S. Carlson and D. M. Hamilton: Crystalline amylose from cultures of a pathogenic yeast (Torula histolytica). J. biol. Chem. 177, 289—293 (1949).

88. Hickey, R. J., C. J. Corum, P. H. Hidy, J. R. Cohen, U. F. B. Nager and E. Kropp: Ascosin, an antifungal antibiotic produced by a streptomycete. Antibiot. and Chemother. (Basel) 2, 472—483 (1952).

89. Hirth, L., et E. Drouhet: Inhibition du virus de la mosaïque du tabac par le polyoside capsulaire de Torulopsis neoformans. Ann. Inst. Pasteur 84, 437—440 (1953).

90. Hobby, G. L., P. P. Regna, N. Dougherty and W. E. Stieg: The antifungal activity of antibiotic XG. J. clin. Invest. 28, 927—933 (1949).

91. Hoff, C. L.: Immunity studies of Cryptococcus hominis (Torula histolytica) in mice. J. Lab. clin. Med. 27, 751—754 (1942).

92. Hoffmeister, W.: Die Torulopsis neoformans-Infektion. Klin. Wschr. 1951, 301—307.

93. Hotchkiss, R. D.: A microchemical reaction resulting in the staining of polysaccharide structures in fixed tissue preparations. Arch. Biochem. 16, 131—141 (1948).

94. Johnson, C. W., J. W. Joyner and R. P. Perry: The inhibitory effect of 4 thiosemicarbazones on Cryptococcus neoformans. Antibiot. and Chemother. (Basel) 2 636—638 (1952).

95. — E. A., and K. L. Burdon: Eumycin — a new antibiotic active against pathogenic fungi and higher bacteria, including bacilli of tuberculosis and diphtheria. J. Bact. 51, 591 (1946).

96. Kaufmann, W.: Die serologische Differenzierung von Torulopsis neoformans und Torulopsis albida mittels der Präzipitation und Agglutination. Zbl. Bakt., II. Abt. 106, 434—437 (1944/45).

97. Keeney, E. L., L. Ajello and E. Lankford: Studies on common pathogenic fungi and on Actinomyces bovis. II. In vitro effect of sulfonamides. Bull. Johns Hopk. Hosp. 75, 393—409 (1944).

98. Kligman, A. M.: Studies of the capsular substance of Torula histolytica and the immunologic properties of Torula cells. J. Immunol. 57, 395—401 (1947).

99. —, and F. D. Weidman: Experimental studies on treatment of human torulosis. Arch. Derm. Syph. (Chicago) 60, 726—741 (1949).

100. —, and E. D. de Lamater: The immunology of the human mycoses. Ann. Rev. Microbiol. 1950, 283—312.

101. — H. Mescon and E. D. de Lamater: Hotchkiss-McManus-Stain for histopathologic diagnosis of fungus diseases. Amer. J. clin. Path. 21, 86—91 (1951).

102. — A. P. Crane and R. F. Norris: Effect of temperature on survival of chick embryos infected intravenously with Cryptococcus neoformans (Torula histolytica). Amer. J. med. Sci. 221, 273—278 (1951).

102 a.—, and F. S. Lewis: In vitro and in vivo activity of candicidin on pathogenic fungi. Proc. Soc. exp. Biol. (N.Y.) 82, 399—404 (1953).

103. KOCH, H.: Experimentelle Untersuchungen über den Einfluß neuerer Sulfonamide auf hautpathogene und andere Pilze. Arch. klin. exp. Derm. **205**, 1—21 (1957).

104. KÖNIGSBAUER, H.: Experimenteller Beitrag zur Behandlung der Torulose mit D 25 (2,2-Dioxy-5,5-dichlordiphenylsulfid). Zbl. Bakt., I. Abt. Orig. **164**, 466—471 (1955).

105. KUHN, L. R.: Growth and viability of Cryptococcus hominis at mouse and rabbit body temperatures. Proc. Soc. exp. Biol. (N.Y.) **41**, 573—574 (1939).

106. — Effect of elevated body temperatures on cryptococcosis in mice. Proc. Soc. exp. Biol. (N.Y.) **71**, 341—343 (1949).

107. LANDY, M., S. B. ROSENMAN and G. H. WARREN: An antibiotic from Bacillus subtilis active against pathogenic fungi. J. Bact. **54**, 24 (1947).

108. LANGERON, M.: Précis de mycologie. Paris: Marion 1945.

109. LARSH, H. W., ST. L. SILBERG and A. HINTON: Use of the tissue culture method in evaluating antifungal agents. Antibiot. Ann. **1956/57**, 918—922.

110. — A. HINTON and ST. L. SILBERG: The use of the tissue culture method in evaluating antifungal agents against systemic fungi. Antibiot. Ann. **1957/58**, 988—991.

111. LITTMAN, M. L., and L. E. ZIMMERMAN: Cryptococcosis. New York and London: Grune & Stratton 1956.

112. — An improved method for detection of urea hydrolysis by fungi. J. infect. Dis. **101**, 51—61 (1957).

113. — Antimycotic effect of chorquinaldol. Trans. N. Y. Acad. Sci. **18**, 161—177 (1955).

114. LODDER, J.: Die anaskosporogenen Hefen, 1. Hälfte. Verh. Kon. Akad. Wet., Amsterdam Afd. Naturk. **32**, 1—256 (1934).

115. — Torulopsis or Cryptococcus. Mycopathologia (Den Haag) **1**, 62—67 (1938).

116. —, and A. DE MINJER: On the biology of the pathogenic Torulopsidoideae. In W. J. NICKERSON, Biology of pathogenic fungi. Waltham, Mass.: Chronica Botanica 1947.

117. —, and N. J. KREGER-VAN RIJ: The yeasts. A taxonomic study, 1, ed., p. 713. Amsterdam: North Holland Publ. Comp. **1952**.

118. LUDWIG, K. A., F. Y. MURRAY, J. K. SMITH, C. R. THOMPSON and H. W. WERNER: Laboratory studies on β-di-ethyleminoethylfencholate, a new antifungal agent. Antibiot. and Chemother. (Basel) **4**, 56—61 (1954).

119. MAGER, J., and M. ASCHNER: Starch reaction as aid in identification of causative agent of European Blastomycosis. Proc. Soc. exp. Biol. (N.Y.) **62**, 71—72.

120. — — Biological studies on capsulated yeasts. J. Bact. **53**, 283—295 (1947).

121. MCMANUS, J. F. A.: Histological and histochemical uses of periodic acid. Stain Technol. **23**, 99—108 (1948).

122. MEZEY, C. M., and R. FOWLER: Cerebrospinal cryptococcosis. J. Amer. med. Ass. **132**, 632—634 (1946).

123. MILLER, J. M., E. B. SCHOENBACH, P. H. LONG, J. S. SCHUTTLEWORK and G. E. SNIDER: Treatment of infections due to Cryptococcus neoformans with stilbamidine, antibiotics and chemotherapy. Antibiot. and Chemother. (Basel) **2**, 444—446 (1952).

124. MOLNAR, J.: Über die dichroitische Färbung der Kapsel des Cryptococcus neoformans. Schweiz. Z. Path. **19**, 82—87 (1956).

125. — Meningo-encephalitis caused by Cryptococcus neoformans. Acta morph. Acad. Sci. hung. **6**, 233—239 (1955).

125a. MOSBERG jr., W. H., and J. A. ALVAREZ-DE CHOUDENS: Torulosis of the central nervous System. Lancet **1951** I, 1259—1260.

125b. —, and J. C. ARNOLD jr.: Torulosis of the central nervous system. Ann. intern. Med. **32**, 1153—1183 (1950).

126. NEGRONI, P., y C. BRIZ DE NEGRONI: Estudios sobre el Cryptococcus neoformans (SANFELICE). II. Aspecto micromorfologico y citologia. Ann. Soc. Ci. argent. **151**, 71—76 (1951).

127. —, y C. A. LANATA: Estudios sobre el Cryptococcus neoformans (SANFELICE). III. La formacion capsular. Ann. Soc. Ci. argent. **113**, 200—211 (1952).

128. NEILL, J. M., C. G. CASTILLO, R. H. SMITH and CH. E. KAPROS: Capsular reactions and soluble antigens of Torula histolytica and of Sporotrichum schenckii. J. exp. Med. **89**, 93—106 (1949).

129. Neill, J. M., and Ch. E. Kapros: Serological tests on soluble antigens from mice infected with Cryptococcus neoformans and Sporotrichum schenckii. Proc. Soc. exp. Biol. (N.Y.) 73, 557—559 (1950).

130. Neill, J. M., I. Abrahams and Ch. E. Kapros: A comparison of the immunogenicity of weekly encpasulated and strongly encapsulated strains of Cryptococcus neoformans. J. Bact. 59, 263—275 (1950).

131. — J. V. Sugg and D. W. McCauley: Serologically reactive material in spinal fluid, blood and urine from a human case of cryptococcosis (torulosis). Proc. Soc. exp. Biol. (N.Y.) 77, 775—778 (1951).

132. Nickerson, W. J.: Studies on film forming yeasts. Mycologia (N.Y.) 36, 224—233 (1944).

133. Niño, F. A.: Contribución al estudio de las blastomicosis en la republica argentina: Cap. V. Granuloma criptococcico. Bol. Inst. Clín. quir. Univ. B. Aires 14, 656—755 (1938).

134. Petru, M., u. M. Vojtěchovská: Zur Frage der sog. Vaginal-Kryptokokken. Zbl. Bakt. I. Abt. Orig. 166, 218—224 (1956).

135. Pounden, W. D., J. M. Amberson and R. F. Jaeger: A severe mastitis problem associated with Cryptococcus neoformans in a large dairy herd. Amer. J. vet. Res. 18, 121—128 (1957).

136. Raper, K. B., and D. F. Alexander: Preservation of molds by the lyophil process. Micologia (N.Y.) 37, 499—525 (1945).

136a Rappaport, B. Z., and B. Kaplan: Generalized Torula mycosis. Arch. Path. (Chicago) 1, 720—741 (1926).

137. Ratcliffe, H. E., and W. R. Cook: Cryptococcosis; review of the literature and report of a case with initial pulmonary findings. U.S. armed Forces med. J. 1, 957—969 (1950).

138. Redaelli, P.: Il problema della Torulopsidaceae e dei loro rapporti con l'uomo e con la pathologica umana studiato particolarmente in Italia. Vi. biol. 13, 171—235 (1931)

139. — R. Ciferri e A. Giordano: Debaryomyces neoformans (Sanfelice) nobis, n. comb., pour les espèces de la groupe Saccharomyces hominis — Cryptococcus neoformans — Torula histolytica. Soz. internaz. Microbiol. Boll. sez. ital. 9, 24—28 (1937).

140. Reid, J. D.: The influence of Vitamin B complex on the growth of Torulopsis (Cryptococcus) neoformans on a synthetic medium. J. Bact. 58, 777—782 (1949).

141. Reilly, H. C., A. Schatz and S. A. Waksman: Antifungal properties of antibiotic substances. J. Bact. 49, 585—594 (1945).

142. Ribi, E., S. B. Salvin and W. R. Brown: Antigens from the yeast phase of Histoplasma capsulatum. I. Morphology of the cell as revealed by the electron microscope. Exp. Cell Res. 10, 394—404 (1956).

143. Rinehart, J. F., and S. K. Abdul-Haj: An improved method for histologic demonstration of acid mucopolysaccharides. Arch. Path. (Chicago) 52, 189—194 (1951).

144. Roberts, C.: Observations on the yeast Lipomyces. Nature (Lond.) 179, 1198—1199 (1957).

145. Salvin, S. B.: Quantitative studies as the serologic relationships of fungi. J. Immunol. 65, 617—626 (1950).

146. — Endotoxins in pathogenic fungi. J. Immunol. 69, 89—99 (1952).

147. Sanfelice, F.: Über eine für Tiere pathogene Sproßpilzart und über die morphologische Übereinstimmung, welche sie bei ihrem Vorkommen in den Geweben mit den vermeintlichen Krebscoccidien zeigt. Zbl. Bakt., I. Abt. Orig. 17, 113—118 (1895).

148. — Über einen neuen pathogenen Blastomyceten, welcher innerhalb der Gewebe unter Bildung kalkartig aussehender Massen degeneriert. Zbl. Bakt., I. Abt. Orig. 18, 521—526 (1895).

149. — Sulla azione patogena dei blastomiceti. Ann. Ist. Ig. Univ. Roma 5, 239—262 (1895).

150. Sasakawa, M.: Zur Systematik pathogener und parasitischer Hefen. Morphologisch-biochemische Studie. Zbl. Bakt., I. Orig. 88, 269—285 (1922).

151. Scherr, G. H., and R. H. Weaver: The dimorphism phenomenon in yeasts. Bact. Rev. 17, 51—92 (1933).

152. Schmidt, E. G., J. A. Alvarez-de Choudens, N. F. McElvain, J. Beardsley and S. A. A. Tawab: A microbiological study of Cryptococcus neoformans. Arch. Biochem. 26, 15—24 (1950).

153. SCHMIDT, W. J.: Dichroitische Färbung tierischer und pflanzlicher Gewebe. In Handbuch der biologischen Arbeitsmethoden, Abt. V, 2/2, S. 1835. Wien: Urban & Schwarzenberg 1932.

154. SEELIGER, H. P. R.: Experimentelle Untersuchungen zur mykologischen Serodiagnostik. Habil.-Schrift Bonn 1954.

155. — Zur Anwendungsmöglichkeit der Präzipitation im Agar-Gel bei der O-Antigenanalyse von Bakterien und Pilzen. Z. Hyg. Infekt.-Kr. 141, 110—121 (1955).

156. — Ein neues Medium zur Pseudomyzelbildung von Candida albicans. Z. Hyg. Infekt.-Kr. 141, 488—494 (1955).

157. — Lungenmykosen und ihre Immunbiologie. Med. Wschr. 1954, 692—698.

158. — Mykologie und Serologie der Pneumomykosen. Tuberk.-Arzt 9, 451—463 (1955).

159. — Use of a urease test for the screening and identification of Cryptococci. J. Bact. 72, 127—131 (1956).

160. — Pilzbefall und Mykosen des Menschen. Röntgen- u. Lab.-Prax. 9, 192—203, 221—233 (1956).

161. — Mykologische Serodiagnostik. Leipzig: Johann Ambrosius Barth 1958.

162. —, u. P. CRIST: Zur Schnelldiagnose der Cryptococcus-Meningitis mittels der Liquorpräzipitation. Mykosen 1, 88—92 (1958).

163. — Über die pilzhemmende Wirkung eines neuen Benzimidazolderivats. Mykosen 1, 162—171 (1958).

164. — Cryptococcose (Torulose). Mitt. der Schering-Werke 19 (1958).

165. — Unveröffentlichte Untersuchungen, 1953—1957.

166. SEGRETAIN, G., et E. DROUHET: L'action de la streptomycine sur Torulopsis histolytica (T. neoformans) in vitro et in vivo. C. R. Soc. Biol. (Paris) 142, 319—320 (1948).

167. — — et F. MARIAT: Essai de mise en évidence à l'état frais de la capsule d'une levure pathogène. Bull. Microscop. appl., Sér. II 2, 17—18 (1952).

168. SENECA, H., J. H. KANE and J. ROCKENBACH: Bactericidal, protozoicidal and fungicidal properties of thiolutin. Antibiot. and Chemother. (Basel) 2, 357—360 (1952).

169. SHEPPE, W. M.: Torula infection in man. Amer. J. med. Sci. 167, 91—108 (1924).

169a. SHIMWELL, J. L.: A simple staining method for the detection of askospores in yeasts. J. Inst. Brewing 35, 474 (1938).

170. SIEGEL, M.: Fungistatic activity of methylparaben and propylparaben. Antibiot. and Chemother. (Basel) 3, 478—480 (1953).

171. SIMON, J.: In vitro inhibition of mixed strains of Cryptococcus neoformans isolated from cattle. Amer. J. vet. Res. 16, 394—396 (1955).

172. SKINNER, CH. E., CH. W. EMMONS and A. M. TSUCHIYA: In Henrici's molds, yeasts and actinomycetes, 2. edit. New York and London: John Wiley & Sons 1947.

173. — Generic name for imperfect yeasts, Cryptococcus and Torulopsis. Amer. Midld. Naturalist 43, 242—250 (1950).

174. SOLOTOROVSKY, M. H., E. J. IRONSON, F. J. GREGORY and S. WINSTEN: Activity of certain diamidines against blastomycosis and Candida infections in mice. Antibiot. and Chemother. (Basel) 4, 165—168 (1954).

175. STANLEY, N. F.: Polysaccharide haptens from Torula histolytica. Amer. J. exp. Biol. Med. Sci. 27, 409—415 (1949).

176. STEWARD, R. A., and K. F. MEYER: Isolation of C. immitis (STILES) from soil. Proc. Soc. exp. Biol. (N. Y.) 29, 937—938 (1932). Zit. nach EMMONS (*60*).

177. STODDARD, J. L., and E. C. CUTLER: Torula infection in man. Monogr. Rockefeller Inst. med. Res. No 6 (1916), 98 S.

178. SYMMERS, W. F.: Torulosis. Lancet 1953 II, 1068—1074.

179. TAGER, M., and T. S. DANOWSKI: Inhibition of the growth of fungi by thiourea derivatives, particularly hydrazine dithiocarbamyl. J. infect. Dis. 82, 126—130 (1948).

180. — A. B. HALES and T. S. DANOWSKI: Studies on the suppression of fungus growth by thiourea. III. The effects of protein fractions, amino acids and thiol compounds. J. infect. Dis. 84, 284—289 (1949).

181. THAYER, J. D., W. GARSON and A. J. WALKER: The in vitro and in vivo effect of circulin on Cryptococcus neoformans. J. Bact. 62, 112 (1951).

182. Todd, R. L., and W. W. Hermann: The life cycle of the organism causing yeast-meningitis. J. Bact. **32**, 89—102 (1936).

183. Tomcsik, J., and S. Guex-Holzer: Demonstration of the bacterial capsule by means of a pH-dependant salt-like combination with protein. J. gen. Microbiol. **10**, 97—109 (1954).

184. Uzman, L. L., H. Rosen and G. E. Foley: Studies on the biology of Cryptococcus. VI. Amino acid composition of the somatic proteins of Cryptococcus neoformans. J. infect. Diss. **98**, 208—210 (1956).

185. Vanbreuseghem, R.: Torulosis et Torulopsis neoformans. Ann. Soc. belge Méd. trop. **23**, 495—501 (1953).

186. Verdun, P.: Précis de parasitologie humaine, parasites animaux et végétaux, les bactéries exceptées. 2. edit. Paris: V. Doin 1912.

187. Vuillemin, P.: Les blastomycètes pathogènes. Rev. gén. Sci **12**, 732—751 (1901).

188. Waksman, S. A.: Outstanding problems in the study of antibacterial and antifungal antibiotics (with special reference to the antibiotics or actinomycetes). In Sternberg and Newcomer, Therapy of fungus diseases, p. 3—12. Boston and Toronto: Little, Brown & Comp. 1955.

189. Walton, R. B., and H. B. Woodruff: A crystalline antifungal agent, mycosubtilin, isolated from Subtilis broth. J. clin. Invest. **28**, 924—926 (1949).

190. Weiss, J. D.: Four pathogenic Torulae (blastomycetes). J. med. Res. **2**, 280—311 (1902).

191. Whiffen, A. J., N. Bohonos and R. L. Emerson: The production of an antifungal antibiotic by Streptomyces griseus. J. Bact. **52**, 610-611 (1946).

192. Wickerham, L. J.: A critical evaluation of the nitrogen assimilation tests commonly used in the classification of yeasts. J. Bact. **52**, 293—301 (1946).

193. —, and K. A. Burton: Carbon assimilation tests for the classification of yeasts. J. Bact. **56**, 363—371 (1948).

194. —, and M. H. Flickinger: Viability of yeasts preserved two years by the lyophil process. Brewers Digest **21**, No 4 (1946).

195. — Recent advances in the taxonomy of yeasts. Ann. Rev. Microbiol. **6**, 317—332 (1952).

196. Wilson, H. M., and A. W. Duryea: Cryptococcus meningitis (Torulosis) treated with a new antibiotic, actidione. Arch. Neurol. Psychiat. (Chicago) **66**, 470—480 (1951).

197. Wolf, F. T.: Inhibition of pathogenic fungi by p-hydroxymethylbenzoate. Mycopathologia (Den Haag) **5**, 117—119 (1950).

198. Wolfram, St., u. F. Zach: Über durch niedere Pilze verursachte Nagelerkrankungen beim Menschen. Arch. Derm. Syph. (Berl.) **170**, 681—694 (1934).

199. Beutmann, W.: Immunbiologische und serologische Studien an Hefen. Arch. Mikrobiol. **29**, 227—256 (1958).

200. Burcik, E., u. W. Beutmann: Zuckerbausteine wasserlöslicher Hefepolysaccharide von Haptencharakter. Naturwiss. **43**, 427—428 (1956).

201. Eveland, W. C., J. D. Marshall and A. M. Silverstein: Rapid identification of pathogenic microorganisms using fluorescent antibodies of contrasting colors. Abstracts of communications, 7. Internat. Congr. for Microbiology, Stockholm, S. 313. 1958.

202. Holtz: Briefl. Mitt., Juli 1958.

203. Klatzo, I., and P. H. Geisler: Demonstration of Cryptococcus neoformans in polarized light. Stain Technol. **33**, 55—56 (1958).

204. Staib, F.: Resistenz von Hefen und hefeartigen Pilzen gegen Schwefelsäure. Z. Hyg. Infekt.-Kr. **145**, 26—33 (1958).

Ätiologie und Epidemiologie der Rickettsiosen des Menschen

Von

F. WEYER [1]

Inhalt

Vorbemerkungen und Stoffabgrenzung

In der Rickettsienforschung zeichnen sich deutlich drei Perioden besonders intensiver und fruchtbarer Arbeit ab, die auf wichtige Einzelbeobachtungen oder ganz bestimmte Fortschritte unserer Kenntnisse zurückgehen. Die erste Periode wird durch die Entdeckung der Erreger des Felsengebirgs- und des klassischen Fleckfiebers eingeleitet, an welcher RICKETTS (1), (2), NICOLLE, CONOR und CONSEIL, ferner DA ROCHA LIMA sowie WOLBACH entscheidenden Anteil hatten. Ein zweiter Impuls ist mit der Abgrenzung des murinen vom klassischen Fleckfieber [NEILL; MOOSER (1)], dem sicheren Nachweis der Erreger in Warmblütern [MAXCY (1), (2), MOOSER (1), WOHLRAB] und mit der Züchtung von Rickettsien

---

[1] Bernhard-Nocht-Institut für Schiffs- und Tropenkrankheiten, Hamburg.

in der Gewebekultur (NIGG und LANDSTEINER) gegeben. Die dritte Forschungs-
periode wird gekennzeichnet durch die Entdeckung des Q-Fiebers [DERRICK (1);
BURNET und FREEMAN; DAVIS und COX], die Entwicklung von Verfahren zur
Gewinnung größerer Rickettsienmengen aus dem Dottersack des Hühnereies
[COX (1); OTTO und WOHLRAB] und durch grundlegende Fortschritte in der serologischen Diagnostik, in der Therapie und Prophylaxe der Rickettsiosen, an
welchen besonders amerikanische Forscher hervorragenden Anteil hatten.

Die letzten Jahre haben die praktischen Ergebnisse bedeutend erweitert. Die
wesentlichsten Forschungsresultate betreffen den Feinbau und Stoffwechsel der
Rickettsien, Virulenzänderungen, das Verhalten der Erreger in Warm- und Kalt-
blütern, das Interferenzphänomen und die Epidemiologie der Rickettsiosen, insbesondere die Epidemiologie des Felsengebirgsfleckfiebers, des Q-Fiebers und
des klassischen Fleckfiebers. Unsere Kenntnisse über die Reservoire, die Übertra-
gungsmechanismen und die natürlichen Überträger der Rickettsiosen erhielten
wichtige Ergänzungen und teilweise neue Aspekte.

Der folgende Bericht berücksichtigt in erster Linie die Forschungsergebnisse aus der
Zeit von 1950 bis etwa 1957[1]. Zur Abrundung der erörterten Fragestellungen ist in manchen
Fällen auch auf wichtigere, vor diesem Zeitpunkt erschienene Arbeiten zurückgegriffen. Diese
Übersicht kann sich auf mehrere, in den letzten Jahren erschienene in- und ausländische
Veröffentlichungen stützen oder an Ergebnisberichte anschließen, in welchen die menschlichen
Rickettsiosen insgesamt oder in Teilgebieten monographisch behandelt wurden. Die betreffen-
den Beiträge sind im Schrifttumsverzeichnis aufgeführt.

Eine erschöpfende Darstellung des seit 1950 erschienenen Schrifttums über Rickettsiosen
kann nicht gegeben werden, es wurde jedoch versucht, ein möglichst vollständiges Bild der
behandelten Themen und Probleme zu vermitteln. Die große Zahl der Veröffentlichungen
macht nicht nur die Beschränkung auf die wichtigsten Arbeiten und Ergebnisse, sondern auch
auf einen *Ausschnitt* aus dem Gesamtgebiet notwendig. Es sollen daher im ersten Teil dieses
Berichtes, welchem eine Erörterung der Systematik, Nomenklatur und Einteilung voran-
gestellt ist, Forschungsergebnisse über Bau, Eigenschaften und Lebensäußerungen der
Rickettsien dargestellt werden, während sich der zweite Teil mit der Übertragung und Epidemiologie der Rickettsiosen befaßt. Im Zusammenhang mit der Epidemiologie muß auch
kurz auf die Verbreitung eingegangen werden. Klinik, Pathologie, Diagnostik und Therapie
der Rickettsiosen sind bereits in Hand- und Lehrbüchern oder an anderen Stellen besprochen.
Die entscheidenden Fortschritte auf diesen Gebieten liegen schon länger zurück und werden
inzwischen praktisch ausgewertet. Auch im epidemiologischen Teil können nicht alle Beobachtungen über Auftreten und Übertragung von Rickettsiosen in verschiedenen Ländern und
über den Erregernachweis bei Mensch und Tier angeführt werden, sondern diese Ergebnisse
sind in einigen charakteristischen Beispielen berücksichtigt.

I. Die Erreger und ihre Eigenschaften

1. Einteilung und Benennung (Klassifikation)

Im Interesse der Verständigung haben sich eine Reihe von Untersuchern
in den letzten Jahren speziell mit Fragen der *Systematik der Rickettsien* beschäftigt. Es ist unerläßlich, genau zu wissen, auf welche Objekte sich die
Beobachtungen beziehen und ob man die gleichen Erreger meint, wenn man
bestimmte Namen benutzt. Taxonomie und Nomenklatur der Rickettsiose-
erreger sind Gegenstand von besonderen Konferenzen und Publikationen gewesen, welche die Situation allerdings nicht endgültig klären konnten. Der von
DA ROCHA LIMA (2), (3) geschaffene, ursprünglich recht enge Rickettsienbegriff

---

[1] Das Manuskript wurde am 1. 7. 58 abgeschlossen.

wurde in der Folge mehrfach erweitert und abgeändert. Zu den Grundeigenschaften der Rickettsien gehört, daß sie sich in den Geweben bestimmter Arthropoden vermehren, einen Teil ihres Lebenscyclus in Arthropoden durchlaufen und bei einem Wirtswechsel von den Arthropoden auf Warmblüter übertragen werden können.

Inzwischen sind durchaus begreifliche Versuche unternommen worden, die Fortschritte unserer Kenntnisse auch in der Klassifikation der Rickettsien zum Ausdruck zu bringen. Die *Systematik* der Rickettsien nimmt auf mehrere Eigenschaften der Organismen Bezug. Da morphologische Kriterien nicht ausreichen, sind andere Merkmale, Färbbarkeit, biochemische Leistungen, antigene Struktur,

Tabelle 1. *Einteilung und Benennung der Rickettsiosen und ihrer Erreger (vergl. S. 76)*

Bezeichnung der Krankheit	Benennung des Erregers
1. Durch Zecken übertragene Rickettsiosen	
Felsengebirgsfleckfieber, Rocky Mountain Spotted Fever	*Rickettsia (Dermacentroxenus) rickettsii (Ixodoxenus rickettsii)*
Amerikanisches (neuweltliches) Fleckfieber oder Zeckenbißfieber	
São Paulo-Fleckfieber, Neotropisches Fleckfieber (Brasilien), Tobia-Fieber (Kolumbien)	*(R.brasiliensis, D.rickettsi* var. *brasiliensis)*
Altweltliche Zeckenfieber oder Zeckenbißfieber	*R. (D.) conori (D. rickettsi* var. *conori, I. conori,*
Boutonneuse-Fieber, Mittelmeerfieber, Marseiller Fieber	*I. conori* var. *mediterraneus)*
Zentralafrikanisches Zeckenbißfieber	*(I. conori* var. *africanus)*
Kenia- und Nigeria-Zeckenbißfieber	
Südafrikanisches Zeckenbißfieber	*(R. megawi* var. *pijperi, R. conori* var. *pijperi, D.rickettsi* var. *pijperi, I.conori* var. *africanus)*
Sibirisches oder asiatisches Zeckenbißfieber, Siberian tick typhus, Ixodo-Rickettsiosis asiatica	*R. conori (D. sibiricus, D.rickettsi* var. *sibirica, I. sibiricus)*
Indisches Zeckenbißfieber	*R. (D.) conori*
Nord-Queensland-Zeckenbißfieber	*R. (D.) australis (I. australis)*
Q-Fieber	*Coxiella burneti*
Derrick-Burnetsche Krankheit	*(C.diaporica, R.burneti, R.diaporica)*
2. Durch Milben übertragene Rickettsiosen	
Rickettsienpocken Gamaso-Rickettsiosis vesiculosa	*R. (D.) akari (Acaroxenus varioloides, Gamasoxenus muris)*
Tsutsugamushi-Fieber, Milbenfleckfieber, Kedani-Krankheit, Scrub typhus	*R. tsutsugamushi, R. orientalis (Zinssera tsutsugamushi, R. akamushi, Trombidixenus orientalis* [var. *nipponica, indica, indonesica, australis] R. tamiyai)*
3. Durch Insekten übertragene Rickettsiosen	
Klassisches (epidemisches) Fleckfieber, Läusefleckfieber, Typhus exanthematicus, Louse-borne typhus	*R. prowazeki (prowazekii)*
Murines (endemisches) Fleckfieber, Flohfleckfieber, Rattenfleckfieber, Flea-borne typhus, Shop typhus	*R. mooseri (R.prowazeki* var. *mooseri, R.typhi, R. muricola, R. manchuriae)*
Wolhynisches Fieber, Febris wolhynica Fünftagefieber, Febris quintana, Trench fever, Schützengrabenfieber	*R. quintana (R.wolhynica, R.weigli, R.pediculi, Wolhynia quintanae, Rochalimaea quintana)*

Beziehungen zu bestimmten Zellen, Verhaltensweise in den Wirten oder in der Kultur, Übertragung, Widerstandsfähigkeit gegen äußere Einflüsse, pathogene Eigenschaften u. a. mit für die Klassifikation herangezogen worden (Wolbach; Pinkerton). Die Schwierigkeiten für die Aufstellung eines einheitlichen Systems liegen darin, daß es keinen objektiven Maßstab für die systematische Bewertung der bisher bekannt gewordenen Eigenschaften gibt. Man darf nicht vergessen, daß wir beim Studium der Rickettsien zunächst doch nur bestimmte, unserer Beobachtung zugängliche Merkmale ermittelt haben. Wir wissen aber nicht, ob es sich hierbei nicht um zufällig bzw. willkürlich gewählte oder ob es sich um *wesentliche* biologische Eigenschaften handelt, aus denen sich etwas über die verwandtschaftlichen Beziehungen der Organismen aussagen läßt. Falls sich die Angaben von Price, Emerson, Nagel, Blumberg und Talmadge bestätigen, daß *R. prowazeki* in vitro in *R. mooseri* umgewandelt werden kann (vgl. S. 105), würden damit die Unzulänglichkeit des gegenwärtigen Rickettsiensystems und die Schwierigkeiten der Systematik besonders dokumentiert werden.

Zur Erleichterung der Orientierung sind in Tabelle 1 die in diesem Bericht benutzten und erwähnten Erregernamen sowie einige andere häufiger im Schrifttum genannte Namen zusammengestellt. Die Begründung für die *gewählte Nomenklatur* ist im Text gegeben. Bei der Aufstellung von Genera und Subgenera ist von den meisten Untersuchern besonderes Gewicht auf die antigene Struktur bzw. das immunbiologische Verhalten der Rickettsien gelegt worden. Aus diesem Grunde wurden die Erreger der Rickettsiosen aus der „Spotted-Fever-Gruppe" im Subgenus *Dermacentroxenus* zusammengefaßt. Als weiteres gemeinsames Merkmal kommt bei diesen Rickettsien noch die gelegentlich in der Kultur und im Überträger oder Warmblüter beobachtete intranucleäre Entwicklung hinzu (Pinkerton und Hass; Hass und Pinkerton), welche für die im Subgenus *Rickettsia* vereinigten Erreger des klassischen und murinen Fleckfiebers nicht gilt.

Philip (1) unterschied unter den menschenpathogenen Rickettsien in der Familie *Rickettsiaceae* das Genus *Rickettsia* mit den Subgenera *Dermacentroxenus* und *Coxiella*. Das Subgenus *Rickettsia* enthält hierbei die Arten *prowazeki* und *orientalis*, das Subgenus *Dermacentroxenus* die Art *rickettsii* und das Subgenus *Coxiella* die Art *burneti*. Innerhalb der Arten wurden mehrere Subspecies unterschieden, wobei der Erreger des murinen Fleckfiebers den Namen *R. prowazeki* var. *typhi* erhielt. Später hat Philip (5) dieses System unter Berücksichtigung neuerer Forschungsergebnisse und Einbeziehung der tierpathogenen Rickettsien ergänzt und abgeändert. Danach bilden die *Rickettsiaceae* eine Familie in der Ordnung *Rickettsiales* und schließen die Genera *Rickettsia*, *Ehrlichia* (mit den Arten *canis, bovis, ovina* und *kurlovi*), *Cowdria* (mit der Art *rumimantium*) und *Coxiella* (mit der Art *burneti*) ein. Das Genus *Rickettsia* enthält das Subgenus *Rickettsia* mit den Arten *prowazeki* und *typhi* (für *mooseri*), das Subgenus *Zinssera* mit der Art *tsutsugamushi* und das Subgenus *Dermacentroxenus* mit den Arten *rickettsii, conori, australis* und *akari*. Zum Subgenus *Dermacentroxenus* wird auch der Erreger des sibirischen Zeckenbißfiebers, *D. sibiricus*, gerechnet. *Rochalimaea* ist ein Subgenus „incertae sedis", in welches der Erreger des Wolhynischen Fiebers einzuordnen wäre. Auch Coles billigt den menschenpathogenen Rickettsien nur die beiden Genera *Rickettsia* und *Coxiella* zu. Die tierpathogenen Rickettsien werden auf die Genera *Cowdria* (*C. rumimantium*) und *Ehrlichia* (*E. canis, E. bovis, E. ovina*) verteilt.

Die russischen Autoren (Sdrodowski und Golinewitsch; Shdanow) unterscheiden, um noch ein anderes System zu erwähnen, in der Familie *Rickettsiaceae* die Genera *Rickettsia*, *Trombidixenus, Coxiella* und *Wolhynia*. Das Genus *Rickettsia* enthält hier das Subgenus *Rickettsia* mit den Arten *prowazeki* und *mooseri*, das Subgenus *Ixodoxenus* mit den Arten *rickettsii, conori, sibiricus* und *australis* und das Subgenus *Gamasoxenus* mit der Art *muris* (= *akari*). Das Genus *Trombidixenus* wird nur durch die Art *orientalis* mit den Varietäten *nipponica, indica, indonesica* und *australis* repräsentiert, das Genus *Coxiella* durch die Art *diaporica* und das Genus *Wolhynia* durch die Art *quintanae* und *rutschkowskyi*. Letztere gilt als Erreger einer nur aus Rußland beschriebenen, durch Zecken (*Ixodes persulcatus*) übertragenen „Ixodo-Rickettsiosis paroxysmalis".

PHILIP (7), (8) wies darauf hin, daß eine Reihe der von den russischen Autoren benutzten Namen auf Grund der Nomenklaturregeln keine Berechtigung haben, sondern nur Synonyme sind. Vorgeschlagen wird von ihm eine neue Klasse „*Mikrotatobiotes*" mit den Ordnungen *Virales* und *Rickettsiales*. Letztere enthält die Familie *Rickettsiaceae*. Weitere Vorschläge beziehen sich auf die Aufstellung von 3 Tribus innerhalb der *Rickettsiaceae: Rickettsieae, Ehrlichieae* und *Wolbachieae*. Zu den *Ehrlichieae* werden die Genera *Ehrlichia, Cowdria* und *Neorickettsia*, zu den *Rickettsieae* die Genera *Rickettsia* und *Coxiella* gerechnet.

Die taxonomischen Schwierigkeiten werden ergänzt durch solche der *Nomenklatur* und Schreibweise. So läuft der Erreger des Tsutsugamushi-Fiebers unter den Namen *Rickettsia tsutsugamushi* und *R. orientalis*, der Erreger des Q-Fiebers unter den Namen *Rickettsia burneti* und *Coxiella burneti* oder *C. diaporica*. Während in diesen Fällen die Verständigung gesichert ist, trifft das z. B. nicht ohne weiteres auf die Bezeichnung *R. typhi* für den Erreger des murinen Fleckfiebers zu, der allgemein unter dem Namen *R. mooseri* bekannt ist. MOOSER (2) hat auch auf die sachliche Berechtigung dieses Namens hingewiesen.

Alle Diskussionen, unter Berücksichtigung von Nomenklaturregeln und historischer Entwicklung, und alle geistvollen Überlegungen zur Klassifikation und Nomenklatur der Rickettsien haben somit bisher weder zu einem allgemein anerkannten Rickettsiensystem noch zu einer gleichartigen Benennung und nicht einmal zu einer einheitlichen Schreibweise geführt. Persönliche Erfahrungen und Anschauungen der Beobachter sowie Festhalten an Gewohnheiten sind hierfür in erster Linie mitbestimmend gewesen. Selbst bei bestem Willen der Beteiligten kommt gewöhnlich nur ein Kompromiß zustande. Zur Aufstellung eines „natürlichen Systems", das auch die phylogenetische Ableitung der Rickettsien berücksichtigt, fehlen uns eben noch wichtige Voraussetzungen [WEYER (16)]. Es erscheint z. B. sehr fraglich, ob die einheitliche systematische Bewertung der Erreger der über die ganze Alte Welt verbreiteten Zeckenbißfieber oder der in ihrem Verhalten und der antigenen Struktur stark divergierenden Erreger des Tsutsugamushi-Fiebers richtig ist. Daher muß im Vordergrund der Taxonomie und Nomenklatur vorerst noch das Prinzip der *Verständigung* stehen. Die Verständigung ist aber bei der relativ kleinen Zahl von Erregern menschlicher Rickettsiosen nicht ernstlich in Frage gestellt und selbst bei der Verwendung des einzigen Genusnamens *Rickettsia* erreicht worden, der von verschiedenen Autoren auch noch für den Erreger des Q-Fiebers benutzt wird (vgl. MARMION).

Einteilung und Benennung der Rickettsiosen werden ebenfalls verschieden gehandhabt, je nachdem, ob man als Einteilungsgrundlage die biologischen Eigenschaften der Erreger, ihre antigene Struktur, das klinische Bild oder die Übertragungswege benutzt. In der Benennung haben außerdem historische Momente und rein lokale Bezeichnungen ihren Niederschlag gefunden. Für eine vergleichende Betrachtung hat sich die Einteilung nach den Überträgern bewährt (vgl. Tabelle 1, S. 75). Man kann danach die durch Zecken, durch Milben und durch Insekten übertragenen oder in der Natur übertragbaren Rickettsiosen unterscheiden, eine Einteilung, die der folgenden, die Epidemiologie besonders berücksichtigenden Darstellung zugrunde gelegt werden soll.

Unter den durch *Zecken* übertragenen oder doch wenigstens an Zecken gebundenen Rickettsiosen steht an erster Stelle das amerikanische *Felsengebirgs-fleckfieber*. Der Erreger, *R. rickettsii*, wird ebenso wie der Erreger der Zeckenbißfieber und der Rickettsienpocken mit Rücksicht auf bestimmte gemeinsame Merkmale in der antigenen Struktur und die Fähigkeit, sich im Kern zu entwickeln, zum Subgenus *Dermacentroxenus* gerechnet, und diese Rickettsiosen werden vielfach als „Spotted-Fever-Gruppe" zusammengefaßt. Den Erreger des in Brasilien verbreiteten Zeckenbißfiebers, das epidemiologisch einige Besonderheiten aufweist, als eigene Art oder Unterart zu bewerten (Tabelle 1), besteht bisher kein ausreichender Grund [vgl. WEYER (18)]. Die Erreger haben die gleiche antigene Struktur und verhalten sich in den Versuchstieren übereinstimmend. Nach PEREIRA und TRAVASSOS lassen sich die brasilianischen Stämme auch serologisch nicht von den nordamerikanischen unterscheiden.

DE MAGALHÃES (2) glaubt jedoch, daß zwischen den Erregern serologische und immunologische Unterschiede bestehen. Während ein Impfstoff aus einem brasilianischen Stamm Meerschweinchen zu 100% gegen eine Infektion schützte, wurden mit dem von PARKER entwickelten Impfstoff aus nordamerikanischen Stämmen nur 34% der Tiere immunisiert. Die Weil-Felix-Reaktion erreicht nach DE MAGALHÃES bei brasilianischem Fleckfieber mit OX 19 und OX 2 einen mittleren Titer von 1:5120, mit OX K von 1:640. Beim amerikanischen Fleckfieber wären die Reaktionen dagegen mit allen Proteus-Stämmen gleichmäßig und nur schwach positiv. Es ist ungewiß, ob diese Unterschiede für alle Stämme gelten, konstant sind und ob sie für die Aufstellung einer besonderen Erregerart oder Unterart ausreichen.

An das Felsengebirgsfleckfieber schließen sich die *altweltlichen Zeckenfieber* bzw. Zeckenbißfieber an, die am Schwarzen Meer und im Mittelmeerraum durch das Boutonneuse-Fieber repräsentiert werden, in Zentralafrika durch das Kenia- und Nigeria-Zeckenbißfieber, in Südafrika durch das südafrikanische, in Indien durch das indische, in Asien durch das sibirische oder asiatische Zeckenbißfieber und in Australien durch das Nordqueensland-Zeckenbißfieber. Verhalten der Erreger in Zecken und Nagern, Immunitätsverhältnisse und Epidemiologie sprechen für eine Verwandtschaft dieser Rickettsiosen mit dem Felsengebirgsfleckfieber. Auffällig ist, daß die Zeckenbißfieber, die klinisch leichter verlaufen als das Felsengebirgsfleckfieber, nur auf die Alte Welt beschränkt sind. Für den Erreger des Zeckenbißfiebers im *Mittelmeergebiet* wurde der Name *R. (D.) conori* gewählt. Der Erreger wird trotz seiner nahen Verwandtschaft mit *R. rickettsii* als eigene Art oder Unterart behandelt. Die intranucleäre Entwicklung sieht man übrigens bei *R. conori* seltener als bei *R. rickettsii*. Es ergeben sich Schwierigkeiten bei der serologischen Abgrenzung gegenüber Felsengebirgsfleckfieber und Rickettsienpocken.

In *Zentralafrika* sind Fälle von Zeckenbißfieber zeitweise auch unter der Bezeichnung „Rotes Kongofieber" gelaufen (DEPOUX und MERVEILLE). Das „Rote Kongofieber" ist aber keine spezifische Krankheit, der Name ist daher zu vermeiden.

Ungeklärt ist, ob alle afrikanischen Zeckenbißfieber sowie die indischen und sibirischen ätiologisch einheitlich sind, wie man das vor allem nach den serologischen Reaktionen anzunehmen geneigt ist. Nach GEAR erreicht jedoch die Weil-Felix-Reaktion beim *südafrikanischen Zeckenbißfieber* fast regelmäßig mit OX 2 einen höheren Titer als mit OX 19. Einige Seren ergaben höhere Titer mit Antigen von *R. akari* als mit Antigen von südafrikanischem Zeckenbißfieber.

Die beobachteten serologischen und immunologischen Unterschiede waren Anlaß, dem Erreger eine Sonderstellung (vgl. Tabelle 1, S. 75) einzuräumen. Diese Ansicht wird gestützt durch Unterschiede im Verhalten in der Kleiderlaus, welche bei Stämmen von Zeckenbißfieber aus Nordafrika, Kenia und Südafrika festgestellt wurden [WEYER (13)]. Während der Erreger des Boutonneuse-Fiebers für die Laus stark pathogen war und die Läuse in kurzer Zeit tötete, ließen sich die Erreger der anderen afrikanischen Zeckenbißfieber mit längeren Passageintervallen leicht in Läusen halten. Bei solchen Beobachtungen muß allerdings daran gedacht werden, daß es sich nur um spezifische Eigenschaften bestimmter *Stämme* handeln könnte [WEYER (18)].

Das *indische Zeckenbißfieber* läßt sich serologisch nicht von den afrikanischen trennen. Ein von PHILIP (4) näher untersuchter, aus Zecken in Kaschmir isolierter Stamm verhielt sich ganz ähnlich wie Stämme von *R. conori* aus dem Mittelmeergebiet und aus Südafrika (vgl. S. 99). Meerschweinchen reagierten mit schwachen oder inapparenten Infektionen. Andere Stämme von indischem Zeckenbißfieber erzeugten beim Meerschweinchen Fieber und nur ausnahmsweise leichte Scrotalschwellungen (RAO). Serologisch waren auch diese Stämme nicht von Boutonneuse-Fieber zu unterscheiden.

Für den Erreger des *sibirischen* oder ostasiatischen *Zeckenbißfiebers* ist von russischen Autoren der Name *Ixodoxenus sibiricus* gewählt worden, obwohl eine sichere Abgrenzung der Krankheit von den anderen Zeckenbißfiebern, insbesondere vom Boutonneuse-Fieber, ebenfalls nicht möglich ist. PHILIP (7) hat daher die Berechtigung eines neuen Artnamens für den Erreger bezweifelt und die Frage der Benennung noch offen gelassen. In Läusen verhielten sich die Erreger ganz ähnlich wie die anderer Zeckenbißfieber [WEYER (18)].

Die in Sibirien isolierten Stämme zeigten antigenetisch und im Tierversuch keine Spezifität. In der Weil-Felix-Reaktion unterschieden sie sich ebenfalls nicht von Boutonneuse-Fieber. In der KBR waren die im ganzen relativ niedrigen Titer etwas höher mit spezifischem Antigen als mit Antigen von Boutonneuse-Fieber. Die Reaktionen wurden am 11. Tag positiv, die Weil-Felix-Reaktion mit OX 19 am 10. oder 12. Tag, manchmal schon am Ende der ersten Krankheitswoche. Mit OX 19 wurden Titerhöhen bis zu 1:6400, mit OX 2 von 1:25 bis 1:400 erreicht. Einige in Amerika von CROCKER, BENNETT, JACKSON, SNYDER, SMADEL, GOULD und GORDON näher untersuchte Stämme von sibirischem Zeckenbißfieber waren serologisch und im Tierversuch nicht von klassischem Fleckfieber zu unterscheiden, so daß die Möglichkeit einer Verwechslung der Stämme diskutiert werden mußte.

Die russischen Beobachter stellten für sibirisches Zeckenbißfieber beim Meerschweinchen eine Inkubationszeit von 5—11 Tagen fest. Das Fieber dauerte 5—6 Tage, das Scrotalphänomen (bei 40% der Tiere) hielt durchschnittlich 3 Tage an. Bei Ratten kam es nach intraperitonealer Inoculation nur zu einer inapparenten Erkrankung. Die Rickettsien waren aber bis zu 140 Tagen im Gehirn der Ratten nachweisbar. Mäuse reagierten bei intranasaler Inoculation mit einer typischen Pneumonie. Nach intraperitonealer Inoculation konnte Peritonitis mit schwachem Rickettsienbefall der Epithelzellen beobachtet werden. Derartige Reaktionen kommen aber auch bei Stämmen von *R. conori* aus Afrika vor [WEYER (13)].

Das *Zeckenbißfieber von Nordqueensland* hat antigenetische Beziehungen zum Felsengebirgsfieber, und es wurde beobachtet, daß der Erreger im Dottersack teilweise intranucleär wächst (PLOTZ, SMADEL, BENNETT, REAGAN und SNYDER). Aus diesem Grunde stellte man ihn in das Subgenus *Dermacentroxenus* und die Krankheit in die Spotted-Fever-Gruppe. Meerschweinchen reagierten mit leichtem Fieber und Scrotalentzündung wie bei einer Infektion mit *R. conori*. Jedoch läßt sich das Zeckenbißfieber von Nordqueensland serologisch von den anderen

Rickettsiosen und insbesondere auch von den Zeckenbißfiebern eindeutig abgrenzen. Daher wurde für den Erreger der Name *R. (D.) australis* gewählt. LACKMAN und PARKER stellten in der KBR mit Antigen von *R. australis* einen Titer von 1:126 fest, mit Antigen von *R. rickettsii* einen Titer von 1:18, mit Antigen von *R. akari* einen Titer von 1:15, mit Antigen von *R. conori* von 1:3 bis 1:10. Die Weil-Felix-Reaktion fällt mit OX K negativ aus, mit OX 19 ist sie positiv, teilweise auch mit OX 2. Wahrscheinlich ist die Krankheit früher häufig mit murinem Fleckfieber verwechselt worden.

Das *Q-Fieber*, ein Name, der für eine definitive Bezeichnung nicht glücklich gewählt ist, sich aber trotzdem längst eingebürgert hat, nimmt bereits klinisch eine *Sonderstellung* ein, da es sich um eine „Pneumo-Rickettsiose" handelt, die ohne Exanthem verläuft. Filtrierbarkeit, Widerstandsfähigkeit, antigene Struktur, fehlende Beziehungen zur Proteus-Gruppe und Übertragungswege unterscheiden den Erreger sehr deutlich von anderen Rickettsien. Die Aufstellung eines eigenen Genus *Coxiella* für diesen Erreger dürfte daher berechtigt oder wenigstens zweckmäßig sein. Die Zugehörigkeit des Erregers zu den Rickettsien ist nicht ernstlich in Zweifel gezogen worden. Die Besprechung des Q-Fiebers im Rahmen der durch Zecken übertragenen Rickettsiosen ist sachlich anfechtbar. Sie erfolgt hier nur aus Gründen der Übersichtlichkeit und weil Zecken als Überträger und Reservoire in der Natur fungieren oder wahrscheinlich beim Übergang der Krankheit vom Wild auf Haustiere beteiligt sind.

Rickettsienpocken und Tsutsugamushi-Fieber haben nur die Übertragung durch parasitische *Milben* gemeinsam, sonst aber *nichts* miteinander zu tun. Erregereigenschaften, Verbreitung und klinisches Bild differieren stark. Auch die Überträger gehören ganz verschiedenen Milbenfamilien an, die sich in ihrer Lebensweise stark unterscheiden. Der Erreger der *Rickettsienpocken, R. (D.) akari*, hat morphologisch Ähnlichkeit mit *R. prowazeki* und *R. mooseri* und besitzt keinerlei antigene Beziehungen zum klassischen Fleckfieber, Tsutsugamushi-Fieber und Q-Fieber. Jedoch besteht eine antigene Verwandtschaft mit *R. rickettsii*. Daher wird die Krankheit vielfach auch zur Spotted-Fever-Gruppe gerechnet und der Erreger mit *R. rickettsii, R. conori* und *R. australis* in das Subgenus *Dermacentroxenus* gestellt.

Die in Nordamerika isolierten Stämme von *R. akari* waren serologisch teilweise nicht von *R. rickettsii* zu unterscheiden. FULLER, MURRAY, AYRES, SNYDER und POTASH fanden zwar in der KBR bei Meerschweinchen, die mit Stämmen von *R. akari* infiziert waren, Unterschiede in der Titerhöhe bei Verwendung von Antigen aus *R. akari* (1:640 und 1:2560) und *R. rickettsii* (1:10 und 1:160), bei einer genauer verfolgten Laboratoriumsinfektion hielt sich der Titer jedoch vom 12.—68. Tag nach Krankheitsbeginn mit beiden Antigenen auf gleicher Höhe (HEISENGER, MURRAY und COHEN). In anderen Fällen ergaben sich auch serologische Beziehungen zu *R. mooseri*.

Von weiteren Eigenschaften des Erregers ist die gelegentlich beobachtete intranucleäre Entwicklung und die fehlende Übertragbarkeit auf den Affen zu erwähnen. Ob die genannten Eigenschaften für die Einreihung in die Spotted-Fever-Gruppe ausreichen und ob die gleichen Eigenschaften und Beziehungen auch für den Erreger der in Rußland nachgewiesenen Rickettsienpocken gelten, ist noch nicht geklärt.

Für den Erreger des auf Ost- und Südostasien beschränkten *Tsutsugamushi-Fiebers* ist der Name *R. tsutsugamushi* oder *R orientalis* gebräuchlich. Als Subgenusname wurde die Bezeichnung *Zinssera* vorgeschlagen (vgl. S. 76). *R. tsutsugamushi* ist sehr empfindlich, neigt zur Autolyse und kann sich im Unterschied zu allen anderen Rickettsien in den Magenzellen der Kleiderlaus nicht entwickeln [WEYER (12), vgl. S. 96]. Dié Fragwürdigkeit einer Einteilung der Rickettsiosen und ihrer Erreger auf antigener Basis wird deutlich, wenn man die beim Tsutsugamushi-Fieber festgestellten auffälligen Stammunterschiede berücksichtigt, die sich auch auf den antigenen Charakter beziehen (vgl. S. 100) und die SHDANOW veranlaßt haben, die Art in mehrere Varietäten aufzugliedern.

KAWAMURA hat auf Hokkaido in Japan aus Wühlmäusen zwei Rickettsienstämme isoliert, die *R. tsutsugamushi* nahestehen, aber ein spezifisches Verhalten zeigten. Für diesen Erreger ist als neuer Artname *R. tamiyai* vorgeschlagen worden. Man nimmt an, daß eine in Hokkaido unter dem Namen „Ezo-Fieber" bekannte Erkrankung des Menschen auf eine Infektion mit *R. tamiyai* zurückgeht. Über die Übertragung ist nichts bekannt. Die aus den Mäusen isolierten Stämme erzeugten bei intracutaner Inoculation von Mensch, Kaninchen und Affe keine Primärläsion, waren für das Kaninchen nur schwach pathogen und immunisierten Meerschweinchen nicht gegenüber *R. tsutsugamushi*. Die Seren reagierten in der KBR nicht mit Antigen aus *R. tsutsugamushi*, sie enthielten Agglutinine für OX 2 und OX K in gleicher Stärke. Seren von mit *R. tsutsugamushi* infizierten Patienten oder Versuchstieren waren in der KBR mit Antigen aus dem neuen Stamm negativ. Die biologischen Besonderheiten der neu benannten Rickettsienart sind von KAWAMURA, TSUNEMATSO, NISHIOKA, YAMANÉ und SAITO beschrieben worden. Eine experimentelle Infektion von Affen unter besonderer Berücksichtigung der anatomisch-pathologischen Veränderungen wurde von KAWAMURA, YAMANÉ und KAWAI und eine experimentelle Infektion des Menschen von HAYASHI, NAKAMURA, KOMINE und KAWAMURA sowie von TAKIGAMI, KAWAMURA, NISHIOKA und IIDA durchgeführt.

Bei der experimentellen Infektion des Menschen betrug die Inkubationszeit 8—12 Tage. Bei einem Patienten entwickelte sich an der Inoculationsstelle eine Primärläsion, beim anderen eine Rötung. In beiden Fällen kam es zu einer Lymphadenitis. Ein teils makulöses, teils makulo-papulöses Exanthem, das am Stamm begann und sich auf die Extremitäten ausbreitete, trat bei einem der Freiwilligen am 12., beim anderen erst am 16. Tage auf. Das Exanthem hielt sich 1 Woche. Aus dem Patientenblut konnte der Erreger isoliert werden. Die KBR erreichte mit Antigen von *R. tamiyai* 5 Wochen nach Krankheitsbeginn einen Titer von 1:640 und 1:1280. Am 14. Tag wurden Stämme von OX 19 in einer Verdünnung von 1:80, Stämme von OX K in einer Verdünnung von 1:40 und Stämme von OX 2 in einer Verdünnung von 1:80 agglutiniert.

Die japanischen Autoren sind der Meinung, daß *R. tamiyai* zur Gruppe der Tsutsugamushi-Fieber gehört. Es besteht daher die Möglichkeit, daß es sich bei dem Erreger nur um eine weitere Variante von *R. tsutsugamushi* handelt. Das Verhalten der Erreger in Mäusen und in Läusen stimmte mit dem verschiedener Stämme von *R. tsutsugamushi* überein [WEYER (17)].

Bei den durch *Insekten* übertragenen Rickettsiosen, die auch als „Typhus-Gruppe" bezeichnet werden, bestehen über den Namen *R. prowazeki* für den Erreger des *klassischen* (epidemischen) *Fleckfiebers* keine diskutablen Meinungsverschiedenheiten (höchstens über die Schreibweise: *prowazeki* oder *prowazekii*!). Die Bezeichnung *R. mooseri* für den Erreger des *murinen* (endemischen) *Fleckfiebers* ist sachlich begründet, unmißverständlich und seit langem gebräuchlich.

Trotz gemeinsamer Merkmale ist die Berechtigung einer Trennung der Erreger in zwei Arten oder Unterarten, die besonders auf Unterschieden in der antigenen Struktur beruht, in der letzten Zeit auch nicht mehr in Zweifel gezogen worden. Allerdings treten ganz neue Gesichtspunkte in dieser Beziehung wie überhaupt in der Systematik der Rickettsien auf, wenn sich die Beobachtungen von PRICE, EMERSON, NAGEL, BLUMBERG und TALMADGE bestätigen, daß sich *R. prowazeki* experimentell in *R. mooseri* verwandeln läßt (vgl. S. 105).

Der Erreger des *Wolhynischen Fiebers* hat den Namen *R. quintana*. Die Krankheit scheint mit den übrigen Rickettsiosen wenig zu tun zu haben. Das Exanthem fehlt, und es fehlen, ebenso wie beim Q-Fieber, alle Beziehungen zur Proteus-Gruppe. Die Erreger sind im Blut des Patienten monatelang und noch nach Verschwinden der klinischen Symptome nachweisbar, sie wachsen im Läusemagen als einzige extracellulär, sind spezialisiert auf die Läuse des Menschen und können auf die für andere Rickettsien empfänglichen Versuchstiere, speziell die Nager, nicht übertragen werden. Diese Tatsachen haben manche Autoren veranlaßt, dem Erreger des Wolhynischen Fiebers den Rickettsiencharakter abzusprechen. Jedoch muß er bei näherer Überlegung auf Grund seiner Morphologie, der färberischen Eigenschaften, der Vermehrungsfähigkeit in der Laus, des fehlenden Wachstums auf künstlichen Nährböden und der Pathogenität für den Menschen nach unseren augenblicklichen Kenntnissen zur Rickettsiengruppe gerechnet werden [WEYER (16)]. Der biologische Fundamentalunterschied, die extracelluläre Entwicklung im Läusemagen, verliert dadurch an grundsätzlicher Bedeutung, daß auch andere Rickettsien unter bestimmten Bedingungen, z.B. in der Hämolymphe der Laus, extracellulär wachsen können (vergl. S. 97).

Wenn bei der Klassifikation der Rickettsien die Aufstellung mehrerer Genera oder Subgenera richtig oder zweckmäßig erscheint, dann hätte der Erreger des Wolhynischen Fiebers wohl den gleichen Anspruch auf ein eigenes Genus oder eine Sonderstellung wie der Erreger des Q-Fiebers. Der Vorschlag von SHDANOW, für diesen Erreger das Genus *Wolhynia* aufzustellen, erscheint daher diskutabel. Die Zuordnung des Wolhynischen Fiebers zu den Rickettsiosen ist jedoch nicht weniger gut begründet als die des Q-Fiebers.

Von GIROUD, ROGER und DUMAS wurde der Name „Neorickettsien" für eine Gruppe von Organismen vorgeschlagen, die den Rickettsien nahestehen sollen und im Rahmen von serologischen Untersuchungen bei Q-Fieber auffielen. Sie werden für eine Reihe von Krankheitssyndromen mit Exanthemen, grippeähnlichen Symptomen, Lungenveränderungen und Gefäßerkrankungen verantwortlich gemacht. Stämme wurden bisher aus Menschen, Kühen, Schafen und Ziegen in Europa, Afrika, Asien und Zentralamerika isoliert. Der Nachweis erfolgte mit einer speziellen Methodik über eine KBR [GIROUD und JADIN (5)]. *Neorickettsia helminthoeka* wurde von PHILIP, HADLOW und HUGHES ein rickettsienähnlicher Mikroorganismus genannt, der als Erreger der „Salmon-Poisoning-disease" bei Hunden an der pazifischen Küste der USA auftritt und durch den Trematoden *Nanophyetus salmincola* übertragen wird. Über Eigenschaften und Übertragung dieses Erregers finden sich nähere Angaben bei PHILIP (6).

2. Morphologie, Feinbau, Wachstum, Vermehrung

Eine einwandfreie Differenzierung der Rickettsienarten auf Grund ihrer färberischen Eigenschaften oder ihrer Morphologie ist, auch unter Zuhilfenahme elektronenmikroskopischer Methoden, nicht möglich. Lediglich der Erreger des

Wolhynischen Fiebers, *R. quintana*, kann durch seine gedrungene Form, die kräftige Tinktionsfähigkeit bei Färbung nach GIEMSA und die schwach entwickelte Polendenfärbung von den übrigen Rickettsien gut abgegrenzt werden. TAKEMORI, HENMI und KITAOKA konnten *R. mooseri* und *R. tsutsugamushi* gleichzeitig über 5 Passagen in Mäusen halten und dabei beide Erreger, die teilweise nebeneinander in derselben Zelle lagen, klar differenzieren. In den Epithelzellen des Mäuseperitoneums erscheint *R. tsutsugamushi*, ähnlich wie *R. quintana*, plump und gedrungen, hat fast stets Diploform, ist dunkler gefärbt und findet sich in einzelnen kleineren Kolonien neben dem Kern, während *R. mooseri* das Cytoplasma der Zelle gleichmäßig und vollständig besiedelt. Die morphologische Unterscheidung der Arten wird durch den Polymorphismus der Rickettsien erschwert, der, besonders ausgeprägt bei *R. prowazeki* und *C. burneti*, in erster Linie Ausdruck von Alters- und Entwicklungsstadien ist und durch Reaktionen der Wirtszellen mit beeinflußt wird. So treten z. B. die langen Erregerketten bei manchen Stämmen von *R. prowazeki* fast regelmäßig zu Beginn der Vermehrung in der Laus auf, während andere Rickettsien, insbesondere *R. mooseri* und *R. quintana*, Kettenformen unter den gleichen Bedingungen ganz vermissen lassen.

Zu den *morphologischen* Studien von WEISS sowie PLOTZ, SMADEL, ANDERSON und CHAMBERS, ferner RIS und FOX u. a. und den ersten elektronenoptischen Untersuchungen, die sich vor allem mit der systematischen Stellung der Rickettsien und ihren Beziehungen zu Bakterien und Viren beschäftigten (BABUDIERI und BOCCIARELLI; WEYER, BERGOLD und FRIEDRICH-FREKSA u. a.), sind weitere Beobachtungen gekommen, die letztlich nur die schon von DA ROCHA LIMA vertretene Ansicht bestätigten, daß die Rickettsien nach Feinbau und Teilungsmechanismus den Bakterien nahestehen (VAN ROOYEN und SCOTT; LIEBERMEISTER und ZEHENDER; KAUSCHE und SHERIS; WEYER und PETERS; URBACH und SPRÖSSIG u. a.). URBACH und SPRÖSSIG fanden fluorescenzmikroskopische Untersuchungen vorteilhaft für den schnellen Nachweis von *C. burneti* in Ausstrichen. VAN ROOYEN und SCOTT untersuchten elektronenoptisch *R. prowazeki*, *R. mooseri*, *R. rickettsii*, *R. akari* und *C. burneti* aus Dottersackkulturen und sahen dabei besondere morphologische Ähnlichkeit zwischen *R. akari* und *R. rickettsii*, während *C. burneti* im Feinbau gewisse Unterschiede zu den anderen Rickettsien erkennen ließ. Bei allen Rickettsien wurde eine verdichtete Innenstruktur festgestellt, die von einer helleren Zone umgeben war. Die mittlere Länge und Breite betrug bei *C. burneti* 0,73 zu 0,32 μ, bei *R. rickettsii* und *R. akari* 1,2 zu 0,6 μ. Die Größen von *R. prowazeki* und *R. mooseri* lagen zwischen diesen Werten.

WEYER und PETERS veröffentlichten die bisher einzige elektronenoptische Aufnahme von *R. quintana*. Der Erreger des Wolhynischen Fiebers stimmt danach in seinem Feinbau mit *R. prowazeki* und *R. mooseri* überein. Die untersuchten Rickettsien stammten aus dem Magen der Kleiderlaus und wurden durch unmittelbares Abstreifen der infizierten, auseinandergebreiteten Mägen auf dem Objektträger gewonnen. An diesem Material wurden auch Größenmessungen vorgenommen. Bei *R. quintana* wurden für Einzelorganismen 0,7—1,4 μ Länge und 0,5—0,8 μ Breite festgestellt, für Teilungsformen bei gleicher Breite Längen von 1,2—1,8 μ. Die Segmente der Vermehrungsketten von *R. prowazeki* erreichten bei einer Breite von 0,5 μ eine Länge bis zu 4,5 μ; einzeln liegende Erreger ohne Zeichen einer Teilung waren ungefähr *5* μ lang.

Stoker, Smith und Fiset untersuchten *C. burneti* in Ultramikrotomschnitten. Als Material dienten gereinigte Dottersackkulturen. Innerhalb der Grenzmembran, die eine Dicke von 5—10 mμ erreichte, konnte neben einer granulären Zone von 25 mμ Dicke — einzelne Granula hatten einen Durchmesser von 5 mμ — ein dichter Zentralkörper ausgemacht werden, der die Form eines unregelmäßig geflochtenen Stranges hatte. Die granuläre Zone erschien weniger dicht als die homologe Zone bei Bakterien und nahm einen entsprechend kleineren Raum des Gesamtvolumens ein. Demgegenüber war der klar umrissene Zentralkörper kompakter und größer.

Bei Schnitten durch den mit *R. mooseri* infizierten Dottersack wurden die Rickettsien nur im Cytoplasma von Mesothelial- und Epithelialzellen gefunden (Wissig, Caro, Jackson und Smadel), wo sie offenbar eine Lysis submikroskopischer Cytoplasmakomponenten verursachten. Eine entzündliche Gewebsreaktion wurde nicht gesehen. Die Rickettsien erschienen als dichte, runde oder stäbchenförmige Körper von etwa 1,2 μ Länge und 0,2 μ Breite. Die Grenzmembranen waren im Unterschied zu den Rickettsien aus gewaschenen Suspensionen bei den in situ geschnittenen überhaupt nicht oder nur selten nachweisbar.

Der Dottersack des Hühnchens dient nicht nur als bequemster Nährboden für die Gewinnung größerer Rickettsienmengen zur Herstellung von Impfstoff und Antigen, sondern wird auch bei Studien über *Vermehrung, Wachstum,* Stoffwechsel, Toxinbildung und therapeutische Effekte benutzt. Das Verfahren wurde zum ersten Mal bei der Kultur von *R. rickettsii* angewandt [Cox (1)] und anschließend auf andere Rickettsienarten ausgedehnt. Die Chorio-Allantoismembran eignet sich ebenfalls als Nährboden, die Methode ist aber der Dottersackkultur unterlegen, besonders wenn es sich um die Züchtung größerer Rickettsienmengen handelt. Die Rickettsienausbeute (*R. mooseri*) kann durch Röntgenbestrahlung, besonders am 6. Tag nach Beimpfung der Eier, erheblich gesteigert werden (Greiff, Powers und Pinkerton). *R. quintana* wächst nicht im Dottersack.

Die Beobachtung, daß sich Rickettsien noch im Dottersack toter Hühnerembryonen vermehren, konnte nicht bestätigt werden. Ormsbee wies nach, daß sich *C. burneti* nicht mehr in totem Gewebe vermehrt und im Hühnerembryo auf den Dottersack beschränkt bleibt. Ormsbee, Lackman und Pickens bestimmten bei *C. burneti* in Dottersackkulturen die Korrelation zwischen der LD 50 und dem KBR-Titer. Die Vermehrung der Rickettsien im Ei und die Infektiosität von Dottersackkulturen wurden mit Hilfe des KBR-Testes im Meerschweinchen geprüft (Ormsbee). Die Wachstumskurve durchlief eine Latenzphase, bevor sie zur Exponentialkurve wurde, auf die eine Periode langsamen Wachstums folgte. Die Kurve zeigte kein klares Maximum und keine stationäre Phase. Kuwata verglich die Vermehrung von *R. tsutsugamushi* und von Ornithose-Virus in der Maus nach intracerebraler Inoculation und stellte fest, daß die Infektiosität von *R. tsutsugamushi* während der ersten 24 Std nach der Inoculation unverändert erhalten blieb, im Gegensatz zum Ornithose-Virus, das zwischen 3 und 12 Std nach der Inoculation einen starken Abfall des Infektionstiters erkennen ließ. Dieser Unterschied wird als spezifisch für die biologische Differenz der beiden Organismen angesehen.

Die Züchtung von Rickettsien im Gewebe wurde mit verbesserten Methoden fortgesetzt. Als besonders geeignet für die Kultur von *C. burneti* erwiesen sich gefäßfreie Entodermzellen viertägiger Hühnerembryonen (Weiss und Pietryk).

Der direkte Nachweis der Rickettsien begünstigt in diesem Fall Wachstumsstudien. In Rolltuben-Kulturen wurden Lymphosarkomzellen der Maus mit Rickettsien (*C. burneti*, *R. prowazeki*, *R. mooseri*, *R. rickettsii* und *R. tsutsugamushi*) aus dem Dottersack beimpft (BOZEMAN, HOPPS, DANAUSKAS, JACKSON und SMADEL). Das rasche Wachstum von *R. tsutsugamushi* (bei 37° C innerhalb von 24 Std Zunahme um das Dreifache) ermöglichte hierbei quantitative Wachstums- und Stoffwechseluntersuchungen und die Prüfung der Wirkung von Antibiotica auf die Rickettsien.

SCHAECHTER, BOZEMAN und SMADEL beobachteten bei *R. rickettsii* und *R. tsutsugamushi* in der Kultur von Rattenfibroblasten in vivo die Teilung der Rickettsien im Phasenkontrastmikroskop. *R. rickettsii* war locker im Cytoplasma verstreut und der direkten Beobachtung leicht zugänglich. Gelegentlich wurden auch Rickettsien im Kern gesehen. *R. tsutsugamushi* wuchs in schwer übersehbaren Ansammlungen in der Nähe des Kerns, wobei Perioden der Zusammenballung und Lockerung abwechselten. Durch cytoplasmatische Prozesse konnten Rickettsien aus der Zelle gedrängt werden, obwohl die Masse erst bei Zerfall der Zelle frei wurde. Mit *R. rickettsii* infizierte Zellen degenerierten in 5—6, mit *R. tsutsugamushi* infizierte in 8—9 Tagen.

Bei Versuchen zur Kultur von Rickettsien in explantiertem Läusegewebe wurde eine Vermehrung der Rickettsien nur dann beobachtet, wenn bereits infizierte Magenzellen zur Explantation gelangten [WEYER (6)]. Vermehrungsfähigkeit und Virulenz der Rickettsien gingen trotz guter Erhaltung der äußeren Form schnell verloren. *R. quintana* blieb bis zum 34. Tag nach der Explantation teilungsfähig, *R. prowazeki* bis zum 4., *R. mooseri* bis zum 16. Tag, sofern zugleich mit dem Läusegewebe auch Warmblütergewebe (Kaninchentestes und -milz) explantiert wurde. In einer Anzahl von Kulturen bewahrten *R. prowazeki* und *R. mooseri* ihre Lebensfähigkeit länger, sie vermehrten sich aber nach Übertragung in den Läusemagen nur extracellulär und waren für Mäuse und Meerschweinchen nicht mehr pathogen.

3. Stoffwechsel, Toxinbildung

Alle Untersuchungen über die chemische Zusammensetzung und den *Stoffwechsel* der Rickettsien sprechen für die Bakteriennatur der Organismen. BOVARNICK und SNYDER stellten in gereinigten Suspensionen lebender Rickettsien (2 Stämme von *R. prowazeki* und 1 Stamm von *R. mooseri*) Stoffwechselaktivität in Form von Sauerstoffverbrauch und Kohlensäureabgabe bei Anwesenheit von Succinat, Pyruvat und besonders Glutaminat fest. Der Sauerstoffverbrauch war direkt proportional der Rickettsienkonzentration, die ihrerseits an Hand der Toxicität für Mäuse geprüft wurde. In Fortführung der Versuche wiesen BOVARNICK und MILLER in *R. prowazeki* eine Transaminase nach. Der Sauerstoffverbrauch stieg bei Zugabe von anorganischem Phosphat und Magnesium. In gereinigten Suspensionen von *R. mooseri* bestand unter optimalen Bedingungen eine annähernde Korrelation zwischen dem Stickstoffgehalt, der Menge von komplementbindendem Antigen, dem Toxintiter und der Oxydation von Glutaminat [WISSEMAN, JACKSON, HAHN, LEY und SMADEL (1), (2)]. KARP bestätigte die Glutaminatoxydation in Dottersacksuspensionen von *R. mooseri*. Die Oxydation konnte auch in Rickettsiensuspensionen aus Mäuselungen beobachtet werden.

RIS und FOX stellten bei *R. prowazeki* färberisch und cytochemisch die Anwesenheit von RNS und DNS fest. SMITH und STOKER wiesen mit Hilfe von Papierchromatographie und Spektrophotometrie in der Trockensubstanz von

C. burneti 9,7% DNS und 4,3% RNS nach. Während die Menge der DNS bei verschiedenen Präparationen konstant blieb, wechselte der RNS-Anteil. In der DNS wurden Purin und Pyrimidin, jedoch nicht 5-Methyl-Cytosin ermittelt. Die DNS im Gewebe von normalen Hühnerembryonen enthielt dagegen 5-Methyl-Cytosin, das bis zu 4,3% des Cytosingehaltes ausmachte. Unter 15 verschiedenen Aminosäuren wurde kein Threonin gefunden, das jedoch COHEN in *R. prowazeki* identifizieren konnte. WYATT und COHEN untersuchten die Purin- und Pyrimidin-Zusammensetzung in der DNS von *C. burneti* und *R. prowazeki* im Vergleich zu den Wirtszellen. Die Zusammensetzung differierte deutlich durch den Anteil von Adenin, Thymin, Guanin und Cytosin in den Wirtszellen. In *R. prowazeki* fehlte ebenso wie in *C. burneti* 5-Methyl-Cytosin. SCHAECHTER, TOUSIMIS, COHN, ROSEN, CAMPBELL und HAHN unterzogen Zellmembranen von Rickettsien einer chemischen Analyse und wiesen in den Membranen 12 Aminosäuren, Polysaccharide (darunter Glucose und Galaktose) und geringe Mengen von RNS nach.

Der erste Nachweis eines *Toxins* in Rickettsien — es handelte sich um *R. mooseri* aus Dottersackkulturen — erfolgte durch GILDEMEISTER und HAAGEN. Seither sind entsprechende Toxine auch in anderen Rickettsien nachgewiesen worden [HAMILTON, ferner BENGTSON, TOPPING und HENDERSON sowie COX (4)]. BELL und PICKENS prüften 10 Stämme von durch Zecken übertragenen Rickettsiosen, darunter verschieden virulente Stämme von Felsengebirgsfleckfieber aus Nordamerika und Brasilien, ferner Stämme von Fièvre boutonneuse sowie südafrikanischem und indischem Zeckenbißfieber. Das Toxin von *R. rickettsii* wurde spezifisch neutralisiert durch Seren von Patienten, die Felsengebirgsfleckfieber überstanden hatten oder gegen diese Krankheit geimpft waren. Mit Hilfe des Toxinneutralisationstestes war eine Differenzierung zwischen *R. rickettsii* und *R. conori* möglich. Toxin wurde auch bei einem Stamm von *R. tsutsugamushi* (Stamm Gilliam) in Malaya nachgewiesen, während 8 andere, in der gleichen Weise geprüfte Stämme nur wenig oder gar kein Toxin enthielten (SMADEL, JACKSON, BENNETT und RIGHTS). Mehrere ausgesprochen schwach toxische Stämme isolierte KOENOKI in Japan aus wilden Ratten.

Das Toxin ist, wie insbesondere Untersuchungen an *R. prowazeki* gezeigt haben, an die Rickettsienkörper gebunden [COX (4)]. Es wird durch Formalin von 0,375% und Äther sowie durch Wärme von 56—60° C innerhalb von 3 min zerstört, bei 50° C weitgehend inaktiviert, bei einer Temperatur von 10—12° C innerhalb von 7 Tagen stark abgeschwächt. Die Hitzelabilität teilt das Toxin mit den immunisierenden bzw. neutralisierenden Antikörpern, während die komplementbindenden Antikörper hitzestabil sind. Das Toxin hält sich in gepufferten Dottersacksuspensionen 72 Std, in Magermilch bei —70° C 3 Monate. Die toxische Substanz kann mit Hilfe des Mäuseschutzversuches zur Standardisierung von Impfstoffen benutzt werden. Im Toxingehalt einzelner Rickettsienstämme der gleichen Art bestehen Differenzen, die Beziehungen zu den Virulenzunterschieden (vgl. S. 101) haben, doch bleibt dadurch die Toxinspezifität unberührt.

Das Toxin wird routinemäßig durch intravenöse Inoculation von weißen Mäusen nachgewiesen. Als toxischer Effekt ist eine innerhalb von längstens 24 Std tödlich wirkende Injektion anzusehen. Baumwollratten sind für das Toxin weniger empfänglich als Mäuse. Eine akut toxische LD 50 für Mäuse ent-

spricht einer 10^1 LD 50 oder einer $10^{6,5}$ ID 50 für Baumwollratten [FULLER (4)].
Die Ratte erwies sich, umgerechnet auf das Körpergewicht, für das Toxin von
R. prowazeki in gleicher Weise empfänglich wie die weiße Maus, aber resistenter
gegen das Toxin von *R. mooseri* [NEVA und SNYDER (1)]. Das Toxin ist in erster
Linie für die histopathologischen Organveränderungen im Warmblüter verantwortlich zu machen (SIEGERT). Die intravenöse Inokulation von Mäusen und
Ratten mit *R. prowazeki* führte zu denselben Gefäßveränderungen, wie sie auch
bei schweren Fleckfiebererkrankungen des Menschen zu beobachten sind. PARKER
und NEVA stellten bei der Ratte nach Verabfolgung des Toxins von *R. prowazeki*
und *R. rickettsii* auch pathologische Veränderungen im Dünndarm fest. Nach
intravenöser Verabfolgung von lebenden *R. prowazeki* und *R. mooseri* wurde bei
Ratten und Mäusen eine Erhöhung des spezifischen Gewichtes des Blutes konstatiert [NEVA und SNYDER (2)]. Der systolische Druck blieb bis kurz vor dem
Tode normal. Der Plasmaverlust im Blut ist wahrscheinlich durch die Permeabili-
tät der Capillaren infolge direkter Verletzung der Endothelzellen bedingt.

CLARKE und FOX beobachteten in vitro eine durch Suspensionen von *R. prowazeki* und *R. mooseri* ausgelöste Hämolyse der roten Blutkörperchen vom Kaninchen und Schaf. Das Phänomen wurde nicht bei Mäusen, Meerschweinchen
und Baumwollratten gesehen und nicht mit Suspensionen von *R. tsutsugamushi*
erreicht. Das *Hämolysin* ist an die Rickettsienkörper gebunden und findet sich
in den Membranen in größerer Menge als im Endoplasma (SCHAECHTER, TOUSIMIS,
COHN, ROSEN, CAMPBELL und HAHN). Es wird zerstört durch 0,5%iges Formol
und durch eine Temperatur von 56^0 C nach einstündiger Einwirkungszeit. Die
Aktivität verläuft parallel zur Toxicität der Rickettsien für Mäuse und der Infektiosität für Baumwollratten. Der hämolytische Faktor ist möglicherweise im
Toxin der Rickettsien enthalten oder teilweise mit ihm identisch. Die Hämolyse
wird durch Hyperkaliämie und einen Anstieg des Kaliumgefälles im Plasma
charakterisiert. Beim Vergleich zweier Stämme von *R. prowazeki* und einem
Stamm von *R. mooseri* bestätigten SNYDER, BOVARNICK, MILLER und CHANG,
daß die hämolytische Wirkung mit einem ganz ähnlichen Titer wie die toxische
Wirkung nach intravenöser Inoculation von weißen Mäusen verläuft. Der
Hämolysintest zeigte sich jedoch unter bestimmten Bedingungen dem Toxicitäts-
test überlegen. Es handelt sich um eine Schnellmethode, die diagnostisch dasselbe leistet wie die Toxintitration durch intravenöse Inoculation von Mäusen.
Man verwendet für den Test am besten Schaf- oder Kaninchenblut. BOVARNICK
und ALLEN (2), (3) konnten den Verlust der Toxicität und der hämolytischen
Aktivität in Suspensionen von *R. prowazeki* unter bestimmten Bedingungen wieder
rückgängig machen (vgl. S. 105). Wurden Fleckfieberrickettsien mehrere Stunden
bei 30^0 C ohne Nährsubstrat gehalten, so verloren sie hämolytische Aktivität
und Toxicität. Eine Behandlung mit Glutaminat führte zu einer partiellen oder
vollständigen Wiederherstellung der verlorenen Eigenschaften. Bei 35^0 C war
die Restitution bedeutend schwächer, obwohl die Rickettsien mehr Sauerstoff
aufnahmen.

Aus Dottersackkulturen von *R. rickettsii*, *R. conori*, *R. akari* und *R. mooseri* wurden von
CHANG, SNYDER und MURRAY Substanzen zur Sensibilisierung von Erythrocyten gewonnen,
wie sie von Bakterien bereits bekannt waren. Die Beobachtung wurde für diagnostische
Zwecke ausgewertet. Nach CHANG, MURRAY und SNYDER wird die Reaktion vom 7.—11. Tag

nach Krankheitsbeginn positiv und ist insbesondere bei Felsengebirgsfieber und Rickettsien-
pocken spezifischer als die Weil-Felix- und die Komplementbindungsreaktion. DOWNS,
FEVURLY und MEYER wiesen in den Geweben und im Urin von mit *R. mooseri* infizierten
Mäusen, WHITMIRE und DOWNS auch bei Hamstern und Baumwollratten, die nach Cortison-
Gaben an der Infektion eingegangen waren, eine hitzestabile Substanz nach, welche die
Agglutination sensibilisierter Erythrocyten aufhob. Auch hierdurch boten sich weitere
diagnostische Möglichkeiten. BARBER fand z. B. den Test beim Nachweis von Q-Fieber
empfindlicher als die gewöhnliche Agglutination.

4. Verhalten gegenüber chemischen und physikalischen Einwirkungen

Die zunächst mehr zufällig festgestellte hohe Widerstandsfähigkeit des
Q-Fieber-Erregers veranlaßte zahlreiche spezielle Untersuchungen über den
Einfluß physikalisch-chemischer Faktoren und bactericider Substanzen auf
C. burneti und andere Rickettsien. Die Resultate sind nicht einheitlich. Die
zum Teil recht erheblichen Unterschiede in den Beobachtungsergebnissen können
nicht allein aus der unterschiedlichen Anwendungs- und Testtechnik erklärt
werden.

Während *R. prowazeki* in vitro durch ein aus höher molekularen Derivaten von Aminosäuren hergestelltes Desinfektionsmittel („Tego 103") innerhalb von 5 min abgetötet wurde,
war der Q-Fieber-Erreger unter denselben Bedingungen selbst nach 24stündiger Einwirkungszeit ungeschädigt [WEYER (4)]. Eine tödliche Wirkung wurde erst nach 48 Std erreicht.
RANSOM und HUEBNER fanden, daß *C. burneti* aus Dottersackkulturen Temperaturen von
63° C für 30—40 min überstehen konnte, während *R. mooseri, R. rickettsii* und *R. akari* bei
50° C innerhalb von 15 min getötet wurden. 0,5%iges Formol tötete *C. burneti* in 96 Std nicht
ab, 1%iges Formol führte erst in 72 Std zu einem tödlichen Effekt. Die übrigen Rickettsien
wurden bereits durch 0,1%iges Formol in 24 Std getötet. 10%ige Suspensionen von *R. tsutsugamushi* waren bei Zusatz von 0,1% Formol nach wenigen Stunden nicht mehr infektiös
(JACKSON und SMADEL). Gegenüber Äther, ultraviolettem Licht, Penicillin und Sulfadiazin
verhielt sich *C. burneti* nicht anders als die übrigen Rickettsien.

Nach KIRBERGER töteten 70%iger Alkohol und Äther *C. burneti* in 1—2 min, 0,2%iges
Formol in 48 Std, 0,5%iges Formol in 24 Std, 0,4%iges Phenol in 48 und 1%iges Phenol in
2 Std. Eine sichere Abtötung wurde auch bei 65° C in 15 min erreicht. MALLOCH und STOKER
ließen bestimmte Chemikalien bei 22° C 3 Std auf Dottersacksuspensionen von *C. burneti*
einwirken und prüften anschließend die Wirkung durch Übertragung der Suspensionen auf
Meerschweinchen, Mäuse und Eier. 0,5%iges Phenol hatte nur eine schwache Wirkung auf
C. burneti, 1%iges Phenol setzte die Infektiosität um das 1000fache herab, führte aber nicht
zu einer Sterilisation der Suspensionen. Bei Verwendung von 1%igem Formol waren noch
2 von 13 Testen positiv. Auch nach Behandlung mit 1%igem Lysol konnten noch lebende
Rickettsien festgestellt werden, wenn auch erst nach Anreicherung durch 2 Passagen. Die
Behandlung mit Lysol bei einer Temperatur von 37° C führte zu einer Abtötung der Rickettsien. Auf Grund dieser Beobachtungen ist es denkbar, daß ein nach den gewöhnlichen Verfahren hergestellter Impfstoff gegen Q-Fieber noch lebende Rickettsien enthält. Tatsächlich
konnten nach MALLOCH und STOKER in einer mit 1%igem Phenol behandelten Pockenvaccine
lebende *C. burneti* nachgewiesen werden.

Erhebliche epidemiologische Bedeutung hat die Widerstandsfähigkeit der mit
Sekreten und Exkreten ausgeschiedenen Q-Fieber-Erreger. Besonders groß ist
die Resistenz trockener Rickettsien. Nach PHILIP (2) blieben Rickettsien in den
Faeces der Zecke *Dermacentor andersoni* wenigstens bis zu 586 Tagen lebend.
Mit Faeces, die 6 Jahre bei Zimmertemperatur aufbewahrt worden waren, konnten
Meerschweinchen noch immunisiert werden, die Erreger waren allerdings nicht
mehr zur Vermehrung zu bringen. Sehr auffällig ist, daß *C. burneti* sich im Wasser
bei einer Temperatur von 20—22° C 160 Tage lebend halten konnte (KULAGIN
und SILITSCH).

WEGENER hat im Zusammenhang mit einer Erörterung der milchhygienischen Bedeutung des Q-Fiebers die Beobachtungen über die Wirkung der gebräuchlichen Pasteurisierungsverfahren auf *C. burneti* zusammengestellt. Nach BINGEL und ENGELHARDT töten die üblichen Pasteurisierungsverfahren (85° C für 7—28 sec) die in der Kuhmilch enthaltenen Q-Fieber-Erreger nicht zuverlässig ab, doch werden dieselben bei diesem Prozeß weitgehend „inaktiviert". In Subpassagen war die Quote der überlebenden Rickettsien so niedrig, als wenn das Ausgangsmaterial um mehr als 1:1000 verdünnt wurde. Wahrscheinlich erleiden die Rickettsien dabei auch qualitative Schäden. Beim Meerschweinchen kam es zu keiner Temperaturerhöhung mehr, wenn sie mit Material beimpft wurden, das 5 sec auf 85° C, 7 oder 14 sec auf 80° C oder 60 sec auf 70° C erhitzt worden war. Die so behandelten Rickettsien sind daher vermutlich auch für den Menschen nicht mehr infektiös. Eine sichere Abtötung der Erreger kam durch eine Temperatur von 100° C innerhalb von 7 sec und von 71° C innerhalb von 80 sec zustande. LENNETTE, CLARK, ABINANTI, BRUNETTI und COVERT untersuchten mit der gleichen Fragestellung die in Kalifornien benutzten Pasteurisierungsverfahren. In 1 von 35 Proben war die Milch nach einer Einwirkung von 61,7° C für 30 min noch infektiös geblieben und in 2 von 42 Proben nach einer Einwirkung von 71,1° C für 15 sec. Nach MARMION, MACCALLUM, ROWLANDS und THIEL behielt künstlich infizierte Milch nach einer Erhitzung auf 70—70,5° C für 15 sec noch ihre Infektiosität, jedoch nicht mehr bei einer Erhitzung auf 71,5° C. Natürlich infizierte Milch war nach einer Einwirkung von 67,8—68,9° C für 15 sec nicht mehr ansteckend. ENRIGHT, SADLER und THOMAS untersuchten den Einfluß von Hitze in Milch, die mit 100000 ID pro 2 cm³ infiziert war. Der Erfolg wurde durch Übertragung der behandelten Milch auf Meerschweinchen und nachträgliche KBR festgestellt. Eine Temperatur von 61,7° C konnte in 30 min nicht alle Rickettsien töten. Dies wurde erst bei einer Temperatur von 62,8° C erreicht. Eine Einwirkungszeit von 15 sec vernichtete die Erreger erst bei 71,7° C, aber noch nicht bei 71,1° C. Die vorschriftsmäßige Pasteurisierung reicht jedenfalls nicht immer zu einer sicheren Abtötung aller Erreger aus. COMBIESCU, DUMITRESCU, ZARNEA, SARAGEA, ESSRIG und JONESCU stellten fest, daß *C. burneti* in steriler Milch wenigstens 45 Tage infektiös bleibt, jedoch in Sauermilch die Infektiosität innerhalb 24 Std verliert.

ALLEN, BOVARNICK und SNYDER behandelten gereinigte Rickettsiensuspensionen von *R. prowazeki* aus Dottersackkulturen mit ultraviolettem Licht. Nach einer Einwirkungszeit von 1 min war die Infektiosität der Suspensionen für den Dottersack und die Baumwollratte auf 0,001% reduziert, die Toxicität auf 50%, die hämolytische Aktivität auf 60%. Gleichzeitig wurde eine geringe Reduktion der respiratorischen Aktivität beobachtet. Nach einer Bestrahlung von 9 min ging die Toxicität auf 9% zurück, die hämolytische Eigenschaft auf 21% und die Atmung auf 62%.

R. prowazeki wurde durch Methylenblau (1:1000) in Kochsalzlösungen innerhalb weniger Minuten zerstört (KRYŃSKI und RADKOWIAK (2))]. Lichteinwirkung beschleunigte den Prozeß. Einen gewissen Effekt entfaltete auch Toluidinblau. Die Infektiosität der Rickettsien blieb in Magermilch und menschlichem Serum (1:1) besser erhalten als in Kochsalzlösung. Milch, die in gleicher Menge mit Wasser verdünnt wurde, erwies sich ebenfalls als schonend für *R. prowazeki* (KRYŃSKI und BECLA). In Pufferlösung blieben die Rickettsien bei 4° C nur bis zu 5 Tagen lebensfähig, in Milch dagegen bis zu 35 Tagen. Die Infektiosität ging in Pufferlösung bei 34° C in 2—7 Std verloren, in Milch hielt sie sich bis zu 72 Std.

Die Untersuchungen zu der praktisch wichtigen Frage, wie lange Fleckfieberrickettsien unter natürlichen Bedingungen in *Faeces* von Läusen und Flöhen lebensfähig bleiben, brachten sehr widerspruchsvolle Ergebnisse. Die auffälligen Unterschiede in der festgestellten Lebensdauer der Rickettsien sind vielleicht auf Eigentümlichkeiten der verwendeten Stämme, auf die Anzahl der in den Faeces enthaltenen Rickettsien und auf die Nachweistechnik zurückzuführen, lassen sich aber auf diese Weise sicher nur zum Teil erklären. Eine eindeutige Antwort ist bis jetzt nicht möglich, und das Problem bedarf weiterer Untersuchung.

Sicher ist, daß nicht nur *C. burneti*, sondern auch *R. prowazeki* und *R. mooseri* in *trockenem Zustand* erheblich widerstandsfähiger sind als in feuchtem. Die ersten präzisen Angaben zu diesem Thema gehen auf Versuche von Starzyk (1) zurück, die feststellte, daß sich *R. prowazeki* bei normaler Trocknung an der Luft und einer Temperatur von 5—7° C 35 Tage in Läusemägen und 66 Tage in Läusefaeces lebend hielt. In einer anderen Versuchsreihe konnte eine Lebensfähigkeit von *R. prowazeki* in Läusefaeces von über 3 Monaten festgestellt werden. Die Faeces wurden bei einer Temperatur von 10—20° C und einer relativen Luftfeuchtigkeit von 30—70% konserviert. Rickettsien in Läusefaeces aus Schaffellen, die unter natürlichen Bedingungen aufbewahrt wurden, blieben nach Starzyk (2) sogar bis zu 12 Monaten und 23 Tagen lebensfähig und virulent.

Lange Überlebenszeiten von Rickettsien beobachteten auch Kitaoka und Shishido. Getrocknete Faeces von Läusen, die an einem Patienten mit klassischem Fleckfieber gesogen hatten, waren bei Zimmertemperatur noch nach 233 Tagen für Meerschweinchen infektiös, jedoch nicht mehr nach 120 Tagen, wenn sie im Eisschrank bei 0—5° C aufbewahrt wurden. *R. mooseri* in Faeces von rectal inoculierten Läusen behielt bei Zimmertemperatur die Lebensfähigkeit bis zu 250 Tagen, war aber bei —20° C innerhalb von 180 Tagen abgestorben. Blanc und Baltazard (1), (2) trockneten Faeces von Flöhen, die mit murinem Fleckfieber infiziert waren, über Calciumchlorid und bewahrten sie in zugeschmolzenen Glasampullen auf. Sie stellten unter diesen Bedingungen eine Lebensdauer der Rickettsien bis zu $4^{1}/_{2}$ Jahren fest. Es ist anzunehmen, daß der Trocknungsprozeß als solcher bei dieser langen Widerstandsfähigkeit der Rickettsien eine wichtige Rolle spielte. Daß im Vakuum getrocknete Rickettsien jahrelang lebensfähig bleiben, ist bekannt und ausreichend belegt.

Es existieren auf der anderen Seite Beobachtungen, daß *R. prowazeki* in Läusefaeces und *R. mooseri* in Flohfaeces schon nach wenigen Tagen ihre Infektiosität verloren hatten. Rickard (3) infizierte etwa 3000 Pestflöhe (*Xenopsylla cheopis*) an Ratten und überführte sie 18 Tage später zur Defäkation in Reagenzgläser. Die abgesetzten Faeces wurden bei Raumtemperatur und einer relativen Luftfeuchtigkeit von 80—100% aufgehoben. Die Rickettsien waren unter diesen Bedingungen nach 48 Std noch lebensfähig, jedoch nicht mehr nach 96 Std. Nach 72 Std fielen nur noch 20% der Übertragungsversuche positiv aus. In eigenen Versuchen blieb *R. mooseri* in Läusefaeces, die bei Zimmertemperatur aufbewahrt wurden, nur rund 10 Tage lebensfähig, *R. prowazeki* 30 Tage. Widerstandsfähiger ist *R. quintana*. Diese Erreger blieben unter den gleichen Bedingungen wenigstens 50 Tage lebend. In beiden Fällen erfolgte der Nachweis lebender Rickettsien durch Übertragung der aufgeschwemmten Faeces in die Kleiderlaus. *R. quintana* besitzt auch eine bemerkenswerte Resistenz gegen höhere Temperaturen. Die Rickettsien wurden in trockenen Läusefaeces durch Wärme von 80°C bei einer Einwirkungszeit von 20 min nicht alle zerstört [WEYER (16)]. Nach SHDANOW überstehen sie in Flüssigkeit eine Temperatur von 60° C für 30 min und bleiben bei Trocknung im Licht 4 Monate lebend.

Die Feuchtigkeit, bei welcher die Rickettsien konserviert werden, ist offenbar für ihre Erhaltung wichtiger als die Temperatur. Bei Zimmertemperatur von 10—20° C und einer relativen Feuchtigkeit von 30—60% waren Rickettsien (*R. prowazeki*) nach 147—190 Tagen nicht mehr infektiös (CHAO). In Läusefaeces, die bei 3—6° C und 60—70% relativer Luftfeuchtigkeit aufbewahrt wurden, konnten virulente Rickettsien noch nach 20 Tagen, aber nicht mehr nach 41 Tagen nachgewiesen werden. Ähnlich waren die Ergebnisse bei einer Temperatur von 25—28° C und einer relativen Luftfeuchtigkeit von 75—85%. Nach 19 Tagen enthielten die Faeces noch lebende Rickettsien, aber nicht mehr nach 25 Tagen. Bei gleicher Temperatur und einer Feuchtigkeit von nur 30—35% waren lebende Rickettsien in den Faeces bis zu 55 Tagen zu finden, aber nicht mehr nach 130 Tagen. Der Nachweis lebender Rickettsien erfolgte durch Übertragung der aufgeschwemmten Faeces auf Meerschweinchen.

Wir müssen aus diesen und anderen Beobachtungen schließen, daß die Lebensdauer der von ihren Überträgern ausgeschiedenen Fleckfieberrickettsien unter natürlichen, insbesondere tropischen und subtropischen Verhältnissen sehr viel kürzer ist als unter den gewählten Laboratoriumsbedingungen. Es steht außerdem fest, daß *R. prowazeki* und *R. mooseri* in den Faeces von Läusen und Flöhen erheblich widerstandsfähiger sind als *R. rickettsii* und *R. conori* in den Faeces von Zecken und experimentell infizierten Läusen.

5. Verhalten in Versuchstieren

a) Verhalten in Warmblütern

Die Rickettsien besitzen ein sehr großes Wirtsspektrum. Überragende Bedeutung als Versuchstiere haben die Nager, speziell Meerschweinchen, Ratten und Mäuse behalten. Differenzierung von Rickettsienarten und -stämmen, Fragen der Rickettsienbiologie und die Gewinnung größerer Rickettsienmengen standen bei diesen Arbeiten als Leitmotiv im Vordergrund. Das erfolgreiche Experimentieren hatte den direkten Nachweis der Rickettsien in den infizierten Nagetieren zur Voraussetzung, der MOOSER (1) auf Grund der bereits von NEILL beschriebenen Scrotalläsion beim Meerschweinchen nach der Infektion mit murinem Fleckfieber gelungen war. Das Scrotalphänomen beobachtet man besonders bei murinem Fleckfieber und Felsengebirgsfleckfieber, es kann aber auch bei Rickettsienpocken und den Zeckenbißfiebern auftreten. Ratten und Mäuse sind die geeignetsten Versuchsobjekte für *R. mooseri*, Mäuse für *R. akari* und *R. tsutsugamushi*. Als günstiges Versuchstier für klassisches Fleckfieber hat sich neben dem Meerschweinchen auch die Baumwollratte (*Sigmodon hispidus*) bewährt (ANDERSON; SNYDER und ANDERSON). Junge Tiere sind empfänglicher als alte.

Es ließ sich nachweisen, daß höchstens 10 Rickettsien ausreichen, um eine Baumwollratte mit klassischem Fleckfieber zu infizieren, so daß hinterher komplementbindende Antikörper nachgewiesen werden können (PRICE, EMERSON, NAGEL,

Blumberg und Talmadge). Mit Hilfe der Toxintitration wurde ermittelt, daß 10^7 Rickettsien einer LD 50 für die Maus entsprechen. Die gleiche Menge von Rickettsien genügt zu einem Nachweis im Hämolysintest. Die intraperitoneale Inoculation von Mäusen mit *R. mooseri*, *R. akari*, *R. tsutsugamushi* oder *C. burneti* führt zu charakteristischen Reaktionen. In den Ausstrichen des Peritoneal-exsudats bzw. der Milz und Leber (*C. burneti*) sind die Rickettsien meist leicht und in größerer Zahl nachzuweisen. Bei anderen Rickettsienarten verläuft die Infektion normalerweise symptomlos, doch beobachtet man gelegentlich auch hier eine Anpassung und Vermehrung der Erreger im Peritonealepithel.

Nach intranasaler Inoculation kommt es gewöhnlich in 2—5 Tagen zu einer tödlichen Pneumonie der Versuchstiere. Die lebhafte und rasche Vermehrung der Rickettsien, besonders von *R. mooseri* und gelegentlich auch von *R. prowazeki*, boten die Möglichkeit zur Herstellung von Antigen und Impfstoff aus Lungen. Die Rickettsien gedeihen nicht nur in der Lunge von Nagetieren, sondern z. B. auch in der Lunge von Hunden, Schafen und Ziegen. Trotz starker pneumonischer Veränderungen ist die Rickettsienausbeute nach intranasaler Inoculation von Mäusen mit den Erregern der durch Zecken und Milben übertragenen Rickettsiosen normalerweise sehr gering. Bei einer intranasalen Inoculation von Mäusen mit *R. tsutsugamushi* entwickelt sich die Pneumonie sehr schleppend. Le Gac, Giroud, Roger, Courmes und Bres sahen bei einem Stamm aus Vietnam erst 20 Tage nach der Inoculation deutliche pneumonische Veränderungen. In eigenen Versuchen mit dem Stamm Kato aus Japan ließen sich anfänglich die ersten schwachen Läsionen in der Lunge nicht vor 6—8 Tagen erkennen. Die längere Haltung des Stammes in Lungenpassagen führte zu einer Anpassung der Rickettsien unter ständiger Vergrößerung der pneumonischen Herde. In den Lungenausstrichen waren nach 1 Woche Rickettsien auch in größerer Zahl nachweisbar. Es fiel auf, daß die Mäuse selten klinische Symptome zeigten, auch wenn die Pneumonie die ganze Lunge ergriffen hatte. Der Tod trat erst nach 15 Tagen ein.

F. Roger und A. Roger fanden bei intranasaler Inoculation mit kleinen Dosen den Hamster besonders empfänglich für *R. mooseri*. Die Tiere erkrankten schnell an einer tödlichen Pneumonie, und in den Lungenausstrichen fanden sich große Mengen von freien Rickettsien. Demgegenüber erzeugte *R. prowazeki* beim Hamster, auch bei Verwendung von massiven Dosen, nur eine flüchtige Pneumonie, und die Weiterführung der Stämme durch Lungenpassagen stieß auf Schwierigkeiten. Die Ausstriche enthielten relativ wenige Rickettsien, die nur im Cytoplasma lagen. Die Reaktionsunterschiede waren so auffällig, daß sie sich für eine Differenzierung von *R. mooseri* und *R. prowazeki* eigneten. Whitmire und Downs fanden bei intraperitonealer Übertragung von *R. mooseri* die Maus am empfänglichsten für die Infektion, danach Baumwollratte, Meerschweinchen und Hamster. Nach Verabfolgung von Cortison verlagerte sich die Empfänglichkeit für eine tödliche Infektion in der Reihenfolge Hamster, Maus und Baumwollratte. Dieselbe Dosis war für Meerschweinchen nicht tödlich. Durch Cortison ließ sich auch die Scrotalreaktion beim Meerschweinchen und Hamster unterdrücken und beim Hamster eine schwache Infektion innerhalb von 24 Std in eine tödliche verwandeln. Weiße Ratten, die normalerweise gegen eine Infektion mit *R. rickettsii* resistent sind, werden durch Injektion von gewaschenen roten Blutkörperchen verschiedener Säuger für die Infektion empfänglich (Harshman, Johnson und Price). Die roten Blutkörperchen enthalten einen hitzelabilen Faktor, der die Rickettsienvermehrung stimuliert. Die chemische Natur dieser Substanz ist noch nicht bekannt.

Das Zusammenspiel zwischen Wirt und Parasit wird nicht nur durch Inoculationsmodus und Infektionsdosis, sondern durch Alter und Konstitution des empfänglichen Wirtes und die Virulenz der Stämme (vgl. S. 101) beeinflußt, wobei jedoch die spezifischen Eigenschaften der Stämme erhalten bleiben. Wojciechowski fand nach längerer paralleler Haltung dreier Stämme von *R. prowazeki* in Läusen, Mäuselungen und Dottersäcken weder morphologische noch antigene Unterschiede zwischen den Stämmen. Auch die Virulenz und Pathogenität für Läuse und Meerschweinchen hatte sich nicht geändert.

Für *R. quintana* konnte außer dem Pavian (Codeleoncini) im Rhesusaffen ein Versuchstier ermittelt werden [Mooser und Weyer (1), (2)]. Rhesusaffen wurden intravenös mit Rickettsiensuspensionen aus Läusemägen inoculiert und reagierten mit unregelmäßigem Fieber und einer charakteristischen Rickettsiämie von wechselnder Dauer. Die Rickettsiämie wurde durch Fütterung von Läusen an den Versuchsaffen sowie durch rectale und intracoelomale Inoculation von Läusen mit dem Blut der Affen genauer verfolgt. Dabei konnte weitgehende Übereinstimmung in der Reaktion der Versuchstiere und des Menschen und im Verhalten der Rickettsien in den beiden Wirten konstatiert werden.

Eine größere Zahl von Untersuchungen betrifft das Verhalten von *C. burneti* in Warmblütern. Im Unterschied zu anderen Rickettsien zeigt dieser Erreger keinen Tropismus zu den Gefäßendothelien, sondern bevorzugt die Zellen des Reticuloendothels. Bock fand die Rickettsien beim Goldhamster in den Endothelzellen der Glomeruli und der intratubulär gelegenen Capillaren sowie in den Epithelzellen der Glomeruli und Tubuli. Meerschweinchen, Hamster und Mäuse scheiden die Erreger längere Zeit mit dem Urin aus (Reczko; Bock; Özbil). Bei intratestaler Impfung von Meerschweinchen sahen Herzberg, Herzberg-Kremmer und Urbach eine besonders lebhafte Vermehrung der Rickettsien und das Auftreten von fädigen Kolonien. Der direkte, auch im Schnittpräparat bestätigte Erregernachweis konnte mit dieser Methode geführt werden, bevor die KBR positiv wurde.

Germer untersuchte auf breiterer Basis (97 Versuchstiere) den Verlauf der Infektion beim Meerschweinchen. Die Inkubation betrug 4—7 Tage, das Fieber dauerte 2—10 Tage und blieb bei 2 Tieren aus. Die Infektion endete für 22 Tiere tödlich. Eine Rickettsiämie blieb bis zu 18 Tagen nach der Entfieberung bestehen. Die KBR erreichte nach 4—8 Tagen einen Titer von 1:16 und ihren Maximalwert mit 1:750 am 25. Tage. 7 Monate nach der Infektion bestand noch ein Titer von 1:32. Die wichtigsten histologischen Veränderungen fanden sich in Milz und Testes. Im Unterschied zu anderen Rickettsiosen kommt es beim Q-Fieber zu keiner generalisierten Vasculitis und Perivasculitis.

Bei Ratten (*R. norvegicus*) verlief die Infektion mit *C. burneti* symptomlos (Syrůček und Sobeslavský). Die Tiere entwickelten eine Rickettsiämie von 7—26 Tagen und schieden Rickettsien mit dem Urin und Kot vom 13.—38. Tage aus. Die KBR wurde in der 2. Woche positiv. Hunde und Meerschweinchen wurden auch peroral infiziert [Ravaioli sowie Babudieri und Moscovici (2)]. Die mit infizierten Dottersäcken gefütterten Hunde enthielten im Kot bis zu 6 Tagen Rickettsien, im Blut bis zu 15 Tagen. Eine perorale Infektion von Meerschweinchen gelang nur bei Verwendung sehr hoher Dosen.

Die Bedeutung von Haustieren und speziell von Schafen als Infektionsquelle (vgl. S. 115) ist durch experimentelle intravenöse und intratracheale Inoculation geprüft worden (Stoenner sowie Lennette, Holmes und Abinanti; Abinanti, Welsh, Lennette und Brunetti). Nach intravenöser Inoculation stellte sich bei Schafen in 7—10 Tagen Fieber ein. Rickettsien waren im Blut längstens bis zu 8 Tagen nachweisbar. Sie konnten aus Leber, Milz und Niere isoliert werden, auch bei Schafen, die im Blut keine Rickettsien enthielten. Urin, Faeces und Mundsekrete waren negativ. Babudieri und Ravaioli beobachteten nach experimenteller subcutaner und intratrachealer Inoculation von Schafen nur geringfügige klinische Symptome. Die Erreger konnten in Blut, Milch und Urin nachgewiesen werden.

Auch Gänse, Enten, Hühner und Tauben ließen sich experimentell mit *C. burneti* infizieren [Babudieri und Moscovici (1), (2)]. Die Vögel blieben nur kurze Zeit seropositiv. Rickettsien konnten jedoch in der Niere von Enten und Tauben noch 6 Wochen nach intraperitonealer Inoculation nachgewiesen werden. Von 12 peroral mit Dottersackmaterial infizierten Tauben enthielten 5 nach 15—26 Tagen komplementbindende Antikörper. Smajewa, Ptschelkina, Mischtschenko und Karulin konnten *C. burneti* durch den Stich der Zecke *Hyalomma marginatum* auch auf Küken übertragen. Syrůček und Raška fanden bei 3 von 8 intraperitoneal infizierten Hühnern erst nach 6 Wochen Antikörper im Blut; 1 Huhn schied aber vom 7.—40. Tag nach der Infektion Rickettsien mit den Exkrementen aus. Sobeślavský inoculierte Hühner subcutan, intraperitoneal und intranasal mit *C. burneti*. Die Hühner schieden den Erreger mit dem Kot bzw. Urin zwischen dem 14. und 42. Tag nach der Inoculation aus. Aus Blut und den inneren Organen einer Henne konnten Erreger nach 96 Tagen isoliert werden. Rickettsien ließen sich auch in 6 von 13 Eiern nachweisen, die zwischen dem 19. und 42. Tag nach der Inoculation abgelegt waren, und in 3 von 4 Küken zwischen dem 2. und 55. Tag nach dem Schlüpfen. Danach scheint bei Hühnern eine transovarielle Weitergabe der Rickettsien auf die folgende Generation möglich zu sein.

Von besonderem theoretischen und praktischen Interesse ist die experimentell nachgewiesene lange *Erregerpersistenz*, die für die meisten Rickettsiosen zu gelten scheint. Nachdem schon Laigret und Jadin *R. mooseri* im Gehirn von Mäusen bis zum 40. Tag nach intraperitonealer Inoculation gefunden hatten, zeigten Lépine und Sautter, daß *R. mooseri* im Gehirn des Hamsters (*Citellus citellus*) in unverändert virulenter Form bis zum 118. Tage nach der Inoculation, unter gelegentlicher Abschwächung der Virulenz, aber unter Wahrung der ursprünglichen Eigenschaften bis zum 226. Tage, in einigen Fällen sogar bis zum 374. Tage gefunden werden konnte. Philip und Parker testeten Blut, Milz, Tunicaexsudat und Gehirn von infizierten Ratten, Mäusen und Meerschweinchen auf ihren Rickettsiengehalt. Die Rickettsien wurden am regelmäßigsten und längsten im Gehirn gefunden. *R. mooseri* war im Gehirn von weißen Ratten 147, 153, 161 und 370 Tage nach der Inoculation nachweisbar, in der Milz nach 125 und 153 Tagen, im Gehirn der Maus bis zum 150. Tag, in der Milz bis zum 132. Tag. Der Erregernachweis war in der Maus weniger regelmäßig als in der Ratte. Aus dem Meerschweinchengehirn wurden die Rickettsien nach 60, 90 und 120 Tagen, aus dem Tunicaexsudat nach 16 Tagen isoliert. *R. rickettsii* fand sich nur einmal nach 30 Tagen im Gehirn der Ratte. In inoculierten Mäusen und Erdhörnchen wurden keine Rickettsien dieser Art nachgewiesen. Versuche mit *R. conori* verliefen negativ. Die Erreger des sibirischen Zeckenbißfiebers fanden sich bis zu 140 Tagen im Gehirn von intraperitoneal inoculierten Ratten (Sdrodowski und Golinewitsch). Parker und Steinhaus isolierten *C. burneti* aus der Milz und Niere des Meerschweinchens bis zum 110. Tage nach Krankheitsbeginn, Reczko fand den Erreger noch nach 526 Tagen in den Organen. Wojciechowski, Lewińska und Mikołajczyk (1) inoculierten Meerschweinchen, zahme und wilde Mäuse (*Microtus arvalis*) sowie zahme und wilde Ratten intraperitoneal mit *R. prowazeki* aus Läusen und fanden die Erreger im Gehirn und in der Milz nach 16—20 Tagen.

Auch *R. tsutsugamushi* persistiert recht lange in den Versuchstieren. Fox (1) konnte die Erreger in Mäusen noch nach 610 Tagen, und zwar bevorzugt in der

Niere, weiterhin im Gehirn und in der Leber nachweisen. Der Urin der Mäuse enthielt keine Erreger. Im Blut und in den Geweben der Baumwollratte fand sich *R. tsutsugamushi* bis zum 102. Tag nach der Inoculation, im Blut allein bis zum 145. Tag, im Gehirn bis zum 269. Tag. Zu dieser Zeit war die Niere immer noch rickettsienhaltig. KOUWENAAR und ESSEVELD isolierten *R. tsutsugamushi* aus Blut und Milz von Meerschweinchen nach 635 Tagen.

b) Verhalten in Arthropoden

Einen großen Raum nehmen die Untersuchungen über Vermehrung und Biologie der Rickettsien in Arthropoden ein, die als die natürlichen Rickettsienwirte anzusehen sind. Die Entwicklungsfähigkeit in Arthropoden kann als typisches Merkmal aller Erreger menschlicher Rickettsiosen gelten. Es hat sich gezeigt, daß sich die meisten in der Natur an bestimmte Überträger gebundenen Rickettsienarten auch in den Wirten anderer Arten vermehren können [WEYER (15)]. Das klassische Versuchstier für Rickettsien ist unter den Arthropoden die *Kleiderlaus*. Die von WEIGL entwickelte Methode der künstlichen Infektion auf rectalem oder intracoelomalem Wege bietet eine einfache Möglichkeit nicht nur zur Kultur von *R. prowazeki* [KRYŃSKI und RADKOWIAK (1)], sondern auch zur Haltung von anderen Rickettsienstämmen und zu einem Vergleich ihrer biologischen Verhaltensweise in der Laus.

Mit der Haltung von *R. prowazeki* und *R. mooseri* in der Kleiderlaus haben sich u. a. SNYDER und WHEELER näher befaßt. FULLER, MURRAY und SNYDER infizierten durch eine aus der Haut 7tägiger Küken hergestellte Membran Läuse mit *R. prowazeki* und *R. mooseri*. FULLER (2) verglich unter Benutzung dieser Infektionsmethode die Empfindlichkeit von Läusen und Baumwollratten für den Nachweis kleiner Rickettsienmengen in Organzerreibungen und fand beide Testmethoden gleichwertig. Die Versuchsläuse wurden nach der Infektion täglich 1mal am Kaninchen gefüttert. Die Herauszüchtung eines an Kaninchenblut adaptierten Kleiderlausstammes [CULPEPPER (1), (2)] hat bis zu einem gewissen Grade die Ernährung der Läuse am Menschen und die künstliche Infektion nach der Methode von WEIGL ersetzt, da sich durch Füttern an intravenös inoculierten Kaninchen auch größere Mengen von Läusen bequem mit Rickettsien infizieren lassen. Doch ermöglicht die künstliche Infektion ein schnelleres und wesentlich exakteres Arbeiten und gestattet ohne einen größeren Zeit- und Materialaufwand das gleichzeitige Experimentieren mit mehreren Stämmen.

FULLER (5) ließ hungrige Larven des 3. Stadiums durch die Membran an rickettsienhaltigen Suspensionen saugen und fütterte die Läuse anschließend am Kaninchen. Er hielt den Wilmington-Stamm von *R. mooseri* auf diese Weise 95 Tage in 19 fortlaufenden Passagen, den Florida-Stamm 43 Tage in 8 Passagen, einen Stamm von *R. prowazeki* (E-Stamm) 79 Tage in 5 Passagen ausschließlich in Läusen. Zwischen den Passagen wurden die Läuse für 3—6 Tage bei 32° C gehalten. Änderungen der Stammeigenschaften traten dabei nicht auf. *R. mooseri* konnte sich auch in Läusen vermehren, die nur mit menschlichem Serum gefüttert wurden. Mit der gleichen Methode stellte FULLER (4) den Einfluß einer Infektion mit *R. prowazeki* auf die Lebensdauer der Läuse fest. Bei einer Temperatur von 25,6° C lebten die infizierten Läuse im Durchschnitt 15,4 Tage (Kontrolle 26,3 Tage), bei 31,9° C 7,4 Tage (Kontrolle 12,5 Tage), bei 35,8° C, ebenso wie die

Läuse der Kontrolle, nur 2,4 Tage. Die Infektionsdosis hatte auf die Lebensdauer nur geringen Einfluß. Bei einer Temperatur von 34,5⁰ C variierte die durchschnittliche Lebensdauer der Läuse, die mit einer bis zum 1000fachen der ID 50 gesteigerten Konzentration infiziert wurden, nur zwischen 5,4 und 3,2 Tagen. Die Läuse der Kontrolle lebten 4,96 Tage.

Bei den nach der Weigl-Methode infizierten und am Menschen gefütterten Läusen unterscheidet KRYŃSKI eine akute Infektion, bei der die Läuse zwischen dem 3. und 6. Tag unter rötlicher Verfärbung sterben, und eine chronische Infektion, bei welcher die Läuse wesentlich länger leben. Benutzt man hohe Infektionsdosen oder sehr virulente Stämme, dann können die Läuse schon innerhalb der ersten 2 Tage an der toxischen Wirkung eingehen. Der generelle Ablauf der Infektion von Läusen mit *R. prowazeki* oder *R. mooseri*, der sich nicht grundsätzlich unterscheidet, ist verschiedenen Variationen unterworfen, die teils an spezifische Eigenschaften der Stämme, teils an Schwankungen der Virulenz beim gleichen Stamm gebunden sind und sich u. a. in der Schnelligkeit der Rickettsienvermehrung und der pathogenen Wirkung für die Laus äußern.

Zu den Eigenschaften der Rickettsien oder der Stämme kommen aber auch von der Laus ausgehende Einflüsse, die als eine Resistenz gegen die Rickettsien gedeutet werden müssen [WEYER (15)]. Bei gleicher Infektionsdosis fanden sich Läuse mit langsamer und stürmischer Rickettsienvermehrung und Läuse, bei denen nur ein Teil der Magenzellen Rickettsien enthielt und die daher durch die Infektion nicht beeinträchtigt wurden. Zu den auffälligsten Beobachtungen gehörte, daß Läuse mit einer kräftigen Infektion, die sich auf das ganze Magenepithel erstreckte, eine normale Lebensdauer zeigen konnten und auch keinerlei anderweitige Störungen erkennen ließen. Die Zellen waren dicht mit Rickettsien besetzt und aufgetrieben, aber nur teilweise zerstört. Offenbar war hier die Rickettsienvermehrung durch Abwehrreaktionen der Laus zum Stillstand gekommen und gleichzeitig die toxische Wirkung paralysiert, ohne daß Vermehrungsfähigkeit oder Virulenz der Rickettsien endgültig aufgehoben waren, wie sich bei der Übertragung dieser Rickettsien in andere Versuchstiere erkennen ließ. Bestimmte Versuchsergebnisse müssen dahin gedeutet werden, daß die Rickettsienvermehrung bei manchen Läusen nicht nur von vornherein unterdrückt oder auf frühen Stadien sistiert wird, sondern daß Rickettsien, die bereits in die Zellen eingedrungen sind und mit der Vermehrung begonnen haben, zugrunde gehen und von den Wirtszellen absorbiert werden [WEYER (15)]. Nicht geklärt ist die Frage, ob es primär gegen Rickettsien resistente Läusepopulationen gibt und ob diese Resistenz an besondere geographische Rassen oder Varianten innerhalb einer Population, z. B. dunkel pigmentierte Formen, gebunden ist.

Außer *R. prowazeki*, *R. mooseri* und *R. quintana* konnten durch rectale Inoculation oder durch Fütterung der Läuse an infizierten Mäusen und Meerschweinchen folgende Rickettsien im Magen der Laus zur Ansiedlung und Vermehrung gebracht werden: *R. rickettsii* (Stamm Michoacan aus Mexiko, 1 Stamm aus Brasilien), *R. conori* (1 Stamm aus Nordafrika, 1 Stamm aus Kenia, 1 Stamm aus Südafrika), *R. akari*, 2 Stämme von sibirischem Zeckenbißfieber [WEYER (18)] und *C. burneti*. Eine Ausnahme machte *R. tsutsugamushi*. Diese Beobachtung gilt für fünf bisher in dieser Richtung geprüfte Stämme: Stamm Karp (Neu-Guinea), Stamm Gilliam (Malaya), Stamm Korea, die Stämme Kato und Hokkaido (Japan) und einen Stamm von *R. tamiyai* aus Japan [WEYER (17)]. Diese Rickettsien werden offenbar gleich nach der Übertragung im Magen der Laus vernichtet oder mit den Faeces ausgeschieden. Die übrigen Rickettsien (mit Ausnahme von *R. quintana*, die nur extracellulär im Magenlumen wächst) vermehrten sich intracellulär in den Zellen der Magenschleimhaut, teilweise in der gleichen Intensität wie *R. prowazeki* oder *R. mooseri*. Gelegentlich kam es bei einigen Arten in vorgeschrittenen Stadien der Infektion nach Schädigung der Schleimhaut zu einem Übergang von Erregern in die Hämolymphe, wobei *C. burneti* auch andere Organe befallen konnte. Die Rickettsien ließen sich in Passagen längere Zeit fortlaufend in Läusen halten, ohne ihre Eigenschaften einschließlich ihrer Virulenz zu ändern.

In der Pathogenität der einzelnen Rickettsienarten oder -stämme für die Laus ergaben sich deutliche Unterschiede. Die pathogene Wirkung basierte nicht so sehr auf einer mechanischen Zerstörung der Magenzellen, wie bei einer Infektion mit *R. mooseri* oder *R. prowazeki*, sondern offenbar auf toxischen Eigenschaften der Erreger. Auch bei schwachem Befall kam es vielfach schon frühzeitig zu einer Degeneration der Zellen oder einer Ablösung des ganzen Epithels und damit zu einer letalen Wirkung für die Laus. Die schwächste pathogene Wirkung hatten *R. akari*, die Stämme von südafrikanischem, Kenia- und sibirischem Zeckenbißfieber, die stärkste ein Stamm von *R. rickettsii* aus Mexiko und ein Stamm von *R. conori* aus Nordafrika. Auffällig waren die Unterschiede zwischen den untersuchten Stämmen von *R. conori* [WEYER (13), (18), vgl. S. 79]. Der Stamm von *R. rickettsii* konnte nur über eine beschränkte Anzahl von Passagen kontinuierlich in der Laus gehalten werden. Ähnlich wie bei einer Infektion mit *R. prowazeki* und *R. mooseri* wurden von den infizierten Läusen virulente Rickettsien mit den Faeces ausgeschieden, jedoch in einem wesentlich begrenzteren Umfang. Auch bei sehr kräftiger Vermehrung von *R. rickettsii*, *R. conori* und *R. akari* in der Laus wurde kein Befall der Kerne gesehen.

Die Versuche über das Verhalten der Rickettsien in der Laus sind durch die Technik der intracölomalen Inoculation weiter ausgebaut worden [WEYER (5)]. Die Hämolymphe der Laus bildet einen ausgezeichneten Nährboden für Rickettsien, in welchem sie in dichter Kultur vorwiegend *extracellulär* wachsen. Das trifft für alle bereits erwähnten Arten und Stämme, einschließlich der Stämme von *R. tsutsugamushi*, zu. Versuche, die an die Hämolymphe angepaßten *R. tsutsugamushi* im Magen der Laus zur Ansiedlung zu bringen, fielen negativ aus. Die Vermehrung in der Hämolymphe war bei *R. tsutsugamushi*, Stamm Karp, im Vergleich zu den anderen Rickettsien schwach, die übrigen Stämme von *R. tsutsugamushi* zeigten aber das gleiche ungestörte Wachstum in der Hämolymphe wie die anderen Rickettsien [WEYER (17)]. Die Stämme konnten durch intracölomale Inoculation fortlaufend in Passagen auf Läusen gehalten werden. *C. burneti* befievon der Hämolymphe aus Magenzellen, Zellen im Fettkörper und in der Hypodermis.

Außer der Kleider- und Kopflaus ist für *R. prowazeki* und *R. quintana* auch die Filzlaus (*Phthirus pubis*) empfänglich [WEYER (9)]. *R. prowazeki* und *R. mooseri* dürften sich ferner in allen blutsaugenden Tierläusen entwickeln können. Bemerkenswert ist, daß sich *R. quintana* nicht auf die Schweinelaus übertragen ließ [WEYER (8)]. Für diesen Erreger scheint außer den Menschenläusen nur die Affenlaus eine gewisse Empfänglichkeit zu besitzen [WEYER (16)].

Experimentell ließ sich eine große Zahl von *Floharten* mit *R. prowazeki* und *R. mooseri* infizieren, wobei sich keine prinzipiellen Unterschiede in der Empfänglichkeit ergaben. Auch im Floh entwickeln sich die Rickettsien in den Magenzellen, jedoch wird der Floh im Unterschied zur Laus durch die Infektion nicht geschädigt, auch seine Lebensdauer ist nicht verkürzt. Die Rickettsien werden wie bei der Laus mit den Faeces ausgeschieden.

Bei verschiedenen Gelegenheiten ist diskutiert worden, ob sich *R. mooseri* auch in der *Rattenmilbe Bdellonyssus bacoti* entwickeln kann und ob diese Milbe für eine Übertragung des murinen Fleckfiebers in Betracht kommt. Die Frage ist bejaht worden auf Grund von älteren Experimenten, die Dove und Shelmire durchführten, die später aber nicht im ganzen Umfang bestätigt wurden.

Dove und Shelmire konnten mit 25 zerriebenen Milben, die an einem infizierten Meerschweinchen gesogen hatten, *R. mooseri* auf ein gesundes Meerschweinchen übertragen. Bei Ratten wurde 10 und 21 Tage nach dem Saugen von Milben eine inapparente Infektion konstatiert. 4 und 7 Tage, nachdem gesunde Milben an infizierten Meerschweinchen gesogen hatten, ließen sich die Erreger durch den Stich auf gesunde Meerschweinchen übertragen, 10 Tage später nicht mehr. Es wird auch von positiven Übertragungsversuchen mit Milbenlarven berichtet, so daß sogar eine transovarielle Übertragung in Erwägung gezogen wurde.

Worth und Rickard (2) unternahmen Versuche mit Milben in Florida, wo *B. bacoti* der häufigste Ektoparasit der Ratten ist. Sie übertrugen Suspensionen von zerriebenen Milben, die 8 Tage vorher an infizierten Ratten gesogen hatten, auf gesunde Ratten; die Ratten erkrankten an murinem Fleckfieber. Alle Übertragungsversuche, die 11—22 Tage nach dem Saugen angesetzt wurden, verliefen jedoch negativ. 535 *B. bacoti* wurden mit 3 frisch infizierten und 3 gesunden Ratten 6 Wochen zusammengehalten. Nur eine der gesunden Ratten hatte sich dabei mit murinem Fleckfieber infiziert.

Diese Beobachtungen sowie ein Teil der Ergebnisse von Dove und Shelmire lassen sich damit erklären, daß die Rickettsien, die beim Saugen aufgenommen werden, eine Zeitlang im Darm der Milben überleben können. In diesem Sinne wird auch die Isolierung von *R. mooseri* aus *B. bacoti* von Ratten in China gedeutet [Pang; Liu (2)]. Strandtmann und Eben fanden in Galvestone keine Rickettsien in *Laelaps nuttalli* und *B. bacoti*. Versuche, murines Fleckfieber von einer Ratte zur anderen durch diese beiden Milben zu übertragen, verliefen negativ, auch alle Versuche, bei denen die Ratten mit zerriebenen Milben inoculiert wurden, die 2 Wochen lang an infizierten Ratten gesogen hatten. Als Wirt für *R. mooseri* kann daher die Rattenmilbe nicht angesehen werden.

Auch die Rückfallfieberzecke *Ornithodorus moubata* und die Larven vom Mehlkäfer, *Tenebrio molitor*, erwiesen sich im Experiment für Rickettsien empfänglich [Weyer (2), (3), (15)]. In der Hämolymphe und im Fettkörper der Mehlkäferlarven wurden *R. prowazeki*, *R. mooseri*, *R. rickettsii*, *R. conori*, *R. akari* und *C. burneti* zur Vermehrung gebracht und unter Wahrung der ursprünglichen Eigenschaften längere Zeit gehalten. Nur *R. quintana* und *R. tsutsugamushi* konnten auf diesem Wege nicht kultiviert werden. Johnson und Traub beobachteten in ähnlichen Versuchen mit *Tenebrio molitor* ein Überleben von *R. tsutsugamushi* bis zu 25 Tagen, jedoch kam es in ihren Experimenten zu keiner Vermehrung der Erreger.

R. prowazeki wurde nach künstlicher intracölomaler Übertragung auf *O. moubata* in den inoculierten *Zecken* bis zu 262 Tagen, in Eiern bis zu 34 Tagen und in Nymphen der F_1-Generation bis zu 263 Tagen nachgewiesen, *R. mooseri* in den beimpften Zecken bis zu 218 Tagen,

in Eiern bis zu 55 Tagen, in Nymphen der F_1-Generation bis zu 237 Tagen [WEYER (2)].
O. moubata schied *R. mooseri* auch mit der Coxalflüssigkeit aus. *R. prowazeki* blieb in *Argas columbae* bis zu 218 Tagen lebensfähig und virulent. Auffällig war in einigen Versuchen, daß die Rickettsien bei der Rückübertragung aus den Zecken in Läusen extracelluläre Lagerung angenommen und damit ihre Infektiosität und Pathogenität für Mäuse und Meerschweinchen eingebüßt hatten. Die Empfänglichkeit der Zecken für Fleckfieberrickettsien dürfte keine praktische Bedeutung haben. Aus der großen Zahl der Versuche zur Isolierung von Rickettsien aus im Freien gesammelten Zecken ist nur ein Fall bekannt, wo in Indien *R. mooseri* in der Zecke *Boophilus australis* nachgewiesen wurde (vgl. S. 129).

Für *R. rickettsii* und *R. conori* ist charakteristisch, daß sie in ihren Wirtszecken nicht nur die Magenzellen, sondern praktisch alle Organe einschließlich der Ovarien befallen können, wobei gelegentlich auch eine intranucleäre Entwicklung gesehen wird (HASS und PINKERTON; vgl. S. 78). Die Empfänglichkeit für diese Rickettsien ist nicht an eine bestimmte Zeckenart oder -gattung gebunden, obwohl es deutliche Unterschiede in der Eignung der Zecken als Wirte gibt. So konnte z. B. ein Stamm von *R. conori* aus Indien nicht auf *Dermacentor andersoni* übertragen werden [PHILIP (4)]. Als geeigneter Wirt und Überträger für das indische Zeckenbißfieber erwies sich die Hundezecke *Rhipicephalus sanguineus*. Die Rickettsien wurden durch den Stich von Larven, Nymphen und Adulten auf Meerschweinchen übertragen, sofern sich die Zecken in einem früheren Entwicklungsstadium durch Saugakt infiziert hatten.

Folgende Zecken übertrugen in den USA zusätzlich zu den auf S. 108 ff. genannten Arten im Experiment *R. rickettsii*: *Amblyomma cajennense, A. striatum, Dermacentor occidentalis, D. parumapertus, D. albipictus, Ornithodorus parkeri, O. hermsi, O. nicollei, O. turicata* und *O. rudis;* in Brasilien außerdem *Amblyomma brasiliensis, A. cooperi, Rhipicephalus sanguineus* und *Ixodes loricatus*, in Kolumbien *Otocentor nitens*. Alle in Südafrika vorkommenden Zecken konnten auch experimentell mit dem Erreger des südafrikanischen Zeckenbißfiebers infiziert werden (*Joint WHO study group on African Rickettsioses*).

KORSCHUNOWA und PETROWA-PIONTKOWSKAJA übertrugen *R. conori* durch Füttern von infizierten *Rhipicephalus sanguineus* und deren F_1-Generation auf Meerschweinchen. Die Rickettsien waren in den Ausstrichen von Magensaft, Speicheldrüsen und Geschlechtsorganen der Weibchen nachweisbar, aber nicht in Männchen. DAVIS infizierte *Ornithodorus venezuelensis* mit einem Stamm von *R. rickettsii* aus Kolumbien. Die Rickettsien konnten mit Zeckenbrei noch nach 343 Tagen auf Meerschweinchen übertragen werden und gingen auch auf die F_1-Generation über. Die Übertragung erfolgte jedoch nicht beim Saugakt. Stämme von *R. rickettsii* aus USA und Brasilien blieben in der betreffenden Zecke für 191 bzw. 243 Tage lebensfähig, in *O. turicata* bis zu 5 Jahren (BRUMPT und DESPORTES). Nach künstlicher Inoculation konnten *R. rickettsii* und *R. conori* auch in *Ornithodorus moubata* zur Vermehrung gebracht werden.

Das größte Wirtsspektrum innerhalb der Arthropoden besitzt der *Q-Fieber-Erreger, C. burneti*. Die Erreger konnten experimentell außer auf eine große Zahl von Schild- und Lederzecken, wie bereits erwähnt, auch auf Läuse und Mehlkäferlarven, ferner auf Bettwanzen, Raubwanzen und Flöhe übertragen und hier zur Vermehrung gebracht werden [WEYER (11)]. Die Zecken und Insekten wurden teils künstlich rectal oder intracölomal, teils auf natürlichem Wege durch Füttern an Mäusen infiziert. Die infizierten Zecken konnten die Rickettsien während des Saugaktes auf Meerschweinchen übertragen, wobei offen blieb, ob die Erreger mit dem Kot oder Speichel ausgeschieden wurden. Die Hämolymphe von Lederzecken (*O. moubata*), Raubwanzen, Flöhen und Mehlkäferlarven bildet ebenso wie die von Läusen einen günstigen Nährboden für *C. burneti*, die sich frei in der Blutflüssigkeit, in erster Linie aber intracellulär in Blutzellen vermehren. Rectal inoculierte Bettwanzen schieden mindestens 101 Tage lang virulente Rickettsien mit dem Kot aus. In Flöhen siedelten sich die Rickettsien nur nach künstlicher intracölomaler Inoculation an, während es bei natürlicher Inoculation durch den Saugakt lediglich zu einem kurzen Aufenthalt der Erreger im Magen oder gelegentlich zum Befall einzelner Magenzellen kam. Blanc und Bruneau (1) konnten Rickettsien nur bis zu 16 Tagen im Flohmagen nachweisen.

Während die Q-Fieber-Erreger bei einer natürlichen Infektion von *O. moubata* durch den Saugakt auf die Magenzellen der Zecke beschränkt blieben, vermehrten sie sich nach künstlicher intracölomaler Übertragung auch in anderen Organen und wurden mit Speichel, Coxalflüssigkeit und Harn ausgeschieden [Weyer (2)]. Infizierte Zecken blieben ohne Schädigung jahrelang (Beobachtung bei 1298 Tagen abgebrochen) Rickettsienträger. Durch Befall der Ovarien kann es zu einer transovariellen Übertragung kommen. Die transovarielle Übertragung von *C. burneti* ist experimentell auch bei anderen Zecken festgestellt worden, z. B. bei Arten der Gattungen *Dermacentor*, *Amblyomma*, *Hyalomma* und *Rhipicephalus*.

Obwohl *C. burneti* im Vergleich zu anderen Rickettsien eine ungewöhnlich große Anpassungsfähigkeit an sehr verschiedene Organe und Gewebe von systematisch weit getrennten Gliederfüßlern besitzt, dürften die experimentell geprüften Wirte unter natürlichen Verhältnissen als Überträger nur eine nebensächliche Rolle spielen. Am ehesten könnten noch die mit dem Kot infizierter Wanzen und Läuse ausgeschiedenen Rickettsien eine Infektionsquelle darstellen. Giroud und Jadin (1) berichten auch über den Nachweis von *C. burneti* in Kleiderläusen in Belgisch-Kongo. Die Frage der praktischen Bedeutung dieser Beobachtung ist anschließend von Giroud, Jadin und Jezierski durch experimentelle Infektion von Läusen geprüft worden. Daß Fliegen als Verschlepper von *C. burneti* in Frage kommen, hat Philip (2) durch Experimente bestätigt.

Rattenmilben (*Bdellonyssus bacoti*) konnten im Versuch virulente Rickettsien wenigstens 10 Tage beherbergen [Weyer (11)]. Semskaja und Ptschelkina stellten fest, daß sich außer *B. bacoti* auch die Hühnermilbe *Dermanyssus gallinae* an Meerschweinchen infizieren läßt. Die Milben übertrugen die Erreger beim Saugakt, auch wenn sie vorher mehrfach an gesunden Tieren gesogen hatten. Die Weibchen blieben bis zu 6 Monaten infektiös.

Arthropoden und Nager werden auch für eine *ätiologische Diagnose* von Rickettsien herangezogen, die trotz aller Fortschritte der serologischen Diagnostik in manchen Fällen von Bedeutung sein kann. Durch den Erregernachweis bietet sich die Möglichkeit, die serologische Diagnose zu sichern oder durch eine Untersuchung der Versuchstiere zu erweitern. Neben der direkten Übertragung von Blut in den Dottersack oder die Versuchstiere spielt bei dem Infektionstest die Xenodiagnose eine gewisse Rolle. Sie kommt vor allem für die Isolierung von Erregern des klassischen Fleckfiebers und des Wolhynischen Fiebers durch die Laus in Frage. Die Xenodiagnose hat sich bei spärlicher Rickettsiämie in manchen Fällen dem Meerschweinchenversuch überlegen gezeigt. Beim Wolhynischen Fieber stellt der Läusetest bisher den einzigen Weg einer sicheren Diagnose dar, er ist allerdings nur bei positivem Ausfall beweisend [Weyer (1)]. Beim Q-Fieber ist für eine Xenodiagnose mit gutem Erfolg auch *O. moubata* benutzt worden [Weyer (11), Burgdorfer; Geigy].

6. Stammunterschiede, Virulenzänderungen

Auffällige Unterschiede in der antigenen Struktur der Rickettsien sind schon frühzeitig bei Stämmen von *R. tsutsugamushi* beschrieben worden. Mit diesen *Stammunterschieden* hängt es zusammen, daß bei Tsutsugamushi-Fieber die serologische Diagnose mit Hilfe von KBR oder Agglutination häufig versagt — man benutzt in erster Linie die Weil-Felix-Reaktion mit OX K-Stämmen — und daß es bisher nicht gelungen ist, einen brauchbaren Impfstoff gegen die Krankheit zu entwickeln. Fox (2) untersuchte 8 Stämme aus Südostasien und kam unter Verwendung des Neutralisationstestes zur Aufstellung von 4—5 Stammgruppen, die sich mit Hilfe der KBR nicht klar differenzieren ließen. In Neu-Guinea wurden auf engem Raum 2 antigenetisch verschiedene Stämme festgestellt. Auch Bennett, Smadel und Gauld schlossen aus der Prüfung von 10 Stämmen, daß jeder Stamm seine eigene Individualität besitzt. Die Neutralisierungsindices der Stämme variierten erheblich. Nur 2 Stämme, einer aus Neu-Guinea und einer von den Philippinen, zeigten in diesem Test eine Schutzkraft, die sich auch auf einen Teil der anderen Stämme erstreckte.

Ley, Smadel, Diercks und Paterson stellten außerdem Unterschiede in der Infektionsdosis verschiedener Stämme für Mensch und Maus fest. Bei manchen Stämmen lag die Infektionsdosis für den Menschen erheblich unter der für Mäuse. Neben oder zusammen mit diesen Differenzen bestehen bei *R. tsutsugamushi* Unterschiede in der Virulenz der Stämme, die sich nicht mit den antigenen Unterschieden decken. Die *Virulenz* ist nach Price (3) die Summe der Eigenschaften eines Parasiten, die die Schwere der Infektion bei einem empfänglichen Wirt bestimmen. Dazu gehören die Eindringungsfähigkeit, die Bevorzugung wichtiger Organe oder Gewebe, die Vermehrungsgeschwindigkeit, die maximale Größe der Population vor einer Vermehrungsbegrenzung und die Produktion von Substanzen, die für die Wirtszellen toxisch sind. Pathogenese ist demgegenüber die Dynamik eines Infektionsprozesses, die sich aus dem Zusammenspiel von Wirt und Parasit ergibt.

In Nord-Queensland wurden von 131 Patienten über die intraperitoneale Inoculation von Mäusen 31 Stämme von *R. tsutsugamushi* isoliert (Carley, Doherty, Derrick, Emanuel und Ross). 13 Stämme erwiesen sich als hochvirulent und töteten fast alle Mäuse, 4 Stämme hatten eine mittlere und 14 eine schwache Virulenz. Einer der milden Stämme, die bei Mäusen nur Milzvergrößerungen verursachten, wurde innerhalb von 2 Jahren über 47 Passagen gehalten, ohne seine Virulenz zu ändern. Bei einem virulenten Stamm betrug die Sterblichkeit der Mäuse, die mit starken pathologischen Veränderungen reagierten, 91%. Bei einem Stamm wurde eine Virulenzsteigerung gesehen. In den ersten 12 Passagen betrug die Letalität der Mäuse 10—36%, von der 22. Passage ab 100%. Jedoch konnten die Ursachen dieser Virulenzänderung nicht geklärt werden. In Japan isolierte Koenoki aus wilden Ratten 4 Stämme, die für Versuchstiere viel weniger toxisch waren als alle bisher bekannten Stämme und die auch beim Menschen nur ganz milde Erkrankungen verursachten.

Virulenzunterschiede sind ganz besonders beim amerikanischen Felsengebirgsfleckfieber bekannt geworden [Parker (1)]. Während in der Waldzecke *Dermacentor andersoni* in erster Linie virulente Stämme von *R. rickettsii* vorkommen, sind in der mit Vorliebe auf Kaninchen und Hasen parasitierenden Zecke *Haemaphysalis leporis-palustris* fast ausschließlich schwach virulente Stämme gefunden worden. Parker, Pickens, Lackman, Bell und Thrailkill isolierten im Bitter-Root-Tal 7 derartige Stämme und verglichen sie mit virulenten. Die mittlere Inkubationszeit für Meerschweinchen war bei diesen Stämmen 2mal so lang wie bei den virulenten, die Fieberdauer weniger als halb so lang. Ein schwaches Scrotalphänomen trat nur bei 56% der Meerschweinchen auf (gegenüber 93% bei virulenten Stämmen), und die Infektion endete nicht letal. Bei den virulenten Stämmen betrug die Letalität 90%. Immunologisch verhielten sich die Stämme wie virulente. Ähnliche Stämme wurden in der gleichen Gegend gelegentlich auch in der Zecke *Dermacentor andersoni* gefunden, aus denen die Mehrzahl der virulenten Stämme herkommt. Im Osten der USA konnten PARKER, BELL, CHALGREN, THRAILKILL und MCKEE aus *Haemaphysalis leporis-palustris* 6 Stämme isolieren, unter denen sich umgekehrt ausnahmsweise auch ein virulenter Stamm befand, der im Meerschweinchen sofort unter charakteristischen Symptomen anging und unter Wahrung seiner Eigenschaften in Passagen gehalten werden konnte. Schwach- und hochvirulente Stämme können im gleichen Gebiet nebeneinander vorkommen.

Starke Virulenzunterschiede sind auch bei Stämmen von Felsengebirgsfieber aus Brasilien bekannt. DE MAGALHÃES (1), (2) unterschied einen selteneren hochvirulenten V.B.-Stamm, der zu schweren, meist letalen Infektionen führt, und zwei relativ häufige V.A.-Stämme, die milde und gelegentlich inapparente Erkrankungen verursachen.

Mit den Virulenzunterschieden bei den nordamerikanischen Stämmen hat sich PRICE (2), (3), (4) eingehender beschäftigt. Er prüfte 4 näher bekannte Stämme von *R. rickettsii* mit den Bezeichnungen R, S, T und U, darunter 3 (R, S und T), die in Zecken in einer avirulenten Phase existieren (S. 103). Stamm R verursachte beim Meerschweinchen schwere Scrotalreaktionen, 8 Tage Fieber und tötete 12 von 20 Tieren. Stamm S löste schwächere Scrotalreaktionen und Fieber von 4,1 Tagen aus. Die Infektion endete meist nicht tödlich. Auf den Stamm T reagierten die Meerschweinchen lediglich mit einem kurzen Fieber ohne Scrotalentzündung, und die Rickettsien des Stammes U führten zu inapparenten, nicht mehr mit Fieber verknüpften Infektionen.

Bei allen Stämmen wurden die gleichen Minimaldosen für die Infektion von Meerschweinchen und für einen letalen Effekt bei Hühnerembryonen benötigt. Die Stämme hatten die gleiche Vermehrungsrate, zeigten keinen Unterschied in der Toxin- und Hämolysinproduktion und erreichten dieselben Maximalwerte bei der Vermehrung in den Organen der Versuchstiere, wobei der Inoculationsmodus keine Rolle spielte. Die Stämme besaßen dieselbe Empfindlichkeit gegenüber kritischen Temperaturen, begannen zum gleichen Zeitpunkt mit der Produktion von Antikörpern und unterschieden sich auch nur geringfügig in immunologischer Beziehung. Der Virulenzgrad der Stämme blieb unter natürlichen Bedingungen wenigstens 3 Jahre erhalten.

PRICE (4) versuchte, die Virulenz der Stämme durch Warmblüter- und Zecken-Passagen zu ändern. Ein Stamm vom Typ R wurde durch 8—13 Milzpassagen in Baumwollschwänzchen (*Sylvilagus nuttalli*) in seiner Virulenz gemildert und verhielt sich wie ein Stamm vom Typ S oder sogar T. Bei Übertragung auf das Meerschweinchen stellte sich aber die ursprüngliche Virulenz wieder ein. Bei einigen Stämmen ließ sich eine Erhöhung der Virulenz nach einer Menschenpassage feststellen, bei einem anderen wurde die Virulenz durch einen längeren Aufenthalt in der Zecke für kurze Zeit abgeschwächt. Soweit überhaupt Änderungen der Virulenz auftraten, waren sie nur vorübergehender Natur. Schwach virulente Stämme aus *Haemaphysalis leporis-palustris* erlangten keine höhere

Virulenz durch Übertragung in den Dottersack, in *Dermacentor andersoni* oder dadurch, daß die infizierten Zecken gefüttert wurden. Umgekehrt setzte die Übertragung virulenter Stämme auf *Haemaphysalis leporis-palustris* die Virulenz nicht herab. Die Virulenz scheint danach fest an Feinbau und Stoffwechsel der Rickettsien gebunden zu sein.

Von besonderem Interesse ist in diesem Zusammenhang das schon lange bekannte Phänomen, daß bestimmte Stämme von *R. rickettsii* in Zecken in einer *avirulenten Phase* auftreten und durch Fütterung der Zecken oder durch Temperaturerhöhung auf 37° C für 48 Std, am schnellsten unter dem Einfluß von Blutnahrung *und* Temperaturerhöhung, reaktiviert und in die virulente Phase überführt werden können (SPENCER und PARKER). Es handelt sich hier also im Unterschied zu den oben geschilderten Eigenschaften verschiedener Stämme um Virulenzschwankungen bei ein und demselben Stamm. Die Rickettsien der avirulenten Phase vermehren sich nicht im Meerschweinchen, wohl aber im Dottersack und erlangen auch auf diesem Wege ihre ursprüngliche Virulenz. In diese Vorgänge haben wir durch neuere Beobachtungen von GILFORD und PRICE einen tieferen Einblick gewonnen, die zu den bemerkenswertesten Fortschritten in der Rickettsienforschung der letzten Jahre gehören. Es gelang nämlich bei Stämmen aus der Zecke *Dermacentor andersoni*, die avirulente Phase in vitro in die virulente zu überführen. Diese Wirkung wurde erreicht durch eine Behandlung der Rickettsien mit Diphosphopyridinnucleotid (DPN) und dem Coenzym A. Umgekehrt führte die Einwirkung von p-Aminobenzoesäure (PABS) zu einem Verlust der Virulenz, d. h. zu einem Übergang der virulenten in die avirulente Phase. Dieser Effekt konnte durch Zugabe von DPN und dem Coenzym A zur Dottersackkultur verhindert werden. Die genannten Stoffe scheinen daher bei den Virulenzänderungen dieselbe Rolle zu spielen wie Stoffwechselprodukte der Zecken, unter denen vielleicht Häutungshormone in Betracht gezogen werden müssen. Es konnte noch gezeigt werden, daß die Virulenzerhöhung mit einer Verstärkung der Absorption von Rickettsien durch Meerschweinchenzellen in vitro verknüpft war.

Während sich Stämme von *C. burneti* trotz ihrer weit getrennten geographischen Provenienz durch eine große Einheitlichkeit auszeichnen, sind Virulenzunterschiede beobachtet, die bei experimenteller Übertragung verschiedener Stämme von *R. prowazeki* und *R. mooseri* auf Warmblüter und Läuse auftraten. Diese Unterschiede sind allerdings nicht genauer erfaßt. Es können auch Schwankungen der Virulenz bei denselben Stämmen auftreten, insbesondere bei der Kultur von *R. prowazeki* in der Laus [WEYER (15)]. Es ist aber nicht bekannt, ob es sich hierbei um ähnliche Phänomene wie bei *R. rickettsii* handelt. Außerdem wurde beobachtet, daß die Vermehrungsintensität der Rickettsien bei der Haltung in Versuchstieren jahreszeitlichen Schwankungen unterworfen ist.

Theoretisches und erhebliches praktisches Interesse hat ein, offenbar durch Mutation entstandener, avirulenter Stamm von *R. prowazeki* gewonnen, der als E(Espagne)-Stamm bekannt geworden ist (PEREZ GALLARDO). CLAVERO und PEREZ GALLARDO (1) beobachteten, daß ein Stamm von *R. prowazeki*, der in Dottersackkulturen gehalten wurde, in der 11. Passage ein abnormales Verhalten zeigte, indem zunächst der Hauttest nach GIROUD negativ ausfiel. Von der 16. Passage ab hatte der Stamm seine Pathogenität für Meerschweinchen verloren, ohne jedoch die immunisierende Eigenschaft einzubüßen. Dieser Stamm ist für die Herstellung eines *Impfstoffes* aus lebenden Rickettsien benutzt worden. Die diesbezüglichen Erfahrungen sollen wegen ihres allgemeinen Interesses erwähnt werden, obwohl sie außerhalb des hier behandelten Themas liegen.

Die ersten Versuche zur Schutzimpfung von Meerschweinchen mit dem Stamm E führten Perez Gallardo und Fox sowie Clavero und Perez Gallardo (2) durch. Bei Versuchen an menschlichen Freiwilligen durch Everitt, Bhatt und Fox wurden zunächst die Beziehungen zwischen Infektionsdosis und Antikörperspiegel bestimmt. Nach der Impfung mit höheren Dosen traten teilweise Primärläsionen, regionäre Lymphadenitis und gelegentlich Fieber auf, Symptome, die bei einer Impfung mit abgetöteten Rickettsien normalerweise ausbleiben. Die serologischen Reaktionen entsprachen denen nach Verwendung sehr hoher Dosen von abgetöteten Rickettsien. Die Vermehrung der Erreger, die bei einer Impfung mit dem E-Stamm in den Körper gelangten, schien recht schwach zu sein. Durch Blutübertragung auf Meerschweinchen oder Fütterung von Läusen an den geimpften Personen konnten keine Rickettsien isoliert werden. Der Impfschutz wurde außerdem durch nachfolgende Infektion der Versuchspersonen mit lebenden Rickettsien eines virulenten Stammes erprobt (Fox, Everitt, Robinson und Conwell).

Der Immunitätsspiegel geimpfter Personen wurde laufend verfolgt und durch Infektion 2 Jahre nach der Impfung geprüft (Fox, Jordan, Conwell und Robinson). Komplementbindungsreaktion und Toxinneutralisationstest zeigten einen starken Rückgang der Antikörper innerhalb der ersten 12 Monate nach der Impfung, wobei die Höhe des Titers in Korrelation zur Impfdosis stand. In den folgenden 12 Monaten fiel der Antikörpergehalt wesentlich langsamer ab. Von 18 Personen, die 3—12 Monate nach der Impfung mit virulenten *R. prowazeki* infiziert wurden, hatte nur eine Person eine schwache fieberhafte Reaktion. 5 Personen. die einige Monate früher immunisiert waren, zeigten im Anschluß an die nachfolgende Infektion keine klinischen Symptome, obwohl sie vorher in der KBR entweder negativ waren oder nur einen ganz niedrigen Titer hatten. Der Verlust der komplementbindenden Antikörper ist somit nicht unbedingt ein Zeichen fehlender Immunität. Von 3 Personen, die mit abgetöteten Rickettsien (Cox-Vaccine) immunisiert waren und 9 Monate später mit lebenden Rickettsien infiziert wurden, reagierten dagegen 2 mit deutlichen klinischen Symptomen. In Fortsetzung dieser Untersuchungen inoculierten Fox, Jordan und Gelfand 8 Personen intradermal mit virulenten Rickettsien, und zwar 48—66 Monate nach der Impfung. Nur 2 Personen reagierten mit einem „typical typhus onset“. Dieselben hatten nach der Impfung keine Antikörper entwickelt. 7 von 8 Kontrollpersonen, die die gleiche Dosis virulenter Rickettsien enthielten, bekamen Fieber. Die Immunkörper bleiben also offenbar viele Jahre im Blut.

Nach den sorgfältigen Laboratoriumsprüfungen und den hierbei gesammelten günstigen Erfahrungen wurde der Impfstoff unter natürlichen Verhältnissen in einem Fleckfiebergebiet in Peru erprobt (Fox, Montoya, Jordan und Espinosa). Über die Ergebnisse und Erfahrungen hat Fox (3) auf Grund der Impfung von über 10000 Personen berichtet. Obwohl die KBR bei den Geimpften im Verlaufe eines Jahres negativ werden konnte, blieben toxinneutralisierende Antikörper und damit die Widerstandsfähigkeit gegen eine nachfolgende Infektion mit virulenten Rickettsien mindestens 2 Jahre bestehen. In Zusammenhang mit den Impfungen kam es weder zu Todesfällen noch zu schweren Erkrankungen. 92% der Geimpften zeigten eine deutliche serologische Reaktion, die für eine Antikörperbildung sprach.

Anläßlich einer zusammenfassenden Darstellung über Beobachtungen an etwa 17000 Personen hat Fox (4) den Impfschutz durch lebende und abgetötete Rickettsien verglichen. Eine solide Immunität bleibt nach einer Vaccination mit lebenden Rickettsien mindestens 3 Jahre, wahrscheinlich erheblich länger bestehen. Die Überlegenheit dieses Impfstoffes über den aus abgetöteten Rickettsien ist danach evident, jedoch sind noch eine Reihe von Fragen, auch technischer Natur, zu klären, bis eine Ablösung der Cox-Vaccine durch den lebenden Impfstoff in Erwägung gezogen werden kann.

Der Stamm E ist auch im Laboratorium einer weiteren Analyse unterzogen worden. Beim Einfrieren (—80° C) und Auftauen von Rickettsien aus Dottersackkulturen in isotonischen Salzlösungen stellten Bovarnick und Allen (1) ein Absinken der Toxicität (für Mäuse), der hämolytischen Aktivität, der Respiration und der Infektiosität für Eier fest. Die ursprünglichen Eigenschaften stellten sich zum Teil wieder ein, wenn die Rickettsien für 2 Std bei 34°C in Gegenwart von DPN und dem Coenzym A gehalten wurden. Die Aktivität wurde hierdurch manchmal bis zum 10fachen gesteigert. Möglicherweise ist das Coenzym auch für das Eindringen der Rickettsien in die Zellen verantwortlich. Der Prozeß der Reaktivierung vollzog sich nicht sofort, sondern allmählich (in 2—3 Std) und verlief bei 0°C sehr viel langsamer als bei 34°C. BOVARNICK und ALLEN (2) stellten in weiteren Versuchen fest, daß der Aktivitätsverlust noch nach 18 Std zu 60—100% aufgehoben werden konnte, wenn die Suspensionen für 3 Std bei 33°C in einem Medium mit DPN, Coenzym A, Mg, Mn, Glutaminat und Albumin gehalten wurden. Der Aktivitätsschwund wird auf einen kombinierten Effekt des DPN-Verlustes und eines Wechsels in der Konzentration anorganischer Ionen zurückgeführt. Es handelt sich hierbei um Vorgänge, die eine deutliche Parallele zu den erwähnten Virulenzänderungen der Stämme von *R. rickettsii* haben (vgl. S. 103).

Der Jahre oder gar Jahrzehnte währende Aufenthalt von *R. prowazeki* im menschlichen Körper, welchen wir bei Fällen Brill-Zinsserscher Krankheit annehmen müssen, hat auf die Virulenz und die sonstigen Stammeigenschaften offenbar keinen Einfluß. Wie entsprechende Untersuchungen zeigten [WOJCIECHOWSKI und MIKOŁAJCZYK (1), (2); WEYER und HORNBOSTEL u.a.], besaßen solche Stämme alle typischen Merkmale von *R. prowazeki*. Das bezieht sich auf das Verhalten der Rickettsien in der Kultur und in Versuchstieren und ebenso auf die serologischen Reaktionen.

Mit plötzlichen, möglicherweise durch Mutation bedingten Änderungen muß jedenfalls bei den Rickettsien gerechnet werden, wie das Beispiel des Stammes E zeigt. ROUX beschreibt einen Stamm von *C. burneti*, der für Meerschweinchen ursprünglich schwach pathogen war und plötzlich eine starke Virulenz entwickelte. Alle mit dem Stamm infizierten Tiere gingen unter den Zeichen einer Encephalitis und Keratitis ein. WEISS, DRESSLER und SUITOR beobachteten bei der Haltung von *R. prowazeki* (Stamm E) in Kombination mit PABS die Entstehung einer Mutante, die im Vergleich zum Originalstamm eine um das 20fache gesteigerte Resistenz gegen PABS aufwies. Die Resistenz machte sich von der 4. Passage ab bemerkbar und hatte in der 8. Passage ihre definitive Form erreicht. Der Stamm änderte sich bis zur 26. Passage nicht mehr. Morphologisch und im Verhalten gegenüber Meerschweinchen stimmten Ausgangsstamm und resistenter Stamm überein.

Erwähnt werden muß in diesem Zusammenhang eine Beobachtung von PRICE, EMERSON, NAGEL, BLUMBERG und TALMADGE, die im Falle ihrer Bestätigung ganz neue Probleme aufwerfen und unsere bisherigen Ansichten über die verwandtschaftlichen Beziehungen der Rickettsien erschüttern würde. Die Autoren glauben, einen Stamm von *R. prowazeki* durch Behandlung mit Extrakten aus *R. mooseri* in einen Stamm verwandelt zu haben, der alle Eigenschaften von *R. mooseri* besaß. Die Änderung war permanent. Ein besonderer Einfluß bei der Umwandlung wird der Anwesenheit von DNS in den Extrakten eingeräumt. Eine genetische Deutung dieses Phänomens scheint vorerst nicht möglich.

7. Interferenz bei Rickettsien

PRICE (1) konnte bei Rickettsien ein Interferenzphänomen (RIP) nachweisen. Er fand, daß Meerschweinchen gegen das Auftreten schwerer klinischer Symptome bei einer Infektion mit hochvirulenten Stämmen von *R. rickettsii* geschützt werden, wenn er die Tiere gleichzeitig oder bis zu 10 Tagen vorher mit schwachvirulenten Stämmen inoculierte. Dasselbe Resultat kam zustande, wenn er anstatt der schwachvirulenten Stämme von *R. rickettsii* Stämme von *R. prowazeki*, *R. mooseri*, *R. tsutsugamushi* oder *C. burneti* verwendete, die ihrerseits bei den Versuchstieren einen spezifischen Infektionsverlauf auslösten. 80—90% der Meerschweinchen, die mit einem Teil eines hochvirulenten und 10—30 Teilen eines schwachvirulenten Stammes inoculiert wurden, reagierten wie bei einer Infektion mit dem schwachvirulenten Stamm. Ähnlich waren die Ergebnisse, wenn Feldmäuse, Baumwollschwänzchen oder Erdhörnchen durch den Stich der Zecke *Dermacentor andersoni* gleichzeitig mit hoch- und schwachvirulenten Stämmen infiziert wurden.

Die Abschwächung gelang auch bei Verwendung von Rickettsien eines homologen virulenten Stammes, die mit ultraviolettem Licht inaktiviert oder durch Ultraschall zerstört waren, aber nicht mit durch Hitze abgetöteten Rickettsien (Price, Johnson, Emerson, Hope und Preston). Wurde die schützende Suspension 3 Std nach der virulenten Suspension verabfolgt, so blieb die Schutzwirkung bereits aus. Die schützende Substanz konnte nach Zentrifugieren von gereinigten, mit Ultraschall behandelten Rickettsiensuspensionen aus der überstehenden Flüssigkeit isoliert werden. Der Schutz war abhängig vom Verhältnis der interferierenden zur Kontrolldosis.

Der Mechanismus der Interferenz, der nicht auf einer einfachen Absättigung der empfänglichen Zellen beruht, ist noch ungeklärt. Es konnte nur ermittelt werden, daß eine Mobilisierung und Phagocytose der weißen Blutkörperchen, Entzündungsvorgänge und Antikörperbildung beim Zustandekommen des RIP keine entscheidende Rolle spielen. Die virulenten Rickettsien breiteten sich in denselben Organen und Geweben und im gleichen Umfang in ungeschützten Versuchstieren wie in geschützten aus, nur war bei letzteren die Vermehrung der virulenten Rickettsien reduziert. Das RIP erwies sich als hochspezifisch und unabhängig von der Inoculationszeit, vorausgesetzt, daß die virulenten Rickettsien nicht später als 10 Tage nach den avirulenten verabfolgt wurden. Die Verwendung von Bakterien und verschiedenen Viren anstatt der Rickettsien war ohne Wirkung.

Mika, Goodlow, Victor und Braun konnten eine gewisse Interferenz zwischen *C. burneti* und *Brucella suis* feststellen. In der gleichen Richtung liegt eine Beobachtung von Salvin und Bell, die eine bemerkenswerte Resistenz von Mäusen gegen *R. mooseri* beobachteten, wenn dieselben mit Histoplasmosis infiziert waren. Während die Infektion mit *R. mooseri* für 94% der Mäuse letal endete, gingen unter dem Schutz der Pilzinfektion nur 19% der Mäuse ein. Auch hier stand die veränderte Reaktion nicht mit der Bildung spezifischer Antikörper in Zusammenhang. Der Effekt trat nur ein, wenn beide Organismen intraperitoneal verabfolgt wurden, und war abhängig vom Zeitpunkt der Inoculation und der Infektionsdosis. Wurden beide Parasiten gleichzeitig gespritzt, so unterblieb die Wirkung. Der maximale Schutz wurde erreicht, wenn die Mäuse eine Dosis von 109 Pilzzellen erhielten und die Inoculation mit *R. mooseri* 7 Tage danach erfolgte. Die Schutzwirkung hielt sich auf annähernd gleicher Höhe noch bis zum 14. Tage, um dann allmählich abzuklingen. *Histoplasma capsulatum* beeinflußte auch die Infektion der Maus mit *R. tsutsugamushi*. Das Wachstum der Pilze spielte bei dem Phänomen keine Rolle, denn ein ähnlicher, wenn auch schwächerer Effekt, wurde mit abgetöteten Pilzzellen erreicht. Dagegen war die Verwendung von 6 weiteren Pilzarten ohne entsprechende Wirkung.

Umgekehrt kann die Infektion mit Rickettsien auch einen Schutz gegen eine Infektion mit Bakterien gewähren. Nach Owen und Larson führte eine Infektion von Mäusen mit

R. mooseri zu einer Reduktion der Letalität und einer Verlängerung der Überlebensdauer gegenüber einer anschließenden Infektion mit *Pasteurella pestis* und *P. tularensis*. Die Interferenz begann rund 16 Std nach der Injektion und wurde innerhalb von 4 Tagen ausgeprägter. Danach erfolgte ein Rückgang der Schutzwirkung, die bei Verabfolgung sehr hoher Bakteriendosen oder hochvirulenter Stämme verdeckt werden konnte. Ein Stamm von *P. pestis* wurde über 10 Passagen zusammen mit *R. mooseri* in Mäusen geführt, ohne eine Resistenz gegen die Schutzwirkung zu entwickeln. Diese Beobachtung spricht gegen die Annahme, daß die Interferenz mit einer antibiotischen Wirkung der Rickettsien zusammenhängt.

II. Epidemiologie

1. Durch Zecken übertragene Rickettsiosen

a) Felsengebirgsfleckfieber

Das Felsengebirgsfleckfieber ist in Kanada (Alberta, Saskatchewan, Britisch-Kolumbien), in den Vereinigten Staaten (mit Ausnahme von Connecticut, Maine, Rhode Island und Vermont), in West- und Zentral-Mexiko, Panama, Brasilien und Kolumbien nachgewiesen. Wahrscheinlich kommt es auch in Venezuela vor.

Die weiteste *Verbreitung* hat die Krankheit in den Vereinigten Staaten. Von 1939—1946 betrug hier die jährliche Morbidität 480 Fälle. Nach CAWLEY und WHEELER wurden von 1945 bis 1954 4517 Erkrankungsfälle gemeldet. In den südöstlichen Staaten sind von 1939 bis 1946 1308 Fälle mit einer Mortalität von 20% registriert worden, in den westlichen Staaten 683 Fälle mit einer Mortalität von 21,5% (McCROAN, RAMSEY, MURPHY und DICK). 1949 bis 1953 kam die Hälfte aller in den USA gemeldeten Fälle aus 10 südöstlichen Staaten. Aus Kanada sind nur Einzelfälle (12 Erkrankungen von 1919—1939) bekannt geworden. Über Fälle aus Panama, bei welchen virulente Stämme isoliert wurden, berichten DE RODANICHE und RODANICHE, ferner CALERO, NÚÑEZ und SILVA-GOYTIA und DE RODANICHE. Untersuchungen über die Verbreitung der Krankheit in Mexiko wurden von SILVA-GOYTIA und ELIZONDO (2), (3) sowie von SILVA-GOYTIA, VASQUEZ CAMPOS und ELIZONDO durchgeführt. Verbreitung und Epidemiologie des „Neotropischen Fleckfiebers" in Brasilien sind ausführlich von DE MAGALHÃES (2), (3) behandelt. In Brasilien wurden von 1929—1942 663 Erkrankungsfälle registriert.

Übertragung und *Epidemiologie* des Felsengebirgsfleckfiebers sind am besten aus den USA bekannt, und unsere Kenntnisse sind hier durch neue wichtige Beobachtungen erweitert worden. Die entscheidende Überträgerrolle spielen Schildzecken. Die Übertragung erfolgt während des Saugaktes mit dem Speichel. Die Rickettsien befallen alle Organe der Zecke und können gelegentlich auch beim Absammeln oder Zerdrücken der Zecken in einen neuen Wirt gelangen. Frische Zeckenfaeces enthalten ebenfalls lebende Rickettsien und bilden eine Infektionsquelle, doch gehen die Erreger in getrockneten Faeces schnell zugrunde. Im Westen der Vereinigten Staaten wird die Krankheit durch *Dermacentor andersoni*, eine Waldzecke mit einem zweijährigen Entwicklungscyclus, übertragen, die im Frühjahr aktiv wird. Die meisten Erkrankungen betreffen Männer (50% über 40 Jahre) und fallen in die Zeit von April bis Mai; in höheren Lagen treten sie bis in den Sommer auf. Im Herbst und Winter kommen nur vereinzelte Fälle vor. Besonders gefährdet sind Wald- und Wegearbeiter, Viehhändler, Jäger, Fischer usw., d. h. Personen, die sich viel im Freien aufhalten. Im Osten ist der Überträger *D. variabilis*, eine Zecke, die ihre Hauptaktivität im Sommer entfaltet. Der Morbiditätsgipfel fällt in den Juli, Erkrankungen kommen von April bis September vor. Zu den Patienten gehören auch Frauen (39,4%) und

Kinder (46,8% unter 15 Jahren), weil die Zecke gern an Hunden saugt und damit leicht in die Wohnungen gelangt. In Texas und Oklahoma ist *Amblyomma americanum* wichtigster Überträger.

Alle Entwicklungsstadien der genannten Zecken, Larven, Nymphen und Imagines, können den Erreger beim Saugakt an einem infizierten Wirt aufnehmen. Larven und Nymphen von *D. andersoni* saugen gern an Nagern und kleineren Raubtieren, die erwachsenen Zecken bevorzugen größere Wild- und Haustiere. Sie saugen auch am Menschen, die Nymphen findet man gelegentlich an Kindern. Die Entwicklungsstadien von *D. variabilis* parasitieren auf kleinen Nagern, vor allem auf bestimmten Mäusen, die Adulten auf Wild- und Haustieren, besonders auf Hunden, und auf dem Menschen. Bei anderen Schildzecken (z. B. *D. occidentalis, Haemaphysalis leporis-palustris, Ixodes dentatus* und *Rhipicephalus sanguineus*) konnten natürliche Infektionen festgestellt werden. Diese Zecken fungieren z. T. als Überträger unter wilden Nagetieren. Am wichtigsten ist in dieser Beziehung *H. leporis-palustris*.

Obwohl die Epidemiologie des Felsengebirgsfiebers in ihren Grundzügen schon recht lange bekannt ist, gelang es erst kürzlich, einen Stamm von *R. rickettsii* aus einem wilden Nager zu isolieren (Gould und Miesse). Unter 65 Wiesenwühlmäusen (*Microtus pennsylvanicus*) in Virginia erwies sich eine Maus als Rickettsienträger. 40 andere kleine Säuger verschiedener Art, die aus der gleichen Umgebung stammten, waren negativ. Die infizierte Maus war von *D. variabilis* befallen.

Price (3), (4) hat die epidemiologische Situation in Maryland genauer untersucht. Wichtigster Überträger ist hier *D. variabilis*. In einem Bezirk wurden jährlich 3000 Zecken dieser Art gesammelt, von denen 0,2—0,3% mit *R. rickettsii* infiziert waren. Hunde und Mäuse sind in Maryland die wichtigsten Zwischenglieder in der Übertragungskette. Die Hunde sind wichtig als sekundäre Rickettsienwirte oder als Blutspender und Transporteure von infizierten Zecken. Von 28 Patienten besaßen 20 Hunde, und in 10 Fällen konnten Stämme aus den auf den Hunden parasitierenden Zecken (*D. variabilis*) isoliert werden. Von 442 *Microtus pennsylvanicus* hatten 49 toxinneutralisierende Antikörper.

M. pennsylvanicus ließ sich experimentell leicht durch Larven und Nymphen von *D. variabilis* infizieren, und die Rickettsien konnten über die Maus von Zecke zu Zecke weitergeleitet werden. In der gleichen Weise wurden die Rickettsien von experimentell infizierten Hunden auf adulte Zecken übertragen. An Mäusen und Hunden konnten sich die Zecken etwa 1 Woche lang infizieren. Die Erreger waren im Blut der Maus 1 Woche und in ihren Organen 1 Monat lang nachweisbar. Komplementbindende und toxinneutralisierende Antikörper wurden in den Mäusen bis zu 1 Jahr nach der Infektion gefunden. Mäuse stellen wahrscheinlich neben den Zecken ein wichtiges Erregerreservoir dar. Baumwollschwänzchen (*Sylvilagus nuttalli*) scheinen als Wirte von Stämmen, die auf den Menschen gehen, keine Rolle zu spielen. An *Microtus* infizierten sich unter optimalen Bedingungen 76% der Larven und 81% der Nymphen von *D. variabilis*, an den Hunden 82% der erwachsenen Zecken. Die adulten Zecken blieben längere Zeit Rickettsienträger, doch scheinen die Rickettsien, wenigstens bei schwachvirulenten Stämmen, unter Bedingungen, die denen in der Natur entsprechen, nach 14—16 Monaten auszusterben. Bis zu diesem Zeitpunkt blieb die Virulenz der

Stämme unverändert. Die Ansicht, daß infizierte Zecken auf Lebenszeit Rickettsienträger bleiben, wird dadurch erschüttert.

PRICE untersuchte auch die transovarielle Übertragung von *R. rickettsii*, betont aber ausdrücklich, daß sich seine Feststellungen nur auf die Stämme mit normaler oder schwacher Virulenz beziehen. Eine transovarielle Übertragung wurde in 30—40% der Versuche erreicht. In 50% der positiven Gruppe war 1 von 5 bis 10 Eiern infiziert, in 15% 1 von 2 bis 4 Eiern, in 35% 1 von 20 bis 40 Eiern. Ähnlich waren die Zahlen bei Larven, die von infizierten Weibchen abstammten. Aus infizierten Eiern entwickelten sich regelmäßig infizierte Larven, Nymphen und Adulte. Die transovarielle Übertragung ist für die Erhaltung der Erreger in der Natur zweifellos von Bedeutung, zur Unterhaltung der Krankheit müssen die Rickettsien aber von infizierten Zecken über einen Warmblüter an empfängliche Zecken weitergegeben werden.

D. variabilis ist auch wichtigster Überträger in Long Island (MILLER). In Alberta überträgt *D. andersoni*. PARKER, BELL, CHALGREN, THRAILKILL und MCKEE untersuchten 191 Gruppen von Zecken, die in Pennsylvania, Virginia, New York, Connecticut, Nordkarolina und Alabama gesammelt waren. Dabei wurden neben mehreren Tularämie-Stämmen 7 Stämme von *R. rickettsii* isoliert, 6 Stämme aus *Haemaphysalis leporis-palustris* (in 91 Testen mit 3168 Zecken) und 1 Stamm aus *Ixodes dentatus* (in 83 Testen mit 2201 Zecken). Bei dieser Gelegenheit wurden zum erstenmal in *H. leporis-palustris* auch virulente Stämme nachgewiesen (vgl. S. 102). Im Bitter-Root-Tal fanden sich an *Sylvilagus nuttalli* alle Entwicklungsstadien von *D. andersoni* (PARKER, PICKENS, LACKMAN und THRAILKILL). Als Reservoir für *R. rickettsii* haben diese Tiere aber wahrscheinlich keine Bedeutung. Von 10 Hasen (*Lepus americanus*) waren 8 serologisch positiv. Aus den Zecken, die an den Hasen parasitierten, konnten 3 Stämme isoliert werden. Insgesamt wurden von 1946—1947 286 Seren von Wildtieren auf komplementbindende Antikörper für *R. rickettsii* untersucht. 6 von 61 Seren, die von *Sylv'lagus nuttalli* stammten, reagierten positiv. Ein Stamm von *R. rickettsii* wurde aus *D. andersoni* isoliert, die von Bergziegen abgesammelt waren, ein weiterer Stamm aus Larven von *D. andersoni*, die am Ziesel (*Citellus lateralis*) gesogen hatten.

Als wichtigster Überträger in Kolumbien und im Osten von Mexiko gilt *Amblyomma cajennense*, im Norden und Nordosten von Mexiko *A. cajennense* und *Rhipicephalus sanguineus*. Die Übertragung durch *A. cajennense* ist nicht saisongebunden. In Panama isolierte DE RODANICHE einen hochvirulenten Stamm (90% Letalität beim Meerschweinchen) aus einer Zerreibung von 11 erwachsenen *A. cajennense*. In Nordmexiko fanden SILVA-GOYTIA und ELIZONDO (1) natürliche Infektionen bei *Rh. sanguineus*, *Ornithodorus nicollei* und *Otobius lagophilus*. Von einem Patienten wurden mit *R. rickettsii* infizierte Kleiderläuse abgesammelt. (Die Vermehrungsfähigkeit von *R. rickettsii* in der Kleiderlaus ist experimentell bestätigt [vgl. S. 96].) Stämme wurden bei 15 von 28 Patienten aus dem Blut isoliert. In der Epidemiegegend hatten 30% der Bevölkerung eine positive KBR. Wahrscheinlich verlaufen zahlreiche Fälle bei Kindern inapparent oder milde.

SILVA-GOYTIA, VASQUEZ CAMPOS und ELIZONDO prüften 3225 Seren von Personen aus 5 mexikanischen Staaten. Der Prozentsatz der positiven Reaktionen schwankte von 5,4—27. Am häufigsten waren positive Reaktionen im Norden des Landes. Unter der Landbevölkerung enthielten 20,3% der Seren komplementbindende Antikörper. Positive Seren wurden auch bei Rindern, Eseln, Ziegen und Hunden gefunden. Den höchsten Prozentsatz positiver Seren hatten Rinder. Es wird angenommen, daß die Haustiere für eine Infektion empfänglich sind, aber nur inapparent erkranken. Sie spielen in erster Linie eine Rolle als Blutspender für die Zecken, die das Hauptreservoir der Krankheit bilden.

Etwas anders scheint die epidemiologische Situation in Brasilien zu liegen. Als Reservoire werden hier Beuteltiere, Hunde, Kaninchen, Meerschweinchen, Katzen, Agutis, Wasserschweine, Ziegen und Rüsselbären angesehen, weil Rickettsien in diesen Tieren gefunden wurden oder weil sie sich leicht mit *R. rickettsii* infizieren ließen. Wichtigster Überträger ist *Amblyomma cajennense*. 1034 Zecken dieser Art enthielten zu 0,5% den Stamm V. B., zu 0,1% den Stamm V. A. [DE MAGALHÃES (1)]. Natürliche Infektionen wurden außerdem bei *A. striatum, A. ovale, A. brasiliensis, A. cooperi* und *Ixodes loricatus* festgestellt. *Rhipicephalus sanguineus* ist Überträger unter den Hunden.

Seit 1933 ist wiederholt darauf hingewiesen worden, daß in Brasilien auch Bettwanzen (*Cimex lectularius* und *C. rotundatus*) maßgeblich an der Übertragung der Krankheit beteiligt sind. DE MAGALH~ES (2) berichtet, daß aus Wanzen, die in Häusern von Patienten gesammelt waren, mehrere Stämme von *R. rickettsii* isoliert werden konnten. Wanzen ließen sich auch durch Saugen am Meerschweinchen und Menschen infizieren. Die Übertragung soll durch den Stich nach 72 Std und durch die Verimpfung von zerriebenen Wanzen nach 48 Std und 10 Tagen gelungen sein. Die Wanzen blieben bis zu 72 Tagen infektiös. Schließlich wird auch ein Übergang der Rickettsien auf Eier und Larven von Wanzen angegeben, die an einem Patienten gesogen hatten. Diese zum Teil sehr auffälligen Beobachtungen bedürfen dringend einer weiteren Nachprüfung. Sicher ist nur, daß *R. rickettsii* einige Tage im Magen der Wanze überleben kann [WEYER (18)].

Es wird angenommen, daß das Felsengebirgsfleckfieber eine lebenslängliche Immunität hinterläßt. PARKER (2) berichtet jedoch von 26 sicheren Zweiterkrankungen, von denen die meisten 10—31 Jahre nach der ersten auftraten. Die Möglichkeit eines Rückfalls ist dabei nicht erwähnt, obwohl sie nicht von der Hand zu weisen ist. Auf die lange Erregerpersistenz im Versuchstier wurde bereits hingewiesen (S. 94). In diesem Zusammenhang ist der von PARKER, MENON, MERIDETH, SNYDER und WOODWARD mitgeteilte Fall von Interesse, wo Rickettsien aus den Lymphknoten eines Patienten noch 1 Jahr nach der Erkrankung isoliert werden konnten.

b) Zeckenbißfieber der Alten Welt

Neuere Beobachtungen über die Zeckenbißfieber betreffen in erster Linie *Verbreitung* und Klinik und enthalten wenig epidemiologische Daten.

QUINTANE hat alle Angaben über die geographische Verbreitung von Fällen des Fièvre boutonneuse im Mittelmeerraum einschließlich der Inseln gesammelt, die seit 1933 zur Beobachtung kamen. In der Provinz Tarragona in Spanien wurden von 1941—1948 112 Fälle von Boutonneuse-Fieber beobachtet (PEREPÉREZ PALAU). Die Erkrankungen traten von Mai bis Oktober auf, 82 Fälle fielen in den August und September. In Portugal hat die Verbreitung des Zeckenbißfiebers nach SAMPAIO und DE M. FAIA (1), (2) stark zugenommen. Im Rahmen dieser Untersuchungen wurden 15 Stämme von *R. conori* isoliert, davon 3 aus Zecken und 12 aus Patientenblut, das zwischen dem 4. und 11. Krankheitstag entnommen war. Erreger wurden gefunden in *Rhipicephalus sanguineus* und *Rh. bursa*, außerdem in einer *Boophilus*-Art (SAMPAIO, DA CRUZ und DE M. FAIA). Autochtone Fälle von Boutonneuse-Fieber traten in Paris (MARTIN und MAROGER) und in Lothringen (MELNOTTE und MENGUS) auf. An der ligurischen Küste, wo von 1942—1952 im ganzen 131 Fälle (31 Erkrankungen im Jahr 1952) von Zeckenbißfieber bekannt wurden, ist *Rhipicephalus sanguineus* Überträger (CAPUTO). Von 1912—1953 wurden in Rumänien 138 Fälle von Zeckenbißfieber beobachtet (COMBIESCU, DUMITRESCU, RUSS und DINULESCU). Auf der Krim wurden Stämme von *R. conori* aus adulten *Rh. sanguineus* isoliert, die als Nymphen an Hunden gesogen hatten (KORSCHUNOWA und PETROWA-PIONTKOWSKAJA).

Wichtigster *Überträger* des Boutonneuse-Fiebers und der übrigen afrikanischen Zeckenbißfieber ist *Rhipicephalus sanguineus*. In Kenia sind Stämme auch aus *Rh. simus, Haemaphysalis leachi* und *Amblyomma variegatum* isoliert worden (HEISCH), in Dakar aus *A. variegatum*, die von einem Schaf abgesammelt worden waren (GIROUD, COLAS-BELCOUR, PFISTER und MOREL). In Nairobi und wahrscheinlich in ganz Kenia sind die wichtigsten Überträger *H. leachi* und *Rh. simus* (HEISCH, MCPHEE und RICKMAN). *Rh. sanguineus* spielt hier als Überträger keine Rolle. Als Überträger des südafrikanischen Zeckenbißfiebers fungieren Zecken, die an Wildtieren saugen. Die meisten Infektionen kommen im Freien zustande. In bestimmten Gegenden sind *Rhipicephalus evertsi* und *Amblyomma hebraeum* die Überträger. Natürliche Infektionen wurden außerdem in *Rh. appendiculatus*, *Hyalomma aegyptium* und *Boophilus decoloratus* nachgewiesen. Die Hundezecke *Haemaphysalis leachi*, aus der ebenfalls Stämme isoliert wurden und bei der eine transovarielle Übertragung beobachtet ist, spielt weniger für die Infektion des Menschen als für die Verbreitung der Krankheit und als Reservoir eine wichtige Rolle (GEAR; GEAR und DE MEILLON).

Die Erkrankungen an Zeckenbißfieber treten das ganze Jahr über auf, doch sind sie in Zusammenhang mit der Aktivität der Zecken in der warmen Jahreszeit am häufigsten. Der Erreger verhält sich nach allem, was bis jetzt bekannt ist, in der übertragenden Zecke grundsätzlich ebenso wie *R. rickettsii* in ihren spezifischen Überträgern, allerdings ist diese Frage nicht genauer untersucht, und es sind hierzu auch in der letzten Zeit keine Beobachtungen gemacht worden. Die einschlägigen Untersuchungen stammen von HASS und PINKERTON. Die Übertragung der Rickettsien erfolgt in erster Linie mit dem Speichel beim Saugakt. CAPUTO berichtet von einer Infektion, die durch Zerdrücken einer auf dem Hunde saugenden Zecke zustande kam.

Ein *Warmblüterreservoir* war bisher weder für das Boutonneuse-Fieber noch für die anderen *afrikanischen Zeckenbißfieber* bekannt, so daß die übertragenden Zecken als die wesentlichsten Quellen angesehen werden, aus denen sich die Krankheit speist. Ob der Hund als einer der wichtigsten Zeckenwirte auch als Reservoir dienen kann, ist nicht geklärt. Hunde lassen sich zwar experimentell infizieren, in der Natur sind aber noch keine Rickettsien aus Hunden isoliert worden. Sie spielen offenbar eine mehr passive Rolle. Einen wichtigen Beitrag zur Frage der natürlichen Reservoire für Zeckenbißfieber haben HEISCH sowie HEISCH, MCPHEE und RICKMAN geliefert. In Nairobi konnten 9 Rickettsienstämme aus *H. leachi*, 1 Stamm aus *Rh. simus* und 2 Stämme aus *A. variegatum* isoliert werden, die bei der Übertragung auf den Menschen teilweise echtes Zeckenbißfieber erzeugten und serologisch nicht von südafrikanischem Zeckenbißfieber zu unterscheiden waren. Bei 5 von 12 Freiwilligen, die mit den Stämmen inoculiert wurden, konnte eine positive Weil-Felix-Reaktion festgestellt werden, die mit OX 19 höher als mit OX 2 ausfiel. Ebenso agglutinierten die Seren von 23 Hunden OX 19 stärker als OX 2. Versuche zur Isolierung von Rickettsien aus Hunden verliefen negativ. Keine Agglutinine für Proteus enthielten die Seren von verschiedenen Muriden: 24 *Arvicanthis abyssinicus*, 18 *Otomys angoniensis*, 2 *Mastomys coucha* und 1 *Lemniscomys striata*. 3 von 12 Seren, die von *Rattus* spec. stammten, agglutinierten OX 19 in der Verdünnung von 1:120 bis 1:240. Für Isolierungsversuche wurden Gehirne und Milzen von 84 *Otomys*, 30 *Arvicanthis* und 3 *Lemniscomys* auf Meerschweinchen übertragen. Auf diesem Wege wurden 1 Stamm aus *Otomys* und 1 Stamm aus *Lemniscomys* isoliert. Die Stämme stimmten mit den aus Zecken isolierten überein oder waren zumindest mit ihnen nahe verwandt. Die Rickettsien verhielten sich wie *R. conori*. Offenbar

handelt es sich hier um den ersten sicheren Nachweis eines Reservoirs für Zeckenbißfieber. *H. leachi* und *Rh. simus* wurden mehrfach in den Löchern von *Arvicanthis* und *Otomys* gefunden. In Südafrika sind als Reservoire für das Zeckenbißfieber *Otomys irroratus* und *Rhadomys pumilio* verdächtigt worden (GEAR).

Die ersten Beobachtungen über ein *Zeckenbißfieber in Indien* stammen von
MEGAW. Bisher sind von hier nur Einzelfälle beschrieben worden. Bei einem
genauer beobachteten Fall erfolgte die Infektion durch den Stich von *Rhipicephalus sanguineus* (RAO). Aus dem Blut des Kranken und aus Zecken der gleichen
Art, die von Hunden aus der näheren Umgebung des Patienten stammten, wurden
durch Inoculation von Meerschweinchen 7 Erregerstämme gewonnen. 3 Hunde,
von denen die infizierten Zecken abgesammelt waren, enthielten keine komplementbindenden Antikörper. Warmblüterreservoire sind bisher auch in Indien
nicht bekannt.

Nach SDRODOWSKI und GOLINEWITSCH sind Überträger des in Zentralasien,
Zentralsibirien und in den Meeresprovinzen von Südostsibirien verbreiteten
asiatischen oder *sibirischen Zeckenbißfiebers* die Schildzecken *Dermacentor silvarum*,
D. nuttalli, *D. marginatus*, *D. pictus*, *Haemaphysalis concinna* und *H. punctata*.
Bei *D. silvarum*, *D. nuttalli* und *H. concinna* wurde eine natürliche Infektion
nachgewiesen. Die Zecken übertrugen die Rickettsien im Experiment auf Meer-
schweinchen, außerdem wurde eine transovarielle Übertragung festgestellt. Als
Erregerreservoir gelten in Zentralsibirien Erdhörnchen. In Ostsibirien wurden
Stämme aus verschiedenen Nagern, darunter Hamster, Ratten und Mäuse, iso-
liert.

Von 62 Erkrankungsfällen, die KIREJEWA in Nordasien beschrieben hat, traten 57 in der
Zeit von Juni bis August auf, während in Sibirien der Mai den Erkrankungsgipfel bringt. Bei
den Patienten handelte es sich durchweg um Leute, die im Walde arbeiteten. Zeckenstiche
und Primärläsionen fanden sich bei 42 Kranken.

Das *Zeckenbißfieber von Nordqueensland* wird seit 1946 als besondere Rickettsiose unterschieden (PLOTZ, SMADEL, BENNETT, REAGAN und SNYDER). Die
Krankheit dürfte auch in Südostqueensland vorkommen, wo POPE durch
Inoculation von Mäusen mit Blut eines Patienten Rickettsien isolierte, die sich
im Tierversuch wie *R. australis* verhielten. Daß diese Rickettsiose durch Zecken
übertragen wird, wurde vermutet auf Grund der Anamnese von Patienten, die
z. T. auch eine Primärläsion hatten. Bei einem kürzlich einwandfrei diagnostizierten und genauer verfolgten Fall (NEILSON) hatte eine Zecke an einem 16jähri-
gen Knaben gesogen. Die Zecke war entfernt worden, bevor sie bestimmt werden
konnte. Der Hauptverdacht fällt auf *Ixodes holocyclus*.

Unklarheit herrscht auch hier über die tierischen Reservoire. Komplementbindende Antikörper (Titer 1:4) hat FENNER bei 8 von 11 Säugerarten aus einem
Endemie-Gebiet festgestellt. Unter den serologisch positiven Tieren waren Fuchs-
kusus, Beutelratten, Mosaikschwanzmäuse und Beuteldachse. Suspensionen aus
den auf den positiven Tieren parasitierenden Zecken (*Ixodes holocyclus* und
Haemaphysalis humerosa), Flöhen und Milben enthielten keine Rickettsien.

c) Q-Fieber

Die große Zahl von Veröffentlichungen über das Q-Fieber, die in erster Linie
Verbreitung und Epidemiologie der Krankheit betreffen, veranlaßte in den letzten

Jahren bereits mehrfach zusammenfassende Darstellungen über den jeweiligen Stand der Forschung, von denen hier nur die Publikationen von BERGE und LENNETTE sowie HENGEL, KAUSCHE, LAUR und RABENSCHLAG erwähnt seien. 1955 war die Krankheit in 51 Ländern bekannt und nur noch nicht in Irland, Holland, Polen, Schweden, Norwegen, Finnland und Neuseeland gefunden worden (KAPLAN und BERTAGNA). Inzwischen wurde sie auch in Polen nachgewiesen.

Nachdem in Deutschland zunächst nur Erkrankungen in Laboratorien bekannt waren (NAUCK und WEYER; KIKUTH und BOCK u. a.), kam es in den folgenden Jahren zu einer Reihe von dörflichen Epidemien, so in Baden (HENGEL, KAUSCHE und SHERIS; SCHUBOTHE, BOCK und WIESMANN), in Württemberg-Hohenzollern (GERMER und GLOCKNER), in der Eifel und bei Aschaffenburg (TERHAAG; NOLDEN; HERZBERG und MAY). In der Schweiz ist die Krankheit bei Gelegenheit mehrerer kleinerer und größerer Epidemien sehr genau verfolgt worden (WIESMANN; WIESMANN, SCHWEIZER und TOBLER u. a.). Eingehende Untersuchungen über Q-Fieber liegen auch aus England vor [STOKER u. Mitarb., MARMION; STOKER und MARMION (1)]. In einem Teil von Kent konnten 16% aller Pneumonien und unklaren fieberhaften Erkrankungen als Q-Fieber diagnostiziert werden. Antikörper wurden bei 2—4%, stellenweise bei 25% der Bevölkerung gefunden. Aus Frankreich sind sporadische Fälle und einzelne kleinere Epidemien bekannt (GIROUD und GOULON; SAUTET).

In Italien sind nach den Schätzungen von BABUDIERI (2) von 1949—1950 etwa 10000 bis 15000 Q-Fieber-Fälle vorgekommen. 1949 wurde die Krankheit in Sizilien nachgewiesen (CASTORINA). Größere Epidemien sind u. a. aufgetreten in Pesaro (BABUDIERI und PAOLUCCI), Domegge di Cadore (DEL CAMPO, DE LOTTO und MAGRÌ) und in der Campagna (DI SAPIO und LECALDANO). In der Provinz Cagliari verteilten sich die in der KBR positiven Seren zu 50% auf Schafe, 40% auf Ziegen und 10% auf Ochsen (CIOGLIA und MEDDA). In jeder Ziegenherde fand sich wenigstens ein positives Tier. Q-Fieber ist auch aus Österreich (LASS), der Tschechoslowakei (RAŠKA, SYRŮČEK und KUBÁSEK; BREZINA und TÁBORSKÁ u. a.), aus Jugoslawien (SIPKA und MILJKOVIC-KREJAKOVIC) und Rumänien (COMBIESCU u. Mitarb.) beschrieben worden. In der Tschechoslowakei waren von 1451 Rindern 542 (37,4%) in der KBR positiv, von 250 Schafen 135 (54,0%), von 93 Ziegen 53 (57,0%), von 21 Pferden 15 und von 13 Hunden 9; in Jugoslawien von 1426 Kühen 0,91%. 11 von 13 positiven Kühen schieden den Erreger in der Milch aus. In Polen ist die Krankheit jetzt ebenfalls nachgewiesen. Von 932 Schlachtkühen enthielten 0,8—1,8% komplementbindende Antikörper, von 119 Zuchtkühen 2,5% (WOJCIECHOWSKI, WNĘK, LEWIŃSKA und FRYGIN). WOJCIECHOWSKI, LEWIŃSKA und MIKOŁAJCZYK fanden bei 5 von 500 Schlachthofarbeitern komplementbindende Antikörper im Serum. 249 Patienten mit atypischer Pneumonie aus Warschau und Umgebung waren serologisch negativ.

Einen Bericht über die Verbreitung des Q-Fiebers in der UdSSR haben LESCHTSCHINSKAJA, POWALISCHINA und DZAGUROW gegeben. Ziemlich häufig ist die Krankheit im Südosten der UdSSR (TSCHUMAKOW). Kleinere Epidemien traten z. B. im Kaukasus auf (KULAGIN und SILITSCH), in Aserbeidschan (STERKHOWA und MIRSOEWA) und in Turkmenien (FEDOROWA, BEKTEMIROW, TARASEWITSCH, KERBABAEW und SEMASCHKO). Im Ural ist Q-Fieber außer bei Menschen auch bei Pferden, Kühen und Ziegen nachgewiesen (BURGANSKI, KAPLINSKI, WYGOWSKI und BERDNIKOW). Auf der Krim fanden BEKTEMIROW, TARASEWITSCH und KARULIN komplementbindende Antikörper bei 48% der ländlichen und 9,1% der städtischen Bevölkerung. Eine positive KBR ergaben 23,9% der Seren von 481 Kühen und 5,7% der Seren von 87 Pferden. In Aserbeidschan fand man Agglutinine gegen Q-Fieber bei 48,6% von 972 Kühen, 23% von 282 Schafen und 7 von 16 Büffeln. Die KBR war bei 9,7% von 289 Kühen und 23,1% von 65 Schafen positiv. Komplementbindende Antikörper hatten 5,9% von 561 Schlachthof- und Landarbeitern, ferner 88 (8,2%) von 1077 Personen, die in den letzten 2 Jahren an einer nicht diagnostizierten fieberhaften Erkrankung gelitten hatten. Bei einem Q-Fieber-Ausbruch auf einer Kolchose im Kaukasus, der ganze Familien erfaßte, hatten 32,6% von 144 gesunden Personen und 33,3% von 72 Kühen komplementbindende Antikörper für Q-Fieber.

In Ägypten stellten TAYLOR, HASSAN und KADER Q-Fieber in Kairo und in 5 verschiedenen Provinzen fest. 22% der Seren von 403 Männern und 33% der Seren von 319 Frauen waren in der KBR positiv. Von 26 Schafen enthielten 16, von 25 Ziegen 15 komplementbindende

Antikörper. TAYLOR, RIZK und ABDEL KADER sahen positive Seren besonders häufig bei Kindern unter 5 Jahren. TAYLOR, MOUNT, HOOGSTRAAL und DRESSLER wiesen in Ägypten Erreger in Zecken und Schafmilch nach. Einige Seren von Füchsen waren in der KBR positiv. ELYAN und DAWOOD fanden positive Seren bei Kühen (16%), Kamelen (13,9%) und Schafen. Von 759 in Marokko untersuchten Personen hatten 37,5% komplementbindende Antikörper für Q-Fieber (GARIN). Von 179 Haustieren waren 54,8% der Ziegen, 44,7% der Rinder und 38% der Schafe positiv. Q-Fieber ist serologisch auch in Tunis bei Menschen, Ziegen, Schafen und Rindern festgestellt worden (MAURIN). Bei Schlachthofarbeitern in Algier waren 18 von 122 Seren für Q-Fieber positiv, ferner 12 Seren von 178 Schafen und 4 Seren von 10 Rindern (LACROIX, SAYAG und DOUARD). In Algier ist es auch zu einer Epidemie beim Menschen gekommen (MINOUN, PIERROU, VASTEL und MARC).

Nachrichten über Vorkommen von Q-Fieber liegen auch aus anderen Teilen Afrikas vor, so aus Äthiopien (GUTFREUND, GELBERG und CHABAUD), wo positive serologische Reaktionen bei Menschen, Schafen, Ziegen und Zebus ermittelt wurden, aus Kenia, besonders dem Rifttal (CRADDOCK und GEAR; BROTHERSTON und COOKE; BROWN), aus Südrhodesien (GELFLAND und BERNEY) und Südafrika (FREYCHE und DEUTSCHMAN). BROWN fand in Kenia Antikörper für Q-Fieber im Blut von Hunden, Pferden, Kamelen, Ziegen und Kühen.

In der Türkei wurde Q-Fieber zuerst 1948 nachgewiesen; diese Beobachtungen sind durch serologische Untersuchungen erweitert worden [PAYZIN (1), (2)]. Von 1590 Patienten mit Pneumonie hatten 357 eine positive KBR für Q-Fieber. Von 1008 Tierseren enthielten 149 komplementbindende Antikörper. Beteiligt waren Schafe zu 16,5, Ziegen zu 13, Kühe zu 16 und Büffel zu 4%. Im Rahmen dieser Untersuchungen wurden mehrere Stämme von *C. burneti* isoliert. Durch serologische Untersuchungen ist das Vorkommen von Q-Fieber in Libanon (GARABEDIAN, DJANIAN und JOHNSTON) und im Iran (RAFYI und MAGHAMI; GADJUSEK und BAHMANYAR) gesichert. In Libanon waren von 450 menschlichen Seren 19,3 % positiv. Bei gefährdeten Personen (Bauern, Schlachthofarbeitern usw.) betrug der Prozentsatz der positiven Reaktionen 39,0. GARABEDIAN und DJANIAN wiesen Q-Fieber-Erreger in Kuh- und Ziegenmilch nach.

Antikörper für Q-Fieber beim Vieh (Büffel, Kühe, Ziegen, Schafe) sind in Ceylon (SCHMID, JAYSUNDERA und VELAUDAPILLAI) und in Indien [KALRA und TANEJA; SOMAN (2)] gefunden worden. Bei einer Erkrankung von Menschen in Nordindien konnte die Diagnose durch den Erregernachweis gesichert werden (ANDERSON und KALRA). Bei den in Vietnam beobachteten Q-Fieber-Fällen (QUINTIN, DESNUES und AUDRAN) ist die Infektion wohl von Frankreich aus eingeschleppt worden. Daß Q-Fieber auch in Peking vorkommt, konnten CHANG, ZIA und CHAO wahrscheinlich machen. Von 37 Patienten mit atypischer Pneumonie reagierten 7 in der KBR für Q-Fieber positiv. In Japan waren unter 754 Seren von Personen, die in Schlachthäusern und Lebensmittelbetrieben arbeiteten, 39 für Q-Fieber positiv (TAKANO, KITAOKA und SHISHIDO). Positive serologische Reaktionen wurden auch beim Vieh festgestellt. Das Fehlen von Q-Fieber in Neuseeland ist durch weitere Untersuchungen bestätigt (SALISBURY; FASTIER). FASTIER fand mit Hilfe des Meerschweinchenversuchs unter 1300 menschlichen Blutproben 3 positive. Das Blut stammte jedoch von Personen, die in den letzten 3 Jahren aus Europa eingewandert waren.

Weit verbreitet ist Q-Fieber in Nordamerika. In Südkalifornien sind vor 1950 etwa 50000 Menschen an Q-Fieber erkrankt gewesen. Auch in Ohio ist die Krankheit kürzlich bei Milchkühen festgestellt worden (REED und WENTWORTH), ferner in Wisconsin (KITZE, HIEMSTRA und MOORE). 1,9% von 22304 Blutproben aus 2130 Rinderherden reagierten in der KBR positiv; die meisten positiven Seren fanden sich im Südosten des Landes. Hier konnten aus 7 von 8 Milchproben positiver Kühe Rickettsien isoliert werden. Außerdem waren 7 von 30 Seren bei Farmern und 92 von 2035 Seren bei Angestellten eines Fleischpackbetriebes positiv. In Texas wurden von 1948—1955 nur 64 Erkrankungen beobachtet (IRONS, EADS und PEAVY). Als Infektionsquelle sind hier vor allem Kühe wichtig. In Kanada kommt Q-Fieber in der Provinz Quebec vor. Sehr eingehende Untersuchungen über Q-Fieber wurden besonders in Kalifornien durchgeführt.

Es liegen auch einige Nachrichten über Q-Fieber in Mexiko (SILVA-GOYTIA), Venezuela (BRICENO IRAGORRY), Uruguay (SALVERAGLIO, BACIGALUPI, SRULEVICH und VIERA), Argentinien (BETTINOTTI) und Brasilien (BRAND O, RIBEIRO DO VALLE und CHRISTOV O; H. RIBEIRO DO VALLE, L. A. RIBEIRO DO VALLE, CHRISTOV O und D'APICE; RIBEIRO DO VALLE,

Martins Castro, Banoi und Ferreira) vor. In São Paulo wurden positive Seren bei 5 von 11 Landarbeitern und 24 von 171 Rindern nachgewiesen. Erregerisolierungen gelangen nicht.

Das besondere Interesse an der anfangs in Dunkel gehüllten *Übertragung und Epidemiologie* des Q-Fiebers hat eine größere Zahl von speziellen Untersuchungen zu diesem Thema veranlaßt, deren Ergebnisse z. T. bereits zusammengefaßt wurden [Sidky; Derrick (2), Weyer (11), Stoker und Marmion (1), (3) u. a.]. Übertragung und Epidemiologie der Krankheit können heute als in den wichtigsten Grundlagen geklärt angesehen werden. Dabei ist zu unterscheiden zwischen dem Q-Fieber des Menschen und dem der Haus- und Wildtiere. Über das Q-Fieber bei Wildtieren wissen wir noch relativ wenig. Sehr viel besser sind wir über das Verhalten des Erregers in Arthropoden orientiert, die entweder in den Entwicklungskreislauf des Erregers eingeschaltet oder bei der Verbreitung und Übertragung der Krankheit unter Wildtieren und von Wild- auf Haustiere beteiligt sind.

Es steht fest, daß die empfänglichsten und für die Übertragung auf den Menschen wichtigsten *Haustiere* Schafe, Kühe und Ziegen sind, und daß die weitaus meisten menschlichen Erkrankungen aus dieser Infektionsquelle gespeist werden. Bei anderen Haustieren, z. B. Kamelen, Dromedaren, Pferden und Hunden, sind in manchen Gegenden Antikörper für Q-Fieber festgestellt worden; nach den bisherigen Erfahrungen spielen diese Tiere aber, abgesehen vielleicht von Pferden, epidemiologisch keine größere Rolle.

Zu erwähnen sind jedoch die Beobachtungen von Mantovani und Benazzi, die *C. burneti* aus dem Blut eines Hundes isolieren konnten, der eine Schafplacenta gefressen hatte. Ebenso wurden die Erreger aus Zecken (*Rhipicephalus sanguineus*) isoliert, die über eine Periode von 2 Monaten von dem betreffenden Hund in 5—7tägigen Abständen abgesammelt wurden. Bei 2 kleineren Epidemien wurden auch 3 mit *C. burneti* infizierte Hunde gefunden. In einem Fall erfolgte die Erregerisolierung beim Hunde 110 Tage nach dem Ausbruch der Epidemie. Die Hunde waren 2—10 Monate nach Beginn der Epidemie in der KBR positiv.

Infizierte Haustiere können den Erreger mit Blut, Speichel, Milch, Urin und Kot ausscheiden. Die Erreger sind bei Kühen u. a. in der Milch, in den Milchdrüsen und Lymphknoten, in der Milz und Leber gefunden worden. Besonders reich an Rickettsien ist das *Fruchtwasser*, so daß menschliche Infektionen gehäuft im Frühjahr zur Zeit des Lammens der Schafe und des Kalbens der Kühe auftreten. Es besteht auch Klarheit darüber, daß die Infektion des Menschen selten durch direkten Kontakt mit den infizierten Haustieren, vielmehr am häufigsten auf aerogenem Wege erfolgt. Daher sind nicht nur die Personenkreise einer Infektion ausgesetzt, die ständig oder unmittelbar mit dem Vieh zu tun haben, sondern mitunter Personen, die nur einmal zufällig Gelegenheit hatten, erregerhaltigen Staub einzuatmen. Die gleichen Gesichtspunkte gelten auch für Laboratoriumsinfektionen. Bei der gegenseitigen Infektion der Haustiere dürfte jedoch der direkte Kontakt mit den erregerhaltigen Ausscheidungen im Vordergrund stehen. Die Milchdrüsen spielen als Eintrittspforte für den Erreger offenbar keine Rolle (Stoenner und Lackman).

Nach Lennette und Clark sind 70% aller Fälle von Q-Fieber in Kalifornien auf den Umgang mit Schafen, Ziegen und Kühen zurückzuführen. In Nordkalifornien spielen Schafe und Ziegen die Hauptrolle, in Südkalifornien Kühe. Am wichtigsten ist die Periode des Lammens und Scherens der Schafe (Abinanti, Lennette, Winn und Welsh). Eine epidemiologische Erhebung bei 350 Personen, die von 1948—1949 erkrankten, ergab, daß die meisten

Fälle auf dem Land in Betrieben mit Milchwirtschaft und Viehzucht auftraten (Clark, Lennette und Romer). 70% erkrankten von März bis April, d. h. in der Zeit, wo die Schafe lammten oder kastriert wurden. Es waren 10mal mehr Männer als Frauen befallen. 93 von den 350 Patienten hatten keinen Kontakt mit dem Vieh. Mit Hilfe der KBR wurden inapparente Fälle in etwa gleicher Zahl wie apparente festgestellt.

Luoto und Huebner wiesen bereits auf die Bedeutung von getrocknetem Placentagewebe als Ansteckungsquelle hin. Sie fanden bei 13 von 33 Kühen die Erreger in der *Placenta*. Placentagewebe war noch in einer Verdünnung von 1:100 Millionen infektiös. Welsh, Lennette, Abinanti und Winn isolierten Erreger aus 21 von 72 Schafplacenten. Von 29 Schafen, die serologisch negativ waren, enthielten 6 (21%) den Erreger in der Placenta. Stoker, Brown, Kett, Collings und Marmion isolierten *C. burneti* in Endemiegebieten aus der Placenta und von der Wolle der Schafe. Abinanti, Welsh, Winn und Lennette wiesen bei 2 von 50 Schafen, die vor 7 Tagen gelammt hatten, Q-Fieber-Erreger an der Wolle der Dammregion nach.

Wiesmann, Schweizer, Fey und Bürki sowie Wiesmann, Schweizer und Tobler fanden in der Nordostschweiz Q-Fieber-Erreger in gesunden Schafen, in abortierenden Schafen und Ziegen, in der Placenta von Rindern und im Urin und Kot von Schafen. Wiesmann und Bürki machten auf die Bedeutung der fieberhaften Aborte für die Verbreitung der Erreger aufmerksam und zeigten, daß das Fell neugeborener Schaflämmer massenhaft Rickettsien enthalten kann. Von besonderer epidemiologischer Bedeutung ist, daß auch äußerlich gesunde Schafe den Erreger im Urin ausscheiden können.

Die mit der *Milch* ausgeschiedenen Rickettsien treten demgegenüber als Infektionsquelle zurück, obwohl der Nachweis von Rickettsien in der Milch von Schafen und Ziegen mehrfach gelungen ist (Jellison, Welsh, Elson und Huebner; Lennette und Welsh; Likar; Sipka und Miljkovic-Krejakovic; Kitze, Hiemstra und Moore; Enright, Sadler und Thomas u. a.). Enright Sadler und Thomas fanden in Kalifornien in 8 von 109 Proben roher Sammelmilch und in Einzelproben bei 18 von 137 Kühen in einer Herde *C. burneti*. Die Anzahl der Erreger variierte bis zu 1000 ID für das Meerschweinchen in 2 cm³ Milch. Eine durch die Zitzen infizierte Kuh enthielt in 2 cm³ Milch 10000 ID Erreger und schied Rickettsien bis zu 405 Tagen aus. Slavin prüfte in England 8343 Milchproben von 2581 Bauernhöfen im Meerschweinchentest. 6,9% der Milchproben enthielten Rickettsien. Die epidemiologische Analyse des Q-Fiebers in einer städtischen Gegend von England durch Marmion und Harvey ergab, daß von 22 Personen mit positiver KBR nur eine Umgang mit Vieh hatte; 91% verbrauchten aber in ihrem Haushalt rohe Kuhmilch, in welcher auch Erreger nachgewiesen werden konnten.

Die Infektion des Menschen erfolgt in den meisten Fällen durch Einatmen der von den Haustieren ausgeschiedenen, *getrockneten* und mit dem *Staub* aufgewirbelten Erreger. Der Zusammenhang mit Schafen ist speziell bei den europäischen Epidemien sehr auffällig. Dafür sprechen eindeutig die Beobachtungen in England, Deutschland, in der Schweiz und in Italien. Sporadische Erkrankungen in England gingen auch auf Kühe zurück, doch konnte bei fast einem Drittel der Fälle die Infektionsquelle nicht ermittelt werden (Marmion, Stoker, McCoy, Malloch und Moore). Bei Erkrankungen von Schlachthofarbeitern in Frank-

reich waren für die Infektion Schafe verantwortlich (SAUTET). In Pesaro (BABU-
DIERI und PAOLUCCI) bildete eine Herde von 20 Schafen die Infektionsquelle bei
einer Epidemie, die 200 Personen erfaßte. Einige Tage vor dem Ausbruch hatten
mehrere Schafe gelammt, eines hatte abortiert. Trockenes Klima und Straßen-
arbeiten begünstigten die Verbreitung des erregerhaltigen Staubes. Außer Erd-
staub können auch Heu und Stroh Infektionsquellen sein, und gefährdet sind
häufig Anwohner von Straßen, auf denen Viehherden durchgetrieben werden.

In der Tschechoslowakei waren Rinder, Schafe und Ziegen die Hauptinfektionsquellen
(RAŠKA, SYRŮČEK und KUBÁSEK; RAŠKA und SYRŮČEK). Eine kleine Q-Fieber-Epidemie in
Polen konnte auf die Einfuhr von Schafen zurückgeführt werden (OLÉS und KURZEJA;
OLÉS, KURZEJA, LEWIŃSKA und FRYGIN). In dem Epidemiegebiet waren von 274 Schafen
73,6% serologisch positiv, während von 311 Schafen aus einem benachbarten Weidegrund
33,7% und von 176 Schafen aus den umliegenden Dörfern nur 0,9% Antikörper enthielten.
WOJCIECHOWSKI und PRZYBYŁA isolierten C. burneti von Schafwolle. Bei einem Q-Fieber-
Ausbruch auf einer Kolchose im Kaukasus wurde Milch als Infektionsquelle festgestellt
(STERKTOWA und MIRSOEWA). BEKTEMIROW, TARASEWITSCH und KARULIN isolierten auf der
Krim C. burneti bei Kühen aus Euter, Leber und Milz, außerdem 3 Stämme aus roher Milch.
Es ist häufig schwierig, die Infektionsquelle zu ermitteln, da die Erreger leicht über weite
Strecken verschleppt werden können. So konnten die Erkrankungen von Personen im euro-
päischen Rußland, die mit der Verarbeitung von aus Zentralasien importierter Baumwolle
beschäftigt waren, erst nach langem Suchen auf Kühe und Schafe zurückgeführt werden, durch
welche die Baumwolle bereits auf den Feldern in Turkmenien verunreinigt worden war (FEDE-
ROWA, BEKTEMIROW, TARASEWITSCH, KERBABAEW und SEMASCHKO). In der Tschechoslowakei
erkrankten 40 Personen in einem Raum, in welchem Baumwollballen aufgebrochen wurden
(BÁRDOŠ, BREZINA, KRAMÁR, SOBEŠLAVSKÝ und ŠIMONÍK). In Mäusen, die auf der Baumwolle
gehalten wurden, konnte später C. burneti nachgewiesen werden. In Krakau kam es zu einer
kleineren Epidemie in Zusammenhang mit der Verarbeitung von Schafwolle aus Rumänien
(LUTYŃSKI, RAGINIS, ZIEMICHÓD und KOŹMIŃSKA).

Daß der Staub Erreger enthalten kann, ist mehrfach bewiesen worden.
DE LAY, LENNETTE und DE OME untersuchten den Staub von Plätzen, auf denen
sich infizierte Kühe und Schafe aufgehalten hatten und konnten in 3 von 5 Versuchen C. burneti isolieren. LENNETTE und WELSH fanden Rickettsien im Luft-
staub eines Stalles, in welchem infizierte Ziegen untergebracht waren. BINGEL,
LAIER und MANNHARDT wiesen die Erreger in einer Mischung von Staub und Korn
aus einem Maissack nach, MARMION und STOKER im Staub aus der Kleidung eines
Schafhirten, BEKTEMIROW, TARASEWITSCH und KARULIN im Staub von Kuh-
ställen.

Der direkte Kontakt mit infektiösen Organen kommt als Infektionsquelle besonders bei
Schlachthofarbeitern in Frage. Die hohe Infektiosität des Q-Fiebers geht u. a. daraus hervor,
daß im Anschluß an eine menschliche Sektion von 18 Ärzten und Studenten 16 innerhalb
kurzer Zeit erkrankten [BABUDIERI (2)]. Es ist daher nicht verwunderlich, daß Patienten
mit starkem Auswurf ihre Umgebung durch Tröpfcheninfektion gefährden können. Über
derartige Fälle berichten u. a. DEUTSCH und PETERSON; SIEGERT, SIMROCK und STRÖDER
sowie QUINTIN, DESNUES und AUDRAN.

Über Spätrückfälle bei Q-Fieber ist bisher nichts bekannt geworden.

Unsere Kenntnisse über die Erregerreservoire des Q-Fiebers unter den Wildtieren sind inzwischen erweitert worden. DERRICK und SMITH hatten bereits in
Australien 3 Stämme von C. burneti aus Isoodon torosus (Großer Streifen-Beuteldachs) isoliert. BLANC, MARTIN und MAURICE fanden in Marokko 2 Stämme in der
Milz von Meriones shawi (Wüstenrennmaus). BLANC und BRUNEAU (4), (5) isolierten
ebenfalls in Marokko Stämme aus Kaninchen, Gerbill (Dipodillus campestris),

Waldmaus (*Apodemus sylvaticus*), Gartenschläfer (*Eliomys mumbianus*) und Temminck (*Eptesicus isabellinus*). Taylor, Mount, Hoogstraal und Dressler wiesen in Ägypten im Blut von Füchsen komplementbindende Antikörper für *C. burneti* nach. Perez Gallardo, Clavero und Hernández fanden in Spanien Q-Fieber-Erreger in Bergkaninchen. 54 von 81 Kaninchenseren reagierten in der KBR für Q-Fieber positiv.

Marmion und Stoker untersuchten in einem Endemiegebiet von Südostengland 229 wilde Kaninchen und 30 andere kleine Säuger, außerdem 166 Wildvögel, ohne Rickettsien zu finden. In der Slowakei wurden komplementbindende Antikörper bei 3 von 5 Hirschen, 2 von 19 Hasen, 14 von 40 Ratten, 2 von 20 Hausmäusen und 3 von 15 Spitzmäusen nachgewiesen (Brezina und Tabórská). An 2 Stellen in West- und Ostböhmen wurden Seren von 574 Nagetieren in der KBR geprüft (Raška und Syrůček). Positive Reaktionen fanden sich bei 14 von 169 Hausmäusen (*Mus musculus*), 11 von 220 Gelbhalsmäusen (*Apodemus flavicollis*) und Waldmäusen (*A. sylvaticus*), bei 1 von 90 Feldmäusen (*Microtus arvalis*), 2 von 67 Rötelmäusen (*Clethrionomys glareolus*), 1 von 6 Schermäusen (*Arvicola terrestris*), 1 von 13 Hasen (*Lepus europaeus*) und 41 von 339 Wanderratten (*Rattus norvegicus*). Die Seren von 126 Spitzmäusen waren negativ. Rickettsien wurden isoliert aus 1 von 20 Wanderratten, 2 von 30 Hausmäusen und einer Rötelmaus. Im gleichen Gebiet fanden Rehn und Ravdan komplementbindende Antikörper bei einem Hirsch. Syrůček, Sobeślavský und Havlík prüften in Nordwestböhmen 104 Seren von Insectivoren mit Hilfe der KBR. Positiv reagierten bei *Sorex araneus* (Waldspitzmaus) 4 von 83 Seren und bei *S. minutus* (Zwergspitzmaus) 1 von 14 Seren. Aus der serologisch positiven Zwergspitzmaus konnte ein Stamm von *C. burneti* isoliert werden. 7 Maulwürfe enthielten keine Antikörper.

Im Süden von Zentralasien wurden Q-Fieber-Erreger in *Rhombomys opimus* (große Rennmaus) und *Spermophilopsis leptodactylus* (Erdhörnchen) gefunden (Smajewa, Ptschelkina, Mischtschenko und Karulin). Bektemirow, Tarasewitsch und Karulin untersuchten auf der Krim eine größere Anzahl von Nagetieren und Vögeln und konnten Q-Fieber-Stämme aus dem Bergsperling und aus dem Streifenhamster (*Cricetulus migratorius*) isolieren.

Vögel sind erst ernstlich als Erregerreservoire in Betracht gezogen worden, nachdem Babudieri (2) und Babudieri und Moscovici (1) in einem Q-Fieber-Gebiet in Italien komplementbindende Antikörper bei einer Gans und mehreren Tauben festgestellt hatten. Unter 32 Tauben waren 3 positiv. Aus einer Taube konnte auch ein Q-Fieber-Stamm isoliert werden. In der Tschechoslowakei wurde der Bedeutung von Vögeln als Virusreservoir besondere Beachtung geschenkt (Syrůček und Raška). In einem Epidemiegebiet war die KBR bei 51 (14,9%) von 342 Hühnern für Q-Fieber positiv, ferner bei 8 von 158 Enten, 2 von 37 Putern, 7 von 59 Gänsen und 8 von 34 Tauben. 28 Kontrollseren von Hühnern aus einem Gebiet ohne Q-Fieber waren negativ. Negative Reaktionen ergaben auch die Seren von 10 Schweinen, 4 Katzen und 6 Kaninchen. Bei einigen Erkrankungen in Turkmenien spielten offenbar Hühner und Hühnereier eine Rolle als Infektionsquelle (Smajewa und Ptschelkina). Rickettsien konnten aus dem Gehirn und der Leber eines Hahnes und von Eierschalen isoliert werden.

Überraschende Ergebnisse brachte die Untersuchung von Singvögeln in der Tschechoslowakei. 27 (15,8%) von 117 Vögeln, die auf einem Hof mit Q-Fieber gefangen wurden, gaben positive Reaktionen in der KBR, ferner 7 (4,3%) von 176 Vögeln eines benachbarten Hofes, während von Vögeln aus anderen Teilen des Landes nur 2 (1,8%) von 117 positiv reagierten. Die positiven Reaktionen verteilten sich auf 8 von 92 Rauchschwalben, 14 von 60 Mehlschwalben, 5 von

35 Haussperlingen, 2 von 23 Rotschwänzchen, 1 von 2 Zaunkönigen, 1 von
8 Kohlmeisen, 1 von 51 Buchfinken, 1 von 31 Goldammern, 1 von 17 weißen
Bachstelzen, 1 von 9 großen Buntspechten und 1 von 9 Grünfinken. Von 35 Versuchen zur Isolierung von Rickettsien aus der Milz und Leber der Singvögel
brachten 4 ein positives Ergebnis. Stämme von *C. burneti* wurden aus Rotschwänzchen, Bachstelzen und Schwalben isoliert.

Der Erregernachweis in Zugvögeln erscheint besonders interessant im Hinblick
darauf, daß die Vögel die Krankheit über sehr weite Räume verschleppen können.
Es kann offenbar von Haustieren aus auch ein Übergang der Erreger auf Wildtiere erfolgen.

Von den zahlreichen Versuchen, im Zusammenhang mit dem Auftreten der
Krankheit die Erreger in *Arthropoden* (Mücken, Fliegen, Flöhen, Läusen, Mallophagen, Käfern, Milben) nachzuweisen, seien nur erwähnt die von CLARK,
BOGUCKI, LENNETTE, DEAN und WALKER; BURGDORFER, GEIGY, GSELL und
WIESMANN; IRONS, EADS, JOHNSON, WALKER und NORRIS sowie IRONS, EADS
und PEAVY. Die letzteren Autoren untersuchten in Texas eine große Zahl
von Ektoparasiten, die von 4000 Tieren (30 Arten), vorwiegend Nagern, abgesammelt waren. Unter den Parasiten waren 15 Zeckenarten einschließlich
Amblyomma americanum, Otobius megnini, Ornithodorus turicata und *O. talaje*,
ferner 5 Milben-, 7 Läuse- und 9 Floharten, außerdem Stechmücken, Stubenfliegen, Stechfliegen, Schmeißfliegen, Bremsen, Raubwanzen und Käfer. Die
Versuche verliefen durchweg negativ. GIROUD und GIROUD und JADIN (1), (4)
glauben, in Belgisch-Kongo *C. burneti* in Kleiderläusen gefunden zu haben und
schließen daraus auf eine Bedeutung der Läuse als Überträger oder Reservoire
der Krankheit. Jedoch konnte diese Annahme durch keine weiteren direkten
Beobachtungen gestützt werden. COMBIESCU fand die Erreger in Pferdelausfliegen (*Hippobosca equina*) und Schaflausfliegen (*Melophagus ovinus*).

Unter natürlichen Verhältnissen kommen als Wirte für *C. burneti* in erster
Linie Zecken in Frage. Die Bedeutung der *Zecken*, aus welchen Q-Fieber-Stämme
in großer Zahl isoliert sind, liegt nicht darin, daß sie die Krankheit auf den Menschen übertragen — es gibt nur ganz vereinzelte Fälle, in denen Zecken als unmittelbare Überträger von Q-Fieber beim Menschen in Betracht gezogen werden
konnten —, als vielmehr in ihrer Eigenschaft als Reservoire und in ihrer noch nicht
ausreichend geklärten Rolle bei der Übertragung unter Wildtieren und von
Wild- auf Haustiere. Von epidemiologischer Bedeutung ist außerdem, daß sich
die Rickettsien im Zeckenkot und damit auch im Fell der Wirtstiere lange Zeit
lebend halten können. Auf diese Infektionsquelle sind z. B. Erkrankungen von
Schlachthofarbeitern zurückzuführen oder Erkrankungen von Personen, die
Rohwolle oder Tierhäute zu verarbeiten hatten. Jedoch ist der Weiterbestand der
Krankheit unter den Haustieren auch ohne Zecken gewährleistet. Das Verhalten
der Rickettsien in den Zecken und insbesondere die Frage der transovariellen
Übertragung sind noch nicht restlos geklärt.

In früheren Zusammenstellungen [WEYER (11), *Advances in the Control of
Zoonoses* 1953, BLANC (2), STOKER und MARMION (2)] sind 18—22 Zeckenarten
aus 8 Genera aufgezählt, aus denen in der Natur Erregerstämme isoliert werden
konnten. Diese Zahlen haben sich besonders durch Beobachtungen in Asien noch
vermehrt. Die ersten Q-Fieber-Stämme in Australien wurden von SMITH und

DERRICK aus *Haemaphysalis humerosa* isoliert. Später fanden CARLEY und POPE 2 Stämme in *Ixodes holocyclus* in Queensland. 423 Zecken von 8 verschiedenen anderen Arten enthielten keine Rickettsien. Natürlich infizierte Zecken sind in Australien offenbar relativ selten. PHILIP und WHITE isolierten an der Mississippi-Golf-Küste Stämme aus *Amblyomma americanum* und *A. maculatum.* In England wurden Rickettsien in *Haemaphysalis punctata* gefunden, ferner in einer Mischung von 1800 *Ixodes ricinus* und 15 *Dermacentor reticulatus* [MARMION, STOKER, McCOY, MALLOCH und MOORE; STOKER und MARMION (2)].

PAYZIN und AKKAY untersuchten in der Türkei in 50 Proben über 500 *Ornithodorus lahorensis* und fanden etwa 4% der Zecken infiziert. BLANC und BRUNEAU (3) entdeckten in Marokko zufällig einen Stamm in *O. erraticus*, nachdem sie schon früher Stämme aus *Hyalomma excavatum* isoliert hatten, die von Wildkaninchen abgesammelt worden waren [BLANC und BRUNEAU (2)]. In Ägypten wurden Q-Fieber-Erreger in *Hyalomma dromedarii* von Kamelen und *H. excavatum* von sudanesischen Bullen gefunden (TAYLOR, MOUNT, HOOGSTRAAL und DRESSLER).

REHN und RAVDAN isolierten in Südostböhmen 2 Stämme von *C. burneti* aus einem Gemisch von 2322 *Ixodes ricinus.* In Zentralasien wurden Stämme isoliert aus *Hyalomma excavatum (anatolicum), H. detritum, Argas persicus* und *A. reflexus,* ferner aus 2 Milben: *Leeuwenhoekia major* und *Dermanyssus passerinus* (TSCHUMAKOW; SMAJEWA, PTSCHELKINA, MISCHTSCHENKO und KARULIN); auf der Krim aus *Hyalomma plumbeum* und *Rhipicephalus bursa*, die von Haus- und Wildtieren abgesammelt waren (BEKTEMIROW, TARASEWITSCH und KARULIN). PTSCHELKINA, SMAJEWA und SUBKOWA fanden den Erreger in *Ixodes crenulatus* vom Iltis, SMAJEWA, MISCHTSCHENKO und PTSCHELKINA in Südkirgisien in *Hyalomma anatolicum* und *Rhipicephalus turanicus* von Pferden und Kühen.

2. Durch Milben übertragene Rickettsiosen

a) Rickettsienpocken

Die Rickettsienpocken sind erst 1946 entdeckt worden. Bis Mitte 1952 wurden 624 Fälle gezählt, die sämtlich aus New York kamen (NICHOLS, RINDGE und RUSSELL). Einzelne Fälle sind auch in Boston aufgetreten (PIKE, COHEN und MURRAY; FRANKLIN, WASSERMANN und FULLER). LA BOCCETTA, ISRAEL, PERRI und SIGEL berichteten über 4 klinisch und serologisch diagnostizierte Fälle aus Pennsylvanien, EUSTIS und FULLER über 7 Fälle in West-Hartford (Connecticut), HOEPRICH, KENT und DINGLE über serologisch diagnostizierte Fälle in Cleveland.

Bei serologischen Untersuchungen in Bosnien und Herzegowina fanden TERZIN und GAON unter 115 Patienten 14,6%, deren Seren mit Antigen von *R. akari* reagierten. Auch in Italien konnte BADIALI unter 124 Seren 2 für *R. akari* positive feststellen. In Europa ist jedoch über Rickettsienpocken bisher nichts bekannt. Die beobachteten Reaktionen sind wahrscheinlich als unspezifisch anzusehen. In Südafrika ergaben mehrere Seren in der KBR mit Antigen von *R. akari* höhere Werte als mit Antigen von *R. conori* (GEAR). Die Angaben über Rickettsienpocken in Französisch-Äquatorialafrika sind kritisch aufzunehmen. LE GAC und GIROUD glauben hier die Krankheit festgestellt zu haben, weil das Serum eines mit dem Blut des verdächtigen Patienten inoculierten Meerschweinchens *R. akari* in einer Verdünnung von 1:20 agglutinierte. Später beschrieben LE GAC, GIROUD, LE HENAFF und BAUP in der gleichen Gegend in einer Familie 12 Fälle von „vesicular rickettsiosis", von denen 5 tödlich ausgingen. 5 von 8 Patienten reagierten in der Agglutination mit *R. akari* in einer

Verdünnung von 1:10 bis 1:20. 7 Seren reagierten in gleicher Stärke mit Antigen von *C. burneti*. LE GAC, GIROUD, ROGER und DUMAN haben noch 3 weitere Fälle mit einem pockenähnlichen Exanthem gesehen. Bei einem Patienten war die KBR mit *R. akari* positiv, außerdem erfolgte eine schwache Reaktion mit Antigen von *C. burneti*. — Das Vorkommen von Rickettsienpocken in Europa und Afrika kann auf Grund dieser Beobachtungen nicht als gesichert angesehen werden, zumal eine Erregerisolierung bisher nicht gelungen ist.

Außerhalb von Amerika sind Rickettsienpocken aus Rußland bekannt (SDRODOWSKI und GOLINEWITSCH; KISELEW und WOLTSCHANETSKAJA). In Korea wurde ein Stamm von *R. akari* aus einer einheimischen Maus (*Microtus fortis pelliceus*) isoliert, ohne daß Krankheitsfälle beim Menschen bekannt geworden wären (FULLER und SMADEL; JACKSON, DANAUSKAS, COALE und SMADEL). Unsere Kenntnisse über die geographische Verbreitung der Rickettsienpocken werden daher möglicherweise noch Erweiterungen erfahren. Es ist nicht ausgeschlossen, daß *R. akari* auch einen natürlichen Cyclus hat, der vom Menschen und von Hausmäusen unabhängig verläuft.

Übertragen werden die Rickettsienpocken nach den Beobachtungen in Amerika und Rußland durch eine blutsaugende Milbe: *Allodermanyssus sanguineus*, die auf der Hausmaus parasitiert. Auf Grund experimenteller Untersuchungen könnte auch die Rattenmilbe *Bdellonyssus bacoti* als Überträger in Frage kommen (PHILIP und HUGHES). Außerdem vermehrt sich *R. akari* lebhaft in Läusen [WEYER (10)], doch wurden unter natürlichen Bedingungen Erregerstämme nur aus den auf Hausmäusen parasitierenden Milben isoliert, die daher bis jetzt die einzig bekannten Überträger darstellen. EUSTIS und FULLER fanden in West-Hartford Rickettsien auch in Milben, die aus dem Keller eines Mietshauses stammten. Außer in Menschen und Milben wurden in Amerika Erreger in Hausmäusen nachgewiesen, die in der Umgebung von Kranken gefangen waren, in Rußland außer in Mäusen auch in *Rattus norvegicus*. FULLER, MURRAY, AYRES, SNYDER und POTASH isolierten in Boston 2 Stämme aus 22 Mäusen an einem Platz, an welchem 7 Monate vorher eine Frau an Rickettsienpocken erkrankt war. Von 8 Mäusen waren 3 in der KBR mit einem Titer von 1:60 positiv. 12 der gefangenen Mäuse hatten Milben, meist *Allodermanyssus sanguineus*; doch enthielten diese Milben keine Rickettsien. In Rußland wurden die meisten Fälle von Rickettsienpocken im Mai und Juni beobachtet (KISELEW und WOLTSCHANETSKAJA).

In New York trat die Krankheit wie in West-Hartford bevorzugt in modernen Wohnungen mit Abfallschächten auf (NICHOLS, RINDGE und RUSSELL). Durch diese Abfallschächte konnten Mäuse in die Küchenräume und zu den im Keller befindlichen Verbrennungsöfen gelangen. Die Außenwände der Öfen boten den Milben günstige Temperatur- und Feuchtigkeitsbedingungen. Die Parasiten fanden sich hier auch in größerer Zahl, außerdem in den Gängen der Mäuse. In Rußland kamen die meisten Erkrankungen ebenfalls in Städten vor.

Die Entwicklung von *Allodermanyssus sanguineus* verläuft über Ei, Larve, Protonymphe und Deutonymphe zur Imago [FULLER (3)]. Bei einer Temperatur von 23—24° C und 80% Feuchtigkeit schlüpften die Larven in 4—5 Tagen. Das Larvenstadium dauerte 3 Tage. Die Larve nimmt keine Nahrung auf. Die Protonymphen sogen 1—2 Tage nach der Häutung Blut — die Blutmahlzeit dauerte 1 Std — und häuteten sich 3—4 Tage später zur Deutonymphe, die im Unterschied zur Rattenmilbe *Bdellonyssus bacoti* ebenfalls Blut saugt. Das 2. Nymphenstadium dauerte 6—10 Tage. Die erwachsenen Milben nehmen wiederholt Blut auf. Die Entwicklungszeit vom Ei bis zur Imago betrug 17—23 Tage. Die Eiablage begann

2—5 Tage nach dem Saugen. Die Protonymphen konnten bis zu 10 Tagen hungern, jedoch setzte nach 1 Woche starke Sterblichkeit ein. Ein von der Hausmaus abgesammeltes Weibchen lebte 4 Wochen.

Über die Entwicklung und das Verhalten von *R. akari* in *Allodermanyssus sanguineus* ist bisher nichts Näheres bekannt geworden. Nach russischen Beobachtungen (KISELEW und WOLTSCHANETSKAJA) kommt eine transovarielle Übertragung vor, da sich der Erreger mit Eiern und Larven von infizierten erwachsenen Milben auf Mäuse übertragen ließ. Die Übertragung erfolgt beim Saugakt.

b) Tsutsugamushi-Fieber

Das Tsutsugamushi-Fieber, das ursprünglich nur aus Japan bekannt war, kommt in ganz Ost- und Südasien vor und ist *verbreitet* von Japan über Formosa, die Philippinen, Indonesien, Neuguinea bis zur Nordostküste von Australien und den angrenzenden Inseln einschließlich Espiritu Santo; in westlicher Richtung über China, Thailand, Malaya, Burma, Pakistan und Indien bis Ceylon und zu den Inseln im Indischen Ozean und im Golf von Bengalen. GIROUD, ferner GIROUD und JADIN (2) sowie LE GAC und GIROUD fanden in Belgisch-Kongo beim Menschen mit Hilfe des Intradermaltestes und der KBR Antikörper gegen *R. tsutsugamushi*. Diese Beobachtungen beweisen allerdings noch nicht, daß Tsutsugamushi-Fieber auch in Afrika vorkommt, sondern zeigen vielleicht nur, daß die betreffenden Reaktionen nicht immer spezifisch sind.

In Ceylon wurde die Krankheit zum erstenmal während des II. Weltkrieges festgestellt. In Jamshedpur in Indien kamen von Oktober 1947 bis November 1949 400 klinische Fälle von Tsutsugamushi-Fieber zur Beobachtung, von denen 11 tödlich verliefen (KHAN). In den Jahren 1948, 1949 und 1950 wurden 207, 200 und 105 Fälle serologisch diagnostiziert (SWAMY und DUTTA). Bei den Patienten konnten mehrere Stämme von *R. tsutsugamushi* isoliert werden. Einzelfälle kamen in Punjab zur Beobachtung (MATHUR und SURI). In Bombay reagierten von 19956 in der Zeit von 1945—1953 untersuchten menschlichen Seren 826 mit Proteus-Stämmen, und zwar 776 mit OX 19, 20 mit OX 2 und 90 mit OX K [SOMAN (3)]. Bei den Patienten konnten insgesamt 22 Stämme von *R. tsutsugamushi* isoliert werden.

Der südlichste Punkt der Erde, an welchem bisher Tsutsugamushi-Fieber gefunden wurde, ist der Distrikt Mackay in Queensland (DERRICK, BERRY, TONGE und BROWN). Die hier schon seit längerer Zeit aufgetretenen Erkrankungen an „Sarino-Fieber" sind erst nachträglich als Tsutsugamushi-Fieber erkannt worden. In Nordqueensland konnten von März 1953 bis November 1954 bei 53 Fällen von Tsutsugamushi-Fieber 45 Erregerstämme via Maus isoliert werden (DOHERTY). Zu kleineren Epidemien kam es in Neuguinea (HOOGERHEIDE und ENSINK) und Singapur (LAWLEY). In Südsumatra wurde in der Nähe eines Ölfeldes in den letzten Jahren eine Zunahme der Krankheit konstatiert (DUMOULIN und BRYAN). Die betreffenden Patienten infizierten sich beim Schlafen im Freien. Der Stamm war wenig virulent. Bei 300 Erkrankungen gab es keine Todesfälle. Mehrere Fälle, bei welchen es zu keinen charakteristischen Symptomen kam, wurden nur serologisch diagnostiziert. Von 1950 bis 1951 traten in Hongkong 10 Fälle von Tsutsugamushi-Fieber auf (STEWART). Auch aus Korea sind Erkrankungen gemeldet worden (MUNRO-FAURE, ANDREW, MISSEN und MACKAY-DICK; FULLER und SMADEL; JACKSON, DANAUSKAS, SMADEL, FULLER, COALE und BOZEMAN). Zwei koreanische Stämme konnten isoliert und im Laboratorium genauer untersucht werden.

In Japan hat die Krankheit nach dem II. Weltkrieg zugenommen, bzw. es sind hier neue Verbreitungsgebiete entdeckt worden (SASA). Ein Herd wurde am Fuß des Fudschijama festgestellt. Tsutsugamushi-Fieber kommt auch auf mehreren Inseln an der Südküste von Honshu vor. Ein weiterer Krankheitsherd befindet sich in der Provinz Mie (FUJITA, SOZUKI und HORIKAWA). Die Abgrenzung einer besonderen Rickettsiose in Japan, deren Erreger den Namen *R. tamiyai* erhielt, wurde bereits erwähnt (S. 81). Die Krankheit kommt in

Hokkaido vor und ist hier auch unter dem Namen „Ezo-Fieber" bekannt (TAKIGAMI, KAWAMURA, NISHIOKA und IIDA).

Neuere Arbeiten über die Epidemiologie des Tsutsugamushi-Fiebers beschäftigen sich speziell mit den Reservoiren, mit der Verbreitung, Systematik und Lebensweise der übertragenden Milben. Die wichtigsten Überträger sind *Trombicula akamushi* (Japan und Neuguinea) und *T. deliensis* (Indien, Burma und Neuguinea). Während die Krankheit in Japan am häufigsten im Sommer ist, bestehen in den Tropen und Subtropen keine eindeutigen Vorzugszeiten. In Assam und Burma tritt die Krankheit besonders bei Beginn und am Ende des Monsuns auf. In Japan und Korea kommen Erkrankungen auch im Winter vor. In Jamshedpur (Indien) fallen sie in die Regenzeit von Juni bis November; das Maximum liegt im September (SWAMY und DUTTA). In dem Endemiegebiet wurden 1951 in den Busch- und Grasstreifen um die Stadt 1418 *Rattus rattus*, 33 *Bandicota bengalensis* (indischer Beuteldachs) und 11 Hausmäuse gefangen. 3,4% der Ratten waren von Milben, darunter *T. deliensis*, parasitiert. In Bombay wurden auf 1442 Ratten 3302 Milben gezählt, die zu 3 Arten von *Trombicula*, jedoch nicht zu *T. deliensis* gehörten [SOMAN (3)]. Rickettsien wurden weder hier noch in Jamshedpur in den Milben gefunden. Dagegen konnten in Kaschmir aus 8 Larvenkollektionen (*T. deliensis*) von Ratten 3 Erregerstämme isoliert werden [KALRA und RAO (3)]. In Jubbulpore fanden KALRA und RAO (2) je einen Stamm von Tsutsugamushi-Fieber in Ratten und Mäusen. 70% der Milben auf Ratten, Feldmäusen und Spitzmäusen waren *T. deliensis*.

Eine wichtige Grundlage für epidemiologische Untersuchungen bilden die systematischen Beiträge von WOMERSLEY sowie AUDY (2), (3), (4) über die Trombiculiden im asiatisch-pazifischen Raum. Nach einer Zusammenstellung von HARRISON und AUDY sind hier 87 Säuger (43 Nagetiere, 32 Muriden, 5 Insectivoren) und eine größere Zahl von Vögeln, besonders auf dem Boden lebende Vögel, z. B. Wachteln, als Wirte von *T. akamushi* und *T. deliensis* ermittelt worden. *R. tsutsugamushi* wurde in 14 verschiedenen Nagerarten, an erster Stelle in Ratten, ferner in 2 *Tupaia*-Arten und in *Isoodon torosus* (großer Streifen-Beuteldachs) nachgewiesen. Die Milben sind vor allem an Populationen von *Rattus rattus* gebunden. Andere Rattenarten, z. B. *R. exulans* in Australien und auf den pazifischen Inseln sowie Spitzmäuse und sonstige Mäuse (z. B. *Microtus, Apodemus*) in Japan, sind wichtige Wirte außerhalb der Verbreitungsgrenzen von *Rattus rattus*.

In Assam ist *T. deliensis* der wichtigste Überträger (TRAUB). Rickettsien konnten hier in 36 Testen aus Milben dieser Art isoliert werden. In Burma und Malaya kommen nach AUDY (1) und AUDY und HARRISON als Überträger *T. akamushi*, die mehr auf den Norden und Osten beschränkt ist, und die über das ganze Gebiet verbreitete *T. deliensis* in Frage. Die Milbenlarven sind überwiegend Ektoparasiten kleiner Säuger. Die Milben kommen in kleinen Populationen auf sog. Milbeninseln in direkter Abhängigkeit von der Verbreitung ihrer Wirte vor. Zu den wichtigsten Wirten, den Ratten, kommen als weitere Wirte in den Dörfern und Städten von Burma der indische Beuteldachs (*Bandicota bengalensis*) und Vögel. Eine ganze Anzahl von Säugern und Vögeln ist außerdem wahrscheinlich an der räumlichen Verbreitung der Milben beteiligt. *Rattus muelleri* ist der wichtigste Wirt für *T. deliensis* in den Waldgebieten. Im freien Gelände wurden jedoch auch größere Mengen von Milben an Ratten aus der Gruppe *R. rattus* gefunden. Das Auftreten der Milbe ist sehr eng an die Luft- und Bodenfeuchtigkeit gebunden. Sie findet sich häufiger im Gras- als im Buschgelände.

In einigen Milbeninseln wurde das Vorkommen von *R. tsutsugamushi* nachgewiesen. Im dichten Wald besteht keine Infektionsgefahr. Doch gibt es Herde an Waldrändern, in Waldlichtungen, auf Grasbänken in Flüssen, auf unbearbeiteten Feldern oder in verlassenen Dörfern und Plantagen. Diese Inseln sind die endemischen Zonen des Milbenfleckfiebers, die durch ökologische Schranken, insbesondere die Verbreitungszentren der Ratten, bestimmt und unmittelbar durch das Eingreifen des Menschen in Form von Abholzungen, Siedlungen usw. beeinflußt werden. Im malaiischen Urwald fanden TRAUB, FRICK und DIERKS *R. tsutsugamushi* in *Rattus muelleri* und *R. edwardsi*. Von besonderem Interesse ist der Nachweis von Erregern in Milben der Art *Euschöngastia indica*, die von 3 Eichhörnchen abgesammelt waren. Es spricht manches dafür, daß sich die Krankheit in beschränktem Umfang in den Wäldern von Südostasien in noch unbekannten Reservoiren in Form eines „Waldmilbenfleckfiebers" hält, das in erster Linie durch *T. akamushi* übertragen wird, für das aber noch andere Arten, die wir bisher nicht kennen, als Überträger in Betracht kommen. Von diesen Herden aus dürfte die Krankheit durch Ratten oder andere Transporteure weiterverschleppt werden. Damit würden beim Tsutsugamushi-Fieber interessante Parallelen zur Epidemiologie des Buschgelbfiebers und der Waldpest bestehen.

In Thailand wurde ein Stamm von *R. tsutsugamushi* aus 3 Beuteldachsen, ein weiterer aus *Rattus rattus* isoliert (TRAUB, PHYLLIS, MIESSE und ELBEL). In Djakarta ist der Überträger der Krankheit *T. akamushi* [GISPEN (1)]. Die Milben werden wahrscheinlich von kleinen Säugern oder Vögeln aus ländlichen Bezirken in die Stadt eingeschleppt. Ein Stamm von *R. tsutsugamushi* konnte aus 4 Hausratten isoliert werden. In Südsumatra fanden sich in einem Gebiet, in welchem Tsutsugamushi-Fieber in den letzten Jahren zugenommen hat, *Rattus r. argentiventer* und *R. r. jalorensis* zu 50% mit Milben (*T. akamushi* und *Euschöngastia indica*) infiziert, die Überträger der Rickettsien unter den Muriden sind (DUMOULIN und BRYAN). Aus einer Ratte konnte *R. tsutsugamushi* isoliert werden, andere Ratten waren serologisch positiv. In Korea waren in einem Gebiet 17% von *Apodemus agrarius* mit *R. tsutsugamushi* infiziert (FULLER und SMADEL; JACKSON, DANAUSKAS, SMADEL, FULLER, COALE und BOZEMAN). Der Erreger wurde außerdem in der Milbe *T. pallida* nachgewiesen, mit welcher auch im Experiment eine transovarielle Übertragung der Rickettsien auf Mäuse gelang. *T. pallida* ist in erster Linie Überträger unter den Nagetieren, nur gelegentlich überträgt sie die Krankheit auch auf den Menschen. *T. scutellaris* ist Überträger auf den Izu-Shichito-Inseln im Herbst und Winter, wenn *T. akamushi* fehlt.

Schon frühzeitig war in Japan bekannt, daß hier mindestens 5 *Trombicula*-Arten als Überträger des Tsutsugamushi-Fiebers in Frage kämen (NAGAYO, MIYAGAWA, MITAMURA, TAMIYA und TENJIN). SASA unterscheidet in Japan nach Epidemiologie, Biologie und Ökologie der Überträger 4 Typen von Tsutsugamushi-Fieber. Der klassische Typ (Niigata-Typ), der eine Letalität von 20—45% hat, wird durch *T. akamushi* übertragen. Das Reservoir bildet die Wühlmaus *Microtus montebelli*. Der Kochi-Typ hat eine Letalität von 50—70%. In dieser Form kommt die Krankheit auf den Inseln Shikoku und Awaji vor (YAMGUCHI, HORIE, MIKI und ONO). Hier wurden unter 37831 Milben von 1297 Wirten 19 verschiedene Arten festgestellt. Der wahrscheinlichste Überträger ist *T. tosa*, das Reservoir bildet *Rattus norvegicus.* Bei den Erkrankungen auf mehreren Inseln an der Südküste von Honshu kommt als Überträger *T. scutellaris* in Frage. *T. scutellaris* ist außerdem der Überträger für den Shichito-Typ, während das Reservoir *Rattus norvegicus* und *Apodemus speciosus* bilden. Der Turumi-Typ ist durch einzelne verstreute Fälle gekennzeichnet. Als Überträger ist *T. pallida* verdächtig, der auch als Überträger am Fuß des Fudschijama in Frage kommt und hier an Wieseln und Feldmäusen parasitiert. In der Provinz Mie isolierten FUJITA, SUZUKI und HORIKAWA in 32 Proben von Milzsuspensionen aus 142 wilden Nagern 3 Stämme von *R. tsutsugamushi*. Die Waldmaus *Apodemus speciosus* wurde infiziert gefunden in Gegenden, in denen auch *T. scutellaris* und *T. pallida* vorkommen, z. B. in dem Herd am Fudschijama (HAYASHI). In Fukien (China) waren von 64 Hauskaninchen 16 mit *R. tsutsugamushi* infiziert (YU, LIN und CH'EN). Die Ansteckung war offenbar durch Milben erfolgt, die mit dem Futtergras in die Ställe gebracht worden waren.

Über das Verhalten der Rickettsien in den übertragenden Milben ist nichts Näheres bekannt geworden, jedoch ist der Übertragungsweg noch nicht als restlos geklärt anzusehen. Auffällig ist z. B. der seltene Nachweis von Rickettsien in Nymphen und Adulten. Möglicherweise ist für eine erfolgreiche Übertragung erst eine Aktivierung der Rickettsien notwendig (AUDY und HARRISON). KRISH-NAN, SMITH, BOSE, NEOGY, ROY und GHOSH konnten experimentell durch Füttern von Larven (*T. deliensis*) der F_1-Generation *R. tsutsugamushi* auf Mäuse über-tragen. In den positiven Fällen hatten an den betreffenden Mäusen 38 bis 78 Lar-ven gesogen.

Besondere Beachtung ist der Biologie der Milben im Freien und im Labo-ratorium geschenkt worden. JONES hat den Saugakt bei *T. autumnalis* näher untersucht. Der Vorgang dürfte bei anderen *Trombicula*-Arten ganz ähnlich verlaufen. Die Larven saugen weder Blut noch Lymphe, sondern nur durch Speichel verflüssigte Zellen. Der Speichel greift besonders die Zellen des Stratum Malpighi an; es kommt hierbei zu einer Entzündung und erhöhten Produktion von Intercellularflüssigkeit. Der Saugakt dauert 3—5 Tage. Auch *T. deliensis* bildet beim Saugen am Menschen ein Stylostom aus [FULLER (1)]. Die Haut-reaktionen auf den Stich nichtinfizierter Milben sind unbedeutend und werden normalerweise nicht beachtet. Die Milbe verursacht auch keine Trombidiose wie andere *Trombicula*-Arten.

MEHTA sah die Larven von *T. deliensis* in Indien an ihren Wirten, meist Ratten, 2 bis 4 Wochen festgesogen. Bevorzugt wurden die Region an den Ohren und Augen und haarlose Stellen des Körpers. Die Häutung erfolgte bei 23—25° C in 15 Tagen. Die Nymphen gediehen in einer Tiefe von 15 cm in feuchter Erde, die natürliche organische Zersetzungsbestandteile enthielt. Die Nymphenzeit dauerte 8 Wochen. Die Adulten lebten im Detritus mehrere Monate. Die ersten Larven der F_1-Generation erschienen nach einem Monat. AUDY und HAR-RISON beobachteten bei *T. deliensis* im Laboratorium eine Entwicklungszeit von 5—8 Wochen. Sie nehmen an, daß der Cyclus unter normalen Bedingungen im Freiland 8—12 Wochen beansprucht. AUDY (3) fand in Malaya *T. deliensis* und *T. akamushi*, die überwiegend Säuger-parasiten sind, häufig an Bodenvögeln, besonders Wachteln. *T. akamushi* parasitierte öfter an Wachteln als an Feldratten. *T. deliensis* hat ein größeres Wirtsspektrum.

Von den in Japan nachgewiesenen rund 60 Trombiculiden konnten 15 wenigstens einige Zeit im Laboratorium gehalten werden, darunter auch *T. akamushi* [SASA und MIURA (1), (2) und SASA]. Nymphen und Erwachsene dieser Art wurden mit Eiern von *Culex pipiens pallens* gefüttert. Der Cyclus von der vollgesogenen Larve bis zur Larve der 2. Generation dauerte 3 Monate. Erwachsene Milben lebten unter günstigen Bedingungen bis zu 238 Tagen. Die Larven von *T. scutellaris*, dem Überträger des Shichito-Typs, sind langsamer als die Larven von *T. akamushi* und besitzen eine schwach negative Geotaxis (SUZUKI). Eine Temperatur von 43—46° C ist für sie tödlich, während sie Temperaturen von 0—2° C für 60 Tage aushalten können. Die Larven finden sich besonders in feuchten Waldungen und in sumpfigem buschreichen Gelände. In den Siedlungen sind sie selten. Weitere Angaben über die Biologie japanischer Laufmilben, darunter *T. scutellaris, T. akamushi* und eine *Neoschöngastia*-Art, finden sich bei Fukuzumi, Obata und Kagiwada, ferner Sasa, Tanaka, Ueno und Miura sowie Sasa, Ueno, Miura und Tanaka.

Eine gewisse epidemiologische Bedeutung kommt der Beobachtung zu, daß die Erreger des Tsutsugamushi-Fiebers nicht nur im Versuchstier (s. S. 94), sondern auch im Menschen längere Zeit persistieren können, so daß theoretisch Rückfälle nach Art der Brill-Zinsserschen Krankheit denkbar sind. Smadel, Ley, Diercks und Cameron konnten in Versuchen mit Freiwilligen, die an Tsutsugamushi-Fieber erkrankt waren, in einem Fall *R. tsutsugamushi* in axillaren Lymphknoten nachweisen. Die Erkrankung lag bereits 15 Monate zurück.

3. Durch Insekten übertragene Rickettsiosen
a) Murines Fleckfieber

Das murine Fleckfieber ist auf tropisches und subtropisches Gebiet beschränkt. Die weiteste *Verbreitung* hat die Krankheit in den USA. Die Anzahl der beim Menschen bekanntgewordenen Fälle wird hier für die Zeit von 1931—1946 mit etwa 42000 angegeben (Snyder). Die Zahl der Erkrankungen stieg von 510 im Jahre 1930 auf 5401 im Jahre 1944 (Pratt und Good). Die Morbidität ging dann stark zurück, in erster Linie durch den Einsatz von DDT zur Bekämpfung der Rattenflöhe. 1949 gab es 983, 1950 686 und 1952 nur noch 186 Fälle. In Georgia sank die Zahl der Kranken von 1256 im Jahre 1943 auf 41 im Jahre 1953 (McCroan, Ramsey, Murphy und Dick).

Aus anderen Teilen der Welt sind im Vergleich zu den USA relativ wenige Erkrankungen an murinem Fleckfieber bekannt geworden. In Uganda wurden von 1947—1950 444 sporadische Fälle serologisch diagnostiziert (Baird und Tonkin). Bei 4 von 33 Patienten konnten Erreger isoliert werden. Eine Anzahl von Fällen wurde in den letzten Jahren in Baku beobachtet (Imamaliew). In Kaschmir kamen innerhalb von 2 Jahren 178 Erkrankungen an murinem Fleckfieber mit einer Mortalität von 8—17% vor, und in Bengalore und Mysore wurden von 1942—1947 145 Patienten mit murinem Fleckfieber im Hospital behandelt [Kalra und Rao (3), (4)]. Murines Fleckfieber wurde auch in der Punjab-Ebene (Mathur) und in Goleonda (Ali und Subrahmanian) nachgewiesen. Auf dem Hochplateau von Zentralannam wurden von 1945—1954 99 Fälle gezählt (Sureau, Rousilhon und Capponi). Für Java sind von 1929—1950 284 Erkrankungen angegeben (Gispen, Gan und Westermann). In Westaustralien wurden von 1927—1952 1332 Erkrankungen an murinem Fleckfieber beobachtet, von denen 50 tödlich endeten (Saint, Drummond und Thorburn). Bis 1942 kamen jährlich im Durchschnitt 30 Fälle vor, in den folgenden Jahren waren es 60—140 Erkrankungen. In Japan entwickelte sich eine kleine Epidemie im Anschluß an eine Rattenbekämpfungsaktion (Kitaoka, Takemori, Shishido und Jo).

Das murine Fleckfieber tritt normalerweise nur in *endemischer* Form auf. Es handelt sich überwiegend um sporadische Fälle, sehr selten kommt es zu Gruppenerkrankungen. Die Bindung an die Nagetiere und deren Flöhe bringt das vermehrte Auftreten in Hafenstädten, in Lebensmittelbetrieben, auf Bauernhöfen und an anderen von Ratten frequentierten Plätzen mit sich. In den Tropen hat die Krankheit keine Vorzugszeiten, in den Subtropen kommen die meisten Fälle in der warmen Jahreszeit vor. Auch in Baku traten das ganze Jahr über Erkrankungen auf. Bevorzugt waren die Monate August bis Februar mit einem Gipfel im Dezember (Imamaliew). Auffällig ist, daß hier 5,7% der Patienten vorher eine Erkrankung an klassischem Fleckfieber durchgemacht haben sollen.

Einige dieser Fälle verliefen schwer. Frauen erkrankten häufiger und schwerer als Männer. In Südwestgeorgia betrafen 70% aller Erkrankungen Farmerfamilien (STEWART und HINES). In den ländlichen Bezirken erkrankten die meisten Personen von Mai bis August, während sich die Fälle in den Städten gleichmäßig über das ganze Jahr verteilten. In Indien traten die meisten Fälle zwischen März und August auf. In Java ist das murine Fleckfieber über die ganze Insel verbreitet und auf dem Lande ebenso häufig wie in den Städten. Bei Tokio wurden 2 Familienepidemien beschrieben. In einem Fall erkrankten alle Familienmitglieder in wenigen Tagen, in einem anderen 10 Personen aus einer Familie innerhalb von 2 Monaten. Auch in Baku wurden in 10 Familien Erkrankungen von 2—3 Mitgliedern beobachtet.

Als *Reservoire* des murinen Fleckfiebers sind aus allen Verbreitungsgebieten in erster Linie Ratten, insbesondere die Wanderratte *Rattus norvegicus*, bekannt. Anfänglich wurde auch der Katze eine gewisse Bedeutung beigemessen, Katzen sind jedoch für den Erreger nur schwach empfänglich und erkranken meist inapparent (LÉPINE und LORANDO). Murines Fleckfieber ist unter natürlichen Verhältnissen nur ausnahmsweise in Katzen nachgewiesen worden. RAYNAL konnte in Shanghai 1942 in einem Jahr, in welchem es ungewöhnlich viel murines Fleckfieber bei Ratten und Menschen gab, in 3 von 25 Katzen *R. mooseri* nachweisen. Auch Mäuse können Rickettsienträger sein. SPARROW isolierte zum erstenmal in Tunis aus 300 Mäusen 2 Stämme von murinem Fleckfieber. BRIGHAM (1) fand *R. mooseri* in einer wilden Oldfield-Maus (*Peromyscus polionotus*) in Alabama. Seitdem sind wiederholt Stämme von murinem Fleckfieber aus Mäusen isoliert worden, z. B. in China (LIU und ZIA (1)]. In Texas wurden 2 Stämme von *R. mooseri* aus in Wohnungen gefangenen Hausmäusen isoliert (KEATON, NASH, MURPHY und IRONS). Alle im Freien gefangenen Hausmäuse waren negativ, auch serologisch. SMITH prüfte in Mississippi die Seren von 995 Hausmäusen auf das Vorhandensein von Antikörpern gegen murines Fleckfieber. Nur ein Test war positiv. Es handelte sich dabei um 2 Mäuse, die mit infizierten Hausratten Kontakt hatten. In Kalifornien wurden unter 126 Hausmäusen 5 seropositive gefunden. Im ganzen ist jedenfalls die Rolle der Mäuse als Reservoir für murines Fleckfieber im Vergleich zu den Ratten gering.

Infizierte *Ratten* waren in den USA am häufigsten in Gebieten mit einer mittleren Januartemperatur von über 8,3° C bei hoher Luftfeuchtigkeit und am seltensten in Gegenden mit einer mittleren Januartemperatur unter 4,4° C (MOHR, GOOD und SCHUBERT). Sie fanden sich auch in den Städten. 1945 hatten etwa 45% der gefangenen Ratten eine positive KBR, 1952 (nach DDT-Einsatz) waren es nur noch 6,7%. Unter 675 in Oklahoma untersuchten Ratten waren 20 in der KBR positiv (MORLAN, UTTERBACK, DENT, WILCOMB, GRIFFITH und ELLIS), in Kalifornien 105 von 941 *R. norvegicus*, 7 von 101 *R. rattus* und 4 von 151 *R. alexandrinus*. In Galvestone (Texas) fanden sich 1945 unter 141 *R. norvegicus* 63% seropositive, 1946 unter 131 Ratten 58,8% und 1947 unter 110 Ratten 32,7% (STRANDTMANN und EBEN). Weitere Beobachtungen über Isolierung von *R. mooseri* aus Ratten (speziell *R. rattus* und *R. norvegicus*) außerhalb der USA liegen u. a. vor aus Brasilien (TRAVASSOS, PEREIRA und VASCONCELOS), Portugiesisch-Guinea (TENDEIRO), aus Baku am Schwarzen Meer (IMAMALIEW), aus Tunis (VERMEIL, CHENE und ZAIBI), Ägypten und dem Sudan (TAYLOR,

Kingston und Rizk), aus Indien [Soman (1), Kalra und Rao (3), (4)] und Indonesien [Gispen (2); Gispen, Gan und Westermann; Gispen und Warsa]. In Baku waren in einigen Bezirken 33% der Ratten infiziert, in Bombay von 323 *R. rattus* 19,2%, von 193 *R. norvegicus* 45% serologisch positiv [Soman (1)].

Rickard (1) und Rickard und Worth beschäftigten sich mit der Rolle von *Baumwoll-ratten* (*Sigmodon hispidus*) als Reservoir des murinen Fleckfiebers im Vergleich zur Haus-ratte (*R. rattus*). Von 635 experimentell infizierten Hausratten waren 6 in der KBR negativ, von 238 Baumwollratten 4. Ein Vergleich von KBR, Agglutination und Weil-Felix-Reaktion ergab, daß die KBR für den Nachweis einer Infektion mit murinem Fleckfieber unter wilden Ratten am günstigsten ist. Bei diesen Tieren ist bereits ein KBR-Titer von 1:8 beweisend, während Titer bis zu 1:64 bei der Baumwollratte noch nichts aussagen.

Diese Beobachtungen wurden durch Freilanduntersuchungen unterbaut. Die Seren von 172 Baumwollratten wurden in Florida in einer Gegend geprüft, in welcher murines Fleckfieber in *R. rattus* und *R. norvegicus* vorkommt. 14,5% der Seren von Baumwollratten waren zwar positiv, jedoch nur mit einem Titer von 1:8 bis 1:64. Bei den anderen Ratten erreichte der Titer 1:128 bis 1:2048. Die experimentelle Infektion der seropositiven Ratten ergab, daß die Haus- und Wanderratten immun waren, die Baumwollratten nicht. Baumwollratten kommt danach als Reservoir des murinen Fleckfiebers keine Bedeutung zu.

Die Suche nach *weiteren tierischen Reservoiren* hatte größtenteils negative Ergebnisse. Morlan, Hill und Schubert untersuchten in einem begrenzten Gebiet von Südwest-Georgia 3202 Seren von 27 Säuger- und 10 Vogelarten auf ihren Gehalt an komplementbindenden Anti-körpern für murines Fleckfieber und fanden 47 positive Reaktionen bei insgesamt 12 Tier-arten, und zwar bei Baumwollschwänzchen, Opossum, Fuchshörnchen, Hausmaus, Reisratte, Baumwollmaus, Oldfield-Maus, Baumwollratte, Hund, Stinktier, Wiesel und einem Häher. In Kalifornien waren die Seren von 51 Erdhörnchen negativ. In Texas wurden 480 Seren von kleinen Säugern, die sich auf 11 Arten verteilten, serologisch untersucht (Keaton, Nash, Murphy und Irons). Positive Reaktionen wurden bei einer Ratte (*Neotoma floridana*), einer Baumwollratte und einem Opossum gefunden. In Bombay waren von 208 Beuteldachsen *Gunomys kok*) 6,2% in der KBR positiv [Soman (1)], in Bengalore wurde ein Stamm von *R. mooseri* aus einem Beuteldachs isoliert, der in einem Fleckfieberhaus gefangen war [Kalra und Rao (3)].

Die Epidemiologie des murinen Fleckfiebers wird einerseits durch die Ver-breitung und Lebensweise der als Reservoire auftretenden Nagetiere, insbesondere der Ratten, bestimmt, andererseits durch die als *Überträger* fungierenden Ekto-parasiten der Ratten. Wichtigster Überträger ist der tropische *Rattenfloh*, *Xenopsylla cheopis*, in welchem auch am häufigsten der Erreger des murinen Fleckfiebers nachgewiesen werden konnte. Über die Ektoparasiten der Ratten und ihre Bedeutung für das murine Fleckfieber sind vor allem in den USA in den letzten Jahren umfangreiche Untersuchungen durchgeführt worden, in welchen u. a. die Parasitenfunde bei über 20000 Hausratten ausgewertet wurden (Morlan, Utterback, Dent, Wilcomb, Griffis und Ellis u. a.). In Oklahoma waren die seropositiven Hausratten durchschnittlich mit 6 *X. cheopis* besetzt, in Kalifornien wurden in 133 Versuchen 5 Stämme von *R. mooseri* aus Rattenflöhen isoliert. Auch in Galvestone ist *X. cheopis* der häufigste Floh auf der Wanderratte (Strandtmann und Eben). In 74 Versuchen mit kleinen Mengen von Flöhen wurde 19mal *R. mooseri* gefunden. In San Antonio (Texas) war *Rattus norvegicus* häufiger und stärker parasitiert als *R. rattus* [Davis (2)]. *X. cheopis* hatte sein Häufigkeitsmaximum von Mai bis August. Die meisten Erkrankungen traten von Juli bis Oktober auf.

In Bengalore wurden 3 Stämme von *R. mooseri* aus einer Kollektion von *X. cheopis*, *X. brasiliensis* und *X. astia* gewonnen, in Mysore ein Stamm aus *X. brasiliensis*, in Kaschmir ein Stamm aus *X. cheopis* und *X. astia*. Alle Flöhe stammten von Ratten [KALRA und RAO (3)]. In Djakarta isolierten GISPEN und WARSA aus 129 *X. cheopis*, die von 37 Ratten abgesammelt worden waren, 5 Stämme von *R. mooseri*.

Andere Flöhe und Ektoparasiten haben nur nebensächliche Bedeutung, obwohl eine große Zahl von Floharten durchaus geeignete Wirte für *R. mooseri* sind. An weiteren Flöhen wurden in den USA auf *Rattus rattus* und *R. norvegicus* der Mäusefloh *Leptopsylla segnis*, der nordische Rattenfloh *Nosopsyllus fasciatus* und der Hühnerkammfloh *Echidnophaga gallinacea* gefunden (PRATT und GOOD). In Galvestone isolierten STRANDTMANN und EBEN aus *Leptopsylla segnis*, der hier auf Ratten viel seltener ist als *X. cheopis*, in 16 Versuchen 2 Stämme von *R. mooseri*. Auch in Peking wurde *R. mooseri* in Mäuseflöhen nachgewiesen [LIU und ZIA (1)]. In Texas konnte ein Stamm von *R. mooseri* aus Katzenflöhen isoliert werden, die von einem Opossum stammten (KEATON, NASH, MURPHY und IRONS). Umfangreiche Untersuchungen an Hunden- und Katzenflöhen in Galvestone brachten nur negative Ergebnisse. Jedoch sind auch aus dem Hühnerkammfloh Stämme von *R. mooseri* isoliert worden, so von BRIGHAM (2) in Georgia und Albany, von STRANDTMANN und EBEN in Galvestone. Die meisten Versuche zum Nachweis von Rickettsien in Rattenmilben (*Bdellonyssus bacoti*, *Laelaps nuttalli*) und Rattenläusen (*Polyplax spinulosa*) verliefen negativ [WORTH und RICKARD (2), GISPEN und WARSA; STRANDTMANN und EBEN). Auffällig sind die negativen Befunde bei der Rattenlaus, da sich dieser Parasit im Experiment leicht infizieren ließ (MOOSER, CASTA EDA und ZINSSER). In China wurde *R. mooseri* in *Bdellonyssus bacoti* nachgewiesen [PANG sowie LIU (1, 2)] (vgl. S. 98).

Durch zahlreiche Versuche ist bekannt, daß sich *R. mooseri* ebensogut in *Läusen* entwickelt wie *R. prowazeki*. KALRA und RAO (4) konnten unter natürlichen Bedingungen in Kaschmir 2 Stämme von murinem Fleckfieber aus Läusen isolieren, die von Patienten abgesammelt waren. LIU und ZIA (2) und LIU (1) haben in China ebenfalls bei isolierten Fällen von Fleckfieber *R. mooseri* in Kleiderläusen gefunden. Bei einer starken Verlausung ist die Übertragung von murinem Fleckfieber durch Kleiderläuse ohne weiteres möglich, zumal dort, wo klassisches und murines Fleckfieber nebeneinander vorkommen, wie z. B. in Mexiko (FREEMANN, VARELA, PLOTZ und ORTIZ MARIOTTE). Die epidemiologische Bedeutung dieser Tatsache dürfte aber gering sein.

Als neuer Zwischenwirt für murines Fleckfieber ist eine *Milbe* bekannt geworden. GISPEN (2) konnte in Java 5 Stämme von *R. mooseri* aus 2598 Milben der Art *Schöngastia indica* isolieren, die auf Ratten parasitiert hatten. 4 Stämme kamen aus Milben, die auf nichtinfizierten Ratten (*R. rattus diardi* und *R. norvegicus*) parasitierten. Die Stämme verhielten sich wie andere javanische und wie amerikanische Stämme von *R. mooseri*. Die betreffende Milbe, die auch zur Familie der *Trombiculidae* gehört und in ihrer Biologie *Trombicula* ähnlich ist, geht nicht an den Menschen, lebt aber sehr häufig auf Ratten. Sie fand sich bei 49,7% der Ratten, und der Milbenindex betrug 15,6. KALRA und RAO (2) konnten in Jubbulpore in Indien einen Stamm von murinem Fleckfieber aus der Zecke *Boophilus australis* isolieren. Dies wäre der erste Nachweis von *R. mooseri* in einer Freilandzecke. GISPEN und WARSA prüften in Java 661 Zecken, die sie von 45 Ratten gesammelt hatten, und fanden keine Rickettsien.

Die seltenen Funde von *R. mooseri* in anderen Wirten als Flöhen haben epidemiologisch wahrscheinlich wenig Bedeutung. Sie besagen auch nichts über die Eignung der betreffenden Tiere als Wirte für *R. mooseri*, zumal wenn die Untersuchung erfolgte, solange die Parasiten frisches Blut ihres Rattenwirtes enthielten;

denn die Rickettsien können auch im Magen ungeeigneter Wirte einige Zeit überleben.

Unklarheit herrscht noch über den Weg, auf welchem die *Infektion* von Ratten und Menschen mit *R. mooseri* zustande kommt. Es wurde angenommen, daß der Mensch sich mit Lebensmitteln infiziert, die durch rickettsienhaltigen Rattenurin verunreinigt sind. Rickettsien sind jedoch im feuchten Zustand nur kurze Zeit lebensfähig. Bei einer experimentellen peroralen Infektion erkrankte von 5 Freiwilligen, die mit *R. mooseri* infizierte Dottersacksuspensionen verspeisten, nur eine Person (Pollard, Wilson, Livesay und Woodland). Rickard (2) versuchte, Hausratten durch Futter zu infizieren, das mit rickettsienhaltigen Flohfaeces verunreinigt war. Keine von den 20 Versuchsratten erkrankte, und die KBR fiel bei allen Tieren negativ aus. Violle hat angegeben, daß er murines Fleckfieber auf Ratten übertragen konnte, wenn er die Tiere mit Brot fütterte, das mit dem Urin von kranken Ratten und Menschen getränkt war. Dem stehen aber mehrere Beobachtungen entgegen, daß infizierte Ratten und Mäuse normalerweise keine Rickettsien mit dem Urin ausscheiden und daß sich *R. mooseri* in Ratten- oder Mäuseurin höchstens 1 Std lebend hält [Worth und Rickard (1), Özbil]. Worth und Rickard hielten parasitenfreie Ratten unter Bedingungen, bei denen die Ratten und ihr Futter für längere Zeit intensiven Kontakt mit dem Urin infizierter Ratten hatten. Auch hierbei fand keine Übertragung statt. Eine Infektion kam in ganz beschränktem Umfang nur durch Beißen und Kannibalismus bei starker Überbevölkerung zustande, wie sie unter natürlichen Bedingungen nicht vorkommt.

Die wesentlichste Bedeutung für die Übertragung haben die Flöhe und die von ihnen ausgeschiedenen rickettsienhaltigen Faeces. Flohfaeces dürften auch die wichtigste Infektionsquelle für den Menschen bilden, wobei die Rickettsien, ähnlich wie beim klassischen Fleckfieber, in Hautläsionen eingerieben werden oder auf die Schleimhäute gelangen. Infektiöser Staub, der eingeatmet wird, könnte noch eher eine Infektionsquelle bilden als Nahrungsmittel, die mit Flohfaeces verunreinigt sind (Imamaliew).

Die enge Verwandtschaft zwischen *R. mooseri* und *R. prowazeki*, die sich u. a. in der antigenen Struktur, dem Verhalten in den Überträgern und der Reaktion der empfänglichen Wirte äußert, hat frühzeitig den Gedanken aufkommen lassen, daß sich das klassische Fleckfieber aus dem murinen herleitet. Mooser (3) hat die Hypothese vom murinen Ursprung des klassischen Fleckfiebers eingehend begründet, aber auch ihre Gegenargumente zur Debatte gestellt. Die Hypothese besagt, daß das klassische Fleckfieber das Produkt einer langdauernden Übertragung des murinen Fleckfiebers im Cyclus Mensch — Laus — Mensch sei. Die verfeinerten Differenzierungsmethoden führten zu dem Schluß, daß sich die Erreger des murinen und klassischen Fleckfiebers wohl biologisch nahestehen und wahrscheinlich auf eine gemeinsame Wurzel zurückgehen, daß sie jetzt aber genetisch getrennte Arten oder Unterarten darstellen [Mooser (4)]. Diese Ansicht wäre erschüttert, wenn sich die Beobachtungen von Price, Emerson, Nagel, Blumberg und Talmadge bestätigen sollten, daß sich *R. prowazeki* unter bestimmten experimentellen Bedingungen in *R. mooseri* verwandeln läßt.

Auch *R. mooseri* persistiert sehr lange in den Organen der infizierten Versuchstiere (S. 94). Spätrückfälle von murinem Fleckfieber beim Menschen sind jedoch bisher nicht bekannt geworden. BENOIST, GIROUD und HÉRAUD diagnostizierten bei einem aus Afrika stammenden, in Paris beschäftigten Arbeiter murines Fleckfieber. Der betreffende Patient hatte 2 Jahre vorher in Nordafrika ein schweres Fleckfieber überstanden. Es blieb ungeklärt, ob die erste Erkrankung ein klassisches, die zweite ein in Paris erworbenes murines Fleckfieber war, oder ob es sich um den endogenen Rückfall eines in Nordafrika durchgemachten murinen Fleckfiebers gehandelt hat.

b) Klassisches Fleckfieber

Seit Einführung einer spezifischen Therapie mit antibiotischen Präparaten und einer vereinfachten erfolgreichen Überträgerbekämpfung durch synthetische Insecticide hat sich das Bild von der geographischen *Verbreitung* des klassischen Fleckfiebers stark geändert. Die letzten größeren Epidemien datieren aus der Zeit nach dem II. Weltkrieg mit ihren starken Menschenfluktuationen. Eingehend ist das Fleckfieber in Italien, speziell in Neapel, beschrieben worden (CHALKE; WHEELER; CRAUFURD-BENSON; SOPER, DAVIS, MARKHAM und RIEHL). Über die gegenwärtige Verbreitung des Fleckfiebers lassen sich keine genauen Angaben machen. Es ist aber damit zu rechnen, daß in den früher wichtigen Verbreitungsgebieten in Ost- und Mittelasien, Osteuropa, Nordafrika, Äthiopien, Mexiko und Südamerika noch kleine Herde zurückgeblieben sind, von denen aus sich die Krankheit unter günstigen Bedingungen erneut weiter ausbreiten könnte. Die Verschiebung des Verbreitungsbildes unter dem Einfluß der Bekämpfung und die Situation nach dem Kriege können einige Beispiele beleuchten.

In Japan wurden 1946 30000 Fälle von klassischem Fleckfieber mit einer Mortalität von 7—10% gezählt (BLANTON und TANI). In der südkoreanischen Zivilbevölkerung traten zu Beginn des Jahres 1951 im Zusammenhang mit den kriegerischen Auseinandersetzungen 3228 Erkrankungen monatlich auf, von denen 473 letal ausgingen (FULLER und SMADEL). Durch Vorbeuge- und Bekämpfungsmaßnahmen war die Zahl in der ersten Hälfte des Jahres 1952 auf durchschnittlich 125 im Monat zurückgegangen. In Kaschmir wurde Fleckfieber wahrscheinlich während des Krieges durch russische Flüchtlinge eingeschleppt. In den Jahren 1943—1944 erkrankten hier 2481 Personen, von denen 730 starben [KALRA und RAO (1)].

Im Distrikt Bialystok in Polen hatte das Fleckfieber bis 1950 noch Epidemiecharakter (KOŁŁOTO). Danach traten im wesentlichen sporadische Fälle auf. 1954 wurden 33 Erkrankungen gezählt, von denen 9 Spätrückfälle waren (vgl. S. 134). In Jugoslawien kam es im Frühjahr 1942 unter den Flüchtlingen zu einer Epidemie mit über 10000 Fällen (DJOURICHITCH), die trotz der schlechten Lebensbedingungen nur eine Letalität von 4% hatte. Eine Epidemie in Barcelona (BORELL) umfaßte 2155 Kranke; die Letalität erreichte bei den Männern 21,6%.

In Mexiko betrug die Morbidität im Jahre 1951 4,8 auf 100000 Einwohner gegenüber 14,4 im Jahre 1943 und 10,4 im Jahre 1947 (ORTIZ MARIOTTE). In Guatemala wurden in einem Teil des Landes bei einer Bevölkerung von 1,2 Millionen folgende Zahlen für klassisches Fleckfieber festgestellt: 1943 1338 Fälle, 1944 2144 Fälle, 1945 2834 Fälle, 1946 (in diesem Jahr setzten größere Bekämpfungsmaßnahmen ein) 1043 Fälle, 1947 251 Fälle, 1948 69 Fälle, 1949 26 Fälle, 1950 10 Fälle und 1951 8 Fälle (CABRERA, McANALLY und MONTOYA).

In den Kriegs- und Nachkriegsjahren nahm klassisches Fleckfieber auch in Nord- und Südafrika zu (FREYCHE und DEUTSCHMAN). 1942 wurden in Nordafrika über 85000 Kranke gezählt, 1943 in Ägypten 40188 Patienten, 1944 in Südafrika 5623 Patienten. Inzwischen ist die Morbidität stark abgesunken. So sind 1949 aus Ägypten nur 186 und aus Südafrika 244 Fälle bekannt geworden. Ein Teil der Erkrankungen, die zwischen 1951 und 1954 in Ägypten und im Sudan auftraten, werden als Rückfälle angesehen (TAYLOR, KINGSTON und RIZK).

Der *Übertragungsmechanismus* beim klassischen Fleckfieber ist ein fester Bestandteil unseres Wissens. Wichtigster Überträger ist die Kleiderlaus *Pediculus humanus humanus* (vgl. MUESBECK), obwohl die Kopflaus (*Pediculus humanus capitis*) und die Filzlaus (*Phthirus pubis*) ebenfalls als Überträger geeignet sind

[Weyer (9)]. Ebenso wie beim murinen Fleckfieber bildet der reichlich mit Rickettsien durchsetzte Kot den Ansteckungsstoff. Der Kot der Läuse wird kurz nach der Abgabe staubförmig. Die dem Kotstaub beigemischten Rickettsien bleiben unter normalen Bedingungen wenigstens einige Tage vermehrungsfähig und virulent (s. S. 90). Eintrittspforten für die Rickettsien sind die Schleimhäute oder kleinste Verletzungen der Haut.

Zu den wesentlichen Fortschritten auf dem Gebiet der Epidemiologie des klassischen Fleckfiebers gehören die neuen Erkenntnisse über das Wesen der isolierten *sporadischen Fleckfieberfälle*, d. h. der Brill-Zinsserschen Krankheit. Es können heute kaum noch Zweifel daran bestehen, daß die Brill-Zinssersche Krankheit der *Spätrückfall* oder die Spätmanifestation eines klassischen Fleckfiebers ist. Wir müssen annehmen, daß sich bei manchen Personen nach Überstehen der Krankheit eine Prämunition unter Verbleiben der Erreger in den inneren Organen einstellt und daß das gewonnene Gleichgewicht zwischen Wirt und Parasit zugunsten des Parasiten sehr viel später durch eine verminderte Resistenz des Wirtsorganismus — Rückfälle treten z.B. im Zusammenhang mit anderen Erkrankungen wie Meningitis, Tuberkulose und Malaria auf — oder aus sonstiger, uns vorerst noch nicht bekannter Ursache, gestört werden kann. Damit kommt es zu einer erneuten Rickettsiämie und einem von klinischen Symptomen begleiteten Rückfall. Daß Rickettsien längere Zeit nach Überstehen einer Erkrankung im Wirtskörper verbleiben können, wird durch eine große Zahl von Tierversuchen gestützt (vgl. S. 94).

Problematik und Kasuistik der Brill-Zinsserschen Krankheit wurden in der letzten Zeit näher behandelt von Worms, von Murray und Snyder (1), (2), Weyer (7), (14), Mooser (4), Jusatz sowie v. Knorre u.a. Fälle von Brill-Zinsserscher Krankheit sind nicht nur aus Nordamerika und Europa, sondern auch aus Südafrika (Gear, Wolstenholme, Harwin und Stakes) und Brasilien (Meira, Jamra und Lodovici) bekannt geworden.

Die Diagnose konnte bei einigen dieser Erkrankungen auch ätiologisch gesichert werden. Soweit eine Erregerisolierung gelang und die Erreger hinterher näher untersucht wurden, besaßen sie alle typischen Merkmale und Eigenschaften von *R. prowazeki* [Murray und Snyder (1), (2), Giroud und Gaillard; Wojciechowski und Mikołajczyk; Weyer und Hornbostel]. Von besonderer epidemiologischer Bedeutung ist, daß die Rickettsiämie bei Brill-Zinsserscher Krankheit für eine Infektion von Läusen ausreicht, so daß solche Fälle in einem verlausten Milieu zum Ausgangspunkt weiterer Erkrankungen und neuer Epidemien werden können [Murray und Snyder (1), Gaon; Weyer und Hornbostel). Damit dürfte auch die alte Frage geklärt sein, wo das Fleckfieber in den interepidemischen Phasen bleibt. Die Theorie der Erregerpersistenz in Läusen oder Läusefaeces kann heute als überholt gelten. Auch inapparente Erkrankungen können die Erhaltung der Rickettsien nicht sichern. Als *Reservoir* für das klassische Fleckfieber müssen wir vielmehr die wenigen menschlichen Dauerträger von Rickettsien ansehen, bei welchen es noch nach Jahrzehnten zu einem echten Rückfall und damit zu einer Rickettsiämie kommen kann, die die Möglichkeit zum Übergreifen des Fleckfiebers auf neue Wirte einschließt. Das klassische Fleckfieber stellt somit neben dem Wolhynischen Fieber die einzige Rickettsiose dar, die auf den Menschen beschränkt und an den Menschen gebunden ist.

Die Ansicht, daß es beim klassischen Fleckfieber Spätrückfälle gibt, fand eine wesentliche Stütze in einer Beobachtung von PRICE (5), der bei 2 von 31 Patienten in Amerika anläßlich von Bauchoperationen Stämme von *R. prowazeki* aus den inguinalen Lymphdrüsen isolieren konnte. Die Patienten waren vor 28 bzw. 42 Jahren aus Rußland eingewandert und hatten mindestens 16 bzw. 19 Jahre vor der Operation keine Gelegenheit zu einer Neuinfektion gehabt. In ihrem Blut wurden komplementbindende und toxinneutralisierende Antikörper nachgewiesen (s. auch PRICE, EMERSON, NAGEL, BLUMBERG und TALMADGE).

Auch serologische Untersuchungen in Amerika haben die Deutung der Brill-Zinsserschen Krankheit als Fleckfieberrezidiv weiter gefestigt. SIGEL, WEISS, BLUMBERG und DOANE stellten bei 12 von 69 in Europa geborenen Personen, die seit 28—53 Jahren in Amerika lebten, komplementbindende Antikörper fest. 80 Kontrollen bei in den USA geborenen Personen fielen negativ aus. MURRAY, COHEN, JAMPOL, OFSTROCK und SNYDER führten entsprechende Untersuchungen in Boston durch. Unter 272 Einwohnern, die aus Fleckfiebergebieten Europas stammten, besaßen 51 komplementbindende und toxinneutralisierende Antikörper im Blut. Bei 11 Personen erreichte die KBR noch einen Titer von 1:80 bis 1:160. Die entsprechenden Kontrollen bei 247 Personen aus USA oder Kanada waren negativ. Ähnliche Beobachtungen machten SCHAEFER, FRIEDMAN und LEWIS. Nach PRICE, EMERSON, NAGEL, BLUMBERG und TALMADGE hatten von 1708 Personen, die in den letzten 25 Jahren aus Ost- und Südosteuropa nach den USA kamen, wenigstens 30% Antikörper gegen Fleckfieber. 201 in den USA geborene Kinder aus der gleichen Personengruppe waren dagegen negativ, ebenso 481 in den Staaten geborene Erwachsene aus derselben Gegend. Als empfindlichster Test für den Nachweis einer längere Zeit zurückliegenden Infektion mit *R. prowazeki* wird die Antikörper-Reaktion auf eine Injektion von Cox-Vaccine benutzt.

Bei einer Patientin mit Brill-Zinsserscher Krankheit, bei welcher eine Erregerisolierung gelang und die Übertragung des Stammes auf den Rhesusaffen eine positive Weil-Felix-Reaktion auslöste, waren die komplementbindenden und neutralisierenden Antikörper kurze Zeit vor Beginn der klinischen Symptome verschwunden. 308 Personen mit komplementbindenden Antikörpern wurden daher 3 Jahre lang in Abständen von 3 Monaten serologisch kontrolliert, wobei die KBR nur geringfügige Schwankungen zeigte. Die neutralisierenden Antikörper verschwanden jedoch bei 8 Patienten teilweise für 1 Jahr. Trotzdem kam es zu keinen Erkrankungen. Es ist durchaus fraglich, ob alle Personen mit komplementbindenden Antikörpern auch Rickettsienträger sind. Unbekannt ist, in welcher Form die Rickettsien im Menschen überleben und eine latente Infektion unterhalten.

Zu der Erklärung der sporadischen Fleckfiebererkrankungen als Spätrückfälle wurde von verschiedenen Seiten kritisch Stellung genommen [v. BORMANN (1), (2), SATROW; SDRODOWSKI; PSCHETNITSCHNOW und RAICHER; TARABAN und KOSOWSKI; JATZIMIRSKAYA-KIRONTOWSKAYA u. Mitarb. u. a.], ohne daß allerdings hierdurch die entscheidenden Argumente entkräftet wären. So nimmt z. B. v. BORMANN an, daß Rickettsien in Läusen 3—4 Monate überleben können. Diese Angabe widerspricht allen diesbezüglichen Erfahrungen. TARABAN und KOSOWSKI fanden keine prinzipiellen epidemiologischen Unterschiede zwischen Erst- und Zweiterkrankungen und halten es für möglich, daß die klinischen Unterschiede zum Teil mit den Eigenschaften der Erreger und einer veränderten Reaktionslage des befallenen Organismus zusammenhängen. JATZIMIRSKAYA-KORONTOWSKAYA u. Mitarb. versuchten vergeblich, aus Sektionsmaterial von 25 Patienten, die in früheren Jahren Fleckfieber überstanden hatten, *R. prowazeki* zu isolieren. Die negativen Ergebnisse können damit zusammenhängen, daß die Technik — Übertragung von Organsuspensionen auf Mäuselungen — für den Nachweis der Erreger nicht empfindlich genug war. Außerdem darf man nicht vergessen, daß die Erregerpersistenz beim Fleckfieber nur einen kleinen Prozentsatz von

Patienten betrifft. Ihre Ermittlung wird daher einen besonderen Zufall darstellen. Auch Price (5) hatte mit seinen Versuchen bei 29 Patienten keinen Erfolg. Die Unterscheidung von Rückfällen und Zweiterkrankungen ist in Gebieten mit endemischem Fleckfieber nicht immer leicht zu treffen. Trotzdem liegen auch aus Jugoslawien (Murray u. Mitarb., Gaon), Rußland, der Tschechoslowakei (Mittermayer und Kratochvíl) und besonders aus Polen [Kostrzewski (2), (3)] zahlreiche Beobachtungen vor, die die Annahme von endogenen Rückfällen nur stützen können.

Im Rahmen von Untersuchungen, welche Kostrzewski (3), (4) an sporadischen Fleckfieberfällen durchführte, die zwischen 1952 und 1955 in Polen auftraten und die 95% aller Fälle ausmachten, wurden 6 Stämme von *R. prowazeki* isoliert, darunter 3 von Patienten mit einem sicheren Rückfall. Es handelte sich um meist milde, oft atypische Erkrankungen von 7—10 Tagen Dauer. Das Exanthem war flüchtig und konnte gelegentlich fehlen. Alle verdächtigen Personen wurden serologisch untersucht. Als positiv gewertet wurden nur Fälle, bei denen die Weil-Felix-Reaktion einen Titer von wenigsten 1:200 oder die KBR einen Titer von 1:50 erreichten. In der Hälfte der insgesamt beobachteten 795 Fälle waren beide Reaktionen positiv.

Die Fälle wurden in 3 Gruppen eingeteilt. Die 1. Gruppe enthielt 74 Patienten, bei welchen eine Infektionsquelle bekannt war und bei denen es sich daher wahrscheinlich um eine Ersterkrankung handelte. 91,8% dieser Patienten hatte eine positive Weil-Felix-Reaktion. Die 2. Gruppe umfaßte 194 Patienten, bei welchen die Infektionsquelle nicht zu eruieren, eine exogene Infektion jedoch nicht zuverlässig auszuschließen war. 61,3% der Seren enthielten Agglutinine für OX 19. Die 3. Gruppe betraf 527 Personen, bei welchen sicher keine Neuinfektion vorlag und bei denen es sich somit aller Wahrscheinlichkeit nach um einen Spätrückfall, also um Brill-Zinssersche Krankheit, handelte. Nur 48,2% dieser Patienten hatten eine positive Weil-Felix-Reaktion. 70,8% aller Fälle von Gruppe 3 wurden 1954, und zwar hauptsächlich in den Sommermonaten, beobachtet. 295 von den 721 Patienten aus den Gruppen 2 und 3 hatten bereits vor 10—20 Jahren ein Fleckfieber durchgemacht. Bei einem Teil der Patienten war die Ersterkrankung an Fleckfieber offenbar inapparent verlaufen. Das Intervall zwischen Erst- und Zweiterkrankung zeigte mit 35% einen Gipfel zwischen 30 und 40 Jahren. Hierbei handelte es sich offenbar um Rückfälle aus der Zeit des I. Weltkrieges. Das längste Intervall betrug 52 Jahre. Kostrzewski, Grużewski und Milewska kommen auf Grund statistischer Erhebungen und Berechnungen zu der Annahme, daß die ersten Rückfälle 6—10 Jahre nach der Erkrankung auftreten und die Relapse am häufigsten nach 18—20 Jahren sind. Die Zahl nimmt dann bis zum 50. Jahr laufend ab. Somit wäre in Polen bis 1960 noch mit einer größeren Zahl von Rückfällen zu rechnen, deren Ersterkrankung im II. Weltkrieg lag. Dazu paßt gut, daß die Zahl der sporadischen Fälle in der Zeit von 1952—1955 von 7 bis auf 61 pro Jahr laufend angestiegen ist.

Niedzielska berichtet über 54 sporadische, klinisch relativ leicht verlaufene Fälle von klassischem Fleckfieber in Lodz, von denen 23 sichere Zweiterkrankungen waren. Bei den übrigen ließ sich die Infektionsquelle nicht ermitteln. Die Patienten waren zwischen 40 und 80 Jahren alt. Das Intervall zur ersten Erkrankung betrug 30 bis 40 Jahre. 15 Patienten hatten kein Exanthem. Die Weil-Felix-Reaktion war nur bei 30 Patienten positiv. Foryś, Lutyński und Raginis fanden bei 34 von 42 Patienten, die zwischen 1918 und 1955 Fleckfieber hatten, noch komplementbindende Antikörper im Serum. Alle Patienten, die in den letzten 4 Jahren erkrankt waren, reagierten positiv.

Obwohl feststeht, daß sich Läuse an Rückfällen infizieren können, muß zugegeben werden, daß bisher keine sicheren, von solchen Erkrankungen ausgehenden Kontaktinfektionen bekannt geworden sind. Lediglich Gaon berichtet von einigen Erkrankungen in Jugoslawien, die wahrscheinlich von einem Fall Brill-Zinssersche Krankheit ihren Ausgang genommen hatten.

Es ist auch daran gedacht worden, *andere Reservoire* als den Menschen für das klassische Fleckfieber in Betracht zu ziehen. REISS-GUTFREUND (1), (2) hat in Addis Abeba in 19 Versuchen einen Stamm von *R. prowazeki* aus einer Ziege und 2 Stämme aus Schafen isoliert. In 92 Versuchen mit 7 verschiedenen Zeckenarten, die von Haustieren abgesammelt worden waren, wurden 2 Stämme von *R. prowazeki* in *Amblyomma variegatum* und 2 Stämme in *Hyalomma rufipes* gefunden. Die Stämme konnten auf Meerschweinchen, Mäuse, Läuse, Schafe und in den Dottersack des Hühnchens übertragen werden und verhielten sich bezüglich ihrer Pathogenität und immunologischen Eigenschaften wie typische Stämme von klassischem Fleckfieber. *R. prowazeki* konnte auch in Zecken (*A. variegatum*, *A. lepidum*, *H. rufipes*, *Rhipicephalus simus*) nachgewiesen werden, die an mit starken Dosen von Rickettsien inoculierten Kaninchen gesogen hatten. Im Hinblick auf die gute Vermehrungsfähigkeit von *R. prowazeki* in Zecken (vgl. S. 98) ist hieran allerdings nichts Auffälliges. Bei einer serologischen Untersuchung mit Hilfe der Agglutinationsprobe wurden in 21% der Seren von 172 Zebus, 14% der Seren von 248 Schafen und 18% der Seren von 229 Ziegen Agglutinine für *R. prowazeki* gefunden.

Die angeführten Befunde könnten eine Bestätigung der Hypothese von GIROUD und JADIN (3) darstellen, die bereits auf Grund von serologischen Untersuchungen Haustiere als Reservoire für klassisches Fleckfieber in Betracht gezogen hatten. In Ruanda Urundi (Belgisch-Kongo) fanden sie Agglutinine für *R. prowazeki* bei 6 von 26 Kühen, 2 von 24 Schafen, 2 von 12 Ziegen, 6 von 17 Schweinen, 3 von 10 Hunden und 8 von 10 Pferden. In der KBR waren nur eine Kuh und ein Pferd positiv. REISS-GUTFREUND (2) nimmt für das klassische Fleckfieber in Äthiopien einen doppelten Kreislauf an; einen Cyclus Haustier — Zecke — Haustier und einen Cyclus Mensch — Laus — Mensch. Bei einer Infektion des Menschen durch Zeckenstich könnten beide Cyclen ineinander übergehen.

Diese Beobachtungen und Schlußfolgerungen müssen mit einem gewissen Vorbehalt angeführt werden. Sie würden im Falle einer Bestätigung ein ganz neues Licht auf die Epidemiologie des klassischen Fleckfiebers werfen. Es wäre dann allerdings überraschend, daß bei der intensiven Bearbeitung des klassischen Fleckfiebers in aller Welt die Rolle von Haustieren als Erregerreservoir noch nicht früher bemerkt worden ist.

c) Wolhynisches Fieber

Übertragung und Epidemiologie des Wolhynischen Fiebers stimmen in allen wesentlichen Punkten mit den Verhältnissen beim klassischen Fleckfieber überein [WEYER (14)]. Das Reservoir für die Krankheit bildet auch hier der Mensch. Tierische Reservoire sind nicht bekannt. Als empfängliches Versuchstier wurde von MOOSER und WEYER (1) der Rhesusaffe ermittelt. Die bei den Affen im Experiment erhobenen Befunde decken sich weitgehend mit den bei menschlichen Erkrankungen gesammelten Erfahrungen. Der Mensch infiziert sich ebenso wie beim Fleckfieber durch erregerhaltigen Läusekot. Verschiedene Beobachtungen sprechen dafür, daß bei gleicher Exposition nicht alle Personen an Wolhynischem Fieber erkranken, sondern daß es Menschen mit einer angeborenen Immunität gegen die Krankheit gibt, oder daß für die Infektion eine besondere Disposition vorhanden sein muß [KOSTRZEWSKI (1)]. Auch die Reaktionen auf einen erfolgten Infekt, bei welchem es zu einer sich gewöhnlich auf mehrere Wochen erstreckenden Rickettsiämie kommt, sind ganz verschieden. Zwischen einer praktisch symptomlosen Infektion und dem typischen Bild des Wolhynischen Fiebers mit einem kräftigen oder mehreren schwächeren Fieberanfällen gibt es eine Reihe von Übergängen.

Empfängliche Personen, die keine oder nur unbedeutende Symptome zeigen, können im Blut durch Xenodiagnose nachweisbare Rickettsien enthalten, die man irrtümlich als apathogen angesehen und daher als besondere Art, *R. pediculi*, bezeichnet hat. Solche „apathogenen" Rickettsien sind nur in Gebieten mit starker Verlausung nachgewiesen worden, in denen auch Wolhynisches Fieber endemisch ist. Bei Kontrollen in anderen Gegenden hat man noch keine apathogenen Rickettsien gefunden. *R. pediculi* ist daher wie *R. weigli* als identisch mit *R. quintana* anzusehen.

Wo sich die Rickettsien im menschlichen Körper vermehren, ob die Vermehrung intra- oder extracellulär abläuft, ob nur im peripheren Blut oder auch in Körperorganen, wissen wir nicht. Im Rahmen der Experimente mit Rhesusaffen konnten MOOSER und WEYER Rickettsien aus Leber, Milz und Knochenmark der Affen isolieren.

Im Verlaufe einer Erkrankung des Menschen kann sich nach einem einmaligen Fieberanfall eine solide Immunität einstellen. Häufiger ist eine zeitlich begrenzte unstabile Immunität. So erkrankten von 67 unter annähernd gleichen Bedingungen stehenden und über längere Zeit kontrollierten Laboratoriumsarbeitern 25 nur 1mal, 25 Personen 2mal, 6 Personen 3mal, eine Person 4mal, eine Person 5mal, je 2 Personen 6mal, 7mal und 8mal und 3 Personen mehr als 8mal [KOSTRZEWSKI (1)]. Die Intervalle zwischen erster und zweiter Erkrankung betrugen 2, 3, 6, 9, 12, 18 Monate und 2, 3, 5 und 6 Jahre.

Es ist sehr wahrscheinlich, daß es auch beim Wolhynischen Fieber Zweiterkrankungen auf Grund von *Spätrückfällen*, vergleichbar der Brill-Zinsserschen Krankheit, gibt. Schwierigkeiten bei der sicheren Beurteilung solcher Zustände ergeben sich vor allem dadurch, daß beim Wolhynischen Fieber eine zuverlässige serologische Diagnostik fehlt. KOSTRZEWSKI (1) hat zwar über gute Erfahrungen mit der Agglutination und der KBR berichtet, das Antigen für die Reaktionen konnte aber nur in kleiner Menge aus Läusefaeces gewonnen werden. Alle anderen Versuche zur Herstellung eines brauchbaren Antigens sind fehlgeschlagen. *R. quintana* wächst auch nicht im Dottersack des Hühnchens. Außerdem stellt der Erreger offenbar nur einen schwachen Reiz zur Antikörperproduktion im Wirtsorganismus dar, so daß insbesondere bei leichteren Fällen nicht ohne weiteres mit einem Antikörpernachweis zu rechnen ist. Die einzige Möglichkeit einer sicheren Diagnose besteht im Läusetest, der allerdings nur im positiven Fall beweisend ist, nämlich dann, wenn die Zahl der Rickettsien im Blut groß genug ist, um eine Ansiedlung im Magen der Laus zu gewährleisten [WEYER (1), (14)]. Der Rickettsienspiegel im Blut schwankt jedoch und reicht nicht immer für einen solchen Nachweis aus.

Eine längere Persistenz von *R. quintana* im Menschen ist die Regel. Die Rickettsien sind im Blut mindestens während der klinischen Symptome und fast immer noch einige Zeit danach zu finden. Bei manchen Patienten konnten Rickettsien fortlaufend bis zu 109 Tagen im Blut nachgewiesen werden, bei einer Laboratoriumsinfektion bis zu 4 Monaten (MOHR und HIRTE). KOSTRZEWSKI (1) erwähnt Erkrankungen, die sich über 1—2 Jahre erstreckten und bei denen sich während der ganzen Zeit Rickettsien im Blut fanden. Bereits im I. Weltkrieg war bei 2 Patienten eine Rickettsienpersistenz von 298 und 443 Tagen festgestellt worden (BYAM und LLOYD). BIELING und OEHLRICHS fanden *R. quintana* im Blut von Läusefütterern, deren Ersterkrankungen oder deren Aufenthalt in Endemiegebieten 27 Monate und 3 Jahre zurücklagen. Mooser, Marti und Leemann stellten bei einem Patienten eine kurze symptomfreie Rickettsiämie nach einem Intervall von fast 4 Monaten fest. Hoenig und Mohr wiesen Rickettsien bei einem Patienten $3^1/_2$ Jahre nach dem Verlassen des Endemiegebietes nach. Der ganze Krankheitsprozeß zog sich in diesem Fall über 10 Jahre hin, wobei allerdings eine exogene Reinfektion nicht sicher auszuschließen war. Kostrzewski (1) berichtet über einen Erregernachweis im Blut nach 4, 5 und ausnahmsweise nach 8 Jahren. Die Rückfälle können ähnlich wie beim klassischen Fleckfieber an eine verminderte Resistenz geknüpft sein, wie sie im Zusammenhang mit anderen Erkrankungen auftritt.

Mögen sich auch manche der erwähnten Fälle mit einer Neuinfektion erklären lassen, so steht eine sehr lange *Erregerpersistenz* beim Wolhynischen Fieber außer jedem Zweifel. Die Rickettsiämie verläuft dabei häufig symptomlos. Auf der anderen Seite ist aber eine Erregerpersistenz, die noch nach Jahren zu einem manifesten Rückfall führt, wesentlich seltener als beim klassischen Fleckfieber. In Hamburg wurden von 1947 bis 1957 über 365 Patienten untersucht, die im II. Weltkrieg Wolhynisches Fieber überstanden oder sich in Endemiegebieten aufgehalten hatten und die mit Fieber oder anderen Symptomen ins Krankenhaus kamen, welche klinisch ohne weiteres als Wolhynisches Fieber gedeutet werden konnten. Bei keinem dieser Patienten ließen sich durch den Läusetest Rickettsien nachweisen (Mohr, sowie Hoenig und Mohr 1954). Ein einwandfreier Rickettsiennachweis ist später bei einem Patienten gelungen, bei dem die Ersterkrankung 13 Jahre und die letzte Infektionsmöglichkeit mindestens 3 Jahre zurücklagen.

Die wenigen Menschen, die die Rickettsien für längere Zeit beherbergen, bilden die *Reservoire*, die den Weiterbestand des Erregers und der Krankheit gewährleisten. Dabei braucht es sich nicht einmal um eine jahrelange Persistenz zu handeln. Beim Vorhandensein von Läusen werden die Rickettsien vom menschlichen Wirt häufig unbemerkt aufgenommen [v. Bormann (2)]. Als Seuche ist das Wolhynische Fieber nur in den beiden Weltkriegen aufgetreten, weil die Zusammenballung großer Menschenmassen auf engem Raum und die mangelnde Hygiene eine starke Zunahme der Verlausung und damit eine erhebliche Rickettsienvermehrung mit sich brachten, wobei die Erreger reichlich neue und zudem nichtimmune Wirte erobern konnten.

Das Wolhynische Fieber ist als endemische und epidemische Krankheit nur aus Ost- und Südeuropa bekannt und besonders während des Ersten Weltkrieges mit den Soldaten auch in weitere Teile Europas verschleppt worden. Alle Angaben über das Auftreten einzelner Fälle in anderen Gegenden, z. B. in Japan, können nicht als gesichert gelten.

Varela, Fournier und Mooser haben in Mexiko Kot von Läusen, die bei der einheimischen Bevölkerung gefunden wurden, auf 2 Freiwillige übertragen. Es kam zu einer fieberhaften Reaktion, wie man sie von Patienten mit Wolhynischem Fieber kennt. Gesunde Läuse, die an den infizierten Personen sogen, enthielten später Organismen vom Typ der *R. quintana*. Falls sich diese Beobachtung bestätigt, wäre damit der erste Nachweis von *R. quintana* in der Neuen Welt erbracht.

Auch eine Beobachtung aus China verdient in diesem Zusammenhang Erwähnung. Liu und Landauer fanden rickettsienähnliche Gebilde in Läusen bei 16 von 25 Rekonvaleszenten, die an Fleckfieber oder einem ungeklärten fieberhaften Infekt erkrankt waren. Die gleichen Organismen wurden bei 74 von 214 Personen nachgewiesen, die mit den Rekonvaleszenten in Kontakt gekommen waren. Sie fanden sich auch in Läusen aus einer Laboratoriumszucht, die an 10 von 19 Rekonvaleszenten und an 12 von 41 Kontaktpersonen gesogen hatten. Die

Versuche wurden an einer Person bis zu 8mal wiederholt. Die Organismen waren teils ständig, teils kurze Zeit, teils intermittierend im Blut vorhanden. Versuche zur Übertragung auf Ratten und Meerschweinchen verliefen negativ. Auf Grund der Beschreibung ist daran zu denken, daß es sich hier vielleicht um *R. quintana* gehandelt hat.

Über die Phylogenie der *R. quintana* und ihre Beziehungen zu den anderen Rickettsien können wir bis jetzt nichts sagen. Weyer (6, 16) hat die Vermutung ausgesprochen, daß extracelluläre Rickettsien aus intracellulären unter Verlust der ursprünglichen Virulenz und Pathogenität hervorgehen können. Es wurde mehrfach eine Umwandlung von intra- in extracelluläre Rickettsien (niemals umgekehrt), z. B. bei Haltung von *R. prowazeki* und *R. mooseri* in Läusen und Zecken sowie in explantierten Läusemägen, beobachtet (siehe S. 85 und 99). Vorerst fehlen aber noch alle Anhaltspunkte dafür, aus welcher Ursache und unter welchen Bedingungen eine solche Wandlung vor sich gehen kann.

Literatur

ABINANTI, F. R., E. H. LENNETTE, J. F. WINN and H. H. WELSH: Q fever studies. XVIII. Presence of Coxiella burnetii in the birth fluids of naturally infected sheep. Amer. J. Hyg. 58, 385—388 (1954).
— H. H. WELSH, E. H. LENNETTE and O. BRUNETTI: Q fever studies. XVI. Some aspects of the experimental infection induced in sheep by the intratracheal route of inoculation. Amer. J. Hyg. 57, 170—184 (1953).
— — J. F. WINN and E. H. LENNETTE: Q fever studies. XIX. Presence and epidemiologic significance of Coxiella burnetii in sheep wool. Amer. J. Hyg. 61, 362—370 (1955).
Advances in the control of zoonoses. Part IV. Q fever. World Health Organ. Monogr. Ser. No 19, p. 155—217. Genf: Palais des Nations 1953.
ALI, M., and M. V. SUBRAHMANIAN: Endemic (flea) typhus in Goleonda. J. Indian med. Ass. 29, 110—111 (1957).
ALLEN, E. G., M. R. BOVARNICK and J. C. SNYDER: The effect of irradiation with ultraviolet light on various properties of typhus rickettsiae. J. Bact. 67, 718—723 (1954).
ANDERSON, C. R.: Experimental typhus infection in the Eastern cotton rat (Sigmodon hispidus hispidus). J. exp. Med. 80, 341—356 (1944).
— R. K., and S. L. KALRA: Q fever studies in India: A case of human Q fever. Indian J. med. Res. 42, 307—314 (1954).
ASCHENBRENNER, R., u. H. EYER: Die Rickettsiosen. In Handbuch der inneren Medizin, 4. Aufl., Bd. I/1, S. 638—761. Berlin: Springer 1952.
AUDY, J. R.: (1) A summary topographical account of scrub typhus 1908—1948. Studies in the distribution and topography of scrub typhus, number 1. Bull. Ind. Med. Res. Fed. Malaya 1949, No 1, 1—84.
— (2) Malaysian parasites. III. A summary review of collection of Trombiculid mites in the Asiatic-Pacific area. Bull. Ind. Med. Res. Fed. Malaya 1953, No 26, 29—44.
— (3) Trombiculid mites infesting birds, reptiles, and Arthropodes in Malaya, with a taxonomic revision, and descriptions of a new genus, two new subgenera, and six new species. Bull. Ruffles Museum 1956, No 28, 27—80.
— (4) A checklist of Trombiculid mites of the Oriental and Australasian regions. Parasitology 47, 217—294 (1957).
—, and J. L. HARRISON: A review of investigations in mite typhus in Burma and Malaya, 1945—1950. Trans. roy. Soc. trop. Med. Hyg. 44, 371—395 (1951).
BABUDIERI, B.: Ricerche sulla febbre Q in Italia. Acta med. ital. 6, 1—9, 29—38 (1951).
—, e D. BOCCIARELLI: Ricerche di microscopia elettronica. I. Studio morphologico di Rickettsia prowazeki. R. C. Ist. sup. Sanità 6, 298—304 (1943).
—, e C. MOSCOVICI: (1) Infezione sperimentale e naturale da Coxiella burneti negli ucelli. R. C. Ist. sup. Sanità 15, 215—219 (1953).
— — (2) Infezione sperimentale da Coxiella burnetii (PHILIP) per via alimentare e per via transcongiuntivale. R. C. Ist. sup. Sanità 18, 65—69 (1955).

BABUDIERI, B., e S. PAOLUCCI: Studio di un epidemico di febbre Q, manifestatosi a S. Lorenzo in Campo (Pesaro). R. C. Ist. sup. Sanità 17, 404—417 (1954).

—, e L. RAVAIOLI: Infezione sperimentale della pecora con Coxiella burneti. R. C. Ist. sup. Sanità 15, 194—214 (1957).

BADIALI, C.: Ricerche sull'eventuale presenza di anticorpi anti-R. akari. Nuovi Ann. Ig. 8, 422—425 (1957).

BAIRD, R. B., and I. M. TONKIN: Endemic typhus in Mengo district, Uganda. E. Afr. med. J. 28, 157—167 (1951).

BARBER, H.: Direct conglutination with Rickettsia burneti, the causal agent of Q fever. J. Hyg. (Lond.) 53, 63—75 (1955).

BÁRDOŠ, V., R. BREZINA, T. KRAMÁR, O. SOBEŠLAWSKÝ and F. ŠIMONIK: An epidemic of Q fever in a factory caused by working with contaminated imported cotton. J. Hyg. Epidemiol., Microbiol. Immunol. 1, 43—48 (1957).

BEKTEMIROW, T. A., S. W. TARASEWITSCH u. B. E. KARULIN: Zur Charakteristik eines endemischen Herdes von Q-Fieber auf der Krim. Ž. Mikrobiol. (Mosk.) 1956, H. 11, 20—26 [Russisch].

BELL, E. J., and E. G. PICKENS: A toxic substance associated with the rickettsias of the spotted fever group. J. Immunol. 70, 461—472 (1953).

—, and C. B. PHILIP: The human rickettsioses. Ann. Rev. Microbiol. 6, 91—118 (1952).

BENGTSON, I. A., N. H. TOPPING and R. G. HENDERSON: Studies of typhus fever. Epidemic typhus: Demonstration of a substance lethal for mice in the yolk sac of eggs infected with Rickettsia prowazeki. Nat. Inst. Hlth. Bull. 1945, No 183, 25—29.

BENNETT, B. L., J. E. SMADEL and R. L. GAULD: Studies on scrub typhus (Tsutsugamushi disease). IV. Heterogenecity of strains of R. tsutsugamushi as demonstrated by cross-neutralization tests. J. Immunol. 62, 453—461 (1949).

BENOIST, F., P. GIROUD et G. HÉRAUD: Typhus murin récurrent. Bull. Soc. méd. Hôp. Paris 5/6, 158—161 (1954).

BERGE, T. O., and E. H. LENNETTE: World distribution of Q fever: Human, animal and arthropod infection. Amer. J. Hyg. 57, 125—143 (1953).

BERGEYS Manual of determinative Bacteriology, 7. edit. (R. S. BREED, E. G. D. MURRAY and N. R. SMITH). Baltimore: Williams & Wilkins Company 1957. 1094 S.

BETTINOTTI, C. M.: Fiebre Q. Anticuerpos fijadores de complemento en la población de Córdoba (R.A.). Sem. méd. (Paris) 108, 125—127 (1956).

BIELING, R., u. L. OELRICHS: Beobachtungen über die Dauer der Infektion mit Rickettsia quintana (pediculi). Z. Hyg. 127, 49—53 (1947).

BINGEL, K. F., u. H. ENGELHARDT: Genügen unsere Pasteurisierungsverfahren zur Unschädlichmachung des Q-Fieber-Erregers in der Kuhmilch? Arch. Hyg. (Berl.) 136, 417—425 (1952).

— R. LAIER u. S. MANNHARDT: Epidemiologische Studien zu 3 Q-Fieberausbrüchen. Z. Hyg. 134, 635—650 (1952).

BLANC, G.: (1) Rickettsia (Coxiella) burneti (Derrick) au Maroc et accessoirement en Algérie et en Afrique équatoriale. Étude expérimentale. Rev. belge Path. 21, 17—36 (1951).

— (2) Épidémiologie de la Q fever (Coxiellose). Maroc. méd. 33, 1021—1025 (1954).

—, et M. BALTAZARD: (1) Longévité du virus de typhus murin dans les déjections de puces infectées. Bull. Soc. Path. exot. 33, 25—32 (1940).

— — (2) Recherches sur le mode de transmission du typhus. II. Le réservoir de virus naturel du typhus, les déjections d'ectoparasites infectés. Arch. Inst. Pasteur Maroc 2, 658—673 (1944).

—, et J. BRUNEAU: (1) Comportement de Rickettsia burnetii chez la puce du rat Xenopsylla cheopis. C. R. Soc. Biol. (Paris) 142, 1334—1335 (1948).

— — (2) Entretien dans la nature de Coxiella burnetii par l'association du lapin de garenne Oryctolagus cuniculus (L) et de la tique Hyalomma excavatum C. L. K. C. R. Acad. Sci. (Paris) 237, 582—584 (1953).

— — (3) Ornithodores et Coxiellose (Q fever). C. R. Acad. Sci. (Paris) 240, 129—131 (1955).

— — (4) Isolement du virus de Q fever de deux rongeurs sauvages provenant de la forêt de Nefifik (Maroc). Bull. Soc. Path. exot. 49, 431—434 (1956).

Blanc, G., et M. Baltazard: (5) Présence chez une chauve-souris du Maroc Eptesicus isabellinus (Temminck) d'un virus de Q fever. Bull. Soc. Path. exot. **50**, 653—656 (1957).

— L. A. Martin et A. Maurice: Le mérion (Meriones Shawi) de la région de Goulimine est un réservoir de virus de la Q fever marocaine. C. R. Acad. Sci. (Paris) **224**, 1673—1674 (1947).

Blanton, F. S., and T. G. Tani: Typhus control at ports in Japan and Korea after world war II. J. econom. Entomol. **44**, 812—813 (1951).

Bock, M.: Experimentelle Untersuchungen zur Epidemiologie des Q-Fiebers. Z. Tropenmed. Parasit. **5**, 348—355 (1954).

Borell, J. R.: Eine Fleckfieber-Epidemie in Barcelona in den Jahren 1942—1943. Z. Hyg. Infekt.-Kr. **133**, 65—82 (1951).

Bormann, F. v.: (1) Die in New York und Boston unter dem Namen „Brillsche Krankheit" bekannten Fleckfiebererkrankungen und die Frage der Spätrückfälle beim klassischen Fleckfieber. Z. Hyg. Infekt.-Kr. **135**, 448—471 (1952).

— (2) Zur Epidemiologie des Wolhynischen Fiebers und des klassischen Fleckfiebers außerhalb der Notzeiten. Med. Klin. **1954**, 140—143.

Bovarnick, M. R., and E. G. Allen: (1) Reversible inactivation of typhus rickettsiae. I. Inactivation by freezing. J. gen. Physiol. **38**, 169—179 (1955).

— — (2) Reversible inactivation of typhus rickettsiae at 0° C. J. Bact. **73**, 56—62 (1957).

— — (3) Reversible inactivation of the toxicity and hemolytic activity of typhus rickettsiae by starvation. J. Bact. **54**, 637—645 (1957).

—, and J. C. Miller: Oxydation and transamination of glutamate by typhus rickettsiae. J. biol. Chem. **184**, 661—676 (1950).

—, and J. C. Snyder: Respiration of typhus rickettsiae. J. exp. Med. **89**, 561—565 (1949).

Bozeman, F. M., H. E. Hopps, J. X. Danauskas, E. B. Jackson and J. E. Smadel: Study on the growth of rickettsiae. I. A tissue culture system for quantitative estimations of Rickettsia tsutsugamushi. J. Immunol. **76**, 475—488 (1956).

Brandão, H., L. A. Ribeiro do Valle e D. de A. Cristovão: Investigações sobre a febre Q en São Paulo. I. Estudo serologica em operarios de um frigorífico. Arch. Fac. Hig. S. Paulo 7, 127—134 (1953).

Brezina, R., u. D. Táborská: (Research on Q fever in Slovakia. 2nd report: Sporadic cases and further investigation of reservoir animals.) Čsl. Epidem. **5**, 152—155 (1956) [Tschechisch].

Briceno Iragorry, L.: La fiebre „Q", su comprobacion en la zona caraquena. Gac. méd. Caracas **60**, 98—114 (1952).

Brigham, G. D.: (1) A strain of endemic typhus fever isolated from a field mouse. Publ. Hlth Rep. (Wash.) **52**, 659—660 (1937).

— (2) Two strains of endemic typhus isolated from naturally infected chicken fleas (Echidnophaga gallinaceae). Publ. Hlth Rep. **56**, 1803—1804 (1941).

Brotherston, J. C., and E. R. N. Cooke: Q fever in Kenya. E. Afr. med. J. **33**, 125—130 (1956).

Brown, R. D.: Q fever — Veterinary aspects. E. Afr. med. J. **33**, 441—445 (1956).

Brumpt, E., et C. Desportes: Grande longévité du virus de la fièvre pourprée des Montagnes Rocheuses et de celui du typhus de São-Paulo chez Ornithodorus turicata. Ann. Parasit. hum. comp. **18**, 145—153 (1941).

Burganski, B. K., M. B. Kaplinski, A. P. Wygowski and I. F. Berdnikow: Q fever in the Urals. J. Microbiol., Epidemiol. Immunobiol. **28**, 349—354 (1957).

Burgdorfer, W.: Ornithodorus moubata als Testobjekt bei Q-Fieberfällen in der Schweiz. Acta trop. (Basel) 8, 44—51 (1951).

— R. Geigy, O. Gsell u. E. Wiesmann: Parasitologische und klinische Beobachtungen an Q-Fieber-Fällen in der Schweiz. Schweiz. med. Wschr. **1951**, 162—166.

Burnet, F. M., and M. Freeman: Experimental studies on the virus of "Q" fever. Med. J. Aust. **1937**, 2, 299—305.

Byam, W., and L. L. Lloyd: Trench fever. Its epidemiology and endemiology. Proc. roy. Soc. Med. **13**, 1—27 (1919).

Cabrera, M. J. I., J. W. McAnally y J. A. Montoya: Informe sobre las actividades en el control del tifo en la República de Guatemala. Bol. Oficina san. panamer. **34**, 225—235 (1953).

CALERO, M. C., J. M. NÚÑEZ and R. SILVA-GOYTIA: Rocky Mountain spotted fever in Panama. Report on two cases. Amer. J. trop. Med. Hyg. 1, 631—636 (1952).

CAPUTO, C.: Contributo all'epidemiologia delle febbri dermotifosimili. La febbre bottonosa in Provincia di Savona durante gli anni 1942—1954. Igiene mod. 50, 45—56 (1957).

CARLEY, J. G., R. L. DOHERTY, E. H. DERRICK, M. L. EMANUEL and C. J. ROSS: The investigation of fevers in North Queensland by mouse inoculation with particular reference to scrub typhus. Aust. Ann. Med. 4, 91—99 (1955).

—, and J. H. POPE: The isolation of Coxiella burneti from the tick Ixodes holocyclus in Queensland. Austr. J. exp. Biol. med. Sci. 31, 613—614 (1953).

CASTORINA, S.: Rassegna sulle rickettsiosi umane della Sicilia. Nuovi Ann. Ig. 2, 161—190 (1951).

CAWLEY, E. P., and C. E. WHEELER: Rocky Mountain spotted fever. J. Amer. med. Ass. 163, 1003—1007 (1957).

CHALKE, H. D.: Typhus: Experiences in the Central Mediterranean forces. Brit. med. J. 1946, 977—980.

CHANG, N. C., S. H. ZIA, F. T. LIU and S. H. CHAO: The possible existence of Q fever in Peking with a brief review on its current knowledge. Chin. med. J. 69, 35—45 (1951).

— R. S. M., E. S. MURRAY and J. C. SNYDER: Erythrocyte-sensitizing substances from rickettsiae of the Rocky Mountain spotted fever group. J. Immunol. 73, 8—15 (1954).

— J. C. SNYDER and E. S. MURRAY: A serologically-active erythrocyte-sensitizing substance from typhus rickettsiae. I. Isolation and titration. II. Serological properties. J. Immunol. 70, 212—221 (1953).

CHAO, S. H.: Die Lebensdauer von Rickettsia prowazeki in Läusefäces. Schweiz. Z. Path. 12, 507—512 (1949).

CIOGLIA, L., e A. MEDDA: Ricerche sulla febbre Q in provincia di Cagliari. Arch. ital. Sci. med. colon. 38, 475—483 (1957).

CLARK, W. H., A. S. BOGUCKI, E. H. LENNETTE, B. H. DEAN and J. R. WALKER: Q fever in California. VI. Description of an epidemic occurring at Davis, California, in 1948. Amer. J. Hyg. 54, 15—24 (1951).

— E. H. LENNETTE and M. S. ROMER: Q fever in California. XI. An epidemiological summary of 350 cases occurring in Northern California during 1948—1949. Amer. J. Hyg. 54, 319—350 (1951).

CLARKE, D. H., and J. P. FOX: The phenomenon of in vitro hemolysis produced by the rickettsiae of typhus fever, with a note on the mechanism of rickettsial toxicity in mice. J. exp. Med. 88, 25—41 (1948).

CLAVERO, G., e F. PÉREZ GALLARDO: (1) Estudio experimental de una cepa apatógena o immunizante de Rickettsia Prowazeki. Cepa E. Rev. Sanid. Hig. públ. (Madr.) 1943, 28 S.

— — (2) Immunización contra el tifus exantematico con vacuna viva, cepa E. Esc. Nac. san. (Madr.) 1949, 1—102.

CODELEONCINI, E.: Sulla presenza in Ethiopia della „Rickettsia Weigli". Considerazioni sul rapporto fra la Rickettsia Weigli, l'agente eziologico della „Febbre delle Trincei" e la Rickettsia pediculi. Boll. Soc. ital. med. igiene trop. 6, 129—151 (1946).

COHEN, S. S.: Studies on commercial typhus vaccines. IV. The chemical composition of the antigens of commercial typhus vaccine. J. Immunol. 65, 475—483 (1950).

COLES, J. D. W. A.: Classification of rickettsiae pathogenic to vertebrates. Ann. N.Y. Acad. Sci. 56, 457—462 (1953).

COMBIESCU, D.: Fièvre "Q" (typhus pulmonaire-rickettsiose pulmonaire). Arch. rom. Path. exp. Microbiol. 16, 37—55 (1957).

— and collaborators: Zit. nach Trop. Dis. Bull. 53, 43 (1956). Clinical, epidemiological, serological and experimental study of Q fever on a collective farm. Stud. Cercet. Inframicrobiol. 5, 229—260 (1954) [Rumänisch].

— N. DUMITRESCU, M. RUSS u. M. DINCULESCU: Epidemiological observations on cases of Boutonneuse fever occurring in Rumania in the last 41 years. Culture and characteristics of a strain of R. conori isolated in a local focus. Stud. Cercet. Inframicrobiol. 4, 99—107 (1953) [Rumänisch].

— — G. ZARNEA, A. SARAGEA, M. ESSRIG u. H. IONESCU: Experimental and epidemiological studies of pulmonary typhus (Q fever). I. Mechanism of transmission of infection of pulmonary typhus (Q fever). Stud. Cercet. Inframicrobiol. 4, 109—134 (1953) [Rumänisch].

Cox, H. R.: (1) Use of yolk sac of developing chick embryo as medium for growing rickettsiac of Rocky Mountain spotted fever and typhus groups. Publ. Hlth Rep. (Wash.) **53**, 2241—2247 (1938).

— (2) Some recent advances and current problems in the field of rickettsial diseases. Arch. Virusforsch. **4**, 518—533 (1951).

— (3) The spotted fever group. In T. M. Rivers, Viral and rickettsial infections of man, 2nd. edit., p. 611—637. Philadelphia: J. B. Lippincott Company 1952.

— (4) Viral and rickettsial toxins. Ann. Rev. Microbiol. **7**, 197—218 (1953).

Craddock, A. L.: Q fever — Human aspects. E. Afr. mcd. J. **33**, 435—439 (1956).

—, and J. Gear: Q fever in Nakuru, Kenya. Lancet 1955 II, 1167—1169.

Craufurd-Benson, H. J.: Naples typhus epidemic, 1942—1943 (as seen by an entomologist). Brit. med. J. **1946**, 559—580.

Crocker, T. T., B. L. Bennett, E. B. Jackson, M. J. Snyder, J. E. Smadel, R. L. Gauld and M. K. Gordon: Siberian tick typhus. Relation of the Russian strains to Rickettsia prowazeki. Publ. Hlth Rep. (Wash.) **65**, 383—394 (1950).

Culpepper, G. H.: (1) The rearing and maintenance of a laboratory colony of the body louse. Amer. J. trop. Med. **24**, 327—329 (1944).

— (2) Rearing and maintaining a laboratory colony of body lice on rabbits. Amer. J. trop. Med. **28**, 499—504 (1948).

Davis, G. E.: (1) The tick Ornithodorus rudis (= venezuelensis) as a host to the rickettsiae of the spotted fevers of Colombia, Brazil, and the United States. Publ. Hlth Rep. (Wash.) **58**, 1016—1020 (1943).

— (2) Observations on rat ectoparasites and typhus fever in San Antonia, Texas. Publ. Hlth Rep. (Wash.) **66**, 1717—1726 (1951).

—, and R. Cox: A filter-passing infectious agent isolated from ticks. I. Isolation from Dermacentor andersoni, reactions in animals, and filtration experiments. Publ. Hlth Rep. (Wash.) **53**, 2259—2267 (1938).

DeLay, P. D., E. H. Lennette and K. B. DeOme: Q fever in California. II. Recovery of Coxiella burneti from naturally-infected air-borne dust. J. Immunol. **65**, 211—220 (1950).

Del Campo, A., E. de Lotto e C. Magrì: L'epidemia di febbre Q di Domegge di Cadore Minerva med. (Torino) **1**, 1—31 (1955).

Depoux, R., et P. Merveille: Sur une petite épidémie de fièvre exanthématique observeé à Brazzaville. Bull. Soc. Path. exot. **48**, 610—615 (1955).

Derrick, E. H.: (1) ''Q'' fever, a new fever entity: Clinical features, diagnosis and laboratory investigation. Med. J. Austr. **1937**, 2, 281—299.

— (2) The epidemiology of ,,Q'' fever: A review. Med. J. Austr. **1953**, 1, 245—253.

— A. H. Berry, J. I. Tonge and H. E. Brown: Fevers of the Mackay district, Queensland. Med. J. Austr. **1953**, 2, 121—129.

—, and D. J. W. Smith: Studies on the epidemiology of Q fever. 2. The isolation of three strains of Rickettsia burneti from the bandicoot, Isoodon torosus. Austr. J. exp. Biol. med. Sci. **18**, 99—102 (1940).

Deutsch, D. L., and E. T. Peterson: Q fever: Transmission from one human being to others. Report of three cases. J. Amer. med. Ass. **143**, 348—350 (1950).

Djourichitch, M.: Quelques caractéristiques de la grande épidemie de typhus survenue, au cours du printemps et de l'été 1942, chez les réfugiés sur les bords de la Drina en Yougoslavie. Bull. Soc. Path. exot. **46**, 286—287 (1953).

Doherty, R. L.: A clinical study of scrub typhus in North Queensland. Med. J. Austr. **1956**, 2, 212—220.

Dove, W., and B. Shelmire: Tropical rat mites, Liponyssus bacoti Hirst, vectors of endemic typhus. J. Amer. med. Ass. **97**, 1506—1510 (1931).

Downs, C. M., J. Fevurly and M. M. Meyer: Studies on hemagglutination inhibition phenomenon. I. the presence of inhibiting antigen in typhus-infected animals. J. Immunol. **75**, 35—42 (1955).

Dumoulin, F. V. G., and J. Bryan, with the collaboration of F. H. Diercks and M. Haccou: Scrub typhus in a South Sumatra oil field. Med. Bull. (New Jersey) **17**, 66—77 (1957).

Elyan, A., and M. M. Dawood: A serological survey of Q fever in Egypt. J. Egypt. publ. Hlth Ass. **29**, 185—190 (1954).

ENRIGHT, J. B., W. W. SADLER and R. C. THOMAS: Pasteurization of milk containing the organism of Q fever. Amer. J. publ. Hlth 47, 695—700 (1957).

EUSTIS, E. B., and H. S. FULLER: Rickettsialpox. II. Recovery of Rickettsia akari from mites, Allodermanyssus sanguineus, from West Hartford, Conn. Proc. Soc. exp. Biol. (N.Y.) 80, 546—549 (1952).

EVERITT, P. N., BHATT and J. P. FOX: Infection of man with avirulent rickettsiae of epidemic typhus (strain E). Amer. J. Hyg. 59, 60—73 (1954).

FASTIER, L. B.: A survey of 1300 human sera for Q fever antibodies. N. Z. med. J. 53, 365—367 (1954).

FEDOROWA, N. I., T. A. BEKTEMIROW, I. V. TARASEWITSCH, E. B. KERBABAEW and L. L. SEMASCHKO: The prevalence of Q fever in cotton workers. Ž. Mikrobiol. (Mosk.) 1956, H. 11, 27—30 [Russisch].

FENNER, F.: The epidemiology of North Queensland tick typhus: Natural mammalian hosts. Med. J. Austr. 1946, 2, 666—668.

FORYŚ, S., R. LUTYŃSKI and Z. RAGINIS: The antibody level in persons convalescent after typhus fever measured by complement-fixation test. Przegl. epidem. 11, 157—162 (1957) [Polnisch].

FOX, J. P.: (1) The long persistence of Rickettsia orientalis in the blood and tissues of infected animals. J. Immunol. 59, 109—114 (1948).

— (2) The neutralization technique in tsutsugamushi diesease (scrub typhus) and the antigenic differentiation of rickettsial strains. J. Immunol. 62, 341—357 (1949).

— (3) A review of experience with an avirulent strain of R. prowazeki (strain E) as a living agent for immunizing man against epidemic typhus. Amer. J. publ. Hlth 45, 1036—1048 (1955).

— (4) Immunization against epidemic typhus. A brief general review and description of the status of living, avirulent R. prowazeki (strain E) as an immunizing agent. Amer. J. trop. Med. Hyg. 5, 464—479 (1956).

— M. G. EVERRITT, T. A. ROBINSON and D. P. CONWELL: Immunization of man against epidimic typhus by infection with avirulent Rickettsia prowazeki (strain E). Observations as to post-vaccination reactions, the relation of serologic response to size and route of infecting dose, and the resistance of challenge with virulent typhus strain. Amer. J. Hyg. 59, 74—88 (1954).

— M. E. JORDAN, D. P. CONWELL and T. A. ROBINSON: Immunization of man against epidemic typhus by infection with avirulent Rickettsia prowazeki (strain E). II. The seroimmune state and resistance to virulent challenge two years after immunization and a note as to the nature of immediate postvaccination reactions. Amer. J. Hyg. 61, 174—182 (1955).

— M. E. JORDAN and H. M. GELFAND: Immunization of man against epidemic typhus by infection with avirulent Rickettsia prowazeki strain E. IV. Persistence of immunity and a note as to differing complement-fixation antigen requirements in post-infection and post-vaccination sera. J. Immunol. 79, 348—354 (1957).

— J. A. MONTOYA, M. E. JORDAN and M. ESPINOSA: Immunization of man against epidemic typhus by infection with avirulent Rickettsia prowazeki (strain E). III. The serologic response and occurrence of post-vaccination reactions in groups vaccinated under field conditions in Peru. Amer. J. Hyg. 61, 183—196 (1955).

FRANKLIN, J., E. WASSERMANN and H. S. FULLER: Rickettsialpox in Boston. Report of a case. New Engl. J. Med. 244, 509—511 (1951).

FREEMAN, G., G. VARELA, H. PLOTZ and C. ORTIZ MARIOTTE: Typhus fever in Mexico: A study of epidemiology by means of complement-fixation. Amer. J. trop. Med. 29, 63—69 (1949).

FREYCHE, M. J., and Z. DEUTSCHMAN: Human rickettsioses in Africa. Epidem. vital. Statist. Rep. 3, 160—201 (1950).

FUJITA, H., T. SUZUKI and T. HORIKAWA: Epidemiological studies on 20 days fever in Mie-prefecture. Part 1: On Rickettsia isolated in Mie-prefecture. Part 2: On Trombiculid mites in Mie-prefecture. Mie med. J. 6, 153—169 (1956).

FUKUZUMI, S., Y. OBATA and R. KAGIWADA: On the trombiculid mites and rickettsia discovered in the M. Fuji foot plain. Kitasato Arch. exp. Med. 23, 11—22 (1951).

Fuller, H. S.: (1) Infestation of man in Burma with Trombiculid mites, with special reference to Trombicula deliensis. Amer. J. Hyg. 45, 363—371 (1947).
— (2) Studies of human body lice, Pediculus humanus corporis. II. Quantitative comparisons of the susceptibility of human body lice and cotton rats to experimental infection with epidemic typhus rickettsiae. Amer. J. Hyg. 58, 188—206 (1953).
— (3) Studies of rickettsialpox. III. Life cycle of the mite vector, Allodermanyssus sanguineus. Amer. J. Hyg. 59, 236—239 (1954).
— (4) Studies of human body lice, Pediculus humanus corporis. III. Initial dosage and ambient temperature as factors influencing the course of infection with Rickettsia prowazeki. Amer. J. Hyg. 59, 140—149 (1954).
— (5) Human body lice. IV. Direct serial passage of typhus rickettsiae by oral infection. Proc. Soc. exp. Biol. (N.Y.) 85, 151—153 (1954).
— (6) Veterinary and medical acarology. Ann. Rev. Entomol. 1, 347—366 (1956).
— E. S. Murray, J. C. Ayres, J. C. Snyder and L. Potash: Studies of rickettsialpox. I. Recovery of the causative agent from house mice in Boston, Massachusetts. Amer. J. Hyg. 54, 82—100 (1951).
—, and J. E. Smadel: Rickettsial diseases and the Korea conflict.
Recent advances in medicine and surgery. Medical Sci. Publ. No 4, Army Medical Service Graduate School, Walter Reed Army Medical Center, Washington, p. 304—310, 1954.
—, and J. C. Snyder: Studies of human body lice, Pediculus humanus corporis. I. A method for feeding lice through a membrane and experimental infection with Rickettsia prowazeki, R. mooseri, and Borrelia novyi. Publ. Hlth Rep. (Wash.) 64, 1287—1292 (1949).
Gadjusek, D. C., and M. Bahmanyar: Sur la Q fever en Iran. Bull. Soc. Path. exot. 48, 31—33 (1955).
Gaon, J.: Recidive Pjegavca (Brill-Zinsserova Bolest). Med. Arh. (Sarajevo) 9, 21—34 (1955) [Jugoslawisch].
Garabedian, G. A., and A. G. Djanian: Q fever in Lebanon (Middle East). II. Attempts to demonstrate by animal inoculations the presence of C. burneti in milk samples. Amer. J. Hyg. 63, 313—318 (1956).
— — and E. A. Johnston: Q fever in Lebanon (Middle East). I. The presence of complement-fixing antibodies in serum samples obtained from residents of Lebanon. Amer. J. Hyg. 63, 308—312 (1956).
Garin, J. P.: La ,,Q" fever au Maroc. Enquête épidémiologique au Tafilalet. Bull. Inst. Hyg. Maroc 11, 185—206 (1951).
Gear, J.: The rickettsial diseases of Southern Africa. S. Afr. J. clin. Sci. 5, 158—175 (1954).
—, and B. de Meillon: The common dog tick Haemaphysalis leachi as a vector of tick typhus. S. Afr. med. J. 13, 815—816 (1939).
— B. Wolstenholme, R. Harwin and F. H. Stakes: Brill's disease (recrudescent epidemic typhus fever): Its occurrence in South Africa. S. Afr. med. J. 26, 566—569 (1952).
Geigy, R.: Ein Zeckentest zum Diagnostizieren des Q-Fiebers. Bull. schweiz. Akad. med. Wiss. 7, 1—8 (1951).
Gelfland, M., and B. Berney: Q fever in Southern Rhodesia. J. trop. Med. Hyg. 56, 184—186 (1953).
Germer, W. D.: Das Q-Fieber beim Meerschweinchen. Ein Beitrag zur allgemeinen Infektionslehre bei einem nur intracellulär gedeihenden Erreger. Arch. Virusforsch. 5, 336—344 (1954).
—, u. B. Glockner: Epidemiologische Q-Fieberstudien in Württemberg-Hohenzollern. Z. Hyg. Infekt.-Kr. 138, 56—65 (1953).
Gildemeister, E., u. E. Haagen: Fleckfieberstudien. I. Mitt. Nachweis eines Toxins in Rickettsien-Eikulturen (Rickettsia mooseri). Dtsch. med. Wschr. 1940, 878—880.
Gilford, J. H., and W. H. Price: Virulent-avirulent conversions of Rickettsia rickettsii in vitro. Proc. Nat. Acad. Sci. (Wash.) 41, 870—873 (1955).
Giroud, P.: Une mission scientifique au Moyen Congo, en Oubangui, au Ruanda-Urundi, au Katanga en Afrique du Sud. Rev. colon. Méd. Chir. 22, 352—558 (1950).
— J. Colas-Belcour, R. Pfister et P. Morel: Amblyomma, Hyalomma, Boophilus, Rhipicephalus d'Afrique sont porteur d'éléments rickettsiens et néorickettsiens et quelquefois des deux types d'agents. Bull. Soc. Path. exot. 50, 529—532 (1957).

GIROUD, P., et J. A. GAILLARD: Les possibilités de diffusion d'une souche de typhus sporadique de type épidémiologique. Bull. Soc. Path. exot. **45**, 350—357 (1952).

—, et M. GOULON: Maladie de Derrick et Burnet ou „fièvre Q" en France. Bull. Inst. nat. Hyg. (Paris) **7**, 877—883 (1952).

—, et J. JADIN: (1) Le pou dans la fièvre Q au Ruanda-Urundi (Congo Belge). Conservation naturelle de l'antigène de la fièvre Q sur le pou. Bull. Soc. Path. exot. **43**, 674—675 (1950).

— — (2) Présence des anticorps vis-à-vis de Rickettsia orientalis chez les indigènes et les asiatiques vivant au Ruanda-Urundi (Congo-Belge). Bull. Soc. Path. exot. **44**, 50—51 (1951).

— — (3) Comportement des animaux domestiques au Ruanda-Urundi (Congo Belge) vis-à-vis de l'antigène épidémique. Bull. Soc. Path. exot. **46**, 870—871 (1953).

— — (4) Infection latente et conservation de Rickettsia burneti chez l'homme; le rôle du pou. Bull. Soc. Path. exot. **47**, 764—765 (1954).

— — (5) Comportement sérologique et isolement de souches néorickettsiennes chez des veaux en allaitement. C. R. Acad. Sci. (Paris) **243**, 721—724 (1956).

— — et A. JEZIERSKI: Le pou peut-il jouer un rôle dans la transmission de la fièvre Q ? C. R. Soc. Biol. (Paris) **145**, 569—570 (1951).

— F. ROGER et N. DUMAS: Contribution à l'étude des néorickettsioses. L'évolution des anticorps au cours des diverses maladies de l'homme et des animaux. Bull. Soc. Path. exot. **48**, 21—24 (1955).

GISPEN, R.: (1) The rôle of rats and mites in the transmission of epidemic scrub typhus in Batavia. Docum. neerl. indones. Morb. trop. **2**, 23—35 (1950).

— (2) The virus of murine typhus in mites (Schöngastia indica, fam. Trombiculidae). Docum. neerl. indones Morb. trop. **2**, 223—230 (1950).

— K. H. GAN and C. D. WESTERMANN: Endemic typhus in Java. I. Incidence and distribution of shop typhus. Docum. neerl. indones Morb. trop. **3**, 60—66 (1951).

—, and R. WARSA: Endemic typhus in Java. II. The natural infection of rats and rat ectoparasites; identity of shop typhus and murine typhus. Docum. neerl. indones Morb. trop. **3**, 155—162 (1951).

GOULD, D. J., and M. L. MIESSE: Recovery of a Rickettsia of the spotted fever group from Microtus pennsylvanicus from Virginia. Proc. Soc. exp. Biol. (N.Y.) **85**, 558—561 (1954).

GREIFF, D., E. L. POWERS and H. PINKERTON: Further studies on the effects of x-radiation on the multiplication of Rickettsia mooseri in embryonated eggs. J. exp. Med. **105**, 217—222 (1957).

GRUMBACH, A., u. W. KIKUTH: Die Infektionskrankheiten des Menschen und ihre Erreger. Stuttgart: Georg Thieme 1958. 1702 S.

GUTFREUND, R.. V. GELBERG et M. A. CHABAUD: Enquête préliminaire sur la fièvre Q en Éthiopie. Bull. Soc. Path. exot. **48**, 451—453 (1955).

HAMILTON, H. L.: Specificity to the toxic factors associated with the epidemic and murine strain of typhus rickettsiae. Amer. J. trop. Med. **25**, 391—395 (1945).

HARRISON, J. L., and J. R. AUDY: Hosts of the mite vector of scrub typhus. I. A check-list of the recorded hosts. II. An analysis of the list of recorded hosts. Ann. trop. Med. Parasit. **45**, 171—194 (1951).

HARSHMAN, S., J. JOHNSON and W. H. PRICE: The effect of red blood cells on experimental infection of white rats with the R strain of Rickettsia rickettsii. Amer. J. Hyg. **67**, 109—117 (1958).

HASS, G. M., and H. PINKERTON: Spotted fever. II. An experimental study of fièvre butonneuse. J. exp. Med. **64**, 601—623 (1936).

HAYASHI, H.: On the causative agent of tsutsugamushi disease isolated from field voles, Apodemus speciosus speciosus, inhabiting the foot area of Mt. Fuji in Japan. Kitasato Arch. exp. Med. **23**, 13—19 (1951).

— K. NAKAMURA, I. KOMINE and A. KAWAMURA: Studies on Rickettsia tamiyai. III. A report on a case of rickettsiosis experimentally infected with R. tamiyai. Jap. J. exp. Med. **24**, 405—409 (1954).

HEISCH, R. B.: Rickettsiae from ticks and rodents in Kenya. Trans. roy. Soc. trop. Med. Hyg. **51**, 287 (1957).

Heisch, R. McPhee and L. R. Rickman: The epidemiology of tick-typhus in Nairobi. E. Afr. med. J. **34**, 459—477 (1957).

Heisenger, M. H., E. S. Murray and S. Cohen: Rickettsialpox case due to laboratory infection. Publ. Hlth Rep. (Wash.) **66**, 311—316 (1951).

Hengel, R., G. A. Kausche, A. Laur u. K. Rabenschlag: Das Q-Fieber. Ergebn. inn. Med. **5**, 220—305 (1954).

— — u. E. Sheris: Über zwei dörfliche Q-Fieber-Epidemien in Baden. Dtsch. med. Wschr. **1950**, 1505—1507.

Herzberg, K., H. Herzberg-Kremmer u. H. Urbach: Weitere Befunde zur Morphologie und Diagnostik des Q-Fieber-Erregers. Zbl. Bakt., I. Abt. Orig. **156**, 14—22 (1950).

—, u. G. May: Untersuchungen an einem Q-Fieber-Ausbruch des Jahres 1954. IV. Mitt. Über europäische Q-Fieber-Antigene. Z. Immun.-Forsch. **112**, 145—161 (1955).

Hoenig, W., u. W. Mohr: Zur Dauer des Wolhynischen Fiebers. Med. Klin. **1954**, 1034—1037.

Hoeprich, P. D., G. T. Kent and J. H. Dingle: Rickettsialpox: Report of a serologically proved case in Cleveland. New Engl. J. Med. **254**, 25—27 (1956).

Hoogerheide, C., and G. J. Ensink: Two small explosions of scrub typhus in Netherlands New Guinea. Docum. Neerl. indones. Morb. trop. **9**, 94—96 (1957).

Imamaliew, S. A.: The clinic-epidemiologic characteristics of endemic (rat) typhus. Ž. Mikrobiol. (Mosk.) **1957**, H. 3, 47—53 [Russisch].

Irons, J. V., R. B. Eads, C. W. Johnson, O. L. Walker and M. A. Norris: Southwest Texas Q fever studies. J. Parasit. **38**, 1—5 (1952).

— — and J. E. Peavy: Q fever in Texas. Texas Rep. Biol. Med. **15**, 896—903 (1957).

Jackson, E. B., J. X. Danauskas, M. C. Coale and J. E. Smadel: Recovery of Rickettsia akari from the Korean vole Microtus fortis pelliceus. Amer. J. Hyg. **66**, 301—308 (1957).

— — J. E. Smadel, H. S. Fuller, M. C. Coale and F. M. Bozeman: Occurrence of Rickettsia tsutsugamushi in Korean rodents and chiggers. Amer. J. Hyg. **66**, 309—320 (1957).

—, and J. E. Smadel: Immunization against scrub typhus. II. Preparation of lyophilized living vaccine. Amer. J. Hyg. **53**, 326—331 (1951).

Jatzimirskaya-Kirontowskaya, M. K., A. F. Bilibin, T. W. Botscharowa, G. A. Sinaiko, E. P. Sawitzkaya u. I. I. Schatrow: Zur Frage der Möglichkeit eines langdauernden Trägertums von R. prowazeki. Ž. Mikrobiol. (Mosk.) **1956**, H. 7, 33—39 [Russisch].

Jellison, W. L., H. H. Welsh, B. E. Elson and R. J. Huebner: Q fever studies in Southern California. XI. Recovery of Coxiella burnetii from milk of sheep. Publ. Hlth Rep. (Wash.) **65**, 395—399 (1950).

Johnson, Ph. T., and R. Traub: Maintenance of Rickettsia tsutsugamushi in mealworm (Tenebrio molitor). J. Parasit. **38**, Suppl., 24—25 (1952).

Joint WHO Study group on African rickettsioses. Report on the first session. World Health Org. techn. Rep. Ser. No 23, 1950, 23 S.

Jones, B. M.: The penetration of the host tissue by the harvest mite, Trombicula autumnalis Shaw. Parasitology **40**, 247—260 (1950).

Jusatz, H. F.: Zunahme der Zweit-Erkrankungen an Fleckfieber? Ärztl. Wschr. **1955**, 499—500.

Kalra, S. L., and K. N. A. Rao: (1) Typhus fevers in Kashmir state. Part I. Epidemic typhus. Indian J. med. Res. **37**, 395—400 (1949).

— — (2) Typhus group of fevers in Jubbulpore area. Indian J. med. Res. **37**, 373—384 (1949).

— — (3) Typhus in Bangalore and Mysore. Indian J. med. Res. **39**, 303—309 (1951).

— — (4) Typhus fevers in Kashmir state. Part II. Murine typhus. Indian J. med. Res. **39**, 297—302 (1951).

— — and B. L. Taneja: Q fever in India: A serological survey. Indian J. med. Res. **42**, 315—318 (1954).

Kaplan, M. M., and P. Bertagna: The geographical distribution of Q fever. Bull. Wld Hlth Org. **13**, 829—860 (1955).

Karp, A.: An immunological purification of typhus rickettsiae. J. Bact. **67**, 450—455 (1954).

Kausche, G. A., u. E. Sheris: Zur Morphologie der Rickettsia burneti. Z. Hyg. Infekt.-Kr. **133**, 148—159 (1951).

KAWAMURA, A.: Notes on the new strain of rickettsia, Rickettsia tamiyai n. sp., isolated from voles in Hokkaido. Jap. J. exp. Med. **24**, 47—50 (1954).

— Y. TSUNEMATSO, K. NISHIOKA, K. YAMANÉ and M. SAITO: Studies on Rickettsia tamiyai. I. Biological characters of Rickettsia tamiyai. Jap. J. exp. Med. **24**, 385—392 (1954).

— K. YAMANÉ and K. KAWAI: Studies on Rickettsia tamiyai. II. On the experimental infection of monkey. With pathological-anatomical description on a fatal case. Jap. J. exp. Med. **24**, 393—403 (1954).

KEATON, R., B. J. NASH, J. N. MURPHY and J. V. IRONS: Complement fixation tests for murine typhus on small animals. Publ. Hlth Rep. (Wash.) **68**, 28—30 (1953).

KHAN, N.: Scrub typhus (as seen in Jamshedpur). Ind. J. med. Sci. **4**, 487—497 (1950).

KIKUTH, W., u. M. BOCK: 23 Fälle von Laborinfektionen mit Q-Fieber. Med. Klin. **1949**, 1056—1060.

KIRBERGER, E.: Die Resistenz der Coxiella burneti in vitro. Z. Tropenmed. Parasit. **3**, 77—86 (1951).

KIREJEWA, R. J.: Clinical features of the tick-borne typhus of Northern Asia. Ž. Mikrobiol. (Mosk.) **1956**, H. 9, 73—77 [Russisch].

KISELEW, R. I., and G. I. WOLTSCHANETSKAJA: The importance of the mite Allodermanyssus sanguineus in the epidemiology of smallpox-like rickettsiosis. In E. N. PAWLOWSKI, P. A. PETRISCHTSCHEWA, D. N. SASUKHIN i N. G. OLSUF'EW, Natural nidi of human diseases and regional epidemiology. Leningrad 1955. 532 S. [Russisch.]

KITAOKA, M., and A. SHISHIDO: Surviving period of Rickettsia in feces from lice infected with epidemic and murine typhus. Jap. med. J. **3**, 265—272 (1950).

— N. TAKEMORI, A. SHISHIDO and K. JO: On epidemiology of murine typhus in Saitama 1948, in Niigata 1949 und multiple occurrence of murine typhus in one family. Jap. med. J. **4**, 143—155 (1951).

KITZE, L. K. (with the assistance of H. C. HIEMSTRA and M. S. MOORE): Q fever in Wisconsin. Serologic evidence of infection in cattle and in human beings and recovery of C. burneti from cattle. Amer. J. Hyg. **65**, 239—247 (1957).

KNORRE, G. v.: Über Fleckfieber-Spätrezidive. Medizinische **1957**, Nr 2, 85—87.

KOENOKI, S.: A study of the toxin of the Tsutsugamushi isolated from wild rats in Kyushu. Igaku Kenkyu **27**, 2105—2125 (1957) [Japanisch].

KOHLS, G. M.: Vectors of rickettsial diseases. In: Rickettsial diseases of man, p. 83—96. Amer. association for the advancement of science. Washington 1948.

KOLLE, W., H. HETSCH u. H. SCHLOSSBERGER: Experimentelle Bakteriologie und Infektionskrankheiten, 11. Aufl., S. 564—602. München u. Berlin: Urban & Schwarzenberg 1953.

KOŁŁOTO, B.: Fleckfieber im Distrikt Bialystok von 1946—1954. Przegl. epidem. **11**, 11—19 (1957) [Polnisch].

KORSCHUNOWA, O. S., and S. P. PETROWA-PIONTKOWSKAJA: On the vector of Marseilles fever. Doklady vsesosjus akad. Nauk **68**, 1151—1153 (1949).

KOSTRZEWSKI, J.: (1) The epidemiology of trench fever. Bull. Acad. pol. Sci. Lettr. **7/10**, 233—263 (1949).

— (2) Epidemiology of sporadic typhus in Poland in 1952—1954. Przegl. epidem. **10**, 1—17 (1956) [Polnisch].

— (3) Typhus exanthématique sporadique en Pologne. Ann. Inst. Pasteur **91**, 15—24 (1956).

— (4) Investigations on the epidemiology of typhus recrudescenses in Poland. J. Hyg. Epidemiol., Microbiol. Immun. **1**, 237—244 (1957).

— A. GRUŻEWSKI and L. MILEWSKA: Further remarks on the possible anticipation of recrudescense of typhus. Przegl. epidem. **11**, 21—26 (1956) [Polnisch].

KOUWENAAR, W., and H. ESSEVELD: The nature of immunity against scrub typhus in guinea pigs. Docum. neerl. indones. Morb. trop. **1**, 34—40 (1949).

KRISHNAN, K. V., R. O. A. SMITH, P. N. BOSE, K. N. NEOGY, B. K. G. ROY and M. GHOSH: Transmission of Rickettsia orientalis by the bite of the larvae of Trombicula deliensis. Indian med. Gaz. **84**, 41—43 (1949).

KRYŃSKI, S.: Forms of Rickettsia prowazeki infection in lice, artificially infected by Weigl's method. Bull. Inst. mar. trop. Med. Gdańsk. **2**, 231—234 (1949).

Kryński, S., and E. Becla: Influence of milk on viability and infectivity of R. prowazeki. Bull. Inst. mar. trop. Med. Gdańsk. **6**, 205—209 (1955).

—, and J. Radkowiak: (1) Principles of cultivation of Rickettsia prowazeki by Weigl's method. Bull. Inst. mar. trop. Med. Gdańsk. **4**, 237—242 (1953).

— — (2) Further studies on the influence of dyes on R. prowazeki. Med. dośw. Mikrobiol. **5**, 449—456 (1953) [Polnisch].

Kulagin, S. M., and V. A. Silitsch: Q fever in the Grozny district. Ž. Mikrobiol. (Mosk.) **1956**, H. 11, 35—39 [Russisch].

Kuwata, T.: Reproduction pattern of Rickettsia tsutsugamushi and ornithosis virus in the mouse brain. Naturwissenschaften **41**, 43—44 (1954).

La Boccetta, A. C., H. L. Israel, A. M. Perri and M. M. Sigel: Rickettsialpox. Report of four apparent cases in Pennsylvania. Amer. J. Med. **13**, 413—422 (1952).

Lackman, D., and R. R. Parker: The serological characterization of North-Queensland tick typhus. Publ. Hlth Rep. (Wash.) **63**, 1624—1628 (1948).

Lacroix, A. C., A. Sayag et Th. Douard: Observations sérologiques sur la Q fever (Burnet-Derrick) dans la région algéroise. Bull. Soc. Path. exot. **49**, 18—21 (1956).

Lass, R.: Qu-Fieber in Österreich. Wien. klin. Wschr. **1952**, 159—162.

Lawley, B. J.: The discovery, investigation and control of scrub typhus in Singapore. Trans. roy. Soc. trop. Med. Hyg. **51**, 56—61 (1957).

Le Gac, P., et P. Giroud: Rickettsiose vésiculeuse en Oubangui-Chari (A.E.F.). Bull. Soc. Path. exot. **44**, 813—815 (1951).

— —, P. Giroud, A. le Henaff et G. Baup: Épidémie familiale de rickettsiose varicelliforme dans un village de l'Oubangui-Chari (A.E.F.). Bull. Soc. Path. exot. **45**, 19—23 (1952).

— — F. Roger, E. Courmes et P. Bres: Étude de quatre souches de Rickettsia orientalis isolées au cours des opérations militaires au Vietnam. Bull. Soc. Path. exot. **49**, 338—345 (1956).

— — — et N. Dumas: Au sujet des infections varicelliformes d'origine rickettsienne. Bull. Soc. Path. exot. **48**, 314—316 (1955).

Laigret, J., et J. Jadin: Sensibilité de la souris blanche aux virus typhiques — passages; conservation du virus dans le cerveau des souris. Arch. Inst. Pasteur Tunis **21**, 381—397 (1933).

Lennette, E. H., and W. H. Clark: Observations on the epidemiology of Q fever in Northern California. J. Amer. med. Ass. **145**, 306—309 (1951).

— — M. M. Abinanti, O. Brunetti and J. M. Covert: Q fever studies. XIII. The effect of pasteurization on Coxiella burneti in naturally infected milk. Amer. J. Hyg. **55**, 246—253 (1953).

— M. A. Holmes and F. R. Abinanti: Q fever studies. XIV. Observations on the pathogenesis of the experimental infection induced in sheep by the intravenous route. Amer. J. Hyg. **55**, 254—267 (1952).

—, and H. H. Welsh: Q fever in California. X. Recovery of Coxiella burneti from the air of premises harboring infected goats. Amer. J. Hyg. **54**, 44—49 (1951).

Lépine, P., et N. Lorando: Le typhus exanthématique du chat. Bull. Soc. Path. exot. **28**, 356—360 (1935).

—, et V. Sautter: Sur la durée de conservation du typhus murin dans l'encéphale du spermophile. Bull. Soc. Path. exot. **29**, 13—16 (1936).

Ley, H. L., J. E. Smadel, F. H. Diercks and P. Y. Paterson: Immunization against scrub typhus. V. The infective dose of Rickettsia tsutsugamushi for men and mice. Amer. J. Hyg. **56**, 313—319 (1952).

Leschtschinskaja, E. V., T. P. Powalischina and S. G. Dzagurow: The distribution of Q fever in the U.S.S.R. Problems Virology **2**, 8—11 (1957).

Liebermeister, K., u. E. Zehender: Zur Morphologie und systematischen Stellung des Q-Fieber-Erregers. Klin. Wschr. **1950**, 276—278.

Likar, M.: The isolation of R. burneti from goat's milk. Zdrav. Vestn. **24**, 216—217 (1955) [Jugoslawisch].

Liu, W. T.: (1) Studies on the murine origin of typhus epidemics in North China. III. Isolation of murine typhus rickettsiae from rats, ratfleas and body-lice of patients during an epidemy in a poorhouse. Chin. med. J. **62**, 119—139 (1944).

Liu, W. T.: (2) Isolation of typhus rickettsiae from rat mites, Liponyssus bacoti, in Peiping. Amer. J. Hyg. **45**, 58—66 (1947).

—, and E. Landauer: Observations on a non-typhus rickettsia-like organism in the human body louse, Pediculus humanus var. corporis. Chin. med. J. **70**, 115—139 (1952).

—, and S. H. Zia: (1) Typhus rickettsiae isolated from mice and mouse-fleas during an epidemic in Peiping. Proc. Soc. exp. Biol. (N.Y.) **45**, 823—826 (1940).

— — (2) Studies on the murine origin of typhus epidemics in North China. I. Murine typhus rickettsia isolated from body lice in the garments of a sporadic case. Amer. J. trop. Med. **21**, 507—523 (1941).

Luoto, L. and R. J. Huebner: Q fever studies in Southern California. IX. Isolations of Q fever organisms from parturient placentas of naturally infected dairy cows. Publ. Hlth Rep. (Wash.) **65**, 541—544 (1950).

Lutyński, R., Z. Raginis, T. Ziemichód and A. Koźmińska: Q fever outbreak in Kraków. Przegl. epidem. **11**, 69—79 (1957) [Polnisch].

Mackie, Th. T., G. W. Hunter and C. B. Worth: A manual of tropical medicine, 2. edit. Philadelphia and London: W. B. Saunders Company 1954. 907 S.

Magalhães, O. de: (1) Percentagem dos carrapatos portadores, na natureza, das reças V.B., V.A.$_1$ e V.A.$_2$ do tifo exantemático neotrópico no Brasil. (Nota prévia.) Brasil-méd. **64**, 85—86 (1950).

— (2) Contribuição ao conhecimento das doenças do grupo tifo exantemático. Monografias do Instituto Oswaldo Cruz Nr 6, Rio de Janeiro 1952, 968 S.

— (3) Contribuição para o conhecimento das doenças do grupo tifo exantemático no Brasil. Mem. Inst. Osw. Cruz **54**, 279—308 (1956).

Malloch, R. A., and M. P. Stoker: Studies on the susceptibility of Rickettsia burneti to chemical disinfectants, and on techniques for detecting small numbers of viral organisms. J. Hyg. (Lond.) **50**, 502—514 (1952).

Mantovani, A., and P. Benazzi: The isolation of Coxiella burneti from Rhipicephalus sanguineus from naturally infected dogs. J. Amer. vet. med. Ass. **122**, 117—118 (1953).

Marmion, B. P.: Q fever. II. Natural history and epidemiology of Q fever in man. Trans. roy. Soc. trop. Med. Hyg. **48**, 197—203 (1954).

—, and M. S. Harvey: The varying epidemiology of Q fever in the South-East region of Great Britain. I. In an urban area. J. Hyg. (Lond.) **54**, 533—546 (1956).

— F. O. MacCallum, A. Rowlands and T. C. Thiel: The effect of pasteurization on milk containing Rickettsia burneti. Monthly Bull. Minist. Hlth Lab. Serv. **10**, 119—128 (1951).

—, and M. G. P. Stoker: The varying epidemiology of Q fever in the South-East region of Great Britain. II. In two rural areas. J. Hyg. (Lond.) **54**, 547—561 (1956).

— — J. H. McCoy, R. A. Malloch and B. Moore: Q fever in Great Britain. An analysis of 69 sporadic cases, with a study of the prevalence of infection in humans and cows. Lancet **1953 I**, 503—510.

Martin, R., et F. Maroger: Premier cas à Paris de fièvre boutonneuse autochtone. Bull. Acad. Méd. (Paris) **136**, 13—15 (1952).

Mathur, T. N.: Murine typhus in Karnal city. Indian med. Gaz. **88**, 623—626 (1953).

—, and A. R. Suri: Scrub typhus in the Karnal district of Punjab (India). Indian med. Gaz. **88**, 621—622 (1953).

Maurin, J.: Recherches sur l'existence de la fièvre Q en Tunisie par la réaction de dérivation de complément. Ann. Inst. Pasteur **86**, 69—75 (1954).

Maxcy, K. F.: (1) An epidemiological study of endemic typhus (Brill's disease) in the Southeastern United States, with special reference to its mode of transmission. Publ. Hlth Rep. (Wash.) **41**, 2967—2995 (1926).

— (2) Endemic typhus fever of the Southeastern United States. Reaction of the guineapig. Publ. Hlth Rep. (Wash.) **44**, 589—600 (1929).

McCroan, J. E., R. L. Ramsey, W. J. Murphy and L. S. Dick: The status of Rocky Mountain spotted fever in the Southeastern United States. Publ. Hlth Rep. (Wash.) **70**, 319—325 (1955).

Megaw, J. W. D.: A typhus-like fever in India, possibly transmitted by ticks. Indian med. Gaz. **56**, 361—371 (1921).

MEHTA, D. R.: Studies on typhus in the Simla Hills. Part IX. On the life history of Trombicula deliensis, Walch, a suspected vector of typhus in the Simla Hills. Indian J. med. Res. **36**, 159—171 (1948).

MEIRA, J. A., M. JAMRA e J. LODOVICI: Moléstia de Brill — (recrudescência de tifo epidêmico). Rev. Hosp. Clin. (S. Paulo) **10**, 237—246 (1955).

MELNOTTE, P., et B. MENGUS: Premier cas autochtone de fièvre boutonneuse en Lorraine. Bull. Soc. Path. exot. **49**, 609—612 (1956).

MIKA, L. A., R. J. GOODLOW, J. VICTOR and W. BRAUN: Studies on mixed infections. I. Brucellosis and Q fever. Proc Soc. exp. Biol. (N.Y.) **87**, 500—507 (1954).

MILLER, J. K.: Rocky Mountain spotted fever in Long Island. Ann. intern. Med. **33**, 1398 bis 1406 (1950).

MINOUN, G., M. PIERROU, G. VASTEL et P. MARC: Enquête sur une épidémie massive de Q fever survenue à Batna (Algérie) chez de jeunes militaires de la Métropole. Bull. Soc. Path. exot. **48**, 590—598 (1955).

MITTERMAYER, T., u. I. KRATOCHVÍL: Brill-Zinsser disease in Košice county. Čas. Lék. čes. **97**, 337—341 (1958) [Tschechisch].

MOHR, C. O., N. E. GOOD and J. H. SCHUBERT: Status of murine typhus infection in domestic rats in the United States, 1952, and relation to infestation by Oriental rat fleas. Amer. J. Publ. Hlth **43**, 1514—1522 (1953).

— W.: Das Wolhynische Fieber. Betrachtungen zur Klinik und zur Frage der Spätfolgen. Med. Mschr. **10**, 220—224 (1956).

—, u. W. HIRTE: Das Wolhynische Fieber. Ergebn. inn. Med. Kinderheilk. **5**, 97—155 (1954).

MOOSER, H.: (1) Reaction of guinea-pigs to Mexican typhus (Tabardillo). Preliminary note on bacteriological observations. J. Amer. med. Ass. **91**, 19—20 (1928).

— (2) On the nomenclature of the agent of murine typhus. Amer. J. trop. Med. **28**, 841—843 (1948).

— (3) Die Beziehungen des murinen Fleckfiebers zum klassischen Fleckfieber. Acta trop. (Basel) Suppl. **4**, 1—87 (1945).

— (4) Die Wandlungen der Auffassungen über die Epidemiologie des epidemischen Fleckfiebers in den letzten 50 Jahren. Schweiz. Z. Path. **16**, 807—920 (1953).

— (5) Die Rickettsien. Rickettsiosen. In A. GRUMBACH u. W. KIKUTH: Die Infektionskrankheiten des Menschen und ihre Erreger, S. 167—197 und S. 1204—1239. Stuttgart: Georg Thieme 1958.

— M. R. CASTAÑEDA and H. ZINSSER: The transmission of the virus of Mexican typhus from rat to rat by Polyplax spinulosus. J. exp. Med. **54**, 567—575 (1931).

— H. R. MARTI u. A. LEEMANN: Beobachtungen an Fünftagefieber. 2. Mitt. Hautläsionen nach kutaner und intrakutaner Inokulation mit Rickettsia quintana. Schweiz. Z. Path. **12**, 476—483 (1949).

— u. F. WEYER: (1) Die Infektion des Rhesusaffen mit Fünftagefieber (Rickettsia quintana). Z. Tropenmed. Parasit. **4**, 513—539 (1953).

— — (2) Experimental infection of Macasus rhesus with Rickettsia quintana (Trench fever). Proc. Soc. exp. Biol. (N.Y.) **83**, 699—701 (1953).

MORLAN, H. B., E. L. HILL and J. H. SCHUBERT: Serological survey for murine typhus infection in Southwest Georgia animals. Publ. Hlth Rep. (Wash.) **65**, 57—63 (1952).

— B. C. UTTERBACK, J. E. DENT, M. J. WILCOMB, M. E. GRIFFITH and L. L. ELLIS: Domestic rats, rat ectoparasites and typhus control. Part I. Domestic rats in relation to murine typhus control. Publ. Health Serv. Publ. No 209, Publ. Health Monogr. No 4, Washington 1952, p. 1—20.

MUESBECK, C. F. B.: Scientific names of the body and head lice. J. econom. Entomol. **46**, 524 (1953).

MUNRO-FAURE, H. D., R. ANDREW, G. A. K. MISSEN and J. MACKAY-DICK: Scrub typhus in Korea. J. roy. Army med. Cps **97**, 227—229 (1951).

MURRAY, E. S., S. COHEN, J. JAMPOL, A. OFSTROCK and J. C. SNYDER: Epidemic typhus antibodies in human subjects in Boston, Massachusetts. New Engl. J. Med. **246**, 355—359 (1952).

Murray, E. S., T. Pšorn, P. Djaković, St. Sielski, V. Broz, F. Ljupsa, J. Gaon, R. Pavlević and J. C. Snyder: Brills disease: IV. Study of 26 cases in Yougoslavia. Amer. J. Publ. Hlth 41, 1359—1369 (1951).

—, and J. C. Snyder: (1) Brill's disease. II. Etiology. Amer. J. Hyg. 53, 22—32 (1951).

— — (2) Brill-Zinsser disease: The interepidemic reservoirs of epidemic louse-borne typhus fever. VI. Congr. Intern. Microbiol. Roma 4, 31—44 (1953).

Nagayo, M., Y. Miyagawa, T. Mitamura, T. Tamiya and S. Tenjin: Five species of tsutsugamushi (the carrier of Japanese river fever) and their relation to the tsutsugamushi disease. Amer. J. Hyg. 1, 569—591 (1921).

Nauck, E. G.: Lehrbuch der Tropenkrankheiten. Stuttgart: Georg Thieme 1956. 432 S.

—, u. F. Weyer: Laboratoriumsinfektionen bei Q-Fieber. Dtsch. med. Wschr. 1949, 198—202.

Neill, M. H.: Experimental typhus fever in guinea pigs. A description of a scrotal lesion in guinea pigs infected with Mexican typhus. Publ. Hlth Rep. (Wash.) 32, 1105—1108 (1917).

Neilson, G. H.: A case of Queensland tick typhus. Med. J. Aust. 1955 (1), 763—764.

Neva, F. A., and J. C. Snyder: (1) Studies on the toxicity of typhus rickettsiae. I. Susceptibility of the white rat, with a note on pathological changes. J. infect. Dis. 91, 72—78 (1952).

— — (2) Studies on the toxicity of typhus rickettsiae. III. Observations on the mechanism of toxic death in white mice and white rats. J. infect. Dis. 97, 73—87 (1955).

Nichols, E., M. E. Rindge and G. G. Russell: The relationship of the habits of the house mouse and the mouse mite (Allodermanyssus sanguineus) to the spread of rickettsialpox. Ann. intern. Med. 39, 92—102 (1953).

Nicolle, C., A. Conor and E. Conseil: Recherches expérimentales sur le typhus exanthématique entreprises à l'Institut Pasteur de Tunis pendant l'année 1910. Ann. Inst. Pasteur 25, 97—144 (1911).

Niedzielska, H.: Clinical analysis of sporadic cases of typhus. Przegł. epidem. 11, 415—420 (1957) [Polnisch].

Nigg, C., and K. Landsteiner: Studies on the cultivation of the typhus fever rickettsia in the presence of live tissues. J. exp. Med. 55, 563—575 (1932).

Nolden, J.: Eine Q-Fieber-Epidemie in der Nordwesteifel. Dtsch. med. Wschr. 1954, 1743—1745.

Özbil, M.: Ein Beitrag zur Frage der Rickettsienausscheidung mit dem Urin. Z. Tropenmed. 6, 453—459 (1955).

Oleś, A., u. K. Kurzeja: Krankheitsverlauf beim Menschen während eines Q-Fieber-Ausbruchs im Distrikt Rzeszów. Przegł. epidem. 11, 81—84 (1957) [Polnisch].

— — Z. Lewińska u. C. Frygin: Serologische Untersuchungen von Haustieren im ersten Q-Fieber-Herd in Polen. Przegł. epidem. 11, 85—89 (1957).

Ormsbee, R. A.: The growth of Coxiella burnetii in embryonated eggs. J. Bact. 63, 73—86 (1952).

— D. B. Lackman and E. G. Pickens: Relationships among complement-fixing values, infectious endpoints, and death curves in experimental Coxiella burnetii infection. J. Immunol. 67, 257—264 (1951).

Ortiz Mariotte, C.: Campaña nacional contra el tifo en México. Bol. Ofic. sanit. panamer. 34, 236—244 (1953).

Otto, R., u. R. Wohlrab: Über neuere Impfstoffe gegen Flecktyphus. Arb. Staatsinst. exp. Ther. Frankf. M. 1940, H. 40, 1—14.

Owen, C. R., and C. L. Larson: Studies on resistance to bacterial infections in animals infected with rickettsiae. J. exp. Med. 103, 753—763 (1956).

Pang, K. H.: Isolation of typhus rickettsia from rat mites during epidemic in an orphanage. Proc. Soc. exp. Biol. (N.Y.) 48, 266—267 (1941).

Parker, F., and F. A. Neva: Studies on the toxicity of typhus rickettsiae. II. Pathologic findings in white rats and white mice. Amer. J. Path. 30, 215—237 (1954).

— R. R.: (1) Transmission of Rocky Mountain spotted fever by the rabbit tick Haemaphysalis leporis-palustris Packard. Amer. J. trop. Med. 3, 39—45 (1923).

— (2) Symptomatology and certain other aspects of Rocky Mountain spotted fever. In: Rickettsial diseases of man. Amer. Ass. for the advancement of science, p. 139—146, Washington 1948.

Parker, R. R., J. F. Bell, W. S. Chalgren, F. B. Thrailkill and M. T. Mckee: The recovery of strains of Rocky Mountain spotted fever and tularemia from ticks of the Eastern United States. J. infect. Dis. **91**, 231—237 (1952).

— E. G. Pickens, D. B. Lackman, E. J. Bell and F. B. Thrailkill: Isolation and characterization of Rocky Mountain spotted fever rickettsiae from the rabbit tick Haemaphysalis leporis-palustris Packard. Publ. Hlth Rep. (Wash.) **66**, 455—463 (1951).

—, and E. A. Steinhaus: American and Australian Q fevers: Persistence of infectious agents in guinea pig tissues after defervescense. Publ. Hlth. Rep. (Wash.) **58**, 523—527 (1943).

— R. T., P. G. Menon, A. M. Merideth, M. J. Snyder and T. E. Woodward: Persistence of Rickettsia rickettsii in a patient recovered from Rocky Mountain spotted fever. J. Immunol. **73**, 383—386 (1954).

Payzin, S.: (1) La Q fever en Turquie. Arch. Inst. Pasteur Maroc 4, 13—18 (1949).

— (2) Epidemiological investigations on Q fever in Turkey. Bull. Wld Hlth Org. **9**, 553—558 (1953).

—, u. N. Akkay: Natural occurrence of Coxiella burneti in Ornithodorus lahorensis ticks. Türk. Ij. tecr. Biyol. Derg. **12**, 17—19 (1952).

Pereira, H. G., and J. Travassos: Estudo comparativo das rickettsias da febre maculosa das Montanhas Rochosas e do „tifo de S. Paulo" pela reaçãon de fixação do complemento. Hospital (Rio) **40**, 87—91 (1951).

Perepérez Palau, F.: Contribución al estudio geográfico y epidemiológico de la fiebre exantemática mediterránea. Rev. Sanid. Hig. públ. (Madr.) **25**, 472—475 (1951).

Pérez Gallardo, F.: Sobre el uso de la cepa E de R. prowazeki como vacuna contra el tifus exantemático. Z. Tropenmed. Parasit. 8, 464—476 (1957).

— G. Clavero y S. Hernández Fernández: Investigaciones sobre la epidemiología de la fiebre „Q" en España. Los conejos de monte y los lirones como reservorios de la Coxiella burneti. Rev. Sanid. Hig. públ. (Madr.) **26**, 81—87 (1952).

—, and J. P. Fox: Infection and immunization of laboratory animals with Rickettsia prowazekii of reduced pathogenicity, strain E. Amer. J. Hyg. 48, 6—21 (1948).

Philip, C. B.: (1) Nomenclature of the pathogenic rickettsiae. Amer. J. Hyg. **37**, 301—309 (1943).

— (2) Observations on experimental Q fever. J. Parasit. **34**, 457—464 (1948).

— (3) The reservoirs of infection in rickettsial diseases of man. In: Rickettsial diseases of man. Amer. ass. for the advancement of science, p. 97—112. Washington 1948.

— (4) Tick transmission of Indian tick typhus and some related rickettsioses. Exp. Parasit. 1, 129—143 (1952).

— (5) Nomenclature of the Rickettsiaceae pathogenic to vertebrates. Ann. N.Y. Acad. Sci. **56**, 484—494 (1953).

— (6) There's always something new under the „parasitological" sun. (The unique story of helminth-borne Salmon poisoning disease.) J. Parasit. **41**, 125—148 (1955).

— (7) Comments on the classification of the order Rickettsiales. Canad. J. Microbiol. **2**, 261—270 (1956).

— (8) Rickettsiales. In R. S. Breed, E. G. D. Murray and N. P. Smith, Bergey's Manual of determinative bacteriology, 7. edit., p. 934—957. Baltimore: Williams & Wilkins Company 1957.

— W. J. Hadlow and L. E. Hughes: Studies on Salmon poisoning disease of Canines. I. The rickettsial relationships and pathogenicity of Neorickettsia helmintheca. Exp. Parasit. **3**, 336—350 (1954).

—, and L. E. Hughes: The tropical rat mite, Liponyssus bacoti, as an experimental vector of rickettsialpox. Amer. J. trop. Med. **28**, 697—705 (1948).

—, and R. R. Parker: The persistence of the viruses of endemic (murine) typhus, Rocky Mountain spotted fever, and boutonneuse fever in tissues of experimental animals. Publ. Hlth Rep. (Wash.) **53**, 1246—1251 (1938).

—, and J. S. White: Disease agents recovered incidental to a tick survey of the Mississippi Golf Coast. J. econ. Entomol. 48, 396—400 (1955).

Pike, G., S. Cohen and E. S. Murray: Rickettsialpox. Report of a serologically proved case occurring in a resident of Boston. New Engl. J. Med. **243**, 913—915 (1950).

PINKERTON, H.: Criteria for the accurate classification of the rickettsial diseases (rickettsioses) and of their etiological agents. Parasitology 28, 172—189 (1936).

—, and G. M. HASS: Spotted fever. I. Intranuclear rickettsiae in spotted fever studied in tissue culture. J. exp. Med. 56, 151—156 (1932).

PLOTZ, H., J. E. SMADEL, T. F. ANDERSON and L. A. CHAMBERS: Morphological structure of rickettsiae. J. exp. Med. 77, 355—358 (1943).

— — B. L. BENNETT, R. L. REAGAN and M. J. SNYDER: North Queensland tick typhus: Studies of the etiological agent and its relation to other rickettsial diseases. Med. J. Aust. 1946 (2), 263—268.

POLLARD, M., D. J. WILSON, A. R. LIVESAY and J. C. WOODLAND: The oral transmission of murine typhus in humans. Tex. Rep. Biol. Med. 4, 446—451 (1946).

POPE, J. H.: The isolation of a rickettsia resembling Rickettsia australis in South-East Queensland. Med. J. Aust. 1955 (1), 761—763.

PRATT, H. D., and N. E. GOOD: Distribution of some common domestic rat ectoparasites in the United States. J. Parasit. 40, 113—129 (1954).

PRICE, W. H.: (1) Interference phenomenon in animal infections with rickettsiae of Rocky Mountain spottet fever. Proc. Soc. exp. Biol. 82, 180—184 (1953).

— (2) A quantitative analysis of the factors involved in the variations in virulence of rickettsiae. Science 118, 49—52 (1953).

— (3) The epidemiology of Rocky Mountain spottet fever. I. The characterization of strain virulence of Rickettsia rickettsii. Amer. J. Hyg. 58, 248—268 (1953).

— (4) The epidemiology of Rocky Mountain spotted fever. II. Studies on the biological survival mechanism of Rickettsia rickettsii. Amer. J. Hyg. 60, 292—319 (1954).

— (5) Studies on the interepidemic survical of louseborne epidemic typhus fever. J. Bact. 69, 106—107 (1955).

— H. EMERSON, H. NAGEL, R. BLUMBERG and S. TALMADGE: Ecologic studies on the interepidemic survival of louseborne epidemic typhus fever. Amer. J. Hyg. 67, 154—178 (1958).

— J. W. JOHNSON, H. EMERSON and C. E. PRESTON: Rickettsial-interference phenomenon: A new protection mechanism. Science 120, 457—459 (1954).

PSCHETNITSCHNOW, A. W., u. B. J. RAICHER: Kritische Bemerkungen aus Anlaß der Veröffentlichung von Prof. TOKAREVIC und seinen Mitarbeitern über das Fleckfieberrezidiv. Ž. Mikrobiol. (Mosk.) 1954, H. 10, 114—122 [Russisch].

PTSCHELKINA, A. A., Z. M. SMAJEWA u. R. I. SUBKOWA: Q-Fieber im nördlichen Kasachstan. Ž. Mikrobiol. (Mosk.) 1956, H. 11, 32—35 [Russisch].

QUINTANE, G.: Répartition géographique actuelle de la fièvre boutonneuse dans le Bassin Méditérranéen. Thesis. Marseille: Impr. Schneider 1954. 67 S.

QUINTIN, DESNUES et AUDRAN: Note sur vingt-cinq cas fièvre Q abservés à bord de certains navires de la marine nationale à Saigon. Rev. Méd. nav. 12, 7—27 (1957).

RAFYI, A., et G. MAGHAMI: Sur la présence du Q fever en Iran. Bull. Soc. Path. exot. 47, 766—768 (1954).

RANSOM, S., and R. J. HUEBNER: Studies on the resistence of Coxiella burneti to physical and chemical agents. Amer. J. Hyg. 53, 110—119 (1951).

RAO, K. N. A.: A case of tick typhus in Srinigar. Indian J. med. Res. 39, 293—296 (1951).

RAŠKA, K., u. L. SYRŮČEK: Ein Beitrag zur Epidemiologie der Q-Rickettsiose. Zbl. Bakt., I. Abt. Orig. 167, 267—280 (1956).

— — and M. KUBÁSEK: Epidemiology of Q fever in Czechoslovakia. Čsl. Hyg. 4, 26—32 (1955) [Tschechisch].

RAVAIOLI, L.: Infezione sperimentale del cane con Coxiella burnetii. R. C. Ist. sup. Sanità 16, 525—530 (1953).

RAYNAL, J. H.: Le chat dans l'épidémiologie du typhus exanthématique murin. Bull. Soc. Path. exot. 40, 367—375 (1947).

RECZKO, E.: Über Ausscheidung und Verweildauer des Q-Fieber-Erregers bei experimentell infizierten Meerschweinchen. Zbl. Bakt., Abt. Orig. 156, 81—84 (1950).

REED, C. F., and B. B. WENTWORTH: Q fever studies in Ohio. J. Amer. vet. med. Ass. 130, 456—461 (1957).

REHN, F., u. R. RAVDAN: Isolierung von Rickettsia burneti aus der Zecke Ixodes ricinus. Čsl. Hyg. 6, 85—88 (1957) [Tschechisch].

Reiss-Gutfreund, R. J.: (1) Isolement de souches de Rickettsia prowazeki à partir du sang des animaux domestiques d'Éthiopie et de leur tiques. Bull. Soc. Path. exot. **48**, 602—606 (1955).

— (2) Un nouveau réservoir de virus pour Rickettsia prowazeki: Les animaux domestiques et leurs tiques. Bull. Soc. Path. exot. **49**, 946—1021 (1956).

Ribeiro, H., L. A. Ribeiro do Valle, D. de A. Cristovão e M. D'Apice: Investigações sobre a febre Q en São Paulo. II. Estudos em tratadores de gado e em bovinos. Arch. Hig. (S. Paulo) **9**, 167—180 (1955).

Ribeiro do Valle, L. A., R. Martins Castro, O. N. Banoi e J. M. Ferreira: Febre Q en São Paulo. Primeiro caso clinico compravado por estudos serologicos. Rev. paul. Med. **46**, 447—456 (1955).

Rickard, E. R.: (1) Postinfection murine typhus antibodies in the sera of rodents. Amer. J. Hyg. **53**, 207—216 (1951).

— (2) Attempted transmission of murine typhus to roof rats by the ingestion of food heavily contaminated with infected flea feces. Amer. J. trop. Med. **31**, 311 (1951).

— (3) The survival time of the rickettsias of murine typhus in infected flea feces. Amer. J. trop. Med. **31**, 306—310 (1951).

—, and C. B. Worth: Complement-fixation tests for murine typhus on the sera of wild-caught cotton rats in Florida. Amer. J. Hyg. **53**, 332—336 (1951).

Ricketts, H. T.: (1) A summary of investigations of the nature and means of transmission of Rocky Mountain spotted fever. Trans. Chicago path. Soc. **7**, 73—82 (1907).

— (2) A micro-organism which apparently has a specific relationsship to Rocky Mountain spotted fever. A preliminary report. J. Amer. med. Ass. **52**, 379—380 (1909).

Ris, H., and J. P. Fox: The cytology of rickettsiae. J. exp. Med. **89**, 681—685 (1949).

Rivers, Th. M.: Viral and rickettsial infections of man, 2. edit. Philadelphia-London-Montreal: J. B. Lippincott Company 1952. 719 S.

Rocha Lima, H. da: (1) Beobachtungen an Flecktyphusläusen. Arch. Schiffs- u. Tropenhyg. **20**, 17—31 (1916).

— (2) Zur Ätiologie des Fleckfiebers. Vorläufige Mitteilung. Berl. klin. Wschr. **1916**, 1—6.

— (3) Rickettsien. In W. Kolle, R. Kraus u. P. Uhlenhuths Handbuch der pathologischen Mikroorganismen, 3. Aufl., Bd. 8/II, S. 1346—1386. Jena: Gustav Fischer, Berlin u. Wien: Urban & Schwarzenberg 1930.

Rodaniche, E. C. de: Natural infection of the tick, Amblyomma cajennense, with Rickettsia rickettsii in Panama. Amer. J. trop. Med. Hyg. **2**, 696—699 (1953).

—, and A. Rodaniche: Spotted fever in Panama; isolation of the etiological agent from a fatal case. Amer. J. trop. Med. Hyg. **30**, 511—517 (1950).

Roger, F., et A. Roger: Différence de comportement des souches rickettsiennes épidémiques et murines dans le poumon du hamster inoculé par voie nasale. Bull. Soc. Path. exot. **49**, 241—242 (1956).

Roux, J.: Le pouvoir pathogène de Rickettsia burneti pour le cobaye: Variations expérimentales. C. R. Soc. Biol. (Paris) **150**, 782—784 (1956).

Rooyen, C. E. van, and G. D. Scott: Electron microscopy of typhus rickettsiae. Canad. J. med. Sci. **27**, 250—253 (1949).

Saint, E. G., A. F. Drummond and I. O. Thorburn: Murine typhus in Western Australia. Med. J. Aust. **1954** (2), 731—737.

Salisbury, R. M.: „Q" fever and its human and animal association. N.Z. med. J. **52**, 260—264 (1953).

Salveraglio, F. J., J. C. Bacigalupi, S. Srulevich y O. Viera: Comprobación epidemiológico y clinica de la fiebre Q en el Uruguay. Ann. Fac. Med. (Montevideo) **41**, 131—138 (1956).

Salvin, S. B., and E. J. Bell: Resistance of mice with experimental histoplasmosis to infection with Rickettsia typhi. J. Immunol. **75**, 57—62 (1955).

Sampaio, A., e M. de M. Faia: (1) Contribuição para o estudo da biologia da Rickettsia conori em Portugal. Bol. Inst. sup. Hig. **7**, 118—124 (1952).

— — (2) Estudo comparativo de algumas reacções sorológicas no diagnóstico da febre escaro-nodular. Bol. Inst. sup. Hig. **7**, 106—117 (1952).

— A. da Cruz e M. de M. Faia: Outro atropodo infectado com a Rickettsia dermacentroxenus conori em Portugal. Bol. Inst. sup. Hig. **7**, 125—130 (1952).

Sapio, G. di, e E. Lecaldano: Sulla diffusione della febbre Q in Campania e sul valore delle reazioni di deviazione del complemento e di agglutinazione microscopica. Riv. Ist.sieroter. ital. **32**, 47—60 (1957).

Sasa, M.: Comparative epidemiology of tsutsugamushi disease in Japan. Jap. J. exp.Med. **24**, 337—361 (1954).

—, and A. Miura: (1) Studies on the life history of tsutsugamushi (Trombiculid mites) of Japan in the laboratory. Jap. J. exper. Med. **23**, 171—185 (1953).

— — (2) Notes on nymphs and adults of Tsutsugamushi of Japan, and their laboratory rearings. (Studies on Tsutsugamushi, part 83.) Jap. J. exp. Med. **25**, 197—209 (1955).

— H. Tanaka, Y. Ueno and A. Miura: Notes on the bionomics of unengorged larvae of Trombicula scutellaris and Trombicula akamushi, with special references to the mechanism of cluster formation and reaction to carbon dioxyde expired by the hosts. (Studies on Tsutsugamushi. Part 90). Jap. J. exp. Med. **27**, 31—43 (1957).

— Y. Ueno, A. Miura and H. Tanaka: Notes on the bionomics of Neoschöngastia spp. and Trombicula hasegwai, the summer-tsutsugamushi at Hachijo islands, together with descriptions of their post-larval stages. (Studies on Tsutsugamushi, Part 96).) Jap. J. exp. Med. **27**, 217—234 (1957).

Satrow, J. J.: Zur Frage des wiederholten (rezidivierenden) Fleckfiebers. Ž. Mikrobiol. (Mosk.) **1953**, H. 6, 63—68; **1954**, H. 7, 19—24 [Russisch].

Sautet, J.: Présence de la maladie de Derrick (fièvre Q) à Marseille. Bull. Acad. Méd. Paris **141**, 174—175 (1957).

Schaechter, M., F. M. Bozemann and J. E. Smadel: Study on the growth of rickettsiae. II. Morphologic observations of living rickettsiae in tissue culture cells. Virology **3**, 160—172 (1957).

— A. J. Tousimis, Z. A. Cohn, H. Rosen, J. Campbell and F. E. Hahn: Morphological, chemical and serological studies of the cell walls of Rickettsia mooseri. J. Bact. **74**, 822—829 (1957).

Schaefer, G. L., M. Friedman and Ch. Lewis: The incidence of epidemic typhus antibodies in individuals born in Eastern Europe. Ann. intern. Med. **42**, 979—982 (1955).

Schmid, E. E., L. B. T. Jaysundera and T. Velaudapillai: Survey of the occurrence of Q fever in Ceylon. Ceylon med. J. **1**, 65—68 (1952).

Schubothe, H., M. Bock u. E. Wiesmann: Q-Fieber in Südbaden. Ein Beitrag zur Diagnose, Klinik und Epidemiologie der Rickettsia-burneti-Erkrankung. Klin. Wschr. **1951**, 569—605.

Sdrodowski, P. F.: Rickettsienprobleme und folgende Aufgaben. Ž. Mikrobiol. (Mosk.) **1954**, H. 7, 3—8 [Russisch].

—, u. E. M. Golinewitsch: Rickettsien und Rickettsiosen, 2. Aufl. Moskau 1956. 492 S. [Russisch.]

Semskaja, A. A., and A. A. Ptschelkina: Experimental infection of the bird mite Dermanyssus gallinae Redi and the rat mite Bdellonyssus bacoti Hirst with the causal agent of Q fever. Dokl. Vses. Akad. Nauk **101**, 391—392 (1955).

Shdanow, W. M.: Viren bei Mensch und Tier. Evolution, Systematik und Bestimmung. Jena: Gustav Fischer 1957. 279 S.

Sidky, M. M.: Epidemiology of Q fever. Bull. Wld Hlth Org. **2**, 563—569 (1950).

Siegert, R.: Das Rickettsientoxin als Ursache histopathologischer Organveränderungen. Z. Hyg. Infekt.-Kr. **128**, 477—485 (1948).

— W. Simrock u. U. Ströder: Über einen epidemischen Ausbruch von Q-Fieber in einem Krankenhaus. Z. Tropenmed. Parasit. **2**, 1—40 (1950).

Sigel, M. M., L. B. Weiss, N. Blumberg and J. C. Doane: Survey of epidemic typhus antibody levels in bloods of individuals born in Eastern and Central Europe and the United States. Amer. J. med. Sci. **223**, 429—432 (1952).

Silva-Goytia, R.: Fiebre Q en Mexico. Nota preliminar. Medicina (Mex.) **30**, 453—455 (1950).

—, y A. Elizondo: (1) Estudios sobre fiebre manchada en México. II. Parásitos hematófagos en contrados naturalmente infectados. Medicina (Mex.) **32**, 278—282 (1952).

— — (2) Estudios sobre fiebre manchada en México. IV. Características epidemiológicas de casos de fiebre manchada ocurridos en la Laguna. Medicina (Mex.) **32**, 569—579 (1952).

Silva-Goytia, R., y A. Elizondo: (3) Estudios sobre fiebre manchada en México. III. Estudio por medio de fijación de complemento de sueros de ciertos animales domésticos de la Comarca Lagunera. Medicina (Mex.) 33, 76—83 (1953).

— H. Vasquez Campos y A. Elizondo: Estudios sobre fiebre manchada en México. V. Incidencia de anticuerpos específicos para Dermacentroxenus rickettsi rickettsi en grupos ocupacionales de diversas áreas geográficas. Medicina (Mex.) 33, 425—435 (1953).

Sipka, M., and V. Miljkovic-Krejakovic: Q fever in milking cattle in Yugoslavia. Acta microbiol. hellenica 2, 155—161 (1957).

Slavin, G.: Q fever: The domestic animal as a source of infection for man. Vet. Rec. 64, 743—750 (1952).

Smadel, J. E.: Q fever. In Th. M. Rivers, Viral and rickettsial infections of man, 2. edit., p. 652—664. Philadelphia: J. B. Lippincott Company 1952.

— Scrub typhus. In Th. M. Rivers, Viral and rickettsial infections of man, 2. edit., p. 638 bis 651. Philadelphia: J. B. Lippincott Company 1952.

— E. B. Jackson, B. L. Bennett and F. L. Rigths: A toxic substance associated with the Gilliam strain of R. orientalis. Proc. Soc. exp. Biol. (N.Y.) 62, 138—140 (1946).

— H. L. Ley, F. H. Diercks and J. A. P. Cameron: Persistence of Rickettsia tsutsugamushi in tissues of patients recovered from scrub typhus. Amer. J. Hyg. 56, 294—302 (1952).

Smajewa, Z. M., N. K. Mischtschenko u. A. A. Ptschelkina: Spontane Infektion von Hyalomma anatolicum Koch mit dem Erreger des Q-Fiebers im südlichen Kirgisien. Ž. Mikrobiol. (Mosk.) 1956, H. 11, 30—31 [Russisch].

—, and A. A. Ptschelkina: Domestic birds as carriers of the Rickettsia of Q fever in the Turkmen S.S.R. J. Microbiol., Epidemiol. and. Immunol. 28, 347—349 (1957).

— A. A. Ptschelkina, N. K. Mischtschenko and B. E. Karulin: The epidemiological importance of birds in a natural focus of Q fever in the South of Central Asia. Dokl. Vses. Akad. Nauk 101, 287—289 (1955) [Russisch].

Smith, D. J. W., and E. H. Derrick: Studies in the epidemiology of Q fever. 1. The isolation of six strains of Rickettsia burneti. Aust. J. exp. Biol. med. Sci. 18, 1—8 (1940).

— J. D., and M. G. P. Stoker: The nucleic acids of Rickettsia burneti. Brit. J. exp. Path. 32, 433—441 (1951).

— W. W.: The house and murine typhus in Mississippi. Publ. Hlth Rep. (Wash.) 69, 591—593 (1954).

Snyder, J. C.: The typhus fevers. In Th. M. Rivers, Viral and rickettsial infections of man, 2. edit., p. 578—610. Philadelphia: J. B. Lippincott Company 1952.

— The rickettsiae: Comments on recent observations on biology and epidemiology. Amer. J. Trop. Med. Hyg. 5, 461—463 (1956).

—, and C. R. Anderson: The susceptibility of the Eastern cotton rat, Sigmodon hispidus hispidus, to European typhus. Science 95, 23 (1942).

— M. R. Bovarnick, J. C. Miller and R. S. Chang: Observations on the hemolytic properties of typhus rickettsiae. J. Bact. 67, 724—730 (1954).

—, and C. M. Wheeler: The experimental infection of the human body louse, Pediculus humanus corporis, with murine and epidemic louse-borne typhus strains. J. exp. Med. 82, 1—19 (1945).

Sobešlavský, O.: Experimental infection of the domestic fowl (Gallus gallus domesticus) with Rickettsia burneti. Čsl. Epidem. 6, 146—151 (1957) [Tschechisch].

Soman, D. A.: (1) The incidence and distribution of murine typhus amongst Bombay rats. Indian med. Gaz. 85, 249—253 (1950).

— (2) Q fever in India. Serological evidence. Indian J. med. Sci. 8, 698—703 (1954).

— (3) Tsutsugamushi disease (scrub typhus) in Bombay city and suburbs. J. Indian med. Ass. 23, 389—394 (1954).

Soper, F. L., W. A. Davis, F. S. Markham and L. A. Riehl: Typhus fever in Italy, 1943—1945, and its control with louse powder. Amer. J. Hyg. 45, 305—334 (1947).

Sparrow, H.: Enquête sur la présence du virus typhique chez les souris de Tunis. Arch. Inst. Pasteur Tunis 24, 435—460 (1935).

Spencer, R. R., and R. R. Parker: Rocky Mountain spotted fever: Infectivity of fasting and recently fed ticks. Publ. Hlth Rep. (Wash.) 38, 333—339 (1923).

STARZYK, H.: (1) Vitalité, toxicité et pourvoir d'immunisation de Rickettsia prowazeki conservées hors de l'organisme du pou, en milieu liquide et en milieu sec. C. R. Soc. Biol. (Paris).**123**, 1221—1225 (1936).

— (2) On endurance properties of R. prowazeki in endemic foci. Przegł. epidem. **2**, 257—259 (1948) [Polnisch].

STERKTOWA, N. N., u. N. M. MIRSOEWA: Q-Fieber in der Aserbeidschanischen S.S.R. Ž. Mikrobiol. (Mosk.) **1956**, H. 12, 84—88 [Russisch].

STEWART, P. D.: Scrub typhus in Hong Kong. J. roy. Army med. Cps. **100**, 121—126 (1954).

— W. H., and V. D. HINES: Murine typhus fever in South-West Georgia, January 1945 to January 1953. Amer. J. trop. Med. Hyg. **3**, 883—889 (1954).

STOENNER, H. G.: Experimental Q fever in cattle. Epizootiologic aspects. J. Amer. vet. med. Ass. **118**, 170—174 (1950).

—, and D. B. LACKMAN: The rôle of the milking process in the intraherd transmission of Q fever among dairy cattle. Amer. J. vet. Res. **13**, 458—465 (1952).

STOKER, M. G. P., R. D. BROWN, F. J. L. KETT, P. C. COLLINGS and B. P. MARMION: Q fever in Britain: Isolation of Rickettsia burneti from placenta and wool of sheeps in an endemic area. J. Hyg. (Lond.) **53**, 313—321 (1955).

—, and B. P. MARMION: (1) Q fever. I. Clinical features and laboratory diagnosis. II. Natural history and epidemiology of Q fever in man. Trans. roy. Soc. trop. Med. Hyg. **48**, 191—207 (1954).

— — (2) Q fever in Britain: Isolation of Rickettsia burneti from the tick Haemaphysalis punctata. J. Hyg. (Lond.) **53**, 323—327 (1955).

— — (3) The spread of Q fever from animals to man. The natural history of a rickettsial disease. Bull. Wld Hlth Org. **13**, 781—806 (1955).

— K. M. SMITH and P. FISET: Internal structure of Rickettsia burnetii as shown by electron microscopy of thin sections. J. gen. Microbiol. **15**, 632—635 (1956).

STRANDTMANN, R. W., and B. J. EBEN: A survey of typhus in rats and rat ectoparasites in Galvestone, Texas. Tex. Rep. Biol. Med. **11**, 144—151 (1953).

SUREAU, P., J. P. ROUSILHON et M. CAPPON: Le typhus murin à Dallat: État actuel de la question. Isolement d'une souche. Bull. Soc. Path. exot. **48**, 599—602 (1955).

SUZUKI, T.: Studies on the bionomics and chemical control of tsutsugamushi (scrub-typhus mites). Part. II. Trombicula scutellaris Nagayo et al. in Southern Kanto of Japan. Jap. J. exp. Med. **24**, 181—197 (1954).

SWAMY, T. V., and B. B. DUTTA: Epidemiology of XK typhus in Jamshedpur. Indian med. Gaz. 88, 522—525 (1953).

SYRŮČEK, L., and K. RAŠKA: Q fever in domestic and wild birds. Bull. Wld Hlth Org. **15**, 329—337 (1956).

—, and O. SOBEŠLAVSKÝ: Experimental infection in rats (Rattus norvegicus) with C. burneti. Čsl. Epidem. **5**, 251—254 (1956) [Tschechisch].

— O. SOBEŠLAVSKÝ and H. HAVLÍK: Isolation of Rickettsia burneti from the pigmy shrew Sorex minutus in a focus of Q-rickettsiosis in North-Western Bohemia. Čsl. Epidem. **6**, 392—395 (1957) [Tschechisch].

TAKANO, K., M. KITAOKA and A. SHISHIDO: Complement fixation in Q fever. III. Serological survey of human sera in Japan. Jap. J. med. Sci. **7**, 417—425 (1954).

TAKEMORI, N., M. HENMI and M. KITAOKA: Simultaneous multiplication of two different rickettsiae in the same cell. Proc. Soc. exp. Biol. **81**, 633—636 (1952).

TAKIGAMI, R., A. KAWAMURA, K. NISHIOKA and H. IIDA: Clinical investigation on „Ezo fever". Jap. J. exp. Med. **25**, 187—195 (1955).

TARABAN, A. S., and Y. Y. KOSOWSKI: The nature of sporadic cases of recurrent typhus. Ž. Mikrobiol. (Mosk.) **1957**, H. 3, 104—105 [Russisch].

TAYLOR, R. M., F. R. HASSAN and M. A. KADER: Q fever in Egypt as revealed by the complement fixation test on human sera. A preliminary report. J. Egypt. publ. Hlth Ass. **27**, 129—140 (1952).

— J. R. KINGSTON and F. RIZK: A note on typhus in Egypt and the Sudan. Amer. J. trop. Med. Hyg. **6**, 563—570 (1957).

— R. A. MOUNT, H. HOOGSTRAAL and H. R. DRESSLER: The presence of Coxiella burneti (Q fever) in Egypt. J. Egypt. publ. Hlth Ass. **27**, 123—128 (1952).

Taylor, R. M., F. H. Rizk and M. A. Kader: The study of epidemiology of Q fever. I. Q fever in Egypt as revealed by the complement fixation test on human sera. A preliminary report. J. Egypt. publ. Hlth Ass. 27, 129—140 (1952).

Tendeiro, J.: Estudos sobre o tifo murino na Guiné Portuguesa. Memórias do Centro de Estudos da Guiné Portuguesa. Nr 13, Bissau 1950, 208 S.

Terhaag, L.: Zur Epidemiologie des Q-Fiebers. Q-Fieber in der Eifel. Arch. Hyg. (Berl.) 137, 247—269 (1953).

Terzin, A. L., and J. Gaon: Some viral and rickettsial infections in Bosnia and Herzegovina. Bull. Wld Hlth Org. 15, 299—316 (1956).

Traub, R.: Observations on tsutsugamushi disease (scrub typhus) in Assam and Burma. The mite, Trombicula deliensis Walch, and its relation to scrub typhus in Assam. Amer. J. Hyg. 50, 361—370 (1949).

— L. P. Frick and F. H. Diercks: Observations on the occurrence of Rickettsia tsutsugamushi in rats and mites in the Malayan jungle. Amer. J. Hyg. 51, 269—273 (1950).

— T. Phyllis, M. L. Miesse and R. E. Elbel: Isolation of Rickettsia tsutsugamushi from rodents from Thailand. Amer. J. trop. Med. Hyg. 3, 356—359 (1954).

Travassos, J., H. G. Pereira and J. V. Vasconcelos: Tifo murino no Rio de Janeiro. II. Identificação da „Rickettsia mooseri" isolada de ratos naturalmente infectados. Hospital (Lond.) 36, 351—362 (1949).

Tschumakow, M. P.: Queensland-Fever- a zoonotic rickettsiosis of man and animals. Vet. Žizn. 31, 26—32 (1954) [Russisch].

Urbach, H., u. M. Spróssig: Die fluoreszensmikroskopische Darstellung der Rickettsia burneti und ihre photographische Wiedergabe. Zbl. Bakt., I. Abt. Orig. 161, 39—44 (1954).

Varela, G., R. Fournier y H. Mooser: Presencia de Rickettsia quintana in piojos Pediculus humanus de la ciudad de México. Inoculatión experimental. Rev. Inst. Salubr. Enferm. trop. (Méx.) 14, 39—42 (1954).

Vermeil, C., R. Chene et A. Zaibi: Virus typhique murin et surveillance murine à Tunis. Arch. Inst. Pasteur (Tunis) 34, 531—535 (1957).

Violle, H.: Expériences sur le virus exanthématique murin. Sensibilité du rat aux urines virulentes d'origine humaine et murine. C. R. Soc. Biol. (Paris) 129, 984—986 (1938).

Wegener, K. H.: Das Q-Fieber und seine milchhygienische Bedeutung. (Literaturstudie mit Beitrag zur Frage des Vorkommens unter den schleswig-holsteinischen Rinderbeständen.) Kieler milchwirtsch. Forsch.-Ber. 9, 509—535 (1957).

Weigl, R.: Untersuchungen und Experimente an Fleckfieberläusen. Die Technik der Rickettsia-Forschung. Beitr. Klin. Inf.-krkh. 8, 353—376 (1920).

Weiss, E., H. R. Dressler and E. C. Suitor: Selection of a mutant strain of Rickettsia prowazeki resistant to p-aminobenzoic acid. J. Bact. 73, 421—430 (1957).

—, and H. C. Pietryk: Growth of Coxiella burnetii in monolayer cultures of chick embryo entodermal cells. J. Bact. 72, 235—241 (1956).

— L. J.: Electron-micrographs of rickettsiae of typhus fever. J. Immunol. 47, 353—357 (1943).

Welsh, H. H., E. H. Lennette, F. R. Abinanti and J. F. Winn: Q Fever in California. IV. Occurrence of Coxiella burnetii in the placenta of naturally infected sheep. Publ. Hlth Rep. (Wash.) 66, 1473—1477 (1951).

Weltseuchenatlas (herausgeg. von E. Rodenwaldt): Karte 27: Endemisches Vorkommen von Läuse-Fleckfieber in Europa 1914—1950. Von F. v. Bormann. 100/III, Text S. I/81. Karte 28: Epidemien des Läuse-Fleckfiebers in Europa 1914—1950. Von F. v. Bormann. 100—a/III. Hamburg: Falk-Verlag 1952ff.

Weyer, F.: (1) Über Rickettsia wolhynica und die Diagnose des Wolhynischen Fiebers durch den Läuseversuch. Zbl. Bakt., I. Abt. Orig. 152, 403—414 (1948).

— (2) Die künstliche Infektion von Zecken mit Rickettsien und anderen Krankheitserregern. Zbl. Bakt., I. Abt. Orig. 152, 449—457 (1948).

— (3) Die Entwicklung von Rickettsien in den Larven vom Mehlkäfer (Tenebrio molitor L.). Schweiz. Z. allg. Path. 13, 478—486 (1950).

— (4) Über die Wirkung von „Tego 103" auf Rickettsien. Zugleich ein Beitrag zur Frage der Empfindlichkeit von Rickettsien. Z. Tropenmed. Parasit. 1, 586—594 (1950).

WEYER, F.: (5) Beobachtungen bei intracoelomaler Infektion von Läusen mit Rickettsien. Z. Tropenmed. Parasit. 2, 40—51 (1950).
— (6) Explantationsversuche bei Läusen in Verbindung mit der Kultur von Rickettsien. Zbl. Bakt., I. Abt. Orig. 159, 13—22 (1952).
— (7) Das Problem der Brillschen Krankheit im L¡cht neuerer Beobachtungen und Forschungsergebnisse. Z. Tropenmed. Parasit. 3, 417—436 (1952).
— (8) Versuche zur künstlichen Infektion der Schweinelaus Haematopinus suis L. mit Rickettsia prowazeki und R. quintana. Schweiz. Z. allg. Path. 15, 203—210 (1952).
— (9) Die experimentelle Infektion der Filzlaus Phthirus pubis L. mit Rickettsia prowazeki und R. quintana. Z. Tropenmed. Parasit. 3, 302—309 (1952).
— (10) The behaviour of Rickettsia akari in the body louse after artificial infection. Amer. J. trop. Med. Hyg. 1, 809—820 (1952).
— (11) Die Beziehungen des Q-Fieber-Erregers (Rickettsia burneti) zu Arthropoden. Z. Tropenmed. Parasit. 4, 344—382 (1953).
— (12) Künstliche Infektion der Kleiderlaus mit Rickettsia tsutsugamushi. Z. Hyg. Infekt.-Kr. 137, 419—428 (1953).
— (13) Unterschiede im Verhalten mehrerer Stämme von Rickettsia conori in der Kleiderlaus. Z. Tropenmed. Parasit. 5, 477—482 (1954).
— (14) Rückfälle bei Fleckfieber und Wolhynischem Fieber. Medizinische 1954, Nr 38 1267—1271.
— (15) Vergleichende Untersuchungen über das Verhalten verschiedener Rickettsien-Arten in der Kleiderlaus. Acta trop. (Basel) 11, 193—221 (1954).
— (16) Eigenschaften und systematische Stellung der Rickettsia quintana mit Bemerkungen zur Systematik und Nomenklatur der Rickettsien. Z. Tropenmed. Parasit. 6, 2—18 (1955).
— (17) Übertragung mehrerer Stämme von Rickettsia tsutsugamushi auf Kleiderläuse. Z. Tropenmed. Parasit. 9, 42—53 (1958).
— (18) Beobachtungen bei der Übertragung von brasilianischem Fleckfieber und sibirischem Zeckenbißfieber auf die Kleiderlaus. Z. Tropenmed. Parasit. 9, 174—193 (1958).
— H. FRIEDRICH-FREKSA u. G. BERGOLD: Die Beziehungen der Rickettsien zu Bakterien und Viren. Naturwissenschaften 32, 361—365 (1944).
— u. H. HORNBOSTEL: Erregernachweis bei einem Fall Brill-Zinsserscher Krankheit in Hamburg. Schweiz. med. Wschr. 1957, 692—695.
—, u. D. PETERS: Untersuchungen zur Rickettsienmorphologie. I. Mitt. Eine einfache und schonende Präparationsmethode für die elektronenmikroskopische Untersuchung von Rickettsien. Z. Naturforsch. 7b, 357—361 (1952).
WHEELER, C. M.: Control of typhus in Italy 1943—1944 by use of DDT. Amer. J. publ. Hlth 36, 119—129 (1946).
WHITEMIRE, C. E., and C. M. DOWNS: Effects of Cortisone on experimental murine typhus. J. Bact. 74, 417—438 (1957).
WIESMANN, E.: Die Q-fever-Forschung in der Schweiz in den Jahren 1947—1951. Z. Tropenmed. Parasit. 3, 297—301 (1952).
—, u. F. BÜRKI: Die veterinär-medizinische Bedeutung der Rickettsia-burneti-Infektion bei Ziege, Schaf und Rind in der Schweiz. Schweiz. Arch. Tierheilk. 97, 569—574 (1955).
— R. SCHWEIZER, H. FEY u. F. BÜRKI: Nachweis von Rickettsia burneti bei Schaf, Ziege und Rind. Schweiz. Z. allg. Path. 18, 1095—1103 (1955).
— — u. H. TOBLER: Q-Fieber in der Nordostschweiz. Schweiz. med. Wschr. 1956, 60—63.
WISSEMAN, C. L., E. B. JACKSON, F. E. HAHN, A. C. LEY and J. E. SMADEL: (1) Metabolic studies of rickettsiae. I. The effects of antimicrobial substances and enzyme inhibitors on the oxydation of glutamate by purified rickettsiae. J. Immunol. 67, 123—136 (1951).
— — — — — (2) Metabolic studies of rickettsiae. II. Studies on the pathway of glutamate oxydation by purified suspensions of Rickettsia mooseri. J. Immunol. 68, 251—264 (1952).
WISSIG, S. L., L. G. CARO, E. B. JACKSON and J. E. SMADEL: Electron microscopic observations on intracellular rickettsiae. Amer. J. Path. 32, 1117—1133 (1956).
WOHLRAB, R.: Die experimentelle Infektion weißer Mäuse mit murinem Fleckfiebervirus. Zbl. Bakt., I. Abt. Orig. 140, 193—201 (1937).

Wojciechowski, E.: The influence of the substrate on the antigenic properties and the virulence of typhus fever rickettsia. Med. dośw. mikrob. 3, 345—359 (1951) [Polnisch].

— Z. Lewińska and E. Mikołajczyk: (1) The persistence of typhus rickettsiae in the organs of experimentally infected rodents. Przegl. epidem. 11, 39—46 (1957) [Polnisch].

— — — (2) Serologische Erhebungen über Q-Fieber in bestimmten Bevölkerungsgruppen von Polen. Przegl. epidem. 11, 59—63 (1957) [Polnisch].

—, and E. Mikołajczyk: (1) Sporadic typhus fever. IV. Studies on etiology. Przegl. epidem. 7, 187—194 (1953) [Polnisch].

— — (2) The comparison of rickettsiae isolated from cases of epidemic and sporadic fever. Med. dośw. mikrob. 5, 103—112 (1953) [Polnisch].

—, and A. Przybyła: Antigenic properties of R. burneti strains isolated from the first Q fever epidemic in Poland. Med. dośw. mikrob. 9, 281—287 (1957) [Polnisch].

— E. Wnęk, S. Lewińska u. C. Frygin: Serologische Untersuchung auf Q-Fieber in einer Gruppe von Schlacht- und Zuchttieren. Przegl. epidem. 11, 65—68 (1957) [Polnisch].

Wolbach, S. B.: The rickettsiae and their relationship to disease. J. Amer. med. Ass. 84, 723—728 (1925).

Womersley, H.: The scrub-typhus and scrub-itch mites (Trombiculidae, Acarina) of the Asiatic-Pacific region. Rec. South Austral. Museum 10, 1—435, 438—673 (1952).

World distribution of rickettsial diseases. I. Louse-borne and flea-borne typhus Pl. 10. II. Tick-borne and mite-borne forms. Pl. 11. III. Ticks and mite vectors. Amer. Geogr. Soc., New York, 1953/54.

Worms, R.: Das Problem des sporadischen Fleckfiebers in Mittel- und Westeuropa. Med. Mschr. 7, 139—145 (1953).

Worth, C. B., and E. R. Rickard: (1) Transmission of murine typhus in roof rats in the absence of ectoparasites. Amer. J. trop. Med. 31, 301—305 (1951).

— — (2) Evaluation of the efficiency of common cotton rat ectoparasites in the transmission of murine typhus. Amer. J. trop. Med. 31, 295—298 (1951).

Wyatt, G. R., and S. S. Cohen: Nucleic acids of rickettsiae. Nature (Lond.) 170, 846—847 (1952).

Yamaguchi, T., N. Horie, T. Miki and Y. Ono: Studies on tsutsugamushi (trombiculid mites) in Shikoku. Part 3. An epidemiological survey of „Shikoku type“ tsutsugamushi disease. Shikoku acta medica 9, 64—77 (1956).

Yu, E.-S., Lin, S.-G. and Ch'en, C.-L.: Natural infection of the domestic rabbits with Rickettsia tsutsugamushi in Fukien. Acta microbiol. Sinica 5, 183—188 (1957) [Chinesisch].

Zur Epidemiologie der Salmonelleninfektion

Von

W. FROMME

Inhalt

Die verbesserte Diagnostik der Erreger der Typhus-, Paratyphus- und Enteritiserkrankungen, die unter dem Sammelbegriff Salmonellen zusammengefaßt werden,
hat zu einer weitgehenden Untergliederung der Ätiologie dieser Erkrankungen
geführt und gestattet, daß zahlreiche Erkrankungen insbesondere des Magen-
Darmkanals nunmehr ätiologisch aufgeklärt und als übertragbar erkannt werden.
Zu Zeiten von SCHOTTMÜLLER und GÄRTNER konnten wir häufig auf der Nähr-
bodenplatte verdächtige Kolonien, die serologisch nicht ansprachen, nicht
weiter verfolgen, mußten also zu einem negativen Ergebnis kommen. So ist ohne
weiteres das häufige Bekanntwerden derartiger Infektionen durch die verfeinerte
Diagnostik zu erklären. Zahlreiche leicht verlaufende, insbesondere Darmerkrankungen, die früher dem epidemiologischen Interesse entgingen, werden heute
als infektiös festgestellt. Es besteht jedoch nach den umfassenden Forschungen
auf dem Gebiete des Salmonellennachweises kein Zweifel, daß gegen früher die
Häufigkeit der Salmonellenbefunde tatsächlich erheblich zugenommen hat und
stetig zunimmt. Hier muß vor allem das von STEINIGER betriebene Studium
der „Freilandbiologie" erwähnt werden, das neue Erkenntnisse über die Verbreitungsweise und das zunehmende Vorkommen von Salmonellen gebracht
hat. Zunächst sei über die verschiedenen Fundorte von Salmonellen berichtet.

Enten- und Hühnereier als Salmonellenträger

Bereits 1931 hatte ich Gelegenheit, in Witten eine Reihe von Enteritiserkrankungen zu beobachten, die auf den Genuß von Enteneiern, insbesondere
von holländischen Kalkenteneiern zurückgeführt werden konnten [FROMME (*35,
36*), WILLFÜHR, FROMME und BRUNS (*192*)]. Es handelte sich um Breslau- und

Gärtnerbakterien, die mit dem Entenei auf Hackfleisch, Paniermehl, Pudding, Milchsuppe übertragen waren. Auch durch Rührei kam eine Gruppe von Erkrankungen zustande. Die Besichtigung einer der größten holländischen Entenzüchtereien ergab damals das Vorkommen von Seuchen bei Jungenten mit einem Absterben bis zu 80%. Bei der bakteriologischen Untersuchung mitgebrachter eingegangener Enten wurden in den Organen Breslaubacillen nachgewiesen. In späteren Untersuchungen gelang die Feststellung dieser Erreger im Eiweiß und Eidotter von Enteneiern, wenn auch selten [Bruns und Fromme (23)]. Löns-Dortmund fand damals im Innern eines auf dem Markt gekauften Enteneies Gärtnerbacillen. Verhältnismäßig häufig ließen sich die Enteritisbakterien an der Außenschale der Eier nachweisen. So konnte z. B. Bruns bei 20 von 40 Eiern dieselben Erreger, mit denen die eierlegenden Enten gefüttert bzw. gespritzt waren, an der Außenschale feststellen. Auch Fürth und Klein (38) wiesen damals in Düsseldorf wiederholt Breslaubacillen auf der weniger oder mehr beschmierten Eischale von Enten nach, während Müller und Rodenkirchen (101) in Mülheim a. d. Ruhr Gärtnerbakterien auch im Innern eines Enteneies feststellen konnten. Es war anzunehmen, daß die Außenschale der Eier vor allem beim Legeakt in der Kloake mit salmonellenhaltigem Kot beschmutzt wird. Unsere Enteneier stammten zumeist von einer bestimmten Entensorte, den Kaki-Campbell-Enten, die wegen ihrer guten Eierausbeute, nämlich bis zu 300 Eiern im Jahr, in den letzten Jahren damals eine weite Verbreitung gefunden hatten. Offenbar handelte es sich um eine überzüchtete Rasse, die für eine Infektion mit Enteritisbakterien im Gegensatz zu den alten Entenrassen besonders anfällig geworden war. Die Übertragung von Salmonellen durch Enteneier wurde dann auch an anderen Orten beobachtet, während Enteritiserkrankungen bei Enten schon seit Jahren bekannt waren. Nach Haffke (47a) soll Scott in England 1926—1930 aus Enteneiweiß Breslaubacillen gezüchtet haben. Beobachtungen über Erkrankungen nach Genuß von Enteneiern in der damaligen Zeit liegen vor von Beller und Reinhardt (10a) 1934, von Miesner und Köser (92a) 1935, von Wesselmann (190a) 1935, von Kathe und Lerche (64) 1936 u. a. In der Zeit von 1933 bis 1939 kamen im deutschen Reich 224 Erkrankungsgruppen mit 1789 Kranken und 25 Toten nach Genuß von Enteneiern den Gesundheitsämtern zur Kenntnis. Bei den 194 bakteriologisch geklärten Erkrankungen fanden sich 131 durch S. typhi murium, 63 durch S. enteritidis verursacht.

Diese Beobachtungen führten dann zu der Verordnung über Enteneier von 1936, die ein mindestens 8 min langes Kochen vorsah, eine Forderung, die durch eine Verordnung aus 1954 auf 10 min verlängert wurde. Nach Schmith (147) ist indessen eine Kochzeit von 15 min erforderlich. Es sei hier übrigens angemerkt, daß die Auflage eines längeren Kochens der Enteneier, also einer Vernichtung der Salmonellen im Innern des Eies, deshalb von geringerer Bedeutung für die Verhütung einer Übertragung ist, weil die Erreger im Ei verhältnismäßig selten vorkommen. Die eigentliche Gefahr besteht in der Übertragung der Erreger, die auf der Außenschale des Eies haften. Wichtiger als das längere Kochen der Eier, das sich auch nur auf den Genuß der Eier in gekochtem Zustande beschränkt, wäre nach Fromme (36) eine Desinfektion der Enteneier in Großgeschäften. Knothe und Budach (69a) empfehlen zur Verhinderung des Eindringens der Keime in das Innere des Eies eine 2 sec lange Desinfektion mit 10%iger heißer Sodalösung. In Holland ist eine Pasteurisierung der Enteneier vorgeschrieben. Daß die Übertragung der auf den Enteneierschalen haftenden Salmonellen z. B. bei der Verwendung in einem Lebensmittelbetrieb trotz der Enteneierverordnung bestehen bleiben kann, beweist die Beobachtung von SCHEIBE (*143*) aus 1953, nach der 40 Erkrankungen und 2 Todesfälle nach Genuß von Backwaren, zu deren Herstellung Enteneier verwendet waren, beobachtet wurden. Auch Berichte von KOLLATH (*69a*) aus 1941, KNOTHE und BUDACH (*68a*) aus 1952, STAAK (*137a*) aus 1958 u. a. sprechen für diese Annahme. Erwähnt sei hier, daß im August 1954 in einem westfälischen Krankenhaus 85 Personen nach Genuß von Eierpfannenkuchen, der mit Enteneiern zubereitet war, an einer Breslau-Enteritis erkrankten [WÜSTENBERG (*195c*)]. Kostproben der Enten der Geflügelfarm enthielten Breslaubakterien. Es handelte sich hier übrigens im Gegensatz zu unseren früheren Beobachtungen nicht um Kaki-Campbell-, sondern um weiße Peking-Enten. Immerhin hat die Verordnung zu einer erheblichen Einschränkung des Verkaufs von Enteneiern und damit auch von Erkrankungen geführt, wohl dadurch mitbedingt, daß außer dem Kochen der Genuß von Enteneiern in anderer Form und Verwendung untersagt war.

Eine Salmonelleninfektion der *Hühner* ist im Gegensatz zu den Entenerkrankungen unter natürlichen Verhältnissen offenbar selten. Meist handelt es sich dabei um S. pullorum, einen Erreger, der nur ausnahmsweise für den Menschen pathogen ist. Immerhin konnte bei einigen Lebensmittelvergiftungen S. pullorum als Krankheitserreger festgestellt werden [FRITZSCHE (*33b*), GERNEZ-RIENSE u. Mitarb. (*45*), SCHAAL (*138*), KREUSCHER und SYLVESTER (*70b*)]. 1933 beschrieb R. MÜLLER (*99*) Infektionen durch S. enteritis in einer Familie nach Genuß von einem mit Hühnerei angerichteten Kartoffelsalat, POPPE (*111*) 1946 nach Genuß von Pudding, dem rohe Hühnereier zugesetzt waren. MÖLLER (*93*) stellte S. typhi murium in Eiweiß von Hühnereiern gelegentlich einer Gruppenerkrankung fest. Bei ruhrartigen Kükenerkrankungen des Hausgeflügels fand BECKER (*10*) als Erreger seltene Salmonellentypen, wahrscheinlich durch ausländische Futtermittel übertragen. CLARENBURG (*24*) berichtete über zunehmende Befunde von S. bareilly beim Geflügel seit 1951. Über S. anatum als Ursache eines Massensterbens in einem Gänsekükenbestand macht SEELE (*151*) Mitteilung. Unter 250 Hühnereiern aus bäuerlichen Kleinbetrieben des Münsterlandes gelang VOGT (*189*) neuerdings in einem Falle der Nachweis von S. pullorum. Nach WINKLE, ROHDE und BISCHOFF (*194*) zeigt der Salmonellenbefund bei Hühnern eine ansteigende Tendenz. Immerhin fanden sie bei einer Untersuchung von 1005 in- und ausländischen Frischeiern Salmonellen in nur 8 Hühnereiern.

Bemerkenswert ist der häufige Befund von Salmonellen in *Eiprodukten*. Im sog. chinesischen Eigelb konnte KNORR (*69*) bereits 1930 Suipestiferkeime in ungeheuren Mengen nachweisen und erwähnt übrigens eine Anmerkung von KOELSCH, daß erhebliche Mengen von Eigelbpulver in der *Handschuhindustrie* Verwendung finden. In der Tat wird auch heute, wie aus dem Bericht einer Lederfabrik in Durlach hervorgeht, flüssiges chinesisches Hühnereigelb, das mit 1% Rosmarin denaturiert und mit Salz konserviert wird, bei der Lederfabrikation verwendet. Das Eipräparat wird aus Hamburg bezogen. Erkrankungen unter

Personen, die sich mit der Verarbeitung des Eiproduktes befassen, sind bisher nicht bekannt geworden. Eigelb findet übrigens auch in anderen Betrieben, z. B. bei der Herstellung von Haarmitteln, Verwendung.

Bereits zu Beginn des zweiten Weltkrieges wies Dack (25) unter 7584 Proben Eipulver aus USA in 9,9% der Proben Salmonellen nach, und zwar 33 verschiedene Typen, von denen in England 22 unbekannt waren. In USA stellten Solowey u. Mitarb. (165) 1943—1945 unter 5498 Eipulverproben in 35% der Proben 58 verschiedene Arten von Salmonellen fest. Über 45 Erkrankungen durch S. tennessee nach Genuß von Rührei aus chinesischem Trockeneipulver berichtet 1949 Zemann (196). Erwähnt seien auch positive Befunde von Felsenfeld, Young und Yoshimura (31). Auf die Gefahr importierter Eiprodukte weist ferner Savage (137) hin. Kelch (67) fand 1956 unter 791 Proben von chinesischem Gefriervollei 69mal = 8,7% Salmonellen 7 verschiedener Typen. Schäfer, Martin und Haas (142) stellten 1956 1,5% der Proben gefrorenen Volleis aus China als infiziert fest, und zwar 4mal durch S. thompson, 3mal durch S. typhi murium und 1mal durch S. newport. In chinesischem Trockenei, dessen Einfuhr nach England wegen seines Salmonellengehaltes abgelehnt war, konnten Rohde und Adam (132) in Hamburg aus 12 Proben 9mal Salmonellen, vor allem S. thompson, feststellen. Kurze Zeit später gelang dann der Nachweis von S. thompson nicht nur in Hamburger Sielwässern, sondern auch in Krankenhäusern und bei Umgebungsuntersuchungen. Anschließend bis März 1956 durchgeführte Untersuchungen von 400 Proben importierter Eikonserven ergaben dann wieder eine Ausbeute von 30 positiven Befunden, vor allem auch von S. thompson. In der Zeit vom 1. 1. 56 bis 28. 2. 57 wurden nach Winkle, Rohde und Bischoff (194) in 68485 Proben von Eiprodukten vor allem chinesischer Herkunft 4382mal = 6,4% 24 verschiedene Arten von Salmonellen, darunter 3100mal S. thompson, nach Zureck (198) 1956 in chinesischem Gefriervollei S. thompson, S. typhi murium, nachgewiesen. Lerche (77) fand im Trockenei amerikanischer Herkunft zahlreiche verschiedenartige Salmonellentypen (S. heidelberg, tennessee, montevideo, infantis, thompson, typhi murium), während bei gleichzeitiger Untersuchung Proben chinesischer Herkunft sich als frei von Salmonellen erwiesen. Demgegenüber fand Albert (5) bei der bakteriologischen Untersuchung von 39502 Proben von Eiprodukten in der Zeit vom 1. 1. bis 30. 11. 56 in 1976 Proben, die zu 94% chinesischer Herkunft waren, Salmonellen also in 5%, und zwar 21 verschiedene Typen, darunter 68% S. thompson, besonders im Gefrierei. Hierbei konnte festgestellt werden, daß sich die Salmonellen überraschend widerstandsfähig gegen Kälteeinwirkung erwiesen. Bei einer monatelangen Einwirkung einer Temperatur von —18 bis —20° C blieben sie voll lebensfähig. Selbst eine Temperatur von —196° C brachte sie nicht zum Absterben. Betreffend Widerstandsfähigkeit der Salmonellen sei hier noch angefügt, daß Aufbewahren in Liquoidblut und in hämolysiertem Blut während 25—29 Monate nach Reimold (122) keine Änderung im serologischen Verhalten bewirkte.

Salmonellenhaltige Eiprodukte lieferten China, USA, Italien, Polen, Jugoslawien, Holland, salmonellenfreie Ware Dänemark, Äthiopien, Frankreich, Argentinien, Japan, Norwegen. Die Bedeutung der zunehmenden Infektionsgefahr gerade durch chinesische Eiprodukte erhellt auch aus der Zunahme des Imports, der nach Rohde und Adam (132) über Hamburg 1948 48 t, 1953 325 t und 1955 8632 t ausmachte.

Bei dem sonst seltenen Nachweis von Salmonellen in Hühnereiern ist, wie auch VOGT (*189*) ausführt, daran zu denken, daß das importierte Eipulver aus Enteneiern hergestellt wird, deren Salmonellengehalt, sei es im Innern, sei es vor allem an der Außenschale, häufig festgestellt wird. Nach RODENWALDT (*128*) ist in Ostasien das Einsammeln der Eier von in riesigen Herden getriebenen Laufenten durch ihren Hirten die übliche Form der Eiergewinnung. Das Entenei hat den Vorrang vor dem Hühnerei. Die Herstellung des Trockeneies geschieht in China durch Ausstreichen des Gelbeies auf eine Leinwand, die in der Sonne getrocknet wird. Ein solches Verfahren veranschaulicht allerdings deutlich die Infektionsmöglichkeiten dieser Eiererzeugnisse. Auch FREITAG (*33a*) führt an, daß in China Enteneier, die mit einer Paste aus schwarzem Tee, Kalk, Salz und Holzasche überzogen, nach monatelanger Aufbewahrung als besondere Delikatesse gelten. Versuche, in chinesischem Kristalleiweiß Hühner- und Enteneiweiß mittels Präcipitation zu differenzieren, sind nach SCHOENHERR (*148*) bisher gescheitert.

Über die *Infektionswege* der *Salmonellen in das Eiinnere* liegen folgende Beobachtungen vor. Der Nachweis von Salmonellen im Dotter bzw. Eiweiß eines Eies läßt die Frage stellen, ob die Krankheitserreger bereits vom Eierstock her bei der Entwicklung des Eies im Dotter bzw. Eiweiß eindringen, oder ob vielmehr nachträglich ein Einwandern durch die Kalkschale erfolgt. BRUNS und FROMME (*23*) fanden bei künstlich mit Enteritisbakterien gefütterten und injizierten Enten, Hühnern und Tauben lediglich bei 2 Hühnern zwei, fünf und sieben Tage nach der Injektion in den von ihnen gelegten Eiern, und zwar sowohl im Dotter wie im Eiereiweiß, Enteritisbakterien. An der Außenseite der Hühnereier konnten jedoch häufig Salmonellen nachgewiesen werden. Schon vorher war die Kongenitalinfektion, d. h. das Hineingelangen mit der Nahrung aufgenommener Salmonellen (S. pullorum) auf dem Blutwege in den Eierstock und auch in den Hoden in USA von den ROMANOFFS (*135*) beschrieben worden. Bekannt ist auch, daß vor der Schalenbildung Fremdkörper in das Ei gelangen können (Wurmteile, Steinchen, Federn), so daß solche Infektionen verständlich sind [RODENWALDT (*127*)]. Häufiger und wohl als Regel ist indes das Einwandern der Salmonellen von außen durch die Eierschale anzunehmen. Die Anwesenheit von Feuchtigkeit ist für die Bakterieneinwanderung wesentlich, weil die Poren der Eierschale in trockenem Zustand durch eine organische Substanz, wahrscheinlich Mucin, verschlossen sind und Mikroorganismen am Einwandern in das Eiinnere hindern [siehe auch SCHWARZ (*150*)]. Nach STEINIGER (*170*) können die Salmonellen erst nach einer Woche oder später und besonders auf feuchter Unterlage und im warmen Raum mit hoher Luftfeuchtigkeit in das Innere einwandern. Für den Schutz gegen eine Infektion des Eiinhaltes ist deshalb trockene Aufbewahrung zu empfehlen. In angetrocknetem Kot halten sich nach LERCHE (*76*) Salmonellen lange lebensfähig und können in das Innere des Eies eindringen, sobald die Bactericidie des Eiweißes geschwunden ist. Das Einlegen in Kalkwasser beschleunigt nach RASCH (*117*) diesen Vorgang. Salmonellen sollen eine beachtliche Affinität zum Eierstock besitzen. Es können infizierte lebensfähige Küken schlüpfen. KATHE und LERCHE (*64*) berichten über Befunde von Breslaubacillen in Eierstocksfollikeln von Enten.

Experimentell konnte LANGE (*72*) 1907 feststellen, daß Coli-, Typhus-, Paratyphus-, Gärtner- und Botulinuskeime die Hühnerschale bis zum Eigelb durchdringen, nachdem die Eier 2—5 Tage in einer entsprechend geimpften Brühe gelegen hatten. Ähnliche Befunde hatte Piorkowski (*105*) bereits 1895 erhoben. Im gleichen Jahre teilt Wilm (*193*) mit, daß auch Choleravibrionen die Eischale durchwandern, z. B. nach Beschmieren der Außenschale mit Cholerastuhl, auch in Berührung mit Hexel und Sägemehl, die mit Cholerastuhl vermischt waren. Experimentell ließen sich nach Vogt (*189*) Hühnereier mit S. pullorum infizieren, und zwar fanden sich diese Erreger im Eidotter, dagegen nicht im Eiklar. Das Eiklar erwies sich wegen seines hohen p_H-Wertes von etwa 9,4 als ausgesprochen entwicklungshemmend gegenüber einem p_H-Wert des Eidotters von 6,1. Andererseits ist die wachstumshemmende Eigenschaft des Eiweißes nach Lodenkämper (*81*) durch bestimmte Stoffe (Lysozym, Prozardin, Inhibin) bekannt. Ergänzend sei noch angefügt, daß Wesselmann (*190a*), Schöneberg (*147a*), Miesner und Köser (*92a*), Beller (*10a*), Schaaf (*137a*) u. a. annehmen, daß sich die Salmonellen primär im Dotter der Enteneier ansiedeln, während andere Autoren die Bedeutung der nachträglichen Infektion durch die Eischalen betonen. So teilt Gordon (*46a*) mit, daß bei 2000 untersuchten Enteneiern im Eiinneren keine Salmonellen gefunden wurden, obwohl aus mehr als 50% der Ovarien dieser Enten Salmonellen gezüchtet werden konnten. Abschließend darf wohl festgestellt werden, daß sowohl eine primäre Infektion der Eier vom Eierstock her durchaus möglich ist, daß aber die nachträgliche Einwanderung der Salmonellen durch die Eischale im Vordergrunde der Überlegung zu treffender Bekämpfungsmaßnahmen zu stehen hat.

Salmonellenbefunde in Gewässern

In den letzten Jahren sind weitere bemerkenswerte Befunde über das Vorkommen von Salmonellen in freien Gewässern gemacht worden. Nachdem 1946 Glatzel (*46*) und 1948 Meyer (*91*) im Husumer Hafen und Müller [s. Steiniger (*168*)] 1947 auch in der Elbemündung regelmäßig im Wasser S. paratyphus B nachweisen konnten, gelang es 1951 Steiniger (*169*), in systematischen Untersuchungen im Meerwasser der schleswig-holsteinischen Küste ebenfalls Paratyphusbakterien, und zwar in auffallend hoher Menge, festzustellen. Unter der anwohnenden Bevölkerung herrschte der Paratyphus B endemisch. Die aus dem Wasser und von Kranken und Ausscheidern gezüchteten Stämme erwiesen sich als identisch. Steiniger berechnet, daß die Zahl der Paratyphusbakterien im Hafenwasser etwa 1000mal größer zu veranschlagen sei als die von Kranken und Ausscheidern ausgeschiedenen Erreger ohne Berücksichtigung der durch eine Desinfektion der Abgänge vernichteten Krankheitserreger. Es müßte also eine Anreicherung außerhalb des menschlichen bzw. tierischen Körpers angenommen werden. Die Salmonellen vermehren sich offenbar im Wasser bei Vorhandensein von Eiweißstoffen. So kann z. B. das Einleiten von blutigen Abgängen aus einem Schlachthof die Salmonellenzahl im Vorfluter stark erhöhen. Steiniger fand im Liter Wasser 10—20 Salmonellen, bei Zufuhr großer Eiweißmengen jedoch Werte von 10—100 im cm³. Bewegung des Wassers an der Küste, der Gezeitenfluß ist günstig für die Vermehrung. In ruhig stehendem Wasser werden die Salmonellen durch die Begleitflora und die Bakterienfresser vernichtet. Auch die Phagen sind offenbar von Bedeutung. Steiniger (*175*) stellte fest, daß in Salz- und Brackwasser Phagen schwer, dagegen Salmonellen häufiger nachzuweisen waren im Gegensatz zu Süßwasser mit umgekehrtem Verhältnis. Der Salzgehalt eines Wassers ist der Entwicklung der Salmonellen nicht hinderlich. LÜTJE und RASCH (*86*) stellten fest, daß sich Salmonellen in Pökellake und Salzfleisch gut vermehren. Sie machten die Beobachtung, daß in Fütterungsversuchen bei Tieren, die nicht erkrankten, durch Zugabe salinischer Mittel eine Erkrankung ausgelöst werden konnte.

Es muß angenommen werden, daß die im Meerwasser festgestellten Salmonellen mit den Abwässern in das Meerwasser gelangen, also primär von Menschen und Tieren stammen. Ihr Vorhandensein deutet also auf vorangegangene Erkrankungen unter der anwohnenden Bevölkerung. Da aber offenbar feststeht, daß sich die Salmonellen auch außerhalb des menschlichen und tierischen Körpers bei Anwesenheit von Eiweißsubstanzen vermehren, besteht auch die Möglichkeit von Neuinfektionen von Menschen und auch Tieren, wenn salmonellenhaltiges Wasser unmittelbar, aber auch durch Vermittlung von Wasservögeln, die, wie weiter unten ausgeführt wird, an einer Salmonellose erkranken können, in die Umgebung des Menschen gelangt.

Die Pathogenität der Salmonellen im Wasser ist offenbar keineswegs erloschen. Es sei an die Husumer Krabbenepidemie 1942 erinnert, bei der 800 Personen an Paratyphus erkrankten mit einer Letalität von 10%, desgleichen an die Makrelenepidemie 1958 auf Nordhorn. Auch bei einer von STEINIGER (*169*) beschriebenen typhös verlaufenen Erkrankung durch S. panama besteht offenbar ein ursächlicher Zusammenhang mit dem Hafenwasser, in dem die gleichen Erreger nachweisbar waren. Andererseits ist bemerkenswert, daß in einem so stark verseuchten Meerwasser, das 10—100 Salmonellen im Kubikzentimeter enthält, Tausende von Menschen badeten, ohne daß Erkrankungen bekannt geworden waren. Wenn, wie STEINIGER (*173*) ausführt, berücksichtigt wird, daß ein guter Schwimmer beim Baden etwa 50 cm³ Wasser und somit bis zu 5000 Salmonellen verschluckt, ist also festzustellen, daß auch bei stärkster Salmonellenvermehrung allein durch Verschlucken des Wassers eine Infektion mit Typhus-Paratyphusbakterien und anderen Salmonellen in der Regel nicht zu befürchten ist. Trotzdem bedeutet, wie auch KRÖGER (*70c*) hervorhebt, die zunehmende Verschmutzung der Vorfluter durch Abgänge infizierter Menschen und Tiere, insbesondere auch infektiöser Schlachtabfälle, eine ernste Gefahr. Die Infektion bei Mensch und Tier kommt zustande durch Aufnahme von Speisen, in denen sich die Salmonellen, durch günstige Umstände bedingt, offenbar millionenfach angereichert haben.

Die an unseren Küsten erhobenen Befunde wurden ebenfalls an anderen Orten bestätigt. So konnte STEINIGER (*178*) 1956 in den Küstengebieten von Barcelona und Mallorca sowohl im Hafenwasser wie auch im Kot besonders von Bartseeschwalben zahlenmäßig häufig Salmonellen, vor allem S. typhi, S. bareilley feststellen. Im Hafenwasser betrug der Titer 5—10 Salmonellen im Kubikzentimeter. Im Kot von Nagetierfressern wie Reihern fand sich S. typhi murium. Auch der Kot von Staren war positiv. Diese Befunde weisen hin auf eine weltweite Verbreitung der Salmonellen.

Die Verhältnisse in der Verteilung der Salmonellen an der schleswig-holsteinischen Westküste haben sich nach STEINIGER (*175*) übrigens insofern geändert, als dort früher fast nur S. paratyphi B, vereinzelt S. typhi im Wasser gefunden wurde, dagegen später auch seltenere Salmonellentypen, so daß man den Salmonellengehalt des Wassers als ein Spiegelbild der Seuchenlage der

Bevölkerung bezeichnen könnte. Wichtig ist hier die Feststellung, daß Salmonellen keineswegs ausschließlich als Krankheitserreger von Mensch und Schlachttier zu gelten haben. Auch Vögel können durch Aufnahme salmonellenhaltiger Substanzen aus dem Wasser erkranken. Selbst S. typhi, die bisher als ein für den Menschen spezifischer Krankheitserreger galt, kann bei freilebenden Vogelarten Infektionen und Ausscheidertum verursachen.

Über weitere Befunde von Salmonellen in Vorflutern und ihre epidemiologische Beurteilung liegen eine Reihe von Mitteilungen vor. Weitgehende Beziehungen bestehen zum Abwasser. Messerschmidt und Wedemeyer (*89*) fanden 1950 in 31% von Abwasserproben in Hannover Salmonellen. Nach der S. bareilly-Epidemie in Norddeutschland wurde nach Richter (*125a*) im Wasser der Unterelbe die gleiche Salmonellenart vorübergehend in zunehmender Häufigkeit festgestellt. S. bareilly konnte von Kröger (*70c*) in Rieselfeld-, Abwasser-, Schlamm- und Gemüseproben im Anschluß an die S. bareilly-Epidemie nachgewiesen werden. Pohl (*107*) fand 1955 in 27 von 29 Proben von Abwässern 14 verschiedene Typen von Salmonellen, deren Herkunft überwiegend auf Menschen zurückgehe. Popp (*108*) stellte in 15 Vorflutern in Braunschweig zum Teil recht seltene Salmonellen fest, am häufigsten S. paratyphi B. In 50% aller Proben sowohl des rohen wie des mechanisch geklärten Abwassers ergaben die weiteren Untersuchungen von Popp (*109*) positive Salmonellenbefunde. Auch kilometerweit unterhalb der Abwassereinlaßstellen wurden Salmonellen gefunden. Die Faulschlammablagerungen erwiesen sich geradezu als Reservoir für Salmonellen. Im Flußschlamm sei die Lebensdauer nahezu unbegrenzt. Man kann von einem ubiquitären Vorkommen der Salmonellen in den Gewässern sprechen. Die Häufigkeit des Vorkommens hängt offenbar von dem Vorhandensein eiweißhaltiger Substanzen ab. Die Bakteriennester liegen nach Lütje (*84*) wie auch Steiniger nur oberflächlich auf dem Boden, während die darunterliegenden Schlammschichten Salmonellen nicht enthalten. Wüstenberg (*1955b*) fand allerdings bei Ausbaggerung von Kanälen in etwa 50% der Schlammschichten Salmonellen. Interessant ist auch die Feststellung von Steiniger (*175*), daß je mehr Colibacillen im Wasser, um so weniger Salmonellen nachgewiesen wurden. Der Colititer als Indicator verliert somit an Stichhaltigkeit, wenn die Salmonellen nicht berücksichtigt werden. Das regelmäßige mengenmäßig wechselnde Vorkommen der Salmonellen in Gewässern und die Beobachtung, daß ein hoher Gehalt von Colibacillen die Zahl der Salmonellen absinken läßt, muß dazu führen, für die Gütebeurteilung eines Wassers außer der Feststellung des Colititers auch einen *Salmonellentiter* zu berücksichtigen. Es sei eine holländische Arbeit von Peters u. Ruys (*103*) erwähnt, in der bereits 1948 auf die quantitative Bedeutung von Salmonellenbefunden im Wasser hingewiesen wird. Popp (*110*) hat dann für die Gütebeurteilung eines Flußwassers den Salmonellenkataster eingeführt.

Mit dem Wasser der Vorfluter besteht die Möglichkeit der Weiterverbreitung der Salmonellen gelegentlich von *Überschwemmungen* auf weite Strecken des Landes, insbesondere auf Wiesen und auf Weidevieh. So beschreibt Lütje (*83*) die Verseuchung einer Weide durch Elbwasser nach einer Sturmflut 1925, die eine S. dublin-Infektion bei 80 Weiderindern zur Folge hatte. Walzberg (*190*) stellte die Verbreitung insbesondere der S. dublin im Raume Weser-Ems als Ursache für Kälbersalmonellose fest. Erkrankungen an Rindersalmonellose an den Flußläufen häufen sich. Ursächliche Zusammenhänge von Salmonellenerkrankungen mit einem verunreinigten Fluß in Oberhessen stellten STRAUCH und MÜNKER (*182*) fest. Auch das Heu wird nach Überschwemmungen infiziert. PIENING (*104*) konnte unter 55 Heuproben in 3 Proben S. dublin nachweisen. Nach HARMSEN (*49b*) gelang der Nachweis im Heu noch nach 11 Monaten. Auch im Erdboden halten sich die Salmonellen lange Zeit. STEINIGER stellte nicht nur in infizierter Gartenerde Salmonellen noch nach 13 Monaten fest, sondern auch in dem angrenzenden Boden. Die Bedeutung hydrogeologischer Verhältnisse geht auch aus Beobachtungen von HOLZ (*57*) und SCHAAL (*140*) u. a. hervor, die eine Massierung positiver Salmonellenbefunde bzw. salmonellenverseuchter Ursprungsgehöfte in Naßbodentypen und lehmiger Bodenart und teilweisen Moorböden im Gegensatz zu Gehöften auf Sandböden feststellen konnten. LÜTJE und RASCH (*85*) wie auch insbesondere RASCH (*114*) und LÜTJE (*84*) sprechen die Vermutung aus, daß sich in einem solchen Gelände die Salmonellen in Wasser, Schlamm bzw. Moor länger halten, weil immer wieder Erkrankungen unter dem Weidevieh beobachtet werden, und zwar auch dann, wenn ein Bestandswechsel erfolgt war (Pensionsweiden), so daß also durchseuchte Tiere als Ausgang der Infektion nicht mehr in Frage kamen. Es ist somit festzustellen, wie auch RASCH (*118*) hervorhebt, daß als Ursprung und Quelle der Salmonellen neben den Dauerausscheidern auch Abwässer eine wichtige Rolle einnehmen, ein Tatbestand, der bei Durchführung der Seuchenbekämpfung zu beachten ist. So weist auch LÜTJE (*84*) darauf hin, daß als Kernpunkt aller dieser Infektionen die Wasserverseuchung der Flußläufe durch Abwasser anzusehen ist. Nicht die primären originalen Erreger, sondern vorwiegend die vegetativen Nachkommen, die eingeschwemmte Eiweißstoffe wie Exkremente, Fischereiabfälle, Schlachthofabgänge usw. besiedeln, bilden den Ausgang der Infektion bei Mensch und Tier.

Salmonellenbefunde bei Vögeln

Über die besondere Bedeutung der *Enten* wie auch der Hühner als Verbreiter der Salmonellen ist bereits ausführlich berichtet. Nach den Feststellungen der letzten Jahre sind aber auch Wassertiere, vor allem Möwen und Wildenten, als Träger und Verschlepper von Salmonellen anzusehen. Ausgedehnte Untersuchungen an der deutschen Nordseeküste, insbesondere durch STEINIGER (*171*) ab 1950, ergaben in jeder 50. Möwenkotprobe Salmonellen. Möwen infizieren sich offenbar an salmonellenhaltigen Eiweißsubstanzen, die besonders an Sielmündungen einfließen, wo sich die Tiere in großen Scharen ansammeln. Ob es sich bei den Möwen um eine passagere oder echte Dauerausscheidung handelt, wurde durch Untersuchungen von Kotproben auf einer Vogelinsel bei Gotland, wo die Möwen sich über den Sommer aufhalten, also fernab von verseuchten Häfen, dahin beantwortet, daß offenbar eine Dauerausscheidung vorliegt. Im Kot halten sich die Erreger lange, wenn er auf grünen Pflanzenteilen abgelegt ist. STEINIGER (*175*) konnte aus Darminhalt einer an enteritischer Krankheit verendeten Flußseeschwalbe S. paratyphi A, aus einer an Flügellähme erkrankten Lachmöwe S. typhi, aus Inhalt und von der Schale fauler Silbermöweneier S. paratyphi A, S. typhi, S. bareilly züchten. Die Erreger dringen nach 7 bis 12 Tagen in das Innere des bebrüteten Eies und bringen den Keimling gewöhnlich zum Absterben. Ein Massensterben von 1200 Jungmöwen, die an Flügellähme und Durchfällen litten, beobachtete Schmidt (*144*). Auch Edwards, Brunner und Moran (*29*) erwähnen, daß verhältnismäßig häufig Vögel als Träger von S. panama festgestellt wurden und anscheinend neben dem menschlichen Keimträger ein Keimreservoir darstellen. Paratyphusbakterien konnten auch aus aufgefundenen Vogelleichen gezüchtet werden. Möwen erkranken also und gehen auch ein. Sie werden auch zu Dauerausscheidern und verstreuen die Krankheitserreger mit ihrem Kot. Eine Infektion der Weiden durch die über Land und entlang der Flüsse fliegenden Vögel ist auf diese Weise möglich. So wurden nach Steiniger (*171*) 1945 im Zisternenwasser im Kreise Husum, das mit auf den Dächern abgelegtem Möwenkot verunreinigtem Regenwasser gespeist wird, Salmonellen wie S. typhi und S. paratyphi B nachgewiesen. Unter den Bewohnern des betreffenden Grundstückes traten Typhuserkrankungen auf. Salmonellen ließen sich bei frischen Eiern nur auf der Außenschale, nicht im Innern nachweisen. Das Möwenei ist nach Wohlrab und Steiniger (*195a*) heute nicht weniger gefahrdrohend als das schon seit Jahrzehnten bekannte besonders infektionsgefährliche Hausentenei.

Auch unter den Taubenbeständen wird über einen in den letzten Jahren zunehmenden Salmonellenbefall berichtet. Beck und Meyer (*9*) teilen Näheres über die Entstehung von Erkrankungen bei Tauben durch Bakterien der Paratyphus-Enteritis-Gruppe mit. Sörin (*164*) beobachtete in Norwegen 1950/51 eine durch S. typhi murium ausgelöste Erkrankung unter mehreren 1000 Tieren. Besonders Jungtauben erkranken und gehen nach 8—14 Tagen ein. Es handelt sich hier um eine Beobachtung ähnlich der bei Jungenten. Bei chronischem Verlauf tritt oft das Symptom der Flügellähme ein. Auf die Mitteilungen von Miesner und Köser (*92a*) sei verwiesen. Nach Schoop und Zettl (*149*) gehört diese Krankheit zu den wichtigsten Taubenerkrankungen überhaupt und ist zur Zeit stark im Anstieg begriffen. Auch sind Fälle von Erkrankungen bei Menschen nach dem Genuß von Taubenfleisch und Taubeneiern beschrieben. Die Taubensalmonellose, vor allem deren chronischer Verlauf mit der Entwicklung zur Dauerausscheidung stellt, wie Helm (*50*) durchaus beizupflichten ist, zumal als Überträger vielfach einer der klassischen Fleischvergifter, das Bacterium enteritidis breslaviensis gefunden wird, eine beachtliche Gefahrenquelle für die menschliche Gesundheit dar. Gefahren der Übertragung auf den Menschen bestehen insbesondere in Lebensmittelbetrieben wie Milch-, Fleisch- und Fischgeschäften, deren Inhaber Taubenzüchter sind.

Salmonellenbefunde bei anderen Tieren

Anschließend sei hervorgehoben, daß auch andere Tiere in zunehmendem Umfang als Verbreiter von Salmonellen anzusehen sind. So fand Steiniger (*180*) in Cuxhaven unter 53 Hausratten 24 Träger von Salmonellen, und zwar 8 verschiedene Typen; 4 Wanderratten waren sämtlich infiziert, darunter zweimal mit S. ratin. Auch Rattenkotproben waren positiv. Salmonellen wurden nachgewiesen in Miesmuscheln, Austern, Krabben, Schnecken, in Schaben und Küchenfliegen. Holz (*58*) fand sie in Schildkröten. In Fischen des Kairoer Fischmarktes wiesen Floyd und Jones (*32b*) ebenfalls Salmonellen nach. Sie wurden festgestellt bei Kolkraben, Seidenreihern, Spießenten [Zureck (*198*)], desgleichen bei Hunden [Messow u. Stoll (*90*)], Katzen [Seele (*151*)], [Gorham u. Garner

(*47*), Schaal u. Schütz (*141*)], Iltis, Bisamratte, Löwen, Waschbär, Krokodil sowie auch bei Sperlingen und Staren. Berkmen (*12*) fand bei Kamelen in Mittelanatolien, Löliger (*81*) beim Nerz Salmonellen. Nach Bachmann und Kast (*8*) ging eine 15jährige Tigerschlange an einer Salmonellose durch S. typhi murium ein. Eine hochgradige Fliegenverseuchung mit Salmonellen stellte Steiniger (*179*) in Tierfuttermittelrohstofflagern fest. Nach experimentellen Untersuchungen von Schaal (*139*) hielt sich S. dublin, mit der Stubenfliegen gefüttert waren, bis zum 7. Tage im Magendarmkanal, bei toten Fliegen bis zum 35. Tage. Die Salmonellen wurden auch mit dem Kot ausgeschieden.

Vorkommen von Salmonellen bei Schlachttieren

In den letzten Jahren häufen sich die Befunde von Salmonellen bei Schlachttieren. Nicht nur durch Salmonellen verursachte Erkrankungen wurden beobachtet. Der Genuß von Fleisch, das wegen Fehlens jeglicher Krankheitserscheinungen des Viehes freigegeben war, löste nach Rohde und Bischoff (*133*) menschliche Erkrankungen aus. In den Mesenteriallymphknoten und anderen Organen gesunder Schweine, Rinder und Pferde fanden sich in zunehmendem Umfang die verschiedensten Arten von Salmonellen, so daß eine alimentäre Infektion angenommen wurde [Rasch (*116*)]. Bemerkenswert ist, daß bereits 1951 über derartige Befunde in Java berichtet wird. Nach Galton, Lowery und Hardy (*41*) wurden in USA bei zum Teil an Durchfällen erkrankten Schweinen verschiedener Farmen 7% positive Befunde erhoben. Rectalabstriche von Schweinen in Schlachthöfen ergaben sogar 78% positive Befunde. Dieselben Verfasser fanden in USA in 23% frischer Schweinewürste und in 12% von Räucherwürsten Salmonellen. Die nordamerikanischen Kleintierzüchter fürchten vor allem die Salmonellen in Dörrfleisch, die das handelsübliche Kleintierfutter zu einer gefährlichen Infektionsquelle machen. Nach Rasch (*117*) waren 1953/54 im norddeutschen Raum von 10695 Proben von Schlachttieren mit Fleischvergiftungsverdacht 466 — 4,5% mit Salmonellen behaftet, und zwar 11 Pferde, 214 Rinder, 221 Kälber, 19 Schweine, 1 Schaf. Schaal (*140*) fand am Niederrhein in etwa 8000 Proben = 1,8% Salmonellen, darunter 83% S. dublin. Unter den Tierarten stand das Kalb mit 84% positiven Befunden an der Spitze. Schweinekotproben waren in 23—25% der Untersuchungen positiv. Die Auswertung der 2250 Salmonellenbefunde in Berlin, die nach Stellmacher (*181*) 58,4% S. dublin ergaben, bedeuten eine wesentliche Zunahme der Salmonellen in Magdeburg, Potsdam und Mecklenburg. In München ergaben nach Meyer und Endress (*92*) die Schlachtviehuntersuchungen in 13736 Proben 0,7% Salmonellenbefunde. Es handelt sich um Typen, die mit Futtermitteln und Eikonserven eingeschleppt wurden. Bei verendeten Ferkeln fand Klein (*68*) S. manhattan, desgleichen in den verfütterten Futtermitteln. Auch Dauerausscheider werden unter den Tieren festgestellt. Rasch (*119*) berichtet, daß selbst bei Dauerausscheidern unter den Rindern bis zu 90% gar keine oder nur spärliche pathologisch-anatomische Veränderungen nachzuweisen sind. Bei 64% war der einzige Fundort die Leber, in 46% die weiblichen Geschlechtsorgane, 4mal fand sich eine septische Verbreitung. Bei Dauerausscheidern wurden nach Pohl (*107*) Salmonellen sowohl im Eutergewebe wie nach Rauch (*121*) auch in Euterlymphknoten festgestellt. Auf die Gefahr der gesunden Keimträger unter den Schlachtviehbeständen, deren

Erfassung lückenhaft bleiben wird, weisen auch Froehner und Grüttner (*34*) hin. In Niedersachsen wurden 1951—1954 bei 10057 bakteriologischen Fleischuntersuchungen 2732mal = 2,7% Salmonellen nachgewiesen. 1955 und 1956 konnten in Hannover durch Untersuchung von Blut- und Kotproben von 2670 Rindern im Anschluß an die bakteriologische Fleischuntersuchung 41 = 1,45% Dauerausscheider ermittelt werden. Bei Dauerausscheidern der Rinder ist nach Lütje und Rasch (*85*) die Gallenblasenschleimhaut der eigentliche Depotplatz der Salmonellen. Nach Jaeschke (*60*) zeigte bei Salmonellenbefunden die Gallenblasenschleimhaut deutliche Veränderungen, zum Teil hämorrhagische Entzündung. Der häufige Befund von Gärtner-Bakterien in den Euterlymphknoten berührt die Frage einer galaktogenen Infektion. Im Speisequark konnte Seifert (*162*) in 4% der 154 Proben S. typhi murium nachweisen. Eine Gärtner-Infektion durch Käse beschreibt 1928 Bermbach (*13*), während Bourmer und Doetsch (*22*) im gleichen Jahre eine Epidemie durch Käse auf einem Gut, verursacht durch S. dublin, beobachteten. Rasch (*119*) spricht sich für eine Intensivierung der Bekämpfung der Rindersalmonellose aus und empfiehlt, daß bei zur Vorzugsmilch bestimmten Kühen periodisch Kotuntersuchungen vorzunehmen sind. Interessant und bedeutungsvoll ist auch die Beobachtung von Struck (*184*), daß bei geschlachteten Kälbern nach 8—14tägiger sachgemäßer Aufbewahrung der Tierkörper bei Kühlhaustemperatur die Salmonellen, die anfangs nur in den Organen nachweisbar waren, auch in der Muskulatur festgestellt werden konnten. In den Mesenteriallymphknoten gesunder Pferde fand Linke (*79*) in Berlin in 2,7% der Untersuchungen S. typhi murium, S. senftenberg, braenderup, infantis, anatum. Neuerdings berichtet Gässlein (*44*) über Salmonellenbefunde bei ungarischen Importschweinen. 42% der Proben enthielten 6 verschiedene Salmonellentypen. In Mittelanatolien wurden nach Berkmen (*12*) bei umfassenden Untersuchungen nur selten Salmonellen gefunden. Kelch (*66*) konnte in kleineren Schlachthöfen in Bayern im Gegensatz zu größeren feststellen, daß von einem an Salmonellose eingegangenen Tier eine Verbreitung durch Kot auf andere Tiere stattgefunden hat. Nach Daigeler und Kotter (*26*) wurden in Bayern bis 1954 vorwiegend die Typen S. typhi murium und S. enteritidis, im Gegensatz zu Norddeutschland S. dublin jedoch nur in 11% nachgewiesen. Rasch und Richter (*120*) stellten als Ausgang von Erkrankungen bei Rindern eine Kuh als Dauerausscheider fest. Über ähnliche Befunde von Salmonellen bei Schlachttieren ist von anderen Autoren berichtet, die hier nicht vollständig aufgeführt werden können. Die Prozentzahl der positiven Befunde steigt an und erreichte nach Zureck (*198*) in Berlin 1956 7%. Angefügt seien noch Mitteilungen von Holz (*56*), Bischoff (*14*), Zureck (*197*), Seidel (*158*), Rasch (*115*), Utojo (*188*), Herter (*52*), Müller (*100*), Daigeler und Kotter (*26*), Abel (*2*), Holz (*57*), Merk (*88*), Schaal und Schütz (*141*), Hunsteger (*59*), Plaschke und Plaschke (*106*), Froehner und Grüttner (*34*). Die vorstehend aufgeführten Salmonellenbefunde bedeuten also eine außerordentliche, offenbar weiter zunehmende Verbreitung der Salmonellen unter den Schlachttieren.

Salmonellenerkrankungen bei Menschen

Eine Klärung über das zunehmende Vorkommen von Salmonellen bahnte sich dann an durch Beobachtungen gelegentlich von zum Teil ausgedehnten

Enteritisepidemien. 1953 erkrankten nach Bonitz (*21*) im norddeutschen Raum nachweislich 6000 Personen nach Genuß von Camembertkäse aus einer großen Molkerei in der Lüneburger Heide. Die Zahl der Erkrankungen dürfte jedoch wahrscheinlich das Vielfache der angegebenen Zahl betragen. Als Erreger wurde S. bareilly festgestellt, der offenbar durch eine Keimträgerin übertragen war, die die Anleimung der Etiketts auf die bereits in Stanniolpapier verpackten Käse besorgte. Dezember 1953 erkrankten in einem Hamburger Großbetrieb etwa 100 Personen an einer durch S. kirkec verursachten Enteritis. Erinnert sei an die Epidemie in Brackwede mit 520 ermittelten Erkrankungen und 3 Todesfällen, hervorgerufen durch den Genuß von Braunschweiger Wurst. Als Erreger fand sich nach Reploh und Rainer (*124b*) S. blockley. Handloser (*48*) schätzt die Zahl der Erkrankungen auf 1500—2000. Kontaktübertragungen wurden kaum beobachtet. Weiter sei hingewiesen auf Salmonelleninfektionen in Krankenhäusern, besonders unter Kindern. Hierüber gibt Rohde (*131*) einen interessanten Bericht. Es handelt sich hier im wesentlichen um eine Kontaktepidemie (Hände, Handtücher) durch S. montevideo mit einer Letalität von 9%. Nach Sinios, Tiling und Hanisch (*163*) erkrankten überwiegend Frühgeborene und Säuglinge des ersten Trimenon. Fast alle erkrankten Kinder wurden zu Dauerausscheidern. 19 gesunde Keimträger erkrankten. Der Keimnachweis im Stuhl gelang oft erst nach Abklingen des akuten Stadiums. Bemerkenswert bei diesen Erkrankungen ist vor allem, daß als Erreger vielfach Salmonellentypen nachgewiesen wurden, die hierzulande bisher unbekannt, dagegen in anderen — besonders überseeischen — Ländern beschrieben waren. Ein Beispiel für die Verbreitungsweise der Salmonellen zeigt, wie Steiniger (*169*) ausführt, das Auftreten der S. panama, die zuerst 1934 bei amerikanischen Soldaten in der Panamazone von Jordan (*62*) festgestellt wurde, dann während des zweiten Weltkrieges weite Verbreitung in USA fand, 1940 in Deutschland von Hohn und Herrmann (*55*) bei enteritischen Erkrankungen, 1941 bei einem Kind aus Berlin von Preuss (*112*), dann gehäuft in Köln von Lempfried (*73*) nachgewiesen wurde. Kreuscher und Sylvester (*70a*) berichten 1950 über 1000 Infizierte durch S. london nach Genuß von Kochschinken, Keil und Haupt (*65*) 1951 über 250 Erkrankungen durch S. typhi murium nach Genuß von Werksküchenessen, Trüb und Schneider (*185*) über Erkrankungen durch S. panama in Nordrhein-Westfalen 1950/52, Merdl und Wolfs (*87*) weiterhin 1953 über Erkrankungen auch mit tödlichem Ausgang durch S. newport, Zureck (*197b*) im gleichen Jahre über 5 Gruppen von Lebensmittelvergiftungen durch S. münchen, S. dublin, S. bareilly, S. newport, Renner (*123*) über 39 Erkrankungen in Ludwigshafen nach Genuß von Mettwurst durch S. anatum, Richter und Pöhlig (*126*) 1954 über Erkrankungen durch S. braenderup und S. newington sowie über eine große Zahl von Keimträgern, Rohde (*129*) über 1000 Erkrankungen nach einem Betriebsessen, Hofmann und Wolle-John (*54*) 1954 über Erkrankungen durch S. saint Paul, Kauker (*63*) über Erkrankungen nach Genuß von Brathähnen durch S. infantis, Schmidt-Lange, Möller, Lederbogen und Rauch (*146*) über tödliche Erkrankungen durch S. blockley nach Wurstgenuß, Rohde und Seitz (*134*) 1955 über Lebensmittelvergiftungen durch S. chester und S. paratyphi B, Wildführ und Hudemann (*191*) über einen neuen Salmonellentyp aus dem Stuhl eines Patienten in Görlitz, Hepp (*51*) über Erkrankungen durch S. typhi murium nach

Genuß von Schweinefleichkonserven jugoslavischer Herkunft mit Nachweis der Erreger auch an der Außenseite der undichten Konservendosen, Ekstams (*30*) über zahlreiche Erkrankungen und Keimträger in einer Schlachterei Südschwedens, Bischoff (*17*) über Erkrankungen nach Genuß von Schweinefleisch durch S. braenderup, Dräger (*27*) über menschliche Erkrankungen, Münchow (*102*) über Erkrankungen durch S. enteritidis nach Genuß von Eis am Stiel, Schmidt-Lange, Joest und Möller (*145*) 1956 über Erkrankungen durch S. osnabrueck, Gärtner (*43*) über zwei Krankheitsfälle mit Nachweis von Salmonella bovis morbifican im Sputum, Primavesi (*113*) über 340 Erkrankungen im Ruhrgebiet nach Genuß von Braunschweiger Streichwurst durch S. heidelberg und über eine große Zahl von Keimträgern, Joest (*61*) 1957 über Erkrankungen durch S. infantis. Reploh (*124a*) bzw. vor dem Esche und Handloser (*30a*) beobachteten in jüngster Zeit 259 Erkrankungen in der Bielefelder Gegend nach Genuß ausschließlich aus Hühnereiern hergestellter Buttercreme. Gaase (*40*) stellte in Bochum Erkrankungen fest, die durch 15 verschiedene Typen seltenen Vorkommens hervorgerufen waren. Erwähnt sei noch, daß nach Saphra (*136*) in New York in der Zeit von April 1939 bis Dezember 1955 7779 menschliche Salmonellenerkrankungen nachgewiesen wurden, darunter 68% mit gastrointestinalen, 9% mit typhösem bzw. septicämischem Verlauf. Der Typ cholerae suis war am stärksten beteiligt. Die Zahl der Todesfälle betrug 4%. Keimträger konnten in 15% der Erkrankungen ermittelt werden. Bekannt ist die Vielgestaltigkeit der klinischen Bilder durch Salmonellenkeime. Im allgemeinen werden die typhös und die als Enteritis verlaufenden Krankheitsbilder unterschieden. Auch enteritische Erkrankungen können mit einer Keiminvasion ins Blut einhergehen und örtliche Erkrankungen durch Keimansiedlungen (Abscesse, Meningitiden, Pneumonien, u. a.) verursachen. Mannweiler und Humbert (*86*) berichten über durch Enteritiserreger der Salmonellagruppe hervorgerufene Bakterienämien. Besonders gefahrvoll sind nach Staak (*167a*) Salmonelleninfektionen für junge und alte, sowie resistenzgeschwächte Personen.

Einschleppung der Salmonellen durch Futtermittel

Es ist nun bekannt, daß in den letzten Jahren in zunehmendem Umfang *Mehlpräparate* zusätzlich bei der Fütterung des Viehs Verwendung finden. Diese Mehle werden besonders aus dem Ausland, vor allem aus Übersee, eingeführt. Es ist von Bedeutung, auf den erheblichen Anstieg der Einfuhrmengen hinzuweisen. Während z. B. 1948 Fischmehl in einer Menge von 394 t eingeführt wurde, betrug diese Zahl 1954 50631 t und stieg 1956 auf 72766 t, allein für den Hamburger Bereich. Die Einfuhr von Fischmehl, Fleischmehl und anderen tierischen Abfällen zur Viehfütterung und zum Düngen in das Bundesgebiet erreichte 1956 insgesamt 101915 t, davon über Hamburg 72492 t. Die bakteriologische Untersuchung dieser Mehle ergab nun vielfach einen bemerkenswerten Befund von Salmonellen verschiedener Art und in wechselnder Menge. In Fischmehlproben z. B. aus Angola gelang nach Rohde und Bischoff (*133*) der Nachweis von Salmonellen bis zu 70% der untersuchten Proben. In einer Probe von 125 g Fischmehl derselben Herkunft konnten 10 verschiedene Typen gezüchtet werden. Ganze Schiffsladungen mit Eipulver oder Ei, mit Fischmehl, Knochenmehl, Fleischmehl und Kadavermehl waren salmonellenhaltig. Die Einfuhr dieser

vielfach salmonellenhaltigen Futtermittel aus dem Ausland und die zunehmende
Verwendung als Beifutter für die Ernährung des Viehs erklärt wohl ohne weiteres
den so häufigen Befund von Salmonellen.

Ein Beispiel für die so weite Verbreitung durch Tierfutter vermag ich (*37*)
aus eigener Erfahrung mitzuteilen. Als Hühnerhalter füttere ich mit einem Voll-
kraftkorn, das u. a. auch Fischmehl, Fleischmehl und Futterknochenschrot enthält. In dem Konzentrat sind alle zur Aufzucht und Ernährung notwendigen
Bestandteile vorhanden, so daß das Verfahren der Fütterung sehr vereinfacht
ist und eine zunehmende Verbreitung gefunden hat. Die Untersuchung einer
Probe dieses Kraftfutters ergab nun den Nachweis der S. indiana, eines bisher
seltenen Erregers, der nach SEELIGER (*157*) auch beim erkrankten Menschen
festgestellt ist. Auch KLEIN (*68*) wies im Futtermittel die gleichen Salmonellen
(S. manhattan) nach wie in den Organen der verendeten Ferkel. Wenn auch
VOGT (*189*) bei der Untersuchung von 185 Hühnerfutterproben aus 22 verschiedenen Verkaufsstellen Salmonellen nicht nachweisen konnte, so dürfte doch auf
Grund der so häufigen Befunde in Mehlprodukten kein Zweifel an der ursäch-
lichen Bedeutung dieser Futtermehle für die Verbreitung der Salmonellen bestehen. BISCHOFF (*15*) fand in indischem Knochenmehl S. heidelberg, S. chester,
S. senftenberg, S. anatum und weist darauf hin, daß das ausländische Material
die inländischen Knochenmehlfabriken und dann weiter Schwein und Mensch
infiziert. In 275 Proben von ausländischem Fisch- und Fleischmehl fanden
BISCHOFF und ROHDE (*19*) 43mal = 15,6% Salmonellen 22 verschiedener Typen.
Sie sind der Ansicht, daß Fische in Flußmündungen und in Küstennähe vor allem
infiziert sind. Nach WINKLE, ROHDE und BISCHOFF (*194*) ließen sich in Hamburg
in 1386 Tierfutterproben 336mal = 17% 47 verschiedene Salmonellenarten nachweisen. Nach einer Mitteilung von ADAM (*4*) enthielten bei Untersuchungen von
Frühjahr 1955 bis August 1956 833 Fischmehlproben in 15,8%, 3 Walmehlproben
in 33%, 22 Fleischmehlproben in 36%, 14 Blutmehlproben in 14%, 52 Knochen-
schrotproben in 11,5% Keime der Salmonellengruppe. HOFMANN und POHL (*53*)
fanden 1957 neue Salmonellentypen in Fischmehl aus Cuxhaven. Es ist also
gar nicht abzusehen, zu welchen Infektionsmöglichkeiten bei Mensch und
Tier diese mit Tiermehlen verschiedener Art aus dem Ausland eingeschleppten
Salmonellen noch weiterhin führen werden. Interessant und beweisend für die
weite Streuung der Salmonellen durch diese infizierten Futtermehle sind die
von STEINIGER (*179*) mitgeteilten Ergebnisse von Fliegenuntersuchungen in verschiedenen Tierfuttermittelrohstofflagern im nordwestdeutschen Raum. Unter
205 Proben waren $^{1}/_{5}$ der Fliegen mit Salmonellen verschiedener Art behaftet,
darunter S. typhi, S. paratyphi A und B. Der Import salmonellenhaltiger Tierfuttermehle erweist sich so, wie WINKLE, ROHDE und BISCHOFF (*194*) mit Recht
ausführen, als noch gefährlicher als die Einfuhr chinesischer Eiprodukte. Abwehr-
maßnahmen sind dringend und beschleunigt erforderlich.

Als *Ergebnis* der vorstehenden Zusammenstellung von Beobachtungen über
den Nachweis von Salmonellen im Wasser, beim Geflügel, bei Schlachttieren,
bei Menschen, in Eiprodukten und Futtermitteln ist also festzustellen, daß als
Ausgang der Ausbreitung die aus dem Ausland eingeführten Eiprodukte, vor
allem aber die Futtermehle anzusehen sind. Die ausgedehnte Verfütterung
der Futtermehle insbesondere führt zu einer weitgehenden Verseuchung des

Schlachtviehes und des Geflügels und damit zu einer großen Infektionsgefahr für den Menschen.

Es ist deshalb auch erklärlich, daß die Zahl der menschlichen Erkrankungen an Salmonellose bis 1956 ansteigt. Im Bundesgebiet erkrankten an bakteriellen Lebensmittelvergiftungen absolut und berechnet auf 10000 Einwohner

1951	2393	= 0,5	1955	3910	= 0,8
1952	1387	= 0,3	1956	4312	= 0,9
1953	3194	= 0,7	1957	2683	= 0,5
1954	2557	= 0,5			

Der Anstieg 1953 auf 3194 Erkrankungen ist auf die Bareilly-Epidemie in Norddeutschland zurückzuführen. 1955 werden 3910 und 1956 als Höchststand 4312 Erkrankungen gemeldet. Dann aber fällt 1957 die Zahl auf 2683 ab. Wieweit dieser Rückgang an Erkrankungen als Auswirkung eingeleiteter Bekämpfungsmaßnahmen anzusehen ist, muß 1958 lehren.

Maßnahmen

Für die Maßnahmen zur Bekämpfung der Salmonelleninfektionen ist zunächst zu berücksichtigen, daß in den letzten Jahren eine weitgehende Verbreitung von Salmonellen bereits stattgefunden hat. Das beweisen die oben angeführten allseits erhobenen Befunde von Salmonellen. Es gilt also einmal, sich dieser im Lande bestehenden Gefahr zu erwehren. Maßgebend sind die gesetzlichen Bestimmungen zur Bekämpfung der übertragbaren Krankheiten. Es dürfte nicht zulässig sein, mit Rücksicht auf den vielfach leichteren Verlauf der Erkrankungen und das häufige Vorkommen stummer Infektionen mit erleichternden Einschränkungen zu verfahren, weil jeder in die Salmonellengruppe einzuordnende Typ grundsätzlich als menschenpathogen angesehen werden muß, wie auch Dräger (27), Rasch (117) u. a. hervorheben.

Voraussetzung wirksamer Durchführung von Maßnahmen ist die Erfassung der Erkrankungen. Bakteriologische Stuhluntersuchungen grundsätzlich bei allen, auch leichten Darmerkrankungen sind immer wieder zu fordern. Zu dieser Übung sollten nicht nur die Krankenhausärzte, sondern auch der praktische Arzt bei ambulanter Behandlung häufiger und immer wieder angehalten werden. Meldung an die Gesundheitsämter, auch der leichtesten Erkrankungen, die bakteriologisch als infektiös bestätigt sind, muß sichergestellt werden, damit die erforderlichen Maßnahmen entsprechend den gesetzlichen Vorschriften in vollem Umfang durchgeführt werden.

Übrigens, das sei hier angeführt, beschränkt sich das Problem der Lebensmittelvergiftungen, wie Seidel (160) mit Recht hervorhebt, keineswegs auf die Ätiologie der Salmonellen. Abgesehen von Erkrankungen durch B. botulinus spielen auch unspezifische Bakterien wie B. proteus, B. coli bei Lebensmittelvergiftungen eine ursächliche Rolle. Die Bezugnahme auf die Arbeiten von Trüb und Wundram (186) sowie Trüb und Reploh (187) aus den Jahren um 1950 muß aber berücksichtigen, daß der ursächliche Zusammenhang dieser zunächst unspezifischen Erkrankungen nicht immer leicht zu führen ist und inzwischen die Diagnose der Salmonellen erhebliche Fortschritte erfahren hat. Die überragende Bedeutung der Salmonellen als Erreger von Lebensmittelvergiftungen bleibt bestehen. Dem Vorschlag SEIDELs, auf dem Gebiete der Lebensmittelvergiftungen eine zentrale Forschungsstelle mit einem entsprechenden Institut einzurichten, wird beigepflichtet.

Da auch das Schlachtvieh im großen Umfang mit Salmonellen bereits infiziert ist und auch erkrankt, sind in den Schlachthöfen, soweit nicht bereits erfolgt, verschärfte Untersuchungen zu fordern. Wieweit darüber hinaus Salmonellenerkrankungen und Keimträger von Salmonellen unter den Viehbeständen festgestellt und entsprechende Maßnahmen getroffen werden können, hängt auch von ihrer Durchführbarkeit ab. Mit Rücksicht auf die starke Verbreitung der Dauerausscheidung von Salmonellen empfiehlt RASCH (*119*), wie erwähnt, bei zur Vorzugsmilchgewinnung bestimmten Kühen periodische Kotuntersuchungen. FROEHNER und GRÜTTNER (*34*) schlagen vor, in den Ursprungsbeständen der bei der bakteriologischen Fleischuntersuchung als salmonellenpositiv ermittelten Tiere die Entnahme von Blut- und Kotproben dreimal auf alle Rinder des Bestandes auszudehnen. Die Herkunft der salmonellenbehafteten Schlachttiere müßte sichergestellt sein. Auf die Notwendigkeit entsprechender Umgebungsuntersuchungen, von Desinfektion der Einrichtungsgegenstände und Gerätschaften, auf das Vorhandensein separater Räume zur Verhütung von Kontaktinfektion in Schlachthäusern weist ZURECK (*198*) hin. Das Salzen der Därme ist anscheinend von besonderer Bedeutung. Ein Kochsalzgehalt von 8—10% beeinträchtigt bzw. hebt nach BERGMANN und SEIDEL (*11*) die Lebensfähigkeit der Salmonellen auf. Als zusätzliche Maßnahmen hält LÜTJE (*84*) systematische Kälberimpfung, Desinfektion, Kotuntersuchungen, Abriegelung der bisherigen Tränken auf Weiden, Ausfällen der Eiweißstoffe in den Vorflutern für notwendig. Jedenfalls ist der Lebenduntersuchung von Schlachtvieh größte Sorgfalt zuzuwenden, wie RASCH (*114*) schon 1953 hervorhebt. Alle Möglichkeiten zur Ermittelung von Dauerausscheidung sind auszuschöpfen. Verstärkte bakteriologische Untersuchungen sind erforderlich. Jede Ausmerzung eines Dauerausscheiders erhöht die Rentabilität der Rindviehhaltung seines Herkunftbestandes und vermindert die Gefahr möglicher Fleischvergiftung. Wichtig ist natürlich verschärfte Überwachung importierter Lebens- und Futtermittel und auch importierten Schlachtviehs. Auch die Geflügelhaltung bedarf der verstärkten Überwachung. ALTERAUGE (*6*) empfiehlt Sperre über infizierte Entenbestände und Tötung von Dauerausscheidern. Den Inhabern von Lebensmittelbetrieben ist das Halten von Tauben möglichst zu untersagen.

Von Bedeutung ist auch bezüglich der Verbreitung der Salmonellen durch Enteneier ein Verbot der Verwendung von Enteneiern in gewerblichen Betrieben, die Lebensmittel verarbeiten, in Krankenhäusern, in Gemeinschaftsküchen von Heimen aller Art, in Gefangenenanstalten, bei Massenunterbringungen und ähnlichen Verhältnissen. Das Niedersächsche Staatsministerium hat bereits unter dem 19. 8. 49 eine entsprechende Verordnung erlassen.

Wichtig ist dann die bakteriologische Überwachung unserer gesamten Lebensmittel- und Futtermittelversorgung durch human- und veterinärmedizinische Untersuchungsstellen. Aufmerksamkeit ist dem Fleischer-, Bäcker- und Konditorgewerbe, gerade auch mit Rücksicht auf infizierte Eiprodukte zuzuwenden. Auch die Fliegenbekämpfung in Lagern und Verarbeitungsstätten von Futtermehlen, in Abdeckereien, Schlachthöfen und Fleischereien ist mit Nachdruck zu betreiben, nachdem festgestellt ist, daß auch Fliegen in solchen Betrieben in beträchtlichem Umfang Träger von Salmonellen sind.

Schließlich muß der Ansammlung salmonelleninfizierter eiweißhaltiger Substanzen in unseren Vorflutern entgegengearbeitet werden. Als besonders gefährdend sind die Abwässer von Fischfabriken und Schlachthäusern anzusehen. Die Schaffung besonderer Kläranlagen ist anzustreben. Der Forderung von Rasch (*118*), „kein ungereinigtes Abwasser in die Flüsse und auf das Land", stehen naturgemäß in der Praxis Schwierigkeiten im Wege. Steiniger (*170*) schlägt eine Fällung der im Abwasser mitgeführten Eiweißsubstanzen in Verbindung mit einer Faulgasgewinnung aus wirtschaftlichen Gründen vor.

Vor allem aber muß mit Beschleunigung und Nachdruck die *Einfuhr* salmonellenhaltiger Lebensmittel und Tierfuttermehle unterbunden werden, um einer zusätzlichen Salmonellenüberschwemmung entgegenzuwirken. Wie Richter (*125 b*) hervorhebt, können die Durchführungsbestimmungen solcher Verordnungen die Wirtschaftsinteressen, sie müssen aber die seuchenhygienischen Interessen befriedigen. Inzwischen ist die Verordnung zum Schutze gegen Infektionen durch Erreger der Salmonellagruppe in Eiprodukten vom 17. 12. 56 erlassen, die am 1. 4. 57 in Kraft getreten ist. Als ausreichende Vorbehandlung im Sinne dieser Verordnung sind Verfahren anzusehen, durch die die Erreger der Salmonellagruppe und die anderen Erreger der Gruppe der Enterobakteriazeen in Eiprodukten abgetötet werden. Die Zahl der anschließend vorzunehmenden Stichproben zur bakteriologischen Untersuchung ist gestaffelt nach Zahl der Packstücke gleichartiger Sendungen und schwankt zwischen 10 und 2 v. H. der Packstücke. Zum Beispiel sind bei Mengen bis zu 60 Packstücken gleicher Art aus mindestens 6 Packstücken, bis zu 1000 Packstücken aus mindestens 5 v. H. aller Packstücke Stichproben zu je 30 g zu entnehmen.

Wenn Lodenkämper (*80*) in einem Gutachten, das der Handelskammer Hamburg im Oktober 1956 zur Verfügung gestellt wurde, die Notwendigkeit besonderer Maßnahmen zum Schutze gegen die Salmonelleneinschleppung durch Eiprodukte bestreitet, so kann den überzeugenden Ausführungen von Winkle, Rohde und Bischoff (*194*) nur beigepflichtet werden. Es ist bedauerlich, daß Argumente, die den tatsächlichen Verhältnissen wohl nicht entsprechen, nun verständlicherweise von interessierter wirtschaftlicher Seite aufgegriffen werden, um dringend notwendige Maßnahmen zum Schutze der Gesundheit von Mensch und Tier zu verhindern. Diese Gefahr besteht insbesondere, wenn jetzt das schwierigere Problem der Einfuhrverhinderung salmonellenhaltiger Tierfuttermittel zu lösen ist. Diese Forderung kann auch dadurch nicht entkräftet werden, daß nach Beobachtungen aus letzter Zeit das in der Verordnung vom 17. 12. 56 vorgeschriebene Verfahren in Einzelfällen nicht ausreichend wirksam gewesen ist. Anz und Wohlrab (*7*) berichten, daß sowohl in Eiprodukten aus Hamburger Bäckereien wie auch in Beständen der Herstellerfirmen in Hannover Salmonellen, und zwar S. montevideo, nachgewiesen werden konnten. Es handelte sich um Kristalleiweiß, das in der Herstellerfirma einem Mahlprozeß unterworfen wird, um eine bessere Vermischung des Mehlgutes zu erreichen. Auch wurde durch das Mahlen die ungleichmäßige Verteilung der Salmonellen wohl weitgehend beseitigt, so daß die bakteriologische Ausbeute an Salmonellen sich erhöhte. Interessant ist übrigens, daß nach dem Mahlprozeß der Staub im Mahlraum

S. montevideo enthielt. Es ist also festzustellen, daß trotz negativer Stichproben importierten Kristalleiweißes Salmonellenfreiheit nicht gewährleistet ist. Die ungleichmäßige Verteilung der Salmonellen im Trockeneiweißprodukt gebietet eine Aufarbeitung vor der Stichprobenentnahme, möglichst am Einfuhrort bzw. am Orte der Firma, die die Aufarbeitung besorgt. Mit Recht weisen ANZ und WOHLRAB darauf hin, daß bereits im Erzeugerland auf die Gewinnung einwand-freien Ausgangsmaterials Wert zu legen ist.

Diese Maßnahmen einer nachträglichen Überprüfung der Eiprodukte auf Salmonellenfreiheit würden sich erübrigen, wenn die Hersteller einwandfreie Produkte lieferten. Heute, im Zeitalter der Weltgesundheitsorganisation dürfte es nicht schwer sein, solche hygienischen Forderungen zur Verhütung von Seuchen-verschleppung, wie auch STUTZ (*183*) hervorhebt, durchzuführen. Es ist deshalb zu begrüßen, daß nach FREITAG (*33a*) in den Ausfuhrländern wie China, USA, Kanada, Argentinien in letzter Zeit Verfahren Verbreitung gefunden haben, die eine Vernichtung der Salmonellen vor der Ausfuhr sicherstellen sollen. Die Chinesen haben nach ihren Angaben neuerdings Pasteurisierungsmaschinen ange-schafft. Wie der Eierprodukten-Einfuhrverband in Hamburg mitteilt, wird das gesamte ausländische Eiprodukteneinfuhrgut, soweit es sich um Trockenei-produkte handelt, einem Hitzeverfahren unterzogen, das als ausreichende Vor-behandlung im Sinne der Verordnung gilt. Die flüssigen Eiprodukte werden im Krausesprühturm mit einer Eingangshitze von 100—110° versprüht. Außerdem findet eine nachfolgende Nacherhitzung statt. Bei Kristalleiweiß wird der Fer-mentierungsprozeß mit Zusatz von Ammoniak auch in China angewandt. Dieses Verfahren wie auch die Einwirkung von entsprechender Hitze beim Kristalli-sierungsprozeß wird angeblich als ausreichende Vorbehandlung im Sinne der Verordnung angesehen. Flüssige Eiprodukte sind mit 6—8% Salz und etwa 1%igem benzoesaurem Natron versetzt. Diese Zusätze genügen angeblich zur Vernichtung der Bakterien.

Schließlich muß auf die Folgerungen hingewiesen werden, die sich aus der Verwendung von Eiprodukten in der Lederindustrie und in anderen Industrien ergeben. Wie oben aufgeführt, macht eine höhere Erhitzung die Eiprodukte für diese Zwecke ungeeignet. Das für technische Zwecke bestimmte flüssige Hühner-eigelb, das zwecks Konservierung mit 10—12% Salz und 2% Borsäure versetzt ist, unterliegt angeblich nicht der Verordnung. Die Ware wird im allgemeinen vor der Einfuhr mit 1°/₀₀ Rosmarinöl denaturiert und somit für den mensch-lichen Genuß unbrauchbar gemacht.

Die aus dem Ausland eingeführten Eiprodukte sind nun ausnahmslos der bakteriologischen Kontrolle zu unterwerfen. Bei Nachweis von Salmonellen bzw. Enterobakteriazeen hat eine erneute Behandlung stattzufinden. Es kommt dabei wesentlich darauf an, Verfahren anzuwenden, die Struktur, Farbe und Geschmack der Rohstoffe nicht verändern. ADAM (*3*) berichtet über günstige Pasteurisierungs-versuche von Eigelbpulver bei einer Temperatur von 70° durch Dampfeinwirkung in evakuiertem Zustand innerhalb 20 min. Auch in USA wird eine Pasteurisie-rung der Eimasse vor der Versprühung salmonellenhaltiger Eiprodukte empfohlen. Bei Kristalleiweiß ist angeblich eine Nachbehandlung praktisch durchführbar, wenn die Kisten in einem Trockenraum etwa 8 Tage vorsichtig nacherhitzt werden. Eine sichere Vernichtung der Salmonellen in Kristalleiweiß stößt aber offenbar auf Schwierigkeiten, wie auch die oben erwähnten Beobachtungen von Anz und Wohlrab beweisen.

Über weitere Versuche einer wirksamen Behandlung von Eiprodukten zur Vernichtung von Salmonellen berichtet Lerche (77). In flüssigem Weißei gelingt die Abtötung von Salmonellen ohne Schwierigkeiten bei Zusatz von 2% einer 25%igen Ammoniaklösung während 6 Std bei 37° C. Das zugesetzte Ammoniak läßt sich durch Trocknen entfernen. Dagegen stößt die sichere Entfernung der Salmonellen im Trockeneiweiß auf Schwierigkeiten. Temperaturen bis zu 76°, auch bis zu 24 Std Einwirkung genügen nicht. Unter gleichzeitiger Einwirkung von Ammoniak, sei es durch Trockenbegasung oder Einbringen von Ammoniaklösung in den Brutraum bei einer Temperatur von 55° für 5—6 Std konnte Lerche jedoch Salmonellenfreiheit erzielen, und zwar ohne Beeinträchtigung der Güte des Weißeis. Voraussetzung ist Schichtung des Trockeneiweißes bis zu 4—5 cm auf mit Gaze bespannten Horden. Kuprianoff (71) berichtet über Versuche in Australien, in denen getrocknetes Eiweiß in Originalblechdosen bei einer Erhitzung von etwa 55° während 7—11 Tagen ausreichend pasteurisiert werden konnte, während Anz und Wohlrab (7) bei Trockeneipasteurisierung und Begasung mit Ätyhlenoxyd unterschiedliche Ergebnisse hatten. Auch nach Kuprianoff hat die Pasteurisierung von Flüssigei in USA keine Gewähr für die Abtötung von Salmonellen ergeben. In Hamburg ist nach Mitteilung von Bischoff (18) eine Pasteurisierungsanlage für Gefriervollei in Betrieb, in der Hitzegrade von 67° für 10 min in Anwendung kommen.

Wieweit die zur Zeit zur Verfügung stehenden Verfahren, Salmonellen in gefrorenem Vollei und Eigelb, in Eipulver und in Trockeneigelb mit Sicherheit abzutöten, ohne Qualitätsverluste herbeizuführen, sich für die Verwendung im Großen eignen, muß noch abgewartet werden. Auf die Bedeutung der Stichprobenuntersuchung wird noch eingegangen. Die oben erwähnten trotz vorangegangener negativer Stichprobenuntersuchung positiven Ergebnisse bei der Nachuntersuchung von Kristalleiweiß beweisen, daß offenbar die nach der Verordnung vorgesehenen Untersuchungmethoden der Stichprobenuntersuchung nicht ausreichen. Sodann wird eine gleichmäßige Aufbereitung des Eiweißes, die das Vorhandensein von Klumpen, die gelegentlich als Bakteriennester anzusehen sind, beseitigt, der Probeentnahme vorangehen müssen.

Für die allgemeine Verbreitung der Salmonellen sind aber gegenüber ihrer Einschleppung durch Eiprodukte in wesentlich größerem Umfang die eingeführten *Futtermittel* verantwortlich zu machen. Nahe liegt die Forderung, mit Salmonellen infiziertes Futtermehl von der Einfuhr rücksichtslos auszuschließen. Dann würden die Länder, in denen Mehle hergestellt werden, gezwungen sein, entsprechende Maßnahmen zur Verhütung bzw. Beseitigung des Befalls mit Salmonellen zu treffen.

Von Interesse ist zunächst eine Erörterung über Herkunft und Infektionswege der Salmonellen, die in den Futtermehlen nachgewiesen werden. Über den Herstellungsgang von Fischmehlen berichtet Adam (4). Die Fische werden an der Luft getrocknet und anschließend gemahlen. Sofern es sich um an den Küsten gefangene Fische handelt, muß mit primär infizierten Fischen, somit mit salmonellenhaltigem Fischmehl, gerechnet werden. In Übersee sind die an der Sonne auf einer Betonbahn getrockneten Fischmassen aber auch der Gefahr einer nachträglichen Infektion durch keimtragende Arbeiter, die das Ausbreiten und Umschaufeln besorgen, ausgesetzt. Es ist naturgemäß schwierig, die eingeborenen Arbeiter zu einem hygienisch einwandfreien Verhalten zu erziehen. Auch mit einer Infektion der Fischmassen durch Seevögel ist zu rechnen. Als tierische Keimträger kommen ferner in Betracht Hunde, Ratten, die das Fischgut infizieren. Eine Untersuchung des Darminhaltes von 1000 Fischen des Kairoer Fischmarktes ergab, wie erwähnt, nach FLOYD und JONES (*32b*) 13mal Salmonellen, 35mal Shigellen. Bei dem üblichen Fabrikationsprozeß der Fischmehlherstellung werden die Fische zur Gewinnung des Fischöls auf 90—100° erhitzt und dann ausgepreßt. Die Preßrückstände werden getrocknet und anschließend zermahlen.

Es kommt also darauf an, soweit die Einfuhr nicht überhaupt unterbunden werden kann, Maßnahmen anzuwenden, die einmal die sichere Vernichtung der Salmonellen erreichen, möglichst ohne wesentliche Schädigung der Mehle. Dabei ist auf die Vermeidung einer nachträglichen Neuinfektion besonderer Wert zu legen. Sodann aber ist auch ein Kontrollverfahren zu entwickeln, das mit ausreichender Sicherheit eine bakteriologische Stichprobenuntersuchung als beweiskräftig gelten läßt für das Freisein von Salmonellen der gesamten Einfuhrware, deren Überprüfung verlangt wird.

Werden durch ein Stichprobenverfahren, das Anspruch auf ausreichende Zuverlässigkeit erheben kann, Salmonellen gefunden, so wäre die Aufnahmeverweigerung der Ware zunächst das sicherste Verfahren zur Fernhaltung der Salmonellen. In England wurde so vorgegangen mit dem Ergebnis, daß das salmonellenhaltige Produkt nach Deutschland weitergeleitet und hier angenommen wurde. In Schweden nahm man nach STEINIGER (*176*) im April 1955 Gefrierfleisch aus Uruguay, in dem S. paratyphi festgestellt war, unter Verschluß. Das Handelsministerium veranlaßte indes Freigabe, da das Gefrierfleisch gekauft und abgenommen werden müsse. Eine Rücksendung der nicht einwandfreien Ware an den Lieferanten würde wahrscheinlich entsprechende Maßnahmen in dem Ursprungsland des Mehles zur Folge haben. Die Importeure dieser Mehle würden indes bei dem so außerordentlich gewachsenen Bedarf an Futtermitteln in eine schwierige Lage geraten, wenn sie die Abnahme verweigerten. Es ist anzunehmen, daß sie es dann vorziehen, selbst kostspielige Behandlungsverfahren des Mehles, in dem Salmonellen nachgewiesen wurden, mit dem Ziele einer sicheren Vernichtung der Salmonellen selber zu unternehmen.

In Westdeutschland besteht die Verordnung vom 11. 6. 46 betr. Einfuhr und Durchfuhr von Knochenmehl und ähnlichen Erzeugnissen sowie Knochen, welche die Einfuhr von Knochenmehl, Fleischmehl, Tiermehl, Tierkörpermehl, Fischmehl pp, in denen Knochenteile und Fleischteile von Säugetieren enthalten sind, in das Zollinland verbietet. Die Verordnung ist erlassen zur Verhütung der Einschleppung von Milzbrand, gibt aber für ein Verbot einer Einschleppung von Salmonellen keine Handhabe. Die Bundesregierung hat nun unter dem 14. 12. 56 eine *Verordnung über die Ein- und Durchfuhr von Futtermitteln tierischer Herkunft aus dem Ausland* erlassen, nach der Futtermittel tierischer Herkunft in das Zollinland nur eingeführt werden, wenn bei der Einfuhr eine Bescheinigung der zuständigen Behörde des Ausfuhrlandes vorgelegt wird, aus der hervorgeht, daß die Ware bei oder nach der Trocknung einem Erhitzungsverfahren unterworfen wurde, durch das etwa vorhandene Salmonellen abgetötet werden. Die Futtermittel unterliegen bei der Einfuhr einer amtlichen Untersuchung durch tierärztliche Sachverständige in einem Staatlichen Veterinäruntersuchungsamt. Sie dürfen erst eingeführt werden, wenn durch bakteriologische Untersuchung festgestellt ist, daß die Ware frei von Salmonellen ist. Für die Untersuchung sind bei gleichartigen Sendungen von 1—100 Säcken aus 5 v.H. der Säcke, von 101—500 Säcken aus 3 v.H. der Säcke und darüber hinaus aus 2 v.H. der Säcke Proben zu entnehmen. Werden Salmonellen festgestellt, sind die Futtermittel nur einfuhrfähig, nachdem sie einem unter ordnungsbehördlicher Aufsicht durchgeführten Erhitzungsverfahren unterworfen wurden, durch das die Salmonellen abgetötet werden. Weiterhin ist vorgeschrieben, daß die Futtermittel tierischer Herkunft nur in unbenutzten Papiersäcken eingeführt werden dürfen. Die entstehenden Kosten fallen den Zollbeteiligten zur Last. Diese Verordnung ist 3 Monate nach ihrer Verkündung in Kraft getreten.

Nordrhein-Westfalen hat dann am 18. 9. 57 eine Viehseuchenverordnung über die Ein- und Durchfuhr von Futtermitteln tierischer Herkunft aus dem Ausland erlassen, die inhaltlich im wesentlichen mit der Bundesverordnung übereinstimmt. Diese Landesverordnung ist am 1. 1. 58 in Kraft getreten. In einem Runderlaß vom 21. 12. 57 sind dazu Ausführungsbestimmungen ergangen. Für die Entnahme der Proben, die bei der Grenzzollstelle unter Zollaufsicht von den Zollbeteiligten zu ziehen und an das Veterinäruntersuchungsamt einzusenden sind, werden geeignete Behältnisse zur Verfügung gestellt. Werden bei der Untersuchung Salmonellen festgestellt und beabsichtigt der Einführende die Futtermittel trotzdem einzuführen, hat die Ordnungsbehörde die Ware dem Empfänger erst dann zur freien Verfügung zu überlassen, wenn die Futtermittel unter ihrer Aufsicht einer Erhitzung von 80° C für die Dauer von wenigstens 15 min ausgesetzt waren. Solange von der Futtermittelindustrie keine eigenen Erhitzungseinrichtungen geschaffen sind, kann diese Erhitzung in den Tierkörperbeseitigungsanstalten durchgeführt werden. In anderen Ländern sind ähnliche Verordnungen nur zum Teil bekannt geworden.

In der DDR werden nach Mitteilung von Hofmann und Pohl (53) durch eine Verordnung des Ministeriums für Land- und Forstwirtschaft vom 5. 11. 54 alle Produkte tierischer Herkunft aus dem Ausland, die tierischen Ernährungszwecken dienen sollen, vor ihrer Verarbeitung zu Mischfutter oder vor ihrer unmittelbaren Verfütterung bakteriologisch auf Salmonellenfreiheit untersucht. Infizierte Futtermittel dürfen erst nach ihrer Entseuchung verarbeitet und verfüttert werden.

Man wird die praktische Auswirkung der Verordnung der Bundesrepublik abwarten. Die bakteriologische Untersuchung gleichartiger Mehlsorten auf 5—2 v. H. Stichproben zu beschränken, erscheint nicht ausreichend. Es wird auf die vorher erwähnten Salmonellenbefunde in Kristalleiweiß hingewiesen, das bei der Einfuhr entsprechend der Verordnung für Eiprodukte behandelt worden war, d. h. eine negative bakteriologische Stichprobenuntersuchung ergeben hatte. Bei den Futtermehlen wird von einer primären Hitzebehandlung abgesehen. Daraus ergibt sich, daß bei der Entnahme von Stichproben ein besonders scharfer Maßstab unbedingt notwendig erscheint.

Dem *Stichprobenverfahren* haften erhebliche Mängel an. Eine Voraussetzung ist die gleichmäßige Verteilung der Keime in den Futtermehlen und Eiprodukten.

Bei Annahme von 250 vermehrungsfähigen Salmonellen in 50 kg Futtermitteln bzw. Eiprodukten würden nach der Poissonschen Regel bei Entnahme von 100 g als Stichprobe 60% der negativen Befunde Fehlurteile sein. Bei Untersuchung von 1000 g beträgt die Prozentzahl der Fehlurteile nur noch 0,64%. Ist die Zahl der Salmonellen geringer, steigt naturgemäß die Häufigkeit der Fehlurteile erheblich an, so daß ein solches Stichprobenverfahren nach Schoenherr (*148*) sinnlos erscheinen muß. Im übrigen ist die Durchführung bakteriologischer Untersuchungen solcher Mengen von Futtermehl und Eiprodukten praktisch nicht möglich. Auch nach Mossel (*94*) sollten die laufenden Betriebsproben in Stufenkontrolle statistisch fundiert entnommen werden. Daß bei der Stichprobenuntersuchung von Fischmehlen Mängel bestehen, besonders auch wegen der ungleichmäßigen Verteilung das Auffinden von Salmonellen von Zufällen abhängt, heben auch Mossel, Eijgelaar und Hensel (*96*) hervor. Sie schlagen deshalb, wie auch Adam (*4*), eine zusätzliche Ergänzung des Nachweises von Salmonellen durch eine Untersuchung auf Enterobakteriazeen, auf Fäkalstreptokokken, auf proteolytische Clostridien sowie die Bestimmungen der Keimzahl vor. Stark fäkal verunreinigte Proben müssen abgelehnt werden. In fäkal nicht oder nur vereinzelt verunreinigten Proben wurden niemals Salmonellen nachgewiesen. Es ist auch bekannt, daß der Nachweis der Salmonellen bei starker Infektion mit sonstigen Enterobakteriazeen erschwert ist, wie Winkle, Rohde und Bischoff (*194*) hervorheben.

Um eine breitere Grundlage des Stichprobenmaterials zu erreichen, wird folgender Vorschlag gemacht. Es dürfte durchführbar sein, Proben aus 30—50 v.H. der Säcke gleichartiger Mehlsorten zu entnehmen, diese Proben zu mischen und aus einer Zahl verschiedener Gemische eine Anzahl von Proben zu untersuchen, z. B. aus einer Sendung von 500 Säcken Entnahmen von Proben aus 250 Säcken zu je einem Gemisch aus 50 Säcken, aus denen je 5 Mischproben auf Salmonellen, möglichst auch auf Enterobakteriazeen und Keimzahl zu untersuchen wären. Das ergäbe die Untersuchung von 25 Proben, die mengenmäßig etwa 30 g zu betragen hätten. Es dürfte sich empfehlen, ein derart verbreitertes Untersuchungsverfahren zumindest zunächst anzuwenden und bei entsprechender Erfahrung später gegebenenfalls auszubauen.

Von seiten der Importeure wird das Verlangen nach Einrichtung von Erhitzungsanlagen sich wahrscheinlich bald verstärkt bemerkbar machen. Es ist deshalb notwendig, daß die zuständigen Stellen entsprechend sicher wirkende und das Mehlgut nicht schädigende Verfahren zur Verfügung stellen. Soweit bekannt, bestehen bisher in Westdeutschland derartige Erhitzungsanlagen noch nicht. Solange das nicht der Fall ist, kann die Einschleppung von Salmonellen durch Futtermehle, wie auch Seeliger (*153*) hervorhebt, nur durch Hitzebehandlung des gesamten Einfuhrgutes verhindert werden.

Als Beispiel der Durchführbarkeit dieser Forderung wird wiederholt auf Dänemark hingewiesen, das angeblich frei ist von Enteritiserkrankungen, die auf Einfuhr ausländischer Mehlprodukte zurückzuführen sind. Alles importierte Fleisch- und Knochenmehl muß in einer dänischen Verwertungsanlage neu sterilisiert werden, bevor es für den Verkauf freigegeben wird. Die eingeführten Produkte werden direkt vom Einfuhrhafen zur Abteilung für „ungereinigtes Material" einer Verwertungsanlage zugeleitet. Die Wiedersterilisation muß in gesättigtem Dampf einer Trocknungsanlage vorgenommen werden, die mit einem Ventil versehen ist, um Dampf direkt in die Mitte der Trommel zu leiten. Nicht mehr als etwa 800 kg dürfen in eine Trommel gefüllt werden, die 400 l faßt. Die Luft wird sorgfältig aus der Trommel und dem mit gesättigtem Dampf behandeltem Material unter ständigem Rühren abgesaugt. In 15 min wird der gesättigte Dampf auf mindestens 1,37 atü (= kg/cm²) gebracht, was einer Temperatur von 125° C entspricht. Der Druck und diese Temperatur müssen während mindestens 20 min aufrechterhalten werden. Von je 25 t neu sterilisierter Produkte sind Proben an das Laboratorium zur bakteriologischen Untersuchung einzusenden. Auf diese Weise werden die Wirksamkeit der Wiedersterilisierung und der allgemeine Stand der Hygiene — besonders die Trennung zwischen den Abteilungen für „gereinigtes" und „ungereinigtes" Material — in den Fabriken, denen das infizierte Fleisch- und Knochenmehl zugeht, unter Kontrolle gehalten. Die Proben werden vom leitenden Tierarzt oder von einem tierärztlich nicht ausgebildeten Fabrikaufseher entnommen, der unter dem Tierarzt arbeitet, der die Sauberkeit in der Fabrik durch mindestens einen Besuch in der Woche überwacht. Leider findet zur Zeit eine ähnliche pflichtmäßige Neusterilisation von importiertem Fischmehl nicht statt. Wie R. Müller (99) jedoch mitteilt, werden auf Grund experimenteller Versuche in einer Trocknungsanlage, wie oben beschrieben, bei einer Temperatur von nur 100° C während 15 min Salmonellabakterien in dem infizierten Fischmehl abgetötet, ohne jedoch das Mehl völlig zu sterilisieren.

In der Abteilung für Veterinärmedizin des Bundesgesundheitsamtes erwies sich an einem Modellversuch mit künstlich infiziertem Fischmehl eine 5 min-Erhitzung von 100—105° C als voll wirksam, während die Erhitzung mit Dampf im Vakuum auf 80° C nicht zur Abtötung der Salmonellen führte. Ob durch die Erhitzung des Fischmehls eine Beeinträchtigung der biologischen Wertigkeit eingetreten ist, wird noch geprüft.

Nach Adam (4) läßt sich Fischmehl in kleineren Mengen in Autoklaven bei 80° C während 20 min entkeimen. Größere Mengen Fischmehl müssen angefeuchtet und unter Umrühren bei 70° C getrocknet oder aber in einem Paddelschneckengang der gleichen Temperatur ausgesetzt werden, um das Ziel der Vernichtung der Salmonellen zu erreichen.

Die Maschinenfabrik Friedrich Haas & Co., G.m.b.H., Remscheid-Lennep, liefert Trockentrommeln zur Trocknung von Fischmehl, Knochenmehl, Knochenschrott usw. Das Mehl wird während der Trocknung im Gleichstromverfahren auf eine Eigentemperatur von etwa 70° C gebracht. Es sei ohne weiteres möglich, jeden erwünschten Erhitzungsgrad zu erreichen, wobei dann die Trocknung im Gegenstromverfahren erfolgen muß. Während der Trocknung befindet sich das Material in ständiger Rieselbewegung. Dabei wird das Gut sehr intensiv von den durchziehenden Heizgasen umspült. Die ausgenutzten Heizgase werden durch einen Ventilator abgesaugt und nach der Reinigung in einem Staubabschneider ins Freie gefördert. Wieweit dieses Verfahren zur sicheren Abtötung der Salmonellen führt, müßte noch untersucht werden.

Derartige Einrichtungen zur Salmonellenvernichtung in eingeführten Mehlprodukten wären zweckmäßig in den Freihäfen bzw. in Nähe der Einfuhrstellen im Lande einzurichten, vielleicht in Anlehnung an Anlagen, die über eine reine und unreine Seite verfügen. Strenge Überwachung durch amtliche Stellen muß sichergestellt sein. Die Einschleppung von Lebensmittelvergiftern durch importierte Lebensmittel und Tierfuttermittel ist jedenfalls ein europäisches Problem, wie auch HARMSEN (*49a*) und SEELIGER mit Recht hervorheben. Es ist deshalb zu begrüßen, daß gelegentlich eines Symposions im Hygienischen Institut der Hansestadt Hamburg am 15. 10. 57 unter WINKLE eine Arbeitsgruppe gebildet wurde, die sich die Erarbeitung von praktischen Vorschlägen für die Gesetzgebung zur Verhütung der Einschleppung von Lebensmittelvergiftungen durch importierte Tierfuttermittel zum Ziele gesetzt hat.

Zusammenfassung

Die Verbreitung von Salmonellen durch Lebensmittel, die als Ausgang einer Infektion anzusehen sind, ist besonders durch die Beobachtungen von enteritischen Erkrankungen nach Genuß von *Enteneiern* bekannt geworden.

Der in den letzten Jahren *zunehmende Befund* von *Salmonellen* verschiedener Typen ist nicht allein auf die verbesserte Diagnostik zurückzuführen. Es handelt sich vielmehr um eine *tatsächliche Zunahme* dieser Krankheitserreger.

Bemerkenswert war in den letzten Jahren immer wieder der Nachweis von Salmonellaarten, die hierzulande bisher nicht beschrieben, dagegen in anderen, besonders überseeischen Ländern festgestellt waren. Es lag deshalb nahe, eine *Einschleppung* vor allem durch *importierte Lebensmittel* anzunehmen.

Die Untersuchungen insbesondere von *chinesischen Eiprodukten* ergaben häufig Befunde unbekannter Salmonellen, die dann auch bei Erkrankten nach Genuß solcher Eiprodukte nachgewiesen wurden. Die hier seit Jahren bekannten Erfahrungen über Verbreitung von Salmonellen durch Enteneier im Gegensatz zu Hühnereiern läßt vermuten, daß die chinesischen Eiprodukte im wesentlichen aus Enteneiern hergestellt werden.

Häufige Befunde von Salmonellen in *Meerwasser* und *Vorflutern* der deutschen Küste erregten dann allgemeine Aufmerksamkeit. Es wurde festgestellt, daß offenbar auch eine *Vermehrung* der *Salmonellen* einschließlich von Typhus- und Paratyphusbakterien im Wasser bei Anwesenheit von Eiweißstoffen stattfindet.

Die große Zahl von Salmonellen, die beim Baden verschluckt werden, ohne Erkrankungen auszulösen, erscheint bedeutsam für die *Beurteilung* der *Grundlage* für das *Zustandekommen einer Infektion*. Es ist dazu offenbar eine Anreicherung von Salmonellen auf einem geeigneten Substrat unter geeigneten Umständen notwendig, die eine millionenfache Vermehrung der Erreger ermöglicht.

Wichtig ist auch die Feststellung, daß sich die Salmonellen in den oberflächlichen Schlammschichten halten und vor allem *durch Filterung nicht zurückgehalten* und in den Vorflutern nachgewiesen werden.

So hat sich der Salmonellennachweis auch in quantitativer Hinsicht für die bakteriologische Beurteilung eines Gewässers als wichtig erwiesen. Es ist heute erforderlich, neben der Bestimmung der Keimzahl und des Colititers auch einen *Salmonellentiter* zugrunde zu legen, der dann durch die Feststellung des Salmonellentyps auch noch von besonderer epidemiologischer Bedeutung ist.

Durch die salmonellenverseuchten Gewässer wird bei Hochwasser und Sturmfluten durch Überschwemmung von Weidegelände eine *Infektion* der *Wiesen* und von *Heu* bewirkt, die sich auf das *Weidevieh* übertragen kann.

Auch die am Wasser lebenden Tiere, besonders die *Möwen* sind anfällig für Infektionen und stumme Feiung. Sie können Salmonellen verschleppen. An Eiern und im Kot dieser Tiere wurden Salmonellen häufig nachgewiesen.

Die immer zahlreicher vereinzelt und in Form von Epidemien auftretenden Erkrankungen beim Menschen, die durch Salmonellen, und zwar hier unbekannter Typen, ausgelöst wurden, bestätigten die Annahme einer *laufenden Einschleppung dieser Salmonellen aus dem Ausland*.

Als nun Salmonellenbefunde bei *Schlachtvieh* ebenfalls in zunehmendem Umfang festgestellt werden konnten, wurde die Aufmerksamkeit auf die *Futtermittel* gerichtet, die als Überträger von Salmonellen in Betracht kamen. Es stellte sich heraus, daß Futtermehle, insbesondere *Fischmehl* in stark steigender Menge aus dem Ausland, insbesondere aus Übersee eingeführt werden und daß die bakteriologischen Untersuchungen dieser Produkte *überraschende Salmonellenbefunde* ergaben. Es fanden sich die verschiedensten Typen oft in der gleichen Mehlprobe.

Wenn berücksichtigt wird, daß die *Beifütterung* dieser Mehle für Vieh und auch Geflügel eine wertvolle Nahrungsbeigabe und außerdem bei der Geflügelfütterung eine Vereinfachung bedeutet, so ist die weite und zunehmende Verbreitung dieser Mehle verständlich, aber auch die *unabsehbare Verstreuung der Salmonellen*.

Die *Vorfluter* werden durch die menschlichen und tierischen Abgänge verseucht. Man denke an die Verbreitung durch *Ratten*, durch *Fliegen* in verseuchten Ställen, die Übertragung auf Lebensmittel, insbesondere auf Milch und Milchprodukte. Eine besondere Gefahr besteht auch für die Umgebung von Futtermittellagern.

Es muß deshalb etwas geschehen, wenn davon ausgegangen wird, daß die verschiedenen Salmonellentypen, wenn auch offenbar in der Mehrzahl der Übertragungen im Körper passager und stumm, *grundsätzlich* als *menschen-* und *tierpathogen* zu beurteilen sind.

Die *Maßnahmen* gegen die Ausbreitung der Salmonellen haben sich einmal gegen die Krankheitserreger im Lande zu richten unter Anwendung der bestehenden Vorschriften über die Bekämpfung übertragbarer Krankheiten vom 1. 12. 38.

Wichtig ist *frühzeitige Erkennung* der oft leicht verlaufenden Erkrankungen. Verstärkte *bakteriologische* und *serologische Untersuchungen* sollten die Ärzte häufiger als bisher veranlassen, insbesondere dann, wenn aus der Vorgeschichte ein ursächlicher Zusammenhang mit verdächtigem Speisegenuß vermutet werden kann. Dabei ist an die häufige Salmonelleninfektion beim Schlachtvieh zu denken.

Die Verwendung von Enteneiern in gewerblichen Lebensmittelbetrieben und Gemeinschaftsküchen jeglicher Art ist zu verbieten.

Bekämpfung der offenen und stummen Salmonellose beim *Schlachtvieh* durch häufige bakteriologische Untersuchungen und enge Zusammenarbeit der Veterinärmedizin mit der Humanmedizin ist wichtig.

Gesundheitsamtliche Überwachung der Betriebe, in denen rohes Eigelb zur industriellen Verarbeitung, insbesondere in der Handschuhfabrikation Verwendung findet, Beaufsichtigung von *Geflügel-*, vor allem von *Enten-* und *Taubenbeständen* in Lebensmittelbetrieben, *Fliegenbekämpfung*, besonders in Lagerstellen von Futtermitteln in schließlich jedem landwirtschaftlichen Betrieb, die *Ver-*

hütung der *Vorfluterverseuchung* durch Abfang eiweißhaltiger Stoffe vor Einlaß sind weitere Forderungen.

Von größter Bedeutung sind aber die *Maßnahmen*, die eine zusätzliche Salmonellenverseuchung durch die *Einfuhr ausländischer Lebens- und Nahrungsmittel zu verhindern* haben.

Alle auf Salmonellen verdächtigen Einfuhren sind, soweit nicht Annahme verweigert werden kann, grundsätzlich einem amtlich überwachten *Hitzeverfahren* zu unterwerfen, das eine Gewähr für vollständige Abtötung von Salmonellen und sonstiger Darmkeime ohne wesentliche biologische Schädigung bietet und einen nachträglich erneuten Befall verhindert.

Inwieweit die *Verordnung* für die Behandlung der *ausländischen Eiprodukte*, die diese Forderung verlangt, ausreichend wirksam ist, bedarf einer besonderen Nachprüfung, nachdem in Eiweißproben, die dem Verfahren unterzogen waren, doch noch Salmonellen nachgewiesen wurden.

Die Verordnung über die *Behandlung ausländischer Futtermehle* sieht im Gegensatz zu der Verordnung über Eiprodukte eine primäre Hitzebehandlung des gesamten Einfuhrgutes nicht vor. Die Hitzebehandlung wird von einem positiven Befund von Stichproben abhängig gemacht, ein Verfahren, das in der vorgeschriebenen Form eine ausreichende Erfassung infizierter Produkte nicht erwarten läßt.

Dem *Stichprobenverfahren* ist besondere Aufmerksamkeit zuzuwenden, insbesondere dann, wenn nicht grundsätzlich eine wirksame Hitzebehandlung der Gesamtware möglich ist.

Vor allem ist bisher ein Hitzebehandlungsverfahren, das in der Lage ist, *große Mengen von Futtermehl* entsprechend zu behandeln, nicht bekannt. Den Importeuren sind beschleunigte Verfahren anzugeben.

Entsprechende *amtliche Überwachung* der *Hitzebehandlung* unter Beachtung von reiner und unreiner Seite und *Verhütung nachträglicher Übertragung* von Salmonellen ist zu fordern.

Ob der zahlenmäßige Rückgang der Erkrankungen an Lebensmittelvergiftungen 1957 auf den Einfluß bereits laufender Bekämpfungsmaßnahmen zurückzuführen ist, bleibt abzuwarten.

Literatur

1. ABEL, G.: Über die Wirksamkeit seuchenhygienischer Maßnahmen bei Massenerkrankungen an Salmonellose und deren bakteriologische Grundlage. Arch. Hyg. (Berl.) **137**, 423 (1953).

2. —Zur Epidemiologie der sog. seltenen Salmonellosen. Zbl. Bakt., I. Abt. Orig. **159**, 433 (1953).

3. ADAM, W.: Pasteurisierungsversuche mit salmonelleninfiziertem Eigelbpulver. Zbl. Bakt., I. Abt. Orig. **167**, 224 (1956).

4. — Über die Herstellung und Pasteurisierung von Fischmehl im Hinblick auf dessen Gehalt an Salmonellabakterien. Berl. Münch. tierärztl. Wschr. **1957**, 49.

5. ALBERT, O. H.: Untersuchungen über das Vorkommen von Salmonellabakterien in Eiprodukten ausländischer Herkunft. Berl. Münch. tierärztl. Wschr. **1957**, 165.

6. ALTERAUGE, W.: Zur Überwachung des Verkaufs von Enteneiern im Rahmen der tierärztlichen Lebensmittelüberwachung und dem Auftreten von Infektionen mit Salmonella typhi murium (Breslaubakterien) beim Menschen. Arch. Lebensmitt.-Hyg. **5**, 271 (1954).

7. ANZ, W., u. R. WOHLRAB: Zur Verbreitung von Salmonellen durch importiertes Kristalleiweiß. Desinf. u. Gesundh.-Wesen **1957**, 156.

8. Bachmann, H., u. H. Kast: Salmonellose bei einer Tigerschlange, ein Beitrag zu den Infektionskrankheiten der Schlange. Berl. Münch. tierärztl. Wschr. **1956**, 304.

9. Beck, u. Meyer: Enzootische Erkrankungen der Tauben durch Bakterien der Paratyphus-Enteritis-Gruppe. Z. Infekt.-Kr. Haustiere **30**, 15 (1927). Zit. nach Helm (*50*).

10. Becker, V. W.: Über Salmonellen beim Hausgeflügel. Berl. Münch. tierärztl. Wschr. **1957**, 168.

10a. Beller, K.: Die Bedeutung des Geflügels für die Entstehung von Lebensmittelvergiftungen. Z. Fleisch- u. Milchhyg. **43**, 365 (1933). Zit. nach vor dem Esche u. Handloser (*30a*).

10b. —, u. Reinhardt: Berl. tierärztl. Wschr. **1934**, 225. Zit. nach Miesner u. Köser (*92a*).

11. Bergmann, G., u. G. Seidel: Über die Wirkung von Kochsalz auf Salmonellen. Arch. Lebensmitt.-Hyg. **8**, 30 (1957).

12. Berkmen, L.: Vorkommen von Salmonellen in Fleisch- und Fleischwaren in Mittelanatolien. Arch. Lebensmitt.-Hyg. **8**, 278 (1957).

13. Bermbach: Erkrankungen von Menschen nach Genuß von Käse aus Milch paratyphuskranker Kühe. Berl. tierärztl. Wschr. **1928, 359.**

14. Bischoff, H.: Ein Salmonella chester-Fund in Proben eines notgeschlachteten Schweines. Lebensmitt.-Tierarzt **1953**, 257.

15. — Über das Vorkommen von Salmonella bareilly im Knochenschrot. Berl. Münch. tierärztl. Wschr. **1955**, 212.

16. — Ein Beitrag zur Klärung nach der Herkunft seltener Salmonellentypen. Berl. Münch. tierärztl. Wschr. **1950**, 306.

17. — Salmonella braenderup als Ursache einer Lebensmittelvergiftung. Berl. Münch. tierärztl. Wschr. **1956**, 293.

18. — Briefliche Mitteilung.

19. Bischoff, J., u. R. Rohde: Salmonellen in Fisch- und Fleischmehl ausländischer Herkunft. Berl. Münch. tierärztl. Wschr. **1956**, 50.

20. Boese, W.: Bakteriologische Speiseeisuntersuchungen und ihre Ergebnisse. Z. ges. Hyg. **137**, 269 (1953).

21. Bonitz, K.: Salmonella-bareilly-Epidemie durch Lebensmittelvergiftung im norddeutschen Raum. Dtsch. med. Wschr. **1953**, 1412.

22. Bourmer, u. Doetsch: Über eine durch den Bazillus enteritidis Gärtner hervorgerufene seuchenhafte Erkrankung in dem Rinderbestande des Gutes Kartheuserhof bei Koblenz und eine durch Käse verursachte Übertragung auf den Menschen. Fleisch- u. Milchhyg. **38**, 387 (1928).

23. Bruns, H., u. W. Fromme: Über Nahrungsmittelerkrankungen durch Enteneier. Münch. med. Wschr. **1934**, 1350.

24. Clarenburg, A.: Salmonella bareilly beim Geflügel. Ihre Beziehungen zur Volksgesundheit. Vortrag auf XV. Internat. Vet. Kongreß in Stockholm. Ref. Lebensmitt.-Tierarzt **1953**, 230.

25. Dack, G. M.: Food Poisoning. Chicago, 135. Zit. nach Rohde u. Adam (*132*).

26. Daigeler, A., u. L. Kotter: Über das Vorkommen von Salmonellen in Bayern während der Jahre 1946—1954. Berl. Münch. tierärztl. Wschr. **1956**, 281.

27. Dräger, H.: Die Bedeutung der Salmonellenbakterien für die Erkrankung des Menschen. Dtsch. Gesundh.-Wesen **40**, 1347 (1956).

28. Edwards, Ph. R.: J. infect. Dis. **45**, 191 (1929). Zit. nach Schaaf (*137a*).

29. — P. R., D. W. Brunner and A. B. Moran: J. infect. Dis. **83**, 220 (1948). Zit. nach Steiniger (*169*).

30. Ekstams, M.: Über eine umfangreiche Lebensmittelvergiftungsepidemie (Salmonella typhi-murium-Intoxikation) in Schweden im Sommer 1953. Medlemsbl. Sveriges Veterinär.-förb. **7**, 73 (1955).

30a. vor dem Esche, P., u. M. Handloser: Zur Frage der Verbreitung von Salmonellen durch Hühnereier. (Im Druck.)

31. Felsenfeld, Young and Yoshimura: J. Amer. vet. med. Ass. **116**, 17 (1950). Zit. nach Vogt (*189*).

32a. Fiedler, C.: Über das Vorkommen von Salmonella bareilly bei Schlachtrindern. Arch. Lebensmitt.-Hyg. **7**, 245 (1956).

32b. FLOYD, TH. M., u. G. B. JONES: Isolierung von Shigellen und Salmonellen aus Nilfischen. Amer. J. trop. Med. Hyg. **1956**, 475. Ref. Zbl. Bakt., I. Abt. Ref. **160**, 56 (1956).

33a. FREITAG, R.: Eikonserven verschiedener Geschmacksrichtung, exotische Delikatesse: Pidan, bebrütete Eier, Balut, Eier in Lehm. Desinf. u. Gesundh.-Wesen **1957**, 160.

33b. FRITZSCHE, K.: Über eine Lebensmittelvergiftung durch Bacterium pullorum in Koblenz. Lebensmitt.-Tierarzt 2, 103 (1951).

34. FROEHNER, H., u. F. GRÜTTNER: Salmonellenbekämpfung in veterinärmedizinischer Sicht. Öff. Gesundh.-Dienst **20**, 12 (1958).

35a. FROMME, W.: Zur Ursache von Nahrungsmittelvergiftungen durch Enteneier. Dtsch. med. Wschr. **1933**, 653.

35b. — Nahrungsmittelvergiftungen nach Genuß von Rührei aus Enteneiern. Dtsch. med. Wschr. **1934**, 1969.

36. — Weitere Beobachtungen über Nahrungsmittelvergiftungen durch Enteneier. Arch. Hyg. (Berl.) **113**, 29 (1934).

37. —, u. A. GAASE: Salmonellen in Hühnervollkraftkornfutter. Desinf. u. Gesundh.-Wesen **1957**, H. 4/5.

38. FÜRTH, E., u. K. KLEIN: Neue Feststellungen über die Entstehungsursache von Enteritis-Gruppenerkrankungen (Enteritis und Entenei). Veröff. Gebiet Med.-Verw. **39**, 363 (1933).

39. GAASE, A.: Zusammenarbeit von Ärzteschaft und Bakt.-Serologischem Institut. Medizinische **1957**, Nr 3/4.

40. — Häufung von seltenen Salmonellentypen. Zbl. Bakt., I. Abt. Orig. **165**, 80 (1956).

41. GALTON, M. M., W. D. LOWERY and I. HARDY: J. infect. Dis. **95**, 232 (1954). Zit. nach SEELIGER (*153*).

42. — SMITH, W. V., u. B. H. ELRATH: J. infect. Dis. **95**, 236 (1954). Zit. nach ROHDE u. BISCHOFF. (*133*).

43. GÄRTNER, H.: Über zwei Krankheitsfälle mit Nachweis von Salmonella bovis morbificans im Sputum. Zbl. Bakt., I. Abt. Orig. **166**, 326 (1956).

44. GÄSSLEIN, F.: Salmonella-Lebensmittelvergiftungen durch ungarische Importschweine. Öff. Gesundh.-Dienst **1957**, 327.

45. GERNEZ-RIENSE, u. Mitarb.: Lebensmittelvergiftung durch Salmonella pullorum. Vet. Bull. (Weybridge) 2264. Ref. Vet. Med. **1952**, 261.

46. GLATZEL,: Zit. nach STEINIGER (*173*).

46a. GORDON: Zit. nach GARSIDE, J. comp. Path. **53**, 80 (1943); **54**, 61 (1944). Zit. nach VOR DEM ESCHE u. HANDLOSER (*30a*).

47. GORHAM, J. R , u. F. M. GARNER: Zit. nach SEELE (*151*).

47a. HAFFKE: Enteritisbakterien bei gesunden und künstlich infizierten Enten. Inaug.-Diss. Hannover 1934. Zit. nach MÖLLER (*93*).

48. HANDLOSER, M.: Epidemiologische Beobachtungen bei einer Masseninfektion durch S. blockley. Arch. Hyg. (Berl.) **140**, 569 (1956).

49a. HARMSEN, H.: Symposion in Hamburg am 15. Okt. 1957. Arch. Lebensmitt.-Hyg. 8, 256 (1957).

49b. — H.: Ist die Abwasserverregnung vom hygienischen Standpunkt aus vertretbar? Umschau 58, 49 (1958).

50. HELM, K.: Salmonella-Infektionen bei Tauben. Arch. Lebensmitt.-Hyg. 8, 170 (1957).

50a. HELMSHORN: Bakteriologisch-serologische und pathologisch-anatomische Untersuchungen im Zusammenhang mit natürlicher und künstlicher Enteritisinfektion beiEnten. Inaug.-Diss. Berlin 1955. Zit. nach MÖLLER (*93*).

51. HEPP, L.: Eine Salmonellose beim Menschen durch den Genuß von Schweinefleischkonserven jugoslawischer Herkunft. Berl. Münch. tierärztl. Wschr. **1955**, 221.

52. HERTER, R.: Salmonella senftenberg bei einem Schlachtrinde. Lebensmitt.-Tierarzt **1953**, 283.

53. HOFMANN, S., u. G. POHL: Zwei neue Salmonellentypen (S. falkensee u. S. seegefeld). Zbl. Bakt., I. Abt. Orig. **167**, 413 (1957).

54. — P., u. R. WOLLE-JOHN: Gehäuftes Vorkommen von Salmonella Saint Paul. Zbl. Bakt., I. Abt. Orig. **162**, 357 (1955).

55. HOHN, W., u. J. HERRMANN: Zit. nach STEINIGER (*169*).

56. Holz, K.: Salmonella morbificans bovis als Ursache von Lebensmittelvergiftungen. Berl. Münch. tierärztl. Wschr. **1950**, 214.

57. — Über die Bedeutung der durch bakteriologische Fleischuntersuchungen ermittelten Salmonellen im Gebiet des Niederrheins. Arch. Lebensmitt.-Hyg. **7**, 121 (1956).

58. — Persönliche Mitteilung.

59. Hunsteger, F.: Salmonella infantis bei einem notgeschlachteten Schwein. Arch. Lebensmitt.-Hyg. **6**, 72 (1955).

60. Jaeschke, H.: Zur Bedeutung der Untersuchung der Gallenblase für die Feststellung von Salmonellen bei Not- und Krankschlachtungen. Inaug.-Diss. tierärztl. Hochschule Hannover, siehe Lebensmitt.-Tierarzt **1953**, 171.

61. Joest, W.: Vorkommen von Salmonellen und Bewertung der Diagnostik. Öff. Gesundh.-Dienst **1957**, 163.

62. Jordan, E. O.: Zit. nach Steiniger (*169*).

63. Kauker, E.: Eine Lebensmittelvergiftung durch Salmonella infantis. Arch. Lebensmitt.-Hyg. **5**, 217 (1954).

64. Kathe, J., u. M. Lerche: Bakterielle Lebensmittelschädigung durch Enteneier. Zbl. Bakt., I. Abt. Orig. **136**, 320 (1936).

65. Keil, R., u. G. Haupt: Untersuchungsergebnisse und epidemiologische Erwägungen bei einer Massenerkrankung, hervorgerufen durch S. typhi murium in Schweinefleisch. Z. Hyg. Infekt.-Kr. **137**, 115 (1953).

66. Kelch, F.: Über die Epidemiologie des postmortalen Befalls von Fleisch- und Fleischwaren mit Salmonellenbakterien. Berl. Münch. tierärztl. Wschr. **1954**, 185.

67. — Über das Vorkommen von Salmonellen in chinesischem Gefriervollei. Berl. Münch. tierärztl. Wschr. **1956**, 307.

68. Klein, H.: Salmonella manhattan in einem Schweinebestand. Arch. Lebensmitt.-Hyg. **8**, 169 (1957).

68a. Knothe, H., u. F. Budach: Bakteriologische Untersuchung der Enteneier. Arch. Hyg. (Berl.) **135**, 574 (1952).

69. Knorr, M.: Über die Herkunft des Paratypus C in Deutschland. Arch. Hyg. (Berl.) **105**, 237 (1936).

69a. Kollath, W.: Paratyphusverhütung und die Khaki-Campbell-Ente. Arch. Hyg. (Berl.) **125**, 127 (1941).

70a. Kreuscher, A., u. W. Sylvester: Salmonella-london-Epidemie im Anschluß an den Genuß von Kochschinken. Zbl. Bakt., I. Abt. Orig. **156**, 485 (1951).

70b. — Über Salmonellenbefunde in zwei Westberliner Bezirken. Zbl. Bakt., I. Abt. Orig. **159**, 176 (1952/53).

70c. Kröger, E.: S. bareilly-Nachweis in Rieselfeld-, Abwasser-, Schlamm- und Gemüseproben nach einer S. bareilly-Epidemie. Z. Hyg. Infekt.-Kr. **139**, 202 (1954).

71. Kuprianoff, J.: Dtsch. Lebensmitt.-Rdsch. **1957**, 223. Zit. nach Anz u. Wohlrab (*7*).

72. Lange, R.: Über das Eindringen von Bakterien in das Hühnerei durch die Eischale. Arch. Hyg. (Berl.) **62**, 201 (1907).

73. Lempfried, H.: Zit. nach Steiniger (*169*).

74. Lennardton: Nord. Vet.-Med. **2**, 193 (1950). Zit. nach Vogt (*189*).

75. Lerche, M.: Die beim Tier vorkommenden Erkrankungen der Bakterien der Paratyphus-Enteritis-Gruppe und ihre Epidemiologie. Zbl. Bakt., I. Abt. Orig. **140**, 1940 (1937).

76. — Paratyphus-Bakterien im Fleisch. Z. Hyg. Infekt.-Kr. **122**, 72 (1940).

77. — Vorkommen und Unschädlichmachung von Salmonellabakterien im Trockeneiweiß. Arch. Lebensmitt.-Hyg. **8**, 267 (1957).

78. — Berl. Münch. tierärztl. Wschr. **1957**, 436.

79. Linke, H.: Über das Vorkommen von Salmonellen in den Mesenteriallymphknoten gesund geschlachteter Pferde. Arch. Lebensmitt.-Hyg. **8**, 244 (1957).

80. Lodenkämper, H.: Zit. nach Winkle, Rohde und Bischoff (*194*).

81. Löliger, H.-Chr.: Über Salmonellose beim Nerz. Berl. Münch. tierärztl. Wschr. **1956**, 31.

82. Lütje, F.: Paratyphuserkrankungen, verursacht durch das B. ent. Gärtneri bei erwachsenen Rindern. Dtsch. tierärztl. Wschr. **1926**, 453.

83. Lütje, F.: Neue Gesichtspunkte auf dem Gebiete der Salmonellose des Kalbes und des Rindes, sowie in bezug auf die Bakterienausscheidung und das vegetative Dasein der Salmonellen in der Umwelt. Berl. Münch. tierärztl. Wschr. 1955, 39.

84. — Zusammenstellung des jüngeren Schrifttums über die Freilandbiologie der Salmonellen, die Salmonellose der Möwenvögel und ihre Beziehungen zum Abwasser und zum Menschen. Berl. Münch. tierärztl. Wschr. 1955, 249.

85. Lütje, F., u. K. Rasch: Beobachtungen über die Salmonellose (Enteritis) des Rindes. Lebensmitt.-Tierarzt 1952, 13, 26, 38, 49, 55.

86. Mannweiler, E., u. R. Humbert: Durch Enteritis-Erreger der Salmonella-Gruppe hervorgerufene Bakteriämien. Dtsch. med. Wschr. 1957, 2215.

87. Merdl, F., u. H. Wolfs: Über septikämische Erkrankungen mit tödlichen Ausgang durch Salmonella new-port. Zbl. Bakt., I. Abt. Orig. 160, 498 (1953).

88. Merk, W.: Salmonella montevideo bei einem Schlachtrind. Arch. Lebensmitt.-Hyg. 7, 194 (1956).

89. Messerschmidt, Th., u. R. Wedemeyer: Der Nachweis von Typhus- und Paratyphusbakterien im städtischen Abwasser und im Vorfluter. Zbl. Bakt., I. Abt. Orig. 156, 189 (1950).

90. Messow, C., u. L. Stoll: Ein Beitrag zur Salmonellose des Hundes. Arch. Lebensmitt.-Hyg. 5, 103 (1954).

91. Meyer: Zit. nach Steiniger (*168*).

92. — M. I., u. Endress: Untersuchungszahl und Ergebnisse der Bakteriologischen Fleischuntersuchungsstelle am Schlachthof München, Zunahme der Salmonellen. Arch. Lebensmitt.-Hyg. 8, 98 (1957).

92a. Miesner, H., u. A. Köser: Lebensmittelvergiftungen und Geflügel. Dtsch. med. Wschr. 1935, 1811.

93. Möller, A.: Breslau-Gruppenerkrankung durch infizierte Hühnereier. Zbl. Bakt., I. Abt. Orig. 164, 535 (1955).

94. Mossel, D. A. A.: Vorschläge zur Verhütung der Verbreitung von Salmonellen durch Lebensmittel bzw. Tierfuttermittel. Vortrag in Hamburg 15. 10. 57. Ref. Arch. Lebensmitt.-Hyg. 8, 257 (1957).

95. — D. A. A., G. Eijgelaar u. H. Hensel: Betrachtungen über die Durchführung der hygienisch-bakteriologischen Beurteilung von Fischmehl. Arch. Lebensmitt.-Hyg. 9, 3 (1958).

96. Müller, J.: Nord. Vet.-Med. 1952, 290. Zit. nach Seeliger (*153*).

97. — Nord. Vet.-Med. 1954, 525. Zit. nach Seeliger (*153*).

98. — Kopenhagen, Statens veterinaere Serumlaboratorium. Briefliche Mitteilung vom 11. Dez. 1957.

99. — R.: Hühnerenteritis- und atypische Gärtnerbakterien durch Gastroenteritiden durch Eierspeisen. Münch. med. Wschr. 1933, 1771.

100. — F.: Beobachtung bei der Salmonellose der Schweine. Berl. Münch. tierärztl. Wschr. 1955, 81.

101. — O., u. J. Rodenkirchen: Über Lebensmittelinfektionen durch Bacillus ent. Gärtner in Mülheim-Ruhr 1930—1932. Veröff. Gebiet Med. Verw. 39, 377 (1933).

102. Münchow, S.: „Eis am Stiel" als Ursache einer Salmonellen-Epidemie. Dtsch. Gesundh.-Wesen 1956, 1341.

103. Peters, H., u. A. Ch. Ruys: Typhus voor Geneeskunde. Ned. T. Geneesk. 1938, 4171. Zit. nach Steiniger (*177*).

104. Piening, C.: Beitrag zur Klärung des sog. Kälberparatyphus. Berl. Münch. tierärztl. Wschr. 1954, 277.

105. Piorkowsky: Über die Einwanderung des Typhusbazillus in das Hühnerei. Arch. Hyg. (Berl.) 25, 145 (1895).

106. Plaschke, I., u. W. Plaschke: Mitteilung über einen Fund von Salmonella blockley im Raum von Berlin. Arch. Lebensmitt.-Hyg. 8, 2 (1957).

107. Pohl, G.: Über das Vorkommen von Salmonellen in geklärten Abwässern, ihren Vorflutern, Rieselfeldabflüssen und Dränagen und Klär- und Faulschlamm. Berl. Münch. tierärztl. Wschr. 1955, 163.

108. Popp, L.: Vorfluter — ein reichhaltiges Salmonellenreservoir. Zbl. Bakt., I. Abt. Orig. **166,** 90 (1956).

109. — Über das Problem der Abwasserunterbringung. Städtehygiene **1956,** 217.

110. — Der Salmonellenkataster eines Flußgebietes. Tagung der Dtsch. Ges. Hyg. u. Mikrobiol. in Berlin 1957.

111. Poppe, H.-H.: Zur Frage der Übertragung von Krankheitserregern durch Hühnereier. Arb. kel. Gesdh.amt **34,** 186 (1910). Zit. nach Vogt (*189*).

112. Preuss, H.: Zit. nach Steiniger (*169*).

113. Primavesi, K., A.: Über eine durch Salmonella Heidelberg hervorgerufene Krankenhausepidemie. Med. Klin. **1956,** 1103.

114. Rasch, K.: Grenzen der Fleischvergiftungsprophylaxe und zur latenten Infektion mit Salmonellen beim Rinde. Lebensmitt.-Tierarzt **1953,** 37.

115. — Salmonella senftenberg bei einem Schlachtrinde. Lebensmitt.-Tierarzt **1953,** 248.

116. — Salmonellen im Knochenschrot. Berl. Münch. tierärztl. Wschr. **1955,** 213.

117. — Das Salmonellenproblem in der Lebensmittelhygiene. Desinf. u. Gesundh.-Wesen **1955,** 97.

118. — Über das Verhalten der Salmonellen in der Außenwelt, ein auch für die Verhütung von Fleischvergiftungen wichtiger Faktor. Arch. Lebensmitt.-Hyg. **6,** 1 (1955).

119. — Zur Bekämpfung der Rindersalmonellose im Blickpunkt der Lebensmittelhygiene. Berl. Münch. tierärztl. Wschr. **1957,** 161.

120. —, u. I. Richter: Epidemiologisches um einen Dauerausscheider von Salmonella heidelberg. Berl. Münch. tierärztl. Wschr. **1956,** 211.

121. Rauch, F.: Zit. nach Rasch (*119*).

122. Reimold, G.: Über die Lebensdauer von Salmonellen. Arch. Hyg. (Berl.) **136,** 479 (1952).

123. Renner, H.: Lebensmittelvergiftung durch Salmonella anatum in Ludwigshafen am Rhein. Lebensmitt.-Tierarzt **1953,** 1.

124a. Reploh, H.: Persönliche Mitteilung.

124b. —, u. A. Rainer: Massenerkrankung an Enteritis durch Wurst. Gesundheitsfürsorge **1955,** 120.

125a. Richter, J.: Städtehygiene **1953,** H. 9.

125b. — Salmonellen in hygienischer und epidemiologischer Sicht unter Berücksichtigung der geplanten gesetzlichen Maßnahmen. Desinf. u. Gesundh.-Wesen **1957,** 50.

126. —, u. W. Pöhlig: Zur Epidemiologie seltener Salmonellen. Öff. Gesundh.-Dienst **1957,** 12.

127. Rodenwaldt, E.: Hygiene und ihre Grundzüge. Stuttgart: Ferdinand Enke 1949.

128. —, u. R. E. Bader: Lehrbuch der Hygiene. 1951.

129. Rohde, R.: Über eine Lebensmittelvergiftung, verursacht durch Salmonella kirkee. Zbl. Bakt., I. Abt. Orig. **162,** 359 (1955).

130. — Über das Auftreten verschiedener Salmonellosen in Hamburg während des Jahres 1954 und die sich daraus ergebenden epidemiologischen Konsequenzen. Arch. Hyg. (Berl.) **1955,** 148.

131. — Über das Auftreten Salmonella montevideo bedingter Hausendemien in Hamburger Kinderkliniken. Zbl. Bakt., I. Abt. Orig. **166,** 67 (1956).

132. —, u. W. Adam: Über die Gefahren bakteriologisch nicht kontrollierter Ei- bzw. Eikonservenimporte. Zbl. Bakt., I. Abt. Orig. **166,** 334 (1956).

133. —, u. J. Bischoff: Die epidemiologische Bedeutung salmonellainfizierter Tierfuttermittel (insbesondere Knochenschrot und Fischmehl) als Quelle verschiedener Lebensmittelvergiftungen. Zbl. Bakt., I. Abt. Ref. **159,** H. 7—9 (1956).

134. —, u. K. Seitz: Über Lebensmittelvergiftungen verursacht durch Salmonella chester und Salmonella paratyphi B unter besonderer Berücksichtigung des abweichenden klinischen Bildes. Dtsch. med. Wschr. **1955,** 116.

135. Romanoff, A. L., and A. I.: The avian egy. New York: John Willy & Sons; Chapman & Hall Limited. London: Zit. nach Schwarz (*150*).

136. Saphra, I., u. I. W. Winter: Clinical manifestations of Salmonellosis in man. New Engl. J. Med. **256,** 1128 (1957). Siehe Arch. Lebensmitt.-Hyg. 8, 256 (1957).

137. Savage, W.: Brit. med. J. **1956,** 317. Zit. nach Vogt (*189*).

137a. SCHAAF, J.: Zur infektiösen Enteritis der Enten. Zbl. Bakt., I. Abt. Orig. **128**, 519 (1933).

138. SCHAAL, E.: Salmonella pullorum und gallinarum bei der bakteriologischen Fleisch-untersuchung eines Kalbes. Arch. Lebensmitt.-Hyg. **7**, 81 (1956).

139. — Über die Lebensfähigkeit von Salmonellen im Magendarmtraktus von Stubenfliegen (Musca domestica) und ihre Ausscheidung durch Kot. Arch. Lebensmitt.-Hyg. **7**, 12 (1956).

140. — Auswertung der am Schlachthof in Duisburg ermittelten Salmonellenfunde der Jahre 1950—1956. Fleischwirtschaft **1957**, 259.

141. —, u. G. SCHÜTZ: Über ein gehäuftes Vorkommen von Salmonella orion im Kot gesunder Schlachtschweine. Arch. Lebensmitt.-Hyg. **7**, 244 (1956).

142. SCHÄFER, W., H. MARTIN u. F. HAAS: Darminfektionen durch salmonellenverseuchte Eikonserven. Med. Klin. **1956**, 1254.

143. SCHEIBE, E.: Über eine bakterielle Lebensmittelvergiftung nach Verwendung von Enten-eiern im Bäckereibetrieb. Dtsch. Gesundh.-Wesen **1953**, 751.

144. SCHMIDT, U.: Massensterben von Möwen infolge einer Bact. enteritidis-Masseninfektion. Zbl. Bakt., I. Abt. Orig. 487 (1955).

145. SCHMIDT-LANGE, W., W. JOEST u. M. MÖLLER: Neue „Salmonella osnabrueck". Zbl. Bakt., I. Abt. Orig. 562 (1957).

146. — M. MÖLLER, K. LEDERBOGEN u. F. RAUCH: Noch unbekannte Salmonella verursacht tödliche Darminfektion. Zbl. Bakt., I. Abt. **164**, 523 (1955).

147. SCHMITH, O.: Über die Gefahr der Übertragung von Salmonelleninfektionen durch Enteneier. Öff. Gesundh.-Dienst **1955**, 293.

147a. SCHÖNEBERG, F.: Über die Infektion von Enteneiern mit Breslaubakterien vom Eileiter aus. Berl. tierärztl. Wschr. **1935**, 474.

148. SCHOENHERR, K.-E.: Briefliche Mitteilung.

149. SCHOOP, G., u. K. ZETTL: Zur Frage der Behandlung der chronischen Salmonellainfektion der Tauben. Tierärztl. Umsch. **1956**, 256.

150. SCHWARZ, L., u. E. KRASEMANN: Beobachtungen an Bruteiern, die vor der Bebrütung einige Zeit gelagert hatten. Arch. Geflügelk. **1956**, 405.

150a. SCOTT: Zit. nach HAFFKE (*47a*).

151. SEELE, W.: Salmonella anatum als Ursache eines Massensterbens in einem Gänse-kükenbestand. Arch. Lebensmitt.-Hyg. **6**, 201 (1955).

152. SEELIGER, H. P. R.: Die praktische Serodiagnostik der pathogenen Eingeweidebakterien. Berl. Münch. tierärztl. Wschr. **1955**, 5.

153. — Aktuelle Probleme der Nahrungsmittelinfektion. Dtsch. med. Wschr. **1956**, 1419.

154. — Salmonellosen in Deutschland. Arch. Hyg. (Berl.) **1956**, 499.

155. — Die Salmonellen in Deutschland. Ärztl. Mitt. (Köln) **1957**, H. 6.

156. — Salmonellen in Deutschland seit 1950. Vortrag in Hamburg 15. Okt. 1957. Ref. Arch. Lebensmitt.-Hyg. **8**, 256 (1957).

157. — Briefliche Mitteilung.

158. SEIDEL, G.: Funde von Salmonella give, S. meleagridis und S. newington anläßlich der bakteriologischen Fleischuntersuchung. Lebensmitt.-Tierarzt **1953**, 124.

159. — Salmonellabefunde im Veterinäruntersuchungsamt und Tiergesundheitsamt Berlin in den Jahren 1950—1956. Arch. Lebensmitt.-Hyg. **8**, 199 (1957).

160. — Über unspezifische Lebensmittelvergiftungen. Arch. Lebensmitt.-Hyg. **8**, 219 (1957).

161. —, u. W. PLASCHKE: Mitteilung über gehäuftes Auftreten von S. braenderup bei Schwei-nen. Arch. Lebensmitt.-Hyg. **7**, 243 (1956).

162. SEIFERT, H.: Nachweis von Breslau-Bakterien (Salmonella typhi murium) in Speise-quark. Lebensmitt.-Tierarzt **1953**, 164.

163. SINIOS, A., E. TILING u. R. HANISCH: Klinik, Pathologie und Epidemiologie der Infektion mit Salmonella montevideo. Dtsch. med. Wschr. **1957**, 1329.

164. SÖRIN, L.: Infektion mit Salmonella typhi murium var. Kopenhagen bei Tauben. Nord. Vet.-Med. **1953**, 385. Zit. nach HELM (*50*).

165. SOLOWEY, M. u. Mitarb.: Amer. J. publ. Hlth **1947**, 976. Zit. nach ROHDE u. ADAM.

166. — SPALDING, u. GORESLINE: Zit. nach ALBERT (*5*).

167. SIELAFF, H.: Salmonella anatum in Würsten. Arch. Lebensmitt.-Hyg. **8**, 3 (1957).

167a. Staak, H.: Bedeutung der Salmonellen-Erkrankungen unter besonderer Berücksichtigung der Verhältnisse in Schleswig-Holstein. Arch. Hyg. (Berl.) **142**, 105 (1958).

168. Steiniger, F.: Paratyphus-B-Bakterien im Nordseewasser. Zbl. Bakt. I. Abt., Orig. **157**, 52 (1951).

169. — Salmonella panama in mit Abwasser verunreinigtem Hafenwasser. Ärztl. Wschr. **1952**, 42.

170. — Aus der Biologie der Typhus-Paratyphus-Enteritis-Keime im Abwasser und in abwasserverunreinigten Vorflutern. Prakt. Desinfektor **1953**, 287.

171. — Der Zyklus — Vogel verunreinigtes Wasser in der Freilandbiologie der Salmonellen. Zbl. Bakt., I. Abt. Orig. **160**, 80 (1953).

172. — Über die Notwendigkeit einer Berücksichtigung freilebender Typhus-Paratyphus-Keime in der Abwassertechnik. Jb. Wasserchem. u. Wasserreinigungstechn. **1954**, 172.

173. — Aus der Freilandbiologie der Salmonellen. Dtsch. med. Wschr. **1954**, 1118.

174. — Paratyphus A im Nordseewasser. Desinf. u. Gesundh.-Wesen **1954**, 161.

175. — Über Verschiebung der Häufigkeit freilebender Salmonellaformen an der Küste der Deutschen Bucht. Zbl. Bakt., I. Abt. Orig. **164**, 440 (1955).

176. — Paratyph. Bakterien im Fleisch. Desinf. u. Gesundh.-Wesen **1955**, 22.

177. — Colititer und Salmonellentiter. Desinf. u. Gesundh.-Wesen **1955**, 102.

178. — Zur Freilandbiologie der Salmonellen im Bereich des westlichen Mittelmeeres. Zbl. Bakt., I. Abt. Orig. **166**, 245 (1956).

179. — Untersuchungen von Fliegen aus Tierfuttermittelrohstoff-Lagern auf Typhus-Paratyphusbakterien. Desinf. u. Gesundh.-Wesen **1957**, 91.

180. — Ratin-Bakterien und andere Salmonellen bei Haus- und Wanderratten in Cuxhaven. Desinf. u. Gesundh.-Wesen **1957**, 160.

181. Stellmacher, W.: Die Auswertung der Salmonellenbefunde des Jahres 1954. Arch. Lebensmitt.-Hyg. **6**, 239 (1955).

182. Strauch, W., u. W. Münker: Bakteriologische Wasseruntersuchungen im Zusammenhang mit in diesem Flußtal aufgetretenen Salmonella-Infektionen bei Haustieren. Berl. Münch. tierärztl. Wschr. **1956**, 205.

182a. Strozzi, P.: Clin. vet. **54**, 927 (1931). Zit. nach Knothe u. Budach (*68a*).

183. Stutz, L.: Zur Verhütung der Salmonelleneinschleppung durch ausländische Eiprodukte und Viehfutter. Desinf. u. Gesundh.-Wesen **1957**, 115.

184. Struck, M.: Kann es bei Tierkörpern, bei denen nur in den Organen Salmonellen festgestellt wurden, nachträglich zur Anreicherung von Salmonellen in der Muskulatur kommen? Berl. Münch. tierärztl. Wschr. **1957**, 163.

185. Trüb, C. L. P., u. P. Schneider: Über das Auftreten gehäufter Einzelerkrankungen und Gruppenerkrankungen durch Salmonella panama und Salmonellentypen der Paratyphus C-Gruppe in Nordrhein-Westfalen 1950/52. Z. Hyg. Infekt.-Kr. **1952**, 121.

186. Trüb, C. L. P., u. G. Wundram: Veröff. Volksgesundh.-Dienst **1942**, 480. Zit. nach Trüb u. Reploh (*187*).

187. Trüb, C. L. P., u. H. Reploh: Erkrankungen durch Lebensmittel in der Nachkriegszeit (1946—1956) im Lande Nordrhein-Westfalen. Arch. Hyg. (Berl.) **1954**, 98.

188. Utojo, R. P.: Büffel, Kühe und Schweine als Salmonellaträger. Zit. Berl. Münch. tierärztl. Wschr. **1955**, 14.

189. Vogt, H.: Untersuchungen über die Verbreitung der Salmonellose durch Hühnereier. Inaug.-Diss. Münster 1957.

190. Walzberg, R.: Zur Verbreitung der Salmonellose im nordwestdeutschen Küstengebiet unter besonderer Berücksichtigung des Raumes Weser-Ems. Arch. Lebensmitt.-Hyg. **6**, 223 (1955).

190a. Wesselmann, A.: Bakteriologische Untersuchungen von Enteneiern, von mit Enteneiern hergestellten Lebensmitteln im Zusammenhang mit bakteriellen Lebensmittelvergiftungen beim Menschen. Inaug.-Diss. Berlin 1935. Zit nach Knothe u. Budach (*68a*).

191. Wildführ, G., u. H. Hudemann: Über einen neuen Salmonellentyp 3,15: eh: 1,2. Z. Hyg. Infekt.-Kr. **1955**, 129.

192. Willführ, W. Fromme u. H. Bruns: Nahrungsmittelerkrankungen durch Enteneier. Veröff. Gebiet Med.-Verw. **39**, H. 3 (1933).

193. WILM: Über die Einwanderung von Choleravibrionen ins Hühnerei. Arch. Hyg.(Berl.) **1895**, 145.

194. WINKLE, S., R. ROHDE u. J. BISCHOFF: Kritische Stellungnahme zu einem Gutachten gegen den Erlaß der „Verordnung zum Schutze gegen Infektion" durch Erreger der Salmonellagruppe in Eiprodukten". Münch. med. Wschr. **1957**, 768.

195a. WOHLRAB, R., u. F. STEINIGER: Zur Ausbreitung von Keimen der typh. Paratyph. Enteritis-Gruppe mit der Tierfütterung. Desinf. u. Gesundh.-Wesen **1956**, 129.

195b. WÜSTENBERG, J.: Zit. nach LÜTJE u. RASCH (*84*).

195c. — Bericht über die Tätigkeit des Hygiene-Instituts des Ruhrgebiets zu Gelsenkirchen über 1954, S. 19.

196. ZEMANN, W.: Eine Paratyphus-C-Gastroenteritis durch den Genuß von Trockeneipulver. Dtsch. med. Wschr. **1949**, 121.

197a. ZURECK, F.: Jahresbericht 1951 des Landesveterinäruntersuchungsamtes Berlin. Lebensmitt.-Tierarzt **1953**, 223, 240, 252.

197b. — Durch Salmonellen verursachte Lebensmittelvergiftungen in Berlin. Lebensmitt.-Tierarzt **1953**, 185.

197c. — Über die im Landesveterinäruntersuchungsamt Berlin in den Jahren 1950—1956 gemachten Salmonellenfunde. Arch. Lebensmitt.-Hyg. 8, 25 (1957).

198. — Über die Notwendigkeit der Vornahme von Umgebungsuntersuchungen zur Feststellung postmortaler Fleischvergiftungen. Arch. Lebensmitt.-Hyg. 8, 99 (1957).

Pasteurella pseudotuberculosis unter besonderer Berücksichtigung ihrer humanmedizinischen Bedeutung

Von

WERNER KNAPP*

Mit 6 Abbildungen

Inhaltsverzeichnis

* Aus dem Hygiene-Institut der Universität Tübingen (Direktor: Prof. Dr. R.-E. BADER).

I. Einleitung

Bis vor wenigen Jahren schienen in der Humanmedizin Infektionen mit Pasteurella pseudotuberculosis, dem Erreger der in der Tierwelt besonders unter Nagetieren und Vögeln weitverbreiteten Pseudotuberkulose, ohne Bedeutung. Nachdem aber die Ätiologie einer besonderen Form der mesenterialen Lymphadenitis des Menschen durch den Nachweis von Past. pseudotuberculosis geklärt wurde und diese bisher unbekannte Infektionskrankheit in zunehmender Zahl im gesamtdeutschen Raum beobachtet wird, verdient die Diagnose der menschlichen Pseudotuberkulose größere Beachtung.

POPPE hat 1928 im deutschen Schrifttum den letzten ausführlichen Bericht über die Pseudotuberkulose und ihren Erreger gegeben. In der vorliegenden Arbeit soll daher in einer zusammenfassenden Darstellung über neue Erkenntnisse und experimentelle Untersuchungen aus der Bakteriologie, Serologie, Klinik und pathologischen Anatomie der Pseudotuberkulose berichtet und auf die noch offenen Fragen in Diagnose, Epidemiologie, Therapie und Prophylaxe hingewiesen werden. Ihre Klärung setzt eine verständnisvolle Zusammenarbeit zwischen Klinikern, Mikrobiologen und Pathologen voraus.

II. Geschichtliches

Kurze Zeit nach der Entdeckung des Erregers der Tuberkulose wurde in zahlreichen Veröffentlichungen über tuberkuloseähnliche Krankheiten berichtet, die besonders in pathologisch-anatomischer Hinsicht der Tuberkulose nahestanden, aber nicht durch Tuberkelbakterien hervorgerufen wurden. Die als Pseudotuberkulosen bezeichneten Erkrankungen traten spontan oder endemisch auf, sie konnten aber auch durch Überimpfung von Organmaterial experimentell ausgelöst werden. So beschrieben MALASSEZ und VIGNAL (1883/84) eine angeblich durch kokkoide Bakterien hervorgerufene Krankheit bei Meerschweinchen, die mit Eiter eines an tuberkulöser Meningitis verstorbenen Kindes infiziert waren, als Tuberculose zoogléique. Das pathologisch-anatomische Bild der Organveränderungen, von denen wir nicht sicher wissen, ob sie spontan oder durch Übertragung des Untersuchungsmaterials entstanden waren (PFEIFFER 1889), entsprach dem einer Tuberkulose. Über ähnliche Beobachtungen berichtete kurze Zeit später EBERTH (1885/86). Auch er sah bei der Sektion von Meerschweinchen und Kaninchen Veränderungen, die denen der Tuberkulose so sehr entsprachen, daß er sie als „Pseudotuberkulose" beschrieb. Es gelang ihm wie den französischen Autoren aber nicht, die Tuberculose zoogléique bzw. Pseudotuberkulose ätiologisch mit einem bestimmten Erreger in Zusammenhang zu bringen. Erst die Untersuchungen von PREISZ (1894/1896) führten zu einer ätiologischen Einengung des Begriffes der Pseudotuberkulose.

PREISZ zeigte, daß die von NOCARD (1885/1889), ZAGARI (1890), PARIETTI (1890) beschriebenen „Pseudotuberkulosen" denselben Erreger und dasselbe pathologisch-anatomische Bild besaßen. Seine Stämme stimmten in Morphologie, Färbbarkeit, kulturellen wie tierexperimentellen Eigenschaften mit dem von PFEIFFER (1889) aus einem Meerschweinchen gezüchteten, als Erreger der Pseudotuberkulose beschriebenen und als Bacillus pseudotuberculosis benannten Stamm überein. Auch NOCARD (1889) sah in vergleichenden Untersuchungen die Identität der von ihm, CHARRIN und ROGER (1888) wie DOR (1888) beschriebenen Stämme. PREISZ (1894/1896) schlug daher vor, diese durch ein wohlcharakterisiertes Bacterium hervorgerufene Erkrankung nach PFEIFFER „Pseudotuberculosis rodentium" und ihren Erreger nach DOR „Streptobacillus pseudotuberculosis" zu nennen. Von der Identität der von MALASSEZ und VIGNAL (1883/84) beschriebenen Tuberculose zoogléique, der Pseudotuberkulose der Meerschweinchen und Kaninchen von EBERTH (1885/86), der Tuberculose zooléique von CHANTEMESSE (1887) oder GRANCHER und LEDOUX-LEBARD (1889/90) und der durch Streptobacillus pseudotuberculosis hervorgerufenen Pseudotuberkulose war PREISZ überzeugt. Zur Unterscheidung der von PREISZ (1891/1894) bzw. KUTSCHER (1894) beobachteten und ätiologisch verschiedenen Pseudotuberkulosen der Schafe und Mäuse wurden dagegen auf Vorschlag von PREISZ die Bezeichnungen *Pseudotuberculosis ovis* und *Pseudotuberculosis murium* gewählt; Bezeichnungen, die sich bis heute gehalten haben. Trotz wiederholter Vorschläge, von Pseudotuberkulosen nur bei Infektionen mit Past. pseudotuberculosis zu sprechen (REIMANN 1932, MANNINGER 1945 u. a.), wird in der Veterinärmedizin die Diagnose Pseudotuberkulose auch heute noch nicht allein bei Infektionen durch Past. pseudotuberculosis gestellt.

III. Nomenklatur und systematische Einordnung

Von seiner ersten Bezeichnung als Bacillus der Pseudotuberkulose durch PFEIFFER (1889) bis zu seiner derzeitigen Benennung als Pasteurella pseudotuberculosis (TOPLEY u. WILSON 1936) hat der Erreger der Pseudotuberkulose die verschiedensten Bezeichnungen erhalten, deren teilweise Abwegigkeit für den Bakteriologen so offenkundig ist (HAUPT 1934), daß sie bei der Aufzählung der Synonyma nicht erwähnt werden.

Von BREED, MURRAY und HITCHENS wurden in Bergey's Manual of Determinative Bacteriology (1948) als Synonyma aufgeführt: Bacillus pseudotuberculosis (EISENBERG 1891), Streptobacillus pseudotuberculosis rodentium (PREISZ 1894), Bacterium bzw. Bacillus pseudotuberculosis rodentium (LEHMANN-NEUMANN 1896, MIGULA 1900), Corynebacterium rodentium (Bergey's Manual 1925; TOPLEY u. WILSON 1929), Corynebacterium pseudotuberculosis rodentium (KELSER 1933) und Malleomyces pseudotuberculosis rodentium (PRIBAM 1933).

Von Cillopasteurella pseudotuberculosis im Gegensatz zu Pasteurella pestis, tularensis und septica sprechen PREVOT (1948) und GOYON (1956).

In der Systematik der Bakterien nimmt der Erreger der Pseudotuberkulose z. Zt. folgende Stellung ein (Bergey's Manual 1957).

Klasse:	Schizomýcetes *Naegeli*
Ordnung IV:	Eubacteriales *Buchanan*
Familie V:	Brucellaceae, *Nom. Nov.*
Gattung I:	Pasteurella *Trevisan*
Art:	Pasteurella pseudotuberculosis *Pfeiffer*

Der Gattung Pasteurella, die in der 6. Ausgabe von Bergey's Manual (1948) noch der Familie IX Parvobacteriaceae zugerechnet wurde, gehörten zuerst die 5 Species: Past. multocida, Past. hemolytica, Past. pestis und Past. pseudotuberculosis an. In der 7. Auflage von Bergey's Manual (1957) werden dagegen der Gattung Pasteurella als zugehörig aufgeführt: Past. multocida, Past. septicaemiae, Past. hemolytica, Past. antipestifer, Past. pestis, Past. pfaffii, *Past. pseudotuberculosis*, Past. tularensis und Past. novicida. Als Gattungsmerkmale werden folgende Eigenschaften herausgestellt: kleine, ellipsoide oder längliche gramnegative, mit Spezialmethoden bipolar färbbare Stäbchen. Aerob oder fakultativ anaerob sich vermehrend, benötigen die Pasteurellen bei der Erstkultur nur geringe Sauerstoffspannung. Die Mehrzahl der Stämme spaltet die verschiedensten Kohlenhydrate unter schwacher Säuerung des Nährmediums, während Milchzucker nicht oder nur schwach fermentiert wird. Keine Gasbildung. Gelatine wird nicht verflüssigt und Milch nicht zur Gerinnung gebracht.

Gegen die in Bergey's Manual (1948/1957) vorgenommene Einordnung der Pseudotuberkulosebakterien in die Gattung Pasteurella innerhalb der Familie Parvobacteriaceae bzw. Brucellaceae wurde von verschiedenen Autoren Stellung genommen (MANNINGER 1945, VAN LOGHEM 1946, GIRARD 1942/43, 1950, 1956, VAN DORSSEN 1951, POLLITZER 1954, THAL 1954). Nach VAN LOGHEM, der für Past.

pseudotuberculosis und pestis die Gattungsbezeichnung „Yersinia" vorschlug, sprechen gegen eine Aufnahme in die Familie Parvobacteriaceae die Größe von Past. pseudotuberculosis und Past. pestis und gegen eine gemeinsame Gattungszugehörigkeit mit Past. multocida Unterschiede in der Morphologie, Färbbarkeit, Kolonieform, im biochemischen Verhalten und in der Antigenzusammensetzung. Aus denselben Gründen nahmen GIRARD (1942/43, 1950) und POLLITZER (1954) gegen die gemeinsame Gattungszugehörigkeit von Past. pseudotuberculosis, Past. pestis und Past. multocida Stellung. GUNNISON, SHEVKY u. Mitarb. (1951) teilten den Standpunkt von VAN LOGHEM (1946) und GIRARD (1950) nicht und lehnten die Aufstellung einer neuen Gattung „Yersinia" ab. GIRARD (1942) kam zudem zu der Feststellung, daß nicht einmal die bipolare Färbbarkeit, die die Einordnung von Past. pseudotuberculosis und Past. pestis in die Gattung Pasteurella maßgeblich bestimmt hatte, ein sicheres, konstantes Merkmal darstellt. VAN DORSSEN (1951) und besonders THAL (1954) wiesen auf die morphologischen, kulturell-biochemischen und antigenen Beziehungen von Past. pseudotuberculosis zu den Enterobacteriaceae hin. Wie VAN LOGHEM hielt THAL (1954) eine Einordnung in die Familie Parvobacteriaceae für nicht richtig und die Aufstellung einer neuen Gattung für erforderlich. HAUPT hatte schon 1934, als er die Einordnung von Bact. pseudotuberculosis in die Gattung Corynebacterium ablehnte, seine Aufnahme in die Gattung Shigella vorgeschlagen. In einer persönlichen Mitteilung schrieb GIRARD (1956) unter erneutem Hinweis auf den Vorschlag von VAN LOGHEM „die Zukunft wird zeigen, ob die Kommision für Nomenklatur seinen Vorschlag annehmen wird und die Gattung ‚Yersinia' zu den Parvobacteriaceae oder zu den Enterobacteriaceae gerechnet werden muß". In der Neuauflage von Bergey's Manual (1957) wurde aber von der Aufstellung einer neuen Gattung abgesehen und die Gattung Pasteurella der Familie *Brucellaceae* zugeordnet.

IV. Bakteriologie

1. Morphologie und Physiologie

Pasteurella pseudotuberculosis ist ein gramnegatives, pleomorphes Stäbchen, das keine Sporen bildet; Kapselbildung im Tierkörper wurde bisher nicht nachgewiesen. Nach MEYER (1952) sollen zwar 22° C-Kulturen bei Färbung mit indischer Tusche eine viscöse Hülle erkennen lassen. In Form und Lagerung von der Nährbodenzusammensetzung, Bebrütungstemperatur und -dauer abhängig, finden sich neben kokkoiden und ovoiden Organismen mit einer Länge und Breite von $0{,}8$—$2{,}0$:$0{,}8\,\mu$ unterschiedlich lange schlanke Stäbchen mit abgerundeten Enden, deren Länge und Breite zwischen $1{,}5$—$6{,}0$:$0{,}4$—$0{,}8\,\mu$ variieren kann, während Past. multocida mit einer Länge und Breite von $0{,}3$—$1{,}5$:$0{,}15$ bis $0{,}25\,\mu$ meist in kokkoider Form vorliegt und selten ovoide oder Stäbchenformen annimmt (GIRARD 1942 u. a.). Die Länge und Breite von Past. pestis ist mit $1{,}5$—$2{,}0$:$0{,}5$—$0{,}8\,\mu$ angegeben (MEYER 1952).

Stäbchenformen finden sich bei Past. pseudotuberculosis auch nach eigenen Beobachtungen fast regelmäßig in flüssigen Nährmedien, während die kokkoiden und ovoiden Formen bei jungen, die ovoiden bis Stäbchenformen bei älteren Kulturen in Peptonwasser vorherrschen. Eine Lagerung der Keime einzeln, paarweise, in kürzeren und längeren Ketten oder Haufen ist uncharakteristisch. Die mehrfach beschriebene Kettenbildung sahen wir bei unseren Stämmen nur selten. Unter ungünstigen Wachstumsbedingungen werden die verschiedensten Degenerationserscheinungen beobachtet. Neben sehr langen, schlanken Stäbchen- und Fadenformen mit 5—7facher Länge der Normalform finden sich verschieden lange, plumpe Stäbchen mit kugel- oder keulenförmig aufgetriebenen Enden, starker Vacuolisierung und unregelmäßiger oder unterschiedlich starker Färbbarkeit. Daneben sind plumpe oder aufgetriebene kokkoide und ovoide Formen nicht selten (POPPE 1928[1]; SCHÜTZE 1929; REIMANN 1932; KAUFFMANN 1933; WEITZENBERG 1935; BOQUET 1937; TRUCHE 1938; TOPPING u. Mitarb. 1938; MOSS u. Mitarb. 1941; HÄSSIG u. Mitarb. 1949; MEYER 1952; FELDMAN 1955; WILSON u. MILES 1955; Bergey's Manual 1957 u. a.).

Färbbarkeit. Past. pseudotuberculosis ist ein gramnegatives Stäbchen. In Ausstrichpräparaten lassen sich die Stäbchen jüngerer Kulturen mit den verschiedenen Anilinfarbstoffen gleichmäßig stark anfärben, während bei älteren Kulturen in größerer Zahl farbblasse oder ungleichmäßig gefärbte Stäbchen nachgewiesen werden (MOSS u. BATTLE 1941; BLAXLAND 1947). Diese Beobachtung gilt in besonderem Maße für Stäbchen mit Degenerationserscheinungen. Die als Gattungsmerkmal herausgestellte stärkere Anfärbung der Pole ist bei Past. pseudotuberculosis nicht konstant, so daß ihre diagnostische und differentialdiagnostische Bedeutung nach GIRARD (1942), VAN LOGHEM (1946) u. a. zweifelhaft ist.

Auch nach REIMANN u. ROSE (1931), MOSS u. BATTLE (1941), HAGAN u. BRUNNER (1951), LEADER u. BAKER (1954) u. a. fehlt bei der Polfärbung oder der von COOK (1952, 1953) für Organausstriche angegebenen modifizierten Ziehl-Neelsen-Färbung die stärkere Anfärbung der Pole nicht selten. In Übereinstimmung mit HÄSSIG u. Mitarb. (1949) sahen wir eine gute Darstellung der Pole häufiger bei jungen Kulturen oder Organausstrichen. Andererseits fanden MOSS u. BATTLE (1941) bipolare Färbbarkeit, Vacuolisierung und andere morphologische Unregelmäßigkeiten besonders bei älteren Kulturen.

Beweglichkeit. Past. pseudotuberculosis wurde lange Zeit als unbeweglich angesehen. Es besteht aber kein Zweifel, daß Kulturen bei Zimmertemperatur beweglich sind, während Brutschranktemperaturen zwischen 30—37° C ihre Beweglichkeit aufheben.

Noch 1928 weist POPPE auf das Fehlen der Eigenbewegung von Past. pseudotuberculosis hin, obwohl KOSSEL und OBERBECK schon 1902 in einer Arbeit, die wenig Beachtung fand, ihre Beweglichkeit beschrieben. Sie berücksichtigten als erste die Züchtungstemperatur und konnten feststellen, daß 3 Stämme von Past. pseudotuberculosis bei 37° C unbeweglich, dagegen bei 22° C beweglich waren. In späteren Arbeiten wurde die Beobachtung der beiden Autoren, sofern bei der Beweglichkeitsprüfung eine Züchtungstemperatur von 18—22 ° C und optimale Wachstumsbedingungen beachtet wurden, von zahlreichen Autoren (ARKWRIGHT 1927; SCHÜTZE 1928; LEVINTHAL 1930; KAUFFMANN 1933; WEITZENBERG 1935; BOQUET 1936/37; PRESTON u. MAITLAND 1952; KNAPP 1956 u. a.) bestätigt. ARKWRIGHT (1927) sah zwar bei einem unter 8 geprüften Stämmen auch bei 37° C Beweglichkeit.

In eingehenden Untersuchungen über die Beweglichkeit und Begeißelung von Past. pseudotuberculosis konnte WEITZENBERG (1935) zeigen, daß allein bei Zimmertemperatur (22° C) die optimale Beweglichkeit des jeweiligen Stammes erreicht wird, während bei 3 und 8° C bewegliche Stäbchen erst nach längerer Zeit auftreten. Nach PRESTON und MAITLAND (1952) liegt die kritische Temperatur bei 30° C. Werden Kulturen, die bei 37° C unbeweglich waren, in 22° C gebracht, so zeigen sich innerhalb weniger Stunden bewegliche Formen (WEITZENBERG 1935). Die Geißelentwicklung soll hierbei ohne gleichzeitige Zellteilung möglich sein (PRESTON u. MAITLAND 1952).

Im Peritonealexudat von Meerschweinchen fand BOQUET (1937) schon 1 Std nach der intraperitonealen Injektion von 4 ml einer gut beweglichen Kultur von Past. pseudotuber-

---

[1] Vor 1928 erschienene Arbeiten werden nur, soweit sie im Rahmen dieser Arbeit von Bedeutung sind, berücksichtigt und im Literaturverzeichnis aufgeführt. Weiterer Literaturnachweis siehe bei POPPE in „Pseudotuberkulose", Handbuch der pathogenen Mikroorganismen, 3. Aufl., Bd. IV/1, S. 413. 1928.

culosis keine beweglichen Keime mehr. Er führte den Beweglichkeitsverlust auf die rasch einsetzende Phagocytose zurück, während Knapp (1956) den die Beweglichkeit hemmenden Einfluß in der Körpertemperatur gegeben sah. Ebenfalls nach Boquet verlieren bei 22° C bewegliche Kulturen auch in vitro bei 37° C schon innerhalb 1—2 Std ihre Beweglichkeit. Preston u. Maitland (1952) gelang es nicht, experimentell einen möglicherweise die Beweglichkeit beeinflussenden Unterschied in der pH-Konzentration der bei 37° C bzw. 22° C bebrüteten Kulturen oder einen nur in 37° C-Kulturen entwickelten encymatischen Stoff, der, mit dem Kulturfiltrat übertragen, zu einer Hemmung der Beweglichkeit bzw. Zerstörung der bei 22° C entwickelten Geißeln führt, nachzuweisen.

Diesen Befunden, die weitgehend an Stämmen tierischer Herkunft erhoben wurden, standen auch in der neueren Literatur Beobachtungen an Stämmen menschlicher Herkunft gegenüber, die bei 22 und 37° C unbeweglich blieben (Neugebauer 1933; Dujardin-Beaumetz 1938; Topping u. Mitarb. 1938; Moss u. Battle 1941; Snyder und Vogel 1943; Piéchaud 1952; Knapp 1954). Doch konnte Knapp (1956) die Beweglichkeit von 14 Stämmen menschlicher Herkunft nachweisen und zeigen, daß ein Unterschied in der Beweglichkeit der bisher in der Literatur als unbeweglich beschriebenen Stämme menschlicher Herkunft und der beweglichen Stämme tierischer Herkunft nicht besteht.

Der Nachweis der Beweglichkeit dieser Stämme gelang nach zahlreichen Versuchen zuerst im U-Kölbchen nach Bader, dessen besondere Bedeutung bei der Beweglichkeitsprüfung anscheinend unbeweglicher, an sich aber begeißelter Stämme damit aufgezeigt wurde (Knapp 1956).

Abimpfungen von gutbeweglichen oder rasch schwärmenden Stammpassagen können von einem plötzlichen, anscheinend vollständigen Verlust der Beweglichkeit und des Schwärmvermögens gefolgt sein (Weitzenberg 1935, Boquet 1936/37, Knapp 1956). Diese Beobachtungen beschränkten sich nicht auf einen Stamm; es hat vielmehr den Anschein, daß alle Kulturen aus unbekannten Gründen die Fähigkeit, bewegliche Formen zu bilden, zeitweise, möglicherweise auch ganz verlieren (Boquet 1936/37, Knapp 1956). Auf die damit verbundenen diagnostischen und besonders auch differentialdiagnostischen Schwierigkeiten gegenüber Past. pestis wurde von Favarissova (1937), Seal (1951) und Pollitzer (1954) u. a. hingewiesen. Auffallend sind auch die starken Unterschiede im Grad der Beweglichkeit der einzelnen Stämme bei gleichen kulturellen Bedingungen (Schütze 1928, Weitzenberg 1935, Boquet 1936, Knapp 1956 u. a.).

Versuche, die Beweglichkeit wenig oder nicht beweglicher Stämme durch häufige Überimpfung oder Wechsel des Nährmediums zu erhöhen, führten bei Weitzenberg (1935) und Boquet (1937) zu keinem verwertbaren Ergebnis.

Die Art der Fortbewegung von Past. pseudotuberculosis im hängenden Tropfen ist ohne besondere Eigenheiten. In stark beweglichen Kulturen zeigen die meisten Keime eine schnelle, geradlinig rotierende oder kurvenförmige Bewegung. Bei stark beweglichen Kulturen fallen unter zahlreichen ruhig liegenden, sich langsam bewegenden oder nur eine starke Molekularbewegung aufweisenden Keimen Stäbchen auf, die sich plötzlich immer rascher drehen oder überschlagen, um dann teils in rotierender, teils in schlängelnder Bewegung das Blickfeld zu verlassen.

Begeißelung: Die Angaben über Anzahl und Anordnung der Geißeln von Past. pseudotuberculosis sind bei den einzelnen Autoren, die sich mit der Geißeldarstellung eingehender befaßten, uneinheitlich und z. T. widersprechend.

Klein (1899) fand mit der van Ermengenschen Silbermethode an einigen Stäbchen einer im hängenden Tropfen unbeweglichen Kultur eine oder zwei endständige Geißeln, und Byloff (1906) mit der Geißelfärbung nach Zettnow nur eine endständige Geißel, die kürzer als der

Bakterienleib war. BURCKHARDT (1914) sah neben Stäbchen mit 1—3 Geißeln gelegentlich auch solche mit vier peritrich angeordneten kurzen Geißeln. PLASAJ (1921) beschrieb im Gegensatz zu BYLOFF und KLEIN das Vorkommen nur eines Geißelfadens mit extrapolarem Ansatz. Im neueren Schrifttum berichtete WEITZENBERG (1935) über Beobachtungen an einem größeren Untersuchungsgut. Bei 25 nach ZETTNOW gefärbten Stämmen beobachtete er an den kokkoiden und ovoiden Formen meist 1—2 und nur vereinzelt 3—6 Geißeln, während längere Stäbchen wiederholt eine peritriche Anordnung von 3—6 Geißeln erkennen ließen. Auf den extrapolaren Abgang der Geißeln wurde von WEITZENBERG besonders hingewiesen. Lophotriche Begeißelung, wie von LEVINTHAL (1930) neben mono- und peritricher Anordnung der Geißeln für Past. pseudotuberculosis beschrieben, glaubte WEITZENBERG (1935) in 2 Fällen gesehen zu haben.

Auch BOQUET (1937) beobachtete wie WEITZENBERG bei 12 von ihm untersuchten Stämmen in der Mehrzahl Stäbchen mit nur einer polaren Geißel und nur selten Keime mit 4—5 extrapolaren Geißeln. Seine Stämme waren zur optimalen Geißelentwicklung auf synthetischem Nährmedium nach SAUTON mit 2% Glucose anstelle von Glycerin gezüchtet und nach der Methode von LEVENSON (1936) gefärbt worden, während WEITZENBERG eine bessere Geißelentwicklung in Kulturen auf festen als in flüssigen Nährmedien, wie z. B. Bouillon, festgestellt und daher Agarmedien für seine Untersuchungen bevorzugt hatte.

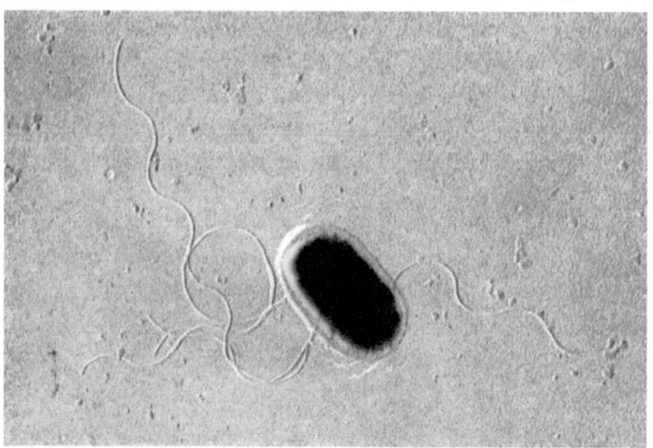

Abb. 1. Peritriche Begeißelung von Past. pseudotuberculosis. Elektronenoptische Vergrößerung 6000×; Gesamtvergrößerung 9000×. Gerät Bosch

Neuere Untersuchungen von PRESTON u. MAITLAND (1952) sowie eigene unveröffentlichte Beobachtungen sprechen aber bei gleichzeitiger kritischer Sichtung der vorliegenden Befunde für eine peritriche Begeißelung von Past. pseudotuberculosis mit durchschnittlich 3—6 Geißeln.

PRESTON u. MAITLAND (1952) untersuchten zuerst bei 22° C bebrütete und nach CONN u. WOLFE (1938) gefärbte Schrägagarkulturen. Sie fanden Stäbchen mit 1—6 peritrich angeordneten Geißeln, doch ließ die größere Zahl der Keime nur 1—2 Geißeln erkennen. In weiteren Versuchen mit Bouillonkulturen konnten dagegen mit derselben Präparations- und Färbetechnik in überwiegender Zahl Keime mit zahlreichen Geißeln nachgewiesen werden. Zu denselben Feststellungen kam KNAPP (1957) bei unveröffentlichten Untersuchungen an 15 Stämmen. Erst bei der Präparation der bei 22° C bebrüteten Bouillonkulturen zeigte die größere Zahl der Keime im Elektronenmikroskop 3—6 peritrich und nur vereinzelt 1—2 parapolar angeordnete Geißeln (Abb. 1 und 2), während bei der Verwendung von Blutagar- oder Schwärmagarkulturen vornehmlich Stäbchen mit 1—2 und nur selten mit mehr Geißeln im Elektronenmikroskop bzw. im Lichtmikroskop nach Anwendung der Färbemethode von SOUS (LEMBACH u. SOUS 1948) gesehen wurden. Die Länge der Geißeln beträgt häufig das 6—9fache des Bakterienleibes. Auffallend ist die ringförmige Lagerung zahlreicher Geißeln.

Elektronenoptische Untersuchungen von PARNAS (1956) ließen bei Past. pseudotuberculosis keine Geißeln erkennen.

Systematische Untersuchungen, inwieweit sich bei einem Wechsel der Umweltbedingungen (Nährmedium, pH-Konzentration, Bebrütungsdauer, Bebrütungstemperaturen zwischen 18—30° C usw.) Anzahl und Anordnung der Geißeln bei Past. pseudotuberculosis ändern, liegen bisher nicht vor. Auf die Bedeutung der Umwelteinflüsse und die Notwendigkeit ihrer Beachtung bei Untersuchungen über die Begeißelung von Bakterien ist von GRIFFIN u. ROBBINS (1944) und STOCKER (1956) in anderem Rahmen hingewiesen worden. Möglicherweise gibt es auch bei Past. pseudotuberculosis neben Stämmen mit peritricher solche mit mono- oder amphitricher Begeißelung, wie von LEIFSON und HUGH (1953) für einige Alcaligenes-Stämme beschrieben.

Von immunobiologischer Bedeutung ist schließlich noch die Darstellung einzelner begeißelter Stäbchen im Elektronenmikroskop bei zwei auf Blutagar bei 37⁰ C gezüchteten Stämmen (KNAPP 1956) und der Nachweis von Geißel-Antikörpern in Sera von Kaninchen, die mit bei 37⁰ C gezüchteten, im Wasserbad bei 56—58⁰ C (BOQUET 1937) oder mit Phenol (KNAPP 1956) abgetöteten Kulturen immunisiert waren. Diese Beobachtungen zeigten ebenso

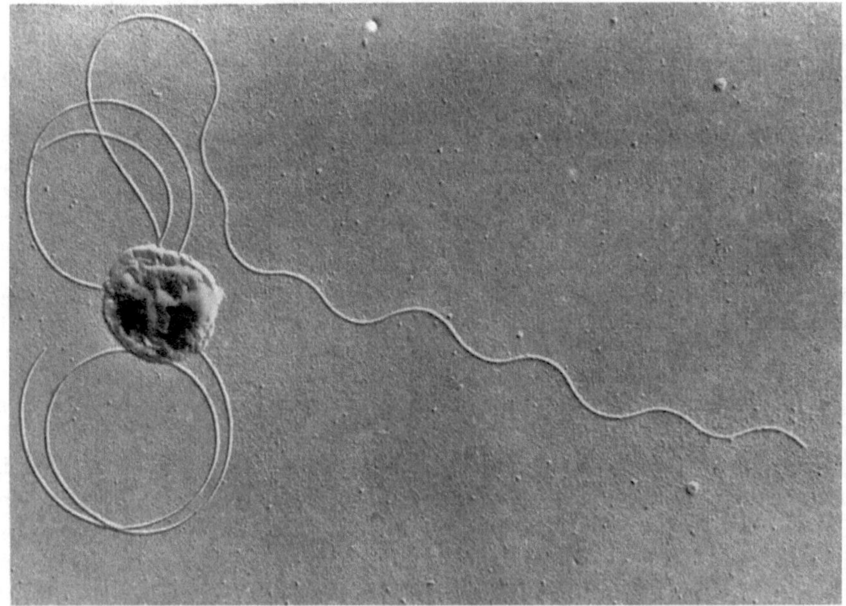

Abb. 2. Peritriche Begeißelung von Past. pseudotuberculosis. Ring- und Spiralformen der Geißeln. Elektronen-optische Vergrößerung 6000 × ; Gesamtvergrößerung 9000 × . Gerät Bosch

wie der Nachweis eines bei 37⁰ C beweglichen bzw. kurze Geißeln besitzenden Stammes (ARKWRIGHT 1927, WEITZENBERG 1935), daß eine Bebrütung der Kulturen von Past. pseudotuberculosis bei 37⁰ C nicht mit absoluter Sicherheit die Geißelbildung hemmt.

Temperaturempfindlichkeit. Die optimale Wachstumstemperatur liegt zwischen 24 und 37⁰ C mit der größten Vermehrungsgeschwindigkeit bei 30⁰ C. Bei Zimmertemperatur (18—22⁰ C) ist das Wachstum etwas verzögert, es erreicht aber nach 2—3 Tagen die gleiche Stärke wie nach 1—2 Tagen bei 30—37⁰ C. Die obere Wachstumsgrenze liegt bei 43⁰ C und die untere bei 4⁰ C (WILSON u. MILES 1955; Bergey's Manual 1957). TUMANSKY u. Mitarb. (1935; zit. nach WLSON u. MILES 1955) sahen Wachstum sogar bei 0⁰ C, wir selbst ab 2⁰ C. Bei der Neigung von Past. pseudotuberculosis, bei 37⁰ C Rauhformen zu bilden und ihre Beweglichkeit zu verlieren, ist die Züchtung der Stämme zur Gewinnung von glatten und beweglichen Kulturen in einem Temperaturbereich von 20—30⁰ C ratsam.

Nach ROSENWALD u. Mitarb. (1944), FEY (1956) und KNAPP (1956) gelingt die Erstzüchtung *einzelner* Stämme aus menschlichem und tierischem Untersuchungsgut mitunter nur bei Zimmertemperatur, so daß auch aus diesem Grunde die kulturelle Erstzüchtung von Past. pseudotuberculosis außer bei 37° C bei 22° C empfohlen wird.

Sauerstoffabhängigkeit. Past. pseudotuberculosis ist ein aerob und fakultativ anaerob wachsender Keim (POPPE 1928 u. a.). Nach KAKEHI (1915) sollen unter anaeroben Bedingungen Wachstum und Vermehrung vollständig gehemmt sein. Diese Beobachtung scheint einmalig gewesen zu sein. Sie fand nur insoweit Bestätigung, als unter anaeroben Bedingungen bei 37 und 22° C auf flüssigen und mehr noch auf festen Nährböden Wachstum und Vermehrung mitunter deutlich, aber nicht vollständig gehemmt sein können. Der Grad der Hemmung ist bei den einzelnen Stämmen verschieden, er ist besonders stark bei Stämmen, die weitgehend in Rauhform vorliegen oder leicht in diese Form dissoziieren. Die Neigung von Past. pseudotuberculosis, Rauhformen zu bilden, ist unter anaeroben Kulturbedingungen größer als unter aeroben.

In eigenen Versuchen vermehrten sich 10 Stämme von Past. pseudotuberculosis Typ I in Tryptose-Bouillon als Nährmedium in aeroben und anaeroben Kulturen bei 37, 22 und 2° gleich stark. Dagegen war in Bouillon mit und ohne Serum- oder Traubenzuckerzusatz die Vermehrung in anaerober Kultur im Vergleich zur aeroben gehemmt. Diese Beobachtungen zeigen zugleich die Notwendigkeit, verschiedene Nährmedien bei der Prüfung der Sauerstoffabhängigkeit zu verwenden.

In einem Fall gelang KNAPP u. MASSHOFF (1954) die Erstzüchtung von Past. pseudotuberculosis Typ I aus einem menschlichen mesenterialen Lymphknoten nur aus einer 5 Tage bei 37° C bebrüteten Leberbouillon. Da Keime im zermörserten Organmaterial nicht gleichmäßig verteilt sein können, besteht die Möglichkeit, daß der fehlende Erregernachweis in fünf gleichzeitig beimpften aeroben Kulturen die Folge ihrer unterschiedlichen Verteilung im Untersuchungsmaterial war.

Abhängigkeit von der Wasserstoffionenkonzentration. Auf festen und flüssigen Nährböden wächst nach eigenen Versuchen Past. pseudotuberculosis in einem p_H-Bereich von 5,2—10,0. Das Vermehrungsoptimum liegt zwischen p_H 6—8. Niedrigere und höhere p_H-Werte führen bei den einzelnen Stämmen zwar unterschiedlich rasch — zuerst auf festen Nährböden — zu einer deutlichen Hemmung oder vollständigen Einstellung der Vermehrung. Zunehmende Säuerung des Nährbodens fördert die Rauhdissoziation und hemmt die Beweglichkeit. Sie ist nach eigenen Beobachtungen bei einzelnen Stämmen einigen Stäbchen noch bei p_H 10 erhalten, fehlt aber allen Stämmen bei p_H 4,6. Das Beweglichkeitsoptimum liegt bei p_H 7—7,3 (BOQUET 1937; PRESTON u. MAITLAND 1952 u. a.).

Wachstum in Stichkulturen. Gelatine wird innerhalb 7 Tagen bei 22° C nicht verflüssigt. Im Gelatineagar-, Nähragar- und Traubenzuckeragarstich ($1^1/_2$ bis 3%iger Agar) zeigt sich längs des Stichkanals ein innerhalb 24—48 Std an Stärke zunehmendes, gleichmäßiges, fadenförmiges, der Spitze zu schwächer werdendes Wachstum, während die Oberfläche langsam von einem grauweißlichen Kulturrasen bedeckt wird. Vermehrungsgeschwindigkeit und -intensität sind im Gelatineagar stärker als im Nähr- und Traubenzuckeragar.

2. Kulturell-biochemische Eigenschaften

a) Kulturelles Wachstum

Wachstum in flüssigen Nährböden. In Fleischwasser-Pepton-Bouillon führt Past. pseudotuberculosis bei 37 und 22° C in 24 Std zu einer schwachen, diffusen Trübung des Nährmediums. Die in den ersten 24 Std eintretende Trübung ist nach gleicher Keimeinsaat bei 37° C Bebrütung der Kulturen häufig stärker als bei 22° C; ein Ausgleich tritt spätestens innerhalb 48—72 Std ein. Nach 2 oder mehr Tagen setzt sich unter gleichzeitiger Klärung des Nährmediums ein bald schleimiger, bald häutchenförmiger oder krümeliger Bodensatz ab, der mitunter nur schwer aufschüttelbar ist. Auch Kahmhautbildung wird beobachtet (POPPE 1928; SCHÜTZE 1929; BOQUET 1937; TOPPING u. Mitarb. 1938; BEAUDETTE 1940; MOSS u. BATTLE 1941; BLAXLAND 1947 u. a.). Rauhformen, die sich rascher und häufiger in 37° C- als in 22° C-Kulturen entwickeln, wachsen ohne Trübung mit klumpigem oder flockigem Bodensatz. Kulturen mit Intermediärformen zeigen zuerst diffuse Trübung und Bodensatz, später verschwindet die Trübung durch völliges Sedimentieren der Keime (BOQUET 1937 u. a.). Der Zusatz von Serum oder Traubenzucker zur Fleischwasser-Pepton-Bouillon wirkt nicht wachstumsfördernd, dagegen werden in Tryptose-Bouillon stärkeres Wachstum und raschere Vermehrung beobachtet.

Peptonwasser mit 0,5% Glucose wird nach OTTEN (1926) von Past. pseudotuberculosis innerhalb 7 Tagen bis zu einem p_H-Bereich von 4,6—4,8 gesäuert, während nach Zugabe von nur 0,05% Glucose die anfängliche Säuerung des Peptonwassers von einer Realkalisierung bis p_H 7,0 oder 7,2 gefolgt ist. Die sekundäre Alkalisierung von Peptonwasser mit geringem Glucosegehalt wurde als differentialdiagnostisches Merkmal gegenüber Past. pestis und Past. multocida angegeben, deren Kulturen eine endgültige Wasserstoffionenkonzentration von p_H 5,1—5,5 bzw. 5,8—6,1 erreichen können. Nach ZLATOGOROFF u. Mitarb. (1928) tritt die Realkalisierung rascher in Kulturen rauher, als glatter Stämme von Past. pseudotuberculosis ein.

In 25—50%iger *Rindergalle* sah CERNAIANU (1928) eine Vermehrungshemmung oder Auflösung der eingeimpften Keime, während BOQUET (1937) über starke Vermehrung in 6%iger und Hemmung in 30%iger Rindergalle berichtete. In eigenen Versuchen konnte in 50%iger Rindergalle während 4wöchiger Bebrütung der Kulturen bei 37° C keine Vermehrung, aber auch keine Reduzierung der Keimzahl nachgewiesen werden.

Nährlösungen mit 3 und 4%igem *Kochsalzzusatz* ermöglichen noch bei 22° C, nicht oder nur schwach bei 37° C Keimwachstum und Vermehrung von Past. pseudotuberculosis. Degenerationserscheinungen wie bizarre, z. T. keulenartig aufgetriebene Faden- oder Spindelformen oder plump aufgetriebene, ovoide und kokkoide Stäbchen mit unregelmäßiger und schwacher Färbbarkeit, werden nur vereinzelt bei 2%, häufiger bei 3% und fast ausschließlich bei 4% Kochsalzzusatz zum Nährmedium beobachtet (SAISAWA 1913; SKORODUMOFF u. Mitarb. 1926; BOQUET 1937; TOPPING u. Mitarb. 1938; MOSS u. BATTLE 1941).

Wachstum auf festen Nährböden. Auf *Nähragar* wächst Past. pseudotuberculosis innerhalb 24 Std in hellen durchsichtigen Kolonien mit einem Durchmesser von etwa 0,25—1,0 mm, die bei weiterer Bebrütung trüb und undurchsichtig werden und eine weißlichgraue bis graugelbliche Farbe aufweisen. Das Zentrum der Kolonien ist meistens erhaben und trüb. Ihr Wachstum ist bei 22° C zunächst langsamer als bei 37° C und wird erst innerhalb 48—72 Std üppiger. Ausgewachsene Kolonien sind bei 22° C von schleimiger oder weicher Konsistenz, im Zentrum leicht erhaben, ihre Oberfläche ist glatt oder leicht granuliert, feuchtspiegelnd, der Kolonierand glatt und regelmäßig. Bei 37° C erscheinen die Kolonien, als Zeichen beginnender Rauhdissoziation, häufig trocken und flach, die deutlich granulierte Oberfläche erscheint mattspiegelnd oder stumpf, während der Kolonierand unregelmäßig begrenzt und saumartig, als Randzone, abgeflacht ist.

Auch auf Serum- und Blutagar sind die Kolonien von Past. pseudotuberculosis nach 24 Std Bebrütung der Kulturen bei 37° C größer als bei 22° C. Sie erreichen am 2. Tag einen

Durchmesser von 2—3 mm. In den ersten 24 Std hell und durchscheinend, werden die Kolonien nach 48—72 Std grauweiß bis gelblich und schließlich undurchsichtig. Hämolyse wurde, von einzelnen nicht nachuntersuchten Beobachtungen abgesehen (KOROBKOVA 1940; MOSS u. BATTLE 1941; HARISIJADES 1953), nicht beobachtet (BOQUET 1937 u. a.).

Eine Verwendung von Serum- oder Blutagarplatten aus kulturellen oder differential-diagnostischen Gründen ist nicht erforderlich.

Nach Zusatz von wachstumsfördernden Stoffen zu einem Gelatine- und vollsynthetischen Aminosäure-Basalmedium sah BERKMAN (1942) bei 5 Stämmen von Past. pseudotuberculosis und Past. pestis im Gegensatz von Past. tularensis und Past. multocida keine Förderung von Wachstum und Vermehrung.

Verschieden sind die Beobachtungen über die Wachstumseigenschaften von Past. pseudotuberculosis auf *Nähragar* mit unterschiedlichem, besonders 3%igem *Kochsalzgehalt*, dessen Verwendung zur differentialdiagnostischen Abgrenzung von Past. pestis verschiedentlich vorgeschlagen wurde (SAISAWA 1913; SKORO-DUMOFF u. SOMOROWITSCH 1926; DIEUDONNÉ u. OTTO 1928; BOQUET 1937; HAAS 1938; TOPPING u. Mitarb. 1938; MOSS u. BATTLE 1941; POLLITZER 1954) (s. auch S. 232).

SAISAWA (1913) sah 5 Stämme auf Nährböden mit 2% Kochsalzgehalt nur kümmerlich und mit 3% Gehalt innerhalb 3 Tagen nicht wachsen. Die auf 2%igem Kochsalzagar ge-wachsenen Keime zeigten Degenerations- und Involutionsformen. Neben sehr langen und spindelförmig aufgetriebenen Stäbchen waren abnorm große und ovale oder runde Formen zu sehen, die aber nicht den eigentümlich aufgequollenen, bald runden, spindelförmigen oder ovalen Involutionsformen, wie sie bei Past. pestis auf 3%igem Kochsalzagar schon innerhalb 24 Std gefunden werden (DIEUDONNÉ u. OTTO 1928; POLLITZER 1954), entsprachen. Nach BOQUET (1937) wächst Past. pseudotuberculosis auf 3%igem Kochsalzagar bei 37° C vor-nehmlich in fadenförmigen Stäbchen, während höhere Kochsalzkonzentrationen das Auftreten kugeliger Formen begünstigen sollen.

Über spärliches Wachstum und starke Pleomorphie der Stäbchen eines Stammes von Past. pseudotuberculosis auf 3%igem Kochsalzagar berichteten TOPPING u. WATTS (1938) sowie MOSS u. BATTLE (1941). Sie sahen neben langen, z. T. fadenförmigen Stäbchen mit Vacuolen und aufgetriebenen Enden kurze plumpe oder große kokkoide bis ovoide Formen, aber keine für Past. pestis charakteristische Pleomorphie. HAAS (1938) fand dagegen bei einem Stamm neben uncharakteristischen Formen dieselben Degenerations- und Involutionsformen, wie sie bei Past. pestis auf 3%igem Kochsalzagar in der Regel innerhalb 24 Std auftreten.

Von Bedeutung ist auch die Berücksichtigung der Bebrütungstemperatur bei der Prüfung der Wachstumseigenschaften von Past. pseudotuberculosis auf Kochsalzagar, wie eigene Versuche an Stämmen von Past. pseudotuberculosis, die bei 22° C und 37° C auf Nähragar-platten mit 2, 3 und 4% Kochsalzzusatz gezüchtet wurden, veranschaulichen.

20 Stämme der Typen I—V, die in ihrem Verhalten keine Unterschiede aufwiesen, wuchsen bei 22° C auf Nähragar mit 2% Kochsalzzusatz ungehemmt und zeigten mit 2 Aus-nahmen keine Degenerations- und Involutionsformen, während bei 37° C 14 Stämme nur stark gehemmt zum Wachstum kamen und im gefärbten Ausstrichpräparat ein Überwiegen von langen, z. T. aufgetriebenen Stäbchen oder langen Fadenformen erkennen ließen. Auf 3%igem Kochsalzagar entwickelten alle Stämme, die bei 22° C eine unterschiedlich starke Wachstumshemmung aufwiesen, neben kokkoiden und Stäbchenformen zahlreiche Degenera-tions- und Involutionsformen. Neben sehr langen, schlanken Stäbchen überwogen Faden-formen, während bei 37° C nur 10 der 20 Stämme noch kümmerlich wuchsen. Im Präparat fanden sich neben großen meist aufgetriebenen Stäbchen zahlreiche Fadenformen mit Knäuelbildung. Auf Nähragar mit 4% Kochsalzgehalt blieb bei 22° C und bei 37° C ein makroskopisch sichtbares Wachstum aus.

Unter den Nährmedien, die bei der selektiven Züchtung von pathogenen und apathogenen Darmkeimen vornehmlich Verwendung finden, wird der *Desoxy-cholat-Citratagar* nach LEIFSON von THAL u. CHEN (1954) und BALTAZARD u. Mitarb. (1956) für die Differentialdiagnose zwischen Past. pseudotuberculosis und Past. pestis besonders empfohlen. Past. pseudotuberculosis soll bei 37⁰ C innerhalb 48 Std unter gelblicher Verfärbung des Nährbodens zu großen, leicht trüben Kolonien auswachsen, während Past. pestis in seinem Wachstum deutlich gehemmt ist. Seine als kümmerlich bezeichneten Kolonien, die nach 48 Std eine punktförmige Rötung aufweisen können, wachsen ohne Verfärbung des Nährbodens. Für die differentialdiagnostischen Untersuchungen soll eine von BALTAZARD u. Mitarb. (1956) angegebene Modifikation des Desoxycholat-Citratagars besonders geeignet sein. Berichte über den Wert dieses in der Salmonella-Diagnostik bewährten Selektivnährbodens bei der Isolierung von Past. pseudotuberculosis aus menschlichem Untersuchungsgut liegen nicht vor.

Eigene Versuche mit Reinkulturen ließen im Vergleich mit der Wachstumsstärke von Past. pseudotuberculosis auf Blut-, Nähr- und Endoagar, auf *Desoxycholat-Citratagar* (BADER 1950) bei 22 und 37⁰ C nur eine schwache Wachstumshemmung erkennen. Die bei 22⁰ C glatten und in den ersten 24 Std hellen, durchsichtigen Kolonien wurden nach 48—72 Std leicht trüb und zeigten wie der Nährboden eine leicht gelblichbraune Farbtönung. Eine deutliche gelbe Verfärbung des Nährbodens wurde nicht beobachtet. Die bei 37⁰ C bebrüteten Kulturen zeigten unterschiedlich rasch eine Rauhdissoziation der Kolonien, die nach unseren Beobachtungen bei 37⁰ C auf Desoxycholat-Citratagar stärker als auf Blut-, Nähr- und Endoagar gefördert wird. Eine Isolierung von Past. pseudotuberculosis aus menschlichem Untersuchungsgut (Lymphknoten, Stuhl, Galle) gelang uns bisher auf diesem Nährboden nicht, doch reicht die Zahl der untersuchten Fälle nicht aus, um schon jetzt ein Urteil über den Wert des Desoxycholat-Citratagar in der kulturellen Diagnostik von menschlichen Pseudotuberkulosen abgeben zu können.

Auf *Wismutsulfitagar* nach WILSON-BLAIR stellte PIÉCHAUD (1952) bei einem aus menschlichem Untersuchungsgut gezüchteten Stamm kein Wachstum fest. Nach eigenen Beobachtungen kann nach starker Beimpfung des Nährbodens und Bebrütung der Kulturen bei 22⁰ C schwaches Wachstum mit grünbräunlicher Verfärbung des Nährmediums eintreten. Bei 37⁰ C wurde innerhalb 48 Std kein makroskopisch sichtbares Wachstum beobachtet.

Auf *Milchzuckersulfitagar* nach ENDO zeigt Past. pseudotuberculosis besonders bei 22⁰ C gutes Wachstum. Die glatten, anfangs hellen und durchsichtigen, später leicht trüben Kolonien erreichen innerhalb 48 Std die gleiche Größe wie auf Nähr- und Blutagarplatten. Die Farbe des Nährmediums bleibt unverändert. Dagegen sind bei 37⁰ C bebrütete Kulturen auf Endoagar deutlich in ihrem Wachstum gehemmt und neigen zur Rauhdissoziation. Die rauhen Kolonien lassen nach 48 Std eine stärkere Granulierung der Oberfläche und eine ausgeprägtere Randsaumbildung mit radiärer Streifung erkennen als auf Nähr- und Blutagar.

In nicht näher beschriebenen Versuchen von ROSENWALD u. DICKINSON (1944) konnte Past. pseudotuberculosis nach Einsaat in *Tetrathionatbrühe* (Difko) leicht auf MacConkey-Agar gezüchtet werden.

Eigene Versuche bestätigten das Überleben von Past. pseudotuberculosis in diesem Anreicherungsmedium, doch zeigte sich, daß eine 12stündige Bebrütung von Reinkulturen von Past. pseudotuberculosis in *Tetrathionatbouillon* nach KAUFFMANN (1936) nicht, aber in der *Selenitbrühe* nach LEIFSON (1936) von einer deutlichen Keimreduzierung gefolgt ist. Die Frage, ob die in der Salmonellabzw. Shigelladiagnostik bewährten Anreicherungsmethoden auch in der Diagnose menschlicher und tierischer Pseudotuberkulosen Eingang finden werden, muß,

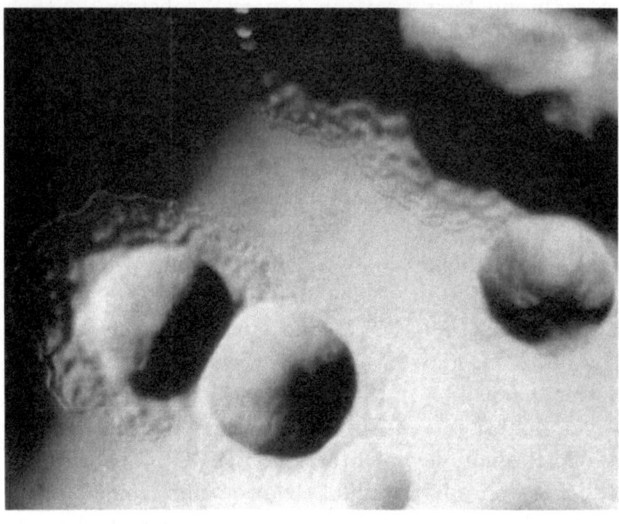

solange keine Erfahrungen an einem größeren Untersuchungsgut vorliegen, unbeantwortet bleiben. Über die Isolierung von Past. pseudotuberculosis aus menschlichem und tierischem Untersuchungsgut liegen keine Veröffentlichungen vor.

Dissoziationsvorgänge. Neben der Glatt- oder S-Form treten besonders leicht bei 37⁰ C-Kulturen Intermediär- und Rauhformen auf (SCHÜTZE 1928; ZLATO-GOROFF u. Mitarb. 1928; POKROWSKAJA 1930; Bo-

Abb. 3. Kolonien von Past. pseudotuberculosis in Intermediärform mit deutlicher Randsaumbildung auf Endoagar nach 48 Std Bebrütung bei 37⁰ C. Vergrößerung 20 ×

QUET 1937 u. a.). Die Rauh- oder R-Form läßt ein deutlich erhabenes, körniges oder geadertes Zentrum, das sich scharf gegen einen zarten, in seiner Begrenzung

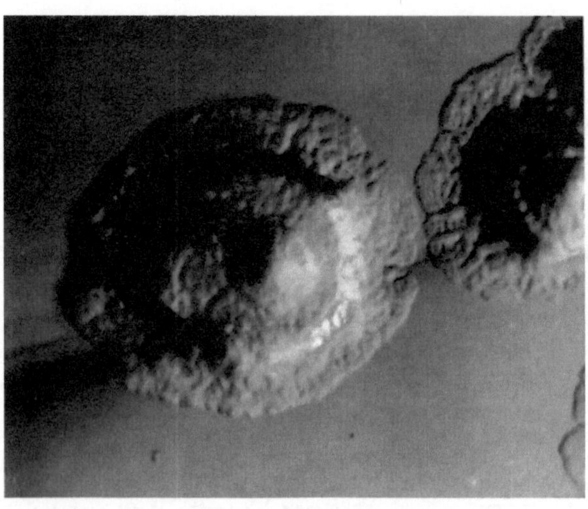

unregelmäßig gezackten, flachen Randsaum absetzt, erkennen. Sie ist undurchsichtig, meist trocken, und ihre rauhe Oberfläche matt spiegelnd oder stumpf. Bei den Intermediär- oder Übergangsformen (I-Formen) überwiegen entweder die charakteristischen Zeichen der S- und R-Form (POKROWSKAJA 1930), oder es entspricht ihre Morphologie weder der einen noch der anderen Form (BOQUET 1937).

BOQUET will auf Grund der großen Unbeständigkeit des morphologischen Aussehens der Glatt- und Rauhkolonien die

Abb. 4. Rauhe Kolonien von Past. pseudotuberculosis auf Endoagar nach 48 Std Bebrütung bei 37⁰ C. Vergrößerung 15 ×

Bezeichnungen glatt und rauh nach dem Verhalten der 37⁰ C-Kulturen in physiologischer Kochsalzlösung durch homogen (inagglutinable S-Form) und agglutinabel (spontan agglutinable R-Form) ersetzt wissen.

Gesetzmäßigkeiten in der Aufpaltung der Rauh- bzw. Intermediärformen ließen sich bisher nicht finden. Spontan agglutinierende R-Formen dissoziieren, solange sie noch einen Glattanteil besitzen, ebenso wie die Intermediärformen in R- und S-Formen. Die Rauhdissoziation wird durch ungünstige Bebrütungstemperatur und Nährbodenzusammensetzung sowie das Alter der Kultur gefördert. Rauhformen, die keinen Glattanteil mehr besitzen, verlieren die Fähigkeit, in Glattform zu wachsen (Lenskaia 1928, Boquet 1937 u. a.) (Abb. 3 und 4).

In eigenen Untersuchungen fiel auf, daß in Kulturen von Stämmen, die leicht in Rauhform dissoziieren, häufig neben normal großen Kolonien glatte und rauhe Mikrokolonien wuchsen, die auch bei mehrtägiger Bebrütung bei 22 und 37⁰ C nicht die normale Größe erreichten. Von Mikrokolonien angelegte Subkulturen zeigten neben normal großen Glatt- und Rauhkolonien wiederum Mikroformen; eine Gesetzmäßigkeit in der Aufspaltung ließ sich auch für diese Wuchsform nicht feststellen.

b) Biochemische Eigenschaften

Past. pseudotuberculosis ist wie Past. pestis im Gegensatz zu Past. multocida und tularensis ein biochemisch aktiver Keim. Seine fermentativen Leistungen gegenüber Zuckern und Alkoholen sind weitgehend konstant, so daß von wenigen Ausnahmen abgesehen die Untersuchungsbefunde in der älteren und neueren Literatur übereinstimmen. Nebenstehende Tabelle 1 gibt unter Berücksichtigung später noch zu besprechender differentialdiagnostischer Belange (siehe S. 231) eine Gegenüberstellung der biochemischen Eigenschaften von Past. pseudotuberculosis, Past. pestis und Past. multocida. Der Zusammenstellung liegen u. a. Arbeiten von Lerche (1927), Poppe (1928), Schütze (1928/29), Haupt (1928), Mørch u. Krogh-Lund (1931), Kauffmann (1933), Topping u. Mitarb. (1938), Rosenbusch u. Merchant (1939), Beaudette (1940), Moss u. Battle (1941), Schipper (1947), Meyer (1928, 1952), van Dorssen (1951), Marthedal u. Velling (1954), Pollitzer (1954), Wilson u. Miles (1955) und eigene Untersuchungsergebnisse an Stämmen von Past. pseudotuberculosis zugrunde.

Tabelle 1. *Biochemische Eigenschaften*

Nährmedium	Past. pseudotuberculosis	Past. pestis	Past. multocida
Adonit	+*	±	−
Arabinose	+	+*	−*
Galaktose	+	±	+
Glycerin	+	±	−*
Glucose	+	+	+
Lävulose	+	+	+*
Maltose	+	±	−*
Mannit	+	+*	+*
Melibiose	+	−	−
Rhamnose	+	−*	−
Salicin	+*	±	−*
Trehalose	+	∓	±
Xylose	+	±	±
Amygdalin	∓	−	−*
Dextrin	±	±	−
Raffinose	∓	−	−*
Sorbit	∓	−*	+
Dulcit	−*	−*	−*
Erythrit	−	−	−*
Inosit	−	−	−
Inulin	−*	−*	−
Lactose	−	−*	−
Saccharose	−*	−*	+*
Lackmusmolke	alkalisch neutral	neutral schwach sauer	neutral
Ammoniak	+	+	+
Harnstoff	+	−	±
Indol	−	−	+*
Nitrat	+	±	+
H_2S	±	−	+
Methylrot	+	±	−
Voges-Proskauer	−	−	−
Methylenblau	+	−	+
KCN	−	−	−

+ = positive Reaktion; ± = positive Reaktion häufiger; ∓ = negative Reaktion häufiger; * = sehr seltene Ausnahmen möglich.

Die Zusammenstellung zeigt, daß Past. pseudotuberculosis Arabinose, Galaktose, Glucose, Glycerin, Lävulose, Maltose, Mannit, Melibiose, Rhamnose, Trehalose und Xylose unter Säuerung, aber ohne Gasbildung spaltet, während Erythrit, Inosit und Lactose nicht angegriffen werden. Ein uneinheitliches fermentatives Verhalten der Stämme ist gegenüber Amygdalin, Dextrin, Raffinose und Sorbit zu beobachten. Das Fehlen der Säuerung von Adonit und Salicin wie die Spaltung von Dulcit, Inulin, Saccharose stellen seltene Ausnahmen dar (POPPE 1928; ZLATOGOROFF u. MOGHILEWSKAJA 1928; HAUPT 1928; GATÉ u. BILLA 1928; SCHÜTZE 1928; TRUCHE u. BAUCHE 1929; MORCH u. KROGH-LUND 1931; URBAIN u. NOUVEL 1937; VERGE u. Mitarb. 1937; TOPPING, WATTS u. LILLIE 1938; GIRARD 1940; BEAUDETTE 1940; MOSS u. BATTLE 1941; ROSENWALD u. DICKINSON 1944; HÄSSIG u. Mitarb. 1949; MEYER 1952; MARTHEDAL u. VELLING 1954, THAL 1954; WILSON u. MILES 1955 u. a.).

Eine Sonderstellung scheinen die von HÄSSIG u. Mitarb. (1949) beschriebenen Stämme (2/15 und 3/24) einzunehmen. Beide aus menschlichem Untersuchungsmaterial gezüchteten, unbeweglichen Stämme spalteten Saccharose, ohne Rhamnose und Xylose anzugreifen. Nach KAUFFMANN (1933), MORCH u. Mitarb. (1931) u. a. wird *Saccharose* von Past. pseudotuberculosis im Gegensatz zu Past. multocida *nie* gespalten. Entgegen stehen weitere Beobachtungen von URBAIN u. NOUVEL (1937), VERGE u. Mitarb. (1937), MOSS u. BATTLE (1941) u. WILSON u. MILES (1955). Nach MOSS u. BATTLE führte der von ihnen aus menschlichem Untersuchungsgut gezüchtete Stamm „New Orleans" in den ersten 2—3 Tagen der Bebrütung zu einer Säuerung des saccharoseenthaltenden Nährbodens, die von einer Realkalisierung innerhalb einer Woche gefolgt war.

Eine Spaltung von *Lactose* 4—5 Wochen nach der Beimpfung des Nährbodens konnten wir bei 4 Stämmen nachweisen. Die biochemische Untersuchung wurde, einer Empfehlung von SCHÜTZE (1929), der einzelne Stämme bei niederen Bebrütungstemperaturen biochemisch leistungsfähiger fand, folgend, bei 22° C und nicht bei 37° C durchgeführt. BOQUET (1937) sah keine Unterschiede in den biochemischen Eigenschaften seiner Stämme bei 37 und 22° C, mit der einen Ausnahme, daß Glycerin bei 18—20° C schwächer und langsamer als bei 37° C abgebaut wurde.

Erwähnenswert ist noch eine Beobachtung von GATÉ u. BILLA (1928). Die Autoren berichteten über einen Meerschweinchenstamm, der zuerst nur *Lävulose* und erst nach mehreren Passagen weitere Zucker und Alkohole mit Ausnahme von Salicin spaltete.

An weiteren, unterschiedlich starken biochemischen Leistungen — im Einzelfall auch einmal fehlend — sind für Past. pseudotuberculosis die rasche Nitrat- und Methylenblaureduktion, Ammoniakbildung, Harnstoffspaltung, positive Methylrot- und Katalasereaktion, sowie eine meist nur schwache H_2S-Bildung zu erwähnen (KAUFFMANN 1933; SNYDER u. VOGEL 1943; BLAXLAND 1947; ISSALY u. Mitarb. 1953; MARTHEDAL u. VELLING 1954 u. a.; s. Tabelle 1, S. 209).

Über ureasenegative Stämme, die sich auch in weiteren biochemischen Eigenschaften atypisch verhielten, berichteten HÄSSIG u. Mitarb. (1949) und FAUCONNIER (1950). THAL u. CHEN (1955) sahen dagegen in der Regelmäßigkeit der Harnstoffspaltung durch Past. pseudotuberculosis eine wertvolle Schnellmethode zur differentialdiagnostischen Abgrenzung der ureasenegativen Past. pestis (weitere Literatur bei POLLITZER 1954).

Auf *Simons-Natriumcitratagar* wuchs ein aus menschlichem Untersuchungsgut gezüchteter Stamm nach PIÉCHAUD (1952) nicht.

Auch wir sahen 14 Stämme von Past. pseudotuberculosis Typ I—V bei einer Bebrütungstemperatur von 37° C auf Natrium-Citratagar nicht wachsen, während bei 22° C 3 Stämme in 10 bis 21 Tagen unter Verwertung von Citrat als einziger Kohlenstoffquelle schwach zum Wachstum kamen.

Schließlich sollen nach ZLATOGOROFF u. Mitarb. (1928) Past. pseudotuberculosis in *Rauhform* Zucker rascher fermentieren als in Glattform; eine Beobachtung, die von BOQUET

(1937) nicht bestätigt wurde. Dagegen wurde nach seiner Beobachtung Peptonwasser mit Glycerinzusatz rascher durch Past. pseudotuberculosis in Rauhform als in Glattform gesäuert.

Sicher ist, daß im Einzelfall von der Norm abweichende Untersuchungsbefunde über die biochemische Leistungsfähigkeit von Past. pseudotuberculosis, wie sie z. B. in der fehlenden Säuerung von Mannit und Maltose bzw. Galaktose und Dextrin von Moss u. BATTLE (1941) und LESBOUYRIES (1934) angegeben wurden (s. Tabelle 1, S. 209), mit einer temporären Variabilität im Fermentapparat der einzelnen Stämme und mit der Verwendung nichteinheitlicher Nährböden oder Bebrütungstemperaturen erklärt werden müssen (GATÉ u. BILLA 1928; SCHÜTZE 1929; KAUFFMANN 1933; BEAUDETTE 1940; SNYDER u. VOGEL 1943; POLLITZER 1954; THAL 1954 u. a.).

V. Serologie
A. Typeneinteilung und serologische Eigenschaften von Past. pseudotuberculosis
1. Typeneinteilung

Die ersten erfolgreichen Versuche einer Antigenanalyse wurden von SCHÜTZE (1928, 1932) durchgeführt. Agglutinationsversuche führten zunächst zur Einteilung von Past. pseudotuberculosis in die drei in ihrem O-, aber nicht in ihrem H-Antigen verschiedenen Gruppen I—III. Durch Absorptionsversuche war eine weitere Unterteilung der Gruppen I und II in die Typen — später als Subtypen bezeichnet — I A, I B und II A, II B möglich. Nachdem M. RHODES (Lister Institut, London, zit. nach SCHÜTZE 1932) einen weiteren, in seinem O-, aber nicht in seinem H-Antigen von den ursprünglich aufgestellten Gruppen I—III verschiedenen Stamm fand, wurde Past. pseudotuberculosis in die 4 Gruppen I—IV eingeteilt. Schließlich konnte SCHÜTZE (1932) in Präcipitations- und Absättigungsversuchen ein allen Gruppen von Past. pseudotuberculosis und Past. pestis gemeinsames Körperantigen, das er als „rough somatic antigen" bezeichnete, nachweisen. Die Befunde von SCHÜTZE (1928, 1932) wurden von KAUFFMANN (1933), BHATNAGAR (1940) und THAL (1954, 1956) durch Agglutinations-, Präcipitations-, Absättigungs- und Immunisierungsversuche bestätigt und erweitert. Schließlich berichtete THAL (1954) über zwei in ihrem O-Antigen verschiedene Stämme, die sich unter 186 von ihm untersuchten Stämmen nicht in die bisher bekannten Gruppen I—III einordnen ließen. Die Aufstellung der weiteren Gruppen IV und V (THAL 1954), die von KNAPP (1955) und GIRARD und CHEVALIER (1955) bestätigt wurden[1], war die Folge.

Ein Vergleich der Stämme Typ IV und V von THAL (1954) und KNAPP (1955) mit dem einzigen Stamm Typ IV von RHODES (zit. nach SCHÜTZE 1932) war bisher nicht möglich, da dieser Stamm im In- und Ausland anscheinend nicht mehr zur Verfügung steht. Typ IV nach THAL unterscheidet sich in seinem O- und H-Antigen, Typ V dagegen nur in seinem O-Antigen von den Typen I—III nach SCHÜTZE. Ihre Identität mit dem von SCHÜTZE beschriebenen Stamm Typ IV, der sich nur in seinem O-, aber nicht in seinem H-Antigen von den Typen I—III unterschied, ist daher auszuschließen. Die von HÄSSIG u. Mitarb. (1949) beschriebenen kulturell-biochemisch atypischen, dem Typ IV nach SCHÜTZE zugehörigen Stämme stimmen

---

[1] Da wir die Einteilung der in ihren O-Antigenen verschiedenen Stämme in Typen und Subtypen, und nicht in Gruppen und Typen, definitionsgemäß für richtiger halten (KNAPP 1956), wird im Rahmen der weiteren Arbeit nur noch von den Typen I—V gesprochen.

mit den von THAL (1954) und KNAPP (1955/56) untersuchten Stämmen in ihrem O- und H-Antigen nicht überein. Ihre Identität mit Past. pseudotuberculosis Typ IV nach SCHÜTZE kann nicht mehr nachgeprüft werden, da, wie erwähnt, der Teststamm und auch das von K. F. MEYER zur Verfügung gestellte Typenserum (zit. nach HÄSSIG u. Mitarb. 1949) nicht mehr vorhanden sind.

Die bisher nachgewiesenen Typen lassen sich, ohne Berücksichtigung der von SCHÜTZE (1932) und HÄSSIG u. Mitarb. (1949) beschriebenen Stämme Typ IV, in Anlehnung und Erweiterung des von KAUFFMANN (1933) für Past. pseudotuberculosis mit Hilfe der Agglutinationsmethode aufgestellten Antigenschemas wie folgt zusammenstellen (KNAPP 1956).

Die geringe Zahl der vorliegenden und serologisch untersuchten Stämme läßt die Frage nach dem Vorkommen von Subtypen innerhalb der Typen III—V, deren Nachweis uns bisher nicht gelang, noch offen. Wahrscheinlich ist die Gesamtzahl der tatsächlich vorkommenden Typen und Subtypen noch nicht bekannt. Die größte Zahl der bisher bei Tieren und Menschen nachgewiesenen Stämme gehört dem Typ I an (THAL 1954, KNAPP 1955/56, GIRARD 1955, GOYON 1956 u. a.).

Tabelle 2. *Antigenstruktur und Typeneinteilung von Past. pseudotuberculosis*

Typ	Subtyp	Körperantigene	Geißelantigene
I	I A	1 2 3	a
	I B	1 2 4	a
II	II A	1 5 6	a
	II B	1 5 7	a
III	—	1 8	a
IV	—	1 9	b
V	—	1 10	a

1 = speciesspezifisches, mit Past. pestis gemeinsames Rauhantigen; 2, 5, 8, 9 und 10 = typenspezifische Körperantigene; 3, 4, 6, 7 = subtypenspezifsche Körperantigene.

Nur in wenigen Fällen wurden aus menschlichem Untersuchungsgut Stämme vom Typ III (JAKOBSTHAL, zit. nach SCHÜTZE 1928; MOSS u. BATTLE 1941) und Typ V (KNAPP 1956; s. Tabelle 2, S. 212) gezüchtet und Infektionen mit Past. pseudotuberculosis Typ II durch Antikörpernachweis in zwei mit Salmonella paratyphi B abgestättigten Patientensera wahrscheinlich gemacht (KNAPP 1956/57). Beim Tier scheint dagegen der Nachweis von Stämmen Typ II und III, im Gegensatz zu Typ IV und V, nicht so selten zu sein (THAL 1954).

2. Art der Antigene

Past. pseudotuberculosis besitzt wie andere Bakterienarten thermostabile O- und thermolabile H-Antigene. Das Vorkommen eines thermolabilen O-Antigens wurde von KNAPP (1956) auf Grund seiner Beobachtungen an abgesättigten Patientensera diskutiert, ohne daß bisher der experimentelle Beweis erbracht wurde. Neben fünf typenspezifischen O-Antigenen der Typen I—V, den stammspezifischen, in Absorptionsversuchen nachgewiesenen O-Antigenen der Subtypen I A, I B, II A und II B und dem allen Typen von Past. pseudotuberculosis mit Past. pestis gemeinsamen thermostabilen, von SCHÜTZE als Rauhantigen bezeichneten O-Antigen (SCHÜTZE 1928/29, 1932; BHATNAGAR 1940; THAL 1954; KNAPP 1955; GIRARD u. CHEVALIER 1955; CRUMPTON u. DAVIES 1957; DAVIES 1958; CRUMPTON, DAVIES u. HUTCHISON 1958 u. a.) wurde bei Typ II von Past. pseudotuberculosis ein mit dem O-Faktor IV der Salmonella B-Gruppe (SCHÜTZE 1932, KAUFFMANN 1933) bzw. bei Typ IV (nach THAL) ein mit dem O-Faktor IX der Salmonella D-Gruppe (KNAPP 1955; TOUCAS u. GIRARD 1956) gemeinsames O-Antigen nachgewiesen und damit die serologischen Beziehungen der Typen II und IV zur Salmonella-Gruppe aufgezeigt.

Nach Untersuchungen von UETAKE u. NAKANO (1949) sollen auch die von SAISAWA (1909) bei einem Soldaten und von KAWASHIMA (1934) bzw. IKEGAKI (1936) bei Affen isolierten nicht typisierten Stämme antigene Beziehungen zur Salmonella D-Gruppe besitzen. Da Angaben über die antigenen Eigenschaften des H-Antigens dieser Stämme und weitere serologische Untersuchungen fehlen, bleibt ihre Zugehörigkeit zu Typ IV nach THAL noch offen.

Die von PIROSKY (1938) auch für Past. multocida beschriebenen antigenen Beziehungen zur Salmonella D-Gruppe ließen keine Partialantigengemeinschaft zwischen Past. multocida und Past. pseudotuberculosis Typ IV auffinden.

Von den zwei bisher nachgewiesenen unterschiedlichen, thermolabilen H-Antigenen ist das von KAUFFMANN (1933) mit dem arabischen Buchstaben a bezeichnete H-Antigen der Typen I—III den Typen I—III und V nach THAL gemeinsam, während Typ IV nach THAL ein mit dem Buchstaben b bezeichnetes unterschiedliches H-Antigen besitzt. Untersuchungen über eine komplexe Natur der H-Antigene liegen nicht vor.

Angaben von STEPHAN (1942) über eine H-Antigengemeinschaft zwischen einzelnen Stämmen von Past. pseudotuberculosis und der Salmonella B-Gruppe blieben unbestätigt.

Untersuchungen über die antigenen Beziehungen zwischen Past. pseudotuberculosis und Past. pestis im Agardiffusionstest nach OUDIN (1952) und OUCHTERLONY (1958), die z. T. zu unterschiedlichen Ergebnissen führten, ermöglichten einen weiteren Einblick in die Antigenstruktur von Past. pseudotuberculosis (CHEN u. MEYER 1955; RANSOM 1956; CRUMPTON u. DAVIES 1956; BHAGAVAN, CHEN u. MEYER 1956; DAVIES 1958; CRUMPTON, DAVIES u. HUTCHISON 1958 u. a.). Sie zeigten, daß Past. pseudotuberculosis und Past. pestis nicht nur das eine von SCHÜTZE (1932) als Rauhantigen beschriebene Körperantigen gemeinsam haben.

BHAGAVAN u. Mitarb. (1956) fanden, daß Past. pseudotuberculosis Typ IV (Stamm 32) mit Past. pestis mindestens 5, und zwar 2 thermostabile und 3 thermolabile Antigene gemeinsam haben müsse, während RANSOM (1956) nur 2 Oberflächen- („envelope antigens") und ein Körperantigen („intracellular antigen") für Past. pseudotuberculosis und Past. pestis als identisch nachwies. Ein für Past. pseudotuberculosis spezifisches, nicht mit Past. pestis gemeinsames Körperantigen fand RANSOM (1956) in seinen nur mit je einem Stamm von Past. pestis (Stamm A 1122) bzw. Past. pseudotuberculosis (Stamm 1, Typ III) und einem Antipest-Serumglobulin durchgeführten Präcipitationsversuchen, im Gegensatz zu SASAKI (1957) und CRUMPTON, DAVIES u. HUTCHISON (1958), die bei Past. pseudotuberculosis typenspezifische, von Past. pestis verschiedene Polysaccharide nachweisen konnten, nicht. CRUMPTON u. DAVIES (1956/57) berichteten über sechs mit Past. pestis gemeinsame Antigene.

Unter ihnen wurde ein aus Protein bestehendes, von avirulenten glatten Peststämmen rasch und von virulenten Stämmen erst nach längerer Bebrütungszeit entwickeltes Antigen („antigen 4") nur bei virulenten, aber nicht bei glatten oder rauhen avirulenten Stämmen von Past. pseudotuberculosis gefunden. Die Autoren diskutierten die Frage, ob dieses Antigen ein „Virulenzantigen" von Past. pseudotuberculosis darstelle. Im Immunisierungsversuch konnte mit ihm kein Impfschutz gegen eine Infektion mit Past. pseudotuberculosis erzielt werden.

Weitere bisher nicht veröffentlichte Versuche im Oudin-Test von KNAPP, CHEN u. MEYER zeigten, daß nicht nur Typ IV (BHAGAVAN u. Mitarb. 1956), sondern alle Typen von Past. pseudotuberculosis mindestens fünf im einzelnen noch nicht näher bestimmte Antigene untereinander und mit Past. pestis gemeinsam haben.

Über die chemische Natur des Körperantigenkomplexes von Past. pseudotuberculosis ist noch wenig bekannt. Versuche, mit der Trichloressigsäure- (BOIVIN u. MESROBEANU 1933; PIROSKY 1938; GIRARD 1941; SCHAR u. THAL 1955) oder Diäthylenglykolmethode in Zellextrakten von Past. pseudotuberculosis und Past. pestis einen Protein-Polysaccharid-Phospholipoid-Komplex und damit den Endotoxincharakter des Körperantigenkomplexes aufzuzeigen, verliefen im Gegensatz zu Untersuchungen bei Past. multocida (PIROSKY 1938, GIRARD 1941) ergebnislos. Von DAVIES (1958) wurden in den mit dem Phenolverfahren nach WESTPHAL, LÜDERITZ u. BISTER (1952) aufgearbeiteten Zellextrakten glatter Kulturen typenspezifische Lipopolysaccharide bestimmt, deren Antigenität, sofern sie nicht mit Lipoproteinen kombiniert wurden, gering war. An Zuckern wurden im Lipopolysaccharidkomplex Glucose, Galaktose, Mannose, Fucose, Tyvelose, Abequose, Glucosamin, Galaktosamin und Heptose bestimmt.

Zu anderen Ergebnissen kam SASAKI (1957). In dem nicht typisierten Past. pseudotuberculosis-Stamm von IKEGAKI (1936) fand SASAKI mit der Trichloressigsäuremethode ebenfalls einen serologisch spezifischen Lipopolysaccharidkomplex, der an Stelle der von DAVIES nachgewiesenen Zucker Tyvelose, Abequose und Aldoheptose die Zucker Xylose, Arabinose und Rhamnose enthielt.

Nach DAVIES (1958) und CRUMPTON u. Mitarb. (1958) sollen sich die in den Lipopolysacchariden von Zellextrakten glatter Kulturen nachweisbaren Polysaccharide von denen rauher Kulturen unterscheiden. Da in den Zellextrakten glatter Kulturen nicht gleichzeitig die Lipopolysaccharide rauher Kulturen nachweisbar sind, nehmen die Autoren an, daß in den glatten Stämmen von Past. pseudotuberculosis das Glatt-O-Antigen nicht ein Rauh-O-Antigen, wie von SCHÜTZE (1928, 1932) auf Grund seiner serologischen Untersuchungsergebnisse zuerst angenommen wurde, überdeckt. Für diese noch nicht ausreichend bewiesene Annahme spricht die Beobachtung, daß Immunsera gegen glatte Stämme von Past. pseudotuberculosis die mit Lipopolysacchariden rauher Stämme sensibilisierten Erythrocyten nicht agglutinierten. Dagegen steht die Feststellung von CRUMPTON u. Mitarb. (1958), daß Immunsera, die mit Lipopolysacchariden glatter Kulturen von Past. pseudotuberculosis hergestellt wurden, die mit Lipopolysacchariden von Past. pestis sensibilisierten Erythrocyten agglutinierten.

Die Typenspezifität der aus glatten Kulturen isolierten Lipopolysaccharide wiesen DAVIES (1958) u. CRUMPTON und Mitarb. (1958) mit Hilfe der Präcipitation und Hämagglutination nach. Immunsera, die nach kombinierter Injektion der Lipopolysaccharide und eines Lipoproteins gewonnen wurden, präcipitierten und hämagglutinierten nur die homologen Lipopolysaccharide und Kulturen. Kreuzreaktionen mit Lipopolysacchariden rauher Kulturen von Past. pseudotuberculosis und von Past. pestis wurden nicht gesehen. Dagegen besitzen die aus Rauhkulturen verschiedener Typen von Past. pseudotuberculosis isolierten Lipopolysaccharide keine Typenspezifität. Sie sind nach CRUMPTON u. Mitarb. (1958) untereinander und mit den Lipopolysacchariden von Past. pestis, denen sie zwar sehr ähnlich sind, nicht identisch.

Über die chemische Natur der von LAZARUS u. GUNNISON (1947) und LAZARUS u. NOZAWA (1948) als Endotoxin beschriebenen toxischen Substanz von Past. pseudotuberculosis Stamm „New Orleans" (Typ III) und Stamm „Saranac" (Typ I) fehlen Angaben.

SCHAR u. THAL (1955) gewannen aus Agarkulturen von Past. pseudotuberculosis Typ III (Stamm 105) eine toxische Substanz („toxic extract"), die sie als Ektotoxin beschrieben. Über ihre chemische Zusammensetzung ist nichts bekannt. Zu ihrer Gewinnung wurden die in physiologischer Kochsalzlösung gewaschenen und in Phosphatpuffer suspendierten Keime zuerst mit 0,5% Toluol abgetötet und anschließend mit 1% Natriumbicarbonat behandelt. Der die toxische Substanz enthaltende Überstand wurde gegen destilliertes Wasser dialysiert.

3. Eigenschaften der Antigene

a) H-Antigene

Die H-Antigene sind an die Geißelsubstanz gebunden, deren optimale Entwicklung bei niederen Temperaturen (18—20° C) erfolgt (ARKWRIGHT 1927, SCHÜTZE 1928, 1932, KAUFFMANN 1933, WEITZENBERG 1935, KNAPP 1956 u. a.). während sie bei Bebrütung der Kulturen bei 37° C ausbleibt oder sehr schwach ist. Daß im Einzelfall bei 37° C bebrütete, unbewegliche Stämme H-Antigen besitzen und im Immunisierungsversuch zur Bildung von H-Antikörpern führen können, wiesen BOQUET (1937) und KNAPP (1956) nach.

Das *H-Antigen* von Past. pseudotuberculosis ist thermolabil. Durch Erhitzen auf 100 oder 120° C gehen seine agglutinablen, agglutininbildenden und agglutininbindenden Eigenschaften verloren. *Phenol* und *Formalin* haben in 0,3 bis 0,5%iger Konzentration keinen Einfluß auf seine antigenen Eigenschaften. Eine Abtötung und Konservierung beweglicher Stämme mit 0,1% Formalin kann dagegen von einer Beeinträchtigung der Agglutinabilität gefolgt sein (BHATNAGAR 1940, KNAPP 1956). Eine Vorbehandlung der Kulturen mit 50% Alkohol führte in unveröffentlichten Versuchen zu einem Verlust der H-Agglutinabilität, aber nicht des H-Agglutininbildungs- und Bindungsvermögens.

Der Typus der H-Agglutination ist nach SCHÜTZE (1928) „grob flockig", nach BOQUET (1937), BHATNAGAR (1940) u. a. „flockig". KNAPP (1956) sah in OH- bzw. H-Sera, die durch Absättigung von Kaninchenimmunsera mit O-Antigenen gewonnen waren, nur selten eine typisch flockige Agglutination von lebenden oder schonend mit Phenol oder Fomalin abgetöteten Kulturen.

In der Regel bildete sich in nur schwach verdünnten menschlichen und tierischen Immunsera ein kompakt schlieriges, der Kuppe des Gläschens anhaftendes und nur schwer aufschüttelbares Agglutinat, während mit zunehmender Serumverdünnung die Aggluninate locker schlierig und leichter aufschüttelbar wurden. Da die Agglutinate von lebenden und schonend abgetöteten Kulturen auch in O-Sera kompakt schlierig sein können, ist eine Deutung als O- oder H-Agglutination häufig mit Schwierigkeiten verbunden oder unmöglich, wenn nicht der Zeitpunkt ihres Auftretens berücksichtigt wird.

Die H-Agglutination lebender oder schonend abgetöteter Kulturen (Wasserbad 52° C) ist in Sera mit hohem Antikörpergehalt frühestens nach 1—2 Std, in Sera mit niederem Titer dagegen erst nach 2—4 Std ablesbar, während die O-Agglutination nach 2—4 bzw. 4—6 Std makroskopisch sichtbar wird.

Diagnostische Bedeutung besitzen nach THAL (1954) die beiden bisher bekannten H-Antigene a und b nicht. Über eine Zusammensetzung der einzelnen H-Antigene aus mehreren Partialantigenen ist nichts bekannt. Ihr Nachweis gelingt leicht in der Objektträger- und Röhrchenagglutination oder durch die Präcipitation (SCHÜTZE 1932 u. a.) in abgesättigten H- oder unabgesättigten OH-Sera, die durch Immunisierung von Kaninchen mit beweglichen Stämmen heterologer Typen gewonnen wurden.

b) O-Antigene

Im Gegensatz zum H-Antigen ist die Entwicklung der O-Antigene nicht an die Bebrütung der Kulturen bei niederen Temperaturen gebunden. Die Erfahrung zeigt aber, daß bei der Neigung von Past. pseudotuberculosis, bei 37° C in Intermediär- oder Rauhformen zu dissoziieren, eine optimale Entwicklung der „Glatt"-O-Antigene bei 22° C sicherer ist als bei 37° C, die die Entwicklung des „Rauh"-O-Antigen fördern. Nur Glattformen ohne Rauhkompenente sind in physiologischer und 3,5%iger Kochsalzlösung stabil und flocken bei Erhitzen auf 100 bzw. 120° C nicht aus.

Durch $2^{1}/_{2}$ bzw. 2stündiges Erhitzen auf 100 bzw. 120° C werden die thermostabilen O-Antigene im Gegensatz zu den thermolabilen H-Antigenen nicht zerstört. Vom H-Antigen befreit, lassen sich mit gekochtem Antigen reine O-Sera herstellen. Auch die Vorbehandlung der Kulturen mit 0,3% Formalin bzw. 0,3—0,5% Phenol oder 50% Alkohol hat keinen Einfluß auf die antigenen

Eigenschaften der O-Antigene. Nach BHATNAGAR (1940) soll die Abtötung lebender Kulturen mit 0,1% Formalin, 0,5% Phenol und 0,02% Silbernitrat von einer starken Abnahme und die Abtötung mit Alkohol oder Quecksilberchlorid (1:10000) von einem Verlust der Agglutinabilität gefolgt sein.

Der Typus der O-Agglutination 2 oder $2^1/_2$ Std auf 100 oder 120° C erhitzter Kulturen ist bei 37° C (Thermostat) und 52° C (Wasserbad) fein- oder grobkörnig. Er unterscheidet sich nicht von dem bei der Agglutination von Salmonella-Stämmen gewohnten Bild. Lebende oder schonend abgetötete Stämme zeigen dagegen in O-Sera häufig kompakt schlierige Agglutinate (s. S. 215).

c) R-Antigen

Mit dem S-R-Formenwechsel ist auch bei Past. pseudotuberculosis ein Wechsel in der Antigenstruktur verbunden. Den Rauhformen können die Glattantigene, die durch das Rauhantigen ersetzt werden, teilweise oder ganz fehlen (BOQUET 1937; THAL 1954; DAVIES 1958; CRUMPTON u. Mitarb. 1958). Die in Rauhform vorliegenden Kulturen flocken in physiologischer bzw. 3,5%iger Kochsalzlösung oder beim Kochen und Autoklavieren aus.

Nach SCHÜTZE (1932) und BHATNAGAR (1940) ist das thermostabile R-Antigen artspezifisch und allen Stämmen von Past. pseudotuberculosis gemeinsam. Es ist mit der Präcipitations- und Komplementbindungsreaktion und unter bestimmten Versuchsbedingungen auch mit der Agglutinationsmethode nachweisbar (SCHÜTZE 1932, BHATNAGAR 1940, THAL 1954 u. a.). THAL (1954) nimmt an, daß das R-Antigen an das O-Antigen gebunden ist und unabhängig davon, ob letzteres stark oder schwach entwickelt ist, nur nach Lösung aus dieser Bindung, z. B. durch Autolyse, als Antigen wirksam wird. In Kulturfiltraten scheint die Menge des gelösten R-Antigens mit dem Alter der Kultur zuzunehmen. Durch $2^1/_2$ Std Kochen soll das R-Antigen nicht autolysierter Kulturen vom hitzestabilen O-Antigen so eingeschlossen werden, daß es weder in der Agglutinationsprobe noch als Agglutinogen wirksam ist.

Nach THAL (1954) präcipitieren mit frisch präparierten nicht autolysierten OH- und O-Antigenen hergestellte Sera im Überschichtungsverfahren nur das typenspezifische O-Antigen, während mit Filtraten alter Kulturen hergestellte Sera sowohl das O- wie auch das R-Antigen erfassen. Bei Stämmen, die nach ihren kulturell-biochemischen Eigenschaften als Past. pseudotuberculosis, nach ihrem serologischen Verhalten aber nicht den 5 Typen zugehörig befunden werden, kann nach THAL (1954, 1956) ihre Artzugehörigkeit durch Bestimmung des R-Antigens geklärt werden. Unser Versuch, die Specieszugehörigkeit der von HÄSSIG u. Mitarb. beschriebenen Stämme auf diesem Weg nachzuweisen, mißlang.

Die aus differentialdiagnostischen Gründen wichtige Agglutination von Past. pseudotuberculosis im Antipestserum wird als R-Antigen-Antikörperreaktion gedeutet, während die Agglutination von Past. pestis im homologen Serum eine Hüllantigen-Antikörperreaktion darstellt (SCHÜTZE 1932, BHATNAGER 1940 u. a.). Die Inagglutinabilität von Past. pestis im Immunserum gegen Past. pseudotuberculosis wird auf das Fehlen von Antikörpern gegen das Hüllantigen zurückgeführt.

4. Toxine

DESSY (1925) und BOQUET (1937) konnten im Gegensatz zu RAMON (1928) keine für nicht sensibilisierte Meerschweinchen toxische Substanz in Kulturen von Past. pseudotuberculosis nachweisen. RAMON führte mit unvorbehandelten

und formalinisierten Kulturfiltraten Immunisierungs- und Neutralisations-
versuche an Pferden durch. Die bei intravenöser Applikation toxische Wirkung
der Kulturfiltrate ließ sich durch Formol abschwächen und durch antitoxisches
Immunserum neutralisieren. LAZARUS u. GUNNISON (1947) und LAZARUS u.
NOZAWA (1948) berichteten über toxische Eigenschaften von 2 unter 27 von ihnen
untersuchten Stämmen von Past. pseudotuberculosis (Stamm „New Orleans",
Typ III, und Stamm „Saranac", Typ I). Sie fanden in Filtraten von Bouillonkulturen, die der lytischen Wirkung eines an Past. pseudotuberculosis adaptierten
Pestphagenstammes ausgesetzt waren, ein als Endotoxin beschriebenes Toxin,
das, ohne charakteristische Organveränderungen auszulösen, bei Kaninchen, Ratten
und weißen Mäusen unter Kreislaufstörungen innerhalb 12 Std zum Tode oder
im Hauttest zu zentraler Nekrose führte. GIRARD (1950) konnte nach einer Untersuchung von 57 Stämmen die Ergebnisse der obengenannten Autoren nur für
diese beiden Stämme von Past. pseudotuberculosis bestätigen. Auch er sah in
der toxischen Substanz, die angeblich nur bei Bakterienzerfall frei wurde, ein
Endotoxin, obwohl ihre Proteinnatur und Thermolabilität, das Fehlen des Gluco-
lipoid-Antigenkomplexes sowie die Möglichkeit ihrer Umwandlung in ein Formol-
toxoid, dem zwar immunogene Eigenschaften fehlen sollen, Eigenschaften eines
Ektotoxins darstellen.

Den Nachweis, daß einzelne Stämme von Past. pseudotuberculosis zwei
Toxine, ein Ektotoxin und ein bis heute noch nicht genügend definiertes Endo-
toxin besitzen, glauben THAL (1954) und SCHAR u. THAL (1955) geführt zu haben.
Ihre Befunde berechtigen zu der Frage, ob es sich nicht auch bei den von RAMON
(1928), LAZARUS u. NOZAWA (1948) und GIRARD (1950) nachgewiesenen toxischen
Substanzen um ein Ektotoxin gehandelt hat.

Unter 186 Stämmen wies THAL nur in der 48stündigen Bouillonkultur der dem serologischen Typ III zugehörigen Stämme ein Toxin nach, das den übrigen Stämmen der serologischen Typen I, II, IV und V auch nach einer Autolyse von 6 Monaten fehlte Das Toxin der
Stämme des serologischen Typ III war wie das Endotoxin von Past. pestis (Lit. bei POLLITZER
1954) thermolabil und mit Formalin leicht in ein immunogen wirkendes Toxoid überzu-
führen. Für Kaninchen, Ratten und Mäuse hoch- und für Meerschweinchen geringer toxisch,
wurde die im Hauttest zur Nekrose führende Toxinwirkung, im Gegensatz zum Toxin von
Past. pestis, das nur bei Ratten und weißen Mäusen toxisch und im Hauttest nekrotisierend
wirkt, aber durch Antitoxin nicht beeinflußt wird, durch antitoxisches Immunserum in vitro
und in vivo nach dem Gesetz der multiplen Proportion neutralisiert. Eine kreuzweise aktive
oder passive Immunität wurde im kreuzweisen Immunisierungsversuch mit beiden Toxinen
bzw. ihren antitoxischen Sera nicht erreicht. Somit scheinen bestimmte serologisch definierte
Stämme von Past. pseudotuberculosis ein Toxin zu bilden, das Eigenschaften eines Ektotoxins (WILSON u. MILES 1955; H. SCHMIDT 1955) besitzt. Die Frage, ob nur Stämme des
serologischen Typ III ein Ektotoxin bilden, ist noch nicht zu beantworten. Weitere Untersuchungen sind an einem größeren Stamm-Material nötig, in die auch die von LAZARUS und
NOZAWA (1948) und GIRARD (1950) als Endotoxinbildner beschriebenen Stämme einzubeziehen
sind. Schließlich ist zu erwähnen, daß schon BHATNAGAR (1940) die besondere Toxicität eines
Stammes von Past. pseudotuberculosis Typ III angab und MOSS u. BATTLE (1941) auf die
hohe Meerschweinchenpathogenität ihres aus menschlichem Untersuchungsmaterial gezüchteten New Orleans-Stammes hinwiesen.

Eine toxische Wirkung (LD_{50}) der Lipopolysaccharide glatter Kulturen von
Past. pseudotuberculosis wurde von DAVIES (1958) an Mäusen in der Konzentration von 500 γ bestimmt. Bei erhitzten Lipopolysaccharidlösungen lag die
LD_{50} bei 200—300 γ.

B. Serologische Untersuchungsmethoden
(Antigen-Antikörperanalyse)

1. Agglutinationsreaktion

Für den Antigen-Antikörpernachweis hielt POPPE noch 1928 die Agglutinationsreaktion wegen der unterschiedlichen Antigenstruktur einzelner Stämme von Past. pseudotuberculosis und methodischer Schwierigkeiten, die hauptsächlich in ihrer Neigung zur Spontanagglutination gesehen wurden, für nicht geeignet. In den folgenden Jahren konnte aber durch die Untersuchungen von SCHÜTZE (1928), TRUCHE u. BAUCHE (1929), KAUFFMANN (1933), BRIGHAM (1935), TUMANSKY (1939), BHATNAGAR (1940) u. a. gezeigt werden, daß bei der Verwendung geeigneter Antigene mit der Agglutinationsmethode ein sicherer Antigen-Antikörpernachweis möglich ist.

Zur Gewinnung stabiler agglutinierender Suspensionen wurden die verschiedensten Verfahren angegeben, von denen nur die Reduzierung des Kochsalzgehaltes der Antigensuspensionen auf 0,2—0,35% (ARKWRIGHT 1927, SCHÜTZE 1928, BOQUET 1937; TOPPING u. Mitarb. 1938), das kurze Schütteln der Suspension in der Hand mit anschließendem Abpipettieren des Überstandes nach 2stündigem Stehen bei Zimmertemperatur (BHATNAGAR 1940), das starke Schütteln der Suspension vor Erhitzung (PRESTON 1952), die mehrfachen Passagen über Schwärmplatten nach SVEN GARD (THAL 1954 u. a.) und die ausschließliche Verwendung bei 22° C gezüchteter lebender oder bei 60° C bzw. z. B. mit 0,25—0,5% Phenol, 0,1—0,25% Formalin schonend abgetöteter Kulturen (BHATNAGAR 1940, KNAPP 1956 u. a.) erwähnt werden.

Antigennachweis. Die Antigenanalyse von Past. pseudotuberculosis kann, sofern die zu untersuchenden Stämme, in Glattform vorliegend, nicht spontan agglutinieren, mit Hilfe der Objektträger- und Röhrchenagglutination durchgeführt werden. Die Technik des O- und H-Antigennachweises entspricht den in der Salmonella-Diagnostik geübten und als bekannt vorauszusetzenden Verfahren.

Als erster hat SCHÜTZE (1928) mit Erfolg die Agglutinationsprobe zur Antigenanalyse von Past. pseudotuberculosis angewandt. Weitere Untersuchungen von SCHÜTZE (1929, 1932), KAUFFMANN (1933), BHATNAGAR (1940), THAL (1954), KNAPP (1955/56), GIRARD (1955) u. a. erbrachten den agglutinatorischen Nachweis der fünf in ihrem Körperantigen verschiedenen Typen (I—V), unter denen allein Typ IV (nach THAL) ein von den anderen Typen verschiedenes H-Antigen besitzt (s. Tabelle 2, S. 212). Kreuzweise Absättigungs- und Agglutinationsversuche zeigten artspezifische antigene Beziehungen zu Past. pestis und typenspezifische Beziehungen zur Salmonella-Gruppe. Sie sind bei Past. pseudotuberculosis Typ II zum O-Faktor IV der Salmonella B-Gruppe (SCHÜTZE 1928, 1932, KAUFFMANN 1933) und von Past. pseudotuberculosis Typ IV zum O-Faktor IX der Salmonella D-Gruppe (KNAPP 1956; TOUCAS, GIRARD u. MINOR 1956) gegeben. Durch diese Befunde wurde die komplexe Natur IV_1 und IV_2 bzw. IX_1 und IX_2 der O-Faktoren IV und IX der Salmonella B- und D-Gruppe aufgezeigt (KAUFFMANN 1933; UETAKE u. NAKANO 1949; KNAPP 1956).

Nach THAL (1954) soll auch der Nachweis des artspezifischen Rauhantigens von Past. pseudotuberculosis mit der Agglutinationsmethode möglich sein. Seine Versuche wurden mit einem O-agglutininfreien Immunserum (R-Serum), das durch Immunisierung von Kaninchen mit einem Rauhstamm (Stamm 74, Typ I) gewonnen wurde, und mit durch 0,3% Formalin abgetöteten Stämmen Typ I—V durchgeführt.

CRUMPTON u. Mitarb. (1958) konnten mit Hilfe der Hämagglutination einerseits die Typenspezifität der von ihnen mit der Phenolmethode nach WESTPHAL u. Mitarb. (1952) aus glatten Kulturen der Typen I—V dargestellten Lipopolysaccharide und zum anderen den Verlust der Typenspezifität der Lipopolysaccharide rauher Kulturen der verschiedenen Typen nachweisen (s. S. 214).

Antikörpernachweis. Unseres Wissens hat LEDOUX-LEBARD (1897) den erstmaligen Nachweis von Agglutininen in tierischen Sera erbracht. Seine Beobachtungen wurden später trotz anfänglicher Bedenken gegen die Zuverlässigkeit der Agglutinationsmethode (POPPE 1928 u. a.) in zahlreichen Arbeiten bestätigt. Agglutinine in menschlichen Sera fanden erstmals ALBRECHT (1910), später TOPPING u. Mitarb. (1938), DUJARDIN u. BEAUMETZ (1938), SNYDER u. VOGEL (1943). Der seltene Agglutininnachweis in Patientensera ist darauf zurückzuführen, daß bei der größeren Zahl der bis 1953 diagnostizierten Fälle menschlicher Pseudotuberkulosen die Diagnose nicht zu Lebzeiten der Patienten gestellt wurde (s. Tabelle 4, S. 240). Zudem ist es sehr wahrscheinlich, daß in wiederholten Fällen bei dem uncharakteristischen Krankheitsbild nicht an die differentialdiagnostische Möglichkeit einer Pseudotuberkulose gedacht und die Untersuchung des Blutes auf Antikörper gegen Past. pseudotuberculosis versäumt wurde. Erst seit der ätiologischen Klärung der von MASSHOFF (1953) als abscedierende reticulocytäre Lymphadenitis beschriebenen besonderen Form der mesenterialen Lymphadenitis, als eine durch Past. pseudotuberculosis verursachte Infektionskrankheit (KNAPP 1954; KNAPP u. MASSHOFF 1954), wurde der Nachweis von Agglutininen in Patientensera häufiger erbracht.

Agglutinierende Antikörper gegen O- und H-Antigene der verschiedenen Typen von Past. pseudotuberculosis wurden in tierischen und menschlichen Immunsera nachgewiesen, sofern bei der mit den üblichen Methoden durchgeführten Immunisierung lebende oder in schonender Weise — ohne Verlust der agglutininbildenden Eigenschaften der Antigene — abgetötete Vollantigene verwendet wurden (THAL 1954, KNAPP 1956 u. a.).

Im Gegensatz zu den Beobachtungen an Immunsera fand KNAPP (1956) in den meisten Sera von Patienten, die an der appendicitischen Verlaufsform der menschlichen Pseudotuberkulose erkrankt, wie von Kaninchen und Affen (unveröffentlichte Versuche), die mit nicht tödlichen Dosen von Past. pseudotuberculosis (Typ I) infiziert waren, keine oder nur in sehr niederen Titern H-Agglutinine. Weiterhin wurden gekochte Antigene, obwohl sie zu einer vollständigen Absättigung dieser Sera führten, nicht oder nur schwach agglutiniert.

Diese serologischen Besonderheiten ließen KNAPP (1956) das Vorhandensein eines experimentell nicht bewiesenen, in seiner Agglutininbildung und Agglutinierbarkeit thermolabilen und in seiner Agglutininbindung thermostabilen O-Antigens zur Diskussion stellen, das im infizierten Organismus unter noch unbekannten Bedingungen die Bildung von Agglutininen gegen das thermostabile O-Antigen vollständig oder zumindest weitgehend hemmen kann. Die fehlende oder sehr schwache H-Antikörperbildung bei der natürlichen Infektion des Menschen bzw. im Infektionsversuch am Kaninchen oder Affen wurde von KNAPP (1956) mit der die Geißelentwicklung stark hemmenden Wirkung der Körpertemperatur zu erklären versucht.

Noch unbeantwortet ist die Frage, ob nicht unabhängig von der Wirkung eines weiteren Körperantigens (Kapselantigen?) Länge und Schwere des Infektionsgeschehens einen Einfluß auf die Bildung der in den Sera von Patienten bzw. experimentell infizierten Tieren nicht oder nur sehr schwach nachweisbaren O- und H-Agglutinine haben, und ob Sera von Schwerkranken nicht dieselben Eigenschaften, wie die durch künstliche Immunisierung mit abgetöteten Antigenen gewonnenen Sera, ausweisen können (KNAPP 1956). Für diese Annahme spricht der Nachweis von O- und H-Agglutininen im Serum eines Patienten mit der septisch-typhösen Verlaufsform der menschlichen Pseudotuberkulose (s. S. 239) bzw. in Sera von Kaninchen und Affen, denen *wiederholt* nicht tödliche steigende Antigenmengen injiziert wurden (KNAPP, CHEN u. MEYER, unveröffentlicht).

Die *Spezifität* der mit den Stämmen Typ I, III und V von Past. pseudotuberculosis beobachteten positiven Agglutinationsreaktionen wurde von KNAPP

(1954, 1956) beschrieben. Dagegen ist bei einer Agglutination der Stämme Typ II und IV in menschlichen und tierischen Sera ihre antigene Beziehung zur Salmonella B- bzw D-Gruppe zu berücksichtigen. Eine Agglutination von Stämmen der Typen II und IV beweist noch nicht das Vorliegen einer Infektion mit Past. pseudotuberculosis. Ihr sicherer serologischer Nachweis setzt die Durchführung eines Absättigungsversuches voraus.

Braun u. Müller (1957) stellten bei 18 von 101 Kindern mit unklaren Magen-Darmbeschwerden unspezifische Agglutinintiter bis 1:40 und in 2 Fällen bis 1:80 der Serumverdünnungen fest. Eine Erklärung für diese unspezifischen Serumreaktionen, die mit einer phenolisierten Schrägagarkulturabschwemmung des Stammes 2^I beobachtet wurden, fehlt. Die Tatsache, daß die Titer nach wenigen Tagen nicht mehr nachweisbar waren, läßt an methodisch bedingte Ursachen denken. Kuhlmann u. Herrmann (1955) sahen bei Patienten mit unklaren Darmerkrankungen Agglutinintiter von 1:40 mit Past. pseudotuberculosis als verdächtig und von 1:80 für eine Pseudotuberkulose als beweisend an. Hecker (1957) berichtete von Titern zwischen 1:20—1:80 bei 7 Patienten, deren mesenteriale Lymphknoten das histologische Bild einer durch Past. pseudotuberculosis verursachten abscedierenden, reticulocytären Lymphadenitis boten. Von Knapp (1956) wurden keine Grenztiter, die das Vorliegen einer Infektion beweisen sollen, festgelegt. Erst das Vorliegen eines großen Untersuchungsgutes wird eine Stellungnahme zu der Frage, ob und welche diagnostische Bedeutung dem Nachweis niederer Agglutinintiter zukommt, ermöglichen.

Methodisch ist bei jeder Widal-Reaktion zu berücksichtigen, daß die Agglutinate der Lebendantigene nicht das beim Salmonella-Widal gewohnte Bild zeigen (s. S. 215) und in physiologischer oder 3,5%iger Kochsalzlösung suspensionsstabile Widal-Stämme in schwach verdünnten Sera spontan agglutinieren können, während bei zunehmender Serumverdünnung die Spontanagglutination ausbleibt. Diese Beobachtungen zeigen, daß jeder Widal mit einer Kochsalz- und Serumkontrolle durchzuführen ist, wobei die Konzentration des Kontrollserums der ersten Verdünnung des Patientenserums entsprechen soll. Nach eigener Erfahrung konnte der Nachweis eines als unspezifisch gedeuteten niederen Agglutinintiters immer auf methodischbedingte Ursachen oder auf unbekannte, die spontane Agglutination verursachende Serumeigenschaften, die einer zweiten, kurze Zeit später entnommenen Serumprobe fehlten, zurückgeführt werden. Eine Wiederholung der Widal-Reaktion mit frischen Antigenen, neuer Kochsalzlösung und, soweit möglich, einer weiteren Serumprobe, ist somit beim Nachweis niederer Titer unerläßlich.

Saisawa (1913), Roman (1916) und später Truche u. Bauche (1929) haben im Gegensatz zu Lerche (1927) eine stärkere Agglutination der homologen Stämme in tierischen Immunsera beschrieben, während Knapp (1956) u. Flamm (1958) dieselben Beobachtungen beim Agglutininnachweis in Patientensera machten. Diese Beobachtungen stellen eine weitere Erklärungsmöglichkeit dar, warum in einzelnen Fällen bei gesicherter Pseudotuberkulose die Widal-Reaktion negativ ausfallen kann oder nur ein sehr niederer Titer bestimmt wird, solange der homologe Stamm nicht zur Verfügung steht.

Bei der unterschiedlich starken Agglutinabilität verschiedener Stämme desselben Typs ist zur sicheren Erfassung der Agglutinine ein Ansatz der Widal-Reaktion mit mehreren Stämmen desselben Typs, die möglichst bei 22° C bebrütet und als Lebendantigene verwendet werden sollen, ratsam (Bhatnagar 1940, Knapp 1956).

Von wenigen Ausnahmen abgesehen, fand Knapp (1956) schon beim Auftreten der ersten Krankheitssymptome der appendicitischen Verlaufsform der menschlichen Pseudotuberkulose (s. S. 243) einen Agglutinintiter gegen Past. pseudotuberculosis Typ I zwischen 1:80 bis 10260 — in der Regel zwischen 1:160 bis 1:640 — der Serumverdünnung, der bei komplikationslosem Verlauf innerhalb 1—3 Monaten wieder negativ wurde.

Nach den bisherigen Beobachtungen spricht der in den meisten Fällen beobachtete rasche Rückgang der klinischen Symptome und des Agglutinintiters für eine Ausheilung der Pseudotuberkulose, während ein Titeranstieg oder fehlender Rückgang auch bei klinischer Symptomenfreiheit als Folge eines latenten Fortbestehens der Infektion gedeutet wird. Über den frühesten Zeitpunkt des Agglutininnachweises lassen sich keine Angaben machen, solange sichere Kenntnisse über Infektionsmodus und Inkubationszeit der menschlichen Pseudotuberkulose fehlen. Ein negativer Ausfall der Agglutinationsprobe in Patientensera muß auf fehlende oder späte Antikörperbildung zurückgeführt werden, sofern nicht in Einzelfällen blockierende Antikörper oder die Verwendung in Patientensera inagglutinabler heterologer Stämme eine ursächliche Rolle spielen (KNAPP 1956).

2. Komplementbindungsreaktion

Als Antigen dienten STEPHAN (1942) auf 60° C erhitzte, mit 50% Alkohol und 0,2% Lecithin versetzte Schrägagarkulturabschwemmungen. THAL (1954) arbeitete mit in Phenolwasser (0,5%) ausgewaschenen und suspendierten Sedimenten von Bouillonkulturen, während KNAPP u. STEUER (1956) Suspensionen lebender, gekochter, beschallter oder mit 0,15—0,3% Formalin bzw. Phenol abgetöteter Schrägagarkulturen, die 48 Std bei 22° C bebrütet waren, zur Komplementbindung und Herstellung von Immunsera verwendeten. Unter den frisch hergestellten Antigenen erreichten die gekochten und beschallten Suspensionen sofort und die übrigen Antigene erst nach einer Lagerung von 2—3 Wochen die endgültige Höhe des Antigentiters.

Antigenanalyse. Die Typenspezifität der O-Antigene von Past. pseudotuberculosis Typ I—V wurde mit der Komplementbindungsreaktion zum ersten Mal von THAL (1954) nachgewiesen und damit erneut die schon durch das Agglutinations- und Präcipitationsverfahren bewiesene Heterogenität des O-Antigens, die zur Aufstellung der fünf verschiedenen Typen von Past. pseudotuberculosis geführt hatte, bestätigt. Vorausgegangene Versuche anderer Autoren, eine Antigenanalyse mit Hilfe der Komplementbindungsreaktion durchzuführen, waren ohne befriedigendes Ergebnis geblieben (SAISAWA 1913; ROMAN 1916; LERCHE 1927; ZLATOGOROFF u. MOGHILEWSKAJA 1928; BOQUET u. DUJARDIN-BEAUMETZ 1929 u. a.).

Antikörpernachweis. Der Nachweis komplementbindender Antikörper *in Sera kranker und immunisierter Tiere* (LERCHE 1927; POPPE 1928; ZLATOGOROFF u. Mitarb. 1928; BOQUET u. DUJARDIN-BEAUMETZ 1929; STEPHAN 1942; GORET u. Mitarb. 1955; KNAPP u. STEUER 1956) war bisher für die Veterinärmedizin ohne größere diagnostische Bedeutung geblieben.

Systematische Untersuchungen über das Vorkommen komplementbindender Antikörper *in menschlichen Sera* wurden zum ersten Male von KNAPP u. STEUER (1956) durchgeführt. Sie zeigten, daß in der Serodiagnostik der menschlichen Pseudotuberkulose die Agglutinationsprobe der Komplementbindungsreaktion überlegen ist. Komplementbindende Antikörper wurden in allen menschlichen und tierischen, durch aktive Immunisierung gewonnenen Sera, wohl als Folge des stärkeren antigenen Reizes, aber nicht in allen Patientensera mit positivem Agglutinintiter nachgewiesen. Sofern der Nachweis komplementbindender Antikörper in Patientensera gelang, erreichte ihr Titer nie die Höhe des Agglutinintiters.

Bei der Auswertung der Komplementbindungsreaktionen sind wie bei der Widal-Reaktion die antigenen Beziehungen zur Salmonellagruppe und zu Past. pestis zu berücksichtigen. Eine schwach-positive Komplementbindungsreaktion zwischen Pestimmunserum und Past. pseudotuberculosis als Antigen beschrieb HAAS (1938) während sie zwischen Past. pestis als Antigen und Immunsera gegen Past. pseudotuberculosis fehlen soll (DAMPEROFF 1910; BOQUET u. Mitarb. 1929; weitere Literatur s. bei POLLITZER 1954).

3. Präcipitationsreaktion

Als Antigene zur Durchführung der Präcipitation und Gewinnung hochwertiger präcipitierender Sera eignen sich Filtrate autolysierter Kulturen, der Überstand oder die Filtrate
$1^1/_2$ Std auf 60° C erhitzter, im Wechsel der Kälte- und Wärmeeinwirkung ausgesetzter, über
Stunden oder Tage im Schüttelapparet bei 22° C mit und ohne Glasperlenzusatz geschüttelter
oder bis zur weitgehenden Zerstörung aller Zellen beschallter Kulturen (POPPE 1928; SCHÜTZE
1932; BOQUET 1937; THAL 1954; KNAPP u. STEUER 1956; KNAPP, CHEN u. MEYER, unveröffentlicht), sofern nicht gekochte oder autoklavierte Antigene zum ausschließlichen
Nachweis von Antikörpern gegen thermostabile präcipitierende Antigene verwendet werden.

Antigenanalyse. Im Überschichtungsverfahren wies SCHÜTZE (1932) bei
Past. pseudotuberculosis neben dem artspezifischen mit Past. pestis gemeinsamen
R-Antigen typenspezifische O-Antigene und ein den von ihm beschriebenen Typen
gemeinsames H-Antigen nach. THAL (1954) bestätigte mit demselben Verfahren
den Nachweis der art- und typenspezifischen Antigene von Past. pseudotuberculosis und die antigenen Beziehungen von Past. pseudotuberculosis Typ II zur
Salmonella B-Bruppe. Nach seinen Beobachtungen ist das artspezifische, mit
Past. pestis gemeinsame R-Antigen von Past. pseudotuberculosis an das O-Antigen, gleichgültig ob dieses stark oder schwach ausgebildet ist, gebunden und in
dieser Form ohne präcipitogene und agglutinogene Wirkung. Zur Gewinnung
hochwertiger präcipitierender R-Sera ist daher die Verwendung von Impfstoffen
mit freiem R-Antigen Voraussetzung.

Die immunogene Inaktivität des gebundenen R-Antigens erklärt THAL (1954) mit seiner
Beobachtung, daß mit frischen, nicht autolysierten Kulturen hergestellte Immunsera nur das
typenspezifische O-Antigen und nicht das artspezifische R-Antigen präcipitieren, während
z. B. mit Filtraten autolysierter Bouillon-Kulturen hergestellte Immunsera sowohl das
R- wie das O-Antigen erfassen. Die Herstellung eines O-Antikörper-freien Immunserums gelang THAL nur mit serologisch reinen R-Stämmen. Versuche, mit Rauhstämmen ein R-Serum
herzustellen, mißlangen, sofern das R-Antigen noch an ein zwar unterentwickeltes O-Antigen
gebunden war.

Präcipitationsversuche im Agardiffusionstest über die antigenen Beziehungen zwischen
Past. pseudotuberculosis und Past. pestis führten zwar zu unterschiedlichen Ergebnissen,
sie zeigten aber einheitlich, daß Past. pseudotuberculosis und Past. pestis nicht nur das eine
von SCHÜTZE (1932) als Rauhantigen beschriebene Antigen gemeinsam haben (CHEN u.
MEYER 1955; BHAGAVAN, CHEN u. MEYER 1956; RANSOM 1956; DAVIES 1958; CRUMPTON u.
Mitarb. 1958 u. a.). KNAPP, CHEN u. MEYER fanden in bisher unveröffentlichten Präcipitationsversuchen, die kreuzweise mit Stämmen und Sera der 5 Typen von Past. pseudotuberculosis sowie Past. pestis (Stamm A 1122) im Oudintest durchgeführt wurden, daß alle
5 Typen von Past. pseudotuberculosis mindestens 5 Antigene untereinander und mit Past.
pestis gemeinsam haben (s. S. 213).

Antikörpernachweis. Für den Antikörpernachweis in Sera kranker Tiere und
Menschen besitzt die Präcipitationsreaktion im Überschichtungs- und Agar-
diffusionsverfahren keine Bedeutung.

Im Oudin-Test konnten KNAPP, CHEN u. MEYER nur in 3 unter 20 Serumproben von
Patienten, die an der appendicitischen Form der Pseudotuberkulose erkrankt waren, mit dem
Überstand einer 8 Tage bei 22° C bebrüteten und geschüttelten Kultur als Antigen, Präcipitine, die jeweils nur zur Bildung eines Präcipitationsringes führten, nachweisen. Die Erfahrung zeigt, daß der präcipitogene Reiz des infizierenden Stammes anscheinend nicht
ausreicht, um während der Erkrankung zur Bildung der in Hyperimmunsera von Kaninchen
nachweisbaren verschiedenen Präcipitine zu führen. Die Gewinnung guter präcipitierender
Immunsera setzt eine Hyperimmunisierung, wie sie nur bei Versuchstieren durchgeführt
werden kann, voraus.

Über die Herstellung präcipitierender menschlicher Immunsera liegen keine Angaben vor.

C. Immunität

1. Aktive Immunität

Über die Immunitätsverhältnisse beim *Menschen* nach Überstehen einer Infektion mit Past. pseudotuberculosis oder nach aktiver Immunisierung mit abgetöteten Bakterienstämmen ist im Gegensatz zum Tier nichts bekannt. Diesbezügliche Untersuchungen beim Menschen waren bisher ohne Interesse. Dagegen wurde mit einer Formolvaccine von Past. pseudotuberculosis auf Madagaskar aktiv gegen Pest immunisiert (Lit. bei POLLITZER 1954). Die Frage, ob das Ausbleiben neuer Pesterkrankungen während der Berichtszeit ursächlich auf eine erfolgreiche Vaccinierung oder auf epidemiologische Gegebenheiten zurückzuführen war, blieb unbeantwortet (BOQUET u. DUJARDIN-BEAUMETZ 1929).

Eine aktive Immunität von *Versuchstieren* ist durch Immunisierung mit lebenden oder abgetöteten Kulturen von Past. pseudotuberculosis zu erzielen (POPPE 1928; SCHÜTZE 1929; BOQUET u. Mitarb. 1929; TRUCHE u. BAUCHE 1929; BOQUET 1937; KUROKAWA 1941; THAL 1954; VAN DORSSEN 1952, 1955; SACHDEVA u. Mitarb. 1956), während Versager (DESSY 1925, LERCHE 1927) mit der Verwendung immunogen unwirksamer Kulturen zu erklären sind (BOQUET 1937, MEYER 1952, THAL 1954).

THAL (1954) stellte bei Meerschweinchen-Immunisierungsversuchen mit einem avirulenten lebenden Stamm von Past. pseudotuberculosis (Stamm 32^{IV}) fest, daß eine aktive Immunität von etwa 5 Monaten Dauer gegen Infektionen mit Stammvertretern der serologisch homologen wie heterologen Typen erreicht wird, wobei er die intraperitoneale Injektion des Impfstoffes, die nach seiner Ansicht zu einer besseren Verteilung der avirulenten lebenden Keime im Organismus führen soll, der subcutanen vorzieht. Der serologischen Heterogenität von Past. pseudotuberculosis steht demnach die immunbiologische Einheitlichkeit gegenüber. Immunisierungserfolge mit abgetöteten Kulturen sind an die wiederholten Injektionen großer Antigenmengen gebunden. Die Beobachtung, daß nur eine kleine Zahl avirulenter lebender Stämme — bei THAL (1954) 2 von 10 geprüften — im Immunisierungsversuch einen Impfschutz verleiht, unterstreicht die Notwendigkeit der Auswahl immunogen sicher wirksamer Stämme. Die Überlegenheit der immunisierenden Wirkung einer avirulenten Lebendvaccine gegenüber einer Formolvaccine scheint durch ältere und neuere Untersuchungen erwiesen zu sein (NOON 1909; THAL 1954; VAN DORSSEN 1955; SACHDEVA u. Mitarb. 1956).

THAL (1954) unterscheidet bei einer mit Past. pseudotuberculosis Typ III ausgelösten Meerschweinchenpseudotuberkulose eine antiinfektiöse und antitoxische Immunität. Seine Beobachtungen, deren Nachuntersuchung und Bestätigung noch aussteht, sprechen dafür, daß durch die Immunisierung mit avirulenten, lebenden, ektotoxinfreien Stämmen nur eine antiinfektiöse, aber keine antitoxische Immunität erreicht wird. Die antitoxische Immunität soll sich gegen die toxische und infektiöse Wirkung toxischer Stämme von Past. pseudotuberculosis Typ III, aber nicht gegen eine Infektion mit virulenten, atoxischen Stämmen richten.

Nach Untersuchungen von BOQUET (1937) und THAL (1954) verbreiten sich im nichtimmunen Organismus *virulente*, subcutan injizierte Keime in weniger als 24 Std über den ganzen Organismus, wobei sie die für eine Pseudotuberkulose typischen Veränderungen auslösen. Die Generalisation *avirulenter* Keime ist dagegen weniger rasch; sie werden in den

Organen, ohne makroskopisch sichtbare Veränderungen hervorzurufen, vernichtet. Bei immunisierten Tieren ist die Verbreitung der virulenten Keime über die Lymph- und Blutbahn gehemmt und ihre rasche Vernichtung in den Organen wie am Ort der Injektion, auf den die manifeste Infektion meist beschränkt bleibt, möglich.

Die durch avirulente Stämme von Past. pseudotuberculosis erreichte spezifische Immunität kann nach BOQUET (1937) von einer unspezifischen Immunität gegen Milzbrand begleitet sein. NICOLLE berichtete schon 1906, daß unter bestimmten Voraussetzungen Infektionen mit Past. pseudotuberculosis auch zu einer Immunität gegen Rotz führen können.

Schon bevor Einzelheiten über die antigenen Beziehungen zwischen Past. pseudotuberculosis und Past. pestis bekannt waren, berichteten zahlreiche Autoren (Lit. bei POPPE 1928; SCHÜTZE 1929) über die Möglichkeit der aktiven Immunisierung von Meerschweinchen und Ratten gegen virulente Pestbakterien durch Past. pseudotuberculosis-Stämme. Meerschweinchen und Ratten können mit hitze-, chloroform- oder formalinabgetöteten Suspensionen von Past. pseudotuberculosis gegen Infektionen mit virulenten Pestbakterien geschützt werden, während umgekehrt gegen Pest immune Tiere für Past. pseudotuberculosis empfänglich sind. Im neueren Schrifttum berichten CHEN u. MEYER (1955 mit weiterer Literaturangabe) über den hohen Impfschutz, der mit Past. pseudotuberculosis gegen Infektionen mit virulenten Pestbakterien erreicht wurde. Auf weitere Arbeiten, die sich mit der Kreuzimmunität befassen, einzugehen, ist in diesem Rahmen nicht notwendig. Doch lassen die Ergebnisse der kreuzweisen Immunisierungsversuche noch manche Fragen offen (Lit. bei POLLITZER 1954).

2. Allergische Reaktion

BACHMANN (1922) sah bei Meerschweinchen, deren Infektion 6—12 Wochen zurücklag, nach intracutaner Injektion des Filtrates von 18 verschiedenen, auf Schrägagar gezüchteten, mit 1 ml Kochsalzlösung abgeschwemmten und bei 60—65° C abgetöteten Stämmen lokale Entzündungserscheinungen, die er als allergische Reaktion der infizierten Tiere deutete. Über entsprechende Beobachtungen berichteten DESSY (1925), SAENZ u. COSTIL (1932), BOQUET (1937), GORET u. Mitarb. (1955). BOQUET (1937) kam zu der Feststellung, daß der Hauttest nach Infektion von Meerschweinchen mit virulenten Stämmen regelmäßig stark, mit schwach oder avirulenten Stämmen dagegen schwach oder fraglich positiv sei. Seine Beobachtungen wurden vom Verfasser nicht bestätigt.

Ein reaktionsloses Verhalten von Meerschweinchen, die 2—6 Wochen nach der Infektion mit virulenten, avirulenten oder ektotoxinbildenden Stämmen (Typ III) infiziert worden waren, wurde bei Durchführung des Hauttestes ebenso beobachtet wie Rötung, Schwellung oder Entwicklung kleiner Nekrosen bei nichtinfizierten Kontrolltieren. Auch bei Patienten mit der appendicitischen Verlaufsform der menschlichen Pseudotuberkulose versagte uns bisher der Hauttest. Hautreaktionen blieben mit verschieden konzentrierten Antigenchargen entweder ganz aus, oder die Patienten zeigten eine gleiche Reaktionsstärke wie gesunde Kontrollpersonen.

BOQUET (1937) gelang es auch, mit einigen Stämmen von Past. pseudotuberculosis beim Kaninchen das Sanarelli-Shwartzman-Phänomen auszulösen.

Die stärksten Reaktionen erhielt er nach intracutaner Injektion von 0,5 ml des Zentrifugates einer 1—3 Monate alten Martin-Bouillon-Kultur, die 1 Std auf 58—60° C erhitzt war. 24 Std später erfolgte die Injektion von 2 ml desselben bei 2° C aufbewahrten Kulturzentrifugates, die innerhalb weniger Stunden zur Entwicklung der typischen hämorrhagischen Reaktion führte.

Die passive Immunisierung von Meerschweinchen gegen eine Infektion mit Past. pseudotuberculosis ist durch Übertragung von Immunserum nicht möglich (DESSY 1925, BOQUET 1937, RUTQVIST u. Mitarb. 1956).

VI. Tierpathogenität

Wenige Bakterienarten sind wohl in der Lage, eine größere Anzahl verschiedenster Tierarten (Einzelheiten s. Abschnitt IX) zu infizieren als Past. pseudotuberculosis (FELDMAN 1955). Unter den bisher bekannten serologischen Typen wird nach einer Zusammenstellung von THAL (1954) Past. pseudotuberculosis Typ I am häufigsten beim infizierten Tier nachgewiesen. Eine Wirtsspezifität der einzelnen Typen besteht nicht. Im Tierversuch an Mäusen, Meerschweinchen und Kaninchen führen die Vertreter der Typen I—III, in der Regel schon in geringen Dosen intravenös, intraperitoneal, intramuskulär, subcutan und auch intracutan appliziert, unter charakteristischen Veränderungen zum Tode der Tiere, während Typ IV für die genannten Tiere nicht und Typ V nur schwach pathogen zu sein scheint. Verlauf und Ausgang der natürlichen und experimentellen Infektion der verschiedenen empfänglichen Tiere sind nach PALLASKE u. MEYN (1932) wesentlich von der Virulenz, der Dosis und dem Infektionsweg des infizierenden Stammes sowie von individuellen Faktoren beim Tier, wie Unterschieden in ihrer Resistenz und Immunität, bestimmt.

Krankheitsverlauf. Die natürliche wie experimentelle Infektion führt bei empfänglichen Tieren zu einem in seiner klinischen Symptomatologie unterschiedlich stark ausgeprägten, uneinheitlichen Krankheitsbild. Neben völliger Symptomenfreiheit bis zum überraschend eintretenden Tod werden als uncharakteristische Prodromalerscheinungen Störungen des Allgemeinbefindens, Freßunlust, Abmagerung, Erbrechen, Schläfrigkeit, Benommenheit oder Unruhe, Sträuben von Fell oder Gefieder, Gehstörungen durch Lähmungen oder Steifheit der Glieder und nicht selten Durchfälle beobachtet. Meist treten diese Symptome — wenn überhaupt — in unterschiedlicher Stärke und Regelmäßigkeit erst wenige Tage vor dem Tode auf (POPPE 1928; GATÉ u. BILLA 1928; BISHOP 1932; TRUCHE u. BAUCHE 1933; VERGE u. Mitarb. 1937; TRUCHE u. ISNARD 1937; BOQUET 1937; TRUCHE 1938; SCHÄFER 1939; KUROKAWA 1939/40; BEAUDETTE 1940; ROSENWALD u. DICKINSON 1944; URBAIN, NOUVEL u. BULLIER 1944; MANNINGER 1945; URBAIN u. NOUVEL 1949; MEYER 1952; COOK 1952; FELDMAN 1955; MAGLIONE u. CERETTO 1956 u. a.). Aus der klinischen Symptomatologie eine Pseudotuberkulose sicher zu diagnostizieren, ist unmöglich.

Wie uncharakteristisch die klinischen Symptome der Pseudotuberkulose des Tieres sind, kann am besten durch eine kurze Zusammenstellung der bei verschiedenen Tierarten wiederholt erhobenen Befunde veranschaulicht werden.

Bei *natürlich infizierten Meerschweinchen* und *Kaninchen* stellen sich bei der klassischen Verlaufsform Abmagerung, Freßunlust und kurze Zeit vor dem Tod ein struppiges Fell ein. Nicht regelmäßig werden Darm- oder Gleichgewichtsstörungen und Benommenheit beobachtet (POPPE 1928; GATÉ u. BILLA 1928; BISHOP 1932; BOQUET 1937; TRUCHE 1938). *Katzen* gehen an der natürlichen Infektion innerhalb 2—3 Wochen zugrunde. Freßunlust, Durchfälle oder Obstipationen, ein aufgetriebener Leib, Erbrechen und gelegentlich Gelbsucht sind die wesentlichsten Symptome (PALLASKE u. MEYN 1932; PALLASKE 1933; TRUCHE 1938 u. a.). Auch bei *Vögeln* sind die Prodromalerscheinungen meist nur uncharakteristisch. Beim *Truthahn* treten während der häufig nur 3—4, selten 8—10 Tage dauernden Krankheit fortschreitende Abmagerung, Appetitlosigkeit, Durchfall, Hängen der Flügel und Entfärbung des Kammes ein. Sehr häufig wird ein Hinken der Tiere schon zu Anfang und selten erst am Ende der Krankheit beobachtet. Unter allgemeinem Marasmus können die Tiere schließlich im Koma zugrunde gehen (TRUCHE 1938; ROSENWALD u. DICKINSON 1944; BLAXLAND 1947; MATHEY u. SIDDLE 1954 u. a.). Besonders empfänglich sind *Kanarienvögel* und *Tauben*, die meist eine akute, schwere Verlaufsform zeigen. Die Tiere werden kurz nach der Infektion apathisch, schläfrig oder benommen, verweigern die Nahrungs- und Flüssigkeitsaufnahme und sterben rasch unter Krämpfen (LESBOUYRIES 1934, TRUCHE 1938). Bei *Hühnern* werden vor allem Appetitlosigkeit, Abmagerung, Steifheit der Glieder, Gehstörungen, Durchfall, zum Schluß auch Krämpfe gesehen (TRUCHE u. ISNARD 1937; TRUCHE 1938; SCHÄFER 1939 u. a.). *Affen* sterben z. T. plötzlich ohne Prodromalstadium, während andere Tiere Appetitlosigkeit, Speichelfluß, Muskelkontraktionen und Gleichgewichtsstörungen aufweisen (URBAIN u. NOUVEL 1949 u. a.). Bei einer *Löwin* wurden als einzige Symptome 3 Wochen vor dem Tod Verlust der Lebhaftigkeit, Appetitlosigkeit und Verdauungsstörungen beobachtet (URBAIN u. Mitarb. 1944), während *Füchse* unter dem klinischen Bild einer leichten Gastroenteritis starben (EIELAND 1947). Junge, nicht richtig ernährte Tiere scheinen für die Infektion besonders empfänglich zu sein (MEYER 1952, FELDMAN 1955).

Entsprechend uncharakteristisch ist auch die Symptomatologie beim *experimentell infizierten* Tier. Erfolgreiche experimentelle Tierversuche wurden außer bei Mäusen, Meerschweinchen und Kaninchen, als den üblicherweise im Laboratorium zu diagnostischen Zwecken verwandten Tieren, auch bei verschiedenen domestizierten und anderen Tierarten durchgeführt. Sie gelangen bei verschiedenen *Geflügelarten* wie z. B. beim Huhn, Leghornhuhn, Kücken, Truthahn (KRAGE u. WEISGERGER 1924; TRUCHE u. BAUCHE 1933; TRUCHE u. ISNARD 1937; SCHÄFER 1939; MARTHEDAL u. Mitarb. 1954 u. a.), beim *Chinchilla* (VERGE u. PLACIDI 1942; CHAPMANN 1948; LEADER u. BAKER 1954), bei *Katzen* (URBAIN u. Mitarb. 1944, mit weiteren Literaturangaben; GORET u. Mitarb. 1955), *Affen* (URBAIN u. NOUVEL 1949 u. a.), *Füchsen* (EIELAND 1947), *Kanarienvögeln* und *Tauben* (PALLASKE 1933; SACHDEVA u. Mitarb. 1956).

Pathologisch-anatomische Veränderungen. Pathologisch-anatomisch sind verschiedene, bei allen empfänglichen Tierarten in sich aber weitgehend einheitliche Verlaufsformen zu erkennen. Grundsätzliche Unterschiede in den Organveränderungen bestehen aber bei den verschiedenen Tierarten nicht (PALLASKE 1933). Allen Tieren, sofern sie nicht kurze Zeit nach der Infektion eingehen, sind verschieden große und alte Herde in den Organen, vornehmlich in Leber und Milz, gemeinsam. Ein Teil der Veränderungen kommt aber erst durch die histologische Untersuchung zum Nachweis. Die Herde zeigen eine weißgraue oder graugelbliche Farbe und trockene Oberfläche. In den größeren Herden werden nicht selten Abszeßbildungen mit zentraler Erweichung gesehen. Bei akutem wie subakutem Verlauf einer natürlichen oder experimentellen Infektion kann die typische Knötchenbildung fehlen. Der Erregernachweis gelingt im akuten Stadium meist aus Blut, Peritonealexsudat oder Organabstrichen und im chronischen Stadium aus vergrößerten mesenterialen Lymphknoten oder den typischen knötchenförmigen Herden. Differentialdiagnostisch läßt der makroskopische Sektionsbefund eine sichere Abtrennung der ebenfalls zu Herdbildung führenden Infektionen mit Mycobacterium tuberculosis, Past. tularensis, Past. pestis, Streptococcus pyogenes (C), Listeria monocytogenes, Corynebacterium ovis und murium, verschiedenen Salmonellaarten und Brucella abortus nicht zu. Erregernachweis und eine histologische Untersuchung veränderter Lymphknoten oder Organe sind unerläßlich.

Für experimentelle Untersuchungen, insbesondere für den Erregernachweis aus menschlichem oder tierischem Untersuchungsgut, sind unter den Laboratoriumstieren *Meerschweinchen* besonders geeignet, da sie mit größerer Regelmäßigkeit — als Mäuse und Kaninchen —

nach der Übertragung des infektiösen Materials eingehen und konstantere Versuchsbedingungen ermöglichen. Mit dem Vorkommen spontaner Infektionen muß aber gerechnet werden, so daß beim Erregernachweis auf weitere Untersuchungen, insbesondere auf das Kulturverfahren, nicht verzichtet werden darf.

Da in den durch Past. pseudotuberculosis verursachten pathologisch-anatomischen und histologischen Veränderungen beim Tier keine grundsätzlichen Unterschiede bestehen (PALLASKE 1933), soll bei der besonderen Bedeutung des Meerschweinchentierversuches in der Diagnose der menschlichen Pseudotuberkulose, nur am Beispiel des Meerschweinchens, kurz über die möglichen klinischen und pathologisch-anatomischen Verlaufsformen und histologischen Befunde referiert werden. Auf die von POPPE (1928), MANNINGER (1945), MEYER (1952), FELDMAN (1955), WILSON u. MILES (1955) und im Abschnitt IX dieser Arbeit zitierten Befunde bzw. Literaturstellen wird ebenso wie auf die neueren tierexperimentellen Untersuchungen von BISHOP (1932), PALLASKE u. MEYN (1932), PALLASKE (1933), BOQUET (1937), OLT (1938), TRUCHE (1938), TOPPING u. Mitarb. (1938), MOSS u. BATTLE (1941), MACCHIAVELLO (1941) hingewiesen.

Pseudotuberkulose des Meerschweinchens

a) Klinische Verlaufsform. Bei der *natürlichen Infektion* der Meerschweinchen lassen sich nach dem klinischen Bild drei verschiedene Verlaufsformen unterscheiden. Die *septicämische Form* führt zum Tod innerhalb 24—48 Std. Die Sektion zeigt entweder keinen makroskopischen pathologisch-anatomischen Befund oder eine deutlich vergrößerte Milz, schwere hämorrhagische Enteritis und Ergüsse in den verschiedenen Körperhöhlen. Bei der *klassischen Verlaufsform* fehlen Krankheitssymptome, oder sie sind uncharakteristisch. Der Tod tritt meistens in 10—30 Tagen ein. Die dritte Verlaufsform, die *Drüsenform*, ist wahrscheinlich eine Folge von Bißverletzungen, die zu einer Entzündung der cervicalen und thorakalen Lymphknoten führen kann (RAMON 1914, MEYER 1952 u. a.).

b) Pathologisch-anatomische Verlaufsform und Veränderungen. Nach RAMON (1914) wird auch heute noch entsprechend den verschiedenen Erscheinungsformen der pathologisch-anatomischen Veränderungen zwischen einer latenten, pleuropulmonalen, abdominellen und hepatischen Verlaufsform der Pseudotuberkulose des Meerschweinchens unterschieden. Die *latente Form* läßt Knötchenbildung vermissen und führt oft nur zu einer leichten Lungenentzündung mit unterschiedlich starker Hyperämie von Milz und Leber. Der Erregernachweis gelingt meist durch Peritoneal- oder Organabstriche. Ausgeprägter im Erscheinungsbild ist die *pleuropulmonale Form.* In der Lunge finden sich zahlreiche Herde verschiedener Größe; die Beteiligung der Pleura manifestiert sich in einer ein- oder doppelseitigen Pleuritis mit Pseudomembranbildung und eitrigem Erguß. An Stelle von Knötchen kann die Lunge auch das Bild einer lobären Pneumonie zeigen. Die am häufigsten zu beobachtende *abdominelle* Verlaufsform ist durch Hyperplasie und Veränderungen der mesenterialen Lymphknoten, insbesondere in der Ileocoecalgegend, gekennzeichnet. Die Lymphknoten können Hasel- bis Walnußgröße erreichen und enthalten im fortgeschrittenen Stadium eine trockene oder nur noch käsige Masse. Bei den meisten Fällen sind Leber bzw. Leber und Milz an der Oberfläche und in der Tiefe des hypertrophischen Parenchyms mit in Größe und Zahl verschiedenen grauweißlichen oder weißgelblichen Herden durchsetzt, während die übrigen Organe keine oder, wie z. B. Lungen und Darm, herdförmige Veränderungen zeigen. Die knötchenförmigen Veränderungen im Darm, die meistens auf die letzten Abschnitte (Ileum, Coecum) beschränkt sind, werden nicht regelmäßig gefunden und treten meist hinter den übrigen Organveränderungen zurück. Nicht selten findet man in den Hoden, gelegentlich auch in den Nieren herdförmige Veränderungen. Ein seröses, z. T. auch hämorrhagisches Peritonealexsudat vervollständigt den Sektionsbefund (POPPE 1928; GATÉ u. BILLA 1928; SCHÜTZE 1928; BISHOP 1932; PALLASKE 1932; VERGE u. Mitarb. 1937; OLT 1938; TRUCHE 1938; TOPPING u. Mitarb. 1938; MOSS u. BATTLE 1941; KNAPP u. MASSHOFF 1954; MAGLIONE u. CERETTO 1956 u. a. m.). Als *hepatische Form* wird der ausschließliche Befall der Leber mit Knötchen aller Entwicklungsstadien beschrieben. In zahlreichen Fällen ist die Gallenblase mitbeteiligt. Auf eine weitere Verlaufsform mit besonderer Beteiligung der cervicalen Lymphknoten, die dieselben Veränderungen wie die mesenterialen Lymphknoten zeigen, hatte ebenfalls schon RAMON (1914) hingewiesen.

Ein Teil der die Infektion überlebenden Tiere wird zu Dauerausscheidern, ohne daß bei der Sektion sichtbare pathologisch-anatomische Veränderungen auf eine vorausgegangene

Infektion hinweisen müssen (BISHOP 1932, MEYER 1951, BERGER 1957, KNAPP unveröffentlichte Beobachtungen).

Bei der *künstlichen Infektion* der Meerschweinchen lassen sich in Abhängigkeit von der Applikationsweise folgende Verlaufsformen herausstellen (RAMON 1914, POPPE 1928, BISHOP 1932, PALLASKE 1933, OLT 1938 u. a. sowie eigene Beobachtungen).

a) Perorale Infektion

Nach Aufnahme infizierten Futters tritt der Tod in etwa 7—15 Tagen ein. Ein Auftreten von Prodromalsymptomen ist selten oder erfolgt erst wenige Tage (1—3) vor dem Tod. Der Sektionsbefund zeigt meist die abdominelle Verlaufsform mit den gleichen Veränderungen wie nach einer natürlichen Infektion. Im Vordergrund stehen miliare Herde in Leber und Milz, eine Lymphadenitis der mesenterialen, vornehmlich ileocöcal gelegenen Lymphknoten und kleine nekrotische Herde in den Peyerschen Plaques des Ileums und Coecums.

b) Intracutane und subcutane Infektion

Nach 1—2 Tagen tritt an der Injektionsstelle der Reinkultur eine lokale Entzündung mit Ödem- und Abszeßbildung ein. Sofern die Infektion nicht lokal begrenzt bleibt, tritt nach 7—15 Tagen — bei Injektionen von Untersuchungsgut nach 2—6 Wochen — der Tod ein. Als Zeichen der Generalisation des Erregers sind wiederum Leber und Milz von miliaren, unterschiedlich großen Herden durchsetzt, während der Darm oft keine makroskopisch sichtbaren Veränderungen aufweist. Die Herde sind klein und schwer zu erkennen, wenn infolge hoher Infektionsdosis oder bei Verwendung toxischer Stämme (Typ III) der Tod rasch eintritt. Sie sind bei langsam fortschreitender Infektion dagegen deutlich entwickelt.

c) Intraperitoneale oder intravenöse Infektion

Nach Injektion von Reinkulturen erliegen die Meerschweinchen meist schon nach 3—8 und von Untersuchungsmaterial erst nach 10—30 Tagen der Infektion. Die Sektion zeigt nach intraperitonealer Infektion entweder eine serofibrinöse Perikarditis und Peritonitis mit entzündlicher Beteiligung des Netzes, das wie Leber und Milz zahlreiche verschieden große, charakteristisch strukturierte, knötchenförmige Herde aufweist, oder nur einzelne Herde in Leber, Lunge und Milz; z. T. werden nur vergrößerte Mesenteriallymphknoten und ein seröses oder hämorrhagisches Exsudat beobachtet. Nach intravenöser Infektion, mit Tod innerhalb 3—4 Tagen, können die makroskopisch sichtbaren Veränderungen so gering sein, daß ihr Erkennen nur bei genauer Durchmusterung der Organe (Lupe) oder durch histologische Untersuchung möglich wird.

d) Konjunktivale und intraoculäre Infektion

Eine Übertragung lebender Keime auf die Conjunctiva von Meerschweinchen und Kaninchen führte in eigenen, mit HAGER u. MASSHOFF durchgeführten unveröffentlichten Beobachtungen innerhalb 3—4 Tagen zu einer Entzündung der Conjunctiva mit starker eitriger Sekretion, Verklebung der Lider, leichter Trübung der Cornea und Bildung leistenförmiger Granulationen und Knötchen. Nach 14tägiger Beobachtung waren auch bei den später der Infektion erlegenen Tieren noch vereinzelte Knötchen bei reizloser Conjunctiva und Cornea nachzuweisen. Das Einbringen der Keime in die Vorderkammer war von einer akuten Entzündung des Auges mit starker Exsudatbildung in der Vorderkammer, diffuser milchiger Trübung der Cornea, akuter Entzündung der Conjunctiva mit starker Granulation und Knötchenbildung und einer Schwellung der regionären Lymphknoten gefolgt. Die pathologisch-anatomische Untersuchung ließ bei den nur z. T. der Infektion erlegenen Meerschweinchen und Kaninchen das bei der abdominellen Verlaufsform charakteristische Bild mit Beteiligung von Leber, Milz, Peritoneum und außer den mesenterialen auch der cervicalen Lymphknoten erkennen. Die Untersuchungen führten zu einer teilweisen Bestätigung der Befunde von DEYL (1896) und GIFFORD u. LAZAR (1930).

e) Histologische Veränderungen

Das histologische Bild der knötchen- oder herdförmigen Veränderungen wurde verschiedentlich eingehend beschrieben (POPPE 1928; PALLASKE u. MEYN 1932; PALLASKE 1933; TOPPING u. Mitarb. 1938; MOSS u. BATTLE 1941; MANNINGER 1945 u. a.). Nach PALLASKE (1933) zeigt die vergleichende Untersuchung nicht nur der makroskopischen, sondern auch der histologischen Veränderungen an den verschiedenen Organen (Leber, Milz, Lunge, Lymphknoten, Appendix) der zahlreichen von ihm untersuchten Tiere bzw. Tierarten (*Meerschweinchen, Kaninchen, Hasen, Maus, Sumpfbiber, Truthühner, Tauben, Kanarienvögel, Katzen*) ein weitgehend einheitliches Bild. Die Veränderungen bestehen aus einem zellreichen Zentrum und einer dieses peripher abgrenzenden Zone. Sie sind Folge einer anfangs gemischten histiocytär-leukocytären, wenn nicht überwiegend histiocytären Zellreaktion, der sich in der weiteren Entwicklungsphase eine überwiegend leukocytäre Reaktion (Abscedierung) anschließt. Als seltene Extreme werden Herdbildungen rein cellulärer oder rein nekrotischer Natur beobachtet (Einzelheiten sind den obengenannten Arbeiten zu entnehmen).

Neuerdings hat MASSHOFF (KNAPP u. MASSHOFF 1954) die Identität der beim Menschen als Folge einer Infektion mit Past. pseudotuberculosis beobachteten Veränderungen der mesenterialen Lymphknoten mit den am lymphoreticulocytären Gewebe des Meerschweinchens erhobenen histologischen Befunde herausgestellt.

Nach MASSHOFF bestehen die wesentlichsten histologischen Merkmale des in seiner Entwicklung in verschiedenen Phasen ablaufenden Entzündungsprozesses in einer zunächst knötchenförmigen, dann konfluierenden und sich diffus ausbreitenden Proliferation der Reticulumzellen, die von einer leukocytären, zur umschriebenen Einschmelzung führenden Infiltration gefolgt ist. Um den Nekroseherd bildet sich schließlich eine Zone von Granulationsgewebe mit vorwiegend histiocytären Zellelementen und Fibroblasten. Charakteristisch ist für den Lymphknoten die herdförmige und für die Kapsel die diffuse Anordnung des Entzündungsprozesses. In den Randzonen des Lymphknotens führt die Entzündung zu herdförmigen Wucherungen von Reticulumzellen, die z. T. von neutro- und eosinophilen Leukocyten durchsetzt sind und deren Zentrum einer langsam fortschreitenden Nekrose unterliegt. Die nicht der Nekrose anheimfallenden Reticulumzellwucherungen weisen, wohl als Zeichen einer chronischen Verlaufsform, eine lebhafte Capillar- und Bindegewebsneubildung auf. Für die charakteristischen histologischen Merkmale dieses beim Menschen stets gleichartig in bestimmter Form ablaufenden Entzündungsprozesses führte MASSHOFF (1953) die Bezeichnung „abscedierende reticulocytäre Lymphadenitis" ein (über weitere histopathologische Untersuchungen s. S. 250).

VII. Widerstandsfähigkeit gegen physikalische und chemische Einflüsse

Nur wenige Autoren untersuchten die Widerstandsfähigkeit von Past. pseudotuberculosis gegen physikalische und chemische Einflüsse. Die vorliegenden Befunde sind z. T. widersprechend und in ihrem Wert schwer zu beurteilen, da systematische Nachuntersuchungen unter einheitlichen Versuchsbedingungen fehlen.

1. Thermoresistenz

Nach SAISAWA (1913) wird Past. pseudotuberculosis bei 60° C innerhalb 30 min, nach MESSERSCHMIDT u. KELLER (1914) bei 56° C schon innerhalb 10 min abgetötet. POPPE (1928) berichtet, daß durch einstündiges Erhitzen auf 60° C nur die Virulenz der Keime und erst durch zweistündiges Erhitzen die Entwicklungsfähigkeit verlorengeht. Dagegen sind in Bergey's Manual (1957) und bei WILSON u. MILES (1955) bzw. bei MEYER (1952) für 60⁰ C Abtötungszeiten von nur 10 bzw. 40 min angegeben. SACHDEVA (1956) gelang es andererseits nicht, eine Vaccine innerhalb 2 Std bei 60⁰ C abzutöten. TRUCHE u. BAUCHE (1929) benötigten 80⁰ C, um Past. pseudotuberculosis in 10 min abzutöten. In eigenen Versuchen wurden in physiologischer Kochsalzlösung suspendierte Kulturen von Past. pseudotuberculosis bei 50—60⁰ C in 180 min nicht regelmäßig und bei 70—80⁰ C erst in 10 bzw. 5 min abgetötet. Eine aus epidemiologischen Überlegungen durchgeführte Prüfung der keimtötenden Wirkung verschiedener bei der Pasteurisierung angewandter Temperaturen zeigte, daß im Suspensions- wie Keimträgerversuch mit den bei der Pasteurisierung von Milch amtlich zugelassenen Temperaturen und Einwirkungszeiten mit und ohne Einschluß der Steige- und Ausgleichszeit im Lang- (30 min 62⁰ C), Hoch- (momentan 85⁰ C) und Kurzerhitzungsverfahren (1 min 71—74⁰ C) keine sichere Abtötung aller Keime gewährleistet ist.

Von 10 in Milch suspendierten oder an Baumwollfäden angetrockneten Stämmen wurden nur zwei bei 60⁰ C in 70 min und kein Stamm bei 70—74⁰ C in 1 min oder bei 85⁰ C in 15 sec abgetötet. Unter Einschluß der Steigezeit töteten 85⁰ C in 1 bzw. 2 min nur 1 bzw. 3 Stämme ab.

In verschlossenen Kulturen bleibt Past. pseudotuberculosis jahrelang (MARLINI 1938, MEYER 1952) und in Organen nach eigenen Versuchen bei + 2⁰ C mindestens 3 Monate, bei 22⁰ C 10—20, bei 37⁰ C 3—5 und bei 45⁰ C 1—2 Tage am Leben. Moss u. BATTLE (1941) gelang im Tierversuch der Erregernachweis auch noch aus Organgewebe, das 16—18 Tage bei 40⁰ C aufbewahrt war. Die Widerstandsfähigkeit gegen Austrocknung ist nur gering (POPPE 1928, SCHÜTZE 1929). In lyophilisiertem Zustand waren eigene Kulturen bisher über eine Beobachtungszeit von 2 Jahren lebensfähig.

2. Widerstandsfähigkeit gegen Desinfektions- und Konservierungsmittel

Die in der neueren Literatur zitierten Befunde (MANNINGER 1945; MEYER 1952; FELDMAN 1955; WILSON u. MILES 1955; Bergey's Manual 1957 u. a.) gehen weitgehend auf die Untersuchungen von SAISAWA (1913), MESSERSCHMIDT u. KELLER (1914), TRUCHE u. BAUCHE (1929) zurück. In 1- bzw. 2%iger Phenollösung sahen SAISAWA (1913) in 5 bzw. 2 min, TRUCHE u. BAUCHE (1929) in 5%iger Phenollösung dagegen erst in 5—10 min eine Abtötung aller Keime. Für eine 1%ige Sublimatlösung gaben TRUCHE u. BAUCHE eine Abtötungszeit von 15 bis 20 sec an, während in einer 4%igen Formalin- bzw. in einer 3- oder 5%igen Eisen- oder Kupfersulfatlösung eine Abtötungszeit von 1 Std benötigt wurde. 40%iger Alkohol soll nach SAISAWA (1913) alle Keime sofort abtöten. Andererseits berichtete MEYER (1952), daß in 60%igem Alkohol Past. pseudotuberculosis innerhalb 5—10 min abgetötet wird.

Eigene Untersuchungen über die Widerstandsfähigkeit von 5 Stämmen von Past. pseudotuberculosis gegenüber chemischen Substanzen, die zur Desinfektion im Laboratorium oder zur Abtötung und Konservierung von Antigenen verwendet werden, sind in Tabelle 3, deren Besprechung sich erübrigt, zusammengestellt (über Widerstandsfähigkeit gegen Antibiotica und Sulfonamide s. S. 256, Therapie).

Tabelle 3. *Widerstandsfähigkeit gegen Desinfektions- und Konservierungsmittel*

Konzen-tration (%)	Abtötungszeit in Minuten				Konzen-tration (%)	Abtötungszeit in Minuten		Konzen-tration (%)	Abtötungszeit in Minuten
	Phenol	For-malin	Sagro-tan	Zephi-rol		Quecksilber-chlorid	Silber-nitrat		Alkohol
5	$^1/_2$	$^1/_2$	$^1/_2$	$^1/_2$	1	sofort	sofort	96	$^1/_2$
2	$^1/_2$—2	5—10	$^1/_2$	$^1/_2$	0,1	$^1/_2$	$^1/_2$—5	75	$^1/_2$
1	5—30	5—10	$^1/_2$—1	$^1/_2$—1	0,01	$^1/_2$—5	60—120	60	$^1/_2$—5
0,5	90—120	5—10	1—2	$^1/_2$—1	0,001	120	120	50	15—30
0,25	3—24 Std	10—20	·/.	·/.	·/.	·/.	·/.	40	15—30
0,1	·/.	30—60	·/.	·/.	·/.	·/.	·/.	·/.	·/.

·/. = nicht untersucht.

VIII. Kulturell-biochemische, serologische und tierexperimentelle Differentialdiagnose innerhalb der Pasteurellagruppe

In Ländern mit endemischem Pestvorkommen kann die differentialdiagnostische Abgrenzung von Infektionen mit Pasteurella pestis oder Past. pseudotuberculosis und der Nachweis von Mischinfektionen von human- und veterinärmedizinischer Bedeutung sein. MEYER u. BATCHELDER (1926) wie MACCHIAVELLO (1941) beschrieben Mischinfektionen bei Nagetieren bzw. Menschen. In pestfreien Ländern steht dagegen, besonders in der Veterinärmedizin, die Differentialdiagnose von Infektionen mit Past. pseudotuberculosis, Past. multocida und Past. tularensis im Vordergrund.

Past. tularensis

POPPE (1928) sah in der *Tularämie* eine besondere Form der Pseudotuberkulose. REIMANN u. ROSE (1931), REIMANN (1932) betonten die Ähnlichkeit der morphologischen, kulturellen und tierexperimentellen Eigenschaften von Past. tularensis und Past. pseudotuberculosis, ohne mit dieser Feststellung allgemeine Anerkennung zu finden. Vielmehr nimmt *Past. tularensis* infolge ihres sehr unterschiedlichen morphologischen, kulturell-biochemischen und serologischen Verhaltens eine Sonderstellung innerhalb der Gattung Pasteurella ein, so daß ihre Einordnung in die Gattung Pasteurella bis heute keine allgemeine Anerkennung gefunden hat (TOPLEY u. WILSON 1929, 1955; MATZKE 1943; HESSELBROCK u. FOSHAY 1945; MEYER 1952; POLLITZER 1954; KNOTHE 1955). Die *differentialdiagnostische Abgrenzung* von *Past. tularensis* ist durch die Feststellung gegeben, daß Past. tularensis im Gegensatz zu Past. pseudotuberculosis bzw. Past. pestis auf einfachen Nährböden (Agar, Gelatine, Bouillon, Kartoffeln, Milch usw.) nicht und auf Spezialnährböden nur sehr langsam wächst, unbeweglich und infolge ihres verschiedenen kulturell-biochemischen und serologischen Verhaltens — keine Partialantigengemeinschaft — leicht von Past. pestis, Past. pseudotuberculosis und Past. multocida zu unterscheiden ist. Der Sektionsbefund mit Past. tularensis infizierter Meerschweinchen zeigt dagegen vielfach große Ähnlichkeit mit dem pathologisch-anatomischen Befund bei Tieren, die mit Past. pestis und pseudotuberculosis oder mit Mycobacterium tuberculosis infiziert sind (REIMANN u. ROSE 1931; REIMANN 1932; BRANDSTETTER 1939 u. a.).

Past. pseudotuberculosis und Past. pestis

Past. pseudotuberculosis und Past. pestis zeigen in ihrem morphologischen, kulturell-biochemischen (s. Tabelle 1, S. 209), serologischen und tierexperimentellen Verhalten besonders nahe Beziehungen zueinander.

Von zahlreichen, zur kulturell-biochemischen Differentialdiagnose wiederholt angegebenen Eigenschaften erwiesen sich die rasche Bildung von Degenerationsformen auf 3%igem NaCl-Agar bei Past. pestis, die Alkalisierung von Pepton-Zuckerwasser, Maltosepeptonwasser, zuckerfreier Bouillon oder Milch durch Past. pseudotuberculosis, die Reduktion von Nitraten zu Nitriten bzw. die Bildung von Nitraten oder salpetriger Säure in Bouillon oder Leberbouillon durch Past. pestis, die Spaltung von Glykogen durch Past. pestis, von Melibiose durch Past. pseudotuberculosis als nicht zuverlässig genug. Entweder zeigten nicht alle Stämme derselben Keimart ein einheitliches Verhalten oder wurde ein der differentialdiagnostisch auszuschließenden Art gleichsinniges Verhalten beobachtet (Lit. bei POLLITZER 1954; BALTAZARD u. Mitarb. 1956).

Fehlende Harnstoffspaltung sah FAUCONNIER (1950) bei 2 Stämmen von Past. pseudotuberculosis und Ureasebildung bei einem Peststamm. Weitere Ausnahmen wurden bei Past. pseudotuberculosis im Ausbleiben und bei Past. pestis im Nachweis der Glycerinspaltung beobachtet (COLAS-BELCOUR 1926; BOQUET 1937; DEVIGNAT u. BOIVIN 1953; GIRARD 1953; POLLITZER 1954). Rhamnosespaltung, die durch Past. pseudotuberculosis rasch erfolgt, tritt auch bei Peststämmen in seltenen Fällen verzögert oder nach mehreren Kulturpassagen auf geeigneten Nährböden ein (CASTELLANI 1939; DEVIGNAT u. BOIVIN 1953/54; POLLITZER 1954; BRYGOO u. COURDURIER 1955; BALTAZARD u. Mitarb. 1956 mit weiterer Schrifttumsangabe). Ob spontane Mutationen zu einer Änderung des biochemischen Verhaltens führen, sei dahingestellt. Den Beobachtungen von BEZSONOVA u. Mitarb. (1937) über eine spontane Mutation von Past. pestis in Past. pseudotuberculosis und von KOROBKOVA (1937) und TUMANSKY (1937) über eine Änderung des Verhaltens von Past. pestis gegen Rhamnose als Folge einer Bakterienphagenbehandlung muß mit äußerster Zurückhaltung begegnet werden, solange sie nicht durch weitere experimentelle Untersuchungen bestätigt sind. ENGLESBERG (1957) berichtete über die Fähigkeit von Past. pestis, Rhamnose als Folge einer experimentell erzeugten, gerichteten Mutation zu oxydieren (weitere Literaturangaben POLLITZER 1954).

Nach GIRARD (1953) sind die Prüfung und der Nachweis der Beweglichkeit, Harnstoffspaltung, Glycerin- und Rhamnosesäuerung die zwar nicht unwidersprochene Voraussetzung einer sicheren Diagnose von Past. pseudotuberculosis (CHEN 1949; POLLITZER 1954, BALTAZARD 1956). DEVIGNAT u. BOIVIN (1953) beschrieben zwei polytrope Nährböden, die das Verhalten der Kulturen gegenüber Glycerin, Rhamnose und Nilblau bzw. Harnstoff und Inulin sowie ihre Beweglichkeit aufzeigen sollen.

Für die Routinepraxis der kulturell-biochemischen Differentialdiagnose zwischen Past. pseudotuberculosis und Past. pestis empfehlen BALTAZARD u. Mitarb. (1956) die Beweglichkeitsprüfung und Züchtung der fraglichen Stämme auf Desoxycholat-Citratagar, Verfahren, die z. T. mit Schwierigkeiten verbunden sind (s. S. 206, 207), die Prüfung der Harnstoffspaltung, der Rhamnose- und Melibiosesäuerung sowie die Durchführung des Methylenblautestes [Einzelheiten: Bull. Wld Hlth Org. 14, 459—509 (1956)].

Infolge ihrer Partialantigengemeinschaft kann auch die serologische Differentialdiagnose zwischen Past. pseudotuberculosis und Past. pestis erschwert sein; Schwierigkeiten, die aber nur in Gebieten mit endemischen Pestvorkommen gegeben sind.

Eine Antigenanalyse fraglicher Pseudotuberkulose- oder Peststämme ist mit einem polyvalenten OH-Serum mit Antikörpern gegen Typ I—V von Past. pseudotuberculosis, typenspezifischen O-Sera von Past. pseudotuberculosis und einem mit Past. pseudotuberculosis abgesättigten Pestserum möglich. Der Agglutininnachweis in Patientensera ist gleichzeitig mit Stämmen von Past. pseudotuberculosis und Past. pestis zu erbringen. Im Serum menschlicher und tierischer Pseudotuberkulosen wird nur Past. pseudotuberculosis agglutiniert, während Sera von Pest- oder mischinfizierten Patienten Past. pseudotuberculosis und Past. pestis agglutinieren können, so daß eine serologisch gesicherte Diagnose nur nach kreuzweiser Absättigung der fraglichen Sera möglich ist.

Die tierexperimentelle Differentialdiagnose wird an Meerschweinchen und weißen Ratten durchgeführt, da Past. pseudotuberculosis nur für Meerschweinchen, Past. pestis dagegen für Meerschweinchen und weiße Ratten pathogen ist.

Eine weitere differentialdiagnostische Unterscheidungsmöglichkeit besteht in der Anwendung von Pestphagen (BALTAZARD u. Mitarb. 1956). Schnellverfahren sind von GUNNISON, LARSON u. LAZARUS (1951) als Plattentest und von CAVANAUGH u. QUAN (1953) als Streifentest angegeben. GUNNISON u. Mitarb. (1951) gelang es, mit dem zuerst an Past. pestis (Stamm A 1122, „P phage") und dann an Past. pseudotuberculosis (Stamm Spokane, „PTB phage") adaptierten Pestphagenstamm von ADVIER (1933) biochemisch typische und atypische virulente und avirulente Stämme von Past. pestis innerhalb 48 Std zu lösen, während 45 Stämme von Past. pseudotuberculosis unbeeinflußt blieben. Verschiedene Autoren sprechen dem Phagenschnelltest, dessen Einzelheiten den Arbeiten von GUNNISON u. Mitarb. (1951) und BALTAZARD (1956) zu entnehmen sind, größere diagnostische Bedeutung als den biochemischen Differenzierungsmethoden zu, doch scheinen die Erfahrungen noch nicht ausreichend genug zu sein, um ein endgültiges Urteil über die Zuverlässigkeit dieses Testes abgeben zu können. So müssen bei der Durchführung des Testes eine genau eingestellte Verdünnung der Phagensuspension und eine Temperatur von etwa 20^0 C eingehalten werden, da zahlreiche Stämme von Past. pseudotuberculosis bei zu hoher Konzentration des adaptierten Phagen bzw. bei Temperaturen von 37^0 C gelöst werden (GUNNISON, SHEVKY, ZION u. ABBOTT 1951).

Eine unspezifische, lytische Wirkung von Pestphagen gegenüber zahlreichen Stämmen von Past. pseudotuberculosis, einzelnen Salmonella- und Shigellastämmen wurde durch Untersuchungen von GIRARD (1942/43), LAZARUS und GUNNISON (1947), GUNNISON und LAZARUS (1948) u. a. bekannt. Die Autoren berichteten über einen Pestphagenstamm, der alle geprüften Stämme von Past. pseudotuberculosis und einzelne Salmonella- und Shigellastämme löste, bei Past. tularensis und Past. multocida aber keine lytische Wirkung zeigte. GUNNISON, SHEVKY u. Mitarb. (1951) wiesen darauf hin, daß einige bei 37^0 C gewachsene Stämme von Past. pseudotuberculosis gegenüber der lytischen Wirkung des P-Phagen verschieden empfindlich sind. Stämme mit starker Neigung zur Rauhdissoziation sowie Intermediärstämme sollen gewöhnlich empfindlicher als glatte Stämme sein. Die stärkere Wirkung des P-Phagenstammes bei 37^0 C auf Rauh- und Intermediärformen wird mit der Tendenz von Past. pseudotuberculosis bei 37^0 C Rauhformen und bei 22^0 C eine Schleimhülle, die zu einer Maskierung der für den P-Phagen empfindlichen Stellen der Bakterienzelle führen soll, zu erklären versucht. Auch temperaturabhängige Unterschiede im Bakterienstoffwechsel, die bei 22^0 C-Bebrütung der Kulturen möglicherweise den Verlust eines für die Phagenwirkung notwendigen CO-Faktors oder die Bildung eines Hemmfaktors zur Folge haben, werden als Ursache diskutiert. Es ist bekannt, daß bakterielle Polysaccharide die lytische Phagenwirkung hemmen können. Über die Lokalisation des für Past. pseudotuberculosis und Past. pestis gemeinsamen der Phagenwirkung zugängigen Faktors ist nichts bekannt. Alle Versuche, Unterschiede in der Antigenstruktur der phagenempfindlichen und -unempfindlichen Stämme von Past. pseudotuberculosis nachzuweisen, verliefen bisher ohne Ergebnis. Eine Übereinstimmung zwischen Agglutinabilität der Stämme von Past. pseudotuberculosis in Pestsera und ihrer Lysis durch den P-Phagen besteht nicht.

GUNNISON, SHEVKY u. Mitarb. (1951) nehmen sogar an, daß überhaupt keine Beziehungen zwischen der Antigenstruktur und der Empfindlichkeit gegenüber P-Phagen bestehen.

Pasteurella multocida

Bei der kulturell-biochemischen Differentialdiagnose von Past. multocida sind Unbeweglichkeit, schlechtes Wachstum auf gewöhnlichen, nicht aber auf eiweißhaltigen Nährböden bei 37°C, fehlendes oder stark gehemmtes Wachstum bei 22°C und die in Tabelle 1 (s. S. 209) aufgeführten biochemischen Eigenschaften zu berücksichtigen.

Im Gegensatz zu Past. pseudotuberculosis ist Past. multocida wie Past. pestis für weiße Ratten pathogen.

Für die auch heute noch mit diagnostischen Schwierigkeiten verbundene serologische Differentialdiagnose von Stämmen menschlicher und tierischer Herkunft wird die Anwendung der Agglutinations-, Hämagglutinations-, Präcipitations- und Komplementbindungsreaktion empfohlen.

Mit der Agglutinationsmethode konnten von LITTLE u. LYON (1943) 3 Typen und mit dem Präcipitations- bzw. Kapsel-Quellungstest von CARTER und BYRNE (1953) vier verschiedene Typen unterschieden werden. In der Hämagglutination sah CARTER (1955) vier typenspezifische Kapselantigene (Typ A—D). Kreuzweise durchgeführte Immunitätsteste ließen ROBERTS (1947) sechs verschiedene Typen aufstellen. Speciesspezifische, konzentrierte Sera sollen auch Past. pseudotuberculosis und Past. pestis agglutinieren, doch fanden WILSON und MILES (1955) mit verdünnten Sera keine übergreifenden Reaktionen. Allem Anschein nach sind die in der Agglutination nachweisbaren 3 Typen nahe verwandt, so daß ein agglutinierendes, unabgesättigtes Typenserum häufig neben den homologen auch die heterologen Typenstämme erfaßt. Einzelheiten über die Methoden des Antigen- und Antikörpernachweises, die nach eigenen Erfahrungen für die Routinediagnostik noch nicht befriedigen und in diesem Rahmen nicht diskutiert werden, sind den Arbeiten von ROSENBUSCH u. MERCHANT (1939), LEVY-BRUHL (1938), LITTLE u. LYON (1943), ROBERTS (1947), SCHIPPER (1947), MEYER (1952), CARTER (1952, 1955, 1956, 1957), CARTER u. ANNAU (1953), CARTER u. BYRNE (1953), BAIN (1954, 1955), WILSON u. MILES (1955) zu entnehmen.

IX. Pseudotuberkulose beim Tier

Von der in der Veterinärmedizin bisher noch weitgehend üblichen Gepflogenheit, von Pseudotuberkulose nicht bei einem ätiologisch fest umrissenen Krankheitsbild, sondern bei mehreren, ätiologisch zwar verschiedenen, in ihrer klinischen und pathologisch-anatomischen Manifestation aber mehr oder minder ähnlichen Infektionskrankheiten zu sprechen, muß zur Vermeidung weiterer begrifflicher Verwirrung abgegangen werden und die Diagnose *Pseudotuberkulose* allein der durch Past. pseudotuberculosis hervorgerufenen Infektionskrankheit vorbehalten bleiben.

Die ätiologische Uneinheitlichkeit der als Pseudotuberkulose diagnostizierten Infektionen und die z. T. sehr mangelhafte Beobachtung der morphologischen und kulturell-biochemischen Eigenschaften der aus den epizootisch, enzootisch und sporadisch erkrankten Tieren gezüchteten Stämme erschwert die Sichtung der Literatur außerordentlich. Sicher spielen bei den als Pseudotuberkulosen beschriebenen Erkrankungen außer Past. pseudotuberculosis, Corynebacterium bovis und murium, Listeria monocytogenes, verschiedene Salmonellaarten, Pilze oder tierische Parasiten ätiologisch eine Rolle.

1. Befall von freilebenden Haus- und Laboratoriumstieren

Epizootischen Charakter hatten wiederholt Infektionen bei *Meerschweinchen* und *Truthähnen* (MEYER u. BATCHELDER 1926; NICOLLE u. SPARROW 1928; POPPE 1928, mit weiterer Literaturangabe; GATÉ u. BILLA 1928; SCHÜTZE 1928;

TRUCHE u. BAUCHE 1929; BISHOP 1932; PALLASKE 1933; BOQUET 1937; ROSEN-
WALD u. DICKINSON 1944; KARLSSON 1945; BLAXLAND 1947; HAGAN 1951;
MEYER 1952; FELDMAN 1955; MATHEY u. SIDDLE 1954; MAGLIONE u. CERETTO
1956 u. a.) oder bei *Feldmäusen* (OLT 1938) und *Maulwürfen* (GOLOVIN 1930),
während enzootische Ausbrüche unter Kaninchen und *Feldhasen* (LERCHE 1927;
POPPE 1928; DE MENDONCA u. Mitarb. 1943; KARLSSON 1945; VAUTRIN 1949;
CLAPHAM 1953 u. a.), *Ratten* (POPPE 1928, CLAUSSEN 1934, HAAS 1938 u. a.),
Katzen (COCU u. Mitarb. 1931; PALLASKE u. MEYN 1932; PALLASKE 1933; HERMS
1936; BOQUET 1937; VERGE u. Mitarb. 1937; STEPHAN 1941; COMMUNAL 1945),
verschiedenen *Geflügel-* und *Vogelarten* wie *Hühnern, Enten* und *Schwänen*
(CHRISTENSEN 1927; TRUCHE u. BAUCHE 1933; TRUCHE u. ISNARD 1937; SCHÄFER
1939; URBAIN u. Mitarb. 1944; KARLSSON 1945) oder *Tauben* (BECK 1928; LES-
BOUYRIES 1934; CECARELLI 1950; VAN DORSSEN 1951; CLAPHAM 1953; MARTHE-
DAL u. VELLING 1954), *Amseln, Dohlen, Sperlingen, Kanarienvögeln* (CHRISTENSEN
1927; HEELSBERGEN 1927; TRUCHE u. BAUCHE 1933; BEAUDETTE 1940; KARLSSON
1945; MANNINGER 1945; MEYER 1952; CLAPHAM 1953; MARTHEDAL u. VELLING
1954; weitere Literaturangabe), *Affen* (POPPE 1928; SAENZ 1930; SAENZ u.
COSTIL 1932; VERGE u. PLACIDI 1942; URBAIN 1942; URBAIN u. NOUVEL 1949;
NOUVEL u. RINJARD 1949), *Widdern* (JAMIESON u. SOLTYS 1947), *Schweinen*
(SCENNIKOV, zit. nach SCHÜTZE 1929); WRAMBY u. Mitarb. 1941; KARLSSON
1945; EIELAND 1947; JACOTOT u. Mitarb. 1950) gesehen wurden.

Sporadische Erkrankungen wurden unter folgenden Tieren bekannt: *Pferd*
(SCHLAFFKE 1921), *Rind* (MAZZINI 1897), *Ziege* (BAUMANN 1927; RAJAGOPALAN
u. Mitarb. 1944), *Schaf* (MURRAY 1932, TRUCHE 1938), *Löwe* (URBAIN, NOUVEL u.
BULLIER 1944; weitere Literaturangabe), *Nerz, Biber* und *Silberfuchs* (RISLAKKI
1942, KARLSSON 1945, EIELAND 1947), *Känguruh* (zit. nach CHRISTENSEN 1927),
Otter (THAL 1954), *Chinchilla* (VERGE u. PLACIDI 1942; CHAPMAN 1948; LEADER
u. BAKER 1954), *Hund* (CLAUSSEN 1938; COLLET u. Mitarb. 1955).

Einen weiteren Überblick über die Vielzahl der für Past. pseudotuberculosis empfänglichen Tiere und die Befallshäufigkeit der einzelnen Tierarten geben die von THAL (1954)
nach Herkunft (Tierart) und serologischer Typenzugehörigkeit untersuchten 186 (davon
119 aus Schweden) Stämme. Allein 93 Stämme stammten von Hasen oder Kaninchen,
23 von Sumpfbibern, 19 von Truthähnen, 10 von Nerzen, 8 von Meerschweinchen, 6 von
Kanarienvögeln, 3 von Affen, je 2 von Finken und Paradiesvögeln und je 1 Stamm von Ratte,
Katze, Ziege, Reh, Fasan, Rebhuhn, Taube, Pfau und Mensch; 10 Stämme waren unbekannter
Herkunft. Serologisch gehörten 126 Stämme Typ I, 44 Typ II, 14 Typ III und je 1 Stamm
Typ IV und Typ V von Past. pseudotuberculosis an. Zu einer entsprechenden Typenverteilung
kam GOYAN (1956).

Erwähnenswert sind schließlich noch aus der neueren Literatur die Beobachtungen von
CLAPHAM (1953) an *blauen Tauben*, die in einem englischen Distrikt in den Wintermonaten
Dezember bis Februar einem seuchenhaften Ausbruch erlagen. Bei 42 Sektionen wurde das
typische Bild einer abdominellen Verlaufsform der Pseudotuberkulose mit Knötchenbildung
in Milz und Leber, aus denen der Erregernachweis gelang, gesehen. Eine Untersuchung aller
im Anschluß an diese Enzootie im gleichen Distrikt tot aufgefundenen Nagetiere und Vögel
führte zum Nachweis der Pseudotuberkulose bei *17 Holztauben, 1 Lerche, 1 Rotkehlchen,
27 Dohlen, 7 Krähen, 3 Eichelhähern, 1 Spatz, 9 Kaninchen, 6 Feldhasen* und *25 Eichhörnchen.*
Zur gleichen Zeit wurden in zwei weiteren Distrikten 1 infiziertes Kaninchen und 2 Tauben
gefunden. MARTHEDAL und VELLING (1954) beobachteten unter Tauben und *Ziervögeln*
(Kanarienvögel, Schneesperling und Seidenschwanz) und URBAIN u. NOUVEL (1949) unter
Affen enzootische Ausbrüche. Über eine durch Past. pseudotuberculosis ausgelöste Hoden-
Nebenhodenentzündung bei 10 von 20 *Widdern* berichteten JAMIESON u. SOLTYS (1947),

während das erste seuchenhafte Auftreten der Pseudotuberkulose bei *Truthähnen* in England von BLAXLAND (1947) beschrieben wurde. Er bestätigte die Beobachtungen von ROSENWALD und DICKINSON (1944) über das erste Auftreten dieser Seuche bei mehreren Herden in Amerika, während in Deutschland und den nordischen Staaten die Pseudotuberkulose der Truthähne schon früher nachgewiesen wurde (POPPE 1928). Weitere Berichte über sporadisches oder enzootisches Vorkommen von Pseudotuberkulose bei Nagern und verschiedenen Vogelarten bestätigten frühere Beobachtungen; sie geben aber nur einen begrenzten Überblick über die geographische Verbreitung und das jahreszeitliche Auftreten der Seuchenausbrüche.

2. Geographische Verbreitung

Die häufigsten Vorkommen wurden bisher aus westeuropäischen und nordischen Ländern, besonders aus Deutschland, Frankreich und Schweden, berichtet. Einschlägige Mitteilungen liegen aber auch aus England, Nord- und Südamerika (Lit. s. Abschnitt IX, 1.) vor, während in Französisch-Äquatorialafrika, Sahara, Belgisch-Kongo und Indochina wie auf Madagaskar die Seuche unbekannt sein soll (GIRARD 1953). Die Berichte sprechen für eine weltweite Verbreitung dieser Tierseuche, der anscheinend auch durch unterschiedliche Klimata keine geographischen oder geomedizinischen Grenzen gesetzt sind. Die scheinbare Häufung ihres Vorkommens in manchen Ländern und innerhalb einzelner Länder in bestimmten Gebieten hängt sicher einmal mit der intensiveren Fahndung nach Herden, zum anderen aber auch mit der regionär unterschiedlichen Verteilung empfänglicher Tierarten, unter denen nach den bisherigen Beobachtungen verschiedene freilebende und zahme Nagetier- wie Vogelarten als natürliche Erregerreservoire vermutet werden, zusammen.

Eine systematische und regelmäßige Untersuchung der interkurrent eingehenden Laboratoriumstiere, insbesondere aber der Meerschweinchen, Kaninchen und Vögel, würde sicher über reine Zufallsbeobachtungen hinaus interessante Rückschlüsse über die Häufigkeit des Vorkommens der Pseudotuberkulose unter Laboratoriumstieren und damit auch über ihre geographische Verbreitung zulassen, sofern die Laboratoriumstiere von verschiedenen Tierzüchtern erworben und nicht in eigener Zucht gehalten werden. Einen Hinweis auf die geographische Verbreitung dieser Krankheit geben auch menschliche Erkrankungsfälle, die in Gegenden nachgewiesen werden, in denen diese Tierseuche bisher noch nicht bekannt oder, von sporadischen Fällen abgesehen, ohne jede Bedeutung ist.

3. Epizootologie

a) Jahreszeitliches Auftreten

Jahreszeitlich scheint das epi- und enzootische Auftreten der Pseudotuberkulose in seinem Maximum an die naßkalten Wintermonate gebunden zu sein. Ihr im neueren Schrifttum von OLT (1938) beschriebenes seuchenhaftes Auftreten (Rodentiose) bei Feldhasen und Feldmäusen oder die von ROSENWALD u. Mitarb. (1944) und BLAXLAND (1947) beobachteten Stallseuchen bei Truthühnern fielen in die Herbst- und Wintermonate, eine Taubenseuche (CLAPHAM 1953) in die Monate Dezember bis Februar und die von GATÉ u. BILLA (1928) untersuchte Meerschweinchenseuche in den Dezember. Nach GATÉ u. BILLA (1928), BOQUET (1937), OLT (1938) u. a. fördern Nässe, Kälte und fehlendes Licht und nach URBAIN u. Mitarb. (1944) ungesunde Ernährung das Angehen einer Infektion. Junge Tiere sollen für die Infektion besonders empfänglich sein. Auch sporadische Erkrankungen scheinen nicht selten in den Wintermonaten oder im Winterhalbjahr vorzukommen (VERGE u. Mitarb. 1937; BEAUDETTE 1940;

URBAIN u. NOUVEL 1949; JACOTOT u. Mitarb. 1950 u. a.), sie werden aber auch unabhängig von erkennbaren jahreszeitlichen und meteorotropen Einflüssen beobachtet.

b) Eintrittspforte und Übertragungsmodus

Über Eintrittspforte und Übertragungsmodus des Erregers wissen wir bis heute nichts Sicheres. Die fast regelmäßige Beobachtung, daß die Organe der Bauchhöhle die stärksten Krankheitserscheinungen zeigen und durch Fütterung infizierte Tiere unter ausgeprägteren Darmerscheinungen sterben als parenteral infizierte, läßt in der peroralen Aufnahme des Erregers den häufigsten Eintrittsweg vermuten. Daß die perorale Aufnahme der Bakterien nicht mit gesetzmäßiger Regelmäßigkeit von einem Haften der Infektion gefolgt ist und neben Schwankungen in der Virulenz des Erregers auch äußere Einflüsse oder individuelle Faktoren, wie Schwankungen in der Resistenz oder Immunität eine wesentliche Rolle spielen müssen, zeigen eigene und von zahlreichen Autoren (BISHOP 1932; PALLASKE u. MEYN 1932; SCHÄFER 1939; EIELAND 1947 u. a.) durchgeführte perorale und parenterale Infektionsversuche, bei denen nicht alle infizierten Tiere klinische und pathologisch-anatomische Zeichen einer stattgehabten Infektion boten.

PFAFF (1905) konnte beispielsweise Kanarienvögel erst nach Reizung des Darmes mit Senfölsamen erfolgreich peroral infizieren, so daß sie nach 5 Tagen krank waren und nach 2 weiteren Tagen eingingen. Bei Spatzen und Tauben verliefen dagegen die Fütterungs- und Infektionsversuche auch nach Reizung mit Senföl erfolglos.

Für einen peroralen Infektionsmodus spricht nach PALLASKE (1933) und OLT (1938) in pathogenetischer Hinsicht auch die von ihnen und anderen Autoren beobachtete besonders starke Reaktion der Magen-, Portal- und Mesenteriallymphknoten sowie die Veränderungen des Duodenums und z. T. auch der Appendix. Die direkte Aufnahme des Erregers durch die Haut, Schleimhaut, vielleicht auch Tonsillen spielt daneben wahrscheinlich nur eine untergeordnete Rolle. Auf die Bedeutung der Tonsillen als Entwicklungsort pseudotuberkulöser Veränderungen wurde von OLT (1914, 1938) hingewiesen. Möglicherweise ist die glanduläre, auf cervicale und thorakale Lymphknoten beschränkte Verlaufsform der Pseudotuberkulose auf die Übertragung des Erregers durch Biß zurückzuführen (MEYER 1952).

Wiederholt wurde die Frage aufgeworfen, ob die perorale Infektion der Tiere nicht durch Genuß faekal-verunreinigten Wassers und Futters oder, besonders bei Nagetieren, durch Fressen infektiösen Fleisches von Fallwild erfolgt (ALBRECHT 1910; LOREY 1911; SAISAWA 1913; FRAENKEL 1924; POPPE 1928; OLT 1938; ROSENWALD u. DICKINSON 1944; URBAIN u. NOUVEL 1944; BLAXLAND 1947; PIÉCHAUD 1952 u. a.). OLT (1938) sah bei Hasen Uterusinfektionen durch Begattung.

Sicher spielen Keimträger oder Dauerausscheider (MEYER 1928; TRUCHE u. BAUCHE 1929; BISHOP 1932; OLT 1938 u. a.), deren prozentualen Anteil wir nicht kennen, und die auch nach experimenteller Infektion in großer Zahl beobachtet werden (OLT 1938, BERGER 1957, eigene unveröffentlichte Beobachtungen), epidemiologisch eine große Rolle. Ob bei der Übertragung des Erregers Zwischenträger beteiligt sind, ist unbekannt. HAAS (1938) sowie BLANC u. BALTHAZARD (1944) gelang es nicht, durch Stiche von Xenopsylla cheopsis Past. pseudotuberculosis, die im Gegensatz zur sehr rasch abgetöteten Past. multocida 35 Tage im Rattenfloh am Leben bleibt, zu übertragen. JAMIESON u. SOLTYS (1947) diskutierten bei einer Stallseuche unter Widdern die Möglichkeit der Übertragung

des Erregers durch Schafzecken (Ixodes ricinus). Eine befriedigende Klärung
der mit der Epizootologie der Pseudotuberkulose zusammenhängenden Fragen
ist unseres Erachtens aber erst nach Auffinden der biologischen Reservoire von
Past. pseudotuberculosis, das nach OLT (1938) für die Nagetiere die Feldmaus
darstellt, möglich.

X. Pseudotuberkulose des Menschen

Die Pseudotuberkulose als menschliche Infektionskrankheit war bis 1953
in der bakteriologischen wie klinischen Diagnostik weitgehend unbekannt.
MEYER stellte 1952 17 durch Past. pseudotuberculosis ausgelöste menschliche
Pseudotuberkulosefälle zusammen. Von KNAPP und MASSHOFF (1954) wurden
zuerst nur 15 von ALBRECHT (1910), LOREY (1911), SAISAWA (1913), ROMAN
(1916), NEUGEBAUER (1933), DUJARDIN-BEAUMETZ u. Mitarb. (1938), TOPPING
u. Mitarb. (1938), MACCHIAVELLO (1941), MOSS u. BATTLE (1941), SNYDER u. VOGEL
(1943), MASON u. MEYER (1948), HÄSSIG u. Mitarb. (2 Fälle 1949), BURIÁNEK
u. Mitarb. (1949) und PIÉCHAUD (1952) beschriebene Fälle als bakteriologisch
eindeutig charakterisiert anerkannt, denen aber noch der von PAUL u. WELTMANN
(1934) beschriebene Krankheitsfall zuzurechnen ist. Der ätiologischen Klärung
der von MASSHOFF (1953), MASSHOFF u. DÖLLE (1953) beschriebenen abscedierenden, reticulocytären Lymphadenitis als eine besondere, durch Past. pseudotuberculosis verursachte Verlaufsform der menschlichen Pseudotuberkulose
durch KNAPP (1954), KNAPP u. MASSHOFF (1954) folgte von klinischer, bakteriologischer und pathologischer Seite die Veröffentlichung weiterer Erkrankungsfälle (s. S. 248, 249).

Bei den von WELTMANN u. FISCHER (1914), BAYER u. HERRENSCHWAND (1919), SCHMORL
(1920) und UMLAUFT (1931) beschriebenen Fällen reichen unseres Erachtens die Angaben über
Erregernachweis und -art nicht aus, um eine sichere ätiologische Klassifizierung vornehmen
zu können. Auch für die von HÄSSIG u. Mitarb. (1949) gezüchteten Stämme, die durch
Säuerung von Saccharose und Sorbit bzw. fehlende Säuerung von Rhamnose und Xylose in
ihren biochemischen Eigenschaften Unterschiede gegenüber dem typischen Verhalten dieser
Species aufweisen und zu den bisher bekannten fünf serologischen Typen keine antigenen Beziehungen zeigen, muß die Frage diskutiert werden, ob sie tatsächlich der Species angehören
und damit zugleich einen weiteren serologischen Typ repräsentieren.

Die bis 1956 durch Erregernachweis gesicherten und in der Weltliteratur veröffentlichten
Erkrankungsfälle menschlicher Pseudotuberkulosen sowie die vom Verfasser bis 1957 kulturell
diagnostizierten oder nachuntersuchten Fälle sind unter Aufführung des Erkrankungs- oder
Veröffentlichungsjahres, Alters der Patienten, der wichtigsten klinischen, operativen und
pathologisch-anatomischen Befunde, des Erregernachweises und Krankheitsverlaufes in
2 Tabellen (Tabelle 4 und 5, S. 240 und 244) zusammengestellt.

1. Häufigkeit und geographische Verbreitung der menschlichen Pseudotuberkulose

Über die tatsächliche Häufigkeit ihres Vorkommens und die geographische
Verbreitung der menschlichen Pseudotuberkulose geben die bisherigen Beobachtungen noch keine befriedigende Auskunft. Doch sprechen die in Europa (Deutschland, Frankreich, Österreich, Schweiz, Tschechoslowakei und Ungarn), Nord-
und Südamerika (USA, Brasilien) und Asien (Japan, Indien) nachgewiesenen
sporadischen Erkrankungsfälle für eine weltweite Verbreitung der menschlichen
Pseudotuberkulose. Ihr Auftreten ist wahrscheinlich an das gleichzeitige Vorkommen der tierischen Pseudotuberkulose gebunden, deren geographische Ver-

breitung ebenfalls noch weitgehend unerforscht und unbekannt ist. So wissen wir heute noch nicht, ob diese Zoonose tatsächlich in europäischen Ländern wie Deutschland, Frankreich, Holland, Österreich, Schweden oder Tschechoslowakei, wie von MEYER (1952) oder FELDMAN (1955) u. a. angenommen, mehr zu Hause ist als in Amerika, England, den Mittelmeerländern, Afrika und Asien, oder ob nicht ihrer Diagnose in den westeuropäischen Ländern des Kontinents größere Beachtung geschenkt wurde. Nach GIRARD (1953) sollen Infektionen mit Past. pseudotuberculosis bisher in Französisch-Nord- und -Äquatorialafrika, Belgisch-Kongo, Indochina und auf Madagaskar unbekannt sein. Es erscheint uns unwahrscheinlich, daß bei der großen Zahl der für Past. pseudotuberculosis empfänglichen Tierarten die Pseudotuberkulose beim Menschen nur in bestimmten geographisch und klimatisch begrenzten Gebieten vorkommen soll. Vielmehr ist anzunehmen, daß die bis heute als Infektionen mit Past. pseudotuberculosis diagnostizierten Erkrankungsfälle nur einen Teil ihrer tatsächlichen Häufigkeit und ihrer geographischen Verbreitung erkennen lassen.

2. Klinik und Pathologie der menschlichen Pseudotuberkulose

Das klinische Bild der menschlichen Pseudotuberkulose ist uncharakteristisch, so daß eine sichere klinische Diagnose und Abgrenzung gegen andere, differentialdiagnostisch in Betracht kommende Erkrankungen nur durch bakteriologisch-serologische Untersuchungen möglich ist. Die klinische Symptomatologie kann einer *septisch-typhösen Erkrankung, akuten-subakuten Appendicitis, akuten Gastroenteritis* oder *subakuten bis chronischen Enteritis* entsprechen. Einzelne Fälle lassen sich in ihrer Symptomatologie auch diesen Krankheitsbildern nur schwer oder nicht zuordnen. Mitunter lassen die ersten Symptome die Verdachtsdiagnose einer akuten Enteritis oder septischen Allgemeininfektion und erst der weitere Verlauf einer Appendicitis stellen.

a) Septisch-typhöse Verlaufsform

Ihr *klinisches Bild* ist durch einen schweren septischen Verlauf gekennzeichnet. Das Prodromalstadium mit meist jähem Temperaturanstieg, Schüttelfrost, z. T. starken Kopf-, Glieder- oder Gelenkschmerzen ist uncharakteristisch. Die Temperatur ist im weiteren Verlauf intermittierend oder septisch. An weiteren, im Einzelfall verschieden stark ausgeprägten Symptomen werden Appetitlosigkeit, rasche Gewichtsabnahme, Erbrechen, Diarrhöe, auch Obstipation beobachtet. Früher oder später stellen sich Schmerzen im Oberbauch, infolge erheblicher Vergrößerung der Leber und Milz oder nur eines der beiden Organe, und Ikterus ein. Auch Ergüsse in den verschiedenen Körperhöhlen, Bronchitis oder Bronchopneumonien, können auftreten und klinische Befunde für eine Miterkrankung von Herz und Nieren sprechen. Der Tod tritt bei den meisten Patienten unter zunehmenden Intoxikationserscheinungen, z. T. im Koma zwischen dem 10. und 20. Tag ein, doch wurde auch eine klinisch wesentlich kürzere oder längere Krankheitsdauer beobachtet.

Eine besondere Ausnahme stellt der von BURIÁNEK u. Mitarb. (1949) beschriebene, chronisch-rezidivierende Krankheitsfall mit einer Krankengeschichte von über $3^{1}/_{2}$ Jahren dar (Tabelle 4, S. 242). Als wesentliche klinische Befunde wurden Milzschwellung, vergrößerte

Tabelle 4. *Bakteriologisch gesicherte menschliche*

Veröffentlichung	Autor	Land	Alter (Jahre)	Wesentliche klinische Symptome	Klinische (Verdachts-) Diagnose
1910	Albrecht	Österreich	15	Symptome einer akuten Appendicitis	Appendicitis
1911	Lorey	Deutschland	53	Diarrhöe, Fieber, Leberschwellung, Ikterus	Typhus abdominalis
1909/13	Saisawa	Japan	21	Tonsillitis, Fieber, Ikterus, Leberschwellung, Bauchdeckenspannung, Verstopfung, dann Diarrhöe	unklarer fieberhafter Infekt
1916	Roman	Tschechoslowakei	46	Bauchbeschwerden, Appetitlosigkeit	Carcinose, Peritonitis
1933	Neugebauer	Tschechoslowakei	45	Leptomeningitis, Tracheobronchitis, Lungenabscesse	unklarer Infekt
1934	Paul u. Weltmann	Österreich	31	Ikterus, Leber- und Milzschwellung	Leberabsceß
1938	Dujardin-Beaumetz	Frankreich	32	Fieber, Diarrhöe, Delirium, Koma	typhöse Erkrankung
1938	Topping u. Mitarb.	USA	29	Fieber, Schüttelfrost, Kopfschmerz mit leichter Nackensteifigkeit, Ikterus	unklarer fieberhafter Infekt, Grippe
1941	Macchiavello	Brasilien	18	Fieber, Schüttelfrost, Ikterus, Diarrhöe	Pest (Mischinfektion)
1941	Moss u. Battle	USA	34	Kopfschmerz, Schüttelfrost, Schmerzen in der rechten Lendengegend, Fieber	unklarer Infekt
1943	Snyder und Vogel	USA	41	Appetitlosigkeit, Bauchdeckenspannung, Obstipation, Fieber u. a.	unklarer Infekt (Sepsis ?)
1948	Mason u. Meyer	USA	?	zit. nach Meyer (1948),	
1949	Hässig[1] u. Mitarb.	Schweiz	60	Fieber, Ikterus	Coma hepaticum, Cholangitis
1949	Hässig u. Mitarb.	Schweiz	69	Diarrhöe, Fieber, Husten, Auswurf	Lungentuberkulose ? Viruspneumonie ?

a. m. = Erregernachweis ante mortem, in allen anderen Fällen post mortem (p. m.);

Pseudotuberkulose der Weltliteratur

Krankheits-dauer (Tage)	Operations-Sektionsbefund	Verlauf	Erregernachweis aus: Agglutinin-Titer	Therapie
4	Lymphknotenpakete in der Ileocoecalgegend; Enteritis follicularis suppurativa	gutartig	Lymphknoten (Tierversuch) 1:160	—
11	miliare Knötchen und Abscesse in der Leber, Mesenteriallymphknoten o. B.	tödlich	Blut (a. m.), Leber, Milz, Gallenblase, Knochenmark	—
11	Leber- und Milzschwellung mit zahlreichen Knötchen und Abscessen, Enteritis follicularis, Schwellung der Peyerschen Plaques und mesenterialen Lymphknoten	tödlich	Blut (a. m.), Perikarderguß	—
11	Leberschwellung, nekrotische Herde in der Leber und vergrößerte Mesenteriallymphknoten	tödlich	Leber, Milz	—
20	Leber- und Milzschwellung mit zahlreichen graugelblichen und nekrotischen Herden, Bronchitis	tödlich	Leber, Milz	—
19	in der Leber zahlreiche graugelbliche, erbsengroße Abscesse	tödlich	Absceßeiter, Milz	—
15	— keine Sektion	tödlich	Blut (a. m.)	—
17	Ascites, Leber- und Milzschwellung. Zahlreiche kleine Abscesse in der Leber, Nekrose am Pankreaskopf	tödlich	(Blut (a. m.) Leber 1:320	—
7	Leberschwellung, nekrotische Herde	tödlich	Leber (Past. pestis und Past. pstbc.)	—
9 Wochen	Ascites, Leber o. B.; Milz vergrößert. Zahlreiche nekrotische Herde. Nekrose der Lymphknoten am Pankreaskopf	tödlich	Blut (a. m.), Blut (p. m.)	—
?	—	gutartig? Klinik gegen ärztl. Rat verlassen	Blut (a. m.)	Sulfathiazol

genaue Angaben fehlen

Krankheits-dauer (Tage)	Operations-Sektionsbefund	Verlauf	Erregernachweis aus: Agglutinin-Titer	Therapie
3—4 Wochen	Pigmentcirrhose der Leber Pseudotuberkulose der Leber. Milzschwellung	tödlich	Leber, Milz	Penicillin, Bilamid Streptomycin
17	pyelophlebitische Leberabscesse. Pigmentcirrhose, Hämosiderose der Leber, Milz und mesenterialen Lymphknoten	tödlich	Leber, Milz	Penicillin

[1] Siehe Besprechung der Fälle S. 210—212.

Tabelle 4

Veröffent-lichungen	Autor	Land	Alter (Jahre)	Wesentliche klinische Symptome	Klinische (Verdachts-) Diagnose
1949	BURIANEK	Tschecho-slowakei	36	Milzschwellung, Schwellung der cervicalen Lymphknoten, Metrorrhagie, Fieber	Sepsis
1952	PIÉCHAUD	Frankreich	? (Kind ?)	Symptome einer akuten Appendicitis	akute Appenticitis
1956	SACHDEVA u. Mitarb.	Indien	22	Fieber, Singultus, Schmerzen im linken Hypochondrium	unklarer fieberhafter Infekt (Enteritis ?)
1956	BERG u. HECKER	Deutschland	10	Schmerzen im Unterbauch, Abwehrspannung	akute Appendicitis

Cervicallymphknoten und Metrorrhagien angegeben. Die Heilung soll durch Streptomycinbehandlung erfolgt sein. Die Diagnose dieses uncharakteristischen Krankheitsbildes als Pseudotuberkulose war durch den Erregernachweis aus Blut möglich, der bisher zu Lebzeiten der Patienten mit einer septisch-typhösen Verlaufsform der menschlichen Pseudotuberkulose gelungen ist (LOREY 1911; SAISAWA 1913; DUJARDIN-BEAUMETZ 1938; TOPPING u. Mitarb. 1938; MOSS u. BATTLE 1941; SNYDER u. VOGEL 1943; BURIANEK u. Mitarb. 1949 und KNAPP 1957). Bei den übrigen Erkrankungsfällen der septisch-typhösen Verlaufsform wurde die Diagnose einer Pseudotuberkulose erst durch den postmortalen Erregernachweis gestellt.

Die bisher veröffentlichten 13 tödlich bzw. die 2 gutartig verlaufenen Fälle sowie ein weiterer von MASON u. MEYER (1952) nicht näher beschriebener Fall der septischen Verlaufsform der menschlichen Pseudotuberkulose lassen kein Urteil über die Häufigkeit ihres Vorkommens zu. Bei der uncharakteristischen Symptomatologie entgehen sicher immer wieder Erkrankungsfälle infolge fehlenden Erreger- oder Antikörpernachweises der ätiologischen Klärung. Erst ein in größerem Rahmen systematisch durchgeführter Erreger- und Antikörpernachweis in Blutproben ätiologisch nicht geklärter Sepsis- oder Typhusverdachtsfälle wird uns sichere Aussagen über die Häufigkeit ihres Vorkommens machen lassen.

Den diagnostischen Wert einer Erweiterung der routinemäßig für die serologische Diagnose von Salmonella-, Shigella- und Brucellainfektionen angewandten Widal-Reaktion auch auf die menschliche Pseudotuberkulose veranschaulicht die auszugsweise Wiedergabe der uns von der Medizinischen Universitätsklinik Tübingen (Direktor: Prof. Dr. BENNHOLD) freundlicherweise überlassenen Krankengeschichte des folgenden, allein durch bakteriologisch-serologische Untersuchung diagnostizierten Falles.

A. B., 26 Jahre, seit 4 Tagen Fieber, starke Kopfschmerzen, Appetitlosigkeit. Bei der Klinikaufnahme 40° C Temperatur, Milz etwas tastbar, Senkung beschleunigt, sonst o. B. Verdachtsdiagnose: Salmonellainfektion. Während die Untersuchung des Patientenblutes auf Salmonella-Keime und -Antikörper negativ verlief, konnten Agglutinine gegen Past. pseudotuberculosis bis zur Serumverdünnung von 1:1280 bzw. 1:10240 am 5. und 8. Tag des Klinikaufenthaltes und Past. pseudotuberculosis (Typ I) in 2 Blutkulturen am 10. und 16. Krankheitstag in Reinkultur nachgewiesen werden. Ohne routinemäßige Untersuchung aller wegen Verdacht auf Salmonellainfektionen eingesandten Blutproben zugleich auch auf Antikörper gegen Past. pseudotuberculosis wäre dieser Fall ätiologisch nicht geklärt worden. Erst der positive Titer ließ uns zahlreiche Blutkulturen anlegen, die auffallenderweise wie bei einem von GRABER u. KNAPP (1955) beschriebenen Fall noch nach Rückgang der akuten Krankheitssymptome und des Fiebers positiv waren.

(Fortsetzung)

Krankheits-dauer (Tage)	Operations-Sektionsbefund	Verlauf	Erregernachweis aus: Agglutinin-Titer	Therapie
über 3 Jahre ?	—	gutartig	Blut	Streptomycin
?	Appendix o. B., vergrößerte Mesenteriallymphknoten	gutartig	Lymphknoten	—
9	Leber- und Milzschwellung Ulcera im Ileum, Schwellung der Peyerschen Plaques und mesenterialen Lymphknoten	tödlich	Lymphknoten	Sulfaguanidin, Penicillin, Aureomycin
14	perforiertes, infiziertes Enterocystom, umschriebene Peritonitis	gutartig	Enterocystom	Achromycin

Die *pathologisch-anatomischen* Veränderungen der septisch-typhösen Verlaufsform sind durch tumorförmige, knotige, abscedierende oder verkäsende, einzeln oder in großer Zahl nachweisbare Herde vornehmlich in der Leber und Milz oder nur einem Organ bestimmt. Weniger häufig werden Ascites, Zeichen einer ulcerierenden Enteritis oder Veränderungen der mesenterialen Lymphknoten beobachtet. Die grauweißen oder geblichen Organherde stellen coagulierte Nekrosen des Parenchyms mit mehr oder weniger starker leukocytärer Durchsetzung dar. Der Erregernachweis gelingt meist ohne Schwierigkeiten. Die makroskopischen Veränderungen entsprechen weitgehend den bei spontaner und experimenteller Infektion von Haus- und Laboratoriumstieren zu beobachtenden Befunden. Sie lassen wie beim infizierten Tier Unterschiede in Zahl, Ausmaß und Anordnung der Organveränderungen erkennen. Auch in ihren histologischen Veränderungen zeigt diese Form der menschlichen Pseudotuberkulose keine Unterschiede gegenüber den beim Tier erhobenen Befunden (Einzelheiten: Abschnitt VI und neuere Arbeiten von TOPPING u. Mitarb. 1938; MOSS u. BATTLE 1941; MACCHIAVELLO 1941; MEYER 1952; FELDMAN 1955; SACHDEVA u. Mitarb. 1956 u. a.).

b) Appendicitische Verlaufsform

MASSHOFF u. DÖLLE (1953) und MASSHOFF (1953) haben von pathologisch-anatomischer Seite in systematischen Untersuchungen das Problem der mesenterialen Lymphadenitis, Lymphadenopathia mesenterialis oder Lymphadenitis mesaraica (HEUSSER 1924) aufgegriffen. Sie fanden in entzündlich veränderten mesenterialen Lymphknoten von Patienten, die klinisch das Bild einer akuten, operativ nicht bestätigten Appendicitis boten, einen bis dahin nicht beachteten Entzündungsprozeß, den sie als *abscedierende, reticulocytäre Lymphadenitis* beschrieben.

Das zuerst von MASSHOFF u. DÖLLE (1953) veröffentlichte Beobachtungsgut betraf 41 einschlägige Fälle. Die Lymphknoten stammten fast ausschließlich von Jugendlichen im Alter von 2—23 Jahren, die wegen Appendicitisverdacht operiert wurden, aber bioptisch und histologisch das Bild einer mesenterialen Lymphadenitis und nicht einer Appendicitis zeigten. Die Mehrzahl der Erkrankten war 10—15 Jahre alt. Ihr Untersuchungsgut enthielt

16*

Tabelle 5. *Weitere bakteriologisch gesicherte*

Anzahl	Stamm Nr. Serologischer Typ	Klinische (Verdachts-) Diagnose	Operationsbefund	Histologische Befunde 1. Appendix 2. Lymphknoten	Alter und Geschlecht des Patienten
1	1^I N. H.	akute Appendicitis	Appendix o. B., entzündlich veränderte mesenteriale Lk., besonders in der Ileocoecalgegend	o. B. a. r. L.	$11^1/_2$ ♂
2	2^I H. A.	chronische rezidivierende Appendicitis	reichlich Exsudat, Appendix o. B. im Ileocoecalwinkel mehrere walnußgroße Lk.	o. B. a. r. L.	17 ♂
3	3^I J. A.	subakute Appendicitis	reichlich Exsudat, Rötung der Serosa der Appendix, Lymphadenitis, mesenteriale, zahlreiche vergrößerte Lk. im Mesenterium ileocoecale	Periappendicitis a. r. L.	$10^1/_2$ ♀
4	4^I Sp. L.	Tumor re. Unterbauch, perityphlitischer Absceß (?)	Konglomerattumor des Coecum und Ileum, vergrößerte Lk. im Mesenterium	a. r. L. a. r. L. (Dr. GRABER, Detmold)	21 ♂
5	9^V St. G.	Leberschwellung, Tumor li. Unterbauch, Mesenterialdrüsen-Tbc ?	Appendix o. B., zahlreiche bis haselnußgroße mesenteriale Lk.	o. B. o. B.	$8^1/_2$ ♀
6	10^V E. B.	akute Appendicitis	Appendix o. B., zahlreiche bis haselnußgroße mesenteriale Lk.	Hyperplasie des lymphatischen Apparates der Mucosa, Sinuskatarrh	3 ♂
7	$11/225^I$ Ha. J.	akute Appendicitis	Appendix o. B., zahlreiche walnußgroße Lk. im Ileocoecalwinkel und Mesenterium	o. B. Sinuskatarrh (Dr. HUSTEN, Essen-Steele)	12 ♂
8	$15/418^I$ Lu. H.	akute Appendicitis	Appendix o. B. im Ileocoecalwinkel wie im Mesenterium zahlreiche walnußgroße Lk.	o. B. a. r. L. (Dr. HUSTEN, Essen-Steele)	12 ♂

a. r. L. = abscedierende reticulocytäre Lymphadenitis; Lk. = Lymphknoten; o.B. = ohne Besonderheiten; ·/. = nicht untersucht oder nicht bekannt; — = ohne Therapie.
[1] Erregernachweis bei Fall 1—14 am Hygiene-Institut Tübingen, bei Fall 15 am Hygiene-Institut Frankfurt (Privatdozent Dr. BRANDIS). [2] In der Arbeit von KÖNIG u. MAURATH blieb

Fälle menschlicher Pseudotuberkulosen [1]

| Krankheits-Dauer | | | | | | | |
Prodromal-erscheinungen (Tage)	Klinik-Aufent-halt (Tage)	Krankheits-Verlauf nach Appendektomie	Einsender	Erregernachweis aus	Höchster Antikörper-titer	Chemotherapie	Veröffent-lichung
seit 2 Tagen Be-schwerden be-sonders im rech-ten Unterbauch	11	o. B.	Chir. Univ.-Klinik Tübingen	Lk.	1:160	4 g Strepto-mycin in 4 Tagen	KNAPP u. MASSHOFF (1954)
seit 1 Woche Beschwerden be-sonders im rech-ten Unterbauch	12	o. B.	Chir. Univ.-Klinik Tübingen	Lk.	·/.	1,5 g Terramycin, 3,0 g Aureomycin	KNAPP u. MASSHOFF (1954) KÖNIG u. MAURATH (1957) [2]
seit 2 Tagen starke Schmer-zen besonders im rechten Unterbauch	10	o. B.	Kreis-krankenhaus Freudenstadt	Lk.	1:1600	—	KNAPP u. MASSHOFF (1954)
seit 2 Tagen plötzlich starke Bauchschmerzen, Fieber	22	o. B.	Krankenhaus Höxter	Blut	1:3200	3,2 g Terramycin in 3 Tagen	GRABER u. KNAPP (1955)
Seit 8 Wochen Bauchbeschwer-den, Leber-schwellung seit 2 Tagen 39° C Fieber, zunehmende Beschwerden	25	o. B.	Landes-krankenhaus Sigmaringen	Lk.	negativ	—	KNAPP (1955)
akute Bauch-beschwerden	8	o. B.	Landes-krankenhaus Sigmaringen	Lk.	negativ	—	KNAPP (1955)
seit 2 Tagen starke Leib-schmerzen mit Brechreiz	13	o. B.	St. Laurentius-hospital Essen-Steele	Lk.	1:1280	—	nicht ver-öffentlicht
seit Wochen Magen- und Bauchbeschwer-den, seit 3 Tagen verstärkt im rechten Unter-bauch	21	o. B.	St. Laurentius-hospital Essen-Steele	Lk.	1:2560	1,5 g Achromycin in 5 Tagen	nicht ver-öffentlicht

Fall 1 unberücksichtigt. Bei Fall 2 und 10 wurden Angaben gemacht, die nicht mit dem bakteriologischen bzw. serologischen Untersuchungsbefund übereinstimmen.

Soweit nicht besonders vermerkt, wurden die histologischen Untersuchungen von Prof. MASSHOFF am Pathologischen Institut der Universität Tübingen (Direktor: Prof. Dr. LETTE-RER) durchgeführt.

Tabelle 5

Anzahl	Stamm-Nr. Serologischer Typ	Klinische (Verdachts-) Diagnose	Operationsbefund	Histologische Befunde 1. Appendix 2. Lymphknoten	Alter und Geschlecht des Patienten
9	16/448[I] B. E.	subakute Appendicitis	Appendix o. B. im Mesenterium einzelne vergrößerte Lk.	rezidivierende Appendicitis, Sinuskatarrh (Prof. Schairer, Ulm)	18 ♀
10	18/498[I] B. P.	akute Appendicitis	Appendix o. B., vergrößere Lk. im Mesenterium	leukocytäre Infiltrate a. r. L.	11 ♂
11	21/548[I] Wy. M.	chronische Appendicitis	Appendix o. B., vergrößerte Lk. im Ileocoecalwinkel	o. B. o. B.	8 ♂
12	22/647[I] Sch. G.	akute Appendicitis	Appendix o. B., im Mesenterium faustgroßer Tumor eines nußgroßen, vereiterten Lk.	entzündliche Infiltrate a. r. L.	$9^{1}/_{2}$ ♂
13	23/675[I] A. B.	Salmonella-infektion, Milz-schwellung	·/.	·/.	26 ♂
14	24/689[I] K. G.	akute Appendicitis, Bronchopneumonie	Appendix gefäßinjiziert, vereinzelt vergrößerte, glasige Lk. im Mesenterium	Rundzellen-infiltrate —	8 ♂
15	17/488[I] G. R.	akute Appendicitis	fixiertes Coecum mit fingerdicker Appendix verbacken. Entzündeter Lk. an lateraler Coecumwand	a. r. L. a. r. L.	15 ♂

weitere 200 Lymphknoten mit uncharakteristischen Veränderungen (Sinuskatarrh, Hyperplasie, Blutresorption, unspezifische Lymph- oder Perilymphadenitis) sowie 82 tuberkulöse, 29 carcinomatöse, 6 sarkomatöse und je 1 Lymphknoten mit aleukämischer Retikulose und Lymphogranulomatose. Die übrigen Lymphknoten aus einer Gesamtzahl von über 800 waren histologisch unverändert.

Affektionen der mesenterialen Lymphknoten haben verschiedene Ursachen, auf die im einzelnen einzugehen hier nicht der Ort ist. Neben der häufig — oft zu Unrecht — angenommenen Tuberkulose kommen in erster Linie, vor allem bei Jugendlichen, uncharakteristische Hyperplasien ohne entzündliche Beteiligung der Serosa in Betracht. Auch Entzündungen des Darmes durch Keime der Typhus-Paratyphus-Gruppe oder Würmern führen zu einer Schwellung mesenterialer Lymphknoten. Schließlich können entzündliche Vergrößerungen durch Tumoren bedingt sein.

(Fortsetzung)

| Krankheits-Dauer | | | | | | | |
Prodromalerscheinungen (Tage)	Klinik-Aufenthalt (Tage)	Krankheits-Verlauf nach Appendektomie	Einsender	Erregernachweis aus	Höchster Antikörpertiter	Chemotherapie	Veröffentlichung
seit einigen Tagen zunehmende Bauchbeschwerden	9	o. B.	Krankenanstalten Ulm	Lk.	negativ (1. Untersuchung 7 Wochen nach Erkrankung)	—	nicht veröffentlicht
vor 14 Tagen Fieber, Erbrechen. Seit 2 Tagen Fieber, Bauchschmerzen	14	o. B.	Chir. Univ.-Klinik Tübingen	Lk.	1:1280	—	KÖNIG u. MAURATH (1957)[2]
seit mehreren Wochen Bauchschmerzen	10	o .B.	Chir. Univ.-Klinik Tübingen	Lk.	negativ	—	KÖNIG u. MAURATH (1957)
seit 8 Tagen Fieber, seit 3 Tagen starke Bauchschmerzen	14	o. B.	Kreiskrankenhaus Balingen	Eiter	1:640	1 g Streptomycin i. p., 1mal täglich 1 Ampulle Supracillin über 12 Tage	nicht veröffentlicht
seit 4 Tagen Kopfschmerzen, Fieber um 40°C, Appetitlosigkeit	30	·/.	Med. Univ.-Klinik Tübingen	Blut	1:10240	1,5 g Aureomycin, 0,25 g Leukomycin, 15 g Steptomycin	s. Seite 242
seit 8 Tagen Husten und Fieber, seit 1 Tag Bauchschmerzen und Übelkeit	11	o. B.	Chir. Univ.-Klinik Tübingen	Lk.	negativ	—	nicht veröffentlicht
·/.	·/.	o. B.	Hygiene-Institut Frankfurt	Lk.	1:320	2 × 100 mg/die Achromycin über 5 Tage	nicht veröffentlicht (Überlassung der Daten durch Doz. Dr. BRANDIS)

Besonders die zu dem *klinischen Bild* einer akuten oder subakuten *Appendicitis* führenden
entzündlichen Veränderungen der mesenterialen Lymphknoten waren immer wieder Anlaß,
durch bakteriologische Untersuchungen die Ätiologie der Erkrankung zu klären. Eine bestimmte Erregerart, die mit einem in seiner klinischen, operativen und histologischen Erscheinungsform wohldefinierten Krankheitsbild ursächlich zusammenhängen konnte, ließ sich nicht
nachweisen. In den Lymphknoten wurden, sofern sie nicht bei der bakteriologischen Untersuchung steril waren, vornehmlich Escherichia coli, Streptococcus faecalis, Aerobacter aerogenes, Proteus vulgaris, Strepto- und Staphylokokken, Oxyuren, Ascariden oder Pilze nachgewiesen. Ein ursächlicher Zusammenhang zwischen den genannten Infektionserregern und
der bioptisch diagnostizierten mesenterialen Lymphadenitis, z. B. durch Nachweis spezifischer
Antikörper im Patientenserum oder durch erfolgreiche Übertragung der kulturell gezüchteten
Bakterien auf Versuchstiere und Auslösung entsprechender Veränderungen an den mesen-

16b

terialen Lymphknoten, wurde in keinem Fall gesehen. Auch der Versuch, durch Übertragung von Drüsenextrakten auf Versuchstiere Viren nachzuweisen, mißlang bisher.

Von manchen Autoren wurden die Veränderungen an den mesenterialen Lymphknoten und Mesenterialblättern, am Ileum oder Coecum als eine entzündliche Reaktion auf mikrobielle und alimentäre Toxine oder als allergisch-hyperergische Reaktion gedeutet (STOPENI 1949, zit. nach FARBSTEIN).

Auf Grund des morphologischen Substrates, das Beziehungen zum Lymphogranuloma inguinale einerseits und zur benignen Viruslymphadenitis (GSELL 1957) andererseits aufweist, wurde von MASSHOFF (1953) zuerst die Virusgenese der abscedierenden reticulocytären Lymphadenitis diskutiert. Untersuchungen von KNAPP (1954), KNAPP u. MASSHOFF (1954) führten aber zur kulturellen und tierexperimentellen Züchtung von Past. pseudotuberculosis aus mesenterialen Lymphknoten und zum Antikörpernachweis im Patientenserum, so daß der schon bekannten septisch-typhösen eine appendicitische Verlaufsform der menschlichen Pseudotuberkulose gegenübergestellt werden konnte.

Das *klinische Bild* dieser Verlaufsform der menschlichen Pseudotuberkulose entspricht dem einer akuten, subakuten oder chronisch-rezidivierenden Appendicitis, ohne daß sich operativ und histologisch die klinische Diagnose bestätigen läßt. Das Krankheitsbild, das vornehmlich bei jugendlichen Patienten beobachtet und differentialdiagnostisch von der echten Appendicitis nicht sicher abzugrenzen ist (KNAPP u. MASSHOFF 1954; KÖNIG u. MAURATH 1957; SANDER 1958), beginnt meist akut mit Schmerzen im Mittel- und rechten Unterbauch, denen sich im Einzelfall die Symptome einer akuten Gastroenteritis mit Erbrechen, Diarrhöe und Schüttelfrost zugesellen. Das Blutbild zeigt Leukocytenvermehrung; die Blutsenkungsgeschwindigkeit ist meist deutlich im Gegensatz zu den Beobachtungen bei der Appendicitis beschleunigt (LENNERT 1957). Bei der *Operation* findet man in der Bauchhöhle in verschiedener Menge ein klares, seröses Exsudat. Die mesenterialen Lymphknoten, besonders im Ileocoecalwinkel, sind entzündlich verändert und geschwollen, und an den Mesenterialblättern zeichnet sich häufig eine diffuse oder auf die Umgebung der veränderten Lymphknoten beschränkte Rötung ab. Schließlich können das distale Ileum und das Coecum hyperämische Schwellung und ein Ödem der Serosa aufweisen. Die Appendix zeigt wie beim Tier entweder keine oder geringfügige entzündliche Veränderungen [MASSHOFF u. DÖLLE 1953; MASSHOFF 1953; KNAPP u. MASSHOFF 1954; GRABER u. KNAPP 1955; LENNERT 1957; HECKER 1957; KÖNIG u. MAURATH 1957; HAENSELT 1957; SANDER 1958; FLAMM u. KOVAC (im Druck)].

KÖNIG u. MAURATH (1957) und SANDER (1958) weisen besonders auf das mannigfaltige klinische Bild der Pseudotuberkulose der Jugendlichen hin. Nach ihrer Meinung kann im Einzelfall eine Appendicitis differentialdiagnostisch nicht frühzeitig genug ausgeschlossen werden, so daß laparotomiert werden muß, obwohl bei dem gutartigen Verlauf dieser Form der mesenterialen Lymphadenitis eine Operation nicht notwendig wäre. Bis zum Vorliegen des serologischen Untersuchungsbefundes sei in den meisten Fällen ein Aufschieben der Operation nicht zu verantworten. CHRISTIANSEN (1957) empfiehlt daher zur Beschleunigung der serologischen Diagnose in 1:10—1:50 verdünntem Patientenserum die Objektträger-Agglutination mit geeigneten Typenstämmen durchzuführen.

Von *klinischer Seite* liegen bisher nur wenige einschlägige Berichte über die *appendicitische* Verlaufsform der menschlichen Pseudotuberkulose vor. BECKER (1954) veröffentlichte 4 Fälle, die nach ihrem klinischen, bioptischen und histologischen Bild die Beobachtungen von MASSHOFF (1953) bestätigten. Bakteriologisch-serologische Untersuchungen fehlten bei den vier männlichen, 10, 12, 18 und 25 Jahre alten Patienten. Einen kasuistischen Beitrag stellt der von SCHMIDT (1955) beschriebene, ebenfalls nur klinisch und histologisch, aber nicht bakteriologisch-serologisch untersuchte Fall dar. BERG u. HECKER (1956) berichteten dagegen über einen neunjährigen Patienten mit einem perforierten, durch Past. pseudotuberculosis infizierten Enterocystom. Die klinischen Symptome waren die einer akuten Appendicitis. Über ein Beobachtungsgut von 8 bzw. 20 Patienten, die klinisch ebenfalls die Symptome einer operativ nicht bestätigten Appendicitis und histologisch 5- bzw. 20mal das Bild einer abscedierenden reticulocytären Lymphadenitis boten, berichteten KÖNIG u. MAURATH (1957) bzw. HECKER (1957). Bakteriologisch-serologische Untersuchungen wurden aber nur bei einem Teil ihrer Patienten durchgeführt. Der Erregernachweis wurde von KÖNIG und MAURATH (s. Tabelle 5, Fußnote) und von HECKER für je 2 und der Agglutininnachweis für 3 bzw. 7 Patienten mitgeteilt. Auffallenderweise wurde bei den Patienten von HECKER (1957) nur ein Agglutinintiter zwischen 1:20—1:80 der Serumverdünnung bestimmt. Auch INGELRANS (1957) gelang der Erregernachweis.

Über ein größeres Beobachtungsgut wurde dagegen von *bakteriologischer* und *pathologischer* Seite berichtet. Den ersten Veröffentlichungen von KNAPP (1954) und KNAPP u. MASSHOFF (1954) folgten weitere Berichte über unter dem klinischen Bild einer akuten Appendicitis verlaufende Infektionen mit Past. pseudotuberculosis von GRABER u. KNAPP (1955), KNAPP (1955, 1956, 1958), PODHRAGYAI u. FODOR (1956), BRANDIS (1957), CHRISTIANSEN (1957), HAENSELT (1957), FLAMM (1958) und HOLCH (1958).

Auch die von ALBRECHT (1910) und PIÉCHAUD (1952) beschriebenen Einzelbeobachtungen müssen heute der appendicitischen Verlaufsform der menschlichen Pseudotuberkulose zugerechnet werden.

Während BRANDIS (1957) und HOLCH (1958) jeweils bei einem Patienten und FLAMM (1958) bei 2 Patienten der Nachweis von Past. pseudotuberculosis aus mesenterialen Lymphknoten gelang, konnte KNAPP (s. Tabelle 5, S. 244) bisher bei insgesamt 14 Patienten Past. pseudotuberculosis, und zwar 11mal aus mesenterialen Lymphknoten, 2mal aus Blut und 1mal aus Eiter, züchten. Von PODHRAGYAI u. FODOR (1957) und HAENSELT (1957) liegen dagegen keine Angaben über Erregernachweise vor.

Unter insgesamt 117 bis 1957 aus dem Untersuchungsgut des Hygiene-Institutes zusammengestellten und ausgewerteten Fällen (KNAPP 1958)[1] war bei 15 die Diagnose einer appendicitischen Verlaufsform der menschlichen Pseudotuberkulose durch den Erregernachweis und bei 94 durch Übereinstimmung der klinischen, bioptischen, serologischen und histologischen Befunde gestellt worden. Bei 7 Patienten mit typischen klinischen, bioptischen und histologischen Befunden wurden keine Antikörper oder auffallenderweise nur in Titern zwischen 1:20—1:40 der Serumverdünnung nachgewiesen. In 1 Fall fehlten bei positivem serologischem Befund die beschriebenen histologischen Veränderungen. Die Höhe der mit der Widal-Reaktion in den Patientenblutproben bestimmten Antikörpertiter, lag von den wenigen Ausnahmen abgesehen, zwischen 1:80—1:12800, vorwiegend zwischen 1:160—1:640 der Serumverdünnung. Bei den 33 von HAENSELT (1957) mitgeteilten, histologisch gesicherten Fällen wurde 26mal die serologische Untersuchung des Patientenblutes

---

[1] Unterschiede in der zahlenmäßigen Aufschlüsselung dieses Beobachtungsgutes gegenüber der Zusammenstellung von KNAPP (1956) sind, soweit sie die histologischen Befunde betreffen, durch erst nachträgliches Bekanntwerden der nunmehr berücksichtigten histologischen Untersuchungsergebnisse bedingt. Bei der Gesamtzahl der ausgewerteten Fälle sind zudem die Fälle, deren histologische Befunde nicht mitgeteilt werden konnten und die Verdachtsfälle nicht mehr berücksichtigt.

am Hygiene-Institut, Tübingen durchgeführt. Sie fiel bei 24 Patienten positiv und bei 2 negativ aus, während sie bei 7 Beobachtungen nicht möglich war. CHRISTIANSEN (1957) sah 3 Fälle. Wie oft die angegebene Keimisolierung gelang, ist nicht bekannt.

Unter den von KNAPP (Tabelle 5, S. 244) aus menschlichem Untersuchungsgut isolierten Stämmen von Past. pseudotuberculosis gehörten 12 dem serologischen Typ I[1] und zwei dem serologischen Typ V an. Die von BRANDIS (1957) und FLAMM (1958) isolierten Stämme waren ebenfalls vom Typ I, während HOLCH (1958) zum ersten Male beim Menschen einen Stamm von Past. pseudotuberculosis Typ II isolieren konnte. Über die Typenzugehörigkeit der von BERG u. HECKER (1956), HECKER (1957) und CHRISTIANSEN (1957) isolierten Stämme liegen keine Angaben vor.

Das Vorkommen von menschlichen Infektionen mit Past. pseudotuberculosis Typ III wurde durch den Erregernachweis von MOSS u. BATTLE (1941) sowie JAKOBSTHAL (zit. nach SCHÜTZE 1928) bewiesen, die jeweils bei einem an der septisch-typhösen Verlaufsform der menschlichen Pseudotuberkulose erkrankten Patienten einen Stamm von Past. pseudotuberculosis Typ III isolierten. Der Nachweis von Past. pseudotuberculosis Typ IV aus menschlichem Untersuchungsgut ist bisher noch nicht gelungen. Nur durch Absättigungsversuche mit Patientensera konnte von KNAPP (1956) bei 2 bzw. 1 Patienten eine Infektion mit Past. pseudotuberculosis Typ II bzw. IV wahrscheinlich gemacht werden. Sicher ist, daß nach den bisherigen bakteriologischen und serologischen Untersuchungsergebnissen Infektionen mit Past. pseudotuberculosis Typ II, III, IV und V beim Menschen äußerst selten vorkommen, und die appendicitische Verlaufsform der menschlichen Pseudotuberkulose in der Regel durch Past. pseudotuberculosis Typ I ausgelöst wird.

Für die Spezifität der Widal-Reaktion mit Past. pseudotuberculosis spricht, daß in Sera von 76 klinisch und operativ gesicherten Appendicitisfällen und in 1601 wegen Verdacht auf Lues- oder Salmonellainfektionen eingesandten Serumproben von KNAPP (1956) keine Agglutinine gegen Past. pseudotuberculosis Typ I, III und V nachgewiesen wurden, während Agglutinationen von Past. pseudotuberculosis Typ II und IV auf die antigenen Beziehungen zur Salmonella B- und D-Gruppe zurückzuführen waren (s. S. 218). Die Frage, welche diagnostische Bedeutung den von KUHLMANN u. HERRMANN (1955), BRAUN u. MÜLLER (1957) und HECKER (1957) bestimmten sehr niederen Agglutinintitern zukommt, läßt sich, solange nicht weitere Beobachtungen an einem größeren Untersuchungsgut von verschiedener Seite vorliegen, nicht beantworten.

Nach den eingehenden *pathologisch-anatomischen* Untersuchungen von MASSHOFF u. DÖLLE (1953) und MASSHOFF (1953) erreichen die entzündlich veränderten *Lymphknoten* mitunter Walnußgröße, sie können isoliert oder als Konglomerate vergrößert sein und dadurch ein tumoriges Aussehen aufweisen. Auf der Schnittfläche findet man neben den meist großen Follikeln kleinste, punktförmige, weißlichgelbliche Flecken oder Abscesse. Im frischen Stadium läßt das histologische Präparat zuerst eine von den Randzonen der Follikel ausgehende, sich zu lockeren Knötchen formierende Proliferation der Reticulumzellen, die nur kurze Zeit rein reticulocytär bleibt, erkennen, während die weiteren Erscheinungsformen durch eine konfluierende und sich diffus ausbreitende reticulocytäre Proliferation und die sich daran anschließende leukocytäre Infiltration mit scharf begrenzten Einschmelzungsherden gekennzeichnet sind (Abb. 5 und 6). Dieses morphologisch einen besonderen Typ der mesenterialen Lymphadenitis darstellende Erscheinungsbild wird als Folge der Reaktion von lymphknoteneigenen und hämatogenen Zellen angesehen. Diese Beobachtungen wurden durch Untersuchungen von MASSHOFF zusammen mit KNAPP (1954) erweitert und fanden durch HÖRSTEBROCK (1954), RANDERATH (1954), SCHOEN (1954), GRABER u.

---

[1] Nachtrag bei der Korrektur: 1957/58 gelang bei weiteren 3 Patienten der Nachweis von Past. pseudotuberculosis Typ I in mesenterialen Lymphknoten.

KNAPP (1955), LENNERT (1956/57), GRABER (1956), PODHRAGYAI u. FODOR (1956), RINIKER (1957), HAENSELT (1957) und FLAMM u. KOVAC (im Druck) eine weitgehende Bestätigung und Ergänzung.

HÖRSTEBROCK (1954) nahm an, daß diese Form der mesenterialen Lymphadenitis, deren sichere Unterscheidung von der benignen Viruslymphadenitis und dem Lymphogranuloma

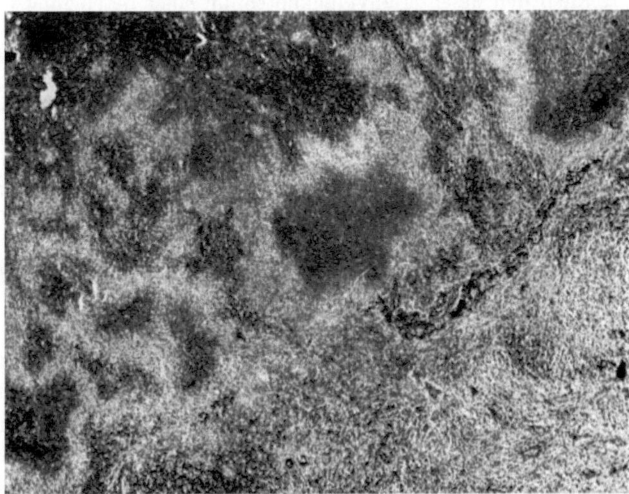

Abb. 5. Typisches Bild der abscedierenden reticulocytären Lymphadenitis. Ausschnitte aus einem kapselnahen Gebiet des Lymphknotens. In der Peripherie des Lymphknotens zahlreiche z. T. konfluierende Herdbildungen. Die einzelnen Herde besitzen einen hellen aus proliferierten Reticulumzellen bestehenden Saum, die dunkleren zentralen Teile bestehen aus zugrunde gegangenen Reticulumzellen und aus Leukocytenmassen [1]

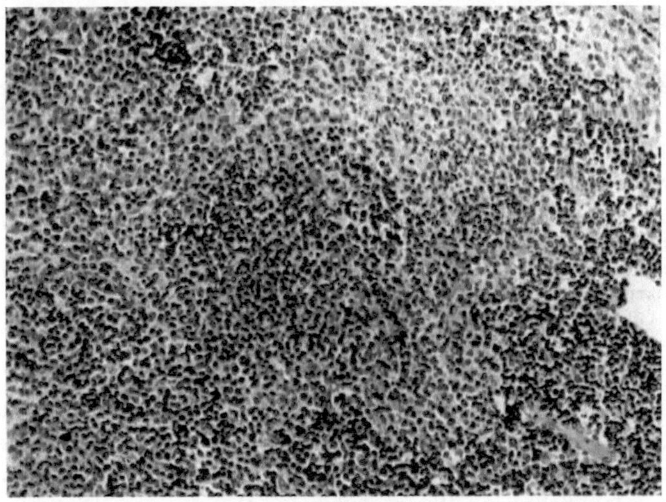

Abb. 6. Abscedierende reticulocytäre Lymphadenitis. Frischer Einzelherd. Starke Reticulumzellwucherung, Durchschwärmung mit Leukocyten, Anreicherung der Leukocyten im Zentrum [1]

inguinale er wie MASSHOFF, LENNERT und andere Autoren für fraglich hielt, die Folge einer Blastomykose sei. Die Frage, ob nicht zumindest die mesenteriale Verlaufsform der sog. benignen Viruslymphadenitis (Katzenkratzlymphadenitis), deren bisher noch nicht nach-

---

[1] Herrn Prof. MASSHOFF sei auch an dieser Stelle für die Überlassung der histologischen Bilder und Befunde gedankt.

gewiesener Erreger zu den entsprechenden morphologischen Veränderungen wie Past. pseudotuberculosis führen kann (HEDINGER 1952, RANDERATH 1954, NORDMANN 1955, LENNERT 1956/57), auch durch Past. pseudotuberculosis ausgelöst wird, wurde von RINIKER (1957) diskutiert.

Von HAENSELT (1957) wurde besonders auf die immer wieder in charakteristischer Form vorliegenden Phasen der Entzündung und auf das gelegentliche Auftreten ausschließlich granulomatöser Herde in den Mesenteriallymphknoten mit einer Beteiligung von neutrophilen Leukocyten hingewiesen. Die histologische Diagnose einer Infektion mit Past. pseudotuberculosis hält HAENSELT unter Berücksichtigung der verschiedenen Phasen der Lymphadenitis, ihrer Lokalisation vornehmlich in der Ileocoecalregion und der weitgehenden Übereinstimmung der histologischen und serologischen Untersuchungsbefunde im Gegensatz zu LENNERT (1956/57) mit großer Wahrscheinlichkeit für möglich. Nach LENNERT ist eine Differentialdiagnose der durch Past. pseudotuberculosis ausgelösten mesenterialen Lymphadenitis von abscedierenden reticulocytären Lymphadenitiden anderer Ätiologie (Tularämie, benigne Viruslymphadenitis, Lymphogranuloma inguinale, Pilzerkrankung u. a.) nur durch bakteriologisch-serologische Untersuchungen möglich, zumal das histologische Substrat der durch Past. pseudotuberculosis ausgelösten abscedierenden, reticulocytären Lymphadenitis nicht als spezifisch, sondern nur als charakteristisch anzusehen ist (MASSHOFF u. DÖLLE 1953; LENNERT 1956/57; HAENSELT 1957 u. a.).

Die *Pathogenese* der morphologischen Veränderungen ist mit größter Wahrscheinlichkeit auf eine enterale Infektion, möglicherweise über das unterste Ileum oder Coecum, zurückzuführen. Während die Entzündung der Lymphknoten im Vordergrund steht und ein Primärherd im Darm kaum zur Beobachtung kommt, sind häufig sekundäre Aussaaten im lymphatischen Apparat des unteren Ileum, Coecum und gelegentlich auch der Appendix nachzuweisen [MASSHOFF 1953; GRABER u. KNAPP 1955; GRABER 1956; LENNERT 1956/57; HAENSELT 1957; RINIKER 1957; FLAMM u. KOVAC (im Druck)]. Die von GRABER (in GRABER u. KNAPP 1955) als Primärinfekte beschriebenen Veränderungen der Darmschleimhaut wurden von MASSHOFF (1953) und RINIKER (1957) als entzündliche, mit Kreislaufschäden verbundene Vorgänge in der Schleimhaut und ihrem lymphatischen Apparat gedeutet, die der Lymphknotenerkrankung parallel gehen oder ihr erst nachfolgen. Bei Ileocoecalresektionen sahen MASSHOFF (1953) und GRABER (1955/56), daß die mesenteriale Lymphadenitis mit ulcerösen Schleimhautveränderungen im terminalen Ileum gekoppelt sein kann.

Das Ergebnis ihrer Fütterungsversuche an weißen Mäusen veranlaßte FLAMM u. KOVAC zu der Folgerung, daß die orale Aufnahme virulenter Kulturen von Past. pseudotuberculosis von der Bildung eines Primärherdes im Darm gefolgt sei und dieser bei genauer Untersuchung stets auch im Darm der Patienten gefunden werden müsse.

c) Enteritische Verlaufsform

Im Vordergrund des *klinischen* Bildes dieser selten beobachteten und bisher nur serologisch nachgewiesenen Verlaufsform stehen neben uncharakteristischen Allgemein- und Prodromalerscheinungen wie Fieber, Kopfschmerzen u. Appetitlosigkeit schwer lokalisierbare Beschwerden im Bauchraum, Diarrhöe und mitunter Erbrechen. Den Symptomen einer akuten, subakuten oder chronischen Enteritis oder Gastroenteritis können klinische Beschwerden, die zu einer Appendicitis-Verdachtsdiagnose führen, vorausgehen oder folgen (KNAPP 1954; KNAPP u. MASSHOFF 1954).

Eine lavierte, unter dem Bild einer subakuten oder chronischen Enteritis verlaufende Form der menschlichen Pseudotuberkulose glaubten KUHLMANN u. HERRMANN (1955) gesehen zu haben. Sie teilten ihr Patientengut nach dem klinischen Bild in 2 Gruppen ein: Patienten mit langjähriger, wenig markant verlaufender, ausgedehnter chronischer Enteritis und Patienten mit umschriebener, subakuter Enteritis.

Bei 8 Patienten (6 Frauen und 2 Männern) der 1. Gruppe blieb das klinische Bild einer chronischen Enteritis mit einem zur Diagnose einer enteritischen chronischen Verlaufsform der menschlichen Pseudotuberkulose führenden Titer gegen Past. pseudotuberculosis zwischen 1:40 und 1:80 der Serumverdünnung, solange nur eine konservative und keine Antibioticabehandlung durchgeführt wurde, über Wochen erhalten. Als Therapieerfolg wurde die klinische Besserung und der völlige Rückgang des Antikörpertiters angesehen. Histologische und bakteriologische Befunde wurden nicht erhoben. Zwei Fälle der 2. Gruppe zeigten einen subakuten schweren Krankheitsverlauf. Während die erste Patientin über Kopfschmerzen, geblähten Leib, Kullern im Bauch, Schweißausbrüche, Kollapsneigung, Durchfälle mit Schleim und Blut und später hochgradiger Verstopfung klagte, standen bei dem zweiten Patienten Schmerzen in der Magengegend, ein Tiefendruckschmerz links unterhalb des Nabels und eine starke Gewichtsabnahme innerhalb 3 Monaten im Vordergrund. Die serologische Untersuchung ergab bei beiden Patienten ebenfalls nur einen sehr niederen Agglutinintiter gegen Past. pseudotuberculosis bis 1:80 der Serumverdünnung. Ein Erregernachweis wurde nicht erbracht. Unter antibiotischer Therapie trat eine wesentliche Besserung der klinischen und röntgenologischen Befunde, die mit konservativer Behandlung nicht zu erreichen war, ein.

Diagnostische Bedeutung sprachen KUHLMANN und ISEBARTH (1954) und KUHLMANN u. HERRMANN (1955) auch der Röntgenuntersuchung zu. Durch eine Reliefdarstellung besonders im Ileumbereich soll sich die spezifische und unspezifische Hyperplasie der größeren in der Darmwand liegenden Lymphfollikel oder Lymphfollikelaggregate nachweisen und durch Andrücken der kontrastreichen Dünndarmschlingen an die Lymphfollikel ein Pelotteneffekt erzielen lassen. Ein für die Lymphadenitis mesenterialis charakteristisches Dünndarmbild wird schließlich in den gedrehten Faltenbergen, die einen drahtspiralenartigen Querverlauf zur Dünndarmachse haben, gesehen. Die befallenen Darmabschnitte sind gasgebläht und die Darmschlingen nur schlecht unter Schmerzen sichtbar.

KUHLMANN und HERRMANN (1955) nehmen an, daß die erst im Erwachsenenalter auftretende subakute wie lavierte Verlaufsform der Pseudotuberkulose die Folge einer Darminfektion im Kindesalter ist. Die Frage, ob diese Annahme zu Recht besteht und die von ihnen veröffentlichten Fälle tatsächlich der menschlichen Pseudotuberkulose zuzurechnen sind, wird erst durch Beobachtungen an einer größeren Patientenzahl mit Erregernachweis oder bioptischen und histologischen Untersuchungen befriedigend zu beantworten sein. Vorläufig fehlt noch ihre Bestätigung durch gesicherte bakteriologisch-serologische Befunde und Untersuchungen über das pathologisch-anatomische Substrat.

KNAPP berichtete 1954 über 4 Patienten, deren Blut wegen Enteritisverdacht zur serologischen Untersuchung eingeschickt wurde. Während die Agglutination mit Keimen der Typhus-Paratyphus-Enteritis-Gruppe negativ verlief, wurden bei allen Patienten Antikörpertiter gegen Past. pseudotuberculosis Typ I zwischen 1:800 und 1:6400 der Serumverdünnung nachgewiesen. Rückfragen bei den behandelnden Ärzten ergaben, daß 3 Patienten wegen Appendicitisverdacht in ein Krankenhaus eingeliefert waren. Der rasche Rückgang der für eine Appendicitis sprechenden klinischen Symptome und das Auftreten starker Durchfälle hatten aber zu einer konservativen Behandlung der Patienten und Einsendung der Blutproben wegen Enteritisverdacht geführt. Der 4. Patient wurde wegen akuter Gastroenteritis mit starken Durchfällen, die nach 3tägiger konservativer Behandlung abklangen, eingewiesen. Am 4. Tag des Krankenhausaufenthaltes stellten sich deutliche Milzschwellung und zunehmender Druckschmerz im Mittelbauch ein, so daß das Vorliegen einer Appendicitis angenommen wurde.

Auch zu diesen Fällen, die möglicherweise der appendicitischen Verlaufsform zuzurechnen sind, wird erst nach Vorliegen eines größeren Beobachtungsgutes endgültig Stellung genommen werden können. Sie zeigen ebenfalls die Notwendigkeit, bei unklaren Darmerkrankungen das Patientenblut auf Agglutinine gegen Past. pseudotuberculosis zu untersuchen und den Erregernachweis, der bisher noch nicht gelungen ist, aus Stuhlproben zu versuchen.

BRAUN u. MÜLLER (1957) prüften durch serologische Untersuchungen die Frage, ob bei Kindern mit unklaren, nicht das klinische Bild einer Appendicitis zeigenden abdominellen oder anderen Krankheitserscheinungen Infektionen mit Past. pseudotuberculosis eine mehr oder weniger große Rolle spielen. Die Autoren fanden bei 66 Kindern mit abdominellen Erscheinungen (20mal akute oder chronische enteritische Beschwerden, 21mal rezidivierende Leibschmerzen und Nabelkoliken, 4mal unklare Brechzustände, 21mal Appendicitis) nur bei einem 2jährigen Kind mit unklaren Brechzuständen einen Titer bis 1:320, der nach Behandlung mit Supracillin innerhalb eines Monats auf 1:10 zurückging. Bei 35 Patienten mit anderen nicht abdominellen Krankheitserscheinungen wurde in 5 Fällen ein als unspezifisch beurteilter Titer bis 1:40 bestimmt. Auch BRAUN u. MÜLLER kamen zu der Feststellung, daß bei Patienten mit unklaren appendicitischen oder enteritischen Beschwerden die Untersuchung des Serums auf Antikörper gegen Past. pseudotuberculosis ratsam sei.

3. Epidemiologie

Über die Epidemiologie der menschlichen Pseudotuberkulose ist nichts Sicheres bekannt. Die weite Verbreitung dieser Infektionskrankheit im Tierreich und der Nachweis, daß die beim Menschen gefundenen Stämme von Past. pseudotuberculosis in ihrem kulturell-biochemischen, serologischen wie tierexperimentellen Verhalten keine Unterschiede gegenüber Stämmen tierischer Herkunft zeigen, läßt die menschliche Infektionsquelle im Tierreich suchen, ohne daß bisher der Übertragungsmodus dieser Infektionskrankheit hätte geklärt werden können. Seine Aufklärung wird, solange nur Einzelerkrankungen und kein en- oder epidemisches Auftreten dieser Infektionskrankheit beim Menschen beobachtet werden, schwierig, wenn nicht sogar unmöglich sein.

Die Tatsache, daß bei der Pseudotuberkulose der Tiere wie der Menschen abdominelle Erscheinungen im Vordergrund stehen und die klinischen wie pathologisch-anatomischen Befunde auf einen enteralen Infektionsweg hinweisen, spricht mit großer Wahrscheinlichkeit für eine perorale Infektion und ein primäres Haften des Erregers im Verdauungstrakt [PIÉCHAUD 1952; MASSHOFF u. DÖLLE 1953; MASSHOFF 1953; GRABER u. KNAPP 1955; GRABER 1956; RINIKER 1957; FLAMM u. KOVAC (im Druck) u. a.] ALBRECHT (1910), LOREY (1911) und SAISAWA (1913) waren die ersten Autoren, die annahmen, daß die menschliche Infektion wie beim Tier über den Verdauungsweg erfolgt, ohne daß bis heute im Einzelfall die Infektionsquelle und die Eintrittspforte des Erregers hätten gefunden werden können. An Infektionsquellen des Menschen wurden der direkte Kontakt mit infizierten Haustieren, besonders Katzen, oder ihren Ausscheidungen (ALBRECHT 1910; PAUL u. WELTMANN 1934), der Genuß von infiziertem Fleisch (LOREY 1911; MOSS u. BATTLE 1941), von Trinkwasser oder verunreinigten Nahrungsmitteln (FRAENKEL 1924) diskutiert, aber bisher in keinem Fall bewiesen. Sicher sind Keimträger und Dauerausscheider unter den empfänglichen Tieren von großer epidemiologischer Bedeutung.

Ob ein von PARIETTI (1890) mit *Milch* gespritztes Kaninchen tatsächlich einer durch die Milchprobe übertragenen und nicht einer spontanen Infektion erlegen ist, kann nach dem vorliegenden Bericht nicht mit ausreichender Sicherheit entschieden werden. Untersuchungen von KNAPP über die Thermoresistenz von Past. pseudotuberculosis in Milchproben (s. S. 230) sowie eine Beobachtung von GRABER u. KNAPP (1955) geben jedenfalls Veranlassung, der Weiterverbreitung der Pseudotuberkulose durch Milch oder Milchprodukte Beachtung zu schenken.

KNAPP (in GRABER u. KNAPP 1955) fand in der Umgebung eines an der appendicitischen Verlaufsform der menschlichen Pseudotuberkulose erkrankten Melkers bei einem Rind mit kurz zuvor ausgeheilter eitriger Euterentzündung bei zweimaliger Untersuchung einen Antikörpertiter gegen Past. pseudotuberculosis Typ I bis 1:200 bzw. 1:320 der Serumverdünnung und bei 4 weiteren von insgesamt 15 untersuchten Rindern desselben Hofgutes einen Titer von 1:10—1:40. Weitere serologische Kontrolluntersuchungen an 131 angeblich gesunden Rindern anderer Höfe ließen bei 21 Tieren einen Titer von 1:10—1:50 bestimmen. Kulturelle und tierexperimentelle Untersuchungen von 34 Sammelmilchproben aus 16 Gemeinden und 2 Gutshöfen führten zu keinem Nachweis von Past. pseudotuberculosis.

Ungeklärt ist bisher auch die an Einzelbeobachtungen wiederholt diskutierte Frage, ob die menschliche Infektion mit Past. pseudotuberculosis einen engen, direkten oder indirekten Kontakt mit Tieren, besonders Haustieren, voraussetzt und ob sie häufiger bei der Land- als bei der Stadtbevölkerung beobachtet wird. Eine Antwort geben die von HECKER (1957) und KNAPP (1958) bisher entsprechend ausgewerteten Fälle nicht.

HECKER fand nach schriftlicher Befragung seiner 20 Patienten bzw. ihrer Angehörigen bei 16 eingegangenen Antworten 11mal den engen Kontakt der erkrankten Großstadtkinder mit Hunden, Katzen, Kaninchen und Hühnern angegeben. Es ist aber nicht bekannt, ob die Tiere gesund oder krank waren.

Von 117 durch KNAPP erfaßten Patienten stammten 55 aus Städten und 62 aus Dörfern oder kleinen Städten mit stark ländlichem Charakter. Bei diesen Patienten bestand möglicherweise ein größerer direkter oder indirekter Kontakt mit Tieren, ohne daß dieser durch Befragung des größeren Teiles der Patienten hätte gesichert werden können.

Wie beim Tier scheint auch beim Menschen die Pseudotuberkulose häufiger in den Herbst- und Wintermonaten aufzutreten. Unter den von KNAPP (1958) beobachteten Patienten erkrankten 61,5% im Winterhalbjahr bzw. 55,5% in den Monaten November bis Februar. Die restlichen 38,5% verteilten sich auf die Monate April bis September. HAENSELT (1957) fand dagegen bei der Auswertung von 33 Fällen einen Frühjahrsgipfel mit Häufung der Erkrankungen in den Monaten April bis Juni, während bei HECKER (1957) von 20 Patienten 16 in den Monaten November bis März und 4 im Monat Juni erkrankten.

Auffallend ist eine eindeutig festzustellende *Alters- und Geschlechtsdisposition* bei Erkrankungen an menschlicher Pseudotuberkulose. Während die septische Verlaufsform bisher fast ausschließlich bei Erwachsenen beobachtet wurde, ist das durch eine abscedierende reticulocytäre Lymphadenitis ausgelöste appendicitische Krankheitsbild vornehmlich bei Jugendlichen im Alter zwischen 2 bis 28 Jahren und nur in Einzelfällen bei älteren Personen zu finden (KNAPP 1954, 1955, 1958; KNAPP u. MASSHOFF 1954; GRABER u. KNAPP 1955; LENNERT 1956/57; HAENSELT 1957; HECKER 1957; KÖNIG u. MAURATH 1957 u. a.). Die Kurve der Altersverteilung zeigt nach MASSHOFF (1954) und LENNERT (1956/57) zwischen dem 10.—15. und nach eigenen Beobachtungen zwischen dem 5.—7. einen kleinen und zwischen dem 10.—17. Lebensjahr einen deutlich ausgeprägten Gipfel.

MASSHOFF u. DÖLLE (1954) haben schon in ihrem ersten Bericht über die abscedierende reticulocytäre Lymphadenitis darauf hingewiesen, daß das männliche Geschlecht häufiger an dieser Infektion erkrankt, als das weibliche. Unter 41 Fällen war das männliche Geschlecht mit 63,5 und das weibliche mit 36,5% vertreten. Entsprechende Unterschiede der Geschlechtsverteilung zeigen die von

HECKER (1957), HAENSELT (1957) und KNAPP (1958) ausgewerteten Erkrankungsfälle. Bei HECKER waren von 20 Jugendlichen 15 männlichen Geschlechtes,
während bei HAENSELT bzw. KNAPP 26 bzw. 96 männlichen nur 2 bzw. 21 weibliche Patienten gegenüberstanden. Ein zahlenmäßiges Überwiegen des männlichen Geschlechtes sahen auch LENNERT (1957) sowie KÖNIG u. MAURATH (1957).

Berichte über *Laboratoriumsinfektionen* mit Past. pseudotuberculosis liegen
bisher nicht vor. Nach eigenen Erfahrungen scheint die Gefahr, daß eine während
der Laboratoriumstätigkeit erfolgte perorale Aufnahme von Past. pseudotuberculosis zu einer klinisch oder bakteriologisch-serologisch nachweisbaren Infektion
führt, geringer zu sein als nach Einbringen der Keime in den Conjunctivalsack.

Drei Institutsangehörige, die virulente Bakteriensuspensionen von Past. pseudotuberculosis Typ I menschlicher Herkunft beim Pipettieren in den Mund bekamen, zeigten keinerlei
Zeichen einer manifesten oder latenten Infektion. Dagegen war bei einer Assistentin die Infektion der Augenbindehaut nach 7 Tagen von Fremdkörpergefühl und einer Schwellung des
Unterlides gefolgt. Die fachärztliche Untersuchung ergab eine Rötung der Conjunctiva tarsi
und bulbi und Follikelschwellungen verschiedener Größe an der oberen und unteren Übergangs-
falte mit kleinen petechialen Blutungsherden. Die präauriculären und submandibulären regionären Lymphknoten waren geschwollen und druckschmerzhaft. Unter Achromycin gingen die
Follikel- und Lymphknotenschwellungen rasch zurück. Ein Erreger- und Antikörpernachweis
gelang wegen des sofortigen Therapiebeginns nicht, dagegen ergab die histologische Untersuchung eines excidierten Follikels denselben histologischen Befund einer abscedierenden
reticulocytären Entzündung, wie sie auch nach der experimentellen Infektion des Conjuncti-
valsackes von Kaninchen und Meerschweinchen gesehen wurde (KNAPP, HAGER u. MASSHOFF
unveröffentlicht; DEYL 1896; GIFFORD u. LAZAR 1930; GRANCINI 1939).

XI. Therapie

Über die Therapie der menschlichen Pseudotuberkulose ist wenig bekannt.
Nur 2 Fälle der septischen Verlaufsform (1 Fall s. S. 242) wurden möglicherweise durch eine rechtzeitig eingeleitete Sulfonamid- bzw. Antibioticatherapie
geheilt. Zwei weitere Patienten starben trotz Antibioticatherapie, während die
übrigen ohne chemotherapeutische Behandlung der Pseudotuberkulose erlagen
(s. Tabelle 4, S. 240). Sichere Aussagen über den Wert der Chemotherapie bei der
septischen Verlaufsform der menschlichen Pseudotuberkulose sind somit unmöglich. Sie setzen Erfahrungen an einem größeren Untersuchungsgut klinisch manifester und bakteriologisch gesicherter Fälle voraus.

Die appendicitische Verlaufsform der Pseudotuberkulose hat sich bisher in
ihrem klinischen Verlauf als gutartig erwiesen. Der in der Regel der Appendekto-
mie folgende komplikationslose Heilungsverlauf schließt in den meisten Fällen
eine chemotherapeutische Behandlung aus oder läßt sie nur als prophylaktische
Maßnahme beginnen. Therapeutische Einzelbeobachtungen an den vom Verfasser
bakteriologisch-serologisch untersuchten Patienten, unter denen nur 1 Fall der
septisch-typhösen Verlaufsform zuzurechnen ist (s. Tabelle 5, S. 244), ermöglichen ebenfalls kein Urteil über die Notwendigkeit und den Wert der chemothera-
peutischen Maßnahmen.

HECKER (1957) behandelte alle Patienten postoperativ mit Supracillin oder Achromycin,
was unseres Erachtens bei dem meist komplikationslosen postoperativen Heilungsverlauf
nicht notwendig ist (KNAPP 1955).

Über einen 12jährigen Jungen, der wegen Appendicitisverdacht operiert wurde und anschließend das Bild der typhösen Verlaufsform mit einem Titer gegen Past. pseudotuberculosis

bis 1:640 zeigte, die mit Leukomycin erfolgreich behandelt wurde, berichteten König u. Maurath (1957).

Sander (1958) behandelte einen Patienten postoperativ bei komplikationslosem Heilungsverlauf mit Supracillin, während bei einem zweiten Patienten, der postoperativ ebenfalls Supracillin bekommen hatte, ein in der 5. Woche nach der Operation auftretendes Rezidiv mit Tetracyclin erfolgreich behandelt wurde.

Gute therapeutische Erfolge glaubten Kuhlmann u. Herrmann (1955) bei 10 Patienten, deren chronische oder subakute Enteritis sie auf eine Infektion mit Past. pseudotuberculosis zurückführten (s. S. 253), durch Behandlung mit Tetracyclin bzw. Hostacyclin (8 Tage 4mal täglich 250 mg) gesehen zu haben, nachdem die vorausgegangene wochenlange symptomatische Therapie erfolglos geblieben war.

Obwohl die Pseudotuberkulose der Tiere eine in der Veterinärmedizin seit langer Zeit bekannte Infektionskrankheit ist, fehlen Mitteilungen über in größerem Rahmen durchgeführte Therapieversuche an spontan infizierten Tieren. Der Versuch von Truche u. Isnard (1937), das enzootische Auftreten der Pseudotuberkulose in einer Geflügelzucht mit Stovarsol zu beherrschen, mißlang. Ebensowenig sahen Urbain u. Nouvel (1949) bei der Behandlung von 15 spontan erkrankten Affen mit Sulfonamiden einen therapeutischen Effekt.

Gewisse Rückschlüsse auf die therapeutische Wirkung von Sulfonamiden und Antibiotica lassen tierexperimentelle Beobachtungen zu.

Von 8 experimentell infizierten Meerschweinchen konnte Vallée (1950) 4 Tiere mit einer Gesamtdosis von 300 mg (6mal 50 mg/die) Streptomycin am Leben erhalten, während 2 Tiere, die über 3 bzw. 6 Tage täglich nur mit 25 mg Streptomycin behandelt wurden, nach 8 bzw. 11 Tagen eingingen. Zwei weitere Tiere, von denen das eine am 21. Tag starb und das zweite überlebte, erhielten über 5 Tage jeweils 50 mg Streptomycin. Mit der Therapie wurde am 3. Tag nach der intraperitonealen Infektion begonnen. Die Kontrolltiere starben am 4. bis 5. Tag. Bei der Sektion der überlebenden, gesund erscheinenden Tiere nach 2 Monaten sah Vallée keine für eine Pseudotuberkulose typischen Veränderungen der Organe. In Versuchen von van Dorssen (1952) starben von 20 Meerschweinchen, die ebenfalls 3 Tage nach der intraperitonealen Infektion über 6 Tage je 50 mg Streptomycin subcutan gespritzt bekamen, 3 Tiere. Mit 25 mg Streptomycin wurde dagegen nur eine Lebensverlängerung bei 9 von 10 Tieren, aber keine Heilung erzielt. Derselbe Erfolg trat auch nach 2maliger Injektion von 25 bzw. 50 mg Streptomycin an 1 oder 2 aufeinanderfolgenden Tagen ein. Entsprechende Sulfonamidversuche (Sulfamethazin) verliefen erfolglos. Nach einer einmaligen Injektion von 500 mg Dihydrostreptomycin 4 Std nach der Infektion blieben 4 intravenös infizierte Kaninchen am Leben, während 6 weitere Tiere nach einer verteilten Dosierung der Gesamtmenge nur 3 Tage überlebten. Eine zweite Infektion überstanden nach van Dorssen (1952), im Gegensatz zu den Beobachtungen von Vallée (1950) bei Meerschweinchen, die mit Streptomycin behandelten Kaninchen nicht und Meerschweinchen nur zum Teil. Harisijades (1953) konnte 13 Tage alte, mit Past. pseudotuberculosis infizierte Hühnerembryonen mit 250—1000 mg Streptomycin pro Ei am Leben erhalten.

In vivo und in vitro prüfte Berger (1957) die Wirkung verschiedener Sulfonamid-(Supronal, Protocid, Aristamid und Diazil) und Antibioticapräparate (Streptomycin, Achromycin, Terramycin, Polymyxin B und Neomycin).

In vivo wurde unter den Sulfonamiden nur mit Aristamid und Protocid in einer Versuchsserie ein geringer lebensverlängernder Effekt erzielt. Bei den behandelten Tieren war nach Ablauf der Beobachtungszeit ein Erregernachweis aus den Organen möglich. Ihre therapeutischen Ergebnisse mit Sulfonamiden entsprachen somit den von Urbain u. Nouvel (1949) bzw. van Dorssen (1952) an Affen bzw. Meerschweinchen erzielten.

Unter den Antibiotica zeigte Penicillin (2000—5000 E Gesamtdosis [G.D.]) eine geringe, nicht sicher verwertbare therapeutische Wirkung. Es führte wie Polymyxin B (0,2—0,5 mg G.D.) nur bei wenigen Tieren zu einer geringen Lebensverlängerung. Ein stärkerer therapeutischer (lebensverlängernder) Effekt war mit Terramycin (1—5 mg G.D.), Neomycin (0,5 bis 1 mg G.D.), Achromycin (1—5 mg G.D.) und Streptomycin (2—10 mg G.D.) zu erzielen. Dagegen wurde mit einer Kombination von Streptomycin und Supronal keine synergistische Wirkungssteigerung erreicht. Eine deutliche Überlegenheit von Streptomycin zeigt die Auswertung der Versuche von BERGER nach der keimtötenden Wirkung der einzelnen Antibiotica. Der Erregernachweis war bei den die Beobachtungszeit überlebenden sezierten Tieren nach Behandlung mit Streptomycin in 6,7%, mit Achromycin in 23,5%, Neomycin in 30,4% und mit Terramycin in 52% möglich.

MAGLIONE (1955) prüfte die Wirkung von Streptomycin, Terramycin und Aureomycin bei der experimentellen Pseudotuberkulose des Meerschweinchens. Einen therapeutischen Effekt sah er nur beim Streptomycin. Bei Versuchen, die Wirkung verschiedener Antibiotica auf die Präcipitinbildung bei Kaninchen zu bestimmen, stellte OLIVO (1953) fest, daß durch Aureomycin der ungünstige Verlauf einer experimentellen Infektion beschleunigt wird. Ein ursächlicher Zusammenhang wurde in der Hemmung der Phagocytose gesehen.

In vitro zeigten Protocid in 10—18,7 mg-%iger und Supronal in 18—50 mg-%iger Konzentration, Aristamid und Diazil dagegen erst in Konzentrationen von 75—100 mg-% eine makroskopisch ablesbare Wachstumshemmung. Die bakteriostatische Wirkung von Penicillin lag zwischen 0,16 und 0,63 E/ml, von Achromycin, Aureomycin, Terramycin, Neomycin, Nebacetin und Polymyxin zwischen 0,19 und 0,78 γ/ml, während sie für Streptomycin mit 0,78—0,25 γ/ml und für Chloramphenicol mit 0,78—6,25 γ/ml bestimmt wurde. Als bactericid wirkende Konzentrationen wurden 6,25—25,0 γ/ml für Streptomycin und 200 bzw. 1000 γ/ml für Chloramphenicol, Aureomycin und Achromycin bzw. Terramycin nachgewiesen. Im Gegensatz zu ihren Beobachtungen im Tierversuch fand BERGER bei der Kombination von Streptomycin und Supronal in vitro eine Potenzierung der Hemmwirkung. Verschiedene Antibioticakombinationen waren, sofern sie nicht in ihrer Wirkung indifferent blieben, von einer additiven Wirkungssteigerung gefolgt.

Diese Befunde stimmen weitgehend mit den von KNAPP (1955) an 24 Stämmen — 11 humaner und 13 tierischer Herkunft — mit 9 Antibiotica- und 6 Sulfonamidpräparaten bestimmten Werten überein. Nur die Hemmwerte von Penicillin (1,0—12,5 γ/ml) und Streptomycin (1,25—10,0 γ/ml) lagen deutlich höher.

Ein Vergleich der Untersuchungsergebnisse von BERGER (1957) und KNAPP (1955) mit früheren, zum Teil an einzelnen Stämmen erhobenen Befunden (QUAN u. Mitarb. 1947; VALLÉE 1950; MEYER 1952; HARISIJADES 1953; BRYGOO u. COURDURIER 1955; IUSCHENKO 1956; SACHDEVA u. Mitarb. 1956) ist dagegen nur mit Vorbehalt möglich. MEYER (1952) berichtet über eine Hemmwirkung von 6,25 γ/ml, VALLÉE (1950) von 5,0 γ/ml, SACHDEVA u. Mitarb. (1956) von 1 bzw. 2 γ/ml Streptomycin. QUAN (1947) sah Tierstämme, die gegen 1,5 OE/ml Penicillin resistent waren. Nach HARISIJADES (1953) zeigten sich im Röhrchentest 7 Stämme von Past. pseudotuberculosis gegen 1,6—3,1 γ/ml Streptomycin und Chloramphenicol, gegen 3,1—6,3 γ/ml Terramycin, gegen 3,1—12,5 γ/ml Aureomycin und gegen 12,5—50 OE/ml Penicillin sensibel. Im Plattentest sahen BRYGOO u. COURDURIER (1955) 6 Stämme durch 10 E/ml Penicillin, 1—10 γ/ml Streptomycin, 10 γ Aureomycin, 10 γ Chloramphenicol und 10—30 γ/ml Terramycin und Polymyxin B im Wachstum gehemmt.

Literatur

ADVIER, M.: Etude d'un bactériophage antipesteux. Bull. Soc. Path. exot. **26**, 94 (1933).

ALBRECHT, H.: Zur Ätiologie der Enteritis follicularis suppurativa. Wien. klin. Wschr. **1910**, 991.

ARKWRIGHT, J. A.: The importance of motility of bacteria classification and diagnosis, with special reference to B. pseudotuberculodsis rodentium. Lancet **1927**, 13.

BACHMANN, W.: Zur Diagnostik der Pseudotuberkulose. Zbl. Bakt., I. Abt. Orig. **87**, 171 (1922).

BADER, R.-E.: Vergleichende Untersuchungen mit einigen neueren Nährböden zur Isolierung von Salmonellen und Shigellen. Z. Hyg. Infekt.-Kr. **131**, 157 (1950).

— Unveröffentlichte Untersuchungen.

BAIN, R. V. S.: Studies on haemorrhagic septicaemia of cattle. II. The detection of naturally aquired immunity. Brit. vet. J. **110**, 519 (1954).

BAIN, R. V. S.: Studies on haemorrhagic septicaemia of cattle. IV. A preliminary examination of antigens of Pasteurella multocida type I. Brit. vet. J. 111, 492 (1955).

BALTAZARD, M. u. Mitarb.: Recommended laboratory methods for the diagnosis of plague. Bull. Wld Htlh Org. 14, 457 (1956).

BAUMANN, R.: Ein Fall von Pseudotuberkulose bei einer jungen Ziege. Z. Infekt.-Kr. Haustiere 31, 141 (1927).

BAYER, G., u. F. v. HERRENSCHWAND: Über die durch Bakterien aus der Gruppe des Bacillus pseudotuberculosis rodentium hervorgerufene Bindehautentzündung (Perinaudsche Conjunctivitis). Albrecht. v. Graefes Arch. Ophtal. 98, 342 (1919).

BEAUDETTE, F. R.: A case of pseudotuberculosis in a blackbird. J. Amer. vet. med. Ass. 97, 151 (1940).

BECK, A.: Die Pseudotuberkulose der Nagetiere und ihre Beziehungen zur Paracholera der Puten, Tauben und Kanarienvögel. Z. Infekt.-Kr. Haustiere 33, 103 (1928).

BECKER, B.: Beitrag zum klinischen Bild der mesenterialen „abscedierenden reticulocytären Lymphadenitis Masshoff". Chirurg 25, 423 (1954).

BERG, H., u. W. CH. HECKER: Perforiertes mit Pasteurella pseudotuberculosis infiziertes Enterokystom. Zbl. Chir. 81, 2483 (1956).

BERGER, K.: Chemotherapieversuche an Pasteurella pseudotuberculosis. Inaug.-Diss. Tübingen 1957.

BERGEY, D. H.: Bergey's Manual of Determinative Bacteriology, 2/4th ed. London: Bailière, Tindall & Cox 1925/1934.

BERKMAN, S.: Accessory growth factor requirement of the members of the genus Pasteurella. J. infect. Dis. 71, 201 (1942).

BEZSONOVA, A., G. LENSKAIA et O. MOLODTZOVA: Quelques cas de transmutation spontanée du B. pestis en B. pseudotuberculosis rodentium Pfeifferi. Off. int. Hyg. publ. 29, 2106 (1937). Zbl. Bakt., I. Abt. Ref. 129, 101 (1938).

BHAGAVAN, N. V., CHEN, T. H. and K. F. MEYER: Further studies of antigenic structure of Pasteurella pestis in gels. Proc. Soc. exp. Biol. (N.Y.) 91, 353 (1956).

BHATNAGAR, S. S.: Bacteriological studies on Pasteurella pestis and Pasteurella pseudotuberculosis, Part. I and II. Ind. J. med. Res. 28, 1, 17 (1940).

BIESTER, H. E., and L. H. SCHWARTE: Diseases of poultry, 2th ed. Ames. Jowa: Jowa State College Press 1948.

BISHOP, L. M.: Study of an outbreak of pseudotuberculosis in guinea-pigs (cavies) due to B. pseudotuberculosis rodentium. Cornell Vet. 22, 1 (1932).

BLANC, G., et M. BALTAZARD: Contribution à l'étude du comportement de microbes pathogènes chez la puce du rat Xenopsylla cheopsis. Le bacille de la pseudotuberculose des rongeurs. C. R. Soc. Biol. (Paris) 138, 811 (1944).

BLAXLAND, J. D.: Pasteurella pseudotuberculosis infection in turkeys. Vet. Rec. 59, 317 (1947).

BOIVIN, A., I. MESROBEANU et L. MESROBEANU: Extraction d'un complexe toxique et antigénique à partir du bacille d'Aërtrycke. C. R. Soc. Biol. (Paris) 114, 307 (1933).

BOQUET, P.: Sur la mobilité du coccobacille de la pseudotuberculose des rongeurs cultivé à 18—20° C. C. R. Soc. Biol. (Paris) 121, 931, (1936).

— Recherches expérimentales sur la pseudotuberculose des rongeurs. Ann. Inst. Pasteur 59, 341 (1937).

—, et ED. DUJARDIN-BEAUMETZ: Sur les relations entre le bacille de la peste et le bacille de la pseudotuberculose des rongeurs. C. R. Soc. Biol. (Paris) 100, 625 (1929).

BRANDIS, H.: Briefliche Mitteilung 1957.

BRANDSTETTER, ST.: Pathologisch-anatomische und histologische Untersuchungen über Tularämie und Pseudotuberkulose (Rodentiose) beim Meerschweinchen. Z. Infekt.-Kr. Haustiere 54, 238 (1939).

BRAUN, O. H., u. K. MÜLLER: Über den Nachweis von Agglutininen gegen Pasteurella pseudotuberculosis bei Kindern. Z. Kinderheilk. 80, 7 (1957).

BREED, R. S., E. G. D. MURRAY and N. R. SMITH: Bergey's Manual of Determinative Bacteriology 7th ed. Baltimore: Williams & Wilkins Company 1957.

BRIGHAM, G. D., and L. F. RETTGER: A systematic study of the Pasteurella genus and certain closely related organisms. J. infect. Dis. 50, 225 (1935).

BRYGOO, E. R., et J. COURDURIER: Action in vitro des antibiotiques sur 101 souches malgaches de Past. pestis. Ann. Inst. Pasteur **89**, 118 (1955).
— — Comportement des souches malgaches de Past. pestis à l'égard du rhamnose. Ann. Inst. Pasteur **89**, 688 (1955).
BURIÁNEK, J., J. MÁLKOVÁ u. F. TION: Pripad sepse pusobene Pasteurellou pseudotuberculosis. Čas. Lék. čes. **88**, 775 (1949).
BURCKHARDT, J. L.: Untersuchungen über die Bewegung und Begeißelung von Bakterien und die Verwertbarkeit dieser Merkmale für die Systematik. Arch. Hyg. (Berl.) **82**, 235 (1914).
BYLOFF, K.: Über eine pestähnliche Erkrankung der Meerschweinchen. Zbl. Bakt., I. Abt. Orig. **41**, 707 (1906); **42**, 5 (1906).
CARTER, G. R.: The type specific capsular antigen of Pasteurella multocida. Canad. J. med. Sci. **30**, 48 (1952).
— Studies of Pasteurella multocida. Amer. J. vet. Res. **16**, 481 (1955).
— A serological study of Pasteurella haemolytica. Canad. J. Microbiol. **2**, 483 (1956).
— Studies on Pasteurella multocida. II. Identification of antigenic characteristics and colonial variants. J. vet. Res. **18**, 210 (1957).
—, and E. ANNAU: Isolation of capsula polysaccharides from colonial variants of Pasteurella multocida. Amer. J. vet. Res. **14**, 475 (1953).
—, and J. L. BYRNE: A serological study of the hemorrhagic septicaemia Pasteurella. Cornell Vet. **43**, 223 (1953).
CASTELLANI, A.: Brief note on a culture medium used in differentiation between Bac. pestis (Past. pestis) and Bac. pseudotuberculosis rodentium (Past. pseudotuberculosis rodentium). J. trop. Med. Hyg. **42**, 158 (1939).
CAVANAUGH, D. C., and S. F. QUAN: Rapid identification of Pasteurella pestis. Amer. J. clin. Path. **23**, 619 (1953).
CECARELLI, A.: Riv. Biol. **42**, 321 (1950). Zit. nach FELDMAN 1955.
CERNAIANU, C.: Sur un nouveau caractère des Pasteurella servant à la différenciation rapide des pasteurella et des paratyphiques. C. R. Soc. Biol. (Paris) **99**, 1180 (1928).
CHANTEMESSE, A.: La tuberculose zoogléique. Ann. Inst. Pasteur **1**, 97 (1887).
CHAPMAN, M.: Pseudotuberculosis rodentium in chincilla. N. Amer. Vet. **29**, 492 (1948).
CHARRIN, et G. H. ROGER: Première note sur une pseudotuberculose bacillaire. C. R. Acad. Sci. (Paris) **106**, 272, 868 (1888).
CHEN, T. H.: The behaviour of Past. pestis in glycerin and rhamnose mediums. J. infect. Dis. **85**, 97 (1949).
—, and K. F. MEYER: Studies on immunisation against plague. J. Immunol. **74**, 501 (1955).
CHRISTENSEN, N. P.: Pseudotuberculose hos fugle. Kgl. Vet. og. Landbohøjsk. /Aarsskr. **50** (1927). Zbl. Bakt., I. Abt. Ref. **87**, 186 (1927).
CHRISTIANSEN, W.: Abscedierende Lymphadenitis mesenterialis. Zbl. Bakt., I. Abt. Ref. **165**, 591 (1957).
CLAPHAM, P. A.: Pseudotuberculosis among Stock Doves in Hampshire. Nature (Lond.) **172**, 353 (1953).
CLAUSSEN, S.: Über Bakteriaemie durch das Bact. pseudotuberculosis rodentium bei der Biberratte. Dtsch. tierärztl. Wschr. **1934**, 21.
— Rodentiose (Pseudotuberkulose) bei Marderhunden. Dtsch. tierärztl. Wschr. **1938**, 262.
COCU, C. TRUCHE et J. BAUCHE: Un cas de pseudotuberculose chez le chat. Bull. Acad. vét. Fr. **4**, 244 (1931).
COLAS-BELCOUR, J.: Valeur du milieu glycériné dans le diagnostic différentiel des cultures des bacilles de la peste et de la pseudotuberculose des rongeurs. C. R. Soc. Biol. (Paris) **94**, 1 (1926).
COLLET, P., L. RENAULT et F. VALENTIN: Pseudotuberculose hépatique à bacille de Malassez et Vignal chez le chien. Bull. Soc. Sci. vét. Lyon **57**, 307 (1955).
COMMUNAL, R.: La pseudotuberculose du chat, S. 122. Paris: Imprimerie R. Foulon 1945.
CONN, H. I., and G. E. WOLFE: Flagella staining as a routine test for bacteria. J. Bact. **36**, 517 (1938).
COOK, R.: A method of demonstrating Pasteurella pseudotuberculosis in smears from animal lesions. J. Path. Bact. **64**, 228 (1952).

Cook, R.: The naturally occuring diseases of laboratory animals and practical measures for the control of the diseases. J. med. Lab. Technol. **11**, 30 (1953).

Crumpton, M. I., and D. A. L. Davies: An antigenic analysis of Pasteurella pestis by diffusion of antigens and antibodies in agar. Proc. Soc. exp. Biol. (N.Y.) **145**, 111 (1956).

— — A protein antigen associated with smooth colony forms of some species of Pasteurella. Nature (Lond.) **180**, 863 (1957).

— — and A. M. Hutchison: The serological specifities of Pasteurella pseudotuberculosis somatic antigens. J. gen. Microbiol. **18**, 129 (1958).

Damperoff, N. J.: Komplementbindungsversuche mit Antipestserum. Zbl. Bakt., 1. Abt. Orig. **55**, 188 (1910).

Davies, D. A. L.: A specific polysaccharide of Pasteurella pestis. Biochem. J. **63**, 105 (1956).

— The smooth and rough somatic antigens of Pasteurella pseudotuberculosis. J. gen. Microbiol. **18**, 118 (1958).

De Mendonca Machado, A., u. J. Pelouro: Pseudotuberculose dos roedores. Repos. Lab. Pat. Vet. Lisboa **5**, 175 (1943).

Dessy, G.: Ricercle sur un bacillo della pseudotuberculosi. Boll. Ist. sieroter. milan. **4**, 123 (1925). Zbl. Bakt., I. Abt. Ref. **83**, 186 (1926).

Devignat, R., et A. Boivin: Deux milieux complexes pour la différenciation des trois variétés de Pasteurella pestis et de Pasteurella pseudotuberculosis. Bull. Soc. Path. exot. **46**, 627 (1953).

— — Comportement biologique et biochimique de P. pestis et de P. pseudotuberculosis. Bull. Wld Hlth Org. **10**, 463 (1954).

—, et A. Chevalier: Des germes pathogènes peuvent-ils nitrifier des protéines? Ann. Inst. Pasteur **82**, 650 (1952).

Deyl, J.: Experimentelle Untersuchungen mit Pseudotuberkulose besonders am Auge. Acad. Sci. de l'empereur François Joseph I, Prag 1894. Zit. nach Eberth, C. J., u. H. Preisz, Ergebn. allg. Path. path. Anat. **1**, 732 (1896).

Dieudonné, A., u. R. Otto: Pest. In Kolle-Krauss-Uhlenhuth, Handbuch der pathogenen Microorganismen, 3. Aufl., Bd. IV/1, 179. 1928.

Doerr, R.: Die Antigene. Wien: Springer 1948.

Dor, L.: Pseudotuberculose bacillaire. C. R. Soc. Biol. (Paris) **106**, 1027 (1888).

Dorssen, C. A. van: Pseudotuberculosis bij duiven. T. Diergeneesk. **76**, 249, 727 (1951).

— Orienteerende proven over therapie en vaccinatie bij pseudotuberculosis van Knaagdieren. T. Diergeneesk. **77**, 235 (1952).

— Enting van caviae tegen Pseudotuberculosis met levende en dode entstoff. T. Diergeneesk. **80**, 718 (1955).

Dujardin-Beaumetz, Ed., B. Ballet et J. Cébron: Pseudotuberculose chez l'homme. Presse méd. **46**, 43 (1938). — Rev. Path. comp. **1938**, 884.

Eberth, C. J.: Zwei Mykosen des Meerschweinchens. Virchows Arch. path. Anat. **100**, 15 (1885).

— Der Bacillus der Pseudotuberkulose des Kaninchens. Fortschr. Med. **3**, 719 (1885). — Virchows Arch. path. Anat. **103**, 488 (1886).

—, u. H. Preisz: Menschliche und tierische Pseudotuberkulose. Ergebn. allg. Path. path. Anat. **1**, 732 (1896).

Eieland, A.: Norsk. Vet. Tidskr. **59**, 1 (1947). Zit. nach Feldman 1955.

Englesberg, E.: Mutation to rhamnose utilization in Pasteurella pestis. J. Bact. **73**, 641 (1957).

Farbstein, M. E.: Die Lymphadenitis mesenterialis. Der heutige Stand der Lehre. Inaug.-Diss. Basel 1955.

Fauconnier, J.: La décomposition de l'urée en milieu synthétique de Ferguson par Pasteurella pseudotuberculosis. Ann. Inst. Pasteur **79**, 104 (1950).

Favarissova, B. I.: Die Beweglichkeit von B. pseudotuberculosis rodentium [Russisch]. Rev. Microbiol. Saratov **16**, 65 (1937). Ref. Bull. Inst. Pasteur **37**, 1169 (1939).

Feldman, W. H.: In Hull, Diseases transmitted from animals to man, 4th ed. Springfield: Ch. C. Thomas 1955.

Fey, H.: Briefliche Mitteilung 1956.

Flamm, H.: Briefliche Mitteilung 1958.

—, u. W. Kovač: Die Pathogenese der pseudotuberkulösen Lymphadenitis ileocaecalis (Manuskript eingesehen). Schweiz. Z. Path. (im Druck).

Fraenkel, E.: Über Pseudotuberkulose des Menschen. Z. Hyg. Infekt.-Kr. **101**, 406 (1924).

Gaté, J., et M. Billa: Etude bactériologique et expérimentale d'une épizootie à manifestations pseudotuberculoses. C. R. Soc. Biol. (Paris) **99**, 814 (1928).

— — A propos d'une épizootie à manifestations pseudotuberculeuses. C. R. Soc. Biol. (Paris) **99**, 812 (1928).

Gifford, S. R., and N. K. Lazar: Inclusion bodies in artificially induced conjunctivitis. Arch. Ophthal. (Chicago) **4**, 468 (1930).

Girard, G.: La réaction des nitrites pour la différenciation du bacille de la peste et du bacille de la pseudotuberculose. C. R. Soc. Biol. (Paris) **133**, 244 (1940).

— Absence d'antigène glucido-lipidique chez le bacille de la peste et le bacille de la pseudotuberculose des rongeurs. C. R. Soc. Biol. (Paris) **135**, 1577 (1941).

— Sur quelques nouveaux caractères différenciant les bacilles de la peste et de la pseudotuberculose des Pasteurella. Ann. Inst. Pasteur **68**, 476 (1942).

— Sensibilité des bacilles pesteux et pseudotuberculeux, d'une part des germes du groupe coli-dysentérique, d'autre part aux bactériophages homologues. Ann. Inst. Pasteur **69**, 52 (1943).

— La toxine de Past. pseudotuberculosis, ses analogies avec la toxine de Past. pestis. Ann. Inst. Pasteur **79**, 33, 105 (1950).

— Méthodes permettant de différencier P. pestis de P. pseudotuberculosis. Bull. Org. mond. Santé **9**, 645 (1953).

— Données récentes sur les infections à bacille de Malassez et Vignal (Past. pseudotuberculosis) en pathologie vétérinaire et humaine. Bull. Acad. vét. Fr. **27**, 497 (1954).

— Briefliche Mitteilung 1956.

—, and A. Chevalier: Classification sérologique de 56 souches de Past. pseudotuberculosis, dont 52 isolées en France. Ann. Inst. Pasteur **88**, 227 (1955).

Golovin, A. D.: Seuchenhaftes Auftreten der Pseudotuberkulose bei Maulwürfen; deutsche Zusammenfassung. Rev. Microbiol. Saratov **9**, 377 (1930).

Goret, P., P. Collet, L. Joubert et C. Pilet: Diagnostic expérimental et pathogénique de la pseudotuberculose du chat. Bull. Soc. Sci. vét. Lyon **57**, 205 (1955).

Goyon, M.: A propos de quelques cas de pseudotuberculose du lièvre dans la Sarthe. Etude bactériologique et sérologique de 10 souches de Cillopasteurella pseudotuberculosis. Rec. Méd. vét. **132**, 539 (1956).

Graber, H.: Der appendizitische Symptomenkomplex als Folge einer enteralen Infektion mit Pasteurella pseudotuberculosis. Chirurg **27**, 401 (1956).

—, u. W. Knapp: Die abscedierende retikulocytäre Lymphadenitis mesenterialis (Masshoff) als Bestandteil eines enteralen Primärkomplexes und Folge einer Infektion mit Pasteurella pseudotuberculosis. Frankf. Z. Path. **66**, 399 (1955).

Grancini, L. E.: Sindrome oculo-glandolare (congiutivite di Parinaud) data dal „Bacterium pseudotuberculosis rodentium (Pfeiffer)". Boll. Oculkt. **18**, 133 (1939).

Grancher et Ledoux-Lebard: Recherches sur la tuberculose zoogléique. Arch. Méd. exp. **2**, 203 (1889); **2**, 588 (1890).

Griffin, A. M., and M. L. Robbins: The flagellation of Listeria monocytogenes. J. Bact. **48**, 114 (1944).

Gsell, O., u. M. Gsell-Burse: Die Katzenkratzkrankheit. Ergebn inn. Med. Kinderheilk. **8**, 76 (1957, ausführliche Literaturangabe).

Gunnison, J. B., A. Larson and A. S. Lazarus: Rapid differentiation between Pasteurella pestis and Pasteurella pseudotuberculosis by action of bacteriophage. J. infect. Dis. **88**, 254 (1951).

—, and A. S. Lazarus: Alteration of Pasteurella pestis bacteriophage following successive transfer on Pasteurella pseudotuberculosis and on Shigella. Proc. Soc. exp. Biol. (N. Y.) **69**, 294 (1948).

— M. C. Shevky, L. V. K. Zion and M. J. Abbott: Lysis of Pasteurella pseudotuberculosis by bacteriophage. J. infect. Dis. **88**, 187 (1951).

HAAS, V. H.: A study of Pseudotuberculosis rodentium recovered from a rat. Publ. Hlth Rep. (Wash.) **53**, 1033 (1938).

HAENSELT, V.: Zur Kenntnis der abscedierenden reticulocytären Lymphadenitis (MASSHOFF). Ärztl. Wschr. **1957**, 509.

HÄSSIG, A., J. KARRER u. F. PUSTERLA: Über Pseudotuberkulose beim Menschen. Schweiz. med. Wschr. **1949**, 1948.

HAGAN, W. A., and D. W. BRUNNER: The infections diseases of domestic animals. 2th ed. Ithaca N. Y.: Cornstock 1951.

HARISIJADES, S. S.: Examination of the action of streptomycin on Pasteurella pseudotuberculosis in embryonated hen's eggs. Acta med. jugosl. **7**, 34 (1953).

— The action in vitro of some antibiotics on Pasteurella pseudotuberculosis. Acta med. jugosl. **7**, 37 (1953).

HAUPT, H.: Ein Beitrag zur kulturellen Unterscheidung des Pfeifferschen Pseudotuberkulosebazillus von ähnlichen vogelpathogenen Bakterienarten (Bact. avicidum u. Bact. gallinarum). Zbl. Bakt., I. Abt. Orig. **109**, 1 (1928).

— Bacterium pfaffi Hadley 1918 = Bacillus pseudotuberculosis Eisenberg 1891. Zbl. Bakt., I. Abt. Orig. **132**, 349 (1934).

— Zur Systematik der Bakterien. Ergebn. Hyg. Bakt. **17**, 175 (1935).

HECKER, W. CH.: Zur Pasteurella-pseudotuberculosis-Erkrankung (Rodentiose) beim Menschen. Arch. Kinderheilk. **156**, 151 (1957).

HEDINGER, C.: Die histologischen Veränderungen bei der sog. Katzenkratzkrankheit, einer benignen Viruslymphadenitis. Virchows Arch. path. Anat. **322**, 159 (1952).

HEELSBERGEN, T. V.: T. Diergeneesk. **54**, 545 (1927). Zit. nach SCHÜTZE 1929.

HERMS: Über die Pseudotuberkulose bei der Katze. Tierärztl. Rdsch. **42**, 324 (1936)

HESSELBROCK, W., and L. FOSHAY: The morphology of Bacterium tularense. J. Bact. **49**, 209 (1945).

HEUSSER, H.: Die Schwellung der mesenterialen Lymphknoten (Lymphadenopathia mesaraica). Bruns' Beitr. klin. Chir. **130**, 85 (1924).

HÖRSTEBROCK, R.: Zur Frage der abscedierenden, reticulocytären Lymphadenitis (MASSHOFF). Zbl. allg. Path. path. Anat. **91**, 221 (1954).

HOLCH, P.: Briefliche Mitteilung 1958.

IKEGAKI, R.: Saikin Gakú-Zasshi **1934**, 751 [Japanisch]. Zit. nach UETAKE u. NAKANO 1949.

INGELRANS, S., et V. POUPARD: Un cas d'adénite mésentérique aiguë à Pasteurella pseudotuberculosis. Lille chir. **12**, 201 (1957).

ISSALY, I., et A. S. ISSALY: Determinación cuantitative de la actividad uréasica de las pasteurellas. Rev. Asoc. bioquím argent. **18**, 154 (1953).

IUSCHENKO, G. V.: Wirkung von Streptomycin bei experimenteller Pseudotuberkulose [Russisch]. Antibiotiki **1**, 46 (1956).

JACOTOT, H., A. VALLÉE et A. LE PRIOL: Sur un cas d'infection du porc par le bacille de Malassez et Vignal. Rev. Path. comp. **58**, 134 (1950).

JAMIESON, S., and M. A. SOLTYS: Infectious epididymo-orchitis of rams associated with Pasteurella pseudotuberculosis. Vet. Rec. **59**, 351 (1947). — Vet. Bull. **18**, 148 (1948).

KAKEHI, S.: A comparison of various strains of Bacillus pseudotuberculosis rodentium (PFEIFFER) with special reference to certain variation phenomena. J. Path. Bact. **20**, 269 (1915).

KARLSSON, K. F.: Pseudotuberculos hos höns Fåglar. Scand. Vet. Bact. **35**, 673 (1945).

KAUFFMANN, F.: Vergleichende Untersuchungen an Pseudotuberkulose-, Paratyphus-, Pasteurella- und Pestbakterien. Z. Hyg. Infekt.-Kr. **114**, 97 (1933).

— Weitere Erfahrungen mit dem kombinierten Anreicherungsverfahren für Salmonellabacillen. Z. Hyg. Infekt.-Kr. **117**, 26 (1936).

KAWASHIMA, K.: Gun'idan Zasshi **71** (1934) [Japanisch]. Zit. nach UETAKE u. NAKANO 1949.

KLEIN, E.: Ein Beitrag zur Kenntnis der Verbreitung des Bacillus pseudotuberculosis. Zbl. Bakt., I. Abt. Orig. **26**, 260 (1899).

KNAPP, W.: Pasteurella pseudotuberculosis als Erreger einer mesenterialen Lymphadenitis beim Menschen. Zbl. Bakt., I. Abt. Orig. **161**, 422 (1954).

— Mesenterial lymphadenitis caused by Pasteurella pseudotuberculosis. J. Amer. med. Ass. **156**, 195 (1954).

KNAPP, W.: Die diagnostische Bedeutung der antigenen Beziehungen zwischen Past. pseudotuberculosis und der Salmonella-Gruppe. Zbl. Bakt., I. Abt. Orig. **164**, 57 (1955).
— Die Bedeutung mikrobiologischer Untersuchungen zur ätiologischen Abgrenzung der mesenterialen Lymphadenitis. Chirurg **26**, 440 (1955).
— Ein Beitrag zur Beweglichkeit von Pasteurella pseudotuberculosis. Z. Hyg. Infekt.-Kr. **142**, 219 (1956).
— Die Agglutinationsreaktion und ihre Besonderheiten in der Serodiagnostik menschlicher Infektionen mit Pasteurella pseudotuberculosis. Z. Hyg. Infekt.-Kr. **143**, 261 (1956).
— Unveröffentlichte Untersuchungen.
— Mesenteric adenitis due to Pasteurella pseudotuberculosis in young people. New Engl. J. Med. **259**, 776 (1958).
— T. H. CHEN u. K. F. MEYER: Unveröffentlichte Untersuchungen.
— H. HAGER u. W. MASSHOFF: Unveröffentlichte Untersuchungen.
KNAPP, W., u. W. MASSHOFF: Zur Aetiologie der abscedierenden reticulocytären Lymphadenitis. Dtsch. med. Wschr. **1954**, 1266.
—, u. W. STEUER: Untersuchungen über den Nachweis komplementbindender und agglutinierender Antikörper gegen Past. pseudotuberculosis in Sera infizierter und immunisierter Menschen und Tiere. Z. Immun.-Forsch. **113**, 370 (1956).
KNOTHE, H.: Über die Epidemiologie der Tularaemie. Beitr. Hyg. Epidem. **1955**.
KÖNIG, P., u. J. MAURATH: Zur Chirurgie der Lymphadenitis mesenterialis durch Pasteurelleninfektion. Chir. Prax. **1**, 165 (1957).
KOROBKOVA, E. I.: Rev. Microbiol. Saratov **16**, 18 (1937); **19**, 3 (1940). Zit. nach POLLITZER 1954.
KOSSEL u. OBERBECK: Bakteriologische Untersuchungen über Pest. Arb. Gesundh.-Amte (Berl.) **18**, 114 (1902).
KRAGE u. WEISGERBER: Eine Putenseuche mit Diplostreptobazillenbefund. Tierärztl. Rdsch. **1924**, 309.
KRAINOVA, A. N.: On the question of the importance of rhamnose for differential diagnosis of B. pestis and B. pseudotuberculosis Pfeiffer. Rev. Microbiol. Saratov **18**, 91 (1939). Zit. nach ENGLESBERG 1957.
KUHLMANN, F., u. W. HERRMANN: Lymphadenitis mesenterialis und Enteritis durch Pasteurella pseudotuberculosis. Med. Klin. **1955**, 1735.
—, u. R. ISEBARTH: Das klinische Bild der Lymphadenitis mesenterialis. Med. Klin. **1954**, 1605.
KURAUCHI, K., and H. HOMMA: Bull. Off. int. Hyg. publ. **28**, 1088 (1936). Zit. nach POLLITZER 1954.
KUROKAWA, M.: A study on the experimental pseudotuberculosis changes by B. pseudotuberculosis rodentium (PFEIFFER) in mice. I. On the behaviour of the bacillus with a special reference to the relation of the infection. Saikin Gaku-Zasshi **526**, 11 (1939). Biol. Abstr. **14**, 662 (1940).
— Studies on the experimental pseudotuberculosis rodentium (PFEIFFER) in mice. II. Chiefly on the infection by intranasal route. Saikin Gaku-Zasshi **527**, 53 (1940). Biol. Abstr. **14**, 1023 (1940).
— Experimental pseudotuberculosis by B. pseudotuberculosis rodentium (PFEIFFER). An experimental on the so-called living bacilli immunity. Saikin Gaku-Zasshi **541**, 62 (1941). Biol. Abstr. **15**, 2040 (1941).
—, and K. MIKAMI: A study on the pseudotuberculosis by B. pseudotuberculosis rodentium (PFEIFFER) in mice. IV. An experiment — concerning the so-called killed bacilli immunity. Saikin Gaku-Zasshi **542**, 35 (1941). Biol. Abstr. **15**, 1687 (1941).
KUTSCHER: Ein Beitrag zur Kenntnis der bacillären Pseudotuberkulose der Nagetiere. Z. Hyg. Infekt.-Kr. **18**, 327 (1894).
LAZARUS, A. S., and J. B. GUNNISON: The action of Pasteurella pestis bacteriophage on strains of Pasteurella, Salmonella and Shigella. J. Bact. **53**, 705 (1947).
—, and M. M. NOZAWA: The endotoxin of Pasteurella pseudotuberculosis. J. Bact. **56**, 187 (1948).
LEADER, R. W., and G. A. BAKER: A report of two cases of Pasteurella pseudotuberculosis infection in the chincilla. Cornell Vet. **44**, 262 (1954).

LEBLOIS, CH.: La pseudo-tuberculose zoogléique chez le chat. Rec. méd. vét. **96**, 307 (1920).

LEDOUX-LEBARD: De l'action du sérum pseudotuberculeux sur le bacille de la pseudotuberculose. Ann. Inst. Pasteur **11**, 909 (1897).

LEIFSON, E.: A method of staining bacteria flagella and capsules together with a study of the origin of flagella. J. Bact. **20**, 203 (1930).

— New selenit enrichment media for isolation of typhoid and paratyphoid (Salmonella) bacilli. Amer. J. Hyg. **24**, 423 (1936).

—, and R. HUGH: Variation in shape and arrangement of bacterial flagella. J. Bact. **65**, 263 (1953).

LEMBACH, K., u. H. SOUS: Geißelfärbung nach SOUS. Zbl. Bakt., I. Abt. Orig. **152**, 445 (1948).

LENNERT, K.: Zur Kenntnis der reticulocytären, abscedierenden Lymphadenitis (MASSHOFF). Vortrag Tagg Norddtsch. Pathologen, Bad Pyrmont, 1956. Ref. Zbl. allg. Path. path. Anat. **96**, 398 (1957).

LENSKAIA, G. N.: Morphological variety of B. pseudotuberculosis rodentium (PFEIFFER) and B. pestis (Englische Zusammenfassung). Rev. Microbiol. Saratov **7**, 254 (1928).

LERCHE, M.: Beobachtungen über eine Putenseuche, die sog. Paracholera. Dtsch. tierärztl. Wschr. **1926**, 405.

— Die „Paracholera" der Puten und ihre Beziehungen zur Pseudotuberkulose der Nagetiere. Zbl. Bakt., I. Abt. Orig. **104**, 493 (1927).

LESBOUYRIES, G.: Pseudotuberculose du pigeon. Bull. Acad. vét. Fr. **7**, 103 (1934).

LEVENSON, S.: Coloration des cils bactériens par une procédé simple. Ann. Inst. Pasteur **56**, 634 (1936).

LEVINTHAL, W.: Eine neue Technik der Beweglichkeitsprüfung an lebenden Bakterien. Z. Hyg. Infekt.-Kr. **111**, 140 (1930).

LÉVY-BRUHL, M.: Les Pasteurelloses humaines. Ann. Méd. **44**, 406 (1938).

LITTLE, P. A., and B. M. LYON: Demonstration of serological types within nonhemolytic Pasteurella. Amer. J. vet. Res. **4**, 110 (1943).

LOGHEM, J. J. VAN: The classification of the plague-bacillus. J. Microbiol. Serol. **10**, 15 (1945).

— La classification du bacille pesteux. Ann. Inst. Pasteur **72**, 975 (1946).

LOREY, A.: Über einen unter dem klinischen Bild des Typhus abdominalis verlaufenden Krankheitsfall, hervorgerufen durch ein anscheinend der Gruppe der Bakterien der Septicaemia haemorrhagica angehörendes Stäbchen. Z. Hyg. Infekt.-Kr. **68**, 49 (1911).

MACCHIAVELLO, A.: Infectión mixta pur Peste y Pasteurella pseudotuberculosis rodentium. Officina Sanitaria Panamericana, Publication Nr. 165, 252. 1941.

MAGLIONE, E.: Contributo allo studio della pseudotuberculosi dei roditori. Ann. Fac. Med. vet. Torino **5**, 249 (1955).

—, e F. CERETTO: Contributo allo studio della pseudotuberculosi dei roditori. Ann. Fac. Med. vet. Torino **5**, 9 (1956).

MALASSEZ, L., et W. VIGNAL: Tuberculose zoogléique. Arch. d. Phys. **2**, 369 (1883).

— — Sur le microorganisme de la tuberculose zoogléique. Arch. d. Phys. **4**, 81 (1884).

MANNINGER, R.: In HUTYRA-MARAK-MANNINGER: Spezielle Pathologie und Therapie der Haustiere, 9. Aufl., S. 560. Berlin: Springer 1945.

MARLINI 1938: Zit. nach K. F. MEYER 1952.

MARTHEDAL, H. E., u. G. VELLING: Pasteurellose og pseudotuberculose hos fjerkrae i Danmark. Nord. Vet.-Med. **6**, 651 (1954).

MASON u. K. F. MEYER: Unveröffentlicht. Zit. nach K. F. MEYER 1952.

MASSHOFF, W.: Eine neuartige Form der mesenterialen Lymphadenitis. Dtsch. med. Wschr. **1953**, 532.

—, u. W. DÖLLE: Über eine besondere Form der sog. mesenterialen Lymphadenitis: „Die abscedierende reticulocytäre Lymphadenopathie." Virchows Arch. path. Anat. **323**, 664 (1953).

MATHEY, W. I., and P. I. SIDDLE: Isolation of Pasteurella pseudotuberculosis from a California turkey. J. Amer. vet. med. Ass. **125**, 482 (1954).

MATZKE, M.: Die Bedeutung der Tularaemie für den europäischen Raum im Spiegel des neueren Schrifttums. Berl. Münch. tierärztl. Wschr. **1943**, 376.

MAZZINI: Pseudotuberculose beim Rind. G. Soc. Acad. Vet. **1897**, 758. Zit. nach POPPE 1928.

MESSERSCHMIDT, TH., u. KELLER: Befunde bei Pseudotuberculose der Nagetiere, verursacht durch den Bac. pseudotuberculosis rodentium. Z. Hyg. Infekt.-Kr. **77**, 289 (1914).

MEYER, K. F.: The newer knowledge of bacteriology and immunology (E. O. JORDAN u. I. S. FALK), p. 614. Chicago: University of Chicago Press 1928.

— The Pasteurella: In DUBOS: Bacterial and mycotic infections of man, 1th and 2th ed. Philadelphia: Lippincott Comp. 1948/1952. (Neuauflage im Druck.)

— Perspective concerning infections in animals transmissible in man. Northw. Med. (Seattle) **50**, 333 (1951).

—, and A. P. BATCHELDER: Selective mediums in the diagnosis of rodents plague. J. infect. Dis. **39**, 370 (1926).

— — A disease in wild rodents caused by Pasteurella muricida. J. infect. Dis. **39**, 386 (1926).

MØRCH, J. R., u. G. KROGH-LUND: Untersuchungen über die Bakterien der Pasteurella-Gruppe. Z. Hyg. Infekt.-Kr. **112**, 471 (1931).

MORETTI, B.: Ein Beitrag zur Pseudotuberkulose. Dtsch. tierärztl. Wschr. **1938**, 35.

MOSS, E. S., and J. D. BATTLE: Human infection with Pasteurella pseudotuberculosis rodentium of Pfeiffer. Amer. J. clin. Path. **11**, 677 (1941).

MURRAY, P. E.: Pseudotuberculosis of sheep due to B. pseudotuberculosis rodentium. Aust. vet. J. **8**, 181 (1932).

NEUGEBAUER, W.: Ein Fall von echter Pseudotuberkulose beim Menschen. Med. Klin. **1933**, 420.

NICOLLE, CH., et H. SPARROW: Sur une épizootie de pseudotuberculose du cobaye observée à Tunis. Arch. Inst. Pasteur Tunis **17**, 338 (1928).

— M.: Etudes sur la morve expérimentale et maladies „spontanées" des cobayes. Ann. Inst. Pasteur **20**, 801 (1906).

NOCARD, E.: Sur une tuberculose zoogléique des oiseaux de basse-cour. Bull. Soc. centr. Med. Vet. **39**, 207 (1885).

— Sur la tuberculose zoogléique. C. R. Soc. Biol. (Paris) **1889**, 608.

NOON, L.: Observations on the evolution of immunity in disease. J. Hyg. (Lond.) **9**, 181 (1909).

NORDMANN, M.: Die Katzenkrankheit. Verh. dtsch. Ges. Path. **38**, 112 (1955).

NOUVEL, I., et I. RINJARD: Pseudotuberculose du singe cynocéphale (Papiopapio) à bacille de Malassez et Vignal. Rev. Path. comp. **1949**, 60.

OLIVO, R.: Aureomicina e recettività des coniglio alla pasteurellosi da P. pseudotuberculosis rodentium. Boll. Soc. ital. Biol. sper. **28**, 1731; **29**, 1032 (1953).

OLT, A.: Über das seuchenhafte Auftreten der Rodentiose unter den Hasen. Z. Infekt.-Kr. Haustiere **52**, 89 (1938).

OTTEN, L.: Die Differentialdiagnose zwischen den Erregern der haemorrhagischen Septikaemien. Zbl. Bakt., I. Abt. Orig. **98**, 484 (1926).

OUCHTERLONY, Ö.: Diffusion-in-gel methods for immunological analysis. In: Progr. Allerg, Vol. 5, p. 1—78. Basel u. New York: S. Karger 1958.

OUDIN, I.: Specific praecipitation in gels and its application to immunochemical analysis. In: Methods in medical research, Vol. 5, p. 335. Chicago: The Year Book Publishers 1952.

PALLASKE, G.: Beitrag zur Patho- und Histogenese der Pseudotuberkulose (Bact. pseudotuberculosis rodentium U der Tiere. Z. Infekt.-Kr. Haustiere **44**, 43 (1933).

—, u. A. MEYN: Über die Pseudotuberkulose (Bact. pseudotub. rod.) bei Katzen. Dtsch. tierärztl. Wschr. **1932**, 576.

PARIETTI, E.: Eine Form der Pseudotuberkulose. Zbl. Bakt., I. Abt. Orig. **8**, 577 (1890).

PARNAS, I.: Utilisation du microscope électronique pour les investigations microbiologiques concernant les bactéries pathogènes. Off. int. Epiz. **45**, 364 (1956).

PAUL, O., u. O. WELTMANN: Pseudotuberkulose beim Menschen. Wien. klin. Wschr. **1934**, 603.

PFAFF, FR.: Eine infektiöse Krankheit der Kanarienvögel. Zbl. Bakt., I. Abt. Orig. **38**, 275 (1905).

PFEIFFER, A.: Über die bacilläre Pseudotuberkulose bei den Nagetieren. Leipzig 1889.

PIÉCHAUD, M.: Un nouveau cas de pseudotuberculose humaine. Ann. Inst. Pasteur **83**, 420 (1952).

PIROSKY, I.: Sur l'antigène glucido-lipidique de Pasteurella. C. R. Soc. Biol. (Paris) **127**, 98, 234, 966 (1938).

— Sur la spécificité des antigènes glucido-lipidiques des Pasteurella et sur leurs affinités sérologiques avec les antigènes glucido-lipidiques des Salmonella. C. R. Soc. Biol. (Paris) **128**, 346, 347 (1938).

PLASAJ, ST.: Zur Morphologie des Bacterium pseudotuberculosis rodentium. Zbl. Bakt., I. Abt. Orig. **86**, 468 (1921).

— Pseudotuberkulose bei Enten. Jugoslav. Vet. Glasnik **9**, 352 (1929).

PODHRAGYAI, L., u. I. FODOR: Abscedálo reticulocytás lymphadenitis. Orv. Hetil. **1956**, 277.

POKROWSKAJA, M.: Über die Dissoziation des Bact. pseudotuberculosis rodentium. Zbl. Bakt., I. Abt. Orig. **116**, 304 (1930).

POLLITZER, R.: Plague. World Health Organisation. Monograph Series No. 22, Genf, 1954.

POPPE, K.: Pseudotuberkulose. In Handbuch pathologische Microorganismen von KRAUSS-UHLENHUTH, 3. Aufl., Bd. IV/1, S. 413. 1928.

PREISZ, H.: Über einen Fall der Pseudotuberkulose beim Schaf und über Pseudotuberkulose im allgemeinen. Zbl. Bakt., I. Abt. Orig. **10**, 568 (1891).

— Recherches comparatives sur les pseudotuberculoses bacillaires et une nouvelle espèce de pseudotuberculose. Ann. Inst. Pasteur **8**, 231 (1894).

— Bacilläre Pseudotuberculose der Tiere. Ergebn. allg. Path. path. Anat. **1**, 733 (1896).

PRESTON, N. W., and H. B. MAITLAND: The influence of temperature on the motility of Pasteurella pseudotuberculosis. J. gen. Microbiol. **7**, 117 (1952).

PRÉVÔT, A. R.: Manuel de classification et de détermination des bactéries. Paris 1948.

QUAN, S. F., L. E. FORSTER, A. LARSON and K. F. MEYER: Streptomycin in experimental plague. Proc. Soc. exp. Biol. (N. Y.) **66**, 528 (1947).

RAJAGOPALAN, V. R., and N. S. SANKARANARAYANAN: A case of pseudotuberculosis in goat. Ind. J. vet. Sci. **14**, 34 (1944).

RAMON, G.: Etudes sur le bacille de Malassez et Vignal. La pseudotuberculose du cobaye. Ann. Inst. Pasteur **28**, 585 (1914).

— Sur la vitesse d'apparition des anticorps. C. R. Soc. Biol. (Paris) **99**, 1295 (1928).

RANDERATH, E.: Beiträge zur Morphologie der sog. Viruslymphadenitis und zu deren Differentialdiagnose. Verh. dtsch. Ges. Path. **38**, 116 (1954).

RANSOM, I. R.: Some aspects of relationship between antigens of Pasteurella pestis and Pasteurella pseudotuberculosis. Proc. Soc. exp. Biol. (N. Y.) **93**, 551 (1956).

REIMANN, H. A.: Further studies on B. pseudotuberculosis. Amer. J. Hyg. **16**, 206 (1932).

—, and W. J. ROSE: The similarity of pseudotuberculosis and tularaemia. Arch. Path. (Chicago) **16**, 584 (1931).

RINIKER, P.: Über die enterale Pseudotuberkulose. Schweiz. Z. Path. **20**, 52 (1957).

RISLAKKI, V.: Pseudotuberculosis in rodents and foxes. Suom Eläinlääk.-L. **48**, 182 (1942).

ROBERTS, R. S.: An immunological study of Past. septica. J. comp. Path. **57**, 261 (1947).

ROMAN, B.: Über einen Fall von bazillärer Pseudotuberkulose beim Menschen. Virchows Arch. path. Anat. **222**, 53 (1916).

ROSENBUSCH, C. T., and I. A. MERCHANT: A study of the haemorrhagic septicemia Pasteurelleae. J. Bact. **37**, 69 (1939).

ROSENWALD, A. S., and E. M. DICKINSON: A report of Pasteurella pseudotuberculosis infection in turkeys. Amer. J. vet. Res. **5**, 246 (1944).

ROWLAND, S.: The relations of pseudotubercle to plague as evidenced by vaccination experiments. J. Hyg. (Lond.) Plague Suppl. **2**, 350 (1912).

RUTQVIST, L., E. THAL and B. ÅBERG: Changes in the serum proteins of guinea-pigs immunized with Pasteurella pseudotuberculosis. Acta path. microbiol. scand. **39**, 94 (1956).

SACHDEVA, L. D., S. L. KALRA and B. L. TANEJA: A case of human infection with Pasteurella pseudotuberculosis rodentium in India. J. med. Sci. **10**, 114 (1956).

SAENZ, A.: Pseudotuberculose spontanée du singe provoquée par le coccobacille de Malassez et Vignal. C. R. Soc. Biol. (Paris) **104**, 1189 (1930).

—, et L. COSTIL: Pseudotuberculose spontanée du singe provoquée par le coccobacille de Malassez et Vignal. C. R. Soc. Biol. (Paris) **110**, 449 (1932).

Saenz, A., et L. Costil: Diagnostic de la pseudotuberculose du cobaye par l'intradermoré-action aux corps microbiens. C. R. Soc. Biol. (Paris) 111, 573 (1932).

Saisawa, K.: Über die Pseudotuberkulose beim Menschen. Z. Hyg. Infekt.-Kr. 73, 353 (1913).

Sander, K.: Beitrag zur abscedierenden reticulocytären Lymphadenitis mesenterialis (Masshoff). Zbl. Chir. 83, 1281 (1958).

Sasaki, T.: Monosaccharides composition of the antigenic polysaccharide of Pasteurella pseudotuberculosis rodentium. Nature (Lond.) 179, 920 (1957).

Schäfer, W.: Das Vorkommen des Bacterium pseudotuberculosis rodentium. Tierärztl. Rdsch. 45, 72 (1939).

Schar, M., and E. Thal: Comparative studies on toxin of Pasteurella pestis and Pasteurella pseudotuberculosis. Proc. Soc. exp. Biol. (N. Y.) 88, 39 (1955).

Schipper, G. J.: Unusual pathogenicity of Pasteurella multocida isolated from the throats of common wild rats. Bull. Johns Hopk. Hosp. 81, 333 (1947).

Schlaffke, K.: Der Bacillus pseudotuberculosis rodentium als Erreger einer rotzähnlichen Erkrankung beim Pferde. Z. Vetkd. 33, 1 (1921).

Schmidt, H.: Fortschritte der Serologie, 2. Aufl. Darmstadt: Steinkopff 1955.

— Kasuistischer Beitrag zur Lymphadenitis mesenterialis Maßhoff. Med. Mschr. 1955, 36.

Schmorl: Demonstration von 3 Fällen allgemeiner Hämochromatose. Münch. med. Wschr. 1920, 913.

Schoen, H.: Eine besondere Form der Lymphadenitis mesaraica. Vortr. Med. Ges. Göttingen, 5. Nov. 1953. Schriftl. Mitt. an Masshoff 1954.

Schütze, H.: Bacterium pseudotuberculosis rodentium. Receptorenanalyse von 18 Stämmen. Arch. Hyg. (Berl.) 100, 181 (1928).

— Pasteurella Trevisan and B. pseudotuberculosis rodentium. A system of bacteriology, vol. 4, p. 446—482. London: His Majesty's Stationery Office 1929.

— Studies on B. pestis antigens: II. The antigenic relationship of B. pestis and B. pseudotuberculosis rodentium. Brit. J. exp. Path. 13, 289 (1932).

Seal, S. C.: Ann. Biochem. 11, 129 (1951). Zit. nach Pollitzer 1954.

Seeliger, H.: Listeriose. Beitr. Hyg. Epid. 2. Aufl. Barth (1958).

Skorodumoff, A. M., et A. D. Somorowitsch: Sur le diagnostic différentiel du B. pseudotuberculosis rodentium et le B. pestis à l'aide des milieux contenant diverses doses de sel. Microb. J. (russ.) 1926, H. 2. Zbl. Bakt., I. Abt. Orig. 84, 439 (1927).

Snyder, G. A., and N. J. Vogel: Human infection by Pasteurella pseudotuberculosis. Northw. Med. (Seattle) 42, 14 (1943).

Stephan, I.: Über die Ausbreitung des Bacillus pseudotuberculosis rodentium Pfeifferei und seine Differenzierung. Tierärztl. Rdsch. 47, 52 (1941).

— Über die serologischen Beziehungen verschiedener Stämme des Bacillus pseudotuberculosis rodentium Pfeifferei untereinander und zu verwandten Bakterien (Receptorenanalyse). In: Medizin und Chemie, S. 465. Berlin: Verlag Chemie 1942.

Stocker, B. A. D.: Bacterial flagella: Morphology, contribution and inheritance. In: Bacterial anatomy, p. 19. Cambridge 1956.

Stopeni: Arch. Atti Soc. ital. Chir. 1949, 213. Zit. nach Farbstein 1955.

Thal, E.: Untersuchungen über Pasteurella pseudotuberculosis. Lund 1954.

— Immunisierung gegen Pasteurella pestis mit einem avirulenten Stamm der Pasteurella pseudotuberculosis. Nord. Vet.-Med. 7, 151 (1955).

— Relations immunologiques entre Pasteurella pestis et Pasteurella pseudotuberculosis. Ann. Inst. Pasteur 91, 68 (1956).

—, and T. H. Chen: Two simple tests for the differentiation of plague and pseudotuberculosis bacilli. J. Bact. 69, 103 (1955).

Topping, N. H., C. E. Watts and R. D. Lillie: A case of human infection with Bact. pseudotuberculosis rodentium. Publ. Hlth (Lond.) 53, 1340 (1938).

Topley, W., and G. Wilson: The Principles of Bacteriology and Immunity, 2th ed. Baltimore: William Wood & Co. 1936 (s. Wilson & Miles 1955).

Toucas, M., G. Girard et L. Le Minor: Relations antigéniques entre les Salmonella du groupe D et Pasteurella pseudotuberculosis Type IV. Ann. Inst. Pasteur 91, 597 (1956).

Truche, C.: La pseudotuberculose chez les oiseaux. Atti del V. Congresso Mondiale di pollicoltura III, 120, 1933. Zit. nach Karlsson 1945.

TRUCHE, C.: Pseudotuberculose du cygne. Bull. Acad. vét. Fr. 8, 278 (1935).

—, et J. BAUCHE: La pseudotuberculose du dindon. Ann. Inst. Pasteur 43, 1081 (1929). — Bull. Acad. vét. Fr. 2, 162 (1929).

— — Le bacille pseudotuberculeux chez la poule et faisan. Bull. Acad. vét. Fr. 6, 43 (1933).

— G., et I. ISNARD: Un nouveau cas de pseudotuberculose chez la poule. Bull. Acad. vét. Fr. 10, 38 (1937).

— M.: La pseudotuberculose chez les animaux. Rev. Path. comp. 38, 874 (1938).

TUMANSKY, V. M.: Rev. Microbiol. Saratov 16, 287 (1939). Zit. nach POLLITZER 1954.

— Relationship of B. pestis and B. pseudotuberculosis rodentium Pfeiffer to rhamnose for differential diagnosis of these microbes. Rev. Microbiol. Saratov 18, 82 (1939). Zit. nach ENGLESBERG 1957.

UETAKE, H., and W. NAKANO: The common antigen between B. pseudotuberculosis rodentium and Salmonella group bacilli. I. Evidence of the O Antigen IX of Salmonella on the Saisawa strain of B. pseudotuberculosis rodentium. Jap. med. J. 2, 109 (1949).

UMLAUFT, W.: Pseudotuberkulose und Haemochromatose. Virchows Arch. path. Anat. 280, 18 (1931).

URBAIN, A.: Au sujet de la pseudotuberculose chez le singe. C. R. Soc. Biol. (Paris) 136, 637 (1942).

—, et J. NOUVEL: Epidémie de pseudotuberculose chez des toucans de cuvier et des toucans ariel. Bull. Acad. vét. Fr. 10, 188 (1937).

— — Epidémie de pseudotuberculose constatée sur des singes patas „Erythrocebus patas" (SCHREBER). Bull. Acad. nat. Méd. (Paris) 133, 299 (1949).

— — et P. BULLIER: Pseudotuberculose du lion. Bull. Acad. vét. Fr. 17, 333 (1944).

VALLÉE, A.: Action de la streptomycine dans la pseudotuberculose expérimentale du cobaye. Ann. Inst. Pasteur 78, 555 (1950).

VAUTRIN, A.: Contribution à l'étude de la pseudotuberculose du lièvre. Paris, Imprimerie R. Foulon 1949, S. 74.

VERGE, J., P. GORET, P. GUILLAUME et F. RUCHOT: Un cas de pseudotuberculose chez le chat. Bull. Acad. vét. Fr. 10, 41 (1937).

—, et L. PLACIDI: Pseudotuberculose chez le singe. C. R. Soc. Biol. (Paris) 136, 482 (1942).

WEITZENBERG, R.: Über die Beweglichkeit des Bac. pseudotuberculosis rodentium (A. PFEIF-FER). Zbl. Bakt., I. Abt. Orig. 133, 343 (1935).

WELTMANN, O., u. R. FISCHER: Nachweis des Bacteriums der Pseudotuberkulose der Nagetiere in einem Fall von Otitis media chronica suppurativa. Z. Hyg. Infekt.-Kr. 78, 447 (1914).

WESTPHAL, O., O. LÜDERITZ u. F. BISTER: Über die Extraktion von Bakterien mit Phenol-Wasser. Z. Naturforsch. 7b, 148 (1952).

WILSON, G. S., and H. A. MILES: Topley and Wilsons Principles of Bacteriology and Immunity 4th ed. London 1955.

WRAMBY, G., and M. FREDRICSON: Ett fall av pseudotuberculos hos svin. Skand. Vet.tidskr. 31, 590 (1941).

ZAGARI: Sulla cosi della tuberculosi zoogleica o pseudotuberculosi. Zbl. Bakt., I. Abt. Orig. 8, 208 (1890).

ZLATOGOROFF, S. J., et B. I. MOGHILEWSKAJA: Constitution des cultures du B. pseudotuberculosis rodentium, leur variabilité et leur parenté avec le B. pestis. Ann. Inst. Pasteur 42, 1615 (1928).

L'immunisation active contre le tétanos*

Par

Robert-H. Regamey

Avec 12 figures

Table des matières

---

* Institut sérothérapique et vaccinal suisse Berne.

Avant-propos

La documentation qui touche au tétanos est extrêmement abondante. En effet, la vaccination active contre cette affection représente l'aboutissement d'une somme de connaissances empruntées à de multiples disciplines; elle offre de nombreuses facettes que les chercheurs ont contemplées ou analysées selon leur optique propre, le plus souvent sans considérer l'ensemble du problème. Aussi existe-t-il quantité de travaux qui ne présentent plus un intérêt d'actualité immédiate. Pour conserver une certaine unité à cet exposé, il fut nécessaire de faire de larges coupes parmi les quelque deux milliers de notes, résumés de travaux, mémoires originaux et monographies consultés.

Une étude sur l'immunisation active contre le tétanos doit nécessairement aborder les problèmes soulevés par la préparation de la toxine et de l'anatoxine, puis ceux qui touchent à l'usage de l'antigène en combinaison avec l'antitoxine ou en association avec d'autres vaccins; ces problèmes, suffisamment vastes pour mériter chacun une revue d'ensemble, ne feront ici que l'objet d'esquisses. La physiopathologie si particulière et encore si méconnue du tétanos n'entre pas non plus dans le cadre de cette exposé; elle n'est abordée ici que dans la mesure où certains de ses aspects s'intègrent parmi les facteurs de l'immunisation passive ou active, en particulier à propos des relations entre la période d'incubation et l'efficacité du sérum ou des vaccins. Le domaine vétérinaire ne fait l'objet que de courtes mentions, mais l'usage de l'anatoxine tétanique dans cette discipline reconnaît les mêmes bases fondamentales que pour son emploi chez l'homme[1].

1. Introduction

11. Incidence du tétanos

La découverte d'un agent infectieux et la connaissance de sa biologie conduisent presque toujours à la mise au point de moyens prophylactiques et thérapeutiques qui contribuent à réduire la morbidité et la mortalité. Le tétanos fait encore exception à cette règle. Le tableau 1 confronte le nombre des décès

---

[1] Les abréviations suivantes ont été employées dans cet exposé: $T.Te$ pour toxine tétanique, $A.Te$ pour anatoxine tétanique. Pour les vaccins associés, Di = diphtérique, Te = tétanique, Per = coquelucheux, TAB = typho-paratyphoïdique A et B, Dys = dysentérique, $Scar$ = scarlatineux, Pol = poliomyélitique.

survenus dans quelques pays à la suite de tétanos, de diphtérie et de fièvre typhoïde; il démontre la régression caractéristique de la mortalité imputable à ces deux dernières maladies, tandis que le tétanos conserve une fréquence relativement élevée.

La mortalité en relation avec la répartition géographique de *Pl. tetani* est fort variable selon les pays ou les régions d'un même pays. Les renseignements que Eckmann (1958) a recueillis récemment confirment les pointages choisis dans le tableau 2.

Malgré les développements considérables qu'ont pris les méthodes thérapeutiques, le tétanos conserve une létalité très élevée. On admet couramment que le 30—35% des tétaniques succombent à la maladie (Kunz 1957 et al.). Mais Lafontaine et Koopmansch (1954) constatent qu'en Belgique la létalité atteint 68%, qu'en France Gauthron (1953) l'estime à 50—70%; Boyer et coll. (1953) arrivent même à 72,3%. Les résultats exceptionnels sont vraisemblablement l'apanage de certaines cliniques particulièrement entraînées dans la thérapie du tétanos: Veronesi (1956), au Brésil, obtient 82,8%, Pinheiro (1957) réalise 76,3%, Christensen et Thurber (1957) au centre Mayo 73% de guérisons. Mais la statistique, pour les U.S.A., de Axnick et Alexander (1957) rappelle à la modestie: pour l'année 1955, la létalité globale est encore de 57,4%.

En Grande-Bretagne, en France, en Suisse, aux U.S.A., par exemple, le tétanos provoque plus de décès que la diphtérie ou la fièvre typhoïde. Kunz (1957) note qu'en Autriche les décès par tétanos sont plus nombreux que ceux par poliomyélite. Le tétanos reste donc une maladie redoutable pour deux raisons principales: 1. la thérapie est peu efficace; 2. la prévention d'urgence offre certaines lacunes. La séroprophylaxie comporte des aléas et connaît des échecs; elle est souvent omise, surtout après des incidents traumatiques bénins.

Par contre, correctement exécutée, l'immunisation active contre le tétanos assure d'éclatants succès, que confirment les expériences massives au cours de la guerre mondiale de 1939—1945. L'anatoxine tétanique s'est révélée d'une effi-

Tableau 1. *Moyennes annuelles du nombre des morts par tétanos, diphtérie et fièvre typhoïde dans quelques pays*
Rapport épidémiologique et démographique de l'O.M.S. 8, 32—33, 140—141, 524—525 (1955).

Pays	Cause des décès	Pour les périodes de		
		1929 à 1933	1939 à 1943	1949 à 1953
Allemagne	Te	—	—	302
	Di	4631	—	636
	Ty	762	—	406
Angleterre + Galles	Te	126	95	71
	Di	2885	2075	24
	Ty	282	107	19
France	Te	—	—	667
	Di	2402	2065*	175
	Ty	1654	1278*	333
Italie	Te	676	747	722
	Di	3261	2619	906
	Ty	6094	5313	1516*
Suisse	Te	18	27	34
	Di	156	76	29
	Ty	38	24	10
U.S.A.	Te	1192	621	372*
	Di	5885	1443	254
	Ty	4970	1177	118*
Egypte	Te	163	305	355*
	Di	837	1498	606
	Ty	762	1061	714*

Te = tétanos, Di = diphtérie, Ty = fièvre typhoïde, * = moyenne calculée sans l'année 1939 ou 1953.

Tableau 2. *Mortalité due au tétanos dans quelques pays*

Mortalité pour 100 000 habitants	Pays	Pour les années	Auteur
0,3	U.S.A.	1947—1955	AXNICK et ALEXANDER 1957
0,2	Allemagne: Brême	1949—1950	HÜBNER et FREUDENBERG 1954
1,6	Allemagne: Bavière	1949—1950	HÜBNER et FREUDENBERG 1954
1,3	Autriche	1950—1954	KUNZ 1957
0,8	Suisse	1949—1953	cf. tableau 1

cacité puissante et durable; elle doit être considérée comme étant, à bien des égards, le vaccin le plus parfait disponible aujourd'hui.

12. Répartition géographique du tétanos

Le bacille tétanique n'est pas ubiquitaire; il offre une répartition géographique capricieuse, qui expliquerait la présence ou l'absence du tétanos dans certaines régions. Des études partielles ont été faites à ce propos.

Aux U.S.A., selon DUBOVSKY et MEYER (1922), les Etats à l'est du Mississipi ont des sols tétanifères, tandis qu'à l'ouest les terres n'hébergent pas de spores tétaniques à l'exception de la Californie. En U.R.S.S., MATVEEV et coll. (1957) notent que la variabilité dans la contamination du sol présente un parallélisme avec la morbidité. Les recherches en particulier de LOEWE (1932) et HINSTORFF (1933) en Allemagne, de KOBLENCS (1935), GROLIER (1952), CUVIER (1955), LAVERGNE et coll. (1949, 1950), BERNARD-GRIFFITHS et GROSLIER (1952), CHAVANNAZ (1953) en France, de LANG (1928), CAMPELL (1929), SAEGESSER (1933), SUTER (1943), ECKMANN et BISAZ (1956) en Suisse ont suscité diverses hypothèses destinées à établir pourquoi certains sols étaient tétanifères et d'autres pas, pourquoi — à contamination égale —, certains terrains étaient plus dangereux que d'autres.

La théorie de VERNEUIL, basée sur la dissémination des spores par les herbivores, n'est guère attrayante, car il existe des terres étendues, riches en bestiaux et privées de tétanos. Pour VAILLARD et son école, *Pl. tetani* est rapidement phagocyté; il n'est pathogène que lorsque la phagocytose est distraite par la présence de bactéries ou de souillures associées. L'existence de souches aérobies, non pathogènes, de bacilles tétaniques (FILDES 1927, D'ANTONA 1934), puis le passage de la forme aérobie atoxinogène à la phase anaérobie toxinogène et la possibilité d'une réversion (D'ANTONA 1951a) impliquent un jeu complexe de facteurs géographiques, atmosphériques et oecologiques.

A *l'altitude*, le tétanos est rare parce que les conditions s'avèrent défavorables à la vie microbienne; les contaminations par les porteurs doivent y être minimes. C'est surtout dans les *terrains calcaires* que le tétanos survient avec une fréquence accrue. L'introduction, chez l'animal, de chlorure de calcium rend possible un tétanos expérimental que les spores seules seraient incapables de déclencher, et l'on peut se demander si, dans les contrées de nature calcaire, la présence d'une combinaison calcique dans la plaie ne favorise pas la multiplication des formes végétatives et, partant, la toxinogenèse (H. SCHMIDT 1952a); les terrains calcaires seraient d'ailleurs, de par leur composition, favorables à l'entretien des formes toxinogènes (LOEWE 1932). L'*humidité* paraît jouer un rôle important. Dans les sols imprégnés d'eau, l'aération est faible; les conditions d'anaérobiose indispensables à la phase pathogène sont plus facilement réalisées. En outre, dans les terrains calcaires, l'eau mal filtrée entraîne avec elle les spores qui se déposent dans les parties déclives: les vallées sont plus riches en tétanos que les hauteurs ou les plateaux.

13. Porteurs de germes et immunité occulte ou naturellement acquise

Considéré le plus souvent comme un saprophyte typique, *Pl. tetani* végète néanmoins chez de nombreux animaux supérieurs en qualité d'épiphyte ou de commensal. Absorbé avec le fourrage, il se rencontre, pathogène ou non, dans le *tractus digestif* et les *fèces*. Les observations discordantes des auteurs sont imputables non seulement à des facteurs relevant de la géographie, mais aussi aux techniques d'isolement et d'identification.

Kerrin (1929) reconnut *Pl. tetani* dans les fèces de cheval, vache, mouton, porc, chien, chat, souris, cobaye, lapin, volaille, jamais chez l'homme. Ommyoji et Shan (1929) furent moins heureux à Moudken; de 67 échantillons prélevés chez 11 espèces, dont l'homme, ils isolèrent trois fois un bacille tétanique: chez un singe, une chèvre et un porc. L'incidence de *Pl. tetani* dans l'intestin du mouton revêt un intérêt particulier pour la stérilisation et le contrôle du catgut (Watts 1938, Savolainen 1950b, 1951, 1952, Cuboni 1957 et al.). Une statistique de Rostock (1950), portant sur 1619 cas, fait ressortir que chez l'homme le bacille tétanique peut être décelé dans le 24% des échantillons de selles examinés. Le contact permanent avec les chevaux augmente le pourcentage des porteurs de germes (Ottino 1927, Scartozzi 1938). Les bacilles rencontrés dans les fèces appartiennent pratiquement à tous les types sérologiques connus (Bauer et Meyer 1926). Cependant certains auteurs n'ont jamais pu identifier de bacilles tétaniques vrais dans les selles: Kerrin (1928) chez 204 enfants et Bandmann (1953) chez 92 ulcéreux.

Le passage de *Pl. tetani* dans le tractus digestif est cliniquement silencieux, car la muqueuse gastro-intestinale intacte est imperméable à la toxine tétanique, même lorsque celle-ci est appliquée en quantités énormes (Regamey 1936b). Le mécanisme de l'inactivation de la toxine reste mal connu; pour Pochon (1936a, b), l'absence de pouvoir pathogène du contenu gastro-intestinal serait imputable surtout à l'action des bactéries cellulolytiques, pour Regamey (1936a), au pouvoir adsorbant des matières alimentaires.

Le bacille tétanique peut engendrer chez son hôte le développement d'anticorps spécifiques. Dejà Römer (1909) avait mis en évidence de l'antitoxine tétanique parmi le tiers des bovidés examinés. Ramon et son école (1936b) ont établi que, parmi les ruminants, les bovidés étaient particulièrement riches en antitoxine sérique, la résorption de l'antigène devant se faire dans la panse, au niveau de la caillette. L'antitoxine «naturelle» peut atteindre le taux de 5 U.I. par cc de sérum; elle est d'autant plus abondante que le sol est plus riche en spores tétaniques (Ramon et Lemétayer 1934, 1935i); elle est décelable aussi, mais plus discrètement, chez d'autres ruminants tels les ovins et les caprins (Ramon et Lemétayer 1933c, e; Valcarenghi et Richou 1933a), buffles (Dima 1937), chameaux, dromadaires et zébus (Ramon 1957a); en sont dépourvus le cheval (Buxton et Glenny 1921 et al.), le porc, le chien (Valcarenghi et Richou 1933b; Richou et Torrisi 1933), le singe, lapin, cobaye, rat, poule et l'homme enfin. Seemüller (1943) n'a jamais trouvé d'antitoxine tétanique dans le sang des 50 taureaux qu'il contrôla; pour Ramon et Lemétayer (1933c), les taureaux sont pauvres en antitoxine parce qu'ils restent à l'étable et n'ont pas l'occasion de s'infecter dans les prairies.

Ramon (1957a) estime que l'homme est incapable de posséder une immunité naturelle acquise. Mais dans leurs études déjà anciennes, Tenbroeck et Bauer (1922, 1923, 1926) relèvent la grande fréquence des porteurs chez les Chinois, dont le sérum renfermerait assez souvent des agglutinines ou de l'antitoxine. Comment s'est déclenchée cette immunité ? Par la résorption de toxine, par une entrée occasionnelle du bacille dans les tissus ? Scartozzi (1938) a trouvé des traces d'antitoxine chez des soldats de cavalerie porteurs de germes soit dans les fèces, soit sur les téguments. Ci et là, on rencontre en Europe des sujets non immunisés et dont le sérum renferme des quantités appréciables d'antitoxine, par exemple de 0,0005 à 0,005 U.I. par cc, mais ces cas restent l'exception. Aux U.S.A., Coleman et Meyer (1926) ont fréquemment décelé des agglutinines contre différents types sérologiques de *Pl. tetani*, mais jamais d'antitoxine.

La maladie elle-même ne laisse pas d'immunité décelable. Une seconde attaque de tétanos peut se produire (BÉRARD et LUMIÈRE 1925; RAMON et ZOELLER 1927d; CHAVANY 1938; MARTIN et coll. 1954; J.A.M.A. 1951a; ECKMANN et BISAZ 1956). Un assistant de Behring (J.A.M.A. 1954b) contracta 3 fois le tétanos et survécut. REGAMEY (1955a) cite le cas d'un équarrisseur, non vacciné, qui fut atteint trois fois à plusieurs années d'intervalle et constate à ce propos qu'il est théoriquement inconcevable qu'une atteinte de tétanos puisse conférer une immunité satisfaisante: si un rescapé de tétanos se sort d'affaire, c'est qu'il a fixé dans ses tissus *moins d'une dose mortelle* de toxine; la réceptivité de l'homme et celle de la souris étant probablement voisines, la dose mortelle pour un homme de 80 kg correspond grosso modo à 0,005 cc d'une toxine dont 0,000001 cc tue la souris de 16 g au 4e jour; transformée en anatoxine, cette petite dose de toxine représente un stimulus nettement insuffisant pour créer un état d'immunité utile: dans la pratique de la vaccination antitétanique, la première injection d'anatoxine brute comporte de 0,2 à 1,0 cc. Si l'infection tétanique devait immuniser, l'injection ultérieure d'A.Te devrait jouer le rôle d'I.R. et provoquer l'apparition rapide d'anticorps, ce qui n'est pas le cas (COOKE et JONES 1943).

Chez l'homme normal, le bacille tétanique ne se rencontre pas seulement dans le tube digestif ou les fèces; il fut identifié par exemple sur la peau (MOLLO 1935, SCARTOZZI 1938), dans la cavité buccale (REITANI 1929) et les expectorations (HALL 1937 et HAUDUROY 1942). Le chirurgien connaît cette variante du tétanos post-opératoire: après l'ouverture d'une cicatrice ancienne, un tétanos se développe, provoqué par des germes «encapsulés» dans les tissus où ils auraient persisté depuis la première intervention (BONNEY et coll. 1938; HAYES 1940; SEIDL et VOGLER 1949; BOYER et coll. 1953; KAISER 1954 et al.). Cette interprétation du tétanos post-opératoire semble devoir être revisée, partiellement du moins, car la contamination pourrait être imputée à une infection de l'air ou du matériel opératoire dans les salles d'hôpitaux (SEVITT 1949, 1953).

En conclusion, le bacille tétanique saprophyte, répandu irrégulièrement sur la surface du globe, peut mener une vie d'épiphyte ou de commensal tant chez l'homme que parmi de nombreux animaux supérieurs. Au contact de *Pl. tetani*, la plupart des ruminants, bovidés surtout, acquièrent une immunité spécifique qui fait défaut dans les autres espèces. D'une façon générale, l'homme ne jouit pas d'une immunité antitétanique d'origine occulte; la maladie tétanique elle-même ne le met pas à l'abri de récidives.

14. La contamination

Il est bien établi que toute plaie, quelle qu'en soit la nature, peut devenir la porte d'entrée du bacille tétanique. Aussi ne seront relevées ci-après que certaines formes d'infection relativement peu courantes.

A Baltimore U.S.A., GILES (1937) poursuivit des études sur les poussières, études les plus précises depuis celles de NICOLAIER en 1884 et BOSSANO en 1889: de 63 échantillons il isola 13 souches de *Pl. tetani*, dont 11 toxinogènes. Les poussières, qui pénètrent partout, expliquent la présence du bacille tétanique dans l'air des salles d'opération (SEVITT loc. cit.); elles créent et entretiennent de véritables endémies tétaniques dans les établissements hospitaliers (LANCET 1956, 1957). Catgut (cf. plus haut), talc (SEIDL et VOGLER 1949), kaolin (Hill et Lederer 1948), blouses, champs opératoires et pansements insuffisamment stérilisés (Hayes 1940) sont des sources classiques d'infection.

Welch et coll. (1942 a, b) ont trouvé des spores tétaniques dans des poudres sulfamidées et observé un tétanos après implantation d'hormones cristallisées, où les spores étaient véhiculées par le support cholestérolé.

Certaines injections médicamenteuses, en particulier la quinine (Ployé 1948, 1949; Casile 1950; Floch et Barrat 1952), mais aussi l'acétylcholine et un vaccin (Casile et Rivierez 1951) furent à l'origine de tétanos. Prévot (1955) estime que, dans ces cas, le matériel d'injection était stérile et que les spores se sont localisées par voie hématogène au lieu de l'injection, locus minoris resistentiae par ischémie. Ce mécanisme, reconnu pour les gangrènes gazeuses post-médicamenteuses à W. perfringens (Regamey 1939) implique à l'origine une bactériémie qui n'est nullement démontrée. Malgré les notes préliminaires de Mayer (1937) et Piringer (1938), l'observation de Savolainen (1949) reste isolée qui, dans un cas mortel de tétanos, cultiva un bacille tétanique du sang, de la rate et de l'utérus. Une bactériémie transitoire implique non seulement la présence d'un foyer, vraisemblablement dans le tube digestif, mais encore une résorption du bacille de Nicolaier au niveau d'une lésion. L'origine exogène du tétanos post-médicamenteux paraît plus probable (cas, par exemple, rapporté par Montant et Mottironi: le bacille tétanique rencontré après une injection de médicament avait son origine dans la terre d'un pot de fleur dont la malade s'occupait).

Interventions obstétricales, médicales ou criminelles (Savolainen 1949, Merz 1955, Guggisberg 1956 et al.), macération du cordon ombilical (Hill et Lederer 1948; Beheyt 1950; Gaud 1950; Lambotte-Legrand 1950 et al.), brûlures (Fasal 1935 et al.), piqûres de tique ou de puce chique (Floch 1949), application vulnéraire de toiles d'araignées (Paoletti 1951), extraction dentaire (Zylka 1957), pansements occlusifs après vaccination antivariolique (Armstrong 1927 et al.), ulcère variqueux (Lafontaine et Koopmansch 1954 et al.) sont autant d'occasions pour qu'un bacille tétanique manifeste sa présence.

Des travaux et statistiques d'un grand nombre d'auteurs, il ressort que la porte d'entrée reste inconnue dans un pourcentage élevé de cas, soit parce qu'au moment où le tétanos se déclare la lésion cutanée, passée inaperçue, n'est plus visible, soit parce que certains modes d'infection nous échappent. Il est indiqué de relever aussi que la maladie peut apparaître dans des régions considérées comme dépourvues de tétanos (H. Schmidt 1937): l'accidenté (skieur, aviateur, usager de la route) porte des spores sur lui.

Quant au rôle de la *flore associée* (Lavergne et coll. 1945) dans l'éclosion et l'évolution du tétanos, il n'existe que peu d'informations à ce sujet. Certaines bactéries d'escorte, suppose-t-on, seraient capables de renforcer les conditions locales d'anaérobiose ou de diminuer les processus normaux de défense, favorisant ainsi le développement et la toxinogenèse du bacille de Nicolaier. Pour Guida et Rodriguez (1946), *Cl. sporogenes* et *Pl. putrificum* ont la propriété de freiner, voire d'inhiber pendant une période déterminée la formation de toxine tétanique: les autres anaérobies sporulés seraient sans action.

Ces quelques remarques démontrent bien les pièges que la contamination tétanique réserve au praticien; elles soulignent aussi les aléas qui grèvent les indications de la séroprophylaxie et, partant, les avantages d'un état préalable d'immunité active.

15. L'immunité passive. Ses échecs

L'emploi si largement répandu de la séroprophylaxie semble consacrer l'efficacité de la méthode. Pourtant l'usage préventif de l'antitoxine soulève des problèmes traités par maints auteurs et résumés en particulier par H. Schmidt (1937) et Regamey (1955a), problèmes qu'il est opportun de reprendre ici à cause de leur signification dans la séro-anatoxiprophylaxie.

A dose préventive, l'antitoxine tétanique se résorbe pratiquement avec la même vitesse, qu'elle soit poussée sous la peau ou dans le muscle (CHRISTENSEN 1952); elle diffuse rapidement et atteint un seuil utile en moins de 9 heures. *Cette protection immédiate est absolument nécessaire.* WOLFF-EISNER et coll. (1927, 1928), NISHIURA (1929), SZÉLYES (1930) ont observé qu'il était difficile de sauver le cobaye et la souris infectés par de la terre tétanifère, même avec des doses énormes et quasi-simultanées de sérum. REGAMEY (non publié) infecte le cobaye avec des fils de soie porteurs de spores tétaniques et note qu'après la 16e heure, il n'est plus à même de sauver l'animal en injectant des doses d'antitoxine correspondant pour l'homme à 1500, 5000, 20000 et 100000 U.I. Compte tenu d'une réactivité particulière des tissus et de la petite taille des animaux d'expérience, il faut admettre que, pour être efficace, le sérum doit être appliqué promptement après la blessure et si possible à dose *plus élevée* qu'on ne le fait d'habitude: 5000 à 10000 U.I. (ABEL et CHALIAN 1938b; J.A.M.A. 1954b; KIND 1957). Cette protection rapide est le seul avantage du sérum, qui offre par contre de nombreux inconvénients.

La durée de la période d'immunité passive est courte, plus courte que d'aucuns l'estiment (MARRI 1933a, J.A.M.A. 1954b). Du diagramme 1, il ressort que la persistance de l'immunité passive est fonction de la quantité d'antitoxine injectée;

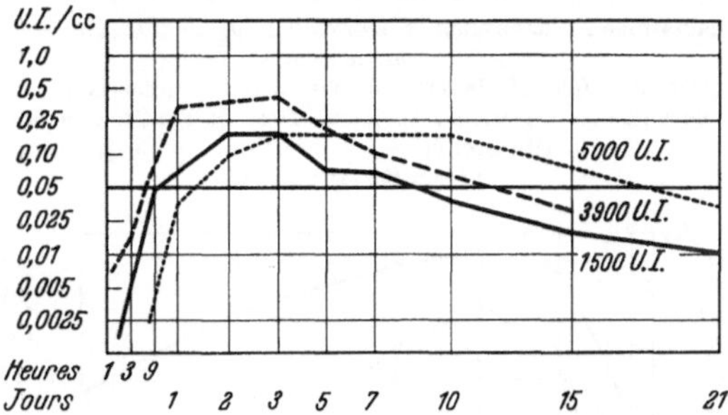

Diagramme 1. *Evolution du taux de l'antitoxine tétanique d'origine équine dans le sérum humain.* Injection sous-cutanée de 1500 U.I. (REGAMEY et SCHLEGEL 1950), 3900 U.I. (CHRISTENSEN 1952), 5000 U.I. (REGAMEY et SCHLEGEL 1950).
(0,05 U.I./cc = seuil empirique — présumé — de sécurité dans le cas de l'immunité passive).

elle varie d'ailleurs grandement d'un sujet à l'autre (BIGLER et WERNER 1941). La dose usuelle de 1500 U.I. procure un répit de 8 à 10 jours, celle de 5000 U.I. un répit de 15 à 21 jours, pour autant que l'on admette 0,05 U.I./cc comme taux de sécurité suffisant dans le cas de l'immunité passive. La valeur de ce seuil est d'ailleurs très empirique. Le choix lui-même de la dose préventive courante (1500 U.I.) ne semble pas basé sur des preuves expérimentales rigoureuses; il est vraisemblable qu'au début on poussait chez le blessé une «injection» de sérum, c'est à dire le contenu d'une seringue normale (10 cc); quand les méthodes d'étalonnage se développèrent et qu'on introduisit la notion de l'unité antitoxique, on conserva tout simplement le nombre des unités que devait contenir la dose originelle d'antitoxine (cf. chiffre 44, p. 316).

La persistance des anticorps d'origine hétérologue se trouve encore réduite si, dans le passé, le blessé a déjà reçu du sérum d'origine animale. Fait curieux, lors d'une réinjection, l'élimination accélérée de l'antitoxine ne se produit pas seulement lorsque la première injection et les injections suivantes sont faites avec du sérum de même espèce animale (H. SCHMIDT 1934; RAMON et FALCHETTI 1935c; FALCHETTI 1937 et al.), mais aussi lorsque les sérums sont d'origine différente (diagrammes 2 et 3).

Les échecs de la séroprophylaxie sont bien connus. Ils ont, comme causes, dans la règle, l'emploi d'un sérum non spécifique ou de mauvaise qualité, l'injection trop tardive du sérum alors que le système nerveux central a déjà inéluctablement fixé la toxine tétanique, l'élaboration persistante de toxine au niveau de foyer infectieux, la disparition accélérée de l'antitoxine, une mauvaise résorption des anticorps au lieu de l'injection, ou encore la fixation trop rapide de l'antitoxine dans les tissus d'un organisme sensibilisé.

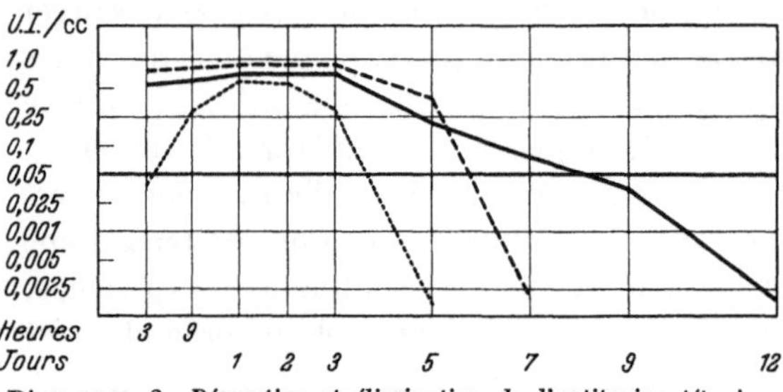

Diagramme 2. *Résorption et élimination de l'antitoxine tétanique chez le lapin*

Injection de 0,5 U.I. de sérum de cheval par gramme-lapin. Intervalle entre les injections: 15 jours. Titre de l'antitoxine dans le sérum après la 1re injection —, 2e injection - - - -, 3e injection.... (Selon Regamey 1944a)

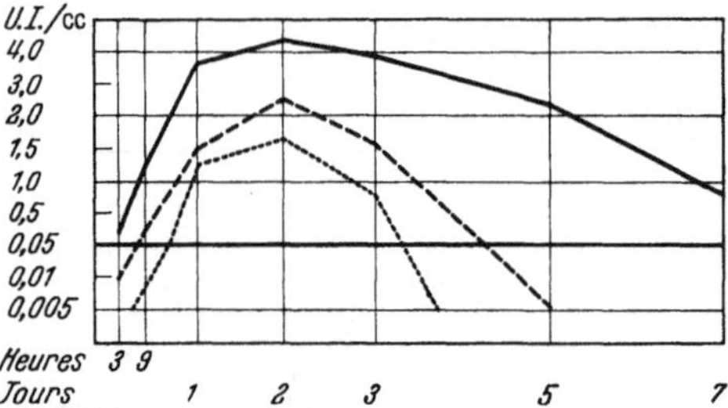

Diagramme 3. *Résorption et élimination de l'antitoxine tétanique chez le lapin*

Injection de 0,1 U.I. de sérum par gramme-lapin. Intervalle entre les injections: 3 semaines. Titre de l'antitoxine dans le sérum (moyenne) pour 3 lapins après la 1re injection (sérum de cobaye) —, 2e injection (sérum d'homme) ---, 3e injection (sérum de cheval) ...... (Selon Regamey 1953)

L'injection d'un sérum d'origine animale comporte toujours le danger de *réactions anaphylactiques,* non seulement chez ceux qui ont reçu préalablement du sérum, mais aussi chez l'hyperallergique, qui peut présenter une réaction mortelle lors d'une *première* injection (Zanniol 1937, H. Schmidt 1952a, p. 90; Regamey: deux cas non publiés). Une enquête auprès de 1327 médecins allemands a révélé 147 réactions anaphylactiques — dont 8 mortelles — consécutives à la séroprophylaxie en l'espace de 2 ans (Mackuth 1935). Les accidents persistent malgré l'emploi de sérums purifiés (Copeman 1951; Swift 1952; Worms et Dreyfus 1953, Littlewood et coll. 1954; Scheibel 1955). On ne parle déjà plus de la « solution d'antitoxine » de Ramon, sérum de haute valeur, mais fortement dilué, formolé et chauffé (Ramon 1939c, Sohier 1939, Ramon et coll. 1940a, b, h). Les antihistaminiques ne méritent qu'une confiance relative (Regamey 1947a, et al.). Les risques d'anaphylaxie sont-ils plus grands que ceux d'un tétanos? Voilà posés des cas de conscience pour le médecin et parfois aussi pour le juge (Hellner 1957, J.A.M.A. 1957c). L'un des seuls progrès que l'on peut envisager dès aujourd'hui réside dans l'usage d'antitoxine d'origine humaine (Winkelbauer 1948, Turner et coll. 1954, Peterson et coll. 1955, Goldsmith 1957).

16. Prophylaxie non spécifique

Ce point, qui n'a guère sa place dans le présent exposé, appelle cependant quelques commentaires.

Les sulfamidés ont une action nette sur la croissance de Pl. tetani (Mayer 1938 et al.), mais leur efficacité dans l'infection expérimentale est fort douteuse (Legroux 1940; Levaditi et Vaisman 1945 et al.). Evans et coll. (1945) ont trouvé que certaines combinaisons du type paraméthylsulfonylbenzamidine exerçaient sur la souris une activité postinfectionelle.

Les antibiotiques tels que pénicilline, érythromycine ou tétracyclines, par exemple, sont capables d'inhiber la croissance et la toxinogenèse non seulement in vitro, mais aussi in vivo (Novak et coll. 1949, Katić 1951a, J.A.M.A. 1951b et al.). La pénicilline est particulièrement appréciée (Diaz-Rivera et coll. 1951, Christensen et Thurber 1957) peut-être à cause de son effet contre les complications pulmonaires (J.A.M.A. 1957). L'antibiotique, comme du reste les autres composés chimiques, ne diffuse pas nécessairement dans le foyer tétanigène. Dans les tissus nécrosés, dévitalisés, le bacille de Nicolaier n'est inhibé que si le médicament atteint un taux suffisant in situ, mais le germe pullule de nouveau et forme de la toxine sitôt que le traitement prophylactique cesse. Aux doses préventives, voire thérapeutiques, les antibiotiques n'ont pas d'action sur la toxine elle-même.

STERN et HUKOVIĆ (1956) ont observé qu'un polypeptide, qu'ils nomment « substance P », jouissait d'un pouvoir préventif; ce corps déprime le système nerveux central et possède en fait les caractères d'un spasmolytique, efficace aussi à l'égard de l'action convulsivante de la strychnine.

La mise au point du Périston par HECHT et WEESE (1943) éveilla certains espoirs. Ce polyvinylpyrrolidon ou « Kollidon », produit de polymérisation à longue chaîne, se fixe électivement aux gammaglobulines et se montre capable de lier ou de déplacer non seulement des colorants du type bleu de Geigy et bleu de trypan, mais aussi certaines toxines microbiennes. En fait, SCHUBERT (1948, 1949) réussit à sauver une partie de ses animaux en injectant des souris simultanément avec de la toxine tétanique et du Kollidon. KRECH (1952), BICK et coll. (1957) ont confirmé dans une certaine mesure les observations de SCHUBERT, tandis que les résultats de BORSELLI (1956) sont négatifs. REGAMEY (non publié) étudia en 1950—1951 — sur les conseils de WEESE — l'influence de divers types de polyvinylpyrrolidon (P.M. 24000—30000, 40000—55000 et 85000) sur des souris et cobayes infectés par des fils de soie tétanifères. Selon BICK et coll. (1957), le polyvinylpyrrolidon (PM 12600) doit être administré en même temps que la T.Te pour exercer une certaine protection. S'il est susceptible de neutraliser une partie de la toxine tétanique, le Kollidon reste sans effet dans l'infection expérimentale; son application chez l'homme instaurée en 1949 par SCHUBERT fut sans lendemain.

Récemment, MÖSE (1957) a reconnu que l'acide usninique à 0,2—0,4%, injecté dans les dix minutes qui suivent l'intoxination de la souris, était capable de neutraliser de faibles quantités de toxine tétanique.

2. La toxine tétanique

L'importance capitale de la tétanospasmine dans l'étiologie du tétanos et la cristallisation de l'intérêt autour de l'anatoxine dans la pratique préventive ont relégué à l'arrière-plan l'analyse des autres éléments antigéniques de *Pl. tetani*. L'habitude qu'on a prise peu à peu de voir dans le tétanos la seule manifestation d'une intoxination, a fait négliger les bénéfices susceptibles d'être retirés d'une meilleure connaissance des caractères antigéniques secondaires.

21. Les antigènes secondaires de Pl. tetani

TULLOCH (1919) classa les bacilles tétaniques qu'il étudia en 4 types sérologiques. Par la suite, d'autres auteurs, en particulier FELIX et ROBERTSON (1928), MacLENNAN (1939) et GUNNISON (1937, 1947) portèrent ce nombre à 10; l'existence d'autres types est d'ailleurs vraisemblable. Comme le *Proteus* et les *Salmonelles*, *Pl. tetani* possède un antigène somatique O thermostable et un antigène flagellaire H thermolabile. Seul le type VI, immobile, est dépourvu d'antigène H. Analysés par les méthodes d'agglutination et de fixation du complément, puis différenciés par les techniques d'adsorption, les antigènes O et H accusent la présence non seulement de facteurs spécifiques de type, mais aussi de facteurs communs soit à plusieurs, soit à tous les types (MacCOY et MacLUNG 1938). Ces antigènes sont de nature encore méconnue, mais l'on admet qu'ils n'ont pas de rapport avec la toxinogenèse: les souches atoxiques peuvent être privées d'antigène O ou l'avoir conservé. Les relations présumées entre la forme des colonies, les caractères antigéniques et la toxinogenèse sont elles aussi mal connues (cf. H. SCHMIDT 1952 a).

Les ferments sécrétés par *Pl. tetani* sont également des antigènes sur lesquels les renseignements sont fragmentaires (SCHULTZE 1942). Ils font pourtant l'objet de nombreux travaux, mais ceux-si s'orientent vers le métabolisme de la bactérie et les conditions du développement de la toxine; leurs caractères immunobiologiques ne sont pas recherchés systématiquement. Comme POPE et coll. (1951) l'on fait pour la toxine diphtérique, RAFYI et coll. (1954), MIR CHAMSY et coll. (1957) ont inauguré l'analyse des antigènes et haptènes tétaniques en utilisant la méthode par diffusion en gel d'*Oudin-Oakley*; leurs premiers résultats ont montré que certaines souches permettaient de reconnaître au moins 18 antigènes sécrétés par le bacille au cours de sa croissance ou libérés par autolyse. La gélatinase tétanique fit l'objet de recherches plus poussées de Ramon (1957 b), qui constata l'absence de parallélisme entre le titre antitoxique et le pouvoir antigélatinolytique du sérum antitétanique.

La protection que confère soit le sérum, soit le vaccin est strictement antitoxique; elle se manifeste par la seule propriété neutralisante à l'égard de la toxine; les germes eux-mêmes ne sont pas influencés. Il est cependant concevable que des antigènes autres que la tétanospasmine jouent un rôle dans l'immunité.

Quelques auteurs ont démontré que le cobaye préparé avec des anticorps anti-O et anti-H survivait à l'injection de bacilles tétaniques lavés, mais non à l'injection de toxine. Ces observations justifient-elles l'étude d'une immunité antibactérienne passive ou active, comme ce fut le cas pour la diphtérie? Il ne le semble pas (Nord 1925).

La *tétanolysine*, reconnue en 1898 par Ehrlich, provisoirement différenciée par Madsen (1899) en proto-, deutéro- et tritotoxine, puis bien étudiée par Reymann (1927), se fixe sur les globules rouges et les lyse. Elle se laisse aisément séparer de la tétanospasmine précisément par sa propriété de se fixer sur les hématies à froid ou par certains processus d'oxydo-réduction (Fleming 1927). Des observations de Regamey (non publié) il ressort que les hématies de 15 espèces animales à sang chaud sont toutes lysées par la toxine. La présence de la lysine dans le bouillon de culture est relativement fugace: Robert (1958) note un effet très précoce avec maximum entre la 12e et la 48e heure, effet qui diminue progressivement pour devenir presque nul vers le 12e jour; la sensibilité des globules rouges varie d'une espèce animale à l'autre pour une toxine issue d'une même souche bactérienne; elle n'est d'ailleurs pas constante: les érythrocytes du rat, par exemple, plus sensibles que ceux du cheval à l'égard de la toxine x sont plus résistants que ceux du cheval à l'égard de la toxine y; les globules rouges de la poule, animal extrêmement réfractaire à la tétanospasmine (Ramon 1957a, p. 443) sont très sensibles à la tétanolysine. L'hémolysine tétanique — tous les auteurs sont d'accord sur ce point — est spécifique; elle n'est pas neutralisée par d'autres antihémolysines actives contre les Clostridies de la gangrène gazeuse, Cl. botulinum, Streptocoque, Staphylocoque, etc. (Prévot 1950).

Antigène beaucoup plus labile que la spasmine, la lysine est particulièrement sensible aux conditions du pH, de la température (Guillaumie et coll. 1944), aux acides, aux oxydants. La filtration sur porcelaine ou membranes d'amiante-cellulose l'inactive en partie, vraisemblablement par des phénomènes d'adsorption. L'hémolysine tétanique se laisse toutefois transformer en analysine par l'action conjuguée du formol et de la chaleur (Lemétayer et Nicol 1945a).

Confirmant les travaux de leurs prédécesseurs, Lemétayer et coll. (1936a, b etc.) mettent en évidence des antihémolysines tétaniques naturelles dans le sérum de plusieurs espèces animales, même chez l'homme, en dehors de toute présence concomitante d'antitétanospasmine. Diverses observations font supposer que ces antihémolysines naturelles ne sont pas immunologiquement spécifiques. Kréguer et Guillaumie (1948) ont rapporté que le pouvoir antihémolytique et antigangréneux des sérums normaux (Robert 1958) était considérablement augmenté par l'immunisation antitétanique, ce qui parle contre une spécificité. Par ailleurs, Kerrin (1930) et al. ont démontré que l'action hémolytique des toxines fournies par les anaérobies sporulés, de *Pl. tetani* en particulier, était neutralisée par la cholestérine et d'une façon générale par les lipoïdes, voire même par certains protéides, ce qui permettrait d'attribuer l'effet antihémolytique non pas à une antihémotoxine spécifique, mais à un constituant banal du sérum. Dans cet ordre d'idées, il serait intéressant de faire des études quantitatives sur les propriétés antihémolytiques polyvalentes des sérums naturels et immuns.

Il importe de relever que chacun s'accorde à confirmer l'observation capitale d'EHRLICH: il n'existe *aucune relation* entre le pouvoir hémolytique et le pouvoir neurotoxique. Les souches dites-atoxiques élaborent autant d'hémolysines que les souches toxinogènes (KERRIN 1930). Propriétés lytiques et propriétés neurotoxiques sont quantitativement dissociées (LEMÉTAYER 1936 b). Pouvoir hémolytique et pouvoir floculant d'une toxine tétanique sont également sans rapports. Au cours de l'immunisation active, les titres antitoxiques et antihémolytiques des sérums n'évoluent pas parallèlement (LEMÉTAYER et NICOL 1945a, b).

La tétanolysine joue certainement un rôle effacé dans la physiopathologie du tétanos. Elle constitue une noxe trop faible pour concurrencer la fraction tétanisante. Son point d'impact, le globule rouge, offre une surface d'attaque et d'absorption relativement plus considérable que les cellules du système nerveux sensibles à la spasmine. De plus, l'organisme se défend mieux contre l'hémotoxine: en passant à travers le foie, la lysine est arrêtée, puis inactivée, tandis que la spasmine ne subit dans cet organe ni fixation, ni transformation (PELLOJA 1951, p. 35).

22. La tétanospasmine

D'une façon générale, lorsqu'on parle de « toxine tétanique », on sous-entend le facteur létal, la tétanospasmine, neurotoxine dont l'effet se manifeste exclusivement sur le tissu nerveux. Il s'agit d'une exotoxine, ce terme étant pris non pas dans son sens strict de poison sécrété par le bacille et présent uniquement dans le milieu ambiant, mais dans celui plus général de toxine-protéine.

La genèse de la T.Te fit l'objet de travaux récents, inaugurés par RAYNAUD (1947, 1951a), puis RAYNAUD et coll. (1949, 1950 et 1952). Ces auteurs ont démontré qu'il était possible, au moyen d'une solution hypertonique ou d'ultrasons, d'extraire la neurotoxine de bacilles lavés, que les germes renfermaient d'autant plus de toxine qu'ils étaient plus jeunes et que la toxinogenèse, plus précisément le passage de la toxine dans le milieu de culture était sans rapport direct avec le métabolisme du croît bactérien. STONE (1952, 1953 c, 1954) étudie l'action du lysozyme sur la lyse des cultures et conclut que la toxine, entièrement produite à l'intérieur de la cellule, n'apparaît pas uniquement à la suite de processus d'autolyse: pendant le développement du bacille, une certaine quantité de toxine diffuse dans le liquide, mais les facteurs qui influencent la perméabilité de la membrane cellulaire sont encore inconnus. Après RAYNAUD, KATIĆ (1951a) puis MIR CHAMSY et SADEGH (1958a) constatent que la pénicilline introduite dans le milieu de culture libère des quantités appréciables de toxine. Une lyse bactérienne réalisée par la trypsine en fin de culture augmente considérablement la récolte de toxine (MIR CHAMSY et NAZARI 1958b). Même la lyse qui survient lorsqu'on place la culture à +4° C pendant quatre à six jours augmente de 25 à 65 % le titre en Lf du filtrat (MIR CHAMSY 1958c).

Déjà au siècle passé la purification du complexe antigénique que représente la T.Te brute, telle qu'elle est issue du bouillon de culture, tenta maints auteurs. En effet, l'obtention de produits purifiés offre un intérêt évident tant pour le développement de nos connaissances sur la physiopathologie du tétanos que pour la préparation de vaccins plus spécifiques. Les recherches, empruntant diverses voies, aboutirent en 1946 à la cristallisation de la T.Te; en voici les dernières phases.

La même année, MASCHMANN (1931), S. SCHMIDT et HANSEN (1931a) puis VELLUZ (1931a) s'adressent aux méthodes d'adsorption. C'est ainsi que MASCHMANN, utilisant comme matériel de départ de la toxine sèche — déjà purifiée par un relarguage avec le sulfate d'ammonium —, obtient un produit 210 à 240 fois plus pur que le liquide de départ: la toxine était adsorbée sur hydroxyde d'aluminium (modification C γ de Willstätter), éluée avec l'ammoniaque à 0,05 %, réadsorbée sur kaolin, puis éluée une nouvelle fois dans la solution ammoniacale. Le produit sec terminé renfermait 25 740 000 DMM par gramme; la réaction du biuret était légèrement positive; les réactions de Millon, tryptophane et sulfure de plomb étaient négatives.

Modern et Ruff (1936) se contentent de l'ultrafiltration. Ils utilisent des bougies imprégnées de collodion acétique à 5,5%. La toxine de départ titrait 0,42 γ, la toxine purifiée 0,056 γ par DMM. La chromatographie en colonne ne donne pas à Ballinari (1941) de résultats sensiblement meilleurs.

Une première étape décisive, où Eaton (1936), Sommer (1937), Boivin et Izard (1937) recourent à l'emploi d'acides (acétique, trichloracétique, phosphotungstique), de cadmium sous forme de $CdCl_2$, du relarguage par les sels neutres, de l'adsorption sur $Al(OH)_3$ aboutit aux observations de Eaton et Gronau (1938 b). Une toxine issue de bouillon de viande de veau fournit à ces auteurs un produit dépourvu du 99% des protéines inactives. Le rapport: N par DMM de la toxine brute/N par DMM de la toxine purifiée est de 125 avec un rendement de 90%; ce quotient peut être porté à 138 par relarguage au sulfate d'ammonium, mais le rendement tombe à 22%. Pour Eaton et Gronau, la T.Te ne serait ni une protéine, ni une substance hydrocarbonée. L'azote de la toxine purifiée provient pour 50% des protéines précipitables par l'acide trichloracétique et en partie de protéoses et peptones: le produit terminé titre 9 000 000 à 18 000 000 DMM par mg N.

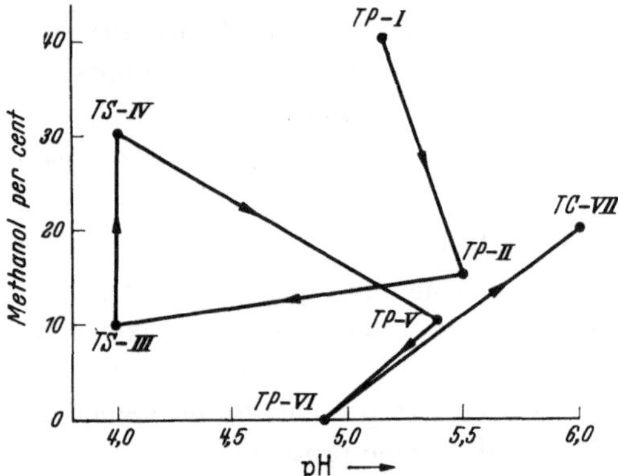

Diagramme 4. *Représentation schématique des conditions de pH et des concentrations de méthanol dans les différentes phases de purification de la toxine tétanique*
TP—I, TP—II etc: précipitation de la toxine; TC—VII: cristallisation de la toxine (températures entre 0 et —5° C). (Selon Pillemer, Wittler, Burrell et Grossberg 1948 c.)

Dans l'étape ultérieure, les expérimentateurs utilisent des toxines provenant de milieux sans protéines, protéoses ou peptones d'origine musculaire (Mueller et Miller 1943). Pickett et coll. (1945), disposant d'une toxine recueillie sur milieu de Mueller et Miller (1945) et recourant au $CdCl_2$ associé au relarguage avec $(NH_4)_2SO_4$ obtiennent une substance 344 fois plus active que la toxine de départ, avec un rendement de 50%; ils sont persuadés qu'un degré supérieur de purification est possible. L'année suivante, Pillemer (1946 a), puis Pillemer, Wittler et Grossberg (1946 c) font connaître leur succès dans la cristallisation de la toxine, cristallisation obtenue à la suite de précipitations successives par le méthanol à basse température. L'ensemble de leurs publications (Pillemer 1948 a, Pillemer et coll. 1948 c, 1949) fait ressortir les points suivants. (La cristallisation de la T.Te fut réalisée selon le diagramme 4.)

Chaque temps de la purification s'accompagne d'une perte de substance active qui, au total, se situe entre 95 et 99,94%!

Les propriétés de la T.Te sont résumées ci-dessous:

Mobilité en tampon de véronal, pH 8,6, force ionique 0,1	$2,8 \times 10^{-5}$
Constante de sédimentation S $_{200}^{1\%}$, W	4,5
Index optique de rotation $[\alpha]_D^{25}$, degrés	—63
Point isoélectrique	$5,1 \pm 0,1$
Lf (unités floculantes) par mg N	3 600
Kf_{50} (vitesse de floculation) en minutes	10
DMM par mg N	$6,6 \times 10^7$

L'analyse élémentaire démontre la présence de 15,7% d'azote, 0,065% de phosphore et 1,04% de soufre. Les hydrates de carbone font défaut. Le poids moléculaire est de 67 286. La T.Te cristallisée apparaît comme une protéine homogène, dans laquelle l'analyse électrophorétique fait ressortir 13 acides amidés, le plus abondant étant l'acide aspartique (15%) et le plus rare étant le tryptophane (<1%). Le nombre de restes aminés par molécule est de: pour le tryptophane 3, l'acide glutamique 47, l'isoleucine 48 et l'acide aspartique 48.

Rien dans la composition ou la disposition des acides aminés n'est susceptible d'expliquer la haute toxicité ou l'affinité pour le tissu nerveux (Dunn, Camien et Pillemer 1949).

La toxine cristallisée est-elle vraiment pure ? Oakley (1954) émet des doutes à ce sujet, estimant qu'on n'a pas jusqu'ici isolé de substance immunologiquement pure. Pillemer et ses collaborateurs ont relevé eux-mêmes les difficultés dues à l'absence de critères d'identification rigoureux. Ils ont observé que la toxine cristallisée, remise en solution et conservée pendant 10 jours à 0° C se transformait spontanément et partiellement en un dimère atoxique, mais floculant, doué d'une constante de sédimentation de 7,0 unités Svedberg (toxine: 4,5) (Pillemer et Moore 1948 b).

La toxine cristallisée déclenche chez la souris un tétanos typique. Même à 500000DMM, elle tue l'animal sans faire apparaître de lésion anatomo-pathologique, qu'elle soit injectée seule (Pillemer et Wartman 1947) ou avec du sérum spécifique (Wartman et Pillemer 1949). Les doses sublétales ne provoquent pas non plus d'altération cellulaire. La durée de la période d'incubation est fonction de la quantité de toxine injectée: pour 500000 DMM, elle est de 30 minutes, la mort survenant en 60 minutes (cf. tableau 3, p. 285). Enfin seuls les sérums antitétaniques sont capables de neutraliser la toxine cristallisée.

Turpin, Raynaud et Rouyer (1952) ont traité la toxine brute de l'Institut Pasteur par la technique de Pillemer et obtenu des résultats irréguliers; ils réalisèrent une seule fois un produit titrant 57000000 DMM par mg N. Deux précipitations successives par l'éthanol ou le méthanol en milieu acide avec redissolution dans l'acétate 0,150 M à pH 6,9 leur livrent en général une toxine purifiée avec 1500—2000 Lf/mg N (chez Pillemer: 3600 Lf/mg N).

Récemment Largier (1956a, 1956b), avec une toxine issue de culture en sac de cellophane (Polson et Sterne 1946) et traitée par électrodécantation sur multimembrane au point isoélectrique, puis fractionnement ultérieur par le sulfate d'ammonium, reçoit un produit qui, d'un poids moléculaire voisin de 68000, semble pur à l'électrophorèse et à la sédimentation. Sa toxine purifée, qui donne une bande unique et étroite au test de diffusion sur gel, et paraît donc exempte d'antigènes contaminants, présente certaines différences avec la substance cristallisée de Pillemer: elle possède d'autres constantes de sédimentation; en outre elle est plus active et titre 4300—4800 Lf par mg N.

Existe-t-il différentes toxines tétaniques ?

Question d'importance capitale, car les vaccins et les sérums antitétaniques sont en général issus ou préparés à partir d'une souche unique et seraient donc « monovalents », c'est à dire incomplets. La fréquence relative des tétanos postsériques et les échecs de la sérothérapie ont suggéré à certains cliniciens l'éventualité d'une pluralité de la toxine tétanique (Haas 1940). Llewellyn Smith (1942/43), constatant que le rapport entre les DMM pour le lapin, le cobaye et la souris varie selon la toxine, émet l'hypothèse de la multiplicité du facteur létal; il n'y aurait par contre pas de différence immunologique, car n'importe quel sérum antitétanique neutralise la toxine, en bloc. L'année suivante, Friedemann et Hollander (1943a, 1943b) étudient 7 T.Te sur le lapin (intracérébral et intramusculaire), le cobaye (intracérébral et intramusculaire) et la souris (intramusculaire). Chacune des méthodes livre des résultats différents: il faut par ex. 100 fois plus d'antitoxine pour bloquer une même quantité de toxine selon que la neutralisation est recherchée in vitro ou in vivo. La dose létale serait déterminée non seulement par le nombre des molécules de toxine par cc, mais aussi par la toxicité propre de chaque molécule individuelle (Friedemann 1947).

Cherchant une explication à ces faits d'apparence discordante, van Heyningen (1950, p. 22) suppose que la molécule de T.Te est de dimensions variables ou qu'elle peut s'accoler à une autre molécule. A des dimensions variables correspondent évidemment des propriétés physico-chimiques différentes. Cette hypothèse s'accorde avec l'observation de Zuger, Hollander et Friedemann (1939) ou de Pelloja (1949a): certaines substances, tissus, extraits d'organes ou produits chimiques sont susceptibles de potentialiser la toxine tétanique. Elle n'est toutefois pas corroborée par les faits que Largier (1956b) rapporte: deux T.Te préparées sur un même milieu, l'une avec la souche non sporulée de Harvard, l'autre avec la souche sporogène Wellcome G. S. 761 fournissent après électrodécantation sur multimembrane deux produits de pureté égale et même supérieure à la toxine cristallisée de Pillemer (1948c); poids et dimensions moléculaires, mobilité électrophorétique, tests de neutralisation et de floculation, constantes de sédimentation sont les mêmes pour les deux toxines purifiées; seule différence: la souche non sporulée de Harvard produit une plus grande quantité de toxine et moins de ballast macromoléculaire.

Il faut reconnaître que, dans l'étal actuel des recherches, l'éventualité d'une pluralité de toxines n'est pas bannie avec certitude. Les faits parlent toutefois pour une identité physico-chimique, sérologique et immunologique. Le doute se concentre plutôt autour des facteurs qui conditionnent l'avidité de la toxine pour le tissu nerveux.

23. A propos de la période d'incubation

Le pronostic du tétanos déclaré est d'autant plus sombre que l'incubation est plus courte. D'une valeur moyenne de 8—12 jours, cette période se ramène parfois à 2—3 jours; elle peut être longue et atteindre plusieurs mois (Bennet et Grodinsky 1935 et al.). Si l'on définit l'incubation comme étant l'intervalle entre l'infection et l'apparition des premiers signes cliniques, on doit reconnaître que les incubations prolongées correspondent à des cas particuliers où la multiplication du bacille, la toxinogenèse ou la propagation de la toxine ont été gênées, par exemple lors de l'usage préventif du sérum: il peut y avoir encapsulation du bacille et/ou neutralisation limitée dans le temps par l'antitoxine présente.

Les phénomènes physiopathologiques qui déterminent l'incubation sont aussi ceux qui conditionnent la pathogénie du tétanos. Dès le début ils ont occupé de nombreux auteurs et suscité des conceptions qui, de Vaillard à Wright, se sont affrontées sans aboutir à une certitude. On en trouvera d'excellentes critiques dans les analyses de d'Antona (1951a), Pelloja (1951), H. Schmidt (1952a, b), Prévot (1955), Wright (1955), Katić (1957), Grumbach (1958).

Ces phénomènes seront envisagés ici sous une forme succincte, dans la mesure seulement où ils touchent l'immunisation antitétanique.

Trois facteurs étrangers à l'organisme doivent être envisagés: la toxine, l'antitoxine et l'anatoxine.

1. On admet que la *toxine tétanique* jouit d'un neurotropisme et que le tissu nerveux possède une affinité particulière pour la toxine. L'apparition et la gravité d'un tétanos sont fonctions de la quantité de toxine qui gagne la cellule nerveuse. On oublie trop souvent que la toxine ne parvient pas toute dans le SNC: une partie, peut-être même la majeure partie du poison diffuse dans d'autres tissus et reste sans effet morbide décelable (Abel et coll. 1936, Pelloja 1950a). Une preuve de cette dissémination relativement abondante de la toxine est

fournie par la réactivation sérologique consécutive au développement de l'infection dans les organismes immunisés (cf. p. 325).

2. Globuline hétérologue, l'*antitoxine* représente pour l'organisme un corps étranger qui doit être détruit au plus vite. Elle est captée dans ce but par les cellules du système réticulo-endothélial et non pas par les cellules nerveuses qui pourraient en avoir besoin. Si elle manifeste une avidité à l'égard de la toxine, l'antitoxine n'a aucune affinité pour la cellule nerveuse. Elle n'est donc utile que dans la mesure où les hasards lui font rencontrer de la toxine. Le poison fixé n'est plus accessible aux anticorps, dont les dimensions moléculaires énormes s'opposent au passage à travers la membrane de la cellule nerveuse. L'antitoxine ne peut donc jouer aucun rôle thérapeutique. Par ailleurs, si la toxine et l'antitoxine manifestent une avidité réciproque, leur réunion n'est pas immédiate; le degré et la vitesse de leur combinaison sont variables d'un échantillon de sérum à l'autre et d'une toxine à l'autre (FRIEDEMANN et HOLLANDER 1943b; CINADER et WEITZ 1950 et al.). Il faut envisager aussi la possibilité d'une compétition pour la toxine entre le pouvoir attractif de la cellule nerveuse et celui de l'antitoxine.

3. *L'anatoxine tétanique* manifeste in vitro une affinité plus grande pour l'antitoxine que la toxine; cette dernière peut être même déplacée (cf. p. 342). Anatoxine et toxine interfèrent donc, mais le phénomène ne doit pas avoir d'importance pratique car, dans la séro-anatoxi-prophylaxie, il y a toujours un excès considérable d'anticorps. L'affinité de l'anatoxine pour la cellule nerveuse n'est pas connue; elle est néanmoins invoquée pour expliquer les cas de protection rapide décrits par KRECH (1949), RAYNAUD et coll. (1951b) et al.

Les buts de la prophylaxie spécifique sont l'interception et la neutralisation de la toxine dès sa formation. Il faut que les anticorps soient en place, en quantité suffisante, dès le moment où la toxine est élaborée et aussi longtemps que du poison émane du foyer tétanigène. L'antitoxine sera donc d'autant plus efficace qu'elle est présente plus tôt, massivement et pendant une période suffisamment prolongée.

Tableau 3. *Rapports entre le nombre de DMM de toxine tétanique cristallisée et le développement du tétanos chez la souris*
(Selon PILLEMER et WARTMAN 1947.)

DMM	N toxinique (en γ)	Début des symptômes tétaniques après ... heures	Mort après ... heures
1	0,000013	30	90—96
10	0,00013	15	30—48
10000	0,13	4	10—15
100000	1,3	1	2
500000	6,4	35/60	1

La rapidité de l'apparition du tétanos est en relation avec la quantité de molécules de toxine capables de se fixer dans les cellules sensibles. Les essais de PILLEMER et WARTMAN (1947), exécutés avec de la T.Te cristallisée, sont concluants à cet égard (tableau 3).

Le raccourcissement de la période d'incubation va de pair avec une accélération de la phase clinique, phénomène étudié en détail par PELLOJA (1950b). Même appliquée à doses considérables, la T.Te ne se manifeste qu'après une certaine latence. PILLEMER et WARTMAN (1947) sont enclins à rejoindre l'ancienne conception de COURMONT et DOYON (1893): la T.Te ne serait qu'un agent intermédiaire dans la production du tétanos; l'élément toxique à proprement parler prendrait naissance chez l'hôte après l'administration de la toxine; la période d'incubation correspondrait au temps nécessaire pour la transformation induite par la toxine. PELLOJA (1949b) a fait l'observation, confirmée par MORGUNOV et KHATUNTSEV (1955), que si l'on injectait des doses sublétales l'animal ne mourrait que lorsque la somme de ces doses atteignait la valeur de la DMM. On peut en conclure que la toxine qui se fixe sitôt en contact avec le tissu nerveux (ABEL et coll. 1938a, WRIGHT 1954), après avoir cheminé des neurones vers les cellules du névraxe dont l'intoxination conduit à la mort (BERNARD et SERVANT 1957): 1. n'immunise pas les cellules nerveuses, fait connu (TEALE et EMBLETON

1919 et al.) et 2. *n'est pas détruite dans le tissu nerveux*, mais qu'elle s'y accumule au contraire pendant une période prolongée sans perdre son efficacité. Et PELLOJA (1951, p. 97) de proposer une interprétation purement biochimique: la durée de la période d'incubation est conditionnée par le temps nécessaire aux molécules de toxine disséminées dans l'organisme pour gagner le S.N.C. — vers lequel les attire leur tropisme particulier —, et pour y atteindre par effet de sommation une concentration assurant le développement des manifestations morbides. H. SCHMIDT (1952 a, p. 76) rappelle les expériences de BROMEIS; cet auteur a déterminé la vitesse de propagation de la T.Te à partir du lieu de l'infection. Cette vitesse est de l'ordre de 1 cm/heure. Dès son arrivée dans le tissu nerveux, le poison ne peut plus être neutralisé par l'antitoxine et continue son chemin jusque dans la moelle épinière.

La cascade des troubles intra- et intercellulaires déclenchés par l'intrusion de la T.Te dans le tissu nerveux n'est pas élucidée, mais certaines conclusions continuent à s'imposer. Les plus importantes dans l'ordre pratique sont les suivantes: après pénétration du bacille tétanique dans l'organisme, aucun signe extérieur ne fait prévoir ni la rapidité, ni l'intensité de la toxinogenèse; l'antitoxine, qu'elle soit d'origine active ou passive, doit être mise à disposition de l'organisme aussi tôt que possible et à dose suffisante.

Si l'axiome « incubation courte, tétanos sévère » se révèle juste, la réciproque n'est pas exacte. Il existe des tétanos tardifs qui sont foudroyants. Leur période d'incubation est camouflée. Dans le foyer traumatisé les germes n'ont pas trouvé, au début, des conditions favorables de développement; ces conditions s'étant réalisées, la toxinogenèse prend un cours accéléré.

24. Titrage de la toxine tétanique

Les T.Te sont conservées soit sous forme liquide, soit à l'état sec. Sous forme *liquide*, à basse température et recouvertes par une couche de toluol, elles montrent une certaine stabilité après avoir subi, au début, une perte d'activité imputable à la transformation d'une partie de la toxine en toxoïde. A l'état *sec*, les toxines obtenues en général par salure au sulfate d'ammonium à saturation, puis desséchées par lyophilisation ou par le vide en présence d'un exsiccant, se gardent intactes pendant des années, même à température du laboratoire. Les titrages précis exigent certaines précautions.

Pour pallier l'instabilité des dilutions de toxine en eau distillée ou en eau physiologique, on recourt à l'usage de *tampons*, de bouillon, de colloïdes protecteurs (ISTRATI 1938, REGAMEY 1941 a). Pour expliquer les titres plus élevés qu'ils observent en ajoutant de la peptone à leurs dilutions de toxine, CONDREA et POENARU (1933 a, b) émettent l'hypothèse que la toxine renferme une substance spécifique inerte, une prototoxine, qui est activée par un principe renfermé dans le bouillon ou la peptone. Mais la peptone n'est qu'un simple protecteur (MUTERMILCH et coll. 1933, BELIN et coll. 1934 et al.), un stabilisateur (BENZONI 1947).

Le pH du diluant doit être légèrement alcalin: 7,2—7,4. LEGROUX et RAMON (1934), puis MUTERMILCH et coll. (1934) ont constaté que la T.Te acide adhérait au verre. Si LEGROUX et RAMON ne changent pas de pipette lors de chaque transport, la toxine diluée à pH 4,7 donne un titre 250 fois trop élevé. Rincée 30 fois à l'eau physiologique, une pipette qui a contenu 1 cc de toxine acidifiée retient sur ses parois 20000 DMM. L'acidification provoque donc une hypertoxicité artificielle, qui diminue et disparaît avec l'élévation du pH. Le phénomène de l'adhérence au verre serait propre à la toxine tétanique.

L'agitation, l'aération (bulles d'air lors des dilutions), les changements de température, la lumière, sont encore des facteurs susceptibles d'influencer le titre d'une toxine (RAMON et coll. 1937 c, 1938 a, 1950 a).

La valeur spécifique d'une T.Te s'apprécie selon des critères multiples, mais qui remontent tous à un test sur l'animal. FRIEDEMANN (1947) fait remarquer que les réactions dites *in vivo* appartiennent le plus souvent, dans la réalité, au groupe des réactions *in vitro*: quand l'expérimentateur injecte à l'animal un mélange de toxine et d'antitoxine, la réaction immunitaire s'est déjà jouée dans l'éprouvette; l'animal n'est qu'un indicateur et le test est un test *in vitro*.

Les différentes modalités de titrage aboutissent à des résultats qui ne sont pas nécessairement concordants. L'amélioration continuelle des techniques de laboratoire n'a pas encore suffisamment évolué pour assurer une exactitude rigoureuse et certaines variables persistent inéluctablement, telles la personnalité de l'expérimentateur, les conditions ambiantes, la sensibilité changeante de l'animal en fonction non seulement de l'espèce, mais aussi de la race, du sexe, du poids, de l'état physiologique et de la saison (HERWICK et coll. 1936), de la température ambiante (IPSEN 1951) etc.

Il y a bien peu de constantes expérimentales que les chercheurs respectent tacitement. Touchant la D.M.M. de la souris par exemple: les uns choisissent des animaux de 16 g, les autres de 20 g; les uns injectent la toxine par voie sous-cutanée, les autres par voie intramusculaire ou intraveineuse; les uns admettent comme point-limite 96 heures, les autres 120 heures. La définition par l'O.M.S., en 1950, de *l'unité internationale d'antitoxine tétanique* (U.I.) fut un premier jalon sur la voie de l'unification; mais les experts de l'O.M.S., qui ont créé l'étalon de sérum antitétanique, n'ont pas encore formulé de proposition définitive sur son mode d'emploi. Voici quelques notions essentielles sur les valeurs fournies par les titrages de la T.Te, notions qui seront complétées plus loin à propos de l'ana-toxine.

L'unité internationale d'antitoxine tétanique: U.I. «L'unité internationale 1950 est l'activité neutralisante spécifique pour la toxine tétanique, contenue dans 0,0003094 g de Préparation Etalon internationale » (Pharmacopoea Internationalis, éd. I, vol. I, 1951, p. 355).

L'U.I. (1950) a remplacé l'ancienne unité internationale de 1928 (cf. O.M.S. 1957, 1225). Les équivalences sont les suivantes:
1 U.I. (1950) = 2 U.I. (1928). U.I.: unité internationale ancienne ou U.A.: unité antitoxique ancienne.

    = 1 I.U.: immunizing unit, unité américaine ou unité Rosenau.

    0,1 U.I. protège pendant 4 jours 100 cobayes de 350 g contre la mort tétanique.

    = 1/82,5 A.E. (antitoxische Einheit) ou B.E. (Behring Einheit).

    1 B.E. protège pendant 4 jours contre la mort tétanique
        40 000 000 g souris (BEHRING et EHRLICH) ou
        28 000 000 g cobaye (ROSENAU et ANDERSON 1908).

    La B.E. est citée ici pour mémoire (cf. H. SCHMIDT 1931, p. 43 et 1952a, p. 23).
Dans cet exposé, les valeurs des toxines, anatoxines et antitoxines tétaniques exprimées par les auteurs ont été *toutes ramenées à l'U.I.* (1950).

L'unité toxique: U.T. (Gifteinheit ou G.E.; cf. H. SCHMIDT 1931), créée par BEHRING, est définie par la quantité de toxine qui, par voie sous-cutanée, tue en 96 heures 2 500 000 souris de 16 g, soit 40 000 000 g souris. L'U.T., reconnue dans les pays d'obédience allemande, tombe peu à peu dans l'oubli; elle offrait certains avantages dans la pratique de l'hyperimmunisation des chevaux.

La dose mortelle minima: D.M.M. est l'équivalent de la d.l.m. ou dosis letalis minima. Il s'agit d'une définition assez élastique (cf. plus haut), et qu'il est toujours indiqué de préciser. D'une façon générale, la D.M.M. représente la quantité de toxine qui, injectée par voie sous-cutanée ou intramusculaire, tue le cobaye de 350 g ou la souris de 16 g en 96 heures, les animaux présentant les symptômes du tétanos (pour ce qui touche à la réceptivité selon la voie d'administration, cf. PELLOJA 1949c, 1950b). A poids égal, le cobaye est 6 fois plus sensible que la souris (Lamont et coll. 1940). On peut faire intervenir dans l'abréviation le point-limite de la mort:

D.M.M.$_4$ = dose minima mortelle tuant l'animal à la fin du 4 ème jour.

L'injection de la D.M.M. ou de ses multiples, employés fréquemment dans les épreuves d'immunité, doit se faire compte tenu du poids de l'animal. Pour simplifier les calculs, on dilue par exemple la toxine de telle sorte que la D.M.M. soit contenue dans 2,5 cc (cobaye) ou 0,5 cc (souris); on injecte alors

	chez le cobaye	chez la souris
pour un poids de	220 g : 2,20 cc	14 g : 0,35 cc
pour un poids de	250 g : 2,50 cc	18 g : 0,45 cc
pour un poids de	315 g : 3,15 cc	20 g : 0,50 cc
	etc.	etc.

La D.M.M.(50). Quand les animaux d'un lot homogène reçoivent la même dose de toxine, tous ne meurent pas en même temps; les uns succombent après 4 jours, d'autres plus tôt ou plus tard, d'autres enfin survivent. Il est alors préférable de recourir à la

D.M.M.$_4$ (50) = dose qui tue le 50% des animaux en 4 jours.

Pour la fixation de cette dose, le calcul selon la méthode de Reed et Muench (1938) rendra d'éminents services.

L'unité floculante: Lf. 1 Lf (limes floculationis) = 1 U.F. ou unité floculante correspond à la quantité d'antigène fixée par 1 unité internationale d'antitoxine tétanique, la réaction exécutée *in vitro* se visualisant par une floculation.

Certaines toxines ne floculent pas. Il est néanmoins possible de les faire réagir en les « entraînant ». On mélange à cet effet la toxine qui ne flocule pas avec un *« helping »*, toxine (ou anatoxine) floculant aisément (Amberg 1946). La floculation qui apparaît renseigne sur la valeur du mélange; un simple calcul permet alors d'en déduire le titre de la toxine revêche. Le procédé du helping, largement répandu, fut déjà esquissé par Renaux (1924), Ramon et Grasset (1926 c) pour le titrage des sérums antidiphtériques qui ne floculaient pas.

On complète souvent le titre floculant par la mention de la Kf, c'est-à-dire de la vitesse de floculation, qui s'exprime en heures ou en minutes. Dans la règle, une toxine flocule d'autant plus vite qu'elle contient plus d'unités floculantes. Mörch et S. Schmidt (1930) ont proposé cette équation:

$$t \cdot a^m = k$$

où t = temps de floculation
a = valeur antigène de la toxine
m = exposant variable et propre à chaque toxine
k = constante.

La vitesse de floculation aurait également une signification sur la vitesse de neutralisation *in vivo* (S. Schmidt 1927 a).

La dose L+ ou «Limes mortis» se rapporte à une mesure indirecte du pouvoir toxique. Comme pour la D.M.M., il est bon de préciser les conditions de l'expérience: espèce animale, durée d'attente avant l'injection du mélange toxine-antitoxine, température pour la phase *in vitro*, mode d'injection, point-limite de la lecture, etc.

Par définition, la L+ est la quantité de toxine dont le mélange avec l'unité internationale d'antitoxine tétanique, injecté à l'animal, provoque la mort par tétanos le quatrième jour après l'injection. Elle est *théoriquement* l'équivalent de 1 Lf + 1 D.M.M. ou, — lors de l'emploi du cobaye de 350 g —, de 1001 D.M.M.$_4$.

Si on utilise la souris, surtout quand il s'agit de déterminer de petites quantités de sérum, on effectue les titrages non pas au seuil L+, mais au niveau L+/10 ou L+/20, puis L+/100 ou L+/200, éventuellement L+/1000 ou L+/2000. Dans le titrage au niveau L+/20, par exemple, la souris reçoit un mélange contenant 1/20 U.I. et la dose d'épreuve de toxine suffisante pour amener la mort en 96 heures.

La dose L$_0$. Selon la définition de la Pharmacopée internationale, Ed. I, Vol. I, 1951, p. 356, la dose L$_0$ représente la plus grande quantité de T.Te qui, mélangée à 1 U.I. d'antitoxine spécifique, a perdu le pouvoir de provoquer les symptômes du tétanos. L$_0$ et ses sous-multiples (L$_{0/5}$, L$_{0/20}$ etc) sont d'usage restreint.

Pour doser la T.Te, quelques auteurs se sont intéressés à d'autres techniques, par exemple à la colorimétrie (NOUREDDINE 1928), à la précipitation (LIBBY et ADAM 1939), à la diffusion sur gel de Oudin (A. RAFYI et coll. 1954, MIR CHAMSY et coll. 1957), à l'activité sur les cellules en culture de tissus (LESFARGUES et DELAUNAY 1946; PENSO et VICARI 1957a, b; LA PLACA 1957), à l'hémagglutination (FULTHORPE 1957, 1958).

Relations entre les valeurs fournies par les différentes méthodes de titrage

Les relations entre D.M.M., Lf et L+ sont-elles constantes ?

La méthode de floculation (NICOLLE et coll. 1920; RAMON 1922a, b, 1923; RAMON et DESCOMBEY 1926b) fournit ce que RAMON appelle la *valeur antigène intrinsèque*, c'est-à-dire le pouvoir antigénique total représenté par la somme des pouvoirs de la toxine proprement dite et de la toxine transformée spontanément en toxoïde. Il s'agit donc d'une mesure « statique » (PRIGGE 1939). Toutefois RAMON et coll. (1937b, 1938c, 1950a) incluent également dans le pouvoir intrinsèque la propriété immunigène, qualité plutôt « énergétique ». Cette extension ne semble pas justifiée (cf. titrage de l'anatoxine tétanique).

RAMON et son école, DELPY et MIR CHAMPSY (1946), MOLONEY et HENNESSY (1944), MUELLER et MILLER (1954) et d'autres auteurs trouvent un parallélisme satisfaisant entre Lf et D.M.M.: 1 Lf de toxine correspond grosso modo à 6000 D.M.M. pour le cobaye et 20000 D.M.M. pour la souris. Mais le quotient Lf/D.M.M. ne saurait être une constante, ce que de nombreux expérimentateurs dont REGAMEY et coll. (1946, 1947b, d) se sont efforcés de démontrer. En effet, la teneur en D.M.M. exprime la valeur de la seule fraction toxique tandis que la Lf associe les pouvoirs antigéniques de la fraction toxinique et de la fraction toxoïdique. C'est aussi à cause de la présence d'une proportion variable de toxoïde dans les différentes toxines que le quotient L+/D.M.M. ne peut pas fournir une constante.

Par contre le rapport Lf/L+, toutes deux valeurs basées sur le pouvoir de combinaison de l'antigène pour l'anticorps — et vice et versa de l'anticorps pour l'antigène — serait susceptible de fournir des indications plus régulières (SURJÁN et GORZÓ 1955) bien que, dans la détermination de la L+, la fraction toxoïdique puisse intervenir avec une certaine ampleur.

La méthode de floculation, basée sur une réaction de zone, n'est pas à l'abri du phénomène de DANYSZ (S. SCHMIDT 1927b; HEIDELBERGER et KENDALL 1930; EAGLE 1937a, b; PAPPENHEIMER et ROBINSON 1937). Elle est souvent troublée par le développement de plusieurs zones dont il est difficile de distinguer celle qui correspond à la valeur antigénique réelle de la toxine (RAMON 1938b, PRÉVOT 1938b, d); la tétanospasmine n'est pas le seul antigène en jeu dans la floculation et l'on a tenté d'expliquer l'apparition de zones secondaires par la présence, dans la toxine, de facteurs antigéniques non spécifiques: antigènes flagellaires (PRÉVOT et POCHON 1938c; PRÉVOT 1938b, d), substances voisines de l'exotoxine (RAMON 1929c, 1938c), antigène thermolabile d'origine non définie (MOLONEY et HENNESSY 1944), protéines somatiques (EISLER 1952). Les zones secondaires peuvent être éliminées par le choix de sérums très spécifiques (RAMON 1940f), par l'emploi d'un sérum de cheval hyperimmunisé avec des toxines obtenues en bouillon sans viande de cheval (KEMÉNY 1954), par la purification préalable de la toxine (MODERN et coll. 1949).

Malgré ses résultats inconstants et souvent difficiles à interpréter, la réaction de floculation a rendu et rend encore des services incontestables et inestimables lorsqu'il s'agit d'évaluer rapidement un antigène — toxine ou anatoxine — tétanique.

25. La toxine tétanique employée comme vaccin

Il est rare de rencontrer des anticorps dans le sérum de malades atteints de tétanos. Noeggerath et Schottelius (1915) signalent qu'ils ont trouvé de faibles quantités d'antitoxine chez des malades et chez des convalescents de tétanos; il n'y avait pas de parallélisme entre le décours de la maladie et le titre sérique. Küster et Martin (1918) ont mis en évidence des agglutinines dans le sérum de malades chroniques. La question de la présence d'antitoxine formée par le malade lui-même au cours de la maladie ne semble pas avoir été approfondie, probablement à cause de la rareté des cas où le tétanique n'a reçu ni sérum, ni anatoxine spécifiques.

Déjà Roux et Yersin (1889) avaient entrevu la possibilité d'immuniser avec la substance toxique, à condition de fixer cette dernière sur un corps qui la rende suffisamment insoluble pour qu'elle ne diffuse que lentement et crée une accoutumance graduelle. Ramon et ses coll. (1935a, 1935b, 1935d, 1935k, 1936a) expérimentent sur le lapin, le cobaye et le mouton l'effet de la T.Te émulsionnée dans un corps gras: lanoline et huile d'olive, seules ou associées. Non seulement l'animal ne présente aucun symptôme de tétanos malgré le nombre élevé de D.M.M. injectées par voie sous-cutanée, intramusculaire ou intraveineuse, mais il développe rapidement une solide immunité. Les auteurs constatent que la toxine n'est pas détruite in situ, puisqu'ils la mettent encore en évidence dans le dépôt extirpé. Tout se passe comme si le corps gras emprisonne la toxine pour ne la laisser diffuser que très lentement et crée sur place des réactions inflammatoires propices à l'immunisation. L'étude de la vaccination par doses progressives et celle de l'hyperimmunisation par doses massives de toxine n'entre pas dans le cadre de cet exposé.

3. L'anatoxine tétanique

31. Historique

En 1909, Löwenstein constate qu'une T.Te additionnée de formaldéhyde, et soumise à l'action d'une lampe de Nernst, perd son pouvoir pathogène mais conserve la propriété de faire apparaître de l'antitoxine. Il reconnaît bientôt avec Eisler (1912) que le rayonnement lumineux ne joue pas un rôle essentiel dans la détoxication: les rayons infrarouges agissent par leur effet calorifique; la transformation de la toxine se produit également dans l'obscurité, même à basse température, mais la chaleur à 36° C accélère notablement le processus. Les tentatives d'appliquer la méthode à l'inactivation de la toxine diphtérique se soldent par des échecs.

Pendant la première guerre mondiale, Eisler (1915) tente avec succès de vacciner ses chevaux destinés à la production d'antitoxine. Des essais portant sur 24 personnes restent cependant négatifs: l'injection de 2,5 ou 10 cc de toxine formolée ne fait pas apparaître d'antitoxine dans le sérum. Ce résultat ne surprend plus aujourd'hui: après un premier stimulus antigénique, il faut une période plus longue que celle observée par Eisler pour développer de l'antitoxine sérique en quantité appréciable. L'auteur, avec Silberstein (1919) précise sa méthode: le bouillon de culture est additionné de 1,5 à 2°/oo de formaline renfermant 40% de formaldéhyde, puis porté au thermostat aussi longtemps que le cobaye ne supporte pas sans incident 3—4 cc, ce qui exige généralement 5 semaines. La «formoltoxine» est injectée aux chevaux à raison de 5 fois 200 cc à 4 jours d'intervalle, puis les animaux reçoivent de la toxine. Deux ans plus tard,

Löwenstein (1921) rend justice à Ehrlich qui, selon une communication personnelle de Paltauf, eut vraisemblablement le premier entre les mains une substance atoxique immunisant contre le tétanos, mais sans jamais avoir pu la reproduire. Il rappelle que la T.Te se caractérise par trois propriétés: elle est toxique (1); elle fait naître de l'antitoxine (2) avec laquelle elle se combine (3); par un procédé chimique, la formolisation, il est possible de dissocier le facteur toxique de la fonction antigénique.

Il est curieux de relever que si Löwenstein et ses collaborateurs échouèrent dans la détoxication de la toxine diphtérique par le formol et que leur premier succès fut réalisé avec la T.Te, l'intérêt de Ramon se porta d'abord sur la toxine diphtérique. En effet, c'est seulement après avoir mis au point l'anatoxine diphtérique que ce chercheur étudia la toxine tétanique. Les travaux entrepris par Descombey (1924) sur l'instigation de Ramon aboutirent à l'élaboration d'un produit qui ne semble d'ailleurs pas beaucoup plus efficace que celui étudié par Eisler en 1915: trois semaines après l'injection sous-cutanée de 1 cc, le cobaye ne supporte qu'un petit nombre de D.M.M. de toxine. Descombey note déjà que l'immunité s'accentue avec le temps, que l'anatoxine flocule avec le sérum spécifique et que de deux anatoxines, celle qui flocule le plus rapidement est aussi celle qui immunise le mieux. Löwenstein fut certes celui qui, le premier, découvrit et utilisa, chez l'animal et chez l'homme, une préparation équivalant à l'A.Te; mais il n'a guère tiré parti de sa trouvaille. S'il ne peut se prévaloir d'une priorité, Ramon a le mérite d'avoir reconnu le bénéfice que l'humanité pouvait trouver dans l'usage des anatoxines; il a complété son oeuvre d'expérimentateur en consacrant un talent incontestable et une énergie inépuisable à répandre l'idée et l'emploi de la vaccination active contre le tétanos au moyen de l'anatoxine spécifique[1].

Outre la question de priorité souvent controversée (Kraus et coll. 1925; H. Schmidt 1925; Weinberg et Ginsbourg 1925; Tréfouel 1945, 1946; Ramon 1957a, p. 35 et 121), les termes d'anatoxine, toxoïde, toxoïde formolé et formoltoxoïde donnèrent lieu à maintes discussions.

Ehrlich entendait par toxoïde une *modification atoxique de la toxine* provoquée par le vieillissement ou le chauffage 30 minutes à 65° C. Dans le toxoïde, le groupe haptophore est conservé tandis que le groupe toxophore, plus sensible aux influences extérieures, est détruit. *Les toxoïdes font apparaître des antitoxines avec lesquelles elles peuvent se combiner.* On retrouve des similitudes avec les agglutinoïdes, les complémentoïdes etc. (Dieudonné et coll. 1925).

L'introduction d'un mot nouveau, « anatoxine », discutable déjà du point de vue étymologique, se justifiait-elle ?

Pour Ramon (1925a, 1957a, p. 40) et ses élèves (Nélis 1925, 1930 et al.), il n'y aurait rien de commun entre les anatoxines et les toxoïdes. Le toxoïde n'a qu'un seul caractère, celui d'être neutralisé par l'antitoxine, tandis que l'anatoxine possède l'innocuité, le pouvoir floculant et le pouvoir immunisant. L'école française a dû se méprendre sur le sens qu'Ehrlich donnait au terme de toxoïde. En effet, l'anatoxine correspond bien aux trois exigences formulées par Ehrlich: atoxique, elle provoque l'apparition d'antitoxine avec laquelle elle se combine. L'anatoxine est un toxoïde (Kraus et coll. 1925, H. Schmidt 1925, Glenny, cité par Nélis 1926). Dans la définition d'Ehrlich, il faut dissocier ce qui touche aux propriétés réelles et ce qui se rapporte à l'interprétation toute théorique d'un mécanisme.

A l'origine, anatoxine et toxoïde ne sont pas des équivalents. Le *toxoïde* possède un sens général et s'applique à n'importe quel antigène complet atoxique, dérivant d'une toxine, quel que soit le mode de détoxication. L'*anatoxine* est de sens plus restrictif; elle désigne un toxoïde obtenu par l'action combinée du formol et de la chaleur; le terme a sa raison d'être puisque celui qui l'ignore doit, pour préciser sa pensée, recourir aux termes de toxoïde formolé ou de formoltoxoïde.

Des nombreuses variantes de toxoïdes rencontrées, décrites ou proposées par les auteurs, seul le toxoïde formolé a acquis droit de cité et s'est largement

---

[1] Les essais sur l'homme entrepris par Vallée et Bazy (1917), confirmés par Okuda (1923) n'ont plus qu'un intérêt historique (toxine iodée).

répandu dans la pratique. Surtout dans la littérature anglo-saxonne, le terme de toxoïde a remplacé peu à peu celui plus long, mais plus exact de formoltoxoïde. Aujourd'hui, *anatoxine tétanique* et *toxoïde tétanique* désignent un produit *identique*.

32. Obtention de la toxine tétanique

Les milieux pour la production de T.Te destinée à l'élaboration d'un vaccin doivent satisfaire à trois impératifs:

1. former une toxine très active, car le pouvoir toxique est certainement l'un des facteurs importants de l'antigénicité ultérieure du vaccin;

2. renfermer le moins possible de protéines étrangères responsables des phénomènes d'escorte d'ordre allergique;

3. être faciles à préparer et d'un rendement économique intéressant.

Un minimum de conditions anaérobies étant réalisées, *Pl. tetani* se multiplie abondamment dans les milieux les plus divers mais, constatation faite déjà plus haut, croît bactérien et toxinogenèse ne vont pas de pair. Les premiers bouillons, constitués par des protéides et des peptones de viande, ont subi une lente évolution pour aboutir à des milieux quasi-synthétiques. Voici quelques formules qui jalonnent le passé relativement court de l'A.Te.

Milieu de Legroux et Ramon (1933). Macération de viande de boeuf avec peptone de laboratoire préparée par digestion peptique (muqueuses d'estomac de porc) du résidu provenant de l'eau de viande. Adjonction de glucose: 1,2%. Le pH de départ est ajusté à 5,8—6,0. Culture en flacons renfermant 5 g de poudre de globules rouges pour 1 litre de milieu; la surface du liquide est en contact avec l'air. Etuve à 37° C pendant 11 jours. La toxine (Ramon 1957a, p. 338) titre par cc 10—15 Lf et 60000—100000 D.M.M. pour le cobaye de 350 g.

Pour des raisons d'économie ou de commodité, la viande de boeuf fut remplacée ci et là par de la viande de cheval ou de veau (d'Antona et Valensin 1934 et al.).

Le procédé fut largement utilisé, pendant de nombreuses années, en Europe surtout. L'Armée suisse, par exemple, est encore aujourd'hui vaccinée avec une anatoxine dérivant du milieu de Legroux et Ramon.

Milieu Vf de Prévot (1939 d). Ce milieu est dérivé de celui de Stickell et Meyer. Il consiste en un hydrolysat peptique de viande et de foie frais de boeuf. Il est ajusté à pH 5,8 et contient environ 0,8% de glucose. La culture, maintenue pendant 11 jours à 33° C, est légèrement agitée dès le troisième jour. La toxine titre de 15—25 Lf/cc. Le taux de l'azote total se situe entre 0,58 et 0,67 mg/cc.

Milieux de Ramon, Amoureux et Pochon (1942 b). Ces milieux s'inspirent de celui que proposa Prévot (1939 d): la viande et le foie de boeuf sont remplacés par du muscle et du foie de cheval. Ils sont également prévus pour l'usage de viande impropre à la consommation (Ramon et Amoureux 1940 i).

L'une des variantes est obtenue par digestion *pepsique*; elle est glucosée à 0,8%. L'autre est un mélange à parties égales de bouillon pepsique et de bouillon *papaïnique* (Ramon et coll. 1943b, c) que fournit l'hydrolyse de viande et de foie par le latex de *Carica papaya*, auquel il est ajouté 4°/₀₀ de glucose et 5°/₀₀ d'extrait de malt. Ces milieux, ajustés à pH 5,8 et dont les caractères chimiques furent étudiés par Amoureux (1943a, b), fournissent des toxines dont la valeur floculante oscille entre 10 et 30 Lf par cc et le pouvoir toxique entre 30000 et 200000 D.M.M. pour le cobaye de 350 g.

Milieu de Taylor (1945). Pour éviter les accidents d'ordre anaphylactique imputés à la présence de peptones dans le bouillon de culture, Taylor propose un milieu — d'ailleurs fort voisin de l'un des milieux de Mueller et Miller (1945) —, avec

1. un autolysat chlorhydrique de panse complète de porc,
2. une infusion de viande.

L'auteur ajoute du $CaCl_2$ de façon à ce que le taux du Fe soit compris entre 0,025 et 0,05 mg/L, puis 0,25% de NaCl, 0,75% de dextrose et 0,2 mg/L d'acide nicotinique. La culture est maintenue 11 jours au thermostat à 34—36° C.

Environ la moitié des lots fournit des titres de toxine oscillant entre 150000 et 225000 D.M.M. cobaye. Une expérience de quatre ans a montré que le vaccin préparé à partir d'une telle toxine ne provoquait aucun incident d'ordre allergique.

Milieux de Mueller et Miller (1940—1954). Le premier milieu de Mueller et Miller (1940) dérive de celui que ces chercheurs avaient mis au point pour l'obtention de la toxine diphtérique; il s'agit d'un hydrolysat de caséine auquel on ajoute du tryptophane, de la glucose, un extrait de foie du commerce et qui ne contient pas plus de 0,3 mg de Fe par litre. Avec Feeney (1943 a, b), les auteurs poursuivent l'étude des facteurs essentiels à la croissance et à la toxinogenèse, leur intérêt se fixant sur les vitamines, les acides aminés et certains ions métalliques. C'est avec une toxine issue d'un milieu de Mueller et Miller (1943, 1945) ensemencé avec la souche de Harvard (1945), que Pillemer et coll. (1948 c) vont réussir la cristallisation de la T.Te. Seuls ou avec des collaborateurs, Mueller et Miller publient au cours des années suivantes une série de travaux consacrés à l'étude des multiples facteurs qui influencent la production de toxine (1947); température (1948 a), phénylalanine (1948 c), gaz produits par la culture elle-même (1948 b), D-sérine (1949 a), glutamine (1949 b) (Lerner et Mueller 1949). Ils notent avec Fisek (1954) le caractère indispensable de l'extrait de muscle. Ils étendent et approfondissent leurs investigations sur les fractions acide, neutre et basique provenant du liquide de digestion pancréatique de la caséine et séparées sur colonne de résine, puis reconnaissent le rôle capital dévolu aux peptides, surtout à certains dérivés de l'histidine (1955, 1956). Bien que de nombreuses modifications aient été apportées aux formules de Mueller et Miller par les laboratoires qui ont adopté les milieux « synthétiques » (Stone 1953 et al.), il paraît indiqué de concrétiser quelques-unes des étapes suivies par les auteurs américains (cf. tableau 4).

Tableau 4. *Comparaison de quelques formules de* Mueller *et* Miller *pour la production de toxine tétanique* (selon J. H. Mueller et P. A. Miller 1954)

Formule de	1947	1951	1954
Digestion pancréatique de caséine . .	15 g	15 g	22,5 g
Infusion de coeur de boeuf	250 cc	37,5 cc	50 cc
Glucose	7,5 g	8,75 g	11 g
NaCl	5 g	2,5 g	2,5 g
$Na_2H\,PO_4$	0,5 g	1,0 g	2 g
$KH_2\,PO_4$	0,17 g	0,1 g	0,15 g
$Mg\,SO_4\,7H_2O$ . . .	0,05 g	0,1 g	0,15 g
Cystine	0,125 g	0,125 g	0,25 g
Tyrosine.	0,125 g	0,5 g	0,5 g
Pantothénate de Ca	—	1,75 mg	1 mg
Uracil	—	0,875 mg	2,5 mg
Acide nicotinique .	—	0,25 mg	—
Thiamine	—	0,25 mg	0,25 mg
Riboflavine	—	0,25 mg	0,25 mg
Pyridoxine.	—	0,25 mg	0,25 mg
Biotine	—	2,5 µg	2,5 µg
Vitamine B_{12} . . .	—	0,5 µg	—
Fer réduit	0,3—0,4 g	0,5 g	0,5 g
Teneur de la toxine en Lf/cc	60—80	80—100	130—150
Teneur de la toxine en D.M.M. souris/cc	$6—8.10^5$	$8—10.10^5$	$13—15.10^5$

Culture en sac de cellophane (1946—1953). Reprenant le principe de MacClean (1937) déjà utilisé d'ailleurs par Polson et Sterne (1946), Laughlin et Brewer (1949), Fredette et Vinet (1952), Koch et Kaplan (1953) décrivent la méthode qu'ils ont suivie pour cultiver *Pl. tetani* en sac de cellophane. Le milieu, extrêmement simple, est un hydrolysat enzymatique de caséine avec NaCl, Na_2HPO_4 et KH_2PO_4; la contre-solution renferme 0,7% de NaCl et 0,05% d'acide thioglycolique neutralisé. Dans les conditions usuelles de culture, le milieu fournit une faible quantité de toxine, qui ne flocule pas; en sac de cellophane (après 9—10 jours de thermostat à 35—35,5° C), la récolte de toxine est 20 fois plus élevée: le produit titre $14,85 \cdot 10^6$ D.M.M. (souris de 20 g) par cc. La toxine est déjà très pure, avec $33 \cdot 10^6$ D.M.M. par mg d'azote, ce qui correspond à 50% de la pureté obtenue par Pillemer et coll. (1948 c) pour la toxine cristallisée.

Schaafsma (1954, 1957), à Johannisbourg, a mis au point la production de grandes quantités de toxine par culture en sac de cellophane. Ses produits titrent de 30 à 100 Lf/cc et même plus; ils offrent une double zone de floculation et donnent de bons résultats dans l'hyperimmunisation du cheval. Pettenella et Sella (1958) viennent de publier une technique simplifiée pour la production industrielle de T.Te en sac de cellophane; utilisant la souche CN 76, de Schaafsma, ils obtiennent des toxines titrant de 95 à 125 Lf/cc avec, par mg N, environ 300 Lf et 300 L +.

Milieu panaché de Surján (1956). Renonçant au milieu de Mueller et Miller, jugé impropre à la production de routine, Surján s'adresse à un milieu hybride renfermant:
1. une infusion de coeur de boeuf,
2. un hydrolysat du résidu de viande obtenu par:
    a) digestion avec la pepsine du commerce, puis,
    b) digestion par la trypsine et l'érepsine préparées extemporanément de pancréas et de muqueuse duodénale de porc,
3. un hydrolysat de caséine préparé comme le précédent.

Le milieu additionné de 0,75% de glucose, ajusté à pH 7,2, est ensemencé avec la souche de Harvard. La culture a lieu pendant 5—6 jours à 34—35° C.

Les 12000 litres préparés jusqu'ici ont fourni des toxines titrant de 12 à 62 Lf par cc. La D.M.M. oscillait entre 0,8 et $15 \cdot 10^{-6}$, la L+ entre 12 et 40 par cc. Le titre de $15 \cdot 10^{-6}$ égale les résultats publiés par Koch et Kaplan (1953) pour la culture en sac de cellophane.

Pedrosa (1948) utilise un milieu présentant une certaine parenté avec celui de Surján, puisqu'il emploie une macération de coeur de boeuf, à laquelle il ajoute une bacto-protéose (Difco). La culture n'est *pas filtrée*, mais formolisée à $4^0/_{00}$. Les anatoxines titrent en moyenne 20 U.I. par cc.

Milieux divers. Voici enfin quelques formules présentant une certaine originalité, mais dont l'usage semble avoir été abandonné ou limité.

Ramon et coll. (1927b) obtiennent sur un milieu contenant 1—7% de bile de boeuf stérilisée un produit peu toxique, mais floculant deux à trois fois plus haut que la toxine issue du bouillon témoin. Ils pensent à la formation d'anatoxine au cours de la toxinogenèse. Cette élévation de l'antigénicité est vraisemblablement factice; on la rencontre aussi lorsque le milieu renferme des extraits de levure (Radvila, non publié). Condrea (1930) utilise un milieu en gélose fluide. Ganslmayer (1935) recourt à un bouillon peptoné et glycériné à 8%, additionné de sang total d'âne; la toxine atteint le titre de 10^6 D.M.M. cobaye par cc; les résultats sont aussi bons sur un bouillon partiellement épuisé par la culture préalable de E. coli (richesse en tryptophane?). Saersoy et Gören (1936), sur un même milieu, ont obtenu des résultats identiques. Hirayama (1950) inaugure un milieu peptoné et glucosé renfermant 0,5% d'extrait de maïs.

La toxinogenèse, pour laquelle on n'a pas encore reconnu le substrat spécifique, pose des problèmes complexes. La toxine n'est certainement pas un produit de catabolisme, car elle peut faire défaut même en présence d'un croît bactérien abondant; est-elle, comme l'a suggéré Knight (1945), l'expression d'une compensation destinée à pallier une déficience ?

Le développement de la toxinogenèse expérimentale a suivi un chemin sinueux, jalonné et concrétisé par l'apparition de certains types de milieux de culture, dont quelques exemples viennent d'être rapportés. L'oeuvre patiente, minutieuse et de longue haleine entreprise par Mueller et Miller ne doit pas faire perdre de vue les contributions de ceux qui ont abordé l'une ou l'autre des multiples facettes du problème de la toxinogenèse: rôle des glucides (Prévot et coll. 1938e, f; Boorsma et coll. 1939; Tasman et Pondman 1941; Ramon et coll. 1942a; André 1948), de la teneur en azote et du métabolisme azoté (Ruggerini 1933; Berthelot et coll. 1939; Prévot et Boorsma 1939c, d; Amoureux et Pochon 1941a, b; Pochon et Amoureux 1941; Amoureux 1943a; Ramon et coll. 1943a; Stone et Berman 1954a; Vinet et Fredette 1958), des vitamines (Kligler et coll. 1938; Prévot et coll. 1939b, 1942; Wildführ 1947), influence des extraits d'organes (Prévot et coll. 1939a, b), du potentiel d'oxydo-réduction propre au milieu (O'Meara 1937, Prévot 1938a), du lysozyme (Stone 1952, 1953c) etc. Pour l'instant du moins, le champ ouvert aux investigations touchant à la toxinogenèse semble encore illimité.

33. Formolisation

L'atténuation de la T.Te fut expérimentée avec un nombre extrêmement élevé de substances. S. Schmidt (1933a) à lui seul étudia plus de deux cents corps. Une revue systématique des agents physiques, chimiques et biologiques, actifs sur la toxine tétanique est présentée dans la monographie de Pelloja (1951, p. 36). La plupart des produits envisagés détruisent non seulement le pouvoir toxique, mais aussi la propriété antigénique.

L'observation de Raubitschek et Russ (1909) sur la qualité immunisante de la toxine traitée par des savons n'est citée ici que pour mémoire, à cause de son ancienneté. Outre l'aldéhyde formique, les agents les plus intéressants parce qu'ils conservent tout ou partie de l'antigène sont assez variés: héxaméthylènetétramine, acétaldéhyde, glucose (Eaton 1937), salicylate de soude (Birkhaug 1931), antipyrine (Vincent 1927), pyrophosphate de soude (Velluz 1933), sulfure de carbone et isothiocyanates (Velluz 1938) qui inactivent la toxine en bloquant certains groupes aminés, comme le fait le cétène (Goldie et Sandor 1938), acides aminés tels la tyrosine (Sbarsky et Jermoljeiva 1927) ou la cystéine (Cowles 1937), la fermentation alcoolique (Comis 1928) etc. Raynaud et coll. (1957) viennent de décrire deux nouvelles formes de toxoïdes: le 2,4-dinitrofluorobenzène toxoïde, et le β-propiolactone toxoïde, combinaisons stables, atoxiques et antigéniques. Mais, dans la pratique, aucun autre corps n'a remplacé jusqu'ici la formaldéhyde choisie par Ramon.

Dans la règle, la formolisation s'effectue avec la formaline du commerce, « solutio formaldehydi » de la Pharmacopoea Internationalis, Ed. I, Vol. I, p.218. La formaline ou soluté de formaldéhyde contient 35,0—36,5% p/p ou 38,0—40,0% p/v de formaldéhyde (CH_2O; P.M. 30,03) et de polymères, avec une quantité variable de méthanol.

La formaldéhyde réagit avec les groupes fonctionnels rencontrés dans les acides aminés et les peptides (French et Edsall 1945). La réaction d'addition la plus usuelle:

$$R - H + CH_2O \rightleftarrows R - CH_2(OH)$$

aboutit à de nombreuses variantes, en particulier à des condensations avec pont méthylénique:

$$R - CH_2(OH + H - R' \rightleftarrows R - CH_2 - R' + H_2O$$

Avec les groupes NH_2 des acides aminés, la réaction classique de Sörensen

$$R - NH_2 + CH_2O \rightleftarrows R - N = CH_2 + H_2O$$

se traduit par une méthylénation que l'on suppose responsable de la détoxication, bien qu'un autre mécanisme soit possible (Eaton 1938a, p. 29).

L'action du formol est sous la dépendance de plusieurs facteurs: température, pH, taux de l'azote total, en particulier de la quantité des restes aminés libres d'origine peptidique. Il existe pour chaque toxine une température favorable qui assure la détoxication la plus rapide et la perte minimum d'antigène. A pH acide, la quantité de formol requise pour une détoxication complète est plus élevée qu'à pH faiblement alcalin. Un excès de formol s'accompagne d'une dégradation de l'antigène, avec augmentation du temps de floculation, perte notable des pouvoirs combinant et immunisant (Cheyroux 1953, 1954 et al.). La totalité du formol n'est pas fixée: si le 67—80% est lié d'une manière irréversible, une partie reste réversible ou libre (Khabas et Tarasova 1939). Après trois ans à 40° C, il reste encore à l'état libre 1 à 2% de l'aldéhyde formique primitivement ajouté à la toxine (Cheyroux loc. cit.).

Certaines réglementations (U.S.A.) tolèrent dans l'anatoxine prête à l'emploi moins de 0,02% de formaldéhyde. Le titrage du formol libre se fait par exemple avec la méthode du diméthyldihydrorésorcinol ou dimédon (cf. Bichsel 1949),

selon la méthode colorimétrique et photométrique de Jacobs et coll. (1951),
qui utilisent le ferrocyanure de K et la phénylhydrazine ou la méthode décrite
récemment par Godfrain et coll. (1957), qui recourent à un procédé colorimétrique basé sur l'emploi de la codéine. D'autres méthodes (Rimini-Shryer, Nash
1953 et al.) donnent également des résultats satisfaisants.

Le mécanisme de l'action de la formaldéhyde est notablement moins bien
étudié pour la T.Te que pour la toxine diphtérique. L'obtention d'une toxine
cristallisée devrait permettre d'aborder le problème avec des chances de succès,
mais les investigations à ce propos étant rares, nos connaissances sont restées
modestes.

Velluz (1929 a, b) admet par exemple que le formol provoque une condensation des
protéines, qui deviennent insolubles et inattaquables par les ferments. Les éléments toxiques
seraient fixés sur ces complexes et inactivés; dans l'organisme, la dislocation lente des complexes formolés ferait réapparaître la propriété antigénique spécifique. Pour Goldie et
Sandor (1938), la toxicité est liée aux groupements aminés primaires superficiels et le pouvoir
antigénique aux groupements aminés centraux de la molécule. La formaldéhyde, trans-
formant la couche superficielle, supprime le pouvoir toxique et respecte le pouvoir antigénique. Pendant la transformation de la toxine en anatoxine, il se produit une diminution de
l'azote aminé libre (S. Schmidt 1933 c), qui suggère la formation de ponts méthylénés entre les
groupes aminés ou la cyclisation de certains acides aminés tels l'histidine, le tryptophane,
la phénylalanine, la tyrosine (Wadsworth et Pangborn 1936; Loiseleur 1942; Fiala 1943;
French et Edsall 1945).

Pour quelques auteurs, la transformation de la toxine en anatoxine évolue en deux étapes
(Wadsworth et Pangborn 1936): au cours de la première étape, l'anatoxine peut récupérer
sa propriété toxique, ce qui n'est plus possible dans la seconde. Pendant la détoxication, la
toxine prend une teinte plus sombre (Stone 1953 a). Malgré quelques rares observations peu
convaincantes de retour à la toxicité (Condrea 1933, Wadsworth et coll. 1937 et al.), un
fait est acquis: *la détoxication par la formaldéhyde constitue un processus irréversible* (Ramon
1933 d, 1957 a, p. 44 et al.). C'est probablement ce caractère d'irréversibilité qui devrait différencier la propriété du toxoïde prise dans le sens large d'Ehrlich et celle du toxoïde prise
dans son sens restreint de toxoïde formolé ou anatoxine de Ramon (S. Schmidt 1933). La
constatation que la toxine tétanique cristallisée se transforme spontanément en un dimère
atoxique (Pillemer et coll. 1948 b, 1950) n'apporte pas de solution au problème que pose la
naissance du toxoïde. Raynaud et coll. (1953 a) soulèvent un doute quant à la détoxication
complète de l'A.Te et discutent l'éventualité d'une « *toxicité résiduelle* » (cf. p. 298).

L'A. Te. a conservé sa spécificité. Elle est un antigène puissant, notablement
plus efficace que les autres anatoxines connues. Elle est plus stable que la toxine
à l'égard du vieillissement, de la chaleur (Blum, 1896/97, avait déjà noté que les
protéines additionnées de formol pouvaient être impunément chauffées) et de
la dialyse contre l'eau (Maschmann 1931, Regamey, non publié). Mais la nature
des changements qui interviennent dans la transformation de la toxine en anatoxine méritent des recherches plus approfondies; ce mécanisme doit se rencontrer
avec des protéines autres que les toxines bactériennes, pour lesquelles la perte de
toxicité représente un critère de transformation intéressant (Pappenheimer 1948).

34. Purification et propriétés physico-chimiques

La purification de l'A.Te présente un double intérêt. Intérêt scientifique
tout d'abord, parce qu'elle va contribuer à la connaissance du processus qui
détermine la « toxoïdation » — ou transformation de la toxine en anatoxine — et
qu'elle va permettre de reconnaître dans quelle mesure la toxoïdation spontanée
(passage spontané du stade de toxine à celui de toxoïde selon la conception

d'EHRLICH) présente une analogie avec la toxoïdation par le formol (selon le procédé de RAMON). Intérêt d'ordre pratique aussi, car il sera possible de mettre entre les mains du médecin un vaccin en grande partie débarrassé d'un ballast inutile et dangereux par ses propriétés allergisantes.

Tentatives de Pillemer et coll. PILLEMER, GROSSBERG et WITTLER (1946 b) partent d'une T.Te. obtenue sur milieu à base d'infusion de viande — protéose — peptone, transformée en anatoxine par l'adjonction de 0,35% de formaline à pH 7,8 et 21 jours d'étuve à 37° C. Recourant à la méthode qui leur avait déjà donné des résultats probants avec l'anatoxine diphtérique, ils opèrent par précipitation à l'alcool méthylique, à basse température, à conditions de pH et de force ionique déterminées. Le produit final, desséché sous vide, perd son efficacité lorsqu'il est filtré après redissolution dans la solution-tampon phosphatée de pH 7,4; il reste actif si la filtration se fait en présence de glycérine 0,3 M. Pour les six lots d'anatoxine examinés, le rendement est bon, car les pertes se situent entre 0 et 5%; le facteur de purification (nombre de Lf par mg N dans l'anatoxine purifiée/nombre de Lf par mg N dans l'anatoxine brute) atteint 300. L'antigène purifié est pratiquement dépourvu de propriétés anaphylactogènes pour le cobaye dans le test de MOLONEY et HENNESSY (1942); il se laisse facilement précipiter par l'alun et jouit de bonnes propriétés immunisantes.

Un peu plus tard, PILLEMER et BENTOFF (1950), partant cette fois d'une anatoxine issue du milieu de MUELLER et MILLER (1945) et affinant leur méthode, réalisent un toxoïde trois fois plus pur que lors des essais précédents; l'antigène purifié titre 1500 Lf/mg. Ailleurs, PILLEMER et coll. (1946 b, p. 216) relatent qu'ils ont obtenu une anatoxine hautement purifiée, contenant plus de 3000 Lf/mg N et se rapprochant des valeurs trouvées pour la toxine. A une petite exception près, les procédés de purification sont identiques pour la toxine et l'anatoxine. La toxine et le dimère atoxique issu spontanément de la toxine ont respectivement une constante de 4,5 S et 7,0 S (PILLEMER et coll. 1948 c), de 3,9 S et 7,5 S (LARGIER 1956 b).

Pour la purification de routine, les producteurs font généralement appel à des procédés classiques, qui permettent de dissocier les diverses fractions protéiniques du complexe que constitue l'anatoxine brute.

Certaines méthodes n'ont pas trouvé d'écho (cf. bibliographie sous NEYROUD 1944). VELLUZ (1931 a, b) ou S. SCHMIDT et HANSEN (1931 a) utilisaient l'adsorption sur un hydroxyde d'aluminium. BALLINARI (1941), REGAMEY et coll. (1943 b) s'adressaient à la chromatographie sur colonne d'adsorbants, ROSS et coll. (1950) à l'adsorption sur une protamine. Les techniques qui suivent ont mieux retenu l'attention.

Procédé à l'acide trichloracétique (Boivin et Izard, 1937). Cette méthode a déjà donné satisfaction à RAMON et coll. (1936 c, 1937 a). La protéine spécifique est précipitée par l'adjonction d'une quantité adéquate de CCl_3COOH, puis centrifugée aussitôt; on redissout le culot dans le soluté de phosphate disodique à 3%. Quand l'opération s'accompagne d'une concentration, le liquide prend une teinte rouge sombre; la décoloration se fait par traitement avec 3—3,5% de charbon animal, sans perte notable d'antigène (REGAMEY, non publié). La récupération, mesurée en Lf, est de l'ordre de 75—95%. L'anatoxine purifiée flocule souvent assez mal et nécessite l'usage d'un helping. JACOBS et BEHAN (1950 a, b), précipitant à pH 4 et redissolvant à pH 8,0 ont un rendement de 83%, avec élimination de 99% de l'azote; l'index de purification atteint 81, voire 197. YOMTOV et SOLOMONOVA (1956) estiment que, comparée aux méthodes de salure par $(NH_4)_2SO_4$ ou de précipitation par l'alcool, la technique inaugurée par BOIVIN et IZARD semble la plus facile à exécuter et la plus rentable pour la production de routine.

DELSAL et MIR CHAMSY (1954) ont combiné la précipitation par CCl_3COOH à la congélation (—15° C) pendant quelques heures; l'anatoxine purifiée titrait 2700 Lf/mg N (milieu de MUELLER) ou 1800—2000 Lf/mg N (milieu Vf).

Procédé du relarguage par les sels neutres. L'adjonction de nombreux sels neutres provoque dans l'anatoxine brute un déséquilibre des protéines en solution, qui se rassemblent en surface. Cette couche superficielle crémeuse est récoltée, puis dialysée contre l'eau courante dans un sac de cellophane; elle fournit une solution de protéines débarrassée d'électrolytes, de protéoses et pauvre en protéines non spécifiques. La purification peut être augmentée par de nouveaux relarguages.

Neyroud (1944), étudiant un système de fractionnement par le sulfate de soude, note que l'antigène le plus pur se recueille lorsque l'anatoxine est précipitée entre 21 et 25% de Na_2SO_4, le 98,8% de l'azote pouvant être déjà éliminé lors du premier relarguage. La méthode, employée sur plus de 20000 litres d'anatoxine tétanique, a permis de revaloriser pour l'immunisation de l'homme ou du cheval des produits bruts pauvres en antigène, avec un rendement de 13 à 60% (Regamey, Neyroud et Calpini 1943a). Avec le sulfate d'ammonium, Levine et Stone (1951) pratiquent une série de précipitations et de dissolutions successives à des taux variables de sel. Ils récupèrent 40—80% de l'antigène présent au début, avec un coefficient de pureté et un pouvoir antigénique satisfaisants. Ramshorst (1957) obtient la meilleure purification et le rendement le plus élevé par salure à 15—25%.

Le procédé du relarguage est souvent utilisé en association avec d'autres techniques, telles l'ultrafiltration ou la précipitation au point isoélectrique.

Procédé de l'ultrafiltration. Empruntée aux techniques de purification en usage pour l'antigène diphtérique (Ramshorst 1951), l'ultrafiltration s'applique

1. soit à la toxine brute, que l'on purifie et concentre avant la formolisation,
2. soit à la toxine formolée, déjà détoxiquée.

L'ultrafiltration s'opère contre l'eau courante, sous pression ou sous vide, à travers une membrane de parlodion à 8% dissous dans l'acide acétique glacial et coiffant un filtre de porcelaine. Elle aboutit à l'élimination des électrolytes, des sucres et d'une quantité importante de l'azote non protéinique. Elle permet aussi de concentrer à volonté le liquide de départ. Le produit final conserve toutefois la totalité de ses protéines, spécifiques ou non. Aussi est-il indiqué de poursuivre la purification par l'emploi d'autres procédés. Tasman et Ramshorst (1952) combinent l'ultrafiltration avec la salure au $(NH_4)_2SO_4$ pour obtenir un antigène 25—40 fois plus pur et plus concentré, les pertes totales ne dépassant pas 15%.

Hendry (1952a, 1953) abandonne le relarguage au $(NH_4)_2SO_4$ à cause de la coloration foncée que prend le précipité lors de l'addition ultérieure d'alun; l'auteur recourt à l'ultrafiltration de l'anatoxine à travers des membranes de papier Schleicher et Schuell no 725, recouvertes d'une couche de parlodion à 4%. Un filtre de Seitz no 20 avec 9 disques laisse passer 35 litres en 8 heures. Le produit de départ, une anatoxine traitée par CCl_3COOH et titrant 700 Lf/mg N, fournit après ultrafiltration un antigène avec 1500—2500 Lf/mg N.

Procédé de l'alcool à basse température. Le traitement par l'acide ou l'alcool d'une anatoxine préalablement concentrée par le vide leur occasionnant trop de pertes, Modern, Ruff et Gatti (1948) s'inspirent de la méthode de Pillemer et coll. (1946b). Une première précipitation par l'éthanol, puis une seconde par l'acide sulfurique au point isoélectrique leur livrent un produit 312 fois plus pur, avec 50% de pertes. Une double précipitation par l'alcool fournit à Modern et coll. (1950, 1950—1953, 1956) une anatoxine 200 fois plus pauvre en azote, les pertes étant cette fois réduites à zéro. L'antigène n'offre plus qu'une seule zone de floculation. A la place d'éthanol, Dresler et Sabaldir (1957) choisissent l'alcool méthylique; la précipitation se fait à —5° C, pH 4,8 et force ionique 0,094. Les résultats irréguliers qu'ils obtiennent sont imputables à la diversité des anatoxines de départ.

Procédé à l'hydroxyde de magnésium et au chlorure de cadmium. Le procédé est celui que Holt (1950a) a décrit pour la purification de l'antigène diphtérique. Un traitement préliminaire est fait avec de l'hydroxyde de magnésium, qui adsorbe et entraîne la plus grande partie des impuretés. Le surnageant est alors traité par le $CdCl_2$. Il se forme un précipité, qui est remis en suspension dans une solution de K_2HPO_4. L'élimination du cadmium, opération délicate, est réalisée par l'adjonction de NaOH, qui précipite le cadmium sous forme de phosphate. On peut d'ailleurs souvent se contenter de la simple purification par le gel d'hydroxyde de magnésium sans recourir à la phase cadmique (Holt 1958).

Procédé à l'acide métaphosphorique. Son étude, faite par Raynaud, Turpin et Lemétayer (1953a) met en évidence un phénomène très particulier. La précipitation par l'acide métaphosphorique, à —15° C, de l'A.Te. à laquelle on ajoute 250 g de chlorure de sodium par litre réalise un rendement de 100%. Les injections massives de cette anatoxine purifiée et concentrée font apparaître chez l'animal un tétanos typique, qu'évite l'injection préalable de sérum antitétanique. Il existe ainsi une *«toxicité résiduelle»* que les auteurs ont aussi trouvée dans les anatoxines ne provenant pas de l'Institut Pasteur. Raynaud et coll. supposent que l'acide métaphosphorique fait réapparaître une partie — faible il est vrai — de la toxicité. Mais on peut également se demander si, de toutes les méthodes de purification, celle à l'acide métaphosphorique ne serait pas la seule qui mette en évidence des traces de toxine non transformée par l'aldéhyde formique, traces que les autres traitements détruisent, traces que les mesures habituelles du contrôle de l'innocuité ne sont pas à même de déceler.

L'antigène contenu dans l'anatoxine de départ et celui du produit purifié ont-ils le même pouvoir immunisant? Pour S. Schmidt et Hansen (1931a), Ramon et coll. (1936c), Wolters et Dehmel (1936), et de nombreux auteurs, la purification n'altère pas la qualité immunigène. L'observation par Megias et Moreno de Vega (1949) d'une anatoxine tétanique purifiée et privée d'antigénicité sur l'animal reste isolée. Stone et coll. (1954a, b) ont constaté qu'à taux égal de Lf, les préparations purifiées issues de milieux, l'un avec et l'autre sans infusion de viande, possédaient exactement le même pouvoir antigénique.

Malheureusement, la plupart des auteurs n'établissent pas de distinction entre pouvoir floculant et propriété immunigène réelle sur l'animal. Le fait que les vaccins purifiés et adsorbés sont de bons antigènes ne constitue qu'une présomption. Des essais plus étendus et plus précis dans la ligne suivie par Stone et Berman (1954a) sont encore indispensables pour établir la relation exacte entre l'antigénicité des anatoxines brutes et celle de leurs dérivés purifiés.

35. Potentialisation de l'anatoxine tétanique

Ramon (1925e, 1926e) reconnut très tôt l'importance d'une série de substances qui, dépourvues de spécificité immunologique, exaltaient le pouvoir antigénique des anatoxines. Il émit aussitôt l'hypothèse que l'augmentation des anticorps se produisait à la faveur de l'inflammation locale qui se développait au lieu de l'injection du vaccin. Ses travaux et ceux de ses collaborateurs sont résumés dans une récente mise au point (Ramon 1957c).

Parmi les corps étudiés en rapport avec l'A.Te., quelques-uns ont trouvé pendant longtemps un large emploi dans l'hyperimmunisation du cheval, surtout le tapioca (Ramon et Descombey 1925d; Ramon 1930, 1938e; Descombey 1930; Ramon et Lemétayer 1931; Tréfouël 1945 et al.). Le chlorure de calcium, l'alun de potassium, l'hydroxyde d'aluminium, le gluconate de calcium, le chlorure de magnésium, soit seuls, soit en combinaison entre eux ou avec le tapioca, eurent leur période de vogue puis furent délaissés (S. Schmidt 1931b; S. Schmidt et Steenberg 1936; Ray et Das 1938a, b; Lahiri 1938; Lemétayer et coll. 1949d; Gören 1951 et al.). D'autres produits, qui furent l'objet d'études plus ou moins approfondies et relatées par Ramon, Lemétayer et Richou (1935h), n'ont pas trouvé d'application pratique: atoxyl, tryparsamide, quinine, benzène, charbon animal, benjoin colloïdal, graisses, huiles végétales, lanoline, bile de boeuf, cholestérine, lécithine, cervelle de cobaye etc., mais ils ont permis de démontrer que la stimulation de la production des anticorps dépendait non seulement de l'antigène, mais aussi de l'espèce animale. Parmi les derniers adjuvants examinés, il y a lieu de citer — corps bactériens mis à part — la saponine (Richou et Thibault 1936), le tannin à 0,6% (Ramon et coll. 1941b), le latex d'Hevea brasiliensis (Jacotot 1947, Ramon et coll. 1948), les antigènes glucido-lipidiques (Ramon et coll. 1950b), le nucléinate de soude (Ramon et coll. 1952a).

Il est curieux de noter qu'en France, où l'étude des stimulants de l'immunité fut activement poussée, l'A.Te. fut et reste utilisée chez l'homme à l'état brut, c'est-à-dire non purifiée et sans adjuvant. L'une des raisons en est peut-être que, dans ce pays, le vaccin tétanique seul est relativement peu employé, les immunisations étant faites dans la règle avec des antigènes associés du type Te TAB ou Di Te TAB. En Allemagne, en Angleterre, aux U.S.A., un phénomène inverse s'est produit: on recourut très tôt aux combinaisons aluminiques et négligea pratiquement l'étude des autres stimulants de l'immunité.

La première observation d'une adsorption de toxine (diphtérique) par l'alun potassique avec élution ultérieure par une solution phosphatée sodique semble remonter à Roux et Yersin (1888). La précipitation de l'anatoxine diphtérique par l'alun fut réalisée par Glenny, Pope, Waddington et Wallace en 1926a: l'adjonction d'une solution d'alun potassique (de fer colloïdal ou d'acide tungstique) au toxoïde formolé fait apparaître un précipité qui entraîne l'antigène avec lui; le mélange anatoxine + alun, injecté à l'animal, engendre une meilleure immunité que l'antigène original (Glenny 1930). L'anatoxine alunée provoque chez l'homme de vives réactions, que l'on peut en grande partie éviter si l'on prend soin de laver le précipité avec un soluté physiologique (Glenny et Barr 1931a). A la place d'alun, Linderström-Lang et S. Schmidt (1930a, b) et S. Schmidt (1930) recourent à l'hydroxyde d'aluminium pour fixer l'antigène; ils obtiennent un vaccin très actif et que S. Schmidt (1931b) propose d'utiliser chez l'homme pour l'immunisation contre le tétanos. Tasman et Ramshorst (1952) introduisent le vaccin tétanique adsorbé sur phosphate d'aluminium, s'inspirant de la technique décrite par Holt (1950a) pour l'antigène diphtérique P.T.A.P. = purified toxoid aluminium phosphate precipitated.

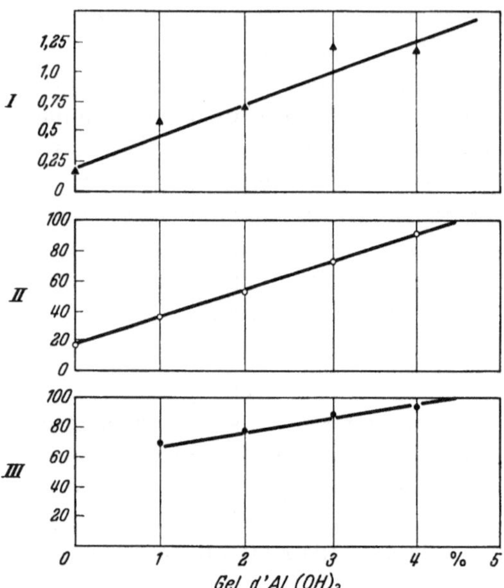

Diagramme 5. ⌜*Moyennes des taux antitoxiques sériques et des pourcentages de survivants pour une anatoxine tétanique adsorbée à des taux variés d'Al(OH)₃*
I Titre antitoxique moyen en U.I./cc
II % des cobayes survivants à la dose d'épreuve (100000 D.M.M.₄ souris)
III % de l'antigène adsorbé sur Al(OH)₃
(Selon Regamey 1947.)

Les vaccins tétaniques du commerce sont de type varié. Les uns sont *liquides*, bruts ou purifiés; les autres sont *précipités* ou *adsorbés*, la fixation sur le composé aluminique étant exécutée soit avec une anatoxine brute, soit avec une anatoxine purifiée au préalable. D'une façon générale, on entend par vaccin *précipité* une anatoxine traitée par l'alun (A.P.T. = alum precipitated toxoid) et par vaccin *adsorbé* une anatoxine fixée sur Al(OH)₃ ou AlPO₄. Les termes de «précipité» et d'«adsorbé» ne correspondent pas à un état physico-chimique déterminé et particulier; ils sont simplement consacrés par l'usage:

— le précipité se forme extemporanément lorsqu'on mélange l'anatoxine avec la solution d'alun;

— l'adsorbat s'obtient dans la régle en introduisant dans l'anatoxine un gel déjà préformé: gel d'Al(OH)₃ du type C, modification γ de Willstätter (Hansen et S. Schmidt 1935) ou gel de AlPO₄ du type Holt (1950a).

Le taux de l'adsorbant ajouté à l'A.Te. est déterminant sur la quantité de l'antigène spécifique entraîné et sur le pouvoir immunisant du mélange (diagramme 5). Mais la quantité d'Al ne saurait être trop élevée à cause des phénomènes réactionnels qui font suite à l'injection; aussi les taux du composé aluminique font-ils ci et là l'objet de prescriptions officielles. Ainsi aux U.S.A., les Minimum Requirements (N.I.H. 1952) exigent l'emploi d'un alun renfermant au minimum 99,5% d'alun potassique $\left(\text{Al K(SO}_4)_2 \cdot 12\,\text{H}_2\text{O}\right)$; le produit terminé ne doit pas contenir plus de 15 mg d'alun par injection pour l'homme. La Pharmacopée Helvétique, Ed. V, Suppl. I, tolère un maximum de 1,2 mg d'aluminium par cc de vaccin précipité ou adsorbé. Dans la règle, les produits du commerce contiennent par cc l'équivalent de 2 à 6 mg d'Al_2O_3.

La potentialisation par un adjuvant peut atteindre 70 fois et plus l'efficacité de l'antigène tétanique original (PRIGGE 1940). Le mécanisme de cette transformation a suscité quelques travaux et quelques théories, mais n'est pas encore élucidé (H. SCHMIDT 1952 a, p. 36, 1955, p. 484; LEVINE et coll. 1955). En fait, les recherches sur le rôle des activateurs se sont développées plutôt à propos de l'anatoxine diphtérique.

En 1925 (d, e) déjà, RAMON, puis RAMON et DESCOMBEY constatent que l'anatoxine tétanique au tapioca poussée sous la peau donne de meilleurs résultats que par voie intramusculaire. La résorption du vaccin est plus rapide dans le muscle que dans le tissu cellulaire sous-cutané, où l'afflux leucocytaire et la multiplication des cellules conjonctives provoqués artificiellement par le tapioca, puis l'action des ferments cellulaires ou plasmatiques sont particulièrement marqués. Les manifestations inflammatoires sont à la base des processus immunitaires. L'injection intradermique fait également une grosse réaction et s'apparente ainsi à l'injection sous-cutanée. L'anatoxine tétanique enrobée dans la lanoline crée un nodule inflammatoire du type paraffinome; la réaction inflammatoire est le fait de la lanoline et non de l'antigène; c'est la persistance de cette réaction qui assure la plus grande immunité (THIBAULT et RICHOU 1936).

De leur côté, GLENNY, BUTTLE et STEVENS (1931 c) formulent une double hypothèse. L'A.Te alunée crée au lieu de l'injection un dépôt d'où l'antigène, élué lentement, excite pendant longtemps les organes producteurs d'anticorps. L'antigène constitue les premiers jours un stimulus primaire et dans la suite un stimulus secondaire. HOLT (1950 a, p. 78) a soumis la théorie du dépôt à une critique expérimentale. Les travaux publiés jusqu'ici permettent de retenir les points suivants:

1. L'antigène brut est en grande partie perdu pour l'organisme, car il s'élimine rapidement (LEMÉTAYER et coll. 1950 a, b, NICOL 1950 et al.). L'adsorbant fixe l'antigène et empêche sa résorption. Les toxines diphtériques ou tétaniques enrobées dans la lanoline (RAMON et coll. 1935 a) ou fixées sur un composé aluminique (OTTENSOOSER 1938, KRECH 1952) ne provoquent plus la mort.

2. EISLER et GOTTDENKER (1937) supposent que l'adsorbant exerce une certaine neutralisation de la toxine. GRASSET (1926) a montré que si la toxine diphtérique adsorbée sur le tapioca perdait son pouvoir nocif pour le cobaye, il suffisait d'ajouter in vitro au mélange toxine + tapioca un ferment amylolytique pour rendre à la toxine son entière activité. Les adjuvants sont donc sans effet propre sur l'antigène, dont ils ne modifient ni le pouvoir antigénique, ni la toxicité (LEMÉTAYER et NICOL 1950).

3. Pour autant qu'il semble justifié de reporter sur l'anatoxine les observations faites à propos de la toxine, l'adjonction d'un adsorbant à la toxine formolée aboutit à la création in situ d'un dépôt renfermant le complexe adsorbant + antigène (FARAGÓ et coll. 1942 et al.).

4. Les phénomènes réactionnels locaux consécutifs à l'injection de l'adsorbat, et dont l'intensité maximum est atteinte après 6 heures déjà (THIBAULT et RICHOU 1936), se concrétisent par la formation d'une barrière de type inflammatoire qui freine la résorption du complexe antigénique et concentre à la porte d'entrée certains moyens de défense de l'organisme (RAMON et LEMÉTAYER 1935 e).

5. S'il n'a pas la possibilité de susciter une réaction tissulaire *locale*, l'adsorbant perd sa propriété activante. Les cobayes traités avec 1 Lf d'A.Te. alunée par voie intracardiaque ont, après 4 semaines, 50 fois moins d'antitoxine dans leur sérum que les témoins préparés avec la même dose par voie sous-cutanée (Holt 1950a, p. 84).

6. L'anatoxine adsorbée et fixée dans les tissus conserve pendant longtemps son pouvoir immunisant. Dans le nodule inflammatoire, l'antigène garde plusieurs jours la propriété de floculer. Le nodule lui-même, prélevé plus de 7 semaines après sa formation dans l'organisme, est encore capable d'immuniser un animal neuf (Faragó 1935, 1941; Harrison 1935). L'immunité optimum est acquise après 30 jours; elle est proportionelle au séjour et dans une certaine mesure, à la quantité absolue de l'adsorbant (A. Schmidt-Burbach et H. Dehmel 1937; Regamey 1947; Holt 1950a).

7. 5—7 injections de très petites doses (0,0033 cc) d'A.Te., poussées chaque jour sous la peau pendant 25 jours, produisent une immunité beaucoup plus considérable si elles sont faites à la même place — où elles créent un foyer inflammatoire — que si elles sont réparties chaque fois en un endroit différent, ou encore que si l'on inocule le total des doses en un temps (Ramon et coll. 1935f). Il est ainsi démontré que la quantité d'antigène ne joue pas le rôle décisif, qui semble dévolu à la présence d'un foyer inflammatoire.

8. Eisler et Eibl (1949) constatent que la puissance de l'activateur est fonction de la dimension de ses particules. Les petites micelles ont une surface relativement plus grandes et retiennent plus d'antigène; elles permettent à celui-ci un meilleur contact avec les cellules productrices d'anticorps. Si elles doivent passer la barrière qui encapsule le dépôt (théorie de Glenny), les petites micelles seront plus facilement résorbées que les grosses; si elles doivent participer à l'élaboration locale d'anticorps (théorie de Freund 1952), elles seront plus aisément manoeuvrées par les cellules.

9. Que le sujet d'expérience soit immun ou non, la rapidité et la nature de la réponse histo-pathologique sont les mêmes (Holt 1950a, p. 99).

10. Les vaccins adsorbés immunisent presque aussi rapidement que l'anatoxine fluide (Gundel et König 1938 et al.).

11. Le pouvoir immunigène de l'adsorbat augmente avec le temps (Vorobiev et coll. 1957b).

En l'absence de plus amples renseignements sur le mécanisme qui détermine le pouvoir activant des composés aluminiques, on concluera provisoirement que l'adsorbant forme avec l'antigène un complexe qui aboutit d'une part à la démultiplication de la surface antigénique et, d'autre part, à la création de phénomènes inflammatoires locaux qui favorisent — selon un processus encore inconnu — la production des anticorps in situ ou ailleurs dans l'organisme.

36. Contrôle des vaccins tétaniques

Il existe une diversité considérable dans les contrôles des produits immuno-biologiques en général et des vaccins tétaniques en particulier (Le Bourdellès et Desbordes 1957a, b; Perry 1958a; Regamey 1958a). Les réglementations gouvernementales varient grandement d'un pays à l'autre. Ici l'Etat se contente

d'accorder une licence de fabrication et ne s'occupe plus de la préparation ultérieure; ailleurs l'Etat se réserve le droit de faire des sondages sur le marché pour
vérifier si les produits mis dans le commerce sont de bon aloi; ailleurs encore
l'Etat, plus sévère, examine chaque lot de vaccin avant d'accorder un permis de
vente; enfin certains gouvernements sont à la fois producteurs et contrôleurs.

Les contrôles qui s'appliquent aux produits finis offrent des lacunes, car les
méthodes de routine pour les épreuves de stérilité et d'innocuité, par exemple,
sont elles-mêmes très déficitaires (REGAMEY 1958 a). Cette carence suggère l'introduction d'un certain nombre de contrôles en cours de fabrication, et l'établissement de « minimum requirements » auxquels doit se soumettre le producteur,
dispositif adopté déjà partout pour le vaccin poliomyélitique. Dans quelques
pays (N.I.H. 1952), les « minimum requirements » interdisent l'usage de certains
produits auxquels on attribue des propriétés allergisantes. Quant aux épreuves
d'efficacité, leur insuffisance est stigmatisée dans une publication de GREEN-
BERG (1955). Cet auteur titra par rapport à l'étalon canadien 47 vaccins tétaniques
provenant de 10 pays et de 16 laboratoires; il nota que le pouvoir antigénique
pouvait, d'un produit à l'autre, varier dans la proportion de 1 à 120!

Les tests décrits ci-après s'appliquent en général aux produits terminés,
mais la plupart sont aussi largement employés en cours de production.

A. Epreuves de stérilité. La recherche de germes aérobies et anaérobies, pathogènes ou
non, se pratique selon différentes méthodes que décrivent de nombreuses pharmacopées.
Malgré les imperfections des techniques actuelles, elle doit garantir une stérilité pratiquement
satisfaisante (BONNEL et RABY 1957; DIDENKO 1957; SCHEIBEL et WEIS BENTZON 1957;
DONY 1958).

B. Epreuves d'identité. Injecté à l'animal ou à l'homme, le produit doit faire apparaître
de l'antitoxine tétanique dans le sang circulant ou augmenter le taux de l'antitoxine éventuellement déjà présente (cf. épreuves d'efficacité).

C. Epreuves de pureté. Les anatoxines brutes, purifiées ou adsorbées ne sont pas des
substances chimiquement pures. La détection de facteurs non spécifiques pour le tétanos,
mais capables de provoquer des phénomènes d'escorte indésirables (protéines étrangères,
peptones, antibiotiques), se heurte à de trop grandes difficultés pour qu'on puisse, du moins
présentement, intégrer des épreuves de pureté dans la routine des contrôles.

D. Epreuves de pyrogénicité. Dans le cas de l'anatoxine tétanique, la recherche de pyrogènes (BERTRAND et QUIVY 1947; VAN GENDEREN 1958) pourrait être interprétée comme une
épreuve de pureté. Le test est décrit par plusieurs pharmacopées, en particulier par la Pharmacopoea internationalis, Ed. I, Vol. II, p. 297; il n'est généralement pas introduit dans la
gamme des contrôles auxquels l'anatoxine tétanique est soumise.

E. Epreuves d'innocuité. Conduites diversement selon les laboratoires, ces épreuves
ont pour but d'établir que le produit

1. ne fait pas apparaître chez l'animal des symptômes, même légers, de tétanos imputables soit à une insuffisance de la détoxication, soit à des spores tétaniques provenant de
la culture originale ou d'une souillure secondaire;

2. ne renferme ni substance toxique, ni bactéries pathogènes.

On injecte simultanément, par voie sous-cutanée, au minimum 2 souris et 2 cobayes
neufs. Les souris reçoivent 0,5—1,0 cc, les cobayes 5—10 fois la dose la plus élevée poussée
en une fois chez l'homme. Les animaux ne doivent présenter ni signe de tétanos, ni
autre signe pathologique grave dans les 10 jours qui suivent. La période d'observation du
cobaye est souvent prolongée à 3, 4 ou 5 semaines afin qu'il soit possible de déceler, par
exemple, des traces de toxine diphtérique dues à quelque erreur de manipulation.

Très souvent les animaux accusent localement une réaction de type inflammatoire qui
disparaît après 2—3 jours. Si le vaccin renferme un adsorbant aluminique, l'inflammation
locale, plus marquée, peut aboutir à des nécroses. En fin d'expérience, les cobayes doivent
être sains et montrer une augmentation de poids.

Les épreuves d'innocuité renseignent en fait sur la tolérance des animaux et non sur celle de l'homme. Certains lots de vaccin, bien supportés chez l'animal, déclenchent chez l'homme des manifestations non spécifiques inattendues et, inversement, des vaccins mal tolérés par le cobaye sont anodins pour l'homme; cette constatation se rapporte surtout aux vaccins associés des types Di Te Per ou Te TAB. Les épreuves usuelles sur l'animal laissent en outre dans l'ombre tous les problèmes qui touchent à l'allergie.

Malgré leurs imperfections, les épreuves d'innocuité fournissent des résultats acceptables pour la routine. Elles offrent peu de renseignements en dehors de la garantie d'une détoxication totale.

F. Epreuves d'efficacité. La grande diversité des appréciations formulées à l'égard des propriétés antigéniques des vaccins tétaniques fut déjà relevée (Greenberg 1955). La création d'un étalon international pour l'anatoxine tétanique (cf. bibliographie sous O.M.S. 1957) n'a pas encore de répercussion appréciable sur les méthodes de titrage. L'*unité internationale* correspond à 0,03 mg de l'anatoxine tétanique étalon de Copenhague, préparation extrêmement stable (Maaløe et Jerne 1952; Jerne et Perry 1956), purifiée à l'alcool, desséchée et contenant du glycocolle.

Selon ce standard,

$$1 \text{ unité internationale (U.I.)} = \frac{420}{833} \text{ Lf, soit 0,5 LF.}$$

L'U.I. est à peu de chose près l'équivalent de la « Schutzeinheit » de Prigge (1939), c'est-à-dire de l'*unité protectrice* ou U.P. (Schutzeinheit ou S.E.), unité allemande en usage à l'«Institut zur experimentellen Therapie » de Francfort a. M. (O.M.S. 1951). L'U.I. correspond à 0,03 mg, l'U.P. à 0,033 mg de l'anatoxine étalon de Copenhague.

Les méthodes destinées à l'évaluation des propriétés immunisantes sont délicates. Il est rare qu'une technique *in vitro* rende superflus les essais sur l'être vivant. L'homme, dans la règle, n'étant pas utilisable comme sujet d'expérience, on recourt à l'animal et introduit ainsi d'emblée une importante variable dans le jeu de l'immunité. En effet, la réponse de l'animal ne peut être reportée sur l'homme qu'avec une certaine réserve. Bien qu'elles aient fait l'objet de nombreux travaux, les méthodes de biométrie appliquées aux antigènes sortent à peine de la période des tâtonnements (Batson 1951; Schulz 1951; Prigge 1953; Miles 1954; Morrell et Greenberg 1954; Holt 1955; Cavalli-Sforza 1956).

Les méthodes les plus couramment utilisées pour le titrage des anatoxines tétaniques sont

a) les épreuves sérologiques

dont la plus répandue est le test de floculation; les autres réactions: précipitation, déviation du complément, diffusion sur gel etc. n'offrent pas d'intérêt pratique;

b) les épreuves mixtes

in vitro et in vivo, du type test de combinaison ou de blocage;

c) les épreuves sur l'animal.

a) Epreuves sérologiques. La réaction de floculation de Ramon sera seule envisagée ici. Cette méthode (Ramon et Richou 1950a) présente pour le titrage de l'anatoxine les mêmes avantages et les mêmes désavantages décrits plus haut à propos de la toxine tétanique, en particulier les phénomènes de double et de triple zone (Goldie et coll. 1942). Rapide, bon marché, elle est devenue indispensable dans les laboratoires mais, contrairement à l'opinion de Thiéry et Richou (1950), elle doit être complétée par des tests sur l'animal. Ramon estime que la floculation mesure le pouvoir antigène intrinsèque de l'anatoxine, c'est-à-dire la propriété immunisante intégrale. Certains doutes s'élèvent à cet égard:

1. La fixation de l'antigène sur l'anticorps ne suit pas nécessairement la loi des proportions multiples, en particulier lorsqu'il s'agit de phénomènes de zone (HEALEY et PINFIELD 1935; EAGLE 1937a, b et al.). La floculation peut avoir son optimum en dehors du point d'exacte neutralisation.

2. Si pour des anatoxines issues dans des conditions identiques d'un même milieu et d'une même souche, la toxicité, le pouvoir floculant et la qualité immunigène peuvent présenter un parallélisme très satisfaisant, il en est tout autrement lorsque les anatoxines proviennent de laboratoires différents. Le travail de GREENBERG (1955) fait ressortir une absence totale de relation entre titre floculant et pouvoir immunisant (les cobayes ont reçu 5,0 cc d'anatoxine et 2 semaines plus tard 20 D.M.M. de toxine):

Titre de l'anatoxine (Lf/cc)	Pouvoir immunigène (en fonction de l'étalon canadien)
5	0,2
8—10	0,6—0,9—0,9—1,7
14—16	0,1—0,2—0,3—0,9
20—30	0,7—0,8—3,3
50	0,07—1,2—1,5
60—65	0,2—0,3

Cette confrontation parle d'elle-même: deux anatoxines titrant par exemple 5 et 60 Lf/cc peuvent avoir le même pouvoir immunisant (cf. également HENDRY 1956).

3. Certains auteurs tels MOLONEY et HENNESSY (1944), NICOL et coll. (1957) ont trouvé un parallélisme entre le pouvoir floculant et le pouvoir combinant de l'anatoxine. D'autres expérimentateurs (SURJÁN et GORZÓ 1955; REGAMEY 1957a) ont fait des constatations contraires.

4. REGAMEY et BERTSCHMANN (1959), étudiant 10 A.Te mises en présence de 10 sérums spécifiques, constatent que les anatoxines fournissent de nombreuses valeurs aberrantes sans relation avec les zones de fausse floculation. Ces aberrations ne sont le propre ni d'une anatoxine, ni d'un sérum déterminé; elles apparaissent dans certains mélanges antigène — anticorps sans qu'il soit possible d'en dégager une loi.

5. Il ne faut pas accorder une grande confiance aux titres inférieurs à 13 Lf/cc (SURJÁN et GORZÓ 1955).

Une chute marquée du titre floculant ou un ralentissement sensible de la vitesse de floculation au cours des processus de détoxication ou de purification doivent faire suspecter une dénaturation. GLENNY et STEVENS (1938) estiment que le titre floculant reste toujours inférieur à 10% lors d'une détoxication correcte. Si une anatoxine ne flocule pas, il est possible — comme pour la toxine — de recourir à l'emploi d'un « helping ».

b) Epreuves mixtes. Dans ces réactions, les deux premiers temps de l'opération, soit la fixation de l'anatoxine sur l'antitoxine, puis la fixation de la toxine sur ce qui reste d'antitoxine, se passent *in vitro*; l'animal joue ensuite le rôle de détecteur de toxine. Le test de combinaison remonte en France à RENAUX (1924), POTTER (1924), HENSEVAL et NÉLIS (1924), en Allemagne à BÄCHER, KRAUS et LÖWENSTEIN (1925, 1926) et dérive en fait du titrage indirect envisagé par EHRLICH pour la recherche des toxoïdes. Bien qu'employée par de nombreux laboratoires, la méthode n'a pas encore fait l'objet d'études systématiques; elle permet de mesurer la quantité d'antitoxine que peut combiner ou bloquer une quantité donnée d'anatoxine dans des conditions déterminées d'expérimentation. Un exemple détaillé de titrage est donné par REGAMEY (1957a), qui propose le terme de U.B. ou unité bloquante (correspondant à la B.E. = Bindungseinheit) pour la quantité d'anatoxine capable de fixer, de bloquer 1 U.I. d'antitoxine tétanique.

Le test du blocage est une épreuve statique qui mesure une quantité d'antigène (NICOL et coll. 1957) comme le test de floculation, dont il diffère cependant par une plus grande spécificité. La floculation visualise un ensemble de processus où la réaction antigène — anticorps tétanique n'est pas seule en jeu; le pouvoir de combinaison n'intéresse que l'antigène et l'antitoxine tétaniques. FISEK (1955a) estime que le test de combinaison permet d'apprécier le pouvoir immunisant d'un vaccin tétanique. Cette constatation mérite d'être approfondie.

HIRAYAMA a proposé en 1949 une épreuve mixte qui utilise le pouvoir que possède l'anatoxine tétanique de déplacer la toxine dans le complexe toxine — antitoxine. Cette méthode ne semble pas avoir été contrôlée ou adoptée par d'autres expérimentateurs.

c) Epreuves sur l'animal. Les titrages d'antigénicité sur l'animal ont emprunté dans leur évolution deux voies distinctes qui ont tendance à se recouper aujourd'hui.

Tableau 5. *Comparaison des méthodes utilisées en Allemagne de l'Ouest, Grande-Bretagne et U.S.A. pour le contrôle de l'antigénicité des vaccins tétaniques*

	Allemagne de l'Ouest Méthode de Francfort, été 1958	Grande-Bretagne British Pharmacopeia Septembre 1958 Immunisation au moyen de 1 stimulus	2 stimulus	U.S.A. Pharmacopeia USP XV Décembre 1955 Anatoxine liquide	Anatoxine alunée ou adsorbée sur Al (OH)$_3$
Animal poids nombre	cobaye 340—380 g 150	cobaye 250—350 g $\geqq 9$	cobaye 250—350 g $\geqq 9$	cobaye 300—400 g $\geqq 10$	cobaye 450—550 g $\geqq 4$
Immunisation mode	1er groupe de 50 cobayes: $^1/_2$ dose d'A.S.** 2e groupe de 50 cobayes: 2 doses d'A.S.** 3e groupe de 50 cobayes: 1 dose d'A ?*	*Anatoxine liquide* (1) 5 fois la dose minimum pour l'homme *Anatoxine alunée* 1 fois la dose minimum pour l'homme ?	*1re injection:* $^1/_{10}$ de la dose minimum pour l'homme Après $\leqq 4$ semaines *2e injection:* même dose comme pour l'anatoxine liquide ?	$\leqq^1/_3$ de la dose totale (qui est de 1,5—3,0 cc) recommandée pour l'homme	$\leqq^1/_2$ de la dose totale (qui est de 1,0—2,0 cc) recommandée pour l'homme
voie	sous-cutanée			sous-cutanée	sous-cutanée
Délai d'attente	4 semaines	$\leqq 6$ semaines	$\leqq 2$ semaines après la seconde injection	$\leqq 6$ semaines	5—6 semaines
Epreuve d'immunité	Injection s.c. de 20 D.M.M. de toxine (volume: 1,75 cc pour 350 g de cobaye)	Recueillir le sérum et titrer l'antitoxine pour chaque animal séparément		Injection s.c. de 10 D.M.M. de toxine	saigner les cobayes, réunir les sérums et titrer le pool
Efficacité minimum exigée	30 U.I./cc	$^2/_3$ des sérums doivent avoir $\geqq 0,05$ U.I./cc *ou* $^1/_3$ des sérums doit titrer $\geqq 0,5$ U.I./cc		Le 80 % des animaux doit rester en vie au moins 10 jours	Le titre du mélange doit être $\geqq 2,0$ U.I./cc

A.S. ** = Antigène standard [anatoxine précipitée par $(NH_4)_2SO_4$ et Na_2SO_4, conservée depuis des années en ampoules sous vide]. En été 1958:

$^1/_2$ dose A.S. = 2,04 mg = 1,2 U.I.,

2 doses A.S. = 8,16 mg = 4,8 U.I.

A. ? * = Antigène à titrer. Les animaux du 3e groupe reçoivent 2,4 unités présumées.

(1) = Le contrôleur est autorisé à panacher un groupe d'animaux: une partie des cobayes reçoit une, les autres deux injections.

En Allemagne, PRIGGE et son école aboutirent relativement vite au choix d'une méthode qui, calbuée sur celle qu'ils avaient mise au point pour les vaccins diphtériques, a fait ses preuves au cours des deux dernières décades (ISTRATI 1938; ISTRATI et coll. 1939, 1940; PRIGGE 1939, 1940, 1953, 1954; CAVALLI-SFORZA 1956).

Dans les pays anglo-saxons, diverses équipes de chercheurs — souvent inspirées par le Comité des experts pour la standardisation biologique de l'O.M.S. — poursuivirent des travaux qui contribuèrent à l'établissement de l'étalon international ainsi qu'à la meilleure connaissance des propriétés immunigènes des A.Te fluides, précipitées ou adsorbéss (LAHIRI 1940, 1942; GREENBERG et coll. 1943, 1945/46, 1953, 1955, 1956; IPSEN et coll. 1953 a, b et al.). Il fut reconnu entre autres qu'il existait une concordance satisfaisante entre l'efficacité d'un vaccin tétanique sur l'homme et celle sur le cobaye ou la couris. Chez le cobaye, les courbes dose-réponse sont de même inclinaison, qu'il s'agisse de vaccin fluide ou adsorbé, ce qui n'est pas le cas pour la souris. Ce dernier animal s'immunise mal avec l'A.Te fluide, mais facilement avec l'A.Te aluminée (ISTRATI et coll. 1957; BARR et coll. 1957a). Un seul étalon de référence suffit donc pour les titrages sur le cobaye, quel que soit le type de vaccin.

On applique néanmoins encore maintenant les méthodes les plus diverses pour le contrôle de l'antigénicité des vaccins tétaniques (FISEK 1955 b). Sauf en Allemagne, il est rare de rencontrer une préparation qui soit évaluée en fonction de l'étalon international de Copenhague. Le tableau 5 démontre ce que l'on voudrait pouvoir appeler des normes de titrage; il tient compte des prescriptions les plus récentes admises en Allemagne, Grande-Bretagne et U.S.A., la France né prévoyant pour l'instant que le test de floculation.

Les contrôles sur l'animal exigent beaucoup de temps. Pour obtenir des renseignements précoces, REGAMEY (1957 b) préconise l'emploi de cobayes jouissant déjà d'une immunité active contre le tétanos. L'injection d'une faible quantité d'antigène déclenche chez eux un bond antitoxique appréciable au bout de quelques jours déjà.

Le titrage de l'antigénicité de la fraction tétanique dans les vaccins associés des types Di Te, Di Te Per, Di Te TAB etc. offre des particularités qui seront évoquées plus loin (cf. p. 331).

37. Conservation et validité

Il existe peu de renseignements précis sur la validité de l'A.Te, vraisemblablement à cause des difficultés inhérentes aux titrages d'antigénicité. RAMON insiste à maintes reprises sur la stabilité du vaccin formolé: des contrôles faits après 7 ans (RAMON 1957 a) ont démontré que l'anatoxine n'avait rien perdu de ses propriétés floculantes, que la vitesse de floculation en particulier n'était pas sensiblement ralentie. Si WILCOX (1934) observe une forte atténuation du pouvoir immunisant sur le cobaye après 38 mois de conservation, c'est parce que sa toxine fut traitée avec un taux trop élevé de formol (1%). Les expériences de OGOBLINA et PONOMAREVA (1944) sont concluantes: l'A.Te congelée à —30°, voire —50° C et décongelée sans interruption pendant 10 jours n'offre aucune altération de son pouvoir immunisant original et n'acquiert aucune propriété nocive pour l'organisme.

Dans les conditions normales de conservation, soit de $+2°$ à $+10°$ C, il ne se produit une perte notable d'antigène que si le produit contient un excès de formol. CHEYROUX (1953, 1954) a démontré qu'après trois ans le taux du formol libre représentait encore le 10% environ du formol initial, ce qui justifie d'une part la neutralisation de l'excès de formol après la détoxication et, d'autre part, l'interdiction dans la pharmacopée américaine d'une quantité supérieure à 0,02% d'aldéhyde formique résiduel libre. Il fut déjà signalé plus haut que les vaccins adsorbés devenaient plus efficaces avec le temps; la « maturation » de l'adsorption multiplierait 40—50 fois le pouvoir antigénique au bout de trois ans (Vorobiev 1957 a).

Le formol, qui assure une certaine stérilité à l'égard des souillures bactériennes, est un préservatif insuffisant contre les moisissures. Certains producteurs ajoutent un agent conservateur sous forme d'éthylthiosalicylate sodique de mercure (Merthiolate, Timerosal) à 0,005—0,01 %, de borate (Merfen) ou d'acétate de phénylmercure à 0,005—0,01 %, de phénol à 0,2—0,5 % etc. Ces produits ne semblent pas affecter les qualités immunisantes du vaccin.

Sous forme desséchée, telle qu'elle fut réalisée déjà en 1929 b par Ruggerini, ou sous forme lyophilisée, l'anatoxine doit acquérir une validité pratiquement illimitée, comparable à celle des préparations étalons (Jerne et Perry 1956).

On conserve l'anatoxine tétanique de préférence entre $+2^0$ et $+10^0$ C, à l'abri de la lumière. On se souviendra que s'ils restent sans effet sur l'anatoxine liquide, les changements de température et la congélation altèrent les vaccins précipités ou asorbés par modification du pouvoir adsorbant des composés aluminiques. Les anatoxines adsorbées sur gel d'aluminium sont très stables pour autant que le pH ne soit pas inférieur à 6,5 (Ramshorst 1954).

Les préparations du commerce portent une indication de validité qui diffère selon les Etats. L'U.S.P. XV, p. 720 n'autorise pas un délai supérieur à 2 ans, calculé à partir de la date de préparation ou de délivrance du produit. La British Pharmacopoeia 1958, p. 666 accorde un délai d'au moins 2 ans à partir de la date de fabrication. Ailleurs, la tolérance s'étend à 3 ans, plus rarement à 5 ans. Il n'existe pas encore de normes pour les produits secs.

4. L'immunisation de base par l'anatoxine tétanique

Dès l'anatoxine mise au point et son efficacité reconnue chez l'animal, puis chez l'homme, Ramon, Descombey, Zoeller et leur école s'efforcèrent de préciser le nombre, le volume et l'espacement des doses d'anatoxine; ils démontrèrent le rôle particulier de l'injection de rappel. D'une façon générale, les observations des auteurs français s'avérèrent exactes, et les multiples travaux qui suivirent ne firent souvent que confirmer des faits déjà solidement établis. Dans l'analyse des processus immunitaires liés à la vaccination antitétanique, il y a lieu de considérer les phases suivantes:

La *stimulation primaire* marque a tout jamais les cellules de l'organisme, mais peut ne pas aboutir à un état de protection suffisant.

Après un intervalle adéquat, la *stimulation secondaire* déclenchée par l'injection du même antigène développe rapidement ses effets. Son étude porte sur l'état d'immunité avant l'injection, l'effet propement dit du second stimulus et l'amortissement progressif du maximum réalisé (Holt 1950 b; MacLeod 1953; Freund 1953; Dixon et Maurer 1955). Les anticorps qui prennent naissance après un second stimulus peuvent avoir une nature différente de ceux consécutifs à la première injection (Foster 1957). Les variations dans la nature de l'antitoxine sont encore plus marquées lors de l'hyperimmunisation (Ramon et Richou 1941 d; Jennings 1956); en effet, le rapport entre l'antitoxine libre et la D.M.M. (50) change au cours de l'immunisation antitétanique (Vorobiev et coll. 1957 b), ce qui fait présumer des anticorps aux propriétés variables.

Stimulus primaire et stimulus secondaire, répété ou non, forment ce que l'on convient d'appeler l'*immunisation de base.*

Une injection ultérieure ayant pour but de prolonger ou de réactiver l'immunité de base porte le nom d'*injection de rappel.* Elle a les caractères d'un stimulus secondaire.

41. Indications de la vaccination contre le tétanos

Plus du tiers des cas de tétanos survenant à la suite de blessures bénignes, il semble indiqué de rendre la vaccination antitétanique universelle. On insistera particulièrement pour immuniser les plus menacés: militaires, agriculteurs, jardiniers, usagers de la route, sportifs et autres. On vaccinera sans inconvénient (GOLD 1937a) les allergiques sensibles au sérum de cheval ou de bœuf (BERGEY et ETRIS 1936), puis tous ceux qui ont déjà reçu une fois ou l'autre un sérum hétérologue.

L'immunisation des futures mères (MAGARA et ADUKATA 1937; COHEN et SCADRON 1946; CIMMINO 1951; MATHES 1955; MÖRL 1956 et al.) doit être recommandée partout où le tétanos du nouveau-né offre une certaine fréquence, en particulier dans la pratique tropicale (FERNAN-NUNEZ 1938 et al.); dans cet ordre d'idées, elle doit même se faire chez la fillette ou l'adolescente (A.P.H.A. 1955). Les premiers travaux de NATTAN-LARRIER et coll. (1926, 1927), puis les belles études de l'école de Garches avec LEMÉTAYER ont jeté les bases d'une méthode de prophylaxie utile. COOKE et JONES (1943) avaient craint une interférence entre le vaccin appliqué au nouveau-né et l'antitoxine d'origine maternelle, mais GREENBERG et FLEMING (1951) ont obtenu, chez des enfants nés de mères immunisées et soumis à l'injection de rappel une année après l'immunisation de base, une bonne réaction chez tous les sujets (titre moyen: 0,67 U.I./cc). EDSALL (1956a) vient de publier des résultats d'ensemble sur le destin des anticorps contenus dans le placenta ou le colostrum.

Les enfants représentent évidemment le contingent le plus indiqué pour la vaccination contre le tétanos. Le nourrisson avant trois mois fournit des titres aussi élevés que le bébé âgé d'un an (PETERSON et CHRISTIE 1951). Les enfants sont plus faciles à immuniser que les adultes (BERGEY et coll. 1939; MILLER et coll. 1949; EDSALL 1957), les filles plus que les garçons (MARVELL et PARISH 1940) et parmi les adultes, les jeunes réagissent à l'immunisation de base par des titres sériques plus élevés que les sujets plus âgés (REGAMEY 1941a). MÉNARD (1955) propose de ne vacciner qu'après la première année, l'enfant étant jusque là relativement peu exposé au tétanos.

Les contre-indications sont envisagées à propos des accidents vaccinaux (cf. p. 320, 326 et 333).

42. Technique de l'immunisation de base

A. Nombre d'injections. L'immunisation de base comporte en principe soit 3 injections d'anatoxine liquide, brute ou purifiée, soit 2 injections d'anatoxine tétanique précipitée ou adsorbée.

L'emploi d'une *seule injection* n'a pas acquis droit de cité. RAMON et ZOELLER (1926f) notent déjà que l'immunité consécutive à la première injection d'anatoxine tétanique est très faible (1 cc de sérum neutralise à peine 1 D.M.M.). L'animal réagit mieux que l'homme. Après 5 cc d'antigène aluminé, le cheval possède déjà 2,5 U.I./cc au bout de 4 semaines (WOLTERS et DEHMEL 1943). Une injection unique chez le cobaye fait apparaître de hauts titres après 1 mois, propriété utilisée pour certains titrages d'efficacité (JONES et MOSS 1936a; WOLTERS et DEHMEL 1936, 1940); les premiers anticorps sont décelables après 12 jours chez la souris, avec maximum au 30ème jour (KOVTOUNOVITCH 1956). Chez l'homme, l'anatoxine fluide commence à manifester ses effets à partir du 11e jour (SACHS 1952), ce qui présente un certain intérêt pour la séro-anatoxiprophylaxie. Les titres présents dans le sérum vers la fin du premier mois sont parfois très faibles (BERGEY et coll. 1934, 1936, 1939; MARRI 1933a; GOLD 1940a; WOLTERS et DEHMEL 1938b, 1942), parfois assez élevés. MacBRYDE (1937) observe que de 94 enfants vaccinés par une injection unique d'anatoxine alunée 73 jours plus tôt, 92 possèdent des titres supérieurs à 0,01 U.I./cc. ERICCSON (1948) se contente même d'une seule injection d'antigène aluminé, recommandant l'injection de rappel dans tous les cas de blessure. En réalité, le stimulus primaire, tel qu'il est pratiqué chez l'homme, ne fait que préparer l'organisme à démultiplier l'effet de la seconde injection de vaccin. Dans le cas du tétanos, on ne saurait compter sur l'effet stimulant de la toxine élaborée au niveau du foyer traumatique, car la quantité si faible d'antigène que représente la dose mortelle de poison n'a pas la force nécessaire pour relancer suffisamment l'immunisation.

L'injection unique d'anatoxine tétanique peut aboutir à une «immunité» précoce. Des faits semblables ont été décrits pour le Pneumocoque, la variole et surtout pour la coqueluche (phénomène d'EVANS et PERKINS 1954a, b, 1955). KRECH (1949), reprenant les tentatives de CREMER, de H. SCHMIDT et SCHOLZ à propos de la diphtérie, note que l'injection d'anatoxine tétanique quelques heures avant l'insertion d'esquilles de bois tétanifères crée chez la souris une protection immédiate, conséquence d'une élaboration instantanée d'anticorps. RAYNAUD et coll. (1951b, 1953b) font des observations analogues: de fortes doses d'anatoxine protègent le cobaye pendant trois jours. L'anatoxine poussée intraveineux prévient la mort à condition d'être en quantité $5 \cdot 10^6$ à $7 \cdot 10^6$ fois plus grande que la toxine. La protection ne saurait être rapportée à une immunité proprement dite, car elle est fugace; les auteurs supposent soit un blocage du système nerveux central, soit un phénomène d'interférence, soit une inhibition compétitive. PLETSITYII et coll. (1956) ont fait apparemment des remarques semblables en U.R.S.S. Si l'on fait appel aux approximations de PELLOJA (1951, p. 94), on peut accorder une certaine probabilité au mécanisme du blocage. Le cobaye possédant $0,2 \cdot 10^9$ neurones et le D.M.M. renfermant $7 \cdot 10^9$ molécules, il faut théoriquement 35 molécules de toxine par neurone pour tuer le cobaye, chiffre certainement trop élevé, car il n'est pas tenu compte de la toxine égarée dans les tissus en dehors du système nerveux central. Le peu d'affinité que manifeste la cellule nerveuse à l'égard de l'anatoxine — comparativement à la toxine — est bousculé par l'énorme excès d'anatoxine (5 à 7 millions de fois plus abondante que la toxine). L'occupation des récepteurs cellulaires serait efficace tant que l'anatoxine n'est pas éliminée. Mais toute tentative d'expliquer le mécanisme de la protection précoce appartient encore au domaine des hypothèses.

Lors du premier stimulus, la même dose d'antigène injectée à la même place ou répartie en 3 ou 6 endroits différents procure des titres d'antitoxine analogues (BORCILA 1935). Pour VOROBIEV (1958) enfin, une dose d'A.Te purifiée et adsorbée est plus efficace que deux doses d'A.Te brute.

Deux injections séparées par un intervalle assez long procurent une bonne immunité de base et une réactivité suffisante lors de l'injection de rappel (cf. diagramme 6, p. 315). Que l'anatoxine soit fluide ou fixée sur un composé aluminique, l'organisme possède après une quinzaine de jours une immunité appréciable (Hegyessy et coll. 1956). Les taux d'antitoxine varient selon les auteurs et selon les vaccins employés. Jones (1936a) rapporte des titres de 0,05 U.I./cc, Boyd (1938) de 0,1 U.I. après 10 jours, MacBryde (1937) de 0,1—5 U.I./cc après deux mois, Regamey (1945) de 0,0004—0,75 U.I./cc après 4 mois, Wolters et Dehmel (1942) de <0,005—0,05 U.I./cc après 2—3 ans. Avec l'anatoxine fluide, la dispersion des titres est plus forte qu'avec l'antigène aluminé. Chacun s'accorde toutefois à reconnaître qu'il vaut mieux, surtout lors de l'emploi d'anatoxine brute, consolider l'immunisation de base par une *troisième injection*. Cette dernière vise également à forcer en quelque sorte la réponse immunitaire des mauvais réacteurs.

L'immunisation de base est complétée par une ou plusieurs injections de rappel (cf. p. 326 et ch. 9, p. 343).

B. Qualité du vaccin. Le choix entre les types suivants de vaccin:
— anatoxine liquide brute ou purifiée,
— anatoxine brute ou purifiée, précipitée par l'alun ou adsorbée soit sur hydroxyde, soit sur phosphate d'aluminium,
— anatoxine brute ou purifiée, fixée ou non sur un composé aluminique et associée à d'autres antigènes (Di Te, Di Te Per, Te TAB, Di Te TAB etc.),
est le plus souvent affaire d'opportunité, d'école, de tradition ou de vogue. L'anatoxine fluide est plus économique (Barr et Sachs 1955). L'avantage d'un antigène activé est manifeste surtout lors du premier stimulus (D'Antona et Piazzi 1956, Vorobiev 1958).

Les produits du commerce portent ou ne portent pas de mention sur la valeur antigénique du vaccin. Le titre est donné en Lf (unités floculantes) ou U.I. (unité immunisante = S.E.: Schutzeinheit); exprimé en Lf, il n'a qu'une signification très relative quant au pouvoir immunisant de l'anatoxine.

C. Quantité de vaccin et volume des doses. Le dosage original de Ramon comportait 1,2 et 2 cc; il a subi maintes modifications. On adopte généralement aujourd'hui, pour des raisons de simple commodité, le schéma de 3 fois 1 cc pour l'anatoxine fluide et celui de 2 fois 0,5 cc. ou 2 fois 1 cc pour le vaccin aluminé. La teneur en substance active relève de la qualité du vaccin: un antigène purifié et puissant pourra être dilué par le producteur, tandis qu'un antigène faible devra faire l'objet d'une concentration et déclenchera plus facilement des phénomènes d'escorte.

Le dosage est le même pour les adultes et pour les enfants.

La schématisation du dosage dans la pratique de la vaccination ne fait pas oublier que les injections ont une signification immunitaire différente selon qu'il s'agit d'un premier, d'un second ou d'un troisième stimulus (Freund et Bonanto 1942; Freund 1947, 1953; Holt 1950a, b; Barr et Llewellyn-Jones 1951 et al.). Il est vraisemblable que les résultats publiés par d'Antona et Piazzi (1956) à la suite d'une étude expérimentale très poussée sur la diphtérie, s'appliquent également au tétanos. Ces deux auteurs démontrèrent que

1. l'on crée une immunité plus élevée et plus durable avec un premier stimulus faible suivi d'un stimulus fort que vice et versa;

2. le premier stimulus est considérablement plus efficace si l'antigène est adsorbé ou, mieux encore, émulsionné dans l'huile;

3. les stimuli ultérieurs fournissent une immunité supérieure lorsque l'antigène est liquide;

4. l'adjonction d'un activateur à l'antigène du premier stimulus est d'une importance capitale pour le développement de l'immunité potentielle.

La quantité absolue d'antigène est en relation avec l'intensité de l'immunité, fait qui constitue la base des méthodes de titrage. Toutefois lors d'une réponse secondaire, la synthèse des anticorps n'est pas toujours en relation directe avec la quantité d'antigène appliquée (Koshland et Engelberger 1957).

D. Mode d'application. Les vaccins tétaniques s'appliquent chez l'homme par voie *sous-cutanée* ou *intramusculaire*. L'injection sous la peau jouit d'une certaine préférence qui se justifie, du point de vue immunologique, par la richesse locale des éléments réticulo-endothéliaux. On recourt à la voie intramusculaire le plus souvent lors de l'emploi de vaccins aluminés, pour éviter à coup sûr la formation d'un nodule réactionnel et peut-être parce que les autres phénomènes d'escorte sont réduits à un minimum (Marvell et Parish 1940; Pletsityii et coll. 1957 et al.). Le vaccinateur doit s'assurer que le vaccin n'est pas poussé intraveineux. L'injection se pratique dans la région sous-claviculaire, dans la fosse sus-épineuse, à la pointe de l'omoplate, à la cuisse ou dans la région deltoïdienne. Le bras est plus particulièrement indiqué chez les sujets allergiques, car il donne la possibilité de placer un tourniquet en cas d'accident anaphylactique aigu (Cooke et coll. 1940).

Depuis les expériences de Grasset (1927) et Borcila (1935), on sait que l'A.Te poussée dans le derme y développe des qualités immunisantes remarquables (J.A.M.A. 1957 d). L'injection *intracutanée* est un peu moins facile à exécuter que les injections sous-cutanées ou intramusculaires; elle permet d'administrer seulement de faibles volumes. On la réserve pour le traitement de certains allergiques, chez lesquels on désire obtenir une résorptieo ralentie.

Que l'on emploie l'anatoxine seule ou mélangée à un excipient, la voie *percutanée* n'aboutit pas à une immunité (Grasset 1927; Kaktine 1931; Tokgöz et coll. 1941; Bilâl 1948 et al.). L'usage d'un « patch », qui crée une asphyxie locale et favorise une certaine résorption transcutanée, engendre une immunité satisfaisante dans le cas de la diphtérie, en particulier lors d'un second stimulus, mais les essais avec l'anatoxine tétanique, purifiée ou non, concentree ou non, se sont soldés par des échecs (d'Antona 1950, 1951 b).

Les résultats ne sont guère meilleurs par *voie buccale*. Ramon et Zoeller (1926 g) relèvent que seule ou précédée par l'ingestion de bile, l'anatoxine tétanique ne déclenche pas d'immunité décelable chez l'homme, alors que le lapin et le cobaye jouissent d'une faible protection, vraisemblablement parce que la bile peut être administrée en proportions beaucoup plus élevées chez l'animal que chez l'homme (Ramon et Grasset 1926 d). Lemétayer (1935) fait ingérer au lapin 940 cc et au cobaye 230 cc d'anatoxine tétanique au cours d'une période de 9 mois sans qu'apparaisse la moindre immunité; par contre la *voie rectale* permet d'immuniser, la résorption de l'antigène se faisant à la marge de l'anus. Katić (1948) rapporte l'observation curieuse suivante: le cheval en voie d'hyperimmunisation fournit considérablement plus d'antitoxine si l'antigène est administré simultanément par voie sous-cutanée et par la bouche. Dans l'état actuel des recherches, la voie gastro-intestinale ne présente aucun intérêt pour la vaccination de l'homme contre le tétanos.

Ramon et Zoeller (1927 c, e) ont constaté que l'anatoxine tétanique concentrée et glycérinée, appliquée seule ou associée avec l'anatoxine diphtérique sous forme d'instillations par *voie nasale*, suscitait une immunité satisfaisante, non pas chez le sujet neuf, mais chez l'individu au bénéfice d'un premier stimulus. Quelques autres expérimentateurs (Gold 1940 b, Said Bilâl 1948 et al.) suggèrent d'envisager la capacité de résorption des muqueuses rhinopharyngées pour réaliser les rappels d'immunité en dehors des injections nécessitées par un traumatisme. L'application d'un antigène tétanique par *voie aérienne* paraît peu efficace (SILBERSCHMIDT 1934; FURBETTA et coll. 1941).

Les voies *intracérébrale* et *méningée* offrent un intérêt théorique. MARIE (1907) reconnaît déjà que la substance cérébrale n'acquiert pas d'immunité et ne forme pas d'anticorps. Pourtant DESCOMBEY (1925a) observe qu'après avoir reçu 0,1 cc d'anatoxine tétanique dans le cerveau, le cobaye possède des anticorps sériques et résiste à l'injection de toxine. De nouvelles expériences, tant sur l'homme que sur l'animal, établissent que l'injection intracérébrale doit être assimilée à une injection sous-cutanée, intramusculaire ou intraveineuse. En effet, il n'y a pas formation d'antitoxine dans la substance cérébrale, qui reste toujours sensible à la toxine tétanique; l'antigène gagne l'économie générale, où les anticorps se forment et d'où ils repassent en faible proportion dans le liquide céphalo-rachidien (DESCOMBEY 1929; MUTERMILCH et SALAMON 1930; RAMON et coll. 1940d, e, g; NÉLIS 1940 et al.).

E. Intervalles entre les injections. L'espacement entre les injections offre une importance considérable (RAMON et DESCOMBEY 1925b; MACBRYDE 1937; BOYD 1938 et al.). En principe les longs intervalles sont plus favorables que les courts, l'optimum pouvant d'ailleurs varier selon le type du vaccin (KESTERMANN et coll. 1939). Chez des sujets traités par l'anatoxine brute, les injections ayant été faites après des intervalles de 11 mois et 4 mois, REGAMEY (1945) a rencontré des titres sériques toujours élevés (1—9 U.I./cc); PESHKIN (1943) également. Dans la pratique, on cherche un compromis pour réaliser une immunisation aussi rapide que possible, compte tenu aussi que les espaces trop longs découragent (J.A.M.A. 1957a). RAUSS et coll. (1958a) estiment qu'il existe une limite inférieure au-dessus de laquelle l'immunité augmente de plus en plus; toutefois, passé un certain degré, la prolongation de l'intervalle n'est suivie que par une élévation modérée de l'efficacité.

Pour l'*anatoxine tétanique liquide*, on sépare la première et la seconde injection par un intervalle de 3—6 semaines, les 3 semaines étant un minimum (RAMON et ZOELLER 1933b). La troisième dose sera poussée au plus tôt 3 semaines après la seconde injection, de préférence entre les 3e et 9e mois.

Pour l'*anatoxine tétanique aluminée*, l'usage prévoit un intervalle de 1 à 2 mois entre le deux premières injections, le plus souvent 6 semaines (WILSON 1941, SACHS 1952). On fera la troisième injection 12 mois après la première (EDSALL 1956b).

43. Efficacité de la vaccination antitétanique

Un test d'efficacité incontestable est celui auquel la vaccination antitétanique fut soumise pendant la guerre mondiale de 1939—1945. Les renseignements fournis par BOYD (1946), LONG et SARTWELL (1947), HALL (1940, 1948), LONG (1948) et SACHS (1952) sont unanimes; ils sont à la base des tableaux 6 et 7.

Tableau 6. *Morbidité tétanique parmi les blessés de guerre*

	Vaccinés	Morbidité pour 1000 blessés
Guerre de 1914—1918		
Allemagne.	non	3,8 (HALL 1948)
Grande-Bretagne . . .	non	1,47 (SACHS 1952)
U.S.A.	non	0,134 (SACHS 1952)
Guerre de 1939—1945		
Grande-Bretagne . . .	oui	0,11 (SACHS 1952)
U.S.A. Army	oui	0,0062 (SACHS 1952)
U.S.A. Navy	oui	0,0044 (HALL 1948)
Japon	non ou mal	env. 1,0 (PARISH 1952)

Tableau 7. *Létalité des cas de tétanos parmi les blessés alliés de la guerre 1939—1945*
(Selon les données de Sachs 1952.)

| | Blessés non vaccinés | | Blessés vaccinés | | |
| | Traitement après la blessure | | | | |
	sérum anti-tétanique	aucun	sérum anti-tétanique	injection de rappel	aucun
Guerre de 1914—1918					
France	66,3	83,3	—	—	—
Grande-Bretagne . .	22,6	53,3	—	—	—
Guerre de 1939—1945					
Grande-Bretagne . .	43,5	48,7	18,2	—	81,8
U.S.A.	50	25	—	50	50

Du tableau 6 il ressort que:

1. Malgré la séroprophylaxie, conduite souvent trop tard, les blessés de 1914—1918 paient un lourd tribut au tétanos.

2. D'une guerre à l'autre la morbidité tétanique a considérablement diminué, tant chez les Anglais que chez les Américains, tous vaccinés.

3. Il existe une grande différence entre la protection des Britanniques et celle de leurs alliés: les premiers jouissaient d'une immunité moins solide (nombre moindre d'injections d'anatoxine) et, lors d'un traumatisme, le blessé recevait non pas de l'anatoxine, mais un sérum antitétanique . . . quand cela était possible (Perry 1940).

4. Chez les Japonais, mal vaccinés, la fréquence du tétanos rappelle celle de l'époque où le sérum était le seul moyen préventif spécifique.

La létalité, par contre, semble être restée dans le même ordre de grandeur pendant les deux guerres (tableau 7). Les valeurs surprenantes pour les U.S.A. sont sans signification statistique (une douzaine de cas). Les *échecs* de la vaccination sont exceptionnels, ils sont commentés page 320.

La réduction considérable des cas de tétanos n'est pas seulement sensible parmi les belligérants; elle l'est aussi pour les civils au bénéfice d'une vaccination active. Regamey (1955a) relève qu'en Suisse, où la morbidité dépasse certainement 0,5 pour 100000, il n'est survenu aucun cas de tétanos chez des vaccinés alors que la population mâle vaccinée à l'Armée aurait dû présenter un minimum de 30—40 infections mortelles.

Quelle est la *durée* de cette immunité? La question a suscité une bibliographie considérable, car sa réponse conditionne et la conduite à suivre en cas de traumatisme chez un sujet vacciné ou en voie d'immunisation, et l'indication de l'I.R. destinée à l'entretien de l'immunité.

Déjà Ramon et Zoeller (1926f) avaient noté qu'une *seule injection* d'anatoxine tétanique était insuffisante; l'immunité qui lui fait suite se produit seulement après des semaines et l'I.R. a peu d'effet (Sneath et Kerslake 1935; Melnik et coll. 1936b) (cf. diagramme 6). A condition que l'intervalle entre les *deux injections* soit adéquat, l'immunité apparaît avec une intensité satisfaisante une quinzaine de jours, voire même 5—7 jours après la seconde dose de vaccin (Saski et Stetkiewicz 1933; J.A.M.A. 1955; Eckmann 1958). 15 jours après la deuxième injection de Te TAB, les 617 enfants traités par Hegyessy et coll. (1956) ont tous un titre antitoxique supérieur à 0,1 U.I./cc; après 9 mois, le 4% possède moins de 0,01, le 86% de 0,01 à 0,1 et le 10% plus de 0,1 U.I./cc.

Une immunisation de base parachevée par une troisième ou même une quatrième injection assure une immunité d'au moins

3—4 ans: BAIRD (1949), MILLER et coll. (1949), WISHART et JACKSON (1951).

8 ans: RAMON (1939a).

10 ans et plus: BIGLER (1951), REGAMEY et SCHLEGEL (1951a), D'ANTONA (1952), SPATH et KÖLE (1952), TURNER et coll. (1954), EDSALL (1955).

7—13 ans: PETERSON et coll. (1955).

16 ans et plus: SCHLEGEL (1956).

20 ans au moins: ECKMANN (1958).

Il fut déjà relevé que le taux de l'antitoxine libre n'était pas l'indicateur réel de l'immunité active, dont la qualité primordiale réside dans la *réactivité* ou pouvoir de réponse à toute nouvelle excitation antigénique suffisante par l'élaboration rapide et abondante de nouveaux anticorps. Or le diagramme 7 démontre clairement que la réactivité, — ou *immunité potentielle* (MAGRASSI 1934) — ne s'émousse pas avec les années: elle conserve la même précocité; elle paraît même s'accentuer avec le temps (WOLTERS et DEHMEL 1936; SPATH et KÖLE 1952).

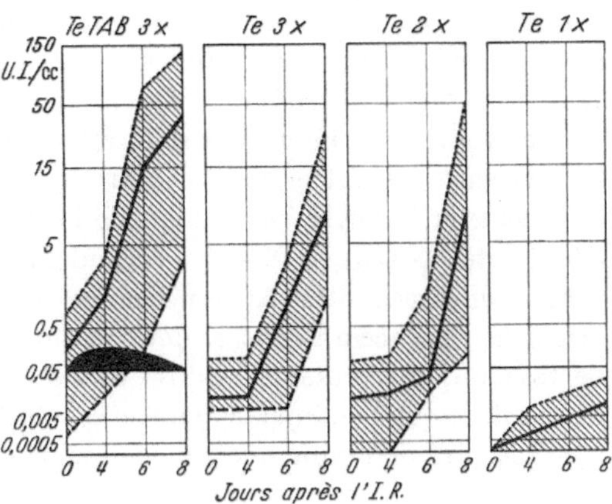

Diagramme 6. *Bond antitoxique après l'injection de 2 cc d'anatoxine tétanique liquide*
Immunisation de base:
Te TAB 3× : 3 injections de vaccin Te TAB
Te 3× (2×, 1×): 3 (2, 1) injections d'anatoxine tétanique liquide. · · · · · titre maximum, - - - - titre moyen,——titre minimum, ▬▬ titre après injection de 1500 U.I. de sérum antitétanique
0,05 U.I./cc: seuil présumé de protection
(Selon REGAMEY 1955a.)

En fait, la durée de l'immunité sera connue au fur et à mesure que parviendront des informations sur les ruptures d'immunité. Pour l'instant, le praticien doit s'en tenir à des normes quelque peu arbitraires, mais offrant une haute sécurité.

Après l'immunisation de base, le titre de l'antitoxine baisse peu à peu dans le sérum du vacciné. Chez certains sujets, les anticorps ne sont plus décelables après un ou deux ans déjà; mais chez la plupart survient une certaine stabilisation (REGAMEY et SCHLEGEL 1951a; STAFFORD et coll. 1954; LOONEY et coll. 1956 et al.). SCHEIBEL (1955) constate qu'avec les années il se forme un équilibre entre la destruction et la formation de la globuline spécifique. Cette recherche d'une constante sérique *en l'absence de toute nouvelle sollicitation antigénique extérieure* n'est pas encore expliquée (BARR et GLENNY 1952, WARD 1957):

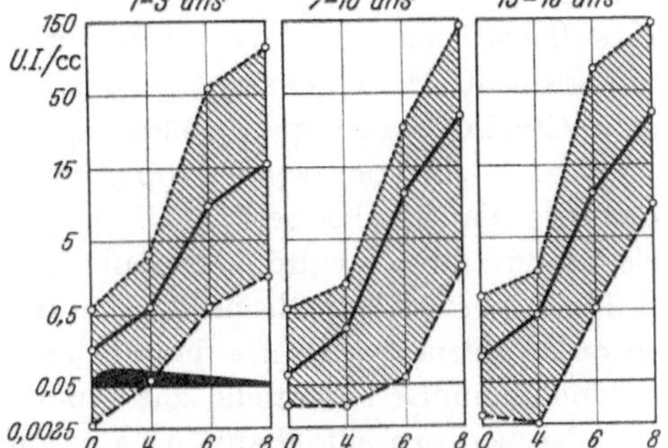

Diagramme 7. *Bond antitoxique après l'injection de 2 cc d'anatoxine tétanique*
Immunisation de base: 3 injections de vaccin Te TAB.
Légende: cf. diagramme 6. (Selon SCHLEGEL 1956)

l'antigène persiste-t-il dans les tissus producteurs d'anticorps, ainsi que COONS le démontre pour le polysaccharide pneumococcique marqué, ou bien s'agit-il d'une altération acquise et transmissible au niveau des cellules élaboratrices de l'antitoxine (Burnet et Fenner 1949)? Dans le cas du tétanos, l'intérêt majeur réside dans la persistance non seulement de l'antitoxine mais aussi et surtout de la *réactivité*, propriété que Roux et Vaillard connaissaient depuis longtemps: «chaque fois que les cellules de l'organisme ont subi l'action de la toxine tétanique, elles conservent longtemps leur résistance vis-à-vis d'elle alors même que le sang ne manifeste pas de propriété antitoxique» (cité selon Lemétayer et coll. 1951).

Les ruptures d'immunité sont relativement fréquentes dans presque toutes les vaccinations autres que le tétanos; aussi l'extrême rareté du tétanos chez le vacciné et la facilité avec laquelle le sujet acquiert une solide protection ont-elles incité Regamey (1944 b) à formuler l'hypothèse suivante: La valeur immunisante d'un vaccin n'est pas fonction de la masse de l'antigène, mais de la toxicité originelle du vaccin avant sa détoxication; en effet, si l'on considère la quantité de toxine que représente la dose totale employée pour l'immunisation de base, on constate que l'on injecte, dans le cas du tétanos, une anatoxine qui correspond à 80 fois plus de toxine que dans le cas de la diphtérie.

44. Seuil antitoxique de protection

Refaisant l'historique à propos du titre antitoxique nécessaire et suffisant pour protéger contre le tétanos, Turner, Stafford et Goldman (1954), puis Looney et coll. (1956) constatent qu'il existe une certaine confusion parce que les taux d'anticorps sériques n'ont pas la même signification chez le sujet passivement immunisé et chez le vacciné. Comme il n'est pas possible d'éprouver l'homme avec des spores ou de la toxine, on en est réduit à l'étude expérimentale chez l'animal. Mais le transfert sur l'homme d'observations touchant le lapin, le cobaye ou la souris suscite des réserves; si les modes d'infection et les conditions de résorption de la toxine peuvent être analogues, elles ne sont cependant pas identiques pour les différentes espèces.

Dans le cas de l'immunité passive, l'injection de 1500 U.I. fait apparaître dans le sérum un titre de 0,1 à 0,2 U.I./cc. Ce taux n'est toutefois *ni indispensable*, *ni suffisant* pour protéger contre l'intoxination ou l'infection tétaniques. Un tétanos peut éclore — exceptionnellement il est vrai — en présence d'une teneur en antitoxine beaucoup plus élevée (cf. tableau 8 et p. 324).

Chez le sujet activement immunisé, la protection totale est fonction de la quantité d'antitoxine sérique libre et — surtout — de l'immunité potentielle, c'est-à-dire de la faculté dont jouit l'organisme vacciné de réagir à toute sollicitation antigénique par la production de nouveaux anticorps. Il semble possible, ce qui d'ailleurs se vérifie dans le cas de la diphtérie (Regamey 1943 c et al.), qu'une immunité potentielle soit si puissante qu'elle assure une protection totale en l'absence apparente d'anticorps circulants. Un extrait des protocoles d'Aegerter (1954) est démonstratif à cet égard (tableau 8). Parmi les quelques 100 cobayes qui ne présentent que peu ou pas d'antitoxine libre après la préparation au moyen d'une faible dose d'anatoxine, l'insertion de fils de soie tétanifères ne tue que les trois quarts des animaux (le 100% des témoins non vaccinés succombant dans les 48 heures); le quart des cobayes survivent avec des symptômes typiques de tétanos. Où placer le seuil de protection? Dès 0,0075 U.I./cc, les animaux ne meurent plus; dès 0,175 U.I./cc, l'infection reste cliniquement muette.

Ces valeurs sont très voisines de celles déjà rapportées par SNEATH et coll. (1937), mais notablement plus favorables que celles de COWLES (1937) (cf. aussi JONES et JAMIESON 1936b; ZUGER et FRIEDEMANN (1938); FRIEDEMANN et coll. 1939 et al.). Mais que penser de ce cobaye qui meurt de tétanos avec 3,0 U.I. d'antitoxine dans son sérum ? Quand l'animal vacciné reçoit en un temps 10000 D.M.M.$_4$ de toxine, une approximation d'un seuil de protection devient impossible.

Tableau 8. *Relation entre le taux de l'antitoxine sérique et l'immunité*
(Selon les observations d'AEGERTER 1954.)

Cobayes infectés avec des fils de soie tétanifères				Titre de l'antitoxine tétanique dans le sérum avant l'épreuve (en U.I./cc)	Intoxination avec 10000 D.M.M.$_4$			
Total	†$_{Te}$	Te	O		Total	†$_{Te}$	Te	O
103	74	29	—	$\leqq$ 0,0025	8	8	—	—
11	5	6	—	0,00375—0,005	1	1	—	—
6	—	3	3	0,0075—0,01	—	—	—	—
8	—	4	4	0,0175—0,025	3	1	—	2
12	—	7	5	0,0375—0,05	2	—	1	1
9	—	1	8	0,075—0,10	4	2	—	2
8	—	—	8	0,175—0,25	6	2	1	3
7	—	—	7	0,375—0,50	5	—	2	3
15	—	—	15	0,75—1,0	14	1	4	9
8	—	1[(A)]	7	1,5—3,5	18	1[(B)]	—	17

†$_{Te}$ = mort de tétanos; Te = symptômes tétaniques, survie; 0 = aucun symptôme tétanique; (A) (B) = 3,0 U.I./cc do sérum.

DESCOMBEY (1925b), puis RAMON et DESCOMBEY (1927g) estimaient qu'un cheval dont le sérum titrait environ 0,001 U.I./cc était capable de résister à l'infection tétanique. PETERSON et CHRISTIE (1951) suggèrent que la présence de 0,001 à 0,003 U.I./cc, accompagnée d'une bonne capacité anamnestique, doit être suffisante pour prévenir la majorité des cas de tétanos survenant « naturellement » chez l'homme. L'auto-expérimentation de WOLTERS et DEHMEL (1942) étayerait cette opinion (cf. p. 325 et diagramme 9).

SNEATH et coll. (1937), BERGEY et coll. (1939), PESHKIN (1941a), FISEK (1957) et al. admettent que 0,01 U.I./cc représente le seuil acceptable, tandis que pour VOLK (1949) 0,02U.I./cc ne protège pas l'enfant. BIGLER (1951) opine pour 0,01—0,1, GOLD (1937a, b), puis PETERSON et coll. (1955) pour 0,1 U.I./cc. BERGEY et coll. (1939) précisent qu'avec un titre d'antitoxine sérique équivalent, le vacciné supporte $2^1/_2$ fois plus de toxine que le sujet immunisé passivement. Cette assertion n'est pas encore confirmée.

La nature même de la réaction immunitaire, si différente selon l'individu, rend arbitraire et aléatoire toute tentative de fixer un seuil minimum de protection dans le cas de la vaccination active contre le tétanos. Le taux de 0,001 à 0,01 U.I./cc qu'admet GREENBERG (1957) est encore trop précis et trop restrictif, malgré son élasticité, car — TENBROEK et BAUER (1926) l'ont bien formulé — la richesse du sang en antitoxine n'est pas en relation directe avec l'immunité antitétanique réelle.

Ainsi dans le cas du tétanos, à moins peut-être de rencontrer des quantités très élevées d'antitoxine chez le vacciné, on ne dispose d'aucun test sûr, sérologique, cutané ou autre, qui permette d'évaluer la réceptivité réelle du sujet.

45. Accidents post-vaccinaux et contre-indications

Les antigènes poussés chez l'homme peuvent faire apparaître chez le vacciné des manifestations d'ordre toxique ou allergique. Les phénomènes *toxiques* surviennent dès la première injection. Quand le sujet possède une hypersensibilité d'origine inconnue, les réactions *allergiques* se produisent aussi après la première application d'antigène; plus souvent elles apparaissent lors d'une réinjection. La coexistence de complications toxiques et allergiques n'est pas rare. Comparés à l'anatoxine diphtérique (Bächer 1927), les vaccins tétaniques ont frappé très tôt par le fait qu'ils étaient bien tolérés: le bacille tétanique étant un hôte rare chez l'homme, les risques d'une sensibilisation occulte sont réduits; sa protéine protoplasmatique est en outre peu allergisante (Gold 1941).

Lors de l'emploi massif de l'A.Te, alunée ou non, au début de la guerre mondiale II, on rencontre quelques réactions nettement anormales, plus sévères que celles publiées par Feierabend (1930), Boyd (1938) et al. Chez un allergique notoire, Cunningham (1940) observe un collapse lors de la seconde injection, malgré l'administration simultanée d'adrénaline. Whittingham (1940) rencontre sur 61042 sujets deux fois vaccinés 651 (1,06%) réactions locales, 12 (0,023%) réactions générales et 2 (0,0033%) graves crises anaphylactiques; tous les incidents surviennent après la seconde injection. Pike (1940) publie un cas de prurit généralisé avec oedème de Quincke, Cooke et coll. (1940) celui d'une urticaire aiguë chez un asthmatique et deux urticaires à retardement, Ralston (1940) l'auto-observation de troubles intestinaux et de spasmes musculaires; Parish et Oakley (1940) décrivent le cas d'une femme avec convulsions et rush scarlatiniforme. Tous ces auteurs avec Auld (1940) et Sachs (1940) imputent les complications à la teneur de l'anatoxine en peptone. En effet, certaines peptones du commerce, préparées à partir de fibrine de bœuf ou de porc, et contenues dans les bouillons de culture (Campbell 1938), doivent être tenues pour responsables de la sensibilisation de l'homme, ce que démontre Gold (1941) en recourant à des séries de tests intracutanés. Brown (1940) propose de n'employer que de l'anatoxine alunée, dont la préparation par lavages répétés du précipité doit assurer dans le produit terminé l'absence absolue de peptone.

Plusieurs analyses systématiques ont par la suite établi la rareté des phénomènes d'escorte imputables à l'anatoxine tétanique, celles par exemple de Marvell et Parish (1940) en Angleterre, de Regamey (1943d, 1944b) en Suisse, de Long et Sartwell (1947) pour les U.S.A. Ces deux derniers auteurs n'ont signalé sur 100000 injections que 60 manifestations allergiques, la plupart bénignes. Plus récemment, Edsall (1956b) mentionne que de 25000 soldats américains vaccinés avec l'anatoxine alunée aucun ne s'est présenté à l'infirmerie pour un motif relevant de la vaccination. Beck (1954) rapporte le cas d'un granulome. Pour diminuer les réactions locales, Schober (1956) mélange l'A.Te à la Novocaïne à 1% āā. Les conclusions de la Société de Pathologie comparée (1957), précisant que l'injection d'A.Te est d'une innocuité absolue et qu'elle n'est suivie d'aucun accident allergique, sont toutefois trop optimistes.

Expérimentalement, l'anatoxine tétanique allergise (Gold 1941, Borissova 1956 et al.). Quoique dépourvue de toxicité primaire, quoique préparée à partir de milieux semi-synthétiques ou sans peptone, quoique hautement purifiée, l'anatoxine tétanique — au même titre qu'un autre vaccin — peut déclencher des manifestations locales, générales et focales. Ces réactions sont précoces, fugaces, rarement graves; elles peuvent aussi survenir tardivement. Voici résumées quelques complications récemment décrites, dont deux mortelles.

Le cas de HOLLÄNDER et WORTMANN (1952) intéresse un homme de 35 ans vacciné sans incident 12 ans plus tôt. L'injection de rappel est suivie immédiatement d'une tuméfaction au lieu de l'injection, avec retentissement ganglionnaire. Au 11e jour, maladie « sérique » typique avec urticaire et arthralgies; rechute 15 jours plus tard. Le test intracutané est fortement positif pour la peptone, négatif pour l'extrait de veau, de porc ou de cheval. En même temps que l'anatoxine, le patient avait reçu de la pénicilline.

Le second cas (REGAMEY 1957, non publié) est celui d'un homme de 44 ans, vacciné contre le tétanos quelques années auparavant. Le blessé, un allergique notoire, reçoit en même temps de la pénicilline et une injection de rappel d'anatoxine tétanique. Collapse et mort dans la demi-heure qui suit. Comme pour l'observation précédente, l'éventualité d'une complication pénicillinique n'est pas exclue.

Le troisième cas (REGAMEY 1954, non publié) est pur. Il s'agit d'un jeune homme de 20 ans, sans passé allergique. 1re injection d'anatoxine brute supportée sans incident. La 2e injection un mois plus tard est suivie d'un malaise vers la huitième heure, avec collapse, dyspnée, crampes; 7,5% d'éosinophiles. La 3e injection, six mois et demi plus tard, se solde dans les deux heures par un collapse avec dyspnée et exitus. A l'autopsie: status après choc anaphylactique aigu, contraction extrême des bronches.

Une complication exceptionnelle est celle que rapporte PIERRET (1954): après l'administration simultanée de sérum antitétanique et d'anatoxine, une fillette de 9 ans accuse presque aussitôt un violent trismus, qui cède aux antispasmolytiques après plusieurs heures seulement.

L'influence de la vaccination antitétanique intéresse aussi l'hématologiste. En effet, de nombreuses peptones, en particulier celles qui sont issues d'estomac de porcs, puis les milieux de culture, les toxines et anatoxines préparées à partir de ces peptones, contiennent des quantités plus ou moins élevées de substance A (OTTENSOOSER 1932). L'injection chez l'homme d'un vaccin contenant la substance A aboutit parfois à une augmentation considérable du titre sérique anti-A, ce qui crée certains dangers lors de l'emploi de donneurs universels (SACHS 1940; HENDRY et SICKLES 1951b; DAUSSET et VIDAL 1951; HOLLÄNDER et WORTMANN 1952; HÄSSIG et coll. 1955 et al.).

Les complications après l'usage de *vaccins associés* et contenant une fraction tétanique sont plus fréquentes que pour l'anatoxine tétanique seule. La part qui revient à cette dernière fraction ne peut pas être évaluée; ce qui précède fait présumer qu'elle doit être faible. Les relations de cause à effet entre les manifestations cliniques d'une part, et l'injection de vaccin d'autre part, sont parfois difficiles à préciser en particulier lorsque la symptomatologie est tardive; elles posent aux tribunaux, aux sociétés d'assurances et aux experts médicaux des problèmes sans cesse renouvelés et dont la solution serait grandement facilitée si toutes les complications anormales étaient publiées.

Voici la liste de quelques travaux concernant les complications survenues après usage de vaccins associés et contenant la fraction tétanique: BELLER (1943), SOHIER et coll. (1944b), MRAVUNAC (1956), MERKLEN et coll. (1956). Des cas mortels ont été décrits par REGAMEY (1947c): il s'agit de trois exitus par choc anaphylactique aigu après la 1re ou la 2e injection de vaccin Te TAB, par EBLEN (1957): fillette de 3½ mois et succombant 18 heures après l'injection de vaccin Di Te Per. La comparaison des cas mortels de REGAMEY après vaccins Te TAB ou Te avec les 2 cas également mortels rapportés par HARVIER (1940) et le cas de SOHIER et ALAIZE (1943), consécutifs à la vaccination TAB, ne permettent pas de tirer des conclusions sur une spécificité des manifestations allergiques propres à l'une ou l'autre fraction du vaccin.

Les accidents survenant après la vaccination antitétanique sont si rares que les *contre-indications* générales aux méthodes d'immunisation active peuvent être allégées et que l'on peut immuniser les allergiques (PARISH et MARVELL 1940). On ne vaccinera pas le petit enfant souffrant de troubles respiratoires, d'infections ou de dentition difficile (MACGUINNESS 1952), les sujets souffrant de troubles rénaux, qu'il faut rechercher au préalable (HUBER et BESSON 1950; DEPARIS 1956). Même les tuberculeux supportent l'A.Te (CHEVALLEY et ZIVY 1939); les quelques activations d'infections latentes (BASSET 1951) peuvent être évitées si, dans tous les cas où l'on suspecte une allergie ou un foyer infectieux latent, on applique le vaccin refracta dosi (cf. GOLD 1937a, COOKE et coll. 1940, EDSALL 1957, SCHLAFER 1957).

Que vaut le test intracutané de ZOELLER-MOLONEY ?

L'injection intracutanée fournit deux types de réactions: l'une, *précoce*, se développe dans les 15 minutes sous la forme d'un érythème de type urticarien; l'autre, *tardive*, avec un acmé après 24 heures, s'accompagne d'un érythème papuleux, avec induration et tuméfaction locale (GOLD 1941). Même si l'antigène est fortement dilué, 1:10, 1:50 ou plus, les réactions cutanées sont extrêmement fréquentes. Or il n'y a pas de parallélisme entre la positivité du test et l'apparition de réactions post-vaccinales: un sujet à test intracutané positif ne présente pas nécessairement une réaction à l'injection sous-cutanée d'anatoxine tétanique, la réciproque étant également vraie. Le test de ZOELLER-MOLONEY doit être réservé à quelques cas particuliers. Lorsqu'il s'agit d'injection de rappel, une attente de 24—48 heures n'est pas tolérable. Par ailleurs le test de scarification proposé par COOKE et coll. (1940) n'a pas été adopté dans la pratique.

Si l'innocuité de l'A.Te et la rareté des phénomènes d'escorte réduisent le nombre des contre-indications, elles ne doivent pas faire oublier les recommandations de CUNNINGHAM (1940), COOKE et coll. (1940), GOLD (1941) et al.: lors d'une injection d'A.Te, tourniquet et adrénaline à 1:1000 doivent être à portée de la main; le patient reste pendant une demi-heure sous contrôle médical.

46. Echecs de l'immunisation active.
Facteurs amortissant la production de l'antitoxine

La guerre mondiale II constitue une expérience massive pendant laquelle l'immunisation active contre le tétanos fit ses preuves. Elle fournit un matériel statistique précieux, mais qui touche pratiquement une seule catégorie d'individus, celle d'hommes adultes. Les travaux sont rares, qui rapportent des incidents inattendus chez la femme, le vieillard ou l'enfant. Les sujets du sexe masculin — du moins les jeunes — s'immunisant plus difficilement que ceux du sexe féminin (MARVELL et PARISH 1940; PARISH 1941), les résultats obtenus chez eux pourront donc être hyperbolés.

L'immunité antitétanique est antitoxique. Elle n'empêche pas *Pl. tetani* d'être présent et de se multiplier dans l'organisme du vacciné. A cet égard, l'observation de JOST (1954) est significative:

Un accidenté par explosion de mine, vacciné cinq ans plus tôt par trois injections d'anatoxine tétanique, reçoit dès son entrée à l'hôpital 1500 U.I. de sérum antitétanique et une injection de rappel avec 2 cc d'anatoxine tétanique. Pendant les trois semaines qui suivent, et bien que le malade ne présente aucun symptôme de tétanos, on identifie dans les sécrétions prélevées sur les plaies un Pl. tetani qui déclenche régulièrement un tétanos mortel chez la souris.

Les auteurs qui commentent les résultats si favorables de la prévention du tétanos par les vaccins tétaniques — au cours de la guerre mondiale II — s'inspirent surtout des travaux déjà cités de Boyd (1946) pour la Grande-Bretagne, de Metcalfe (1955) pour l'Australie, de Wishart (1947) pour le Canada, de Hall (1940, 1948), Long (1944, 1948), Long et Sartwell (1947) pour les Etats-Unis (cf. également Ramon 1951). La U.S. Navy eut 90000 blessés et 4 cas de tétanos survenus chez des blessés en dehors de combats; 4 tétanos compliquèrent des accidents bénins. La U.S. Army enregistra 2700000 blessés et 12 cas de tétanos. L'Armée anglaise eut au minimum 103 cas de tétanos (Barr et Sachs 1955), dont 35 à la suite de blessures de combat, 18 après d'autres accidents et 50 parmi les troupes de partisans ou les prisonniers. Les Australiens rapportent 13 cas, la plupart légers, et les Canadiens 3 cas, dont 1 mortel (Hansen 1958). Les résultats connus pour la Grande-Bretagne et les Etat-Unis sont résumés dans le tableau 9.

Le tableau 9 appelle quelques commentaires. L'Armée anglaise était certainement moins bien immunisée que les troupes américaines (cf. p. 343). L'introduction d'une troisième inoculation de base fit tomber la mortalité chez les blessés de 0,43 à 0,01⁰/₀₀ (Sachs 1952). Ce qui frappe dans le tableau 9, c'est la haute létalité des sujets correctement vaccinés mais sans

Tableau 9. *Répartition des cas de tétanos dans l'Armée anglaise (G.B.) et la U.S. Army (U.S.A.) pendant la guerre mondiale 1939—1945* (Selon Sachs 1952.)

	Nombre de cas	Décès
Immunisation de base complète		
Après la blessure:		
Injection d'anatoxine		
G.B.	—	—
U.S.A.	4	2
Injection de sérum		
G.B.	11	2
U.S.A.	—	—
Pas de traitement spécifique		
G.B.	11	9
U.S.A.	2	1
Immunisation de base incomplète, douteuse ou nulle		
Après la blessure:		
Injection de sérum		
G.B.	26	12
U.S.A.	2	1
Pas de traitement spécifique		
G.B.	54	24
U.S.A.	4	1

— signifie que le procédé n'a pas été appliqué.

le bénéfice de l'injection de rappel après la blessure, létalité qui est même plus élevée que pour les sujets mal ou pas du tout vaccinés. Les Russes ont eu également quelques cas de tétanos chez des vaccinés (Čertkova et Šain 1957).

En dehors des expériences de la guerre, la majorité des échecs de la prévention du tétanos par l'anatoxine intéresse aussi des blessés chez lesquels l'*injection de rappel fit défaut* (Simon et Patey 1940; Boyd et MacLennan 1942; Norman 1943; Lewin 1944; Hedrick 1953; Eckmann 1955; Christensen et Thurber 1957; J.A.M.A. 1957b). Le tableau 10, qui récapitule les observations de Boyer et coll. (1953), est particulièrement démonstratif à cet égard.

Un tétanos peut aussi se développer chez des blessés immunisés et soumis à une injection de rappel posttraumatique (tableau 9); il s'agit toutefois d'un phénomène rare, qui n'apparaît que chez le 0,0015⁰/₀₀ des blessés vaccinés et correctement traités.

Différents facteurs ont été évoqués pour expliquer les déficiences de l'immunité antitétanique (Edsall 1957).

Observation valable d'ailleurs pour chaque type de vaccination, il existe de mauvais « réacteurs » qui, sans cause apparente, n'élaborent pas d'antitoxine sérique (SNEATH 1934; CARLOS et NETO 1948 et al.). Il reste évident toutefois que l'absence d'antitoxine titrable dans le sérum ne doit pas faire conclure à la nullité de la réaction immunitaire, car il est souvent difficile de mettre en évidence de petites quantités d'antitoxine et les méthodes usuelles ne permettent pas de mesurer l'immunité cellulaire ou potentielle.

La production d'antitoxine tétanique peut faire totalement ou partiellement défaut dans les syndromes d'hypo- ou d'agammaglobulinémie (BARRETT et VOLWILER 1957), lorsqu'il y a une tare dans la maturation immunologique (EGDAHL 1958), et tout particulièrement dans les « syndromes par manque d'anticorps » (BARANDUN et coll. 1958, ISLIKER 1957, RIVA et coll. 1958 et al.). La propriété que possède l'organisme de former des gammaglobulines n'assure cependant pas à coup sûr la synthèse d'anticorps: REGAMEY (non publié) a examiné l'évolution de l'antitoxine tétanique chez des malades souffrant d'un syndrome cirrhotique, néphrotique ou hépato-néphrotique; il a constaté que malgré la présence de gammaglobulines en quantité parfois exagérée, les sérums pouvaient être anormalement pauvres en anticorps spécifiques [HAVENS et coll. (1957) ont signalé le phénomène opposé: les cirrhotiques hépatiques réagiraient à l'I. R. par le formation exagérée d'antitoxine]. D'autres déficiences organiques, héréditaires ou acquises, s'accompagnent de tares immunitaires. SIEGEL (1948) signale que chez vingt mongoloïdes traités par un vaccin Te Ty, la réponse immunitaire était plus lente et moins prononcée que chez les témoins. MARVELL et PARISH (1940) ont noté que le cobaye malade s'immunisait

Tableau 10. *Cas de tétanos survenus chez des sujets vaccinés* (Selon BOYER et coll. 1953.)

Sujets (âge)	Vaccin	Nombre d'injections	L'immunisation de base remonte à	Injection de rappel
♂ 11	Di Te TAB	2	5 ans	0
♂ 46	Di Te TAB	2	8 ans	0
♂ 42	Di Te TAB	2	4 ans	0
♀ 35	Te	3	10 ans	0
♂ 50	Di Te TAB	3	9 ans	0
♂ ?	Di Te TAB	1	11 ans	0
♂ ?	Di Te TAB	3	4 ans	0
♂ 58	?	?	3 ans	0
♂ 43	Di Te TAB	2	13 ans	?
♂ 10	Di Te	3	8 ans	0

moins bien (0,020 U.I./cc) que le cobaye sain (0,142 U.I./cc). Un organisme épuisé par des vaccinations antérieures (LEMÉTAYER et coll. 1946), tel le cheval hyperimmunisé, réagit plus faiblement à l'anatoxine tétanique que l'organisme neuf. Le sommeil médicamenteux et l'hypothermie exercent une action dépressive sur la formation de l'antitoxine chez le lapin (ZDRODOWSKI 1957).

Les radiations gamma du cobalt-60 (HALE et STONER 1956) jouent sur la formation des anticorps un effet dépresseur qui est fonction de l'intensité du rayonnement, puis de l'intervalle entre le traitement actinique et l'injection d'A.Te. SILVERMAN et CHIN (1955, 1956) ont fait une observation analogue après avoir irradié la souris avec les rayons X. L'influence était moins marquée chez le cobaye (HENDRY 1952b). L'augmentation ou la diminution de l'activité mitotique offre un certain parallélisme avec la production des anticorps; les substances qui ralentissent la formation des mitoses, adrénaline, colchicine, sanomycine, thyroxine, ACTH, cortisone freinent la réponse immunitaire (HADNAGY et KREPSZ 1957).

L'influence de l'ACTH et de la cortisone sur le développement des processus immunitaires a fait déjà l'objet de nombreuses recherches. Tandis que, à propos du tétanos, les auteurs se sont plutôt orientés vers l'étude de l'action de ces substances sur la réceptivité de l'organisme pour la toxine tétanique, BEUREY et coll. 1953 établissent que, chez le lapin, l'ACTH et la cortisone ralentissent la production et diminuent la quantité de l'antitoxine. Il y a lieu de rapprocher ici l'influence dépressive, directe ou indirecte, de la surrénale sur la production de l'antitoxine tétanique après une saignée exagérée (GUIDOLIN et CORRÉA 1950; NICOL 1955 et al.), puis chez les grands stressés, constatation d'importance à cause des indications de la prophylaxie spécifique posttraumatique chez les vaccinés. L'homme serait néanmoins plus résistant à la cortisone que le rat, la souris, le lapin ou le furet (SHEWELL et LONG 1956; LONG 1957), et l'influence de l'ACTH pourrait être neutralisée par l'hormone de croissance (HAYASHIDA et LI 1957) (cf. p. 326, la répercussion de l'état général sur l'indication de l'I.R.).

Phase négative, paralysie immunologique ou blocage immunitaire ont été étudiés à propos des antigènes les plus divers. En 1929 b déjà, RAMON relate que si l'injection au cheval de grosses doses de toxine ou d'anatoxine diphtériques (400—500 cc) est suivie d'un abaissement passager du titre antitoxique, il n'y a pas de phase négative à proprement parler. Les anticorps se concentrent momentanément soit dans l'oedème local après l'injection sous-cutanée, soit dans le foie fortement congestionné après l'injection intraveineuse; le sang circulant est dilué par un appel d'eau en provenance de l'intestin. Les études minutieuses de MADSEN, JENSEN et IPSEN (1937) sur l'homme et le lapin établissent que, dans l'organisme immunisé, chaque injection d'antigène s'accompagne d'une chute immédiate, mais faible et transitoire de l'antitoxine libre; cette chute

passagère est sans signification pratique pour la protection réelle du sujet. LEMÉTAYER et coll. (1955) viennent de décrire une « phase négative » dans un groupe de 36 chevaux: l'injection du cheval immunisé avec 10 cc d'anatoxine tétanique s'accompagne, pendant 4—6 jours, d'une diminution énorme de l'antitoxine chez la moitié des chevaux et de la disparition totale des anticorps chez le tiers des animaux. RAMON (1957 a, p. 130) n'a jamais observé un phénomène analogue ni chez l'homme, ni chez le cheval.

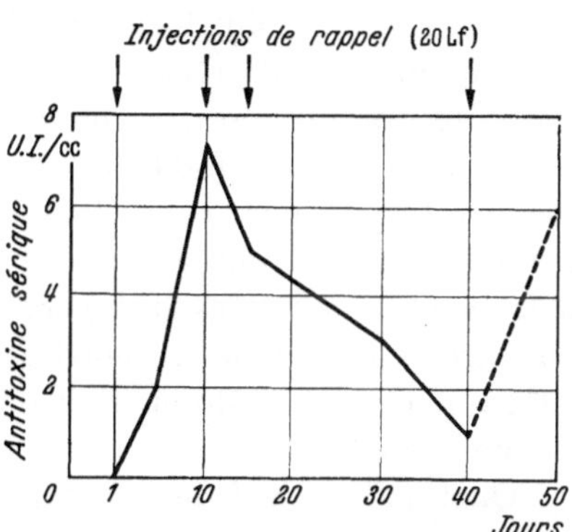

Diagramme 8. *Courbe de l'antitoxine tétanique avec phénomène d'inhibition immunitaire après injections trop rapprochées de l'antigène* (Selon HALAPINE, cité par ZDRODOWSKI 1957)

En 1956, voulant hyperimmuniser des volontaires dans le but d'obtenir une gamma-globuline antitétanique pour la prophylaxie passive des hyperallergiques au sérum hété-rologue, REGAMEY, MARTIN DU PAN et SCHLEGEL (non publié) ont constaté que les titres déjà satisfaisants après la première ou la seconde injection de rappel accusaient une chute massive lorsqu'ils continuaient les injections. Des faits semblables ont été signalés par quelques auteurs. SÉDALLIAN et CLAVEL (1944/45) décrivent chez le lapin en voie d'immunisation des périodes de hausse et de baisse; injectés en période de baisse, les animaux ne réagissent pas manifeste-ment à l'antigène. HOLT (1950 a, p. 80) mentionne aussi un état réfractaire chez le cobaye soumis à un stimulus trop tôt après une forte réponse immunitaire ou lorsque les excitations antigéniques sont nombreuses et rapprochées. L'existence d'un *phénomène d'inhibition immunologique*, en particulier lors de la vaccination contre le tétanos, fut également étudiée par l'école russe (ZDRODOWSKI 1957). Le diagramme 8 démontre la dépression consécutive à l'injection trop rapprochée d'une nouvelle dose d'antigène. L'organisme ne répond pas à chaque injection de rappel par une nouvelle décharge d'anticorps: il a besoin d'un certain temps pour épuiser une sollicitation antigénique. CINADER et DUBERT (1956) supposent que le blocage de la réponse antigénique résulte d'une accumulation d'anticorps à l'intérieur des cellules: la synthèse de nouveaux anticorps se trouve empêchée, car l'antigène nouveau venu est aussitôt neutralisé. Il reste à expliquer pour-quoi le stimulus prématuré fait baisser le titre antitoxique avec une telle intensité.

21*

On a maintes fois relevé ici que des animaux possédant des quantités relativement élevées, 0,5 U.I. par exemple, d'antitoxine tétanique dans leur sérum, succombaient au tétanos lorsqu'ils étaient éprouvés avec de la toxine ou des spores tétaniques. Marvell et Parish (1940) rappellent l'observation de Glenny: un cheval hyperimmunisé meurt de tétanos quand on lui injecte de grosses doses de toxine; une quantité suffisante de toxine se fixe avant d'avoir été neutralisée. C'est en partie à cause de cette éventualité que l'Armée anglaise traita le blessé avec du sérum et non avec de l'anatoxine pendant la guerre de 1939 à 1945. Cette apparition et cette évolution d'un tétanos sous le couvert de l'immunité passive sont d'intérêt et soulèvent la question d'une distinction entre la notion de *l'immunité* et celle de la *protection:*

L'organisme passivement immunisé se défend avec la seule antitoxine circulante. Celui qui est vacciné activement dispose d'une immunité plus complexe, *humorale* par la présence d'anticorps déversés en permanence dans le sang, *potentielle* — ou cellulaire — par la propriété qu'ont acquise les organes producteurs d'anticorps de réagir promptement à tout nouveau stimulus antigénique. La protection que confère une immunité active — qui semble solide — présente pourtant ci et là des faiblesses qui, l'unicité de la toxine tétanique étant admise, font envisager diverses possibilités: 1. l'antitoxine sérique ne possède pas une avidité suffisante pour la toxine et ne l'empêche pas de gagner les centres nerveux; 2. les conditions locales du foyer infectieux permettent l'élaboration d'une toxine qui atteint les cellules réceptrices sans avoir rencontré l'antitoxine (d'Antona 1951a, p. 150); 3. l'intoxination tétanique se produit ci et là selon des processus encore inconnus.

Quoi qu'il en soit de la nature des échecs de la vaccination antitétanique, ces échecs sont si rares en proportion du nombre des vaccinés qu'ils constituent de véritables curiosités immunologiques, et peuvent être pratiquement ignorés (Regamey 1955a).

5. L'injection de rappel

L'I.R., déjà introduite au chapitre précédent, a pour but de réactiver l'immunité dans un organisme préalablement vacciné. Elle représente par définition et par essence un stimulus de type secondaire, n'ayant de valeur que chez le sujet jouissant d'une *immunité potentielle* (Magrassi 1934). La réactivation est *spécifique* (Ramon et Lemétayer 1933a; van der Scheer et Wyckoff 1940; Freund 1953; Dixon et Maurer 1955). Les quelques mouvements du titre antitoxique qui font suite à l'injection de substances étrangères à l'antigène tétanique (Billard 1926, Ruggerini 1929a, 1930 et al.) sont apparemment dûs au déversement dans la circulation d'anticorps préformés et lâchés par les cellules que sollicite un nouvel antigène. Barr et Glenny (1952) supposent que, dans un organisme préalablement immunisé avec un antigène X, l'injection d'un antigène Y excite la formation générale des globulines et par-là même l'élaboration temporaire de globulines anti-X.

Quelques essais de réactivation réflexe ont été faits en Russie (Zdrodowski 1957), mais n'ont pas donné de résultats positifs.

51. Indications de l'injection de rappel

L'I.R. poursuit un double but: celui de réactiver l'immunité chez le blessé et celui d'entretenir l'état réfractaire en dehors de toute menace d'infection.

Injection de rappel posttraumatique. Il semble logique de compter sur l'immunité dont jouit le vacciné pour pallier une intoxination posttraumatique. Le plus souvent, en effet, la quantité d'antitoxine circulant dans les humeurs est telle, qu'elle égale ou dépasse le taux qui fait suite à l'injection préventive de sérum antitétanique. Mais le médecin n'a aucune possibilité d'évaluer rapidement la protection spécifique d'un blessé, même correctement vacciné. Une réactivation d'immunité s'impose, car la plupart des échecs de la vaccination contre le tétanos s'observent chez des traumatisés qui n'ont pas reçu d'I.R. (cf. p. 321 et tableau 10).

La toxine qui prend naissance dans la plaie est vraisemblablement capable de mettre en mouvement le mécanisme de l'immunité potentielle, mais l'effet sera tardif: il y aura d'abord toxinogenèse, puis sollicitation des organes producteurs d'anticorps et enfin réponse immunitaire. Le décalage entre le moment où la toxine se forme et celui de la mise à disposition d'anticorps frais ou suffisants constitue une phase dangereuse.

Chez l'animal, Jones et Jamieson (1936 b), puis Jaulmes et Jude (1939) n'ont pas noté de réponse sérique lorsqu'ils inséraient des spores chez des cobayes immunisés; ils supposaient que le stimulus antigénique faisait défaut, la toxine devant être immédiatement neutralisée par l'antitoxine présente. Wolters et Dehmel (1938 a, 1940), Zuger et coll. (1940, 1942), Katić (1956) ont au contraire observé que le cobaye immunisé réagissait à l'infection par une production abondante d'anticorps.

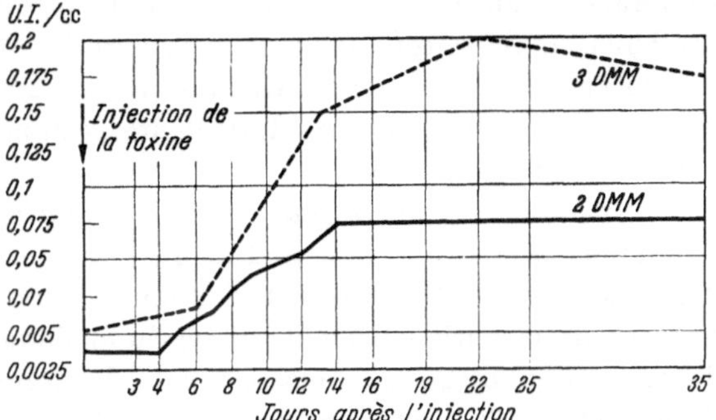

Diagramme 9. *Réponse sérologique chez l'homme activement immunisé et soumis à l'injection de 2 et 3 D.M.M. de toxine tétanique* (Selon Wolters et Dehmel 1942)

Moreno de Vega (1957) voit chez ses cobayes vaccinés 10 mois plus tôt par deux injections d'A.Te fluide les titres passer de 0,01 à 42,5 U.I./cc. Voici encore une observation intéressante: les cobayes d'Aegerter (1954), préparés avec 0,25 Lf d'A.Te fluide et chez lesquels on insère 35 jours plus tard des fils de soie tétanifères, répondent au bout de 10 jours par des titres qui ont passé de < 0,0025—0,5 U.I./cc avant l'infection à 0,175—6,0 U.I./cc; le 30% des animaux a présenté des signes légers, le 40% des symptômes graves de tétanos, mais tous ont survécu.

Une information existe à ce propos pour l'homme. Wolters et Dehmel (1942), activement immunisés, se sont injectés l'un 2 D.M.M. et l'autre 3 D.M.M. de toxine tétanique. Le bond antitoxique débuta entre les 4e et 6e jours. Les titres de l'antitoxine tétanique passèrent de 0,0035 à 0,075 U.I./cc (avec 2 D.M.M.) et de 0,005 à 0,2 U.I./cc (avec 3 D.M M.). L'exemple personnel de deux auteurs allemands démontre bien, d'une part, l'effet stimulant de doses extrêmement faibles d'antigène et, d'autre part, l'absence de relation entre la précocité du bond et l'intensité du stimulus (diagramme 9).

D'une façon générale, on recommande l'I.R. dans tous les cas de traumatisme chez les vaccinés, après des lésions par brûlures ou engelures, avant l'accouchement, lors d'une intervention sur un foyer cicatrisé etc. Pourtant l'absolutisme ou le schéma d'indications trop simple (Kleinschmidt 1953, par exemple) ne sauraient être de rigueur. La fréquence si grande du tétanos après des blessures insignifiantes est l'apanage des non-vaccinés; le praticien qui traite un sujet immunisé doit prendre ses responsabilités et renoncer parfois à l'I.R., en particulier lorsque le rappel précédent remonte à quelques mois seulement ou lorsque les conditions de la blessure ne laissent aucun doute sur l'exclusion d'une infection tétanique. Peterson, Christie et Williams (1955) estiment même qu'en cas de traumatisme, il est inutile de faire une I.R. dans les trois ans qui suivent la précédente réactivation de l'immunité.

La question de l'I.R. chez les grands shockés, lors de gros délabrements ou de fortes hémorragies, en cas de débilité physique, n'est pas entièrement élucidée. Marvell et Parish (1940), constatant que le cobaye en mauvaise santé s'immunise mal, en déduisent que l'homme en état de shock sévère pourrait ne pas fournir une réponse adéquate à l'I.R. Par contre Edsall (1954a) émet des doutes sur l'indication invoquée des affections débilitantes. Aussi de nombreux auteurs recourent-ils à la combinaison *anatoxine tétanique* + *sérum antitétanique* que l'on applique à raison de 1500, ou mieux 5000-10000 U.I. (J.A.M.A. 1952, Peterson et coll. 1955, Looney et coll. 1956, Christensen et Thurber 1957 et al.). Miller et coll. (1949, 1950), qui ont fait des études systématiques sur l'application simultanée d'antitoxine et d'anatoxine tétaniques comme I.R., proposent cette combinaison dans tous les cas où la dernière injection de vaccin remonte à 4 ans ou plus. En accord avec Edsall (1954), Eckmann (1958), dont l'expérience porte sur plusieurs centaines d'accidentés, estime par contre que le blessé vacciné n'a nul besoin de sérum.

L'I.R. doit être faite le plus vite possible après le traumatisme afin de réactiver l'immunité pendant la période d'incubation. Si les circonstances ont fait reculer l'I.R. de plus de 24 heures et si l'immunisation de base est équivoque, il est prudent de recourir églement au sérum antitétanique (J.A.M.A. 1954a, 1957b), de même que lorsque le vacciné a subi des lésions actiniques (Eckmann 1958). J. Mérieux (1957) propose de faire systématiquement l'I. R. d' A.Te associée à la première dose d'antibiotique que reçoit le blessé.

Les échecs sont ceux de la vaccination antitétanique (cf. p. 320). L'anamnèse doit être considérée avec un esprit critique (Baumann 1951 et al.). Schürmann (1954) relate un cas de tétanos mortel survenu chez un homme vacciné trois fois au service militaire et qui reçut une I.R. antitétanique; l'enquête post mortem établit qu'il s'agissait d'une vaccination de base contre le typhus exanthématique. L'établissement de carnets de vaccination, réclamé par Gardner et coll. (1958), s'est déjà développé dans plusieurs pays.

Injection de rappel d'entretien. Les informations sur l'état de l'immunité chez le vacciné se complètent chaque année. En 1956, Schlegel reconnaissait que la validité de l'immunisation de base atteignait 16 ans. En 1958, Eckmann l'estime à 20 ans au moins. Si les titres sériques sont faibles, l'organisme a conservé intacte sa faculté de réagir promptement et abondamment à l'I.R.

Néanmoins le clinicien reste prudent et propose des I.R. faites

tous les 3—4 ans: Yeazell et Deamer (1943), Peshkin (1944, 1945), Ménard (1955), surtout chez les enfants, puis Looney et coll. (1956) chez l'adulte,

tous les 5 ans: Barr et Sachs (1955), Turner et coll. (1954),

tous les 5 ans, éventuellement tous les 10 ans: Peterson et coll. (1955), Christensen (1957), Eckmann (1958),

tous les 10 ans: Bigler (1951), Regamey (1955a).

Chez les enfants, Regamey (1955a) préconise une première I.R., un an après l'immunisation de base, puis d'autres I.R. tous les 3—5 ans. Scheibel (1957a) estime que pour être vraiment complète, la protection doit être assurée, chez tous les sujets, par une I.R. une année après l'immunisation de base. Dans certaines circonstances, selon les occupations du sujet, son âge ou son état de santé, on pourra recourir à des injections annuelles de 0,1 cc (J.A.M.A. 1952).

Les contres-indications sont celles de la vaccination de base.

52. L'antigène et son application

A. Antigène liquide ou aluminé? Vaccin associé? Théoriquement, l'antigène liquide, vite résorbé, doit procurer un réveil plus massif et plus rapide de l'immunité (REGAMEY 1941 b). JONES et MOSS (1937 d) le préfèrent parce qu'il est mieux supporté, mais le choix doit être guidé par des arguments d'ordre immunologique.

Avec l'anatoxine brute, la totalité de l'antigène est disponible le premier jour; avec le vaccin-dépôt, l'antigène est libéré peu à peu, mais dès le début la quantité directement disponible crée une excitation suffisante pour être utile. La plupart des auteurs recommandent l'anatoxine liquide pour l'injection de rappel parce qu'elle engendre un bond antitoxique plus hâtif (SCHLEGEL 1951, TASMAN 1955, PETERSON et coll. 1955); le bond apparaît 1 jour, voire 2—3 jours plus tôt qu'avec le vaccin aluminé (JONES et MOSS 1937 a, GOLD 1940 a, MILLER et coll. 1943, 1949, EDSALL 1956 b). Comparant les propriétés réactivantes de 3 vaccins, chez 90 enfants, MILLER et coll. (1949) constatent un bond antitoxique identique pour l'anatoxine fluide et celle adsorbée sur Al (OH)$_3$; ce dernier vaccin s'avère meilleur que l'A.Te alunée. A propos de la diphtérie, D'ANTONA et PIAZZI (1956) ont nettement démontré l'intérêt du vaccin liquide. A long terme, après un mois par exemple, l'effet des différents types de vaccin s'égalise.

La réserve d'antigène que renferme la dose proposée pour l'I.R. est si grande que, dans la pratique, l'usage d'un vaccin dépôt ne constitue nullement une contre-indication (PETERSON et coll. 1955).

Lorsque les conditions feront préférer l'emploi d'un vaccin associé Di Te ou Di Te Per, par exemple, afin de réactiver en un temps l'immunité contre plusieurs infections, on n'aura aucune crainte quant au réveil de l'antitoxine tétanique (SACQUÉPÉE, PILOD et JUDE 1938). Lors de l'injection de rappel, le vaccin associé se montre aussi bon, si ce n'est meilleur que l'anatoxine seule (RAUSS et coll. 1958 b et al.).

B. Quantité d'antigène. RAMON et son école s'en tenaient au début à 2 cc d'anatoxine tétanique liquide, dose relativement élevée. On sait aujourd'hui (J.A.M.A. 1957 b) qu'en fait 0,1 cc d'antigène suffit et que si l'on conseille 0,5 ou 1,0 cc, c'est pour éviter des erreurs de mesure, simplifier l'application et tenir compte des mauvais réacteurs. WISHART et JACKSON (1951) rapportent que 1 Lf d'anatoxine fluide, dose en général inférieure à 0,1 cc, constitue un stimulus suffisant. HENDRY et CLARK (1951 a), opérant avec un vaccin aluné, notent que 0,1 cc procure un rappel aussi efficace que 0,5 ou 1,0 cc du même vaccin. Avec l'anatoxine fluide par contre, le bond antitoxique est en relation avec la quantité d'antigène: REGAMEY (1941 b), examinant 109 adultes vaccinés, constate que du 4e au 10e jour après l'injection de rappel les titres moyens passent, pour 1 cc d'anatoxine, de 0,17 à 1,35 U.I./cc (augmentation de 668%), pour 2 cc d'anatoxine, de 0,18 à 2,79 U.I./cc (augmentation de 1500%).

L'usage actuel consacre pour l'I.R. la dose soit de 1,0 cc d'A.Te fluide, soit de 0,5 cc de vaccin aluminé; des doses inférieures peuvent se justifier surtout chez les allergiques. Il reste néanmoins certain que le volume du vaccin n'est pas seul en jeu; sa qualité immunigène revêt une importance primordiale (FISEK 1957 et al.).

C. Mode d'application de l'antigène. Ce qui fut dit à propos de l'immunisation de base peut être répété ici. Dans la pratique, seules s'avèrent utilisables sur une large échelle les voies *sous-cutanée* ou *intramusculaire*. Les voies *nasale* ou *oculaire* donnent des résultats irréguliers (GOLD 1939; WISHART et JACKSON

1950); elles pourraient faire l'objet d'études à l'intention de certains allergiques (cf. p. 312 et 320).

La quantité d'antigène (0,1 cc) appliquée par voie *intracutanée* dans le test de ZOELLER-MOLONEY est suffisamment active pour jouer le rôle d'I.R. Cette voie est recommandée ci et là pour réduire la fréquence des réactions désagréables (J.A.M.A. 1953).

53. Analyse de l'injection de rappel

L'étude de la réponse à l'I.R. est celle de l'influence du titre de l'antitoxine avant l'I.R., de la précocité et de la hauteur du bond antitoxique, de la persistance de l'immunité consécutive à la réactivation.

A. Influence du titre avant l'injection de rappel. Toutes autres conditions étant égales chez le vacciné, la teneur du sérum en antitoxine relève de l' « immunisabilité » (IPSEN 1954a) du sujet — caractère essentiellement individuel — et du temps écoulé entre l'immunisation de base et le moment du contrôle sérologique. Il fut déjà relevé que le titre antitoxique ne donnait pas une image fidèle de l'immunité; ce titre ne permet de présumer, avant l'I.R., ni la rapidité, ni la hauteur du bond antitoxique. Les études de REGAMEY (1944b), puis de PETERSON et coll. (1955) en particulier, établissent que l'immunité potentielle, dont l'une des réponses à l'I.R. se traduit par le bond antitoxique, n'est pas conditionnée par la seule présence d'antitoxine libre. Un sujet riche en anticorps apparents avant l'I.R. peut ne présenter qu'une faible réactivation tandis que, au contraire, l'I.R. peut déclencher un bond antitoxique considérable chez un individu primitivement pauvre en anticorps libres.

Le tableau 11, emprunté à LOONEY et coll. (1956) concerne une centaine de sujets. Il démontre qu'avec un titre faible, le bond antitoxique relatif est bien plus considérable que lorsque le titre avant l'I.R. est élevé. Vers le 21e jour, les taux ont même tendance à s'égaliser pour tous les groupes. Par contre, après 1 an, les sujets pauvres avant l'I.R. sont aussi ceux qui accusent le recul le plus sensible.

B. Précocité du bon antitoxique. L'I.R., en cas de blessure, n'a d'utilité que si elle renforce l'immunité pendant la période d'incubation. Il est vraisemblable que les organes producteurs d'anticorps réagissent sitôt que leur parvient l'excitation antigénique; mais ce qui se passe dans la cellule échappe au contrôle et l'on ne perçoit que la manifestation *tardive* de la réactivation, celle du déversement des anticorps libres dans la circulation.

PLETSITYII et coll. (1956) ont cru voir que chez le singe le titre antitoxique augmentait déjà au bout de 5 heures pour atteindre 250% à la fin de la première journée. Cette observation reste isolée; l'antitoxine apparaît dans la règle plus tard. SACQUÉPÉE et JUDE (1937), KESTERMANN et VOGT (1940), puis BIGLER (1951) et MacLEOD (1953) signalent un bond déjà sensible après 3 jours. Le démarrage se fait le plus souvent dès la fin du 4e jour (REDI 1935, COWLES 1937, BOYD 1946, TURNER et coll. 1954) ou du 5e (BANTON et MILLER 1949; HENDRY et CLARK 1951a; SEIBOLD et BACHMANN 1955; J. MÉRIEUX 1957) et plus rarement du 6e jour (LOONEY et coll. 1953). REGAMEY (1941b) rapporte que le bond débute à la fin du 4e jour chez les $^2/_3$ de ses patients; au 6e jour tous ont répondu. Ci et là quelques réactivations sont tardives et ne s'expriment qu'après 7 jours (GOLD 1940a, MILLER et coll. 1949). La précocité du bond antitoxique en fonction de la nature de l'immunisation de base ressort des diagrammes 6 et 7. La rapidité de la formation de l'antitoxine après l'I.R. fut bien étudiée par MILLER et HUMBER (1943); d'une façon générale, elle dépend du type de l'immunisation de base, de l'intervalle depuis la dernière injection, de la qualité du vaccin utilisé pour le rappel, de la quantité de ce vaccin et de facteurs individuels.

L'I.R. n'obtient donc pas de réponse sérologique avant le 4e — 6e jour. Est-ce trop tard lors d'une infection tétanique de courte incubation (Banton et Miller 1949)? Si une réponse catégorique est trop osée, on tiendra compte néanmoins des considérations qui suivent:

1. L'immunité du vacciné est souvent suffisante sans qu'il faille recourir à l'I.R.

2. Pendant la guerre de 1939—1945, la grande majorité des cas de tétanos surviennent chez des vaccinés (Grande-Bretagne et U.S.A.) après une incubation de 7 jours et plus (Sachs 1952).

3. Le fait que 30% environ des tétanos apparaissent après une incubation de moins de 7 jours (Rostock 1950, p. 51) s'oppose à l'extrême rareté de la maladie et à la fréquence des longues incubations chez le vacciné: ce paradoxe ne s'explique que par l'efficacité de l'immunité de base ou/et de l'I.R.

4. L'absence d'antitoxine nouvelle pendant les premiers jours n'élimine pas la vraisemblance d'une réactivité immunitaire cellulaire. Si de nombreux auteurs admettent que dans le tétanos seule l'immunité humorale compte, rien ne permet de refuser l'hypothèse soit d'une immunité « sessile » due à l'action d'anticorps en voie de maturation et n'ayant pas encore quitté les tissus, soit même d'une immunité tissulaire en dehors du S.N.C. lui-même.

La rapidité du bond n'est pas influencée par l'injection concomitante de hyaluronidase (Miller et Ryan 1950).

Tableau 11. *Répercussion de la teneur du sérum en antitoxine sur l'effet de l'injection de rappel.* (Selon Looney, Edsall, Ipsen et Chasen 1956)

Titre avant l'I.R.*	Titre moyen . . . après l'I.R.			
	6 jours	14 jours	21 jours	1 an
$<0,025$	0,12	(2,3)	13,0	0,17
0,045	0,9	10,7	12,4	0,95
0,14	3,8	13,9	47,0	1,36
0,45	1,4	23,7	33,0	1,9
1,4	4,7	23,7	(55,0)	3,3
4,5	7,1	13,4	31,0	3,2
14,0	(5,6)	(18,0)	(55,0)	(8,8)

* = moyenne géométrique en U.I./cc; ( ) = valeurs concernant seulement 1,2 ou 3 sérums.

C. Hauteur du bond antitoxique. Avec l'A.Te. fluide, l'acmé de la production d'antitoxine se situe entre les 8e et 12e jours (Regamey 1944b et al.). Avec le vaccin aluné, les titres les plus élevés apparaissent plus tard, parfois même à la fin du premier mois. L'augmentation du taux des anticorps est rapide. 109 patients répartis en deux groupes, traités les uns avec 1 cc et les autres avec 2 cc d'anatoxine tétanique fluide (Regamey 1941b) ont présenté les augmentations moyennes de

	pour 1 cc d'anatoxine	pour 2 cc d'anatoxine
après 4 jours	0,35 fois	0,55 fois
6 jours	5,6 fois	15,4 fois
8 jours	18,0 fois	25,7 fois
10 jours	—	26 fois
20 jours	14,0 fois	—

Des bonds de 50 à 1000 fois ont été signalés (Ramon et Zoeller 1929a; Sacquépée et coll. 1937; Gold 1940a). Des titres très élevés ont été atteints: 75 U.I./cc (Hendry 1956), 235 U.I./cc (Regamey, non publié). Le bond est plus accentué quand l'immunité de base comporte l'injection d'un vaccin associé (Schlegel 1951, Ikić 1957b) (cf. diagramme 6). Il est aussi plus marqué lorsque l'immunité de base fut renforcée par une ou plusieurs I.R. (Schlegel 1951). Mais les réponses sérologiques les plus faibles sont celles qui présentent le plus d'intérêt, à cause des échecs qu'elles peuvent susciter. L'immunisation de base ne conduit pas toujours à des titres que d'aucuns considèrent comme indicateurs d'une immunité suffisante (Bergey et coll. 1939 et al.). Or l'I.R. supprime tout souci à cet égard. MacBryde et Poston (1946), puis Miller et coll. (1949) relèvent que des enfants vaccinés 4—7 ans plus tôt ont tous 0,1 U.I./cc — ou plus — 7 jours après l'I.R. Parmi 573 enfants réactivés avec 1 cc de vaccin à l'Al(OH)$_3$, 96% acquièrent un titre supérieur à 0,1 U.I./cc; aucun taux ne reste inférieur à 0,003 U.I./cc (Peterson et Christie 1951). Chez 617 enfants examinés 2 semaines après l'I.R., Hegyessy et coll. (1956) constatent les pourcentages suivants: < 0,1 U.I./cc: 2%. 0,1—1,0 U.I./cc: 20%. 1,0—10 U.I./cc: 68% et > 10 U.I./cc: 10%.

L'I.R. fait apparaître des titres antitoxiques notablement plus élevés que ceux de l'immunisation de base; ces titres persistent aussi plus longtemps (Peshkin 1945). Le bond est d'autant plus élevé que l'immunité est plus ancienne et que le nombre des injections antérieures est plus grand (Spath et Köle 1952 et al.); il est plus marqué chez les sujets qui possèdent des titres très bas au moment de l'injection de rappel (Regamey 1944b, Looney et coll. 1956).

D. Persistance de l'immunité après la réactivation. Les renseignements à ce propos sont encore fragmentaires. On connaît l'effet de l'immunisation de base combinée aux I.R. dans les armées vaccinées de 1939 à 1945 et l'on peut en déduire que la durée de l'immunité doit être plus longue qu'après la simple immunisation de base. Les expériences de Wishart et Jackson (1951) qui portent sur des étudiants contrôlés pendant les trois années suivant l'I.R., et celles de Looney et coll. (1953, 1956) sont intéressantes à cet égard; elles invitent à la conclusion pratique suivante: si une première I.R. est faite dans les cinq années qui suivent l'immunisation de base, il y a lieu de répéter l'I.R. d'entretien après dix ans.

54. L'injection de rappel combinée à l'antitoxine

Chez le blessé insuffisamment vacciné, ou chez le traumatisé dont l'immunisation de base est très ancienne, la prophylaxie spécifique d'urgence réclame l'administration immédiate de sérum antitétanique. Peut-on faire simultanément une injection d'anatoxine ?

Les bases expérimentales propres à ce cas particulier sont encore fragiles (Eckmann 1958 et al.) et le problème qui se pose entre dans le cadre de la séro-anatoxiprévention (cf. plus loin, p. 341). L'emploi simultané d'antitoxine et d'anatoxine semble justifié. En effet, la protection quasi-instantanée qu'apporte le sérum n'est pas sensiblement diminuée par la présence du vaccin, et la fraction d'anatoxine qui restera active sera susceptible de réveiller ou d'accentuer les processus immunitaires actifs.

Le blessé correctement vacciné ne reçoit pas de sérum (cf. Bürkle de la Camp 1957, Haas 1957 et al., cités pas Eckmann 1958; cf. aussi ch. 51, p. 324).

6. Les vaccins tétaniques associés

L'emploi des vaccins à antigènes multiples est envisagé ici sous l'angle de la composante tétanique. Ce sont Ramon et Zoeller (1926a) qui ont expérimenté pour la première fois les associations de Di TAB, Te TAB et Di Te; ces auteurs notent qu'avec le vaccin associé l'immunité antitétanique semble supérieure à celle que l'on obtient avec l'A.Te seule. Les vaccinations associées ont pris une extension considérable, car elles permettent d'immuniser contre plusieurs maladies à la fois, d'où économie et gain de temps.

61. Nature des vaccins tétaniques associés

Les combinaisons les plus diverses ont été expérimentées et beaucoup d'entre elles sont employées dans la pratique. L'A.Te est incorporée au mélange sous forme brute, purifiée, concentrée, fixée ou non sur un composé aluminique. Elle voisine avec une autre anatoxine (diphtérique, scarlatineuse, staphylococcique, dysentérique, botulinique, gangréneuse). avec une suspension ou un extrait de bactéries (Salmonelles, Shigelles, H. pertussis, Streptocoques scarlatineux, V. cholerae, B. rhusiopathiae, B. anthracis), avec des virus (poliomyélite, Rickettsies).

Les premières associations étaient faites extemporanément par mélange des vaccins dans la seringue. La préparation pour l'usage à long terme des vaccins multivalents a posé de nombreux problèmes, en particulier celui de l'interaction in vitro des différents antigènes. Les anatoxines brutes ont un reste de formol libre et capable de réagir avec un corps bactérien ou un virus; ce point revêt une grande importance quand une fraction poliomyélitique est en jeu.

La teneur en antigène varie considérablement d'un vaccin, d'un pays ou d'un producteur à l'autre. Les recettes de fabrication ressortent de l'arbitraire au début, puis l'expérience sur l'animal et sur l'homme imposent certaines corrections. Les formules adoptées dans la pratique ne sont pas toujours les meilleures; elles sont parfois imprégnées d'un dogmatisme périmé, que pourront seules conjurer des études systématiques et la mise au point de procédés rigoureux de contrôle.

Dans les vaccins associés les plus usuels, la *fraction diphtérique* est généralement de présentation analogue à celle de l'A.Te; c'est elle qui occasionne les phénomènes d'escorte de nature allergique dont la fréquence augmente avec l'âge du sujet. La *fraction TAB*, presque toujours douée d'une certaine toxicité, comporte à l'ordinaire une suspension de germes tués; on rencontre de plus en plus des extraits salmonelliques du type GRASSET, bruts ou traités par le formol, adsorbés ou non (HALLAUER et coll. 1941; LOVREKOVICH et RAUSS 1942; REGA-MEY 1943d; MAZETTI et coll. 1954; MEISLOWA et coll. 1956), plus rarement des germes traités à l'acétone (BENZONI et LA ROCA 1955) ou desséchés (KLUEVA 1957; PONTECORVO et SOPRANO 1957). La *fraction coquelucheuse* est dans la règle une suspension de H. pertussis tués, la *fraction poliomyélitique*, une suspension multivalente de virus tué (SALK).

Le contrôle de l'antigénicité des vaccins associés n'est pas au point. Les méthodes adoptées pour l'examen de l'anatoxine isolée ne conviennent pas nécessairement pour l'évaluation de la fraction tétanique d'un vaccin multivalent (BARR 1957b, HOLT 1957, IKIĆ 1957a, REGAMEY 1957, RÉTHY et coll. 1957, SCHEIBEL 1957b, JOÓ et RÉTHY 1957).

62. Types principaux de vaccins tétaniques associés

Vaccin Di Te. Cette association assez répandue est obligatoire en France pour le petit enfant entre le douzième et le dix-huitième mois (loi du 7 septembre 1948, BONNEFOI et coll. 1958). Elle a donné de bons résultats épidémiologiques (POULAIN 1948). Le vaccin Di Te procure généralement une immunité plus solide que l'antigène isolé (Ramon 1935 g, Melnik et coll. 1936, Regamey 1948 et al.). Chez l'enfant, trois doses sont indiquées (Peshkin 1941 b). Chez l'adulte, le vaccin Di Te engendre des réactions assez sévères (Jones et Moss 1937 c). Aussi essaie-t-on de réduire la proportion de l'antigène diphtérique en recourant à 1 Lf seulement par dose, mais en multipliant le nombre des injections (Edsall 1952, Edsall et coll. 1954 b, Ipsen 1954 b). Jones et Moss (1937 c) ont observé que les titres de l'antitoxine diphtérique étaient sensiblement plus élevés chez les sujets qui avaient présenté de fortes réactions vaccinales. Cette meilleure réponse sérologique doit être mise sur la présence d'une immunité d'origine occulte qui sensibilise l'organisme et transforme l'immunisation de base en une suite de stimuli secondaires. Bousfield et Holt (1957) ont reconnu que les proportions les plus favorables étaient 30 Lf de vaccin Di et 12 Lf de vaccin Te.

Vaccin Te TAB. L'une des plus anciennes associations introduites par Ramon et Zoeller (1926 a, 1929 a), et probablement la plus utilisée avec le vaccin Di Te Per. La posologie du début: 2; 2,5 et 2,5 cc à 15 et 8 jours d'intervalle fut réduite à 1; 2 et 2 cc, voire même à 3 fois 1 cc à 2—4 semaines d'intervalle. La fraction salmonellique contient généralement un milliard de germes tués par cc. Son action antigénique est puissante (Melnik et coll. 1936 b et al., Petrović 1956), nettement plus efficace que celle des vaccins Di Te (Regamey 1948) ou Te (Regamey 1944 b, Greenberg 1955, Ikić 1957 b). La protection contre le tétanos est déjà manifeste quinze jours après la seconde injection (d'Antona et coll. 1938). Un mois après l'immunisation de base, la moyenne des titres est de 2,5 U.I./cc, après 9 mois de 0,5 U.I./cc et aucun sérum n'est inférieur à 0,05 U.I./cc (Regamey 1941 b). L'état d'immunité antitétanique justifierait une immunisation par deux injections (MacLean et Holt 1940), surtout si l'anatoxine est adsorbée (Réthy et coll. 1957, Pontecorvo et coll. 1957). L'A.Te injectée en même temps que le TAB acquiert les propriétés d'un antigène aluné (Chen 1933). Les faibles titres rencontrés par Neto (1947) semblent constituer une exception.

Les réactions consécutives à l'injection de vaccin Te TAB sont parfois vives, surtout lorsque la fraction TAB n'a pas encore vieilli. Elles sont indépendantes de la voie d'inoculation — sous-cutanée ou intramusculaire —, et des groupes sanguins (Knox et Stamm 1957). Récemment, Robertson et Leonard (1956), puis Fuà (1956) ont attiré l'attention sur les réactions tardives, survenant après 2 semaines, avec retentissement articulaire, cardite rhumatismale et répercussions au niveau du système réticulo-endothélial, du foie, de la rate, du système lymphatique et de la moelle osseuse (cf. Beller 1943 et p. 319). Des cas de mort ont été rapportés (Regamey 1947 c).

Vaccin Di Te TAB. Employé a l'origine avec un dosage de 2, 3 et 3 cc à quinze jours d'intervalle, ce vaccin donne à Pilod (1938) un titre très élevé d'antitoxine tétanique chez le 99,7 % des vaccinés. Par la suite, le dosage fut ramené à 1; 2 et 2 cc. Ramon et coll. (1941 a, c) constatent que le vaccin Di Te TAB engendre moins de phénomènes d'escorte que le vaccin TAB seul et mettent cette bénignité sur le compte de la présence du formol en excès dans les anatoxines, la formaldéhyde libre détoxiquant partiellement les Salmonelles.

Sur 22600 vaccinés, Pilod observe 3 % de réactions fortes, 7 % de réactions bénignes. Sohier (1944 a) signale une urticaire et une albuminurie parmi 331 enfants; les réactions fébriles sont plus marquées dès l'âge de 7 ans; elles sont aussi plus fréquentes chez l'adulte (Besson et Giraud 1944). Récemment André et coll. (1958) ont constaté une leucose aiguë chez un vacciné et relèvent dans les antécédents immédiats d'une trentaine de leucoses la fréquence relative d'une troisième injection de vaccin ou d'un rappel.

Les résultats sérologiques sont estimés satisfaisants (Melnotte et coll. 1948, Sellers et coll. 1950, Jude et Masse 1951), malgré l'absence d'anticorps Vi.

Vaccin Di Te Per. Destiné à l'enfant, ce vaccin fut l'objet de nombreuses études (Sant' Agnese 1949, 1950; Comb et Trafton 1950; Peterson et Christie 1951; Vahlquist et coll. 1954; Ipsen et Bowen 1955 et al.). Le vaccin contient, par dose, 10—30 milliards de H. pertussis tués. Les anatoxines sont le plus souvent purifiées et adsorbées; dans cet état, elles sont plus antigéniques que simplement mélangées aux bactéries à l'état liquide (Chevé et coll. 1957). On pousse le vaccin par voie sous-cutanée ou intramusculaire, de préférence dans la fosse sus-épineuse (Bradford et coll. 1949): 3 injections à 4—6 semaines d'intervalle (J.A.M.A. 1957 a). Volk et coll. (1953) proposent 2—3 injections de base dès le troisième mois de la vie; une injection de rappel de 0,2 cc seulement assure en moins de quinze jours un titre supérieur à 0,05 U.I./cc; les anticorps persistent plus longtemps après l'injection de rappel qu'après l'immunisation de base. Sauer et Tucker (1950) protègent contre le tétanos le 96% des nourrissons âgés de 3—6 mois avec trois injections de vaccin Di Te Per, le 100% avec quatre injections. Mais les titres baissent assez rapidement; après 3 ans, le 35% des vaccinés a moins de 0,02 U.I./cc, et Volk (1949) estime qu'une injection de rappel s'impose après 2—5 ans. Fleming et coll. (1948, 1950) recommandent aussi la vaccination Di Te Per dès le 3e mois de la vie; un an plus tard, la première injection de rappel développe en moyenne 0,55 U.I./cc.

Des réactions fébriles postvaccinales surviennent dans le 22% des cas (Hesselvik et coll. 1954). Des manifestations plus graves sont rares; elles sont dues à la fraction coquelucheuse (Bradford et coll. 1949), mais en partie aussi aux anatoxines dans les vaccins non purifiés (Murray 1950). Elles sont plus fréquentes lors de la 3e injection et lors de l'injection de rappel (Volk et coll. 1953). Les accidents poliomyélitiques consécutifs à l'injection du vaccin Di Te Per ont été imputés à la fraction coquelucheuse en combinaison avec l'alumine (Cockburn 1955, M.R.C. 1956); ils ne seraient plus à craindre par l'usage d'anatoxines purifiées non aluminées (Holt 1958, Perry 1958b). Pour réduire la fréquence des réactions locales, Hesselvik et Ericcson (1954) incorporent à leur vaccin un anesthésique local, qui réduit la douleur sans atténuer le pouvoir immunisant. Chez le petit enfant, on reconnaîtra comme contre-indications: les troubles respiratoires, les infections aiguës, les manifestations dentaires et l'albuminurie; on prêtera également une attention particulière aux réactions exagérées lors de la première injection (Bradford et coll. 1949).

Les germes coquelucheux augmentent le pouvoir immunigène des anatoxines (Ungar 1952, 1954, Faragó et coll. 1949). L'immunisation coquelucheuse ne bénéficie pas de l'association. Pour Ikić (1957), seule l'immunité antidiphtérique serait favorablement influencée par la présence des autres antigènes.

Vaccin TePer. Cette variante, d'usage plus restreint, appelle les mêmes commentaires que le Di Te Per (Miller et coll. 1944, Ungar 1956 et al.).

Vaccin Di Te Per Pol. Il est naturel, surtout dans la pratique pédiatrique, de chercher à associer la vaccination classique par le Di Te Per avec l'immunisation contre la poliomyélite. Les essais sur l'animal (Kendrick et Brown 1957; Levine et Wyman 1957) suggèrent que les différentes espèces ne répondent pas avec la même intensité aux divers constituants du vaccin; l'antitoxine tétanique en particulier semble affaiblie chez le cobaye. Les premières publications sur la vaccination de l'enfant et de l'adulte sont dues à la plume de Batson et coll. (1958), de J. Mérieux (1957), de C. Mérieux (1958), de Barrett et coll. (1958); une réponse immunitaire jugée satisfaisante fut constatée à l'égard des quatre fractions du vaccin Di Te Per Pol.

63. Posologie et action des vaccins tétaniques associés

Lapin (1949) a fait une excellente revue des travaux sur les vaccinations associées, étude que complètent de nombreuses publications récentes émanant de cliniciens, épidémiologistes et hommes de laboratoire (Ramon 1952b, Teichmann 1952 et al.). Il ne sera tenu compte ici que de quelques points particuliers à la fraction tétanique.

On sait que *l'intervalle* entre les injections d'un vaccin exerce une influence considérable sur l'intensité de l'immunité. Pour l'emploi des vaccins associés, il faut tenir compte à la fois du long espace propice à l'A.Te et des intervalles rapprochés indiqués pour la fraction TAB, par exemple. L'espace entre la première et la seconde injection devrait être si possible d'un mois au minimum (Réthy et coll. 1957). Volk (1948) recommande la *voie intramusculaire* plutôt que la voie sous-cutanée, parce que les réactions locales ou générales sont moindres. Les *contre-indications* sont plus sévères que pour l'anatoxine seule; elles sont celles des antigènes associés à l'anatoxine tétanique; les mêmes remarques s'imposent pour les *phénomènes d'escorte*.

Les associations de vaccins permettent de réduire soit le nombre des injections, soit la quantité absolue des antigènes. En effet, contrairement à l'opinion longtemps exprimée par l'école allemande, il n'y a pas lieu de redouter une concurrence des antigènes (cf. plus loin). Au contraire les vaccins mélangés s'entr'aident mutuellement.

Déjà le mélange de l'A.Te avec l'anatoxine diphtérique (MELNIK et coll. 1936 b) ou staphylococcique (RAMON et coll. 1940 c) conduit à une potentialisation de l'effet antigénique. Cet effet, qui a d'ailleurs échappé à d'autres auteurs (JONES et Moss 1937 b et al.), s'explique par le développement de phénomènes inflammatoires plus intenses en lieu de l'injection ou par une sommation de l'excitation antigénique; cette dernière hypothèse paraît confirmée par l'observation de EICHHORN et RICHOU (1936): les anatoxines tétanique et staphylococcique, injectées mélangées ou séparément en deux endroits différents, procurent une immunité de même valeur.

Quand l'A.Te est associée à des vaccins bactériens, l'effet de potentialisation sur la réponse antitétanique est considérable, surtout avec le TAB (RODRIGUEZ et NETO 1947; KATIĆ 1953; IKIĆ 1957 b); deux injections de vaccin Te TAB suffisent alors pour déclencher une immunité antitétanique solide (MACLEAN et HOLT 1940). Pour RAMON (1937 d), les bactéries se comportent comme le tapioca et l'activation trouve son origine dans les réactions locales plus marquées qu'après l'injection d'anatoxine seule. On a tenté aussi de rapporter l'élévation de l'immunité à une adsorption de l'antigène sur la bactérie, qui démultiplierait la surface antigénique de l'anatoxine. Les expériences de MERCIER et NICOL (1939) établissent toutefois que les anatoxines ne se fixent pas de façon sensible sur les bactéries. Avec un vaccin associé A.Te + Rickettsies tuées, du type Castañeda, JUDE (1950) constate que l'immunité antitétanique se développe avec un certain retard et met en cause une adsorption de l'antigène non sur le virus, mais sur les débris pulmonaires.

Il y a lieu de noter en passant qu'après vaccination au moyen d'un antigène associé du type Di Te Staph., c'est l'immunité antitétanique qui persiste le plus longtemps (RAMON et coll. 1940), puis que KESTERMAN et VOGT (1940) ont fait d'intéressants essais de vaccination associée contre le tétanos et la gangrène gazeuse.

En médecine vétérinaire, l'anatoxine tétanique fut incorporée à des vaccins divers, contre le charbon bactéridien, le rouget et les clostridioses (DELPY et MIR CHAMSY 1951; KATIĆ 1951, 1953; SZÉLYES et coll. 1952). Chez leurs chevaux fournisseurs de sérums, TOKGÖZ et et BILÂL (1937) commencent l'immunisation avec l'association Di Te et continuent avec l'antigène qui fournit la meilleure réponse immunitaire. BARR et GLENNY (1950) ont mentionné que la réaction favorable à l'A.Te pouvait être interprétée chez le cheval comme un test d'immunisabilité à l'égard de n'importe quel autre antigène.

64. Phénomènes d'interférence

Si l'on admet que, d'une façon générale, les composants d'un vaccin associé se potentialisent mutuellement, certaines observations créent une note discordante. GLENNY et WADDINGTON (1926 b) relatent que le mélange de plusieurs antigènes peut déprimer l'action immunisante de l'un d'eux. RAINSFORD (1941), constatant incidemment qu'un vaccin Te TAB fournissait de mauvais titres antisalmonelliques, attribuait la perte de l'antigène à l'action délétère du formol libre sur les antigènes bactériens. BARR et LLEWELLYN-JONES (1953 a, b) choisissent des cobayes préimmunisés contre la diphtérie et leur injectent des vaccins Di Te; dans ces vaccins la fraction diphtérique est constante, tandis que la quantité d'antigène tétanique est variable. Les auteurs constatent que la réponse antidiphtérique décroît à mesure que la dose d'anatoxine augmente. Ce n'est pas la rapidité, mais bien l'amplitude du bond antitoxique qui offre ce phénomène d'interférence.

BARR (1956) fait une observation analogue chez le cobaye traité avec deux doses d'antigène Te TAB: des quantités faibles de vaccin ont un effet synergique, des quantités élevées exercent une action freinatrice. La qualité antigénique du toxoïde joue un rôle important pour déterminer s'il y aura potentialisation ou réponse réduite à l'injection du vaccin associé.

L'équipe chinoise de Chen (1956, 1957 a, b) a recherché et étudié chez l'enfant les interférences inhérentes à l'emploi du vaccin Di Te Per. Elle a reconnu que lors de la vaccination primaire, une immunité antidiphtérique préalable influençait la rapidité de la réponse antitétanique et l'amplitude de la formation des anticorps coquelucheux, probablement par des processus d'encombrement (crowding out). C'est le degré de l'immunité antidiphtérique présente qui conditionne la réponse aux fractions tétanique et coquelucheuse. Chen et ses collaborateurs tentent d'analyser mathématiquement le phénomène à propos de l'injection de rappel.

Les constatations de Barr et Chen s'accordent ainsi avec ce qui fut déjà relevé à propos de l'injection de rappel dans le cas de l'immunisation antitétanique isolée: le bond antitoxique est d'autant plus élevé que le titre de l'antitoxine au départ est plus bas. Il est possible aussi que l'affaiblissement relatif de l'immunité antitétanique décrit par Holt (1957) chez le cobaye traité avec des doses élevées de Di Te Per soit apparenté au phénomène décrit par Barr et Llewellyn-Jones. Hegyessy et coll. (1956), Hazza et Habbu (1956), Ikić (1957 a), Rauss et coll. (1958) ont cherché — sans succès —, à détecter l'« effet de Barr » après usage de vaccins associés renfermant les fractions Te, TAB ou Dys. (Cf. également Barr et Glenny 1952, Bousfield et Holt 1957 et al.).

Les phénomènes d'interférence relevés ci-dessus ne constituent vraisemblablement pas des cas particuliers de l'immunité. Ils doivent avoir leur place dans le cadre d'une loi générale. On peut déjà les rapprocher des processus de blocage ou d'interférence avec les immunoglobulines hétérologues (Alder 1957), puis avec les polysaccharides des Pneumocoques (Felton 1949, cité par Ungar 1957), qui n'aboutissent pas à la formation d'anticorps chez la souris lorsqu'ils sont injectés en excès. Peut-être ont-ils une signification dans le développement de l'immunité active chez le tout jeune enfant qui a conservé des anticorps maternels, ou encore chez le sujet soumis à la séro-anatoxiprophylaxie ?

7. La séro-vaccination antitétanique

La controverse que suscite encore le bien-fondé de l'emploi simultané de l'A.Te et de l'antitoxine a créé un sentiment d'insécurité. La versatilité des opinions parmi les chirurgiens allemands au cours des années 1956—1957, par exemple, est une manifestation de ce malaise. La cause doit en être recherchée dans le fait que la méthode de séro-anatoxiprophylaxie fut introduite dans la pratique sans bénéficier de bases expérimentales suffisantes.

71. Bases théoriques

Préconisée par Ramon et Laffaille en 1925 c et réalisée pour la première fois chez l'homme par Zoeller et Ramon en 1926, la séro-vaccination a pour but de procurer au blessé les avantages réunis qu'offrent la vaccination par l'anatoxine et la séroprophylaxie. L'immunité active, lente à paraître, doit se développer sous le couvert de l'immunité passive, la première devant faire suite à la seconde sans interruption. Cette hypothèse comporte deux paradoxes:

1. Chez le blessé, on injecte le sérum dans l'espoir que toute trace de toxine sera neutralisée. Lors de la séro-vaccination, l'antitoxine doit donc fixer toute l'anatoxine et la rendre inefficace.

2. L'antitoxine hétérologue disparaît rapidement de l'organisme; l'antitoxine endogène n'apparaît qu'après la seconde injection de vaccin, donc trop tard. Il y a, dans la continuité de l'immunité, un « trou », une période de carence pendant laquelle l'organisme n'est pas protégé (diagramme 10).

Les deux substances, antitoxine et A.Te doivent pouvoir agir séparément. C'est pourquoi elles sont injectées en deux endroits différents, le sérum étant poussé un peu plus tard que le vaccin.

Quelques considérations retiennent l'attention.

Les 1500 U.I. que comporte la dose usuelle d'antitoxine ne souffrent guère de leur rencontre avec les quelque 10—20 Lf de l'A.Te (MADSEN, JENSEN et IPSEN 1937), même si l'on tient compte d'un phénomène de DANYSZ ou d'une neutralisation inégale au détriment de l'antitoxine.

L'inverse, la neutralisation de l'A.Te par l'excès d'anticorps, constitue un réel paradoxe. Mais le sérum se répartit dans *tout* l'organisme et le rapport 1500 U.I. de sérum/10—20 Lf d'A.Te se trouve considérable-

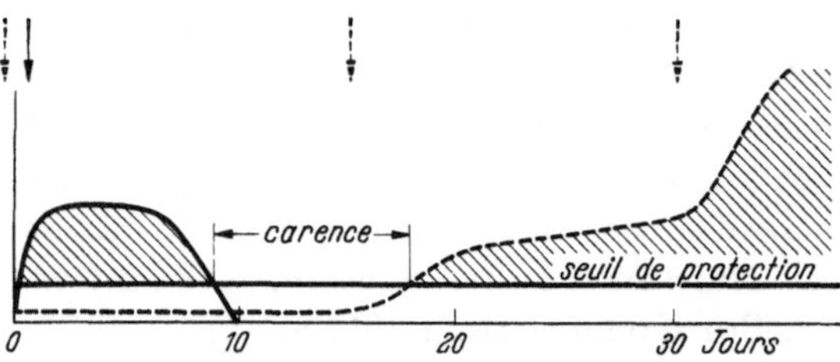

Diagramme 10 (schématique). *Période de carence entre les deux périodes d'immunité passive et active*
Flèche pleine: injection de 1500 U.I. de sérum; ▬ immunité passive consécutive à cette injection; flèche pointillée: injection d'anatoxine; - - - immunité active (Selon REGAMEY 1955 a.)

ment réduit au lieu de l'injection de l'A.Te, si bien qu'une partie de l'antigène échappe vraisemblablement à l'antitoxine et réalise son but immunigène. Quant à l'A.Te, neutralisée par l'anticorps, que devient-elle? Les complexes moléculaires A.Te + antitoxine doivent être retenus quelque part dans le système réticulo-endothélial. L'A.Te y acquiert-elle la propriété d'un antigène-dépôt? Lors des processus cataboliques, l'antigène est-il remis tout ou partie en liberté?

Des processus d'interférence ne sont pas exclus. L'antitoxine tétanique est un corps étranger qui se comporte comme un antigène. Sa masse, comparée à celle de l'A.Te injectée simultanément, exerce une sollicitation considérablement plus puissante sur les organes producteurs d'anticorps. L'effet de l'A.Te en est-il affecté?

72. Bases expérimentales

Contrairement à l'immunisation active par l'A.Te, la séro-vaccination n'a suscité qu'un nombre restreint de travaux expérimentaux sur l'animal. Aussi les informations restent-elles fragmentaires.

Il est par ailleurs curieux de constater que la première communication de RAMON et LAFFAILLE (1925 c) inclut déjà la quasi-totalité des conclusions qui forment le cadre des conceptions actuelles (tableau 12). Les essais intéressent le cobaye.

Ces essais démontrent en effet que, chez le cobaye:

1. l'immunité passive disparaît avant le 16e jour,

2. l'A.Te isolée développe une solide immunité en moins de deux semaines,

3. l'antigène et l'A.Te mélangés dans la seringue n'engendrent pas d'immunité décelable, même si l'on pratique un rappel avec l'A.Te seule,

4. l'antitoxine et l'A.Te injectées séparément créent un stimulus primaire, appréciable après un rappel.

RAMON et LAFFAILLE relèvent expressément que chez le cobaye l'action immunisante de l'A.Te est gênée par la présence de l'antitoxine et que, chez l'homme, on emploie proportionnellement moins d'A.Te, mais que « la résistance à la maladie naturelle n'exige pas un tel degré d'immunité ».

Tableau 12. *Influence de l'antitoxine tétanique sur le développement de l'immunité active chez le cobaye*
(Selon RAMON et LAFFAILLE 1925 c.)

Lots	Traitement de départ	16e jour	20e jour	27e jour
		l'animal supporte	réactivation avec	l'animal supporte
I	A.Te	> 60 D.M.M.		
	A.Te		A.Te	> 200 D.M.M.
II	A.Te + S (mélangés)	< 2 D.M.M.		
	A.Te + S (mélangés)		A.Te	< 5 D.M.M.
III	A.Te + S (séparés)	< 2 D.M.M.		
	A.Te + S (séparés)		A.Te	10—50 D.M.M.
IV	S	< 2 D.M.M.		
	S		A.Te	< 5 D.M.M.

A.Te = 1,0 cc d'anatoxine tétanique liquide; S = 0,5 cc de sérum antitétanique (2250 U.I.)

Deux éléments majeurs vont empêcher de reporter sans restriction sur l'homme les observations recueillies chez l'animal. L'un réside dans le fait que les sujets réagissent différemment selon l'espèce : le cobaye élimine l'antitoxine hétérologue beaucoup plus vite que l'homme (MARRI 1935 et al.); le lapin s'immunise mieux que le cobaye avec les mélanges : sérum + A.Te etc. L'autre caractérise une erreur fréquente de certains expérimentateurs, qui recourent à des doses soit d'antitoxine, soit d'A.Te beaucoup trop élevées, faussant ainsi le jeu de l'immunité. Les globulines font des blocages partiels du système réticulo-endothélial; il se produit des « effets de BARR » (cf. p. 334) entre la globuline hétérologue et l'A.Te, qui se comportent toutes deux comme des antigènes. C'est ainsi que s'expliquerait l'observation de D'ANTONA (1935) : le cobaye ne forme pas d'anticorps si l'A.Te est injectée en même temps que du sérum antitétanique de cheval; l'installation de l'immunité active n'est par contre pas inhibée si l'antitoxine est issue du cobaye.

D'une façon générale, on retire des travaux de laboratoire un faisceau de conclusions plus ou moins concordantes :

1. Le sérum antitoxique injecté en même temps que l'A.Te conserve son pouvoir immunisant (RITOSSA 1934; WOLTERS et DEHMEL 1937). Le dépression de la protection passive rencontrée par RICHTER (1955) et MÖRL (1956) ne trouve pas encore d'explication.

2. L'A.Te mélangée au sérum dans la seringue s'avère un mauvais antigène (MARRI 1933, 1935, RITOSSA 1934, D'ANTONA 1935 et al.). Par contre REGAMEY, SIMON et WANTZ (1955b) obtiennent une immunité aussi solide avec un mélange fait extemporanément. Les premiers auteurs ont étudié l'immunité après le seul traitement par l'A.Te et l'antitoxine, tandis

que Regamey et coll. ont éprouvé leurs cobayes après une série de 3 injections, les deux dernières étant faites avec l'A.Te seule. Le mélange constitue donc un stimulus primaire suffisant, ce que devait faire présumer l'usage d'ailleurs peu répandu de « floculates », vaccin constitué par le précipité que donne l'A.Te avec l'antitoxine tétanique.

3. Au contraire de l'A.Te seule, l'injection simultanée d'A.Te et de sérum, en deux endroits différents, ne déclenche pas d'immunité active apparente; elle constitue néanmoins un premier stimulus. Si l'immunité s'installe, elle ne se manifeste qu'avec retard (Marri 1933; Ramon et coll. 1937f, 1938g; Otten et Hennemann 1940; Aegerter 1954 et al.). Pontano (1935a, b, c) relate que, pour être pleinement effectives, les injections d'antitoxine et d'A.Te doivent être séparées par un intervalle d'au moins 7 jours. Otten et Hennemann (1939, 1940) constatent l'existence d'une phase de non-protection qui sépare la présence effective des deux immunités passive et active, phase qui se situe entre les 3e et 4e semaines après l'injection simultanée. Pour Wolters et Dehmel (1937), qui recourent à l'A.Te aluminée, l'A.Te n'est pas notoirement inhibée par l'anticorps.

4. L'injection simultanée d'antitoxine et d'A.Te suivie d'une ou deux injections d'A.Te seule engendre une bonne immunité, le plus souvent aussi solide que celle qui fait suite à 3 injections d'A.Te (Marri 1933; Ritossa 1934; Wolters et Dehmel 1951; Regamey et coll. 1955b; Teichmann 1958 et al.). La répétition de doses combinées de sérum et d'A.Te est hautement défavorable (Ambrosioni et Murgia 1946; Aegerter 1954 et al.).

73. Expériences cliniques

La séro-vaccination s'est largement répandue dans les pays d'obédience française, malgré la rareté et l'insuffisance des études systématiques chez l'homme. Les expériences sur l'homme sont d'ailleurs difficiles à interpréter, parce qu'il n'existe que deux méthodes d'investigation. La première consiste à suivre l'évolution des titres sériques de l'antitoxine, procédé qui renseigne insuffisamment sur la valeur réelle de l'immunité active, et la seconde à rechercher l'efficacité sur le blessé contaminé. Or les risques de fausses conclusions sont ici considérables, car il est impossible d'évaluer la proportion des foyers traumatiques infectés: le succès de la séro-vaccination posttraumatique peut sembler assuré parce qu'il n'existe en réalité aucune infection tétanique (Scherrer 1956)!

Les essais fondamentaux ne portaient en réalité que sur un nombre restreint de patients (Zoeller et Ramon 1926, Ramon et Zoeller 1927f). Leurs résultats sont reconnus avec objectivité: « l'injection de sérum déprime chez certains sujets la valeur de l'immunité active ». « Dix jours après une injection de 10 cc de sérum, il arrive que l'immunité passive ait disparu; or, à ce moment, l'immunité active n'existe pas encore; il faut attendre quelques jours avant qu'une seconde injection ... en provoque l'apparition. Il y a donc, entre l'immunité passive et l'immunité active un écart variable selon les sujets. » Le problème touchant à l'efficacité de la séro-vaccination était nettement posé.

De nombreux praticiens ont adopté la méthode de séro-anatoxiprophylaxie en se rapportant aux résultats obtenus par Clavel et Clavel (1933). Ces auteurs ont traité 30 blessés par la séro-vaccination, dont 8 furent contrôlés sérologiquement 1—18 mois après la seconde injection d'A.Te; 6 d'entre eux avaient un titre inférieur à 0,01 U.I./cc; 1 malade possédait 0,025 et l'autre, traité avec plusieurs injections d'A.Te, 0,3 U.I./cc.

Les observations de Sacquépée (1933), Sacquépée et Jude (1937) ne fournissent pas d'indication sur la phase critique; elles suggèrent par contre que la séro-vaccination en trois temps: Sérum + A.Te et deux injections ultérieures d'A.Te aboutit à une immunité très satisfaisante et que, selon les auteurs, la réactivité à l'I.R. est la même, que le sujet ait reçu la première injection d'A.Te associée ou non au sérum. 6 mois après l'immunité de base, 127 sérums titrent 0,1—0,5 U.I./cc, 110 ont 0,01—0,035, et 3 seulement moins de 0,01 U.I./cc.

Une I.R. faite un an plus tard laisse après 6 mois des taux d'antitoxine qui s'échelonnent entre 0,005 et 25 U.I./cc, le 96% des sujets ayant un titre égal ou supérieur à 0,17 U.I./cc.

Une étude déjà plus nuancée, mais ne portant que sur 35 sujets, est celle de GOLD et BACHERS (1943), étude dont le tableau 13 résume l'essentiel.

Le tableau 13 démontre en particulier que ni l'intensité, ni la durée de l'immunité passive ne sont influencées par le vaccin, puis que l'A.Te combinée au sérum lors de la première injection est moins efficace seulement si l'immunisation de base ne comporte que deux injections d'antigène.

La plupart des auteurs se rallient aux conclusions de CLAVEL et CLAVEL (1933) (OTTO 1934, CINO 1936 et al.) ou de GOLD et BACHERS (1943) (ERICCSON 1948, TZAMALOUKAS 1950, BAUMANN 1958 et al.). Même lorsque le sujet n'a reçu qu'une première injection d'A.Te associée à l'antitoxine, la réponse ultérieure à l'I.R. est remarquable (REDDI 1935, CARO et LILLO 1955, SCHOBER 1956, MÖRL 1956, ECKMANN 1957b, 1958). Au bout de 2—3 ans, les patients d'ERICCSON (1948) au seul bénéfice d'une injection d'A.Te associée au sérum répondent à l'I.R. par des titres qui dépassent parfois 2,0 U.I./cc; ceux d'ECKMANN (1957 h) (cf. diagrammes 11 et 12) démontrent que la remontée du titre antitoxique est un peu plus lente, comparée à celle des sujets qui jouissent d'une immunisation de base plus riche; le maximum du bond est atteint après 16—20 jours (au lieu de 8—12 jours), mais la hauteur du bond est tout aussi élevée.

La question la plus grave du problème de la séro-vaccination, celle qui touche à la période de carence, n'a pas trouvé de solution. Les partisans de la seule séroprophylaxie chez le blessé non vacciné sont encore nombreux (MÖRL 1955, RICHTER 1955, SCHLEGEL 1956 et al.); ils ne tolèrent pas le compromis et séparent dans le temps les immunisations passive et active. Par contre, les

Tableau 13. *Effet de l'injection simultanée de sérum antitétanique et de différents vaccins tétaniques en fonction du nombre des doses d'antigène et de l'intervalle entre les injections.* (Extrait des résultats de GOLD et BACHERS 1943)

Titres en U.I./cc après

Groupe	Traitement			0	1 jour	1 sem.	2 sem.	4 sem.	5 sem.	8 sem.	9 sem.	10 sem.	12 sem.	14 sem.	16 sem.	20 sem.	24 sem.
1	S			< 0,004	0,18	0,30	0,01	< 0,004									
2	S + Ap		Ap	< 0,004	0,16	0,21	0,01	0,04					> 0,007 ■	0,6	0,6		0,14
3	Ap		Ap	< 0,004									< 0,004	0,34	1,3		0,7
4	S + Af		Af	< 0,004	0,15	0,17	0,05	0,005		< 0,004 ■	0,004	0,06	0,003			0,002	
5	Af		Af	< 0,004						< 0,004	< 0,01	0,25	0,19			0,009	
6	S + Af	Af	Af	< 0,004	0,19	0,18	0,02	0,002 ■	0,002	0,003 ■	0,15	0,4	0,42			0,24	
7	Af	Af	Af	< 0,004					0,003	0,04	0,5	0,75	0,41			0,43	

S = 1500 U.I. d'antitoxine tétanique; Ap = 1 cc d'anatoxine tétanique alunée; Af = 1 cc d'anatoxine tétanique fluide; ■ = injection de l'antigène.

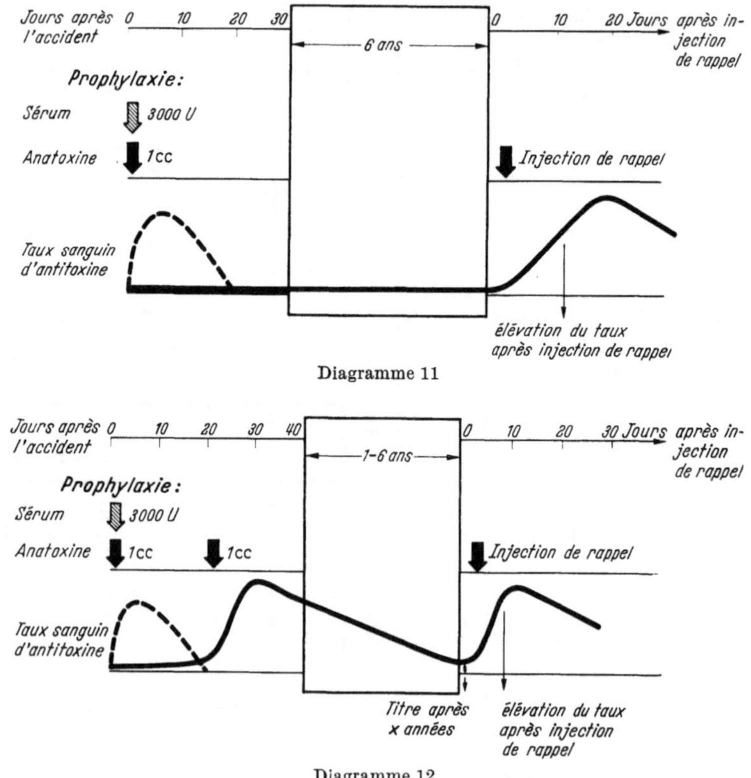

Diagramme 11

Diagramme 12

Diagrammes 11 et 12. *Réactivité à l'injection de rappel de sujets ayant reçu autrefois:*
diagramme 11: 1500 U.I. d'antitoxine tétanique et 1 cc d'anatoxine tétanique,
diagramme 12: même traitement que dans le diagramme 11 mais, en plus, 2ᵉ injection d'anatoxine après 3
semaines.
— — — Courbe d'immunité passive. — Courbe d'immunité active . (Selon Eckmann 1957 b)

partisans de la séro-vaccination invoquent des arguments qui ne sont pas sans valeur:

— l'administration simultanée de sérum et de vaccin ne gêne pas notablement l'effet de l'antitoxine;

— même si l'A.Te associée au sérum ne développe pas tout son effet lors de la première injection, elle n'empêche pas l'établissement ultérieur de l'immunité active excitée par d'autres injections d'antigène;

— si l'injection simultanée contribue à l'apparition d'une protection si minime soit-elle, c'est autant de gagné.

Il n'y pas lieu de prêter à la période de carence une signification que l'on est peut-être tenté d'exagérer. Ce que l'on sait sur le minimum d'immunité nécessaire pour résister à l'infection tétanique naturelle est lacunaire: le sérum n'aurait-il pas une efficacité plus prolongée que les mesures sérologiques ne le laissent supposer? Et l'immunité active n'a-t-elle pas une utilité plus précoce que celle décelée par les méthodes conventionelles? A première vue et en dehors de toute statistique assurée, il semble que le nombre considérable de traitements séro-vaccinaux ait rencontré peu d'insuccès [un cas cité par Scherrer (1956) chez un séro-vacciné ayant reçu deux injections d'A.Te aluminée, après persistance

dans la plaie d'un fragment tétanifère; un autre cas (REGAMEY, non publié)
chez un enfant de 3 ans, ayant reçu 1 cc d'A.Te et 275 U.I. de sérum antitétanique
lors d'une opération sur un ancien foyer d'ostéomyélite; apparition d'un tétanos
4 semaines plus tard; malgré anatoxi- et sérothérapie, exitus].

Si le type de vaccin le plus propice n'est pas encore déterminé, VIGODTCHIKOV (1956)
recommande une anatoxine purifiée, concentrée et adsorbée. En Allemagne, seules les A.Te
aluminées sont utilisées; en France, on n'emploie encore que l'A.Te brute. Des recherches
systématiques chez l'homme sur la possibilité de mélanger le sérum et le vaccin dans la même
seringue n'ont pas été faites. Cette pratique simplifierait le traitement.

La quantité d'antitoxine injectée en même temps que l'A.Te a son importance
sur le freinage de l'immunité active (COOKE et JONES 1943; J.A.M.A. 1954 b;
ECKMANN 1957 a, 1958 et al.). On ne sait pas toutefois si l'interférence est due
à l'action spécifique de l'antitoxine, ou si elle est imputable au support protéinique
de l'anticorps. Dans la deuxième alternative, qui semble justifiée par les cons-
tations de D'ANTONA (1935), on pourrait réduire notablement le rôle inhibiteur
de la globuline en recourant à des sérums puissants et hautement purifiés.

Le problème de la séro-vaccination se cristallise autour du premier temps du
traitement, celui de l'administration simultanée de l'antitoxine et de l'A.Te.
Pour le reste, il entre dans le cadre de l'immunisation active contre le tétanos.

8. La séro-anatoxithérapie antitétanique

Dans sa première note sur la séro-anatoxithérapie, RAMON (1937 e) propose
l'usage de l'A.Te au cours des toxi-infections en évolution pour réduire la quantité
et le nombre des injections de sérum antitétanique, puis pour éviter les rechutes
précoces ou les récidives lointaines de la maladie. Les expériences sur le lapin
(RAMON et coll. 1938 d) démontrent que les injections d'A.Te poussées dans un
organisme passivement hyperimmunisé créent en effet une immunité active;
toutefois les doses choisies d'A.Te sont énormes et correspondent à un volume de
75 — 100 cc chez l'homme. Les premiers essais sur cinq sujets (RAMON et coll.
1938 f) démontrent que l'immunité passive conférée par de grosses doses de sérum
se continue après des temps variables, 50—150 jours, par l'immunité active.
Dans une étude sur 13 malades (RAMON et coll. 1939 d) notent toutefois que, dans
un organisme hyperimmunisé, les injections d'A.Te ne semblent pas jouer le
rôle habituel dévolu à l'I.R. (l'A.Te est poussée aux doses de 2 cc au début,
puis 3 cc le sixième jour, 4 cc le treizième jour, 5 cc le vingtième jour etc).
RAMON (1939 b, 1940 k) précise sa méthode et recommande une injection unique
de 75000 U.I. de sérum antitétanique en même temps que 2 cc d'A.Te, puis à
5—6 jours d'intervalle 2, 4 et 6 cc d'A.Te. La nouvelle thérapie est aussi introduite en médecine vétérinaire (RICHOU 1947).

La séro-anatoxithérapie suscita un grand intérêt dans les pays de langue latine.
Les quelque vingt publications parues de 1939 à 1950 rapportent des résultats
étonnants, la plupart 100% de guérisons. Mais le nombre peu élevé des cas
traités (1—13 par auteur) pourrait suggérer que seules les affections à décours
favorable ont retenu l'attention. Les travaux plus récents paraissent par contre
objectifs.

Compte tenu que la séro-anatoxithérapie fut conduite en même temps que
d'autres méthodes nouvelles qui, elles aussi, ont pu jouer un rôle favorable dans

la statistique, on peut se demander si l'usage de la séro-anatoxiprophylaxie offre un avantage réel.

Quelques études expérimentales consacrées au mécanisme de la séro-anatoxithérapie ont été entreprises ces dernières années par quatre groupes de chercheurs.

— Pour Giberti et Ponzoni (1951), Ambrosioni et Giberti (1951), Ponzoni et Giberti (1951), l'A. Te injectée au cobaye avant ou après la toxine, ou encore simultanément, exerce une action thérapeutique indiscutable; elle a une efficacité comparable à celle d'un sérum antitoxique spécifique. Les auteurs supposent que le vaccin déclenche un stress qui provoque une décharge de 11-oxy-corticostéroïdes et favorise ainsi la résolution des phénomènes morbides.

— Wolters et Dehmel (1951) constatent que le cobaye infecté avec des spores tétaniques peut être sauvé par l'injection d'A.Te aluminée.

— Regamey et Aegerter (1951b) et Aegerter (1954) n'observent, chez le cobaye également, qu'un effet thérapeutique extrêmement discret et seulement si les injections de vaccin sont répétées plusieurs jours de suite; par ailleurs, l'injection simultanée d'A.Te et de sérum affaiblit notablement le pouvoir thérapeutique de l'antitoxine. Ces auteurs sont donc en désaccord avec les chercheurs précédents, ce que peut expliquer le choix des doses de vaccin. Regamey et Aegerter ont utilisé des doses faibles, calculées en fonction de la posologie chez l'homme et ramenées au poids du cobaye, tandis que les auteurs italiens et allemands ont poussé des quantités d'A.Te correspondant à 150—500 cc pour l'homme!

Auteurs	Nombre de cas	Pourcentages de guérison avec la séro-anatoxithérapie %	avant %
Lavergne (1943)	294	50	40
Winkelbauer (1948)	54	78	66
Siméon et coll. (1950)	296	65—75	55—60
Köle (1951)	80	78	65
Tzamaloukas (1954)	19	84	?

— Lemétayer et son équipe de Garches ont fait en 1949 (b, c, e, f, h, i) une série de travaux sur l'action thérapeutique de l'A.Te. Ils ont constaté en particulier que si l'A.Te n'avait pas d'action *in vitro* sur le complexe cerveau-toxine tétanique, elle était par contre capable de rendre la liberté à la toxine fixée par l'antitoxine. Le déplacement est manifeste lorsqu'on injecte de l'anatoxine au cobaye traité préalablement par un mélange neutre de toxine et d'antitoxine: l'animal succombe au tétanos (phénomène décrit par Kjaer en 1931). L'A.Te n'a aucun pouvoir thérapeutique sur le tétanos local.

Pour autant que la séro-anatoxithérapie ait une efficacité réelle, H. Schmidt (1952a, p. 106) envisage deux explications possibles:

1. L'A.Te, en bloquant les récepteurs de la cellule nerveuse, empêcherait la toxine de s'y ancrer; elle pourrait même déplacer la toxine déjà fixée. Cette hypothèse trouve un certain appui dans les constatations, rapportées ci-dessus, de Lemétayer et de son école. Elle semble également corroborée par les travaux de d'Antona et Valensin (1934), de Wolters et Fischoeder (1954), de Fulthorpe (1956) et al., qui ont examiné la capacité de fixation du tissu nerveux et reconnu les possibilités d'un blocage par l'A.Te. Mais cette supposition se heurte à l'interprétation de Pons (1939), selon laquelle le déplacement obéit avant tout à une loi de masse et non à des processus d'avidité; l'A.Te répartie dans tout l'organisme n'atteindrait jamais, aux doses usuelles, la concentration nécessaire pour répondre aux conditions de cette loi.

2. L'A.Te arrive rapidement à déclencher des processus immunitaires (cf. Krech 1949 et al., p. 310). Les anticorps seraient sessiles au début, donc non décelables par des recherches sérologiques. Cette hypothèse paraît peu probable, car il faudrait imaginer des anticorps paralysés pendant une période prolongée, ce qui ne correspondrait pas aux phénomènes immunitaires connus.

Une troisième hypothèse pourrait être retenue, celle d'une réaction apparentée aux phénomènes d'interférences décrits par Evans et Perkins (1954a, b, 1955). Le vaccin coquelucheux rend en l'espace de quelques heures — et pour une dizaine de jours — la souris quasi-insensible à l'infection par H. pertussis vivant, aucune manifestation de type immunitaire ne semblant être en jeu. L'A.Te rend également pour une courte période l'animal moins réceptif à l'égard de la toxine tétanique. Cet effet, qui pourrait s'exercer en dehors de tout processus immunitaire, est-il propre à l'A.Te ? D'autres substances, organiques ou non, biologiques ou non, semblent posséder des propriétés analogues.

Les bases expérimentables de la séro-anatoxithérapie sont encore chancelantes. Elles méritent d'être approfondies, car le traitement du tétanos est resté décevant et toute méthode susceptible d'augmenter les chances de guérison doit retenir l'attention.

9. Calendriers de vaccination antitétanique

Grande-Bretagne

(Barr et Sachs 1955)

Pendant la guerre de 1939—1945, les Anglais recoururent à différents modes de prévention caractérisés par *l'absence d'I.R. en cas de blessure.*

  a) De 1938 à janvier 1941
    2 × 1 cc d'A.Te à 6 semaines d'intervalle.

  b) Dès janvier 1941
    3 × 1 cc d'A.Te à 6 semaines d'intervalle.

  c) Dès novembre 1942
    I.R. annuelle avec 1 cc d'A.Te
      En cas de blessure:
    chez le vacciné: 1500 U.I. d'antitoxine tétanique,
    chez le non-vacciné: 3 fois 1500 U.I. à 1 semaine d'intervalle.

Procédure actuelle (valable pour hommes, femmes et enfants)
2 × 1 cc d'A.Te à 6—12 semaines d'intervalle, sous-cutané.
1 cc d'A.Te 6 à 12 mois plus tard.
I.R. = 1 cc d'A.Te tous les 5 ans.

En cas de blessure:
sujets immuns: 1 cc d'A.Te,
sujets non immuns: 1 cc d'A.Te et 1500 U.I. d'antitoxine tétanique, sous-cutané, à répéter
                  2 ou 3 fois à 7 jours d'intervalle, selon l'appréciation du médecin.
Sont considérés comme *immuns* ceux qui ont reçu *2 injections* d'A.Te selon le schéma précédent, dans les 6 mois qui suivent la seconde injection, et ceux qui ont reçu *3 injections* dans les 5 ans qui suivent la troisième injection.
Tous les autres cas sont considérés comme étant *non immuns.*

France

(Deparis 1956)

3 injections d'A.Te de 1; 2 et 2 cc, à 2 ou 3 semaines d'intervalle.
I.R. de 2 cc d'A.Te, après 1 an.

En cas de blessure:
chez le vacciné: I.R. de 2 cc, d' A. Te,
chez le non-vacciné: 2 cc d'A.Te et 1500 U.I. d'antitoxine tétanique, ensuite 2 injections
                  de 2 cc d'A.Te à 15 jours d'intervalle.

Suisse

A. Dans l'Armée (REGAMEY 1947 c, 1955 a):
  a) Dès janvier 1940
        3 injections de 1; 2 et 2 cc de vaccin Te TAB sous-cutané, à 2—3 se maines d'intervalle.
  b) Dès l'été 1940
        3 injections de 1; 2 et 2 cc d'A.Te à 4 semaines d'intervalle.
  c) Dès 1956
        3 × 1 cc d'A.Te à 4 semaines d'intervalle.
        I.R. prévue seulement en cas de traumatisme: 1 cc d'A.Te.

B. Schéma de SCHLEGEL (1956), Zürich:
        3 injections de 0,5; 1 et 1 cc d'A.Te aluminée ou
        3 injections de 1; 2 et 2 cc d'A.Te fluide à 6—12 semaines d'intervalle.
        I.R.: 2 cc d'A.Te fluide.

En cas de blessure:
Blessé avec au moins 2 injections de vaccin antitétanique, associé ou non, la dernière injection remontant à 15 jours au moins et 6 mois au plus: I.R.
Blessé non immun: 3000 U.I. ou, si le blessé a déjà reçu une fois du sérum, 5000 U.I. d'antitoxine tétanique. L'immunisation active est instituée 6—8 semaines plus tard.

C. Schéma de ECKMANN (1958), Bâle:
Immunisation de base selon recommandations du producteur de vaccin.

En cas de blessure:
Sont considérés comme *immuns* les sujets
a) qui au cours des 6 derniers mois ont reçu *2 injections* d'A.Te séparées par un intervalle d'au moins 3 semaines, la dernière injection devant remonter à 7 jours au moins,
b) qui ont reçu au cours des 18 mois précédents *plus de 2 injections* à 20 jours d'intervalle au moins.
Tous les autres sujets sont considérés comme étant *non immuns.*
Les sujets immuns ne reçoivent aucun traitement spécifique. Les sujets non immuns reçoivent tous une injection de 0,5 cc d'A.Te dépôt. En outre,
a) les blessés ayant eu au préalable *1 injection* d'A.Te au moins 20 jours plus tôt reçoivent dans la règle 1000 U.I. de sérum antitétanique. I.R. faite après 3—5 mois;
b) les blessés n'ayant jamais eu d'A.Te bénéficient de 1500—3000 U.I. de sérum. Injections d'A.Te après 1, puis 3—4 mois;
c) les blessés qui ont déjà reçu du sérum autrefois doivent disposer de 5000-10000 U.I. de sérum.

Etats-Unis

(LONG et SARTWELL 1947)

Dans l'Armée de terre
  Dès 1941
        3 × 1 cc d'A.Te *fluide* à 3 semaines d'intervalle.
        I.R. de 1 cc d'A.Te après 1 an.
  Dès avril 1941 à septembre 1944
        I.R. avant le départ au front pour autant que l'injection précédente remonte à plus de 6 mois.
  En cas de blessure, brûlures et avant une opération:
        I.R. de 1 cc.

Dans la Marine
        2 × 0,5 cc d'A. Te *alunée* à 5—8 semaines d'intervalle.
        I.R.: 0,5 cc d'A.Te alunée après 1 an.
        I.R.: 0,5 cc après 4 ans ou lors du départ pour le front.
  En cas de blessure:
        I.R.: 0,5 cc.
  Dès 1949, *pour toute l'Armée* (LOONEY et coll. 1956)
        2 × 0,5 cc d'A.Te *alunée* à 1 mois d'intervalle.
        I.R.: 0,5 cc d'A.Te alunée environ 1 an plus tard, puis tous les 4 ans ou lors de traumatisme.

Propositions pour les I.R. d'entretien
(après l'immunisation de base)

TURNER et coll. (1954):
I.R. tous les 5 ans, bien que le rappel tous les 10 ans devrait suffire chez la plupart des vaccinés.

REGAMEY (1955a):
enfant: I.R. 1 an après le premier cycle de vaccination, puis tous les 3—5 ans,
adulte: I.R. tous les 10—12 ans.

PETERSON et coll. (1955):
enfant: I.R. après 1 an,
I.R. à l'âge scolaire, puis tous les 10 ans,
adulte: I.R. après 5 ans, puis tous les 10 ans.

Indications sur la conduite à tenir chez les allergiques

Cf. H. SCHMIDT (1952, p. 90—91)
EDSALL (1956b)
CHRISTENSEN (Mayo clinic) (1957).

Bibliographie

Les titres de quelques publications antérieures et possédant le caractère d'un exposé général sur l'immunisation active contre le tétanos sont imprimés en *italiques*. Le nom de l'auteur est en outre précédé d'un *astérisque*.

ABEL, J. J., E. A. EVANS and B. HAMPIL: Researches on tetanus. V. Distribution and fate of tetanus toxin in the body. Bull. Johns Hopk. Hosp. **59**, 307—392 (1936).

— B. HAMPIL, A. F. JONAS and W. CHALIAN: Researches on tetanus. VII. (1) The time required for the fixation of a fatal quantity of tetanus toxin; (2) the return passage of toxin by way of the lymphatic capillaries to the cardiovascular system; (3) the return passage as the basis of a method for the approximate determination of the volume of lymph in the closed lymphatic system. Bull. Johns Hopk. Hosp. **62**, 522—563 (1938a).

—, and W. CHALIAN: Researches on tetanus. VIII. At what point in the course of tetanus does antitetanic serum fail to save life? Bull. Johns Hopk. Hosp. **62**, 610—633 (1938b).

AEGERTER, F.: L'immunisation simultanée, active et passive, contre le tétanos. Etude sur le cobaye. Thèse de doctorat, Berne 1954.

ALDER, F. L.: Antibody formation after injection of heterologous immune globulin. II. Competition of antigens. J. Immunol. **78**, 201—210 (1957).

AMBERG, B.: Die Flockungsreaktion nach RAMON und ihre Anwendung zum Auswerten von Diphtherie- und Tetanus-Serumfraktionen. Thèse de doctorat, Berne 1946.

AMBROSIONI, P., e A. GIBERTI: Ricerche sperimentali sull'azione della anatossina e del siero antitossico nel trattamento preventivo e curativo dell'intossicazione tetanica. Riv. Ist. sieroter. ital. **26**, 85—111 (1951).

—, e A. MURGIA: Ricerche sperimentali sul valore immunitario della siero-anatossiprofilassi e della siero-anatossiterapia nel tetano. Notiz. Ammin. sanit. **7**, fasc. III (1946).

AMOUREUX, G.: Composition chimique du milieu de culture et formation de la toxine tétanique. C. R. Soc. Biol. (Paris) **137**, 351—353 (1943a).

— Comparaison entre les digestions pepsique et papaïnique. Application à la préparation de la toxine tétanique. Rev. Immunol. (Paris) **8**, 119—124 (1943b).

—, et J. POCHON: Dégradation des substances protéiques des milieux de cultures et toxinogenèse tétanique. C. R. Soc. Biol. (Paris) **135**, 1060—1062 (1941a).

— — Recherches sur le métabolisme d'une souche toxigène du bacille tétanique. C. R. Soc. Biol. (Paris) **135**, 1169—1172 (1941b).

ANDRÉ, L., J. MARTY, R. RENNER, P. GIUDICELLI et G. DESMOULINS: Leucose aiguë survenue après une vaccination T.A.B.D.T. Presse méd. **1958**, 689.

— P. M. G.: Etude sur «Plectridium tetani». I. Utilisation des glucides contrôlée par la méthode de Somogyi; II. Toxinogenèse en rapport avec la glucidolyse. Thèse de doctorat, Paris 1948.

D'ANTONA, D.: Varianti aerobiche ed anaerobiche del bacillo tetanico. G. Batt. Immun. **13**, 965—992 (1934).
— Il problema della immunizzazione antitossica con siero ed anatossina. G. Batt. Immun. **14**, 305—332 (1935).
— Une méthode pour provoquer la résorption des antigènes ou des toxines à travers la peau: Intoxication générale aiguë et mort des animaux par application percutanée de toxine diphtérique ou tétanique. Rev. Immunol. (Paris) **14**, 101—122 (1950).
*— *Le tétanos. Synthèse théorique et pratique de nos principales connaissances.* Rev. Immunol. (Paris) **15**, 93—157 (1951a).
— Vaccination percutanée. 4ème réunion d'Unisérum. Siena 31 mai — 2 juin 1951b.
*— *La vaccination contre le tétanos.* Rev. Immunol. (Paris) **16**, 1—85 (1952).
—, et S. PIAZZI: Effets de la variation quantitative et qualitative des stimuli antigéniques primaires et secondaires. Mécanisme d'action des adjuvants de l'immunité. Rev. Immunol. (Paris) **20**, 317—344 (1956).
—, et M. VALENSIN: Préparation de la toxine tétanique par la méthode de R. Legroux et G. Ramon. C. R. Soc. Biol. (Paris) **115**, 1209—1211 (1934).
— — e D. CENTINI: Richerche sulla vaccinazione associata antitetanica — antitifica — paratifica. Dati che depongono per la superiorità dell'immunità antitetanica attiva sulla immunità passiva. Minerva med. (Torino) **1938**, 221. Rif. Bull. Inst. Pasteur **37**, 980 (1939).
A.P.H.A.: Control of communicable diseases in man, p. 183. New York: American Publ. Health Association 1955.
ARMSTRONG, C.: Tetanus following vaccination against small pox, and its prevention with special reference to the use of vaccination shields and dressing. Publ. Hlth Rep. (Wash.) **42**, 3061—3071 (1927).
AULD, A. G.: Anaphylaxis after injection of tetanus toxoid. Brit. med. J. **1**, 368 (1940).
AXNICK, N. W., and E. R. ALEXANDER: Tetanus in the United States: A review of the problem. Amer. J. Publ. Hlth **47**, 1493—1501 (1957).
BÄCHER: Sensibilisierende Eigenschaften der Toxoide (Anatoxine). 12. Tagg der Dtsch. Ver.igg für Microbiologie in Wien. Zbl. Bakt., I. Abt. Orig. **104**, 150*—152* (1927).
— — R. KRAUSS u. E. LÖWENSTEIN: Zur Frage der aktiven Schutzimpfung gegen Diphtherie. Z. Immun.-Forsch. **42**, 350—368 (1925); **45**, 86—92 (1926).
BAIRD, H. W.: The duration of immunity to tetanus. Yale J. Biol. Med. **21**, 385—390 (1949).
BALLINARI, R.: Studien über die Anwendung der strömenden Adsorption (Chromatographie) zur Reinigung und Konzentrierung bakterieller Toxine. Thèse de doctorat, Berne 1941.
BANDMANN, F.: Zum Nachweis von Tetanusbacillen im Darm von Ulcus- und Carcinomträgern. Z. Hyg. Infekt.-Kr. **136**, 559—567 (1953).
BANTON, H. J., and P. A. MILLER: An observation of antitoxin titers after booster doses of tetanus toxoid. New Engl. J. Med. **240**, 13—14 (1949).
BARANDUN, S., H. J. HUSER u. A. HÄSSIG: Klinische Erscheinungsformen des Antikörpermangelsyndroms. Schweiz. med. Wschr. **1958**, 78—82.
BARR, M.: The effect of typhoid-paratyphoid A and B vaccine on the antigenic efficiency of tetanus toxoids of different quality. Brit. J. exp. Path. **37**, 11—19 (1956).
— The preparation, composition and control of combined vaccines. Combined vaccines containing tetanus toxoid. Troisième rencontre internationale de standardisation biologique. Opatijà 2—6 septembre 1957b.
— A. J. FULTHORPE and M. LLEWELLYN-JONES: The immunity responses of mice to injections of diphtheria and tetanus prophylactics. Brit. J. exp. Path. **38**, 312—318 (1957a).
—, and A. T. GLENNY: The relation between antitoxic responses to prophylactic injections of different antigens. Brit. J. exp. Path. **31**, 779—783 (1950).
— The effect of certain non-specific factors on the production of antitoxin. Brit. J. exp. Path. **33**, 543—561 (1952).
—, and M. LLEWELLYN-JONES: Factors influencing the development of potential immunity and the character of the secondary response. Brit. J. exp. Path. **32**, 231—245 (1951).
— — Some factors influencing the response of animals to immunization with combined prophylactics. Brit. J. exp. Path. **34**, 12—22 (1953a).

BARR, M., and LLEWELLYN-JONES: Interference with antitoxic responses in immunization with combined prophylactics. Brit. J. exp. Path. 34, 233—240 (1953b).

—, and A. SACHS: Report on the investigation into the prevention of tetanus in the British Army. The War Office 11 262, London 1955.

BARRETT, C. D., E. A. TIMM, J. G. MOLNER, B. I. WILNER, C. P. ANDERSON, H. E. CARNES and I. W. MACLEAN: Multiple antigen for immunization against poliomyelitis, diphtheria, pertussis and tetanus. J. Amer. med. Ass. 167, 1103—1207 (1958).

—, and W. VOLWILER: Agammaglobulinemia and hypogammaglobulinemia. The first five years. J. Amer. med. Ass. 164, 866—870 (1957).

BASSET, J.: Vaccinations obligatoires collectives et accidents de vaccination. Vaccination contre la diphtérie, le tétanos et les fièvres typho-paratyphoïdes. Rev. méd. Vét. 102, 613—628 (1951).

BATSON, H. C.: Statistical methods in immunology. J. Immunol. 66, 737—756 (1951).

— A. CHRISTIE, B. MAZUR and J. H. BARRICK: Response of the young infant to poliomyelitis vaccine given separately and combined with other antigens. Pediatrics 21, 1—7 (1958).

BAUER, J. H., and K. F. MEYER: Human intestinal carriers of tetanus spores in California. J. infect. Dis. 38, 295—305 (1926).

BAUMANN, E.: Aussprache zum Vortrag A. HÜBNER. XV. Jahrestagung der dtsch. Ges. für Unfallheilkunde, Versicherungs- und Versorgungsmedizin. 21. Okt. 1951.

— Diskussionsvotum. Hefte zur Unfallmedizin Nr 56, S. 32—33, 1957.

BECK, W.: Aluminium hydroxide granuloma after immunization against tetanus. Medizinische 1954, 363—365. Ref. J. Amer. med. Ass. 155, 795 (1954).

BEHEYT, P.: La recrudescence du tétanos à Léopoldville. Ann. Soc. belg. Méd. trop. 30, 341—348 (1950).

BELIN, M., S. MUTERMILCH et E. SALAMON: Du rôle du pH dans la protection de la toxine tétanique. Action d'un tampon, le glycocolle. C. R. Soc. Biol. (Paris) 115, 1476—1478 (1934).

BELLER, A.: Untersuchung von 51 Krankheitsfällen, die in den Jahren 1940/41 als T.P.T.-Impfschäden bei der E.M.V. angemeldet worden sind, auf ihren ursprünglichen Zusammenhang mit der T.P.T.-Impfung. Thèse de doctorat, Berne 1943.

BENNET, A. E., and M. GRODINSKY: Chronic and delayed tetanus. Unusually prolonged incubation period in case of the tetanus. Neb. St. med. J. 20, 161 (1935). Réf. Bull. Inst. Pasteur 33, 1022 (1935).

BENZONI, G.: L'azione stabilizzante del peptone sulla tossina tetanica diluita. Riv. Ist. sieroter. ital. 22, 70—79 (1947).

—, e V. LA ROCA: Esperienze comparative di vaccinazione umana contro il tifo, i paratifi ed il tetano con vaccini T.A.B. all'acetone, all'alcool ed al formolo in dose unica e duplice, associati ad anatossina. Boll. Ist. sieroter milan. 34, 114—164 (1955).

BÉRARD, L., et A. LUMIÈRE: Rechutes et récidives de tétanos. Presse méd. 1925, 993—994.

BERGEY, D. H., C. P. BROWN and S. ETRIS: Immunization against tetanus with alum-precipitated tetanus toxoid. Amer. J. publ. Hlth 29, 334—336 (1939).

—, and S. ETRIS: Immunization of humans with alum precipitated tetanus toxoid. Amer. J. publ. Hlth 24, 582—586 (1934).

— — Active immunization against tetanus infection with refined tetanus toxoid. J. Immunol. 31, 363—371 (1936).

BERNARD, J. G., et P. SERVANT: Agression toxinique intracérébrale et immunité antitétanique chez le lapin. Rev. Immunol. (Paris) 21, 188—196 (1957).

BERNARD-GRIFFITHS, G., et GROSLIER: Les terrains tétanigènes dans le Puy-de-Dôme. Concours méd. 74, 4157—4158 (1952).

BERTHELOT, A., A. R. PRÉVOT, G. AMOUREUX et H. J. BOORSMA: Relations entre les teneurs en azote des divers constituants du milieu de culture et la toxinogenèse tétanique. C. R. Soc. Biol. (Paris) 131, 1133—1134 (1939).

BERTRAND, I., et D. QUIVY: Le problème des pyrogènes: son état actuel. Rev. Hémat. 2, 234—240 (1947).

BESSON, A., et P. GIRAUD: Sur les vaccinations effectuées à l'aide du vaccin triple associé (antidiphtérique, antitétanique, antityphoparatyphoïdique). Leur inocuité. Bull. Acad. Méd. (Paris) 128, 441—444 (1944).

BEUREY, J., M. PIERSON, E. DE LAVERGNE et J. C. BURDIN: Effet de l'ACTH et de la cortisone sur l'apparition de l'antitoxine tétanique chez le lapin. Action lors de l'injection de rappel. C. R. Soc. Biol. (Paris) 147, 849—852 (1953).

BICHSEL, G.: Über die Eignung verschiedener Desinfektionsmittel für die Konservierung von antitoxischen Seren und Impfstoffen. Thèse de doctorat, Berne 1949.

BICK, H., R. S. ABERNATHY and W. W. SPINK: Protection of mice with polyvinylpyrrolidone against lethal doses of tetanus toxin. Proc. Soc. exp. Biol. (N.Y.) 94, 385—388 (1957).

BIGLER, J. A.: Tetanus immunization. A ten year study. Amer. J. Dis. Child. 81, 226—232 (1951).

—, and M. WERNER: Active immunization against tetanus in infants and children. J. Amer. med. Ass. 116, 2355—2366 (1941).

BILLARD, G.: Immunisation non spécifique de certaines neurotoxines entre elles. C. R. Soc. Biol. (Paris) 95, 546—548 (1926).

BIRKHAUG, K. E.: Detoxifying and desinfecting properties of sodium salicylate. J. infect. Dis. 48, 212—225 (1931).

BLUM, F.: Über eine neue Klasse von Verbindungen der Eiweißkörper. Z. physiol. Chem. 22, 127—131 (1896/97).

BOIVIN, A., et Y. IZARD: Méthode pour la purification, à l'acide trichloracétique, des toxines et anatoxines diphtériques, tétaniques et staphylococciques. C. R. Soc. Biol. (Paris) 124, 25—28 (1937).

BONNEFOI, A., A. LAFFAILLE et R. PANTHIER: Indications sur l'ordre chronologique des vaccinations obligatoires ou recommandées. Rev. Prat. (Paris) 8, 53—56 (1958).

BONNEL, P. H., et C. RABY: Le contrôle de la stérilité des produits biologiques au moyen de milieux semi-liquides permettant la culture simultanée des aérobies et des anaérobies. Troisième rencontre internationale de standardisation biologique. Opatijà 2—6 septembre 1957.

BONNEY, V., C. BOX et T. MACLENNAN: Tetanus bacillus recovered from the scar ten years after an attack of post-operative tetanus. Brit. med. J. 1, 10—11 (1938).

BOORSMA, A. J., A. R. PRÉVOT et R. VEILLON: A propos de la fermentation du glucose par «Pl. tetani». C. R. Soc. Biol. (Paris) 131, 1137—1140 (1939).

BORCILA, J.: Immunisation du cobaye par l'anatoxine tétanique; essais comparatifs sur divers modes d'injection de l'antigène. C. R. Soc. Biol. (Paris) 119, 915—917 (1935).

BORISSOVA, L. V.: Sur les propriétés anaphylactogènes des anatoxines diphtériques et tétaniques. J. Microbiol. Epidemiol. Immunobiol. 27, No 4, 55—59 (1956). Réf. Bull. Inst. Pasteur 55, 1521 (1957).

BORSELLI, L.: Influenza del Periston N sull' intossicazione tetanica delle cavie. Riv. Clin. pediat. 57, 682 (1956).

BOURDELLÈS, B. LE, et J. DESBORDES: Etude comparative des dispositions réglementaires relatives aux techniques de contrôle des préparations immuno-biologiques dans quelques pays d'Europe. Troisième rencontre internationale de standardisation biologique. Opatijà 2—6 septembre 1957a.

— — L'organisation du contrôle des préparations immuno-microbiologiques dans quelques pays d'Europe. Rev. Hyg. Méd. soc. 5, 3—25 (1957b).

BOUSFIELD, G., and L. B. HOLT: Combined prophylactics. Variations in response, in children, to diphtheria toxoid produced by addition of tetanus toxoid and H. pertussis vaccine. Brit. med. J. 2, 1213—1215 (1957).

BOYD, J. S. K.: Active immunization against tetanus. J. roy. Army med. Cps 70, 289—307 (1938).

— Tetanus in the African and European theatres of war. 1939—1945. Lancet 1946 I, 113—119.

—, and J. D. MACLENNAN: Tetanus in the Middle East. Effects of active immunization. Lancet 1942. II, 745—749.

BOYER, J., L. CORRE-HURST, H. SAPIN-JALOUSTRE et M. TISSIER: Le tétanos en milieu urbain. Conditions d'apparition. Déductions prophylactiques. Presse méd. 1953, 701—703.

BRADFORD, W. L., E. DAY and F. MARTIN: Humoral antibody formation in infants aged one to three months injected with a triple (diphtheria-tetanus-pertussis) alum- precipitated antigen. Pediatrics 4, 711—718 (1949).

Brown, H. H.: Anaphylaxis after injection of tetanus toxoid. Brit. med. J. 2, 683 (1940).

Burnet, F. M., and F. Fenner: The production of antibodies. Melbourne: Ed. MacMillian & Co. 1949.

Buxton, J. B., and A. T. Glenny: The active immunisation of horses against tetanus. Lancet 1921 II, 1109.

Campbell, J. G. C.: The preparation of formol toxoid and alum precipitated toxoid. J. roy. Army med. Cps 70, 311 (1938).

Campell, R.: Tetanusprophylaxe in der Praxis mit besonderer Berücksichtigung der Verhältnisse im Kanton Graubünden. Schweiz. med. Wschr. 1929, 813—818.

Carlos, J., e F. Neto: Novas observações sòbre e formaçao de antitoxina tetânica em individuos vaccinados. Arch. Inst. Biol. Exerc. (Rio de J.) 9, 76—83 (1948). Ref. Bull. Inst. Pasteur 49, 1054 (1951).

Caro, M. de, e L. Lillo: Siero-vaccinazione antitetanica profilattica nei non vaccinati. R.C. Ist. sup. Sanità 18, 261—269 (1955).

Casile, M.: Un nouveau cas de tétanos après injections de quinine observé à Cayenne. Bull. Soc. Path. exot. Paris 43, 662—666 (1950).

—, et E. Rivierez: Tétanos consécutifs à des injections médicamenteuses diverses, autres que la quinine. Bull. Soc. Path. exot. Paris 44, 143—147 (1951).

Cavalli-Sforza, L. L.: Biometry of immune response. Arb. Staatsinst. exp. Ther. Frankfurt 52, 22—38 (1956).

Čertkova, F. A., u. E. S. Šain: Zur Frage der Tetanus-Revakzinierung. Z. Microbiol. Epidemiol. Immunol. 2, 50—54 (1957). Ref. Zbl. Bakt., I. Abt. Ref. 165, 296 (1957).

Chavannaz, J.: Les zones tétanigènes en France. Mém. Acad. Chir. 79, 46—47 (1953).

Chavany, J. A.: Les tétanos d'apparence bénigne à manifestations céphaliques prédominantes. Possibilités de reprises tardives à évolution mortelle. Nécessité d'une immunisation prolongée par la vaccination anatoxinique. Gaz. méd. Fr. 45, 383—385 (1938).

Chen, B.-L., C.-T. Chou, C.-T.Huang, Y.-T. Wang, H.-H. Ko and W.-C. Huang: Studies on diphtheria-pertussis-tetanus combined immunization in children. I. Heterologous interference of pertussis agglutinin and tetanus antitoxin response by pre-existing latent diphtheria immunity. J. Immunol. 77, 144—155 (1956).

— — — — — — and C.-F. Ling: Studies on diphtheria-pertussis-tetanus combined immunization in children. II. Immune responses after the primary vaccination. J. Immunol. 79, 39—45 (1957a).

— — — — — — Studies on diphtheria-pertussis-tetanus combined immunization in children. III. Immune responses after booster vaccination. J. Immunol. 79, 393—400 (1957b).

— W. K.: Stimulating effect of alum and T.A.B. vaccine in tetanus prophylaxis. Proc. Soc. exp. Biol. (N.Y.) 31, 334—336 (1933).

Chevalley, M., et P. G. Zivy: Vaccination par l'anatoxine mixte (diphtérique et tétanique) des enfants tuberculeux hospitalisés à Brévannes. Bull. Soc. Pédiat. Paris 37, 26—39 (1939).

Chevé, J., L. Nicol, Melle Amoureux, O. Girard et J. Zourbas: Etude comparative de vaccins triples antidiphtérique, antitétanique et anticoquelucheux simples et adsorbés. Troisième rencontre internationale de standardisation biologique. Opatijà 2—6 septembre 1957.

Cheyroux, M.: Sur les rapports entre la chute de toxicité et la fixation du formol pendant la transformation de la toxine tétanique en anatoxine. C. R. Soc. Biol. (Paris) 147, 988—990 (1953).

— Toxine tétanique et formol. Ann. Inst. Pasteur 86, 356—369 (1954).

Christensen, N. A.: Present concepts regarding prophylaxis of tetanus. Proc. Mayo Clin. 32, 160—166 (1957).

—, and D. L. Thurber: Clinical experience with tetanus: 91 cases. Proc. Mayo Clin. 32, 146—158 (1957).

— P. E.: Comparative studies on the rate of absorption after subcutaneous and intramuscular injection of tetanus antitoxin. Acta path. microbiol. scand. 31, 262—274 (1952).

Cimmino, A.: Nuovi orientamenti della profilassi antitetanica. Nuovi Ann. Ig. 2, 241—257 (1951).

CINADER, B., and J. M. DUBERT: Specific inhibition of response to purified protein antigens. Proc. roy. Soc. B **146**, 18—33 (1956).

—, and B. WEITZ: Beta- and gamma-globulin tetanus antitoxin of the hyperimmune horse. Nature (Lond.) **166**, 785 (1950).

CINO, J. A.: Per una razionale profilassi del tetano. Rif. med. **52**, 900—901 (1936).

CLAVEL, C., et Mme C. CLAVEL: Combinaison de la vaccinothérapie à la sérothérapie dans le traitement préventif du tétanos. Presse méd. **1933**, 1683—1684.

COCKBURN, W. C.: The use of mixed diphtheria-pertussis-tetanus antigens. Bull. Org. mond. Santé **13**, 409—422 (1955).

COHEN, P., and S. J. SCADRON: The effects of active immunization of the mother upon the offspring. J. Pediat. **29**, 609—619 (1946).

COLEMAN, G. E., and K. F. MEYER: Study of tetanus agglutinins and antitoxin in human serums. J. infect. Dis. **39**, 332—336 (1926).

COMIS, A.: Action de certaines levures et plus particulièrement de la fermentation alcoolique sur les toxines. C. R. Soc. Biol. (Paris) **98**, 1091—1093 (1928).

CONDREA, P.: Nouvelle méthode pour la préparation d'une toxine tétanique de titre élevé. C. R. Soc. Biol. (Paris) **103**, 1044—1046 (1930).

-- Toxicité spécifique de l'anatoxine tétanique. C. R. Soc. Biol. (Paris) **112**, 1499—1501 (1933).

—, et H. POENARU: L'influence des dilutions sur la toxicité de la toxine tétanique. C. R. Soc. Biol. (Paris) **112**, 1482—1484 (1933a).

— — Recherches sur la nature de la toxine tétanique. Influence de la peptone sur la toxicité de la toxine tétanique. C. R. Soc. Biol. (Paris) **112**, 1484—1487 (1933b).

COOKE, J. V., et F. G. JONES: The duration of passive tetanus immunity and its effect on active immunization with tetanus toxoid. J. Amer. med. Ass. **121**, 1201—1209 (1943).

— R. A., S. HAMPTON, W. B. SHERMAN and A. STULL: Allergy induced by immunization with tetanus toxoid. J. Amer. med. Ass. **114**, 1854—1858 (1940).

COPEMAN, W. J.: Tetanus reaction. Canad. med. Ass. J. **65**, 471 (1951).

COURMONT, J., et M. DOYON: La substance toxique qui engendre le tétanos résulte de l'action sur l'organisme récepteur d'un ferment soluble fabriqué par le bacille de Nicolaïer. C. R. Soc. Biol. (Paris) **45**, 294—298 (1893).

COWLES, P. B.: Immunization against tetanus. J. Bact. **33**, 113—114 (1937).

CUBONI, E.: Il bacillo del tetano nella corda di budello o catgut grezzo. Boll. Ist. sieroter. milan. **36**, 1—14 (1957).

CUNNINGHAM, A. A.: Anaphylaxis after injection of tetanus toxoid. Brit. med. J. **2**, 522—523 (1940).

CUVIER, G.: Rapport sur l'étiologie du tétanos en milieu rural. Cinquième congrès national de médecine rurale. Reims 12 juin 1955.

DAUSSET, J., et G. VIDAL: Accidents de la transfusion chez des receveurs de groupe A ayant reçu du sang de groupe 0. — Rôle de la vaccination par l'anatoxine diphtérique et tétanique. Sang **22**, 478—489 (1951).

DELPY, L. P., et H. MIR CHAMSY: Relation entre le pouvoir toxique et le pouvoir antigène des toxines diphtériques et tétaniques. Arch. Inst. Hessarek 2, 23—30 (1946).

— — Immunisation des chevaux contre la fièvre charbonneuse et le tétanos, par vaccins associés. Bull. Acad. vét. Fr. **24**, 62—65 (1951). Réf. Bull. Inst. Pasteur **50**, 705 (1952).

DELSAL, J. L., et H. MIR CHAMSY: La cryo-dénaturation sélective: méthode de purification des anatoxines tétaniques. C. R. Acad. Sci. (Paris) **239**, 600—602 (1954).

DEPARIS, M.: Comment prescrire vaccins et sérums. Paris: Expansion scientifique française, 1956.

DESCOMBEY, P.: L'anatoxine tétanique. C. R. Soc. Biol. (Paris) **91**, 239—241 (1924).

— Vaccination du cobaye par injection intra-cérébrale d'anatoxine tétanique. C. R. Soc. Biol. (Paris) **92**, 482—483 (1925a).

— Vaccination du cheval par l'anatoxine tétanique. Ann. Inst. Pasteur **39**, 485—504 (1925b).

— Sur la vaccination du cobaye contre le tétanos par injection intracérébrale d'anatoxine tétanique. Ann. Inst. Pasteur **43**, 634—643 (1929).

— Sur une technique de production de l'antitoxine tétanique; ses résultats. C. R. Soc. Biol. (Paris) **104**, 1157—1158 (1930).

DIAZ-RIVERA, R. S., E. RAMIREZ, E. R. PONS jr. and M. V. TORREGROSA: Management of
tetanus. Effect of penicillin on «Clostridium tetani», «in vivo». J. Amer. med. Ass.
147, 1635—1641 (1951).

DIDENKO, S. I.: Theses to the report «control of sera and vaccines sterility in the USSR».
Troisième rencontre internationale de standardisation biologique. Opatijà 2—6 septembre
1957.

DIEUDONNÉ, A., u. W. WEICHARDT: Immunität, Schutzimpfung und Serumtherapie. Leipzig:
Johann Ambrosius Barth 1925.

DIMA, G.: Immunitatea ocultà in tetanus la om si animali. Thèse vét. (Bucarest) 1937. Réf.
Bull. Inst. Pasteur. 37, 734 (1939).

DIXON, F. J., and P. H. MAURER: Specificity of the secondary response to protein antigens.
J. Immunol. 74, 418—431 (1955).

DONY, J.: Considérations sur les épreuves de stérilité. Pharm. Acta Helv. 33, 10—31
(1958).

DRESLER, K. G., et A. G. SABALDIR: Méthode de purification de l'anatoxine tétanique par
l'alcool. Oukr. biokhim. J. (ukrainien) 26, 74—79 (1957). Réf. Bull. Inst. Pasteur 55,
2465 (1957).

DUBOVSKY, B. J., and K. F. MEYER: The occurrence of B. tetani in soil and on vegetables.
J. infect. Dis. 31, 614—616 (1922).

DUNN, M. S., M. N. CAMIEN and L. PILLEMER: The amino acid composition of tetanal toxin.
Arch. Biochem. 22, 374—376 (1949).

EAGLE, A.: On the mutual multivalence of toxin and antitoxin. J. Immunol. 32, 119—127
(1937b).

— On the mutual multivalence of toxin and antitoxin. J. Bact. 33, 65 (1937a).

EATON, M. D.: Purification of tetanic toxin. Proc. Soc. exp. Biol. (N.Y.) 35, 16—19 (1936).

— The action of aldehydes on purified diphtheria toxin. J. Bact. 33, 52—53 (1937).

— Recent chemical investigations of bacterial toxins. Bact. Rev. 2, 3—45 (1938a).

—, et A. GRONAU: Comparative studies on the purification of tetanus and diphtheria toxins.
J. Bact. 36, 423—432 (1938b).

EBLEN, J. G.: Fatal reaction following triple immunization. J. Amer. med. Ass. 163, 1416
(1957).

ECKMANN, L.: Tetanus bei einem aktiv immunisierten Verwundeten. Communication per-
sonelle 29. 4. 1955.

— Zur Tetanusprophylaxe. Dtsch. med. Wschr. 1957a, 435—438.

— A propos de la prophylaxie du tétanos. Méd. et Hyg. (Genève) 15, 360 (1957b).

*— Die Prophylaxe des Tetanus unter besonderer Berücksichtigung der gleichzeitigen Anwendung
von Antitoxin und Toxoid. Thèse d'habilitation, Bâle 1959 (sous presse).

—, u. E. BISAZ: Tetanusprobleme in der Schweiz. Schweiz. med. Wschr. 1956, 641—645.

EDSALL, G.: Immunization of adults against diphtheria and tetanus. Amer. J. publ. Hlth 42,
393—400 (1952).

— Tetanus immunization. J. Amer. med. Ass. 156, 1350 (1954a).

— Immunization. Ann. Rev. Microbiol. 9, 347—368 (1955).

— Active and passive immunity of the infant. Ann. N.Y. Acad. Sci. 66, 32—43 (1956a).

— Tetanus immunization. J. Amer. med. Ass. 160, 1186 (1956b).

— The nature of the antigen and the basic characteristics of the immune response. J. Allergy
28, 1—17 (1957).

— J. S. ALTMAN and A. J. GASPAR: Combined tetanus-diphtheria immunization of adults:
use of small doses of diphtheria toxoid. Amer. J. publ. Hlth 44, 1537—1545 (1954b).

EGDAHL, R. H.: Immunological maturation and defects in immunological capacity. Int.
Arch. Allergy 12, 305—321 (1958).

EICHHORN, E., et R. RICHOU: Recherches expérimentales sur les «vaccinations associées»
(antitétanique et antistaphylococcique). C. R. Soc. Biol. (Paris) 121, 309—311 (1936).

EISLER, M.: Über Immunisierung mit durch Formaldehyd verändertem Tetanustoxin. Wien.
klin. Wschr. 1915, 1223—1225.

— Direkte und indirekte Giftwerte sowie Bakterieneiweiß in Bouillonkulturen von Tetanusbacillen. Z. Hyg. Infekt.-Kr. 135, 577—587 (1952).

—, u. H. EIBL: Über die Wirkungsweise der Aluminium-Diphtherieimpfstoffe. Klin. Med.
(Wien) 4, 441—453 (1949).

EISLER, M., u. GOTTDENKER: Über die Entgiftung des Diphtherietoxins durch Lanolin und Sterine, sowie über die Beeinflussung seines Immunisierungsvermögens durch Cholesterin. I. Mitt. Z. Immun.-Forsch. **90**, 427—451 (1937).

—, u. E. LÖWENSTEIN: Über Formalinwirkung auf Tetanustoxin und andere Bakterientoxine. Zbl. Bakt., I. Abt. Orig. **61**, 271—288 (1912).

—, u. F. SILBERSTEIN: Ein Beitrag zur Gewinnung von Tetanusserum. Z. Hyg. Infekt.-Kr. **89**, 29—47 (1919).

ERICCSON, H.: Studies on tetanus prophylaxis. J. clin. Path. **1**, 306—310 (1948).

EVANS, D. G., A. T. FULLER et J. WALKER: Chemotherapy in experimental tetanus. Lancet **1945 II**, 336—338.

—, and F. T. PERKINS: The ability of pertussis vaccine to produce in mice specific immunity of a type not associated with antibody production. Brit. J. exp. Path. **35**, 322—330 (1954a).

— — Interference immunity produced by pertussis vaccine to pertussis infection in mice. Brit. J. exp. Path. **35**, 603—608 (1954b).

— — The production of both interference and antibody immunity by pertussis vaccine to pertussis infection in mice. Brit. J. exp. Path. **36**, 391—401 (1955).

FALCHETTI, E.: Sur le sort de l'antitoxine lors des réinjections de sérum antitétanique par voie sous-cutanée. C. R. Soc. Biol. (Paris) **124**, 96—98 (1937).

FARAGÓ, F.: Über das Schicksal und die Wirksamkeit des Anatoxin-Praezipitatdepots im Organismus. Z. Immun.-Forsch. **86**, 191—204 (1935).

— Die Bedeutung unspezifischer Adsorbentien auf dem Gebiete der Immunitätslehre. Klin. Wschr. **1941**, 815—817.

—, and S. PUSZTAI: An investigation into the effect of combined diphtheria — tetanus — pertussis prophylactic. Brit. J. exp. Path. **30**, 572—581 (1949).

—, u. K. UJHELYI: Neuere Untersuchungen über die Depotwirkung des niederschlaghaltigen Impfstoffes. Z. Immun.-Forsch. **101**, 178—173 (1942).

FASAL, P.: Zur Frage der Tetanusprophylaxe bei Verbrennungen. Wien. klin. Wschr. **1935**, 181—182.

FEENEY, R. E., J. H. MUELLER and P. A. MILLER: Growth requirements of Clostridium tetani. II. Factors exhausted by growth of the organism. J. Bact. **46**, 559—562 (1943a).

— — — Growth requirements of Clostridium tetani. III. A « synthetic » medium. J. Bact. **46**, 563—571 (1943b).

FEIERABEND, B.: Active immunization in the guinea-pigs, horses and man with tetanus anatoxin. Čas. Lek. čes. **69**, 1457—1462 (1930). Ref. Amer. J. Dis. Child. **46**, 631 (1933).

FELIX, A., and M. ROBERTSON: Serological studies in the group of the spore-bearing anaërobes. I. The qualitative analysis of the bacterial antigens of «B. oedematis maligni» (vibrion septique) and «B. tetani». Brit. J. exp. Path. **9**, 6—18 (1928).

FERNAN-NUNEZ, M.: Tetanus toxoid: active immunization against tetanus. Wis. med. J. **37**, 21—24 (1938).

FIALA, S.: Über die Art der Einwirkung des Formaldehyds auf die Eiweißkörper. Naturwiss. **31**, 370 (1943). Ref. Bull. Inst. Pasteur **43**, 157 (1945).

FILDES, P.: Tetanus. III. Non-toxic variants of «B. tetani». Brit. J. exp. Path. **8**, 219—226 (1927).

FISEK, N. H.: The application of the neutralization inhibition phenomenon to the quantitative determination of toxoids. Première rencontre européenne de standardisation biologique. Lyon 21—25 juin 1955a.

— Tetanus toxoid. Première rencontre européenne de standardisation biologique. Lyon 22—25 juin 1955b.

— The immune response of men to fluid tetanus toxoids. Troisième rencontre internationale de standardisation biologique. Opatijà 2—6 septembre 1957.

— J. H. MUELLER and P. A. MILLER: Muscle extractives in the production of tetanus toxin. J. Bact. **67**, 329—334 (1954).

FLEMING, D. S., and L. GREENBERG: The use of combined antigens in the immunization of infants. Canad. med. Ass. J. **62**, 146—248 (1950).

— — and E. M. BEITH: The use of combined antigens in the immunization of infants. Canad. med. Ass. J. **59**, 101—105 (1948).

— W. L.: Studies on the oxidation and reduction of immunological substances. VII. The differentiation of tetanolysin and tetanospasmin. J. exp. Med. **46**, 279—290 (1927).

FLOCH, H.: Tétanos post-quinique et vaccination antitétanique. Bull. Acad. nat. Méd. (Paris) **133**, 419—422 (1949).

—, et R. BARRAT: Le tétanos post-quininique. Bull. Soc. Path. exot. Paris **45**, 329—335 (1952).

FOSTER, W. D.: Observations on a qualitative difference between the antibody of primary and secondary response. J. Immunol. **78**, 291—295 (1957).

FREDETTE, V., and G. VINET: Production en sac de cellophane de toxines tétaniques renfermant au moins 600000 doses mortelles (cobaye) par millilitre. Canad. J. med. Sci. **30**, 155—156 (1952).

FRENCH, D., and J. T. EDSALL: The reactions of formaldehyde with aminoacids and proteins. Advanc. Protein Chem. **2**, 277—335 (1945).

FREUND, J.: Some aspects of active immunization. Ann. Rev. Microbiol. **1**, 291—308 (1947).

— The response of immunized animals to specific and non-specific stimuli. In A. M. PAPPENHEIMER jr., The nature and significance of the antibody response. New York: Columbia University Press 1953.

—, and M. V. BONANTO: The effect of the amount of antigen on antitoxin-formation during the primary and secondary immunizations. J. Immunol. **45**, 71—78 (1942).

— E. M. SCHRYVER, M. B. McGUINESS and M. B. GEITNER: Diphtheria antitoxin formation in the horse at site of injection of toxoid and adjuvants. Proc. Soc. exp. Biol. (N.Y.) **81**, 657—658 (1952).

FRIEDEMANN, U.: Dynamics and mechanism of immunity reactions in vivo. Bact. Rev. **11**, 275—302 (1947).

—, and A. HOLLANDER: Studies on tetanal toxin. I. Qualitative differences among various toxins revealed by bioassays in different species and by different routes of injection. J. Immunol. **47**, 23—28 (1943a).

— — Studies on tetanal toxin. II. The antitoxin requirements of tetanal toxin in the direct and indirect intraventricular tests. J. Immunol. **47**, 29—33 (1943b).

— B. ZUGER and A. HOLLANDER: Investigations on the pathogenesis of tetanus. I. The permeability of the C. N. S. barrier to tetanal toxin. Passive immunity against toxin introduced by various routes. J. Immunol. **36**, 473—484 (1939).

FUÀ, C.: Cardite acuta benigna successiva a vaccinazione TAB Te. Minerva med. (Torino) **47**II, 2073—2076 (1956).

FULTHORPE, A. J.: Adsorption of tetanus toxin by brain tissue. J. Hyg. (Lond.) **54**, 315—327 (1956).

— Tetanus antitoxin titration by haemagglutination. J. Hyg. (Lond.) **55**, 382—401 (1957).

— Tetanus antitoxin titration by haemagglutination at a low level of test. J. Hyg. (Lond.) **56**, 183—189 (1958).

FURBETTA, F., M. ADORNI-BRACESI e G. PIAZZA: Ricerche sperimentali di vaccinoprofilassi antitetanica per inalazione della anatossina. Boll. Ist. sieroter. milan. **20**, 187—198 (1941).

GANSLMAYER, I.: Über Tetanustoxine. Zbl. Bakt., I. Abt. Orig. **133**, 152—154 (1935).

GARDNER, P. A., A. K. McIVER, M. ELLIS and H. W. JONES: Records of immunization. Lancet **1958** I, 1176—1177.

GAUD, J.: Le tétanos ombilical au Maroc. Bull. Inst. Hyg. Maroc **10**, 319—322 (1950).

GENDEREN, H. VAN: Pyrogens: their mechanism of action. Quatrième congrès de standardisation biologique. Bruxelles 24—30 juillet 1958.

GIBERTI, A., e R. PONZONI: Moderna concezione sul meccanismo terapeutico della siero — vaccinazione antitetanica. Boll. Ist. sieroter. milan. **30**, 660—666 (1951).

GILES, E. C.: The isolation of tetanus bacilli from street dust. Its bearing on surgical practice. J. Amer. med. Ass. **109**, 484—486 (1937).

GLENNY, A. T.: Insoluble precipitates in diphtheria and tetanus immunization. Brit. med. J. **2**, 244—245 (1930).

—, and M. BARR: Alum-toxoid precipitates as antigens. J. Path. Bact. **34**, 118—119 (1931a).

— G. A. H. BUTTLE and M. F. STEVENS: Rate of disappearence of diphtheria toxoid injected into rabbits and guinea-pigs: toxoid precipitated with alum. J. Path. Bact. **34**, 267—275 (1931c).

— C. G. POPE, H. WADDINGTON and U. WALLACE: Immunological notes. XXIII. The antigenic value of toxoid precipitated by potassium alum. J. Path. Bact. **29**, 38—39 (1926a).

GLENNY, A. T., and M. F. STEVENS: The laboratory control of tetanus prophylaxis. J. roy. Army med. Cps 70, 308—310 (1938).

—, and H. WADDINGTON: Combined Schick test and diphtheria prophylactic; combined diphtheria-scarlet-fever prophylactic. J. Path. Bact. 29, 118—122 (1926 b).

GODFRAIN, J. C., P. BERTRAND et L. LIANDIER: Recherche et essais de dosage de l'aldéhyde formique dans les milieux biologiques, anatoxines, sérums, vaccins. Ann. Inst. Pasteur 93, 525—531 (1957).

GÖREN, S.: Rôle adjoint de l'hydroxyde d'aluminium dans l'accroissement de l'immunité par l'anatoxine tétanique. Türk. Ij. tecr. Biyol. Derg. 11, 51—60 (1951). Réf. Bull. Inst. Pasteur 50, 1162 (1952).

GOLD, H.: Studies on tetanus toxoid. I. Active immunization of allergic individuals with tetanus toxoid, alum precipitated, refined. J. Allergy 8, 230—245 (1937 a).

— Studies on tetanus toxoid. II. Active immunization of normal persons with tetanus toxoid, alum precipitated, refined. J. Amer. med. Ass. 109, 481—484 (1937 b).

— Active immunization against tetanus. Ann. intern. Med. 13, 768—782 (1939).

— On the value of a «repeat» injection of tetanus toxoid (secondary stimulus) in active immunization against tetanus. J. Lab. clin. Med. 25, 506—511 (1940 a).

— Active immunization against tetanus by the combined subcutaneous and intranasal routes. Amer. J. Surg. 48, 359—375 (1940 b).

— Sensitization induced by tetanus toxoid, alum precipitated. J. Lab. clin. Med. 27, 26—36 (1941).

—, and H. BACHERS: Combined active-passive immunization against tetanus. J. Immunol. 47, 335—344 (1943).

GOLDIE, H., C. H. PARSONS and M. S. BOWERS: Titration of tetanal toxins, toxoids and antitoxins with the flocculative test. J. infect. Dis. 71, 212—219 (1942).

—, et G. SANDOR: Inactivation de la toxine tétanique par le cétène. C. R. Soc. Biol. (Paris) 129, 454—457 (1938).

GOLDSMITH, S.: Traitement prophylactique du tétanos avec le sérum antitétanique d'origine humaine. Thèse de doctorat, Genève 1957.

GRABAR, P.: Review of the microbiological and immunological literature published in 1956 in the U.S.S.R. Ann. Rev. Microbiol. 11, 43—76 (1957).

GRASSET, E.: Sur la résistance du cobaye aux toxines additionnées de tapioca et sur le rôle du tapioca. C. R. Soc. Biol. (Paris) 94, 260—262 (1926).

— Essais de vaccination intra-cutanée et cutanée, chez le cobaye, par l'anatoxine tétanique. C. R. Soc. Biol. (Paris) 96, 783—785 (1927).

GREENBERG, L.: International standard for tetanus toxoid. Bull. Org. mond. Santé 9, 837—842 (1953).

— The relative immunizing efficiency of tetanus toxoid preparations. Bull. Org. mond. Santé 12, 761—768 (1955).

— An immunological study of Canadian Indian. Canad. med. Ass. J. 77, 211—216 (1957).

—, and R. BENOIT: Control of potency and the dosage of diphtheria and tetanus toxoids. J. Amer. med. Ass. 160, 108—113 (1956).

—, and D. S. FLEMING: The effect of inherited antibodies on the active immunization of infants. II. Duration of immunity. J. Pediat. 39, 672—676 (1951).

— J. GIBBARD et C. A. MORRELL: Le titrage biologique et le contrôle de l'anatoxine tétanique. Bull. Org. Hyg. S. D. N. 12, 385—405 (1945/46).

— C. A. MORRELL and J. GIBBARD: The biological assay of tetanus toxoid. J. Immunol. 46, 333—340 (1943).

GROLIER, B. C. G.: Le tétanos dans le Puy-de-Dôme. Problème des régions tétanigènes. Thèse de doctorat, Paris 1952.

*GRUMBACH, A.: Der Tetanus. In A. GRUMBACH u. W. KIKUTH, Die Infektionskrankheiten der Menschen und ihre Erreger, Bd. II, S. 998—1034. Stuttgart: Georg Thieme 1958.

GUGGISBERG, W.: Kasuistischer Beitrag zum Tetanus puerperalis. Praxis 31, 717—719 (1956).

GUIDA, V., and C. RODRIGUEZ: On the influence of «Cl. sporogenes» upon tetanus toxin production. Arch. Inst. Biol. (S. Paulo) 17, 73 (1946). Réf. Bull. Inst. Pasteur 48, 363 (1950).

GUIDOLIN, R., e A. CORRÉA: Queda da imunidade anti-tetànica em cavalos, apòs sangrias. An. Inst. Pinheiros **13**, 153—159 (1950). Réf. Bull. Inst. Pasteur **50**, 1163 (1952).

GUILLAUMIE, M., A. KREGUER et M. FABRE: Activité hémolytique des toxines « perfringens », tétanique et histolytique après chauffage à différentes températures. C. R. Soc. Biol. (Paris) **138**, 302—303 (1944).

GUNDEL, M., u. F. KÖNIG: Experimentelle Untersuchungen zur aktiv-passiven Immunisierung gegen Diphtherie. Z. Immun.-Forsch. **92**, 235—252 (1938).

GUNNISON, J. B.: Agglutination reactions of the heat stable antigens of Clostridium tetani. J. Immunol. **32**, 63—74 (1937).

— Complement fixation reactions of the heat labile and the heat stable antigens of «Clostridium tetani». J. Immunol. **57**, 67—70 (1947).

HAAS, W.: Wundstarrkrampf trotz Wundausschneidung und Serumgabe. Bemerkungen zu der Arbeit von E. KÖNIG. Chirurg **1940**, 430—431.

HADNAGY, Cs., u. I. KREPSZ: Mitotische Zellteilung und Antikörperbildung. Z. Immun.-Forsch. **114**, 493—501 (1957).

HÄSSIG, A., R. MONTANDON u. H. R. PLÜSS: Über Iso- und Heteroimmunisierungen im ABO-Blutgruppensystem im Hinblick auf die Frage des gefährlichen 0-Spenders. Klin. Wschr. **1955**, 259—264.

HALE, W. M., and R. D. STONER: The effect of cobalt-60 gamma radiation on tetanus antitoxin formation in mice. J. Immunol. **77**, 410—417 (1956).

HALL, W. E. B.: Tetanus with total hemolysis with report of case. Penn. med. J. **41**, 16—22 (1937).

— W. W.: Tetanus toxoid immunization in the United States Navy. Ann. intern. Med. **14**, 565—582 (1940).

— The U.S. Navy's war record with tetanus toxoid. Ann. intern. Med. **28**, 298—308 (1948).

HALLAUER, C., u. R. H. REGAMEY: Über einen modifizierten Typhus-Paratyphus-Tetanus-(T.P.T.)-Impfstoff. Schweiz. Z. allg. Path. **4**, 350—356 (1941).

HANSEN, A., et S. SCHMIDT: Préparation de l'hydroxyde d'aluminium destiné à l'adsorption des toxines (anatoxines) et d'ultravirus. C. R. Soc. Biol. (Paris) **120**, 1150—1152 (1935).

*— F.: *Die Tetanusschutzimpfung.* In H. SPIESS, Schutzimpfungen. Stuttgart: Georg Thieme 1958.

HARRISON, W. T.: Some observations on the use of alum precipitated diphtheria toxoid. Amer. J. publ. Hlth **25**, 298—300 (1935).

HARVIER, P.: Accidents de choc mortels après vaccination antitypho-paratyphique. Bull. Acad. Méd. (Paris) **123**, 771—714 (1940).

HAUDUROY, P.: Présence de « Clostridium tetani » dans les expectorations. C. R. Soc. Biol. (Paris) **136**, 294—295 (1942).

HAVENS jr., W. P., R. M. MYERSON and J. KLATCHKO: Production of tetanus antitoxin by patients with hepatic cirrhosis. New Engl. J. Med. **257**, 637—643 (1957).

HAYASHIDA, T., and C. H. LI: The influence of adrenocorticotropic and growth hormones on antibody formation. J. exp. Med. **105**, 93—98 (1957).

HAYES, S. N.: Imperfect sterilization of dressings as a probable cause of post-operative tetanus. Brit. med. J. **2**, 825—827 (1940).

HAZZA, A. K., and M. K. HABBU: Studies on combined immunization. Combined prophylactic of typhoid vaccine and tetanus toxoid. Indian J. med. Res. **44**, 185—190 (1956).

HEALEY, M., and S. PINFIELD: An in-vitro investigation of the reaction between diphtheria toxin and antitoxin. Brit. J. exp. Path. **16**, 535—553 (1935).

HECHT, G., u. H. WEESE: Periston, ein neuer Blutflüssigkeitsersatz. Münch. med. Wschr. **1943**, 11—15.

HEDRICK, E. C.: Tetanus: Two cases in «immunized» persons. Calif. Med. J. **79**, 49—50 (1953). Ref. Therapeutical Notes (Parke, Davis and Co.), novembre 1953, p. 298.

HEGYESSY, G., S. BOZSÓKY and E. SCHULEK jr.: A study of the immunity against tetanus-toxin following the use of a combined typhoid-tetanus vaccine. Brit. J. exp. Path. **37**, 300—305 (1956).

HEIDELBERGER, M., et F. E. KENDALL: Le déplacement de la toxine des mélanges neutralisés de toxine-anatoxine par le « toxoïde » ou anatoxine. C. R. Soc. Biol. (Paris) **104**, 37—39 (1930).

HELLNER, H.: Vergessene Tetanusserumverabreichung — fahrlässige Tötung? Med. Klin. 1957, 294—299.

HENDRY, J. L.: Tetanus toxins and toxoids. A. R. Div. Lab. Res. N.Y.S. Dep. Health, Albany 1952a, p. 48.

— Effect of X-irradiation on the anamnestic response in guinea pigs to tetanus toxoid. A.R. Div. Lab. Res. N.Y.S. Dep. Health, Albany 1952b, p. 48—49.

— Diphtheria and tetanus toxoids. A.R. Div. Lab. Res. N.Y.S. Dep. Health, Albany 1953, p. 44—45.

— Standardisation of toxoids. A.R. Div. Lab. Res. N.Y.S. Dep. Health, Albany 1956, p. 62.

—, and R. F. CLARK: Tetanus antitoxin and toxoid. A.R. Div. Lab. Res. N.Y.S. Dep. Health, Albany 1951a, p. 44—45.

—, and G. R. SICKLES: Anti-A antibody in the sera of group-0 persons following injections of tetanus toxoid. A.R. Div. Lab. Res. N.Y.S. Dep. Health, Albany 1951b, p. 45—46.

HENSEVAL, M., et P. NÉLIS: Recherches sur la vaccination antidiphtérique à l'aide de l'anatoxine. C. R. Soc. Biol. (Paris) 91, 902—904 (1924).

HERWICK, R. P., E. F. WEIR and A. C. TATUM: Seasonal variation in susceptility of animals to tetanus toxin. Proc. Soc. exp. Biol. (N.Y.) 35, 256—257 (1936).

HESSELVIK, L., and H. ERICCSON: Active basal immunity and its application to epidemiology. IV. Initial pain in diphtheria-tetanus-pertussis vaccination. Acta paediat. (Uppsala) 43, 22—26 (1954).

HEYNINGEN, W. E. VAN: Bacterial toxins. Oxford: Blackwell, Scientific publications 1950.

HILL, C. A. ST., and H. LEDERER: Two cases of tetanus neonatorum. Brit. med. J. 1, 980—981 (1948).

HINSTORFF, D.: Die Abhängigkeit der Tetanusprophylaxe von der Verschiedenheit des regionalen Vorkommens der Erkrankung. Chirurg 5, 9—16 (1933).

HIRAYAMA, H.: Studies on the procedures for potency test of tetanus toxoid. Kitasato Arch. exp. Med. 22, No 2/3, 1—10 (1949).

— T.: Studies on a new culture medium for tetanus bacilli for mass production. Kitasato Arch. exp. Med. 23, No 1, 1—3 (1950).

HOLLÄNDER, L., and F. WORTMANN: Serumkrankheit nach Tetanus-Anatoxin-Injektion mit gleichzeitiger Bildung von blutgruppenspezifischen Immun-Antikörpern. Praxis 1952, 1047—1048.

HOLT, L. B.: Developments in diphtheria prophylaxis. London: William Heinemann, Medical books 1950a.

— Quantitative studies in diphtheria prophylaxis: The second response. Brit. J. exp. Path. 31, 233—241 (1950b).

— Quantitative studies in diphtheria prophylaxis: an attempt to derive a mathematical caracterization of the antigenicity of diphtheria prophylactic. Biometrics 11, 83—94 (1955).

— Some pitfalls in the biometry of mixed antigens. Troisième rencontre internationale de standardisation biologique. Opatijà 2—6 septembre 1957.

— Communication personnelle 1958.

HUBER, J., et A. BESSON: Des injections de rappel d'anatoxine diphthérique et tétanique au début et au cours de la période scolaire. Bull. Acad. Méd. (Paris) 134, 233—234 (1950).

HÜBNER, A., et K. FREUDENBERG: A propos de la vaccination antitétanique. Statistique des cas de tétanos en Allemagne. Rev. Immunol. (Paris) 18, 344—355 (1954).

IKIĆ, D.: Standardization of combined antigens. Troisième rencontre internationale de standardisation biologique. Opatijà 2—6 septembre 1957a.

— Amount of A.U. in humans one year after vaccination with two anti-tetanus preparations of different potency. Troisième rencontre internationale de standardisation biologique. Opatijà 2—6 septembre 1957b.

IPSEN jr., J.: The effect of environmental temperature on the reaction of mice to tetanus toxin. J. Immunol. 66, 687—694 (1951).

— Bio-assay of four tetanus toxoids (aluminium-precipitated) in mice, guinea pigs and humans. J. Immunol. 70, 426—434 (1953a).

— Inherent immunizability to tetanus toxoid, based on studies in pure inbred mice. J. Immunol. 72, 243—247 (1954a).

IPSEN jr., J.: Immunization of adults against diphtheria and tetanus. New Engl. J. Med. 251, 459—466 (1954 b).
—, and H. E. BOWEN: Effects of routine immunization of children with triple vaccine (diphtheria-tetanus-pertussis). Amer. J. publ. Hlth 45, 312—318 (1955).
— J. H. BROWN, E. G. GERWE, L. GREENBERG, J. L. HENDRY, J. HOHENADEL, P. MASUCCI, C. NEWMAN and E. M. TAYLOR: Precision of potency assay of alum-precipitated tetanus toxoid in mice. An inter-institutional study. J. Immunol. 70, 171—180 (1953 b).
ISLIKER, H. C.: The chemical nature of antibodies. Advanc. Protein Chem. 12, 387—463 (1957).
ISTRATI, G.: Experimentelle Untersuchungen über aktive Tetanus-Immunität. I. Mitt. Eine Methode zur Messung kleiner Antitoxinmengen. Zbl. Bakt., I. Abt. Orig. 143, 106—119 (1938).
— L. KICKSCH u. R. PRIGGE: Experimentelle Untersuchungen über aktive Tetanus-Immunität. II. Mitt. Die Beziehung zwischen dem antigenen Reiz und der Antitoxinbildung. Zbl. Bakt., I. Abt. Orig. 145, 19—24 (1939).
— — — Experimentelle Untersuchungen über aktive Tetanus-Immunität. III. Mitt. Die Messung der Wirksamkeit von Tetanus-Impfstoffen. Zbl. Bakt., I. Abt. Orig. 145, 233—240 (1940).
— V. STEFANESCO, M. ISTRATI, N. MUNTIU et T. ARDELEANU: Recherches expérimentales sur l'immunité active antitétanique. 1ère communication. Titrage de l'activité immunogène de l'anatoxine tétanique. Arch. roum. Path. exp. Microbiol. 16, 349—363 (1957).
JACOBS, M. B.: Purification of tetanus toxoid. J. Amer. pharm. Ass. 39, 469—471 (1950a).
—, and M. A. BEHAN: Concentration of tetanus toxoid. J. Amer. pharm. Ass. 39, 466—469 (1950b).
— E. L. EASTMAN and R. L. SHEPARD: Rapid colorimetric determination of formaldehyde in toxoids. J. Amer. pharm. Ass. 40, 365—367 (1951).
JACOTOT, H.: Le latex d' « Hevea brasiliensis » peut être employé comme adjuvant de certains vaccins pour stimuler l'immunité qu'ils engendrent. Rev. Immunol. (Paris) 11, 113—124 (1947).
J.A.M.A.: Second attack of tetanus. J. Amer. med. Ass. 145, 1222 (1951a).
— Tetanus prophylaxis with Penicillin. J. Amer. med. Ass. 145, 1092—1093 (1951b).
— Tetanus wonded boosters. J. Amer. med. Ass. 148, 160 (1952).
— Booster tetanus injection. J. Amer. med. Ass. 152, 1084 (1953).
— Use of tetanus toxoid after injuries. J. Amer. med. Ass. 154, 879 (1954a).
— Tetanus immunization. J. Amer. med. Ass. 156, 501—502 (1954b).
— Tetanus immunization. J. Amer. med. Ass. 157, 1663 (1955).
— Intervals between DPT immunization. J. Amer. med. Ass. 163, 1095—1096 (1957a).
— Treatment of tetanus. J. Amer. med. Ass. 164, 358 (1957b).
— Is neglect of antitetanic serum defensible? J. Amer. med. Ass. 165, 627 (1957c).
— Various routes of immunization. J. Amer. med. Ass. 164, 1626 (1957d).
JAULMES, CH., et A. JUDE: Immunité antitétanique et inoculation tétanigène chez le cobaye. Rev. Immunol. (Paris) 5, 451—461 (1939).
JENNINGS, R. K.: A misleading «reaction of identity» in agar diffusion precipitin studies of tetanus antitoxin. J. Immunol. 77, 156—164 (1956).
JERNE, N. K., and W. L. M. PERRY: The stability of biological standards. Bull. Org. mond. Santé 14, 167—182 (1956).
JONES, F. G., and W. A. JAMIESON: Studies on tetanus toxoid. III. Antitoxic response in guinea pigs immunized with tetanus alum-precipitated toxoid followed by tetanus spores. J. Bact. 32, 33—40 (1936b).
—, and J. M. MOSS: Studies on tetanus toxoid. I. The antitoxic titer of human subjects following immunization with tetanus toxoid and tetanus alum precipitated toxoid. J. Immunol. 30, 115—125 (1936a).
— — Studies on tetanus toxoid. II. The response of human subjects to an injection of tetanus toxoid or tetanus alum precipitated toxoid one year after immunization. J. Bact. 33, 53 (1937a).
— — The antitoxic titers of human subjects following immunization with combined diphtheria and tetanus toxoid, alum precipitated. J. Bact. 33, 53 (1937b).

Jones, F. G., and J. M. Moss: The antitoxic titers of human subjects following immunization with combined diphtheria and tetanus toxoids, alum precipitated. J. Immunol. **33**, 173—181 (1937c).

— — Studies on tetanus toxoid. The response of human subjects to an injection of tetanus toxoid or tetanus alum precipitated toxoid one year after immunization. J. Immunol. **33**, 183—190 (1937 d).

Joó, I., and L. Réthy: Preparation, standardization and control of combined diphtheria-tetanus-pertussis vaccines in Hungary. Troisième rencontre internationale de standardisation biologique. Opatijà, 2—6 septembre 1957.

Jost, A.: Infection tétanique chez un vacciné. Communication personnelle 28. 9. 54.

Jude, A.: Immunisation de l'homme par un vaccin combiné antityphoparatyphoïdique A et B, antidiphtérique, antitétanique et antirickettsies (vaccination T.A.B.D.T.R.). Rev. Immunol. (Paris) **14**, 215—232 (1950).

—, et A. Masse: Antitoxine tétanique, agglutinines T.A.B. et test de séroprotection chez des sujets anciennement vaccinés par le vaccin antityphoparatyphoïdique A et B, antidiphtérique et antitétanique (vaccin T.A.B.D.T.). Rev. Immunol. (Paris) **15**, 20—30 (1951).

Kaiser, M.: Der postoperative Tetanus. Wien. klin. Wschr. **1954**, 727—730.

Kaktine, A.: L'immunisation active des cobayes contre le tétanos. C. R. Soc. Biol. (Paris) **108**, 738—739 (1931).

Katić, R. V.: Immunisation perorale et parentérale des chevaux en vue de l'obtention rapide et intensive de l'antitoxine tétanique. Srpski Arhiv celok. Lek. **46**, 928—929 (1948).

— Action de la pénicilline sur « B. tetani » et sa toxine. Rev. Immunol. (Paris) **15**, 371—381 (1951 a).

— Une possibilité d'emploi du vaccin associé contre le rouget du porc et le tétanos. Bull. off. int. Épiz. **35**, 232—237 (1951 b).

— Action synergique de l'antigène dans l'immunisation avec des antigènes polyvalents. Rev. Immunol. (Paris) **17**, 30—40 (1953).

— Untersuchungen der Immunität gegen Tetanus auf Grund eines biologischen Versuches am Meerschweinchen. Schweiz. Arch. Tierheilk. **98**, 290—293 (1956).

*— *Tétanos*. Monographie tome 281 de l'Académie serbe des Sciences. Belgrade 1957.

Kemény, L.: Die Anwendung der Flocculationsprobe bei Tetanus. Acta vet. Acad. Sci. hung. **1**, 163—170 (1951). Réf. Bull. Inst. Pasteur **52**, 154 (1954).

Kendrick, P. L., and G. C. Brown: Immune response in guinea pigs and monkeys to individual components of combined diphtheria-pertussis-tetanus antigen plus poliomyelitis vaccine. Amer. J. publ. Hlth **47**, 473—483 (1957).

Kerrin, J. C.: The incidence of «B. tetani» in human faeces. Brit. J. exp. Path. **9**, 69—71 (1928).

— The distribution of « B. tetani » in the intestines of animals. Brit. J. exp. Path. **10**, 370—373 (1929).

— Studies on the hemolysin produced by atoxic strains of «B. tetani ». Brit. J. exp. Path. **11**, 153—157 (1930).

Kestermann, E., T. Schleining u. K. E. Vogt: Aktive Immunisierung gegen Tetanus. Klin. Wschr. **1939**, 1553—1555.

—, u. K. E. Vogt: Zur Frage der aktiven Schutzimpfung gegen Gasödem und Tetanus. Klin. Wschr. **1940**, 1009—1010.

— — Nachuntersuchungen bei aktiv gegen Tetanus immunisierten Menschen. Klin. Wschr. **1940**, 1129—1130.

Khabas, I., and M. Tarasova: «Free» and «reversibly combined » formaldehyde in anatoxins. J. Microbiol. Epidemiol. Immunol. **1939**, No 11/12, 127—135. Réf. Zbl. ges. Hyg. **46**, 385 (1940).

Kind, L. S.: Prophylactic value of tetanus antitoxin in mice previously injected with horse serum. Brit. med. J. **1**, 498—499 (1957).

Kjaer, T.: Dissociation du complexe toxine-antitoxine tétanique au moyen de l'anatoxine tétanique. C. R. Soc. Biol. (Paris) **107**, 333—336 (1931).

Kleinschmidt, H.: Diphtherie- und Tetanusbehandlung bei aktiv immunisierten Kindern. Medizinische **1953**, 927.

KLIGLER, I. J., K. GUGGENHEIM and F. M. WARBURG: Influence of ascorbic acid on the growth and toxin production of «Cl. tetani» and on the detoxication of tetanus toxin. J. Path. Bact. **46**, 619—629 (1938).

KLUEVA, N.: Les vaccins combinés, leur composition, préparation et contrôle. Troisième rencontre internationale de standardisation biologique. Opatijà 2—6 septembre 1957.

KNIGHT, B. C. J. G.: La formation des toxines par les bactéries. Bull. Inst. Pasteur **43**, 257—267 (1945).

KNOX, J. D. E., et W. P. STAMM: Morbidity from T.A.B.T. inoculation in R.A.F. recruits. Brit. med. J. **2**, 1462—1464 (1957).

KOBLENCS, Z.: Le tétanos en Meurthe-et-Moselle. Contribution à l'étude des régions tétanigènes et de la prophylaxie du tétanos. Thèse de doctorat, Nancy 1934/35.

KOCH, W., and D. KAPLAN: A simple method for obtaining highly potent tetanus toxin. J. Immunol. **70**, 1—5 (1953).

KÖLE, W.: Über die kombinierte Therapie des Wundstarrkrampfes und die damit erzielten Ergebnisse. Wien. klin. Wschr. **1951**, 393—398.

KOSHLAND, M. E., and F. ENGELBERGER: Mechanism of antibody formation. II. Rate of diphtheria antitoxin formation in the booster response. J. Immunol. **79**, 172—180 (1957).

KOVTOUNOVITCH, L. G.: Rôle du facteur temps dans le développement de l'immunité spécifique. J. Microbiol. Epidemiol. Immunbiol. **27**, No 5, 59—63 (1956). Réf. Bull. Inst. Pasteur **55**, 2467 (1957).

KRAUS, R., T. AWOKI u. N. KOVÁCS: Über aktive Immunisierung mit atoxischen Kulturfiltraten. Dysenteriebazillus Shiga-Kruse. Z. Immun.-Forsch. **45**, 42—48 (1925).

KRECH, U.: Behandlung der experimentellen Tetanusinfektion mit Toxoid. Z. Immun.-Forsch. **106**, 241—248 (1949).

— Die unspezifische Bindung und Inaktivierung bakterieller Toxine «in vivo» und «in vitro». Z. Immun.-Forsch. **109**, 177—183 (1952).

KRÉGUER, A., et M. GUILLAUMIE: Activité antihémolytique des sérums antitétaniques. C. R. Soc. Biol. (Paris) **142**, 813—815 (1948).

KÜSTER u. E. MARTIN: Erfahrungen über chronischen Tetanus. Beitr. klin. Chir. **112**, 268—283 (1918).

KUNZ, H.: Zur Frage der Notwendigkeit der aktiven Immunisierung gegen Wundstarrkrampf. Zbl. Bakt., I. Abt. Ref. **163**, 259—262 (1957).

LAFONTAINE, A., et W. KOOPMANSCH: Sur la fréquence persistante du tétanos en Belgique. Brux. méd. **34**, 111—114 (1954).

LAHIRI, D. C.: Use of non-specific substances in the production of tetanus and diphtheria antitoxin. Indian J. med. Res. **26**, 311—316 (1938).

— Studies on relationship of flocculation value of tetanus-formol-toxoid determinated by Ramon's method and its antigenic value determinated on experimental animals. Indian J. med. Res. **27**, 651—656 (1940).

— A method of measuring antigenicity of tetanal toxoid in the mouse. Indian J. med. Res. **33**, 371—379 (1942).

LAMBOTTE-LEGRAND, J. et C.: Le tétanos chez le nouveau-né indigène à Léopoldville. Ann. Soc. belge Méd. trop. **30**, 541—546 (1950).

LAMONT, A., W. M. FIROR and H. B. SHUMACKER jr.: The «lethal dose» of toxin in experimental tetanus. Bull. Johns Hopk. Hosp. **67**, 25—40 (1940).

Lancet: Tetanus in the theatre? Lancet **1956 II**, 29.

— Tetanus after operation. Lancet **1957 I**, 575, 634, 796.

LANG, W. J.: Über die Verbreitung des «Bacillus tetani» Nicolaier im Erdboden der Schweiz unter spezieller Berücksichtigung der Kantone Graubünden, Waadt und Wallis und der Höhenlagenverbreitung. Thèse de doctorat, Lausanne 1928.

LAPIN, J. H.: Combined immunizations. Advanc. Pediat. **4**, 145—230 (1949).

LA PLACA, M.: Osservazioni sull'azione delle tossine difterica e tetanica nei confronti di cellule carcinomatose coltivate in vitro (stipite KB). Riv. Ist. sieroter. ital. **32**, 350—358 (1957).

LARGIER, J. F.: Purification of tetanus toxin. Biochim. biophys. Acta **21**, 433—438 (1956a).

— Investigation of the tetanus toxin from two different strains of Clostridium tetani. J. Immunol. **76**, 393—398 (1956b).

Laughlin and Brewer: Soc. amer. Bact. Abstracts of Papers 1949 (May) p. 89. Cité par W. Koch et D. Kaplan 1952.

Lavergne, V. de: Sur le traitement du tétanos d'après une statistique hospitalière de 294 cas (tétanos de guerre exceptés). Bull. Acad. Méd. (Paris) 127, 90—92 (1943).

—, et J. R. Helluy: Recherches sur le tétanos. Rec. Inst. nat. Hyg. (Paris) 4, 397—402 (1950).

— — et G. Faivre: Complexité de la flore microbienne anaérobie sporulée dans les plaies tétanigènes. C. R. Soc. Biol. (Paris) 139, 1151—1152 (1945).

— — — Contribution à l'étude morphologique et biologique de « Clostridium tetani ». Rev. Immunol. (Paris) 13, 315—324 (1949).

Legroux, R.: Chimioprévention de l'infection bactérienne des plaies de guerre. Mém. Acad. Chir. 66, 415—420 (1940).

—, et G. Ramon: Sur la production de la toxine tétanique. C. R. Soc. Biol. (Paris) 113, 861—864 (1933).

— — Sur les propriétés de la toxine tétanique rendue hypertoxique (Hypertoxine). C. R. Acad. Sci. (Paris) 198, 620—622 (1934).

Lemétayer, E.: Essais d'immunisation active du lapin et du cobaye par administration d'anatoxine tétanique par les voies buccale et rectale. C. R. Soc. Biol. (Paris) 119, 53—56 (1935).

— Neutralisation de l'hémolysine de la toxine tétanique par les sérums normaux. C. R. Soc. Biol. (Paris) 123, 745—747 (1936a).

— Hémolysine de la toxine tétanique et antihémolysine. C. R. Soc. Biol. (Paris) 123, 742—745 (1936b).

— Mécanisme de l'action des adjuvants de l'immunisation. Bull. Acad. vét. Fr. 23, 59—68 (1950a).

—, et L. Nicol: Développement de la production de l'antitétanolysine sous l'influence d'injections d'anatoxine tétanique. Analysine tétanique. C. R. Soc. Biol. (Paris) 139, 666—667 (1945a).

— — Développement comparatif de la production de l'antitétanolysine et de l'antitoxine spécifique chez les chevaux producteurs de sérum antitétanique. C. R. Soc. Biol. (Paris) 139, 865—867 (1945b).

— — Mécanisme de l'action des adjuvants de l'immunisation. Antigène retard. Bull. Acad. vét. Fr. 22, 251—343 (1949a).

— — De l'emploi des substances dites « adjuvantes et stimulantes de l'immunisation et l'hyperimmunisation actives ». Bull. Acad. vét. Fr. 23, 47—57 (1950b).

— — O. Girard et R. Corvazier: Essais sur l'aptitude réduite à la production des antitoxines des chevaux ayant déjà été utilisés comme producteurs d'un sérum spécifique. C. R. Soc. Biol. (Paris) 140, 727—729 (1946).

— — — — et M. Cheyroux: Recherches sur le pouvoir dissociant « in vitro » de l'anatoxine tétanique vis-à-vis du complexe cerveau-toxine spécifique. C. R. Soc. Biol. (Paris) 143, 675—676 (1949b).

— — — — — Recherches sur le pouvoir dissociant « in vitro » de l'anatoxine spécifique vis-à-vis du complexe nerf-toxine tétanique. C. R. Soc. Biol. (Paris) 143, 676—677 (1949c).

— — — — — Recherches sur la valeur immunisante comparative de l'anatoxine tétanique additionnée de tapioca et de l'anatoxine tétanique précipitée par l'alun. C. R. Soc. Biol. (Paris) 143, 169—170 (1949d).

— — — — — Recherches sur l'action préventive et curative de l'anatoxine spécifique dans l'intoxication tétanique chez le cobaye. C. R. Soc. Biol. (Paris) 143, 1355—1357 (1949e).

— — — — — Influence des injections d'anatoxine tétanique chez des cobayes ayant reçu de la toxine puis du sérum antitétanique. C. R. Soc. Biol. (Paris) 143, 1507—1508 (1949f).

— — — — — De l'immunité potentielle; sa persistance. Bull. Acad. vét. Fr. 24, 197—202 (1951).

— — — — — et A. Gursel: Nouvelles recherches sur l'action de l'anatoxine spécifique dans l'intoxication tétanique du cobaye. C. R. Soc. Biol. (Paris) 143, 1446—1447 (1949g).

— — — — — — Essais de dissociation « in vivo » du complexe toxine -antitoxine tétaniques par l'injection d'anatoxine spécifique. C. R. Soc. Biol. (Paris) 143, 1505—1507 (1949h).

LEMÉTAYER, E., L. NICOL, O. GIRARD, R. CORVAZIER, M. CHEYROUX, A. GURSEL et P. RECULARD: Quelques considérations sur l'injection de rappel dans la vaccination anti-tétanique chez le cheval. Existence possible d'une « Phase négative ». Bull. Acad. vét. Fr. **28**, 425—436 (1955).

— — — — — et C. SIBELLE: Essai de dissociation « in vivo » du complexe toxine tétanique + nerf par l'injection d'anatoxine spécifique. C. R. Soc. Biol. (Paris) **143**, 673—674 (1949i).

LERNER, E. M., and J. H. MUELLER: The role of glutamine in the glucose metabolism of «Clostridium tetani». J. biol. chem. **181**, 43—45 (1949).

LESFARGUES, E., et A. DELAUNAY: Cultures de tissus appliquées à la solution de problèmes immunologiques. I. Etude du pouvoir nécrosant des toxines microbiennes « in vitro ». Ann. Inst. Pasteur **72**, 38—43 (1946).

LEVADITI, C., et A. VAISMAN: La pénicilline dans l'infection tétanique de la souris. C. R. Soc. Biol. (Paris) **139**, 1079—1080 (1945).

LEVINE, L., and J. L. STONE: The purification of tetanus toxoid by ammonium sulfate fractionation. J. Immunol. **67**, 235—242 (1951).

— — and L. WYMAN: Factors affecting the efficiency of the aluminium adjuvant in diphtheria and tetanus toxoids. J. Immunol. **75**, 301—307 (1955).

—, and L. WYMAN: Factors affecting the efficiency of combined prophylactics: effect of poliomyelitis vaccine on diphtheria toxoid. J. Immunol. **79**, 89—93 (1957).

LEWIN, W.: Tetanus after head injury in an immunized subject. Brit. med. J. **2**, 11—12 (1944).

LIBBY, R. L., and J. N. ADAM: A precipitative method for the titration of tetanus toxin and antitoxin. Amer. J. publ. Hlth **29**, 615—618 (1939).

LINDERSTRÖM-LANG, K., et S. SCHMIDT: Sur la purification de l'antitoxine diphtérique. C. R. Soc. Biol. (Paris) **103**, 618—620 (1930a).

— — Sur la purification de la toxine diphtérique. C. R. Soc. Biol. (Paris) **103**, 620—625 (1930b).

LITTLEWOOD, A. H. M., A. K. MANT and G. P. WRIGHT: Fatal tetanus in a boy after prophylactic tetanus antitoxin. Brit. med. J. **2**, 444—445 (1954).

LOEWE, O.: Die Tetanusprophylaxe in ihrer Abhängigkeit von geographischen und geologischen Verhältnissen. Med. Welt **1932**, 1813—1816.

LÖWENSTEIN, E.: Über aktive Schutzimpfung bei Tetanus durch Toxoide. Z. Hyg. Infekt.-Kr. **62**, 491—508 (1909).

— Über Immunisierung mit atoxischen Toxinen. Dtsch. med. Wschr. **1921**, 833—834.

LOISELEUR, J.: Sur la valeur antigène des protéines formolées. Ann. Inst. Pasteur **68**, 439—443 (1942).

LONG, A. P.: Immunizations in the United States Army. Amer. J. publ. Hlth **34**, 27—33 (1944).

— Current status of immunization procedures. Tetanus and exotic diseases of military importance. Amer. J. publ. Hlth **38**, 485—489 (1948).

—, and P. E. SARTWELL: Tetanus in the United States Army in World War II. Bull. U.S. Army med. Dep. **7**, 371—385 (1947).

— D. A.: The influence of corticosteroids on immunological responses to bacterial infections. Int. Arch. Allergy **10**, 5—12 (1957).

LOONEY, J. M., G. EDSALL and W. H. CHASEN: Effect of a booster dose of tetanus toxoid after 5 or more years. Fed. Proc. **12**, 452 (1953).

— — J. IPSEN jr. and W. H. CHASEN: Persistence of antitoxin levels after tetanus-toxoid inoculation in adults, and effect of a booster dose after various intervals. New Engl. J. Med. **254**, 6—12 (1956).

LOVREKOVICH, I., u. K. RAUSS: Schutzimpfung gegen Abdominaltyphus durch die einmalige Injektion eines neuen, präzipitierten Impfstoffes. Z. Immun.-Forsch. **101**, 194—210 (1942).

MAALØE, O., and N. K. JERNE: The standardization of immunological substances. Ann. Rev. Microbiol. **6**, 349—366 (1952).

MACBRYDE, A.: Tetanus immunization with alum precipitated toxoid. Sth. med. J. (Bgham, Ala.) **30**, 565—567 (1937).

MacBryde, A., and M. A. Poston: Immunization with tetanus toxoid. The persistence of antitoxin and the effect of stimulating doses of alum precipitated tetanus toxoid after five years period. J. Pediat. **28**, 692—696 (1946).

MacClean, D.: Staphylococcus toxin: factors which control its production in a fluid medium. J. Path. Bact. **44**, 47—70 (1937).

MacComb, J. A., and M. Z. Trafton: Immune responses and reactions to diphtheria and tetanus toxoids with pertussis vaccine aluminium phosphate precipitated. New Engl. J. Med. **243**, 442—444 (1950).

MacCoy, E., and L. S. MacClung: Serological relations among spore-forming anaerobic bacteria. Bact. Rev. **2**, 47—97 (1938).

MacGuinness, A. C.: Review of current trends in active and passive immunization. J. Amer. med. Ass. **148**, 261—265 (1952).

Mackuth, E.: Über Serumanaphylaxie nach Tetanusantitoxininjektionen. Münch. med. Wschr. **1935**, 1392—1393.

MacLean, I. H., and L. B. Holt: Combined immunization with tetanus toxoid and T.A.B. Lancet **1940 II**, 581—583.

MacLennan, J. D.: The serological identification of «Cl. tetani». Brit. J. exp. Path. **20**, 371—376 (1939).

MacLeod, C. M.: Relations of the incubation period and the secondary immune response to lasting immunity to infectious diseases. J. Immunol. **70**, 421—425 (1953).

Madsen, T.: Über Tetanolysin. Z. Hyg. Infekt.-Kr. **32**, 214—238 (1899).

— C. Jensen and J. Ipsen: Problems in active and passive immunity. Bull. Johns Hopk. Hosp. **61**, 221—245 (1937).

Magara, M., et K. Adukata: L'immunisation antitétanique et antidiphtérique du nouveau-né et de l'enfant par la vaccination des femmes enceintes, au moyen de l'anatoxine. C. R. Soc. Biol. (Paris) **125**, 782—785 (1937).

Magrassi, F.: Contributo sperimentale allo studio dell'immunità locale antitossica. I. Immunità antidifterica. Boll. Ist. sieroter. milan. **13**, 600—627, 953—992 (1934).

Marie, A.: Sensibilité des cellules cérébrales à la toxine tétanique. C. R. Soc. Biol. (Paris) **62**, 1164—1166 (1907).

Marri, P.: Ricerche sull'immunità antitetanica. Nota prima. Durata e valore dell'immunità antitossica in soggetti iniettati con siero antitetanico, con anatossina tetanica e con siero ed anatossina simultaneamente. Pathologica **25**, 649—656 (1933a).

— Ricerche sull'immunità antitetanica. Nota seconda. Sul grado di reattività specifica all'anatossina tetanica di soggetti precedentemente inoculati con anatossina e siero antitetanico. Pathologica **25**, 729—735 (1933).

— Sulla immunizzazione antitetanica associata attiva e passiva. Minerva med. (Torino) **26 II**, 286—289 (1935).

Martin, H. L., and F. MacDowell: Recurrent tetanus: Report of a case. Ann. intern. Med. **41**, 159—163 (1954).

Marvell, D. M., and H. J. Parish: Tetanus prophylaxis and circulating antitoxin in men and women. Brit. med. J. **2**, 891—895 (1940).

Maschmann, E.: Über Tetanustoxin. Hoppe-Seylers Z. physiol. Chem. **201**, 219—254 (1931).

Mathes, C. J.: Tetanus in Dade County: Ten years survey. J. Fla med. Ass. **41**, 847—849 (1955).

Matveev, K. I., S. V. Soloviev and Z. M. Volkova: Contamination of the soil with «Cl. tetani» and tetanus morbidity. J. Microbiol. Epidemiol. Immunobiol. **28**, 54—78 (1957). Réf. Bull. Inst. Pasteur **55**, 2461 (1957).

Mayer, J. B.: Über den Nachweis und die Verbreitungsweise der Tetanusbazillen im menschlichen und tierischen Organismus. Zbl. Bakt. I. Abt. Orig. **139**, 137—151 (1937).

—, R. L.: Essais de chimiothérapie du tétanos. Note préliminaire sur l'action préventive des dérivés sulfamidés sur le développement du tétanos expérimental de la souris. Bull. Acad. Méd. (Paris) **120**, 277—285 (1938).

Mazetti, G., G. Piazza e S. Freni: Il nuovo metodo di vaccinazione antitifo-paratifo-tetano in corso di esperimento nell'esercito. Minerva med. (Torino) **45 (II)**, 1260—1266 (1954).

Megias, J., y F. Moreno de Vega: Estudio experimental acerca de las vacunas antitetánicas de alto poder inmunizante. An. Inst. Llorente **2**, 1—24 (1949).

MEISLOWA, P., A. RYZEWSKA and Z. SPORZYŃSKA: Studies on the activity of a combined triple vaccine (typhoid endotoxin, tetanus toxoid, diphtheria toxoid). Med. dośw. Microbiol. 8, 317—334 (1956). Réf. Bull. Inst. Pasteur 55, 2414 (1957).

MELNIK, M. I., et G. M. STAROBINETZ: Recherches sur la vaccination associée par une seule injection. III. Immunisation antidiphtérique et antitétanique. Ann. Inst. Metschnikoff 5, 23 (1936 b). Réf. Bull. Inst. Pasteur 35, 538 (1937).

— — et M. M. GOLCO: Immunisation active associée contre la fièvre typhoïde et le tétanos. Ann. Inst. Metchnikoff 5, 1 (1936 a). Ref. Bull. Inst. Pasteur 35, 537 (1937).

MELNOTTE, P., C. JAULMES, R. BOLZINGER et L. GIRIER: Séro-diagnostic antigénique dissocié (O, H et Vi) avant et après vaccination (TABDT). Rev. Immunol. (Paris) 12, 97—115 (1948).

MÉNARD, E.: Aperçu prophylactique du tétanos. Cinquième congrès national de médecine rurale. Reims 12 juin 1955.

MERCIER, P., et L. NICOL: Sur l'immunité conférée par les vaccins associés. Résultats obtenus avec le mélange des antigènes et avec chacun d'eux pris isolément. C. R. Soc. Biol. (Paris) 130, 710—712 (1939).

MÉRIEUX, C.: Sur l'aasociation des anatoxines au vaccin antipoliomyélitique et la simplification du calendrier des vaccinations. Quatrième congrès international de standardisation biologique. Bruxelles 24—30 juillet 1958.

— J.: Nouvelles techniques pour la mise en oeuvre de l'injection de rappel d'anatoxine tétanique. Thèse de doctorat, Lyon 1957.

MERKLEN, F. P., R. CABARROU, G. R. MELKI et P. HARTER: Erythrodermie généralisée survenue après vaccination triple associée (antidiphtérique, antitétanique et antityphoparatyphique). Bull. Soc. franç. Derm. Syph. 63, 130 (1956).

MERZ, W. R.: Quatre cas de tétanos après opération césarienne. Méd. et Hyg. (Genève) 1955, 307.

METCALFE: 1955. Cité par G. RAMON 1951 et 1957 a.

MILES, A. A.: Problems in the measurement of immunity and of the potency of immunizing agents. Fed. Proc. 13, 799—807 (1954).

MILLER jr., J. J., and J. B. HUMBER: Observations on tetanus immunization; the dosage of alum-precipitated toxoid and the use of fluid toxoid after trauma. J. Pediat. 23, 516—521 (1943).

— — and J. O. DOWRIE: Immunization with combined diphtheria and tetanus toxoids (aluminium hydroxide adsorbed) containing Hemophilus pertussis vaccine. J. Pediat. 24, 281—289 (1944).

—, and M. L. RYAN: Combined active-passive re-immunization against tetanus in previously immunized individuals: experimental and clinical evidence. J. Immunol. 65, 143—154 (1950).

— — and R. R. BEARD: The speed of the secondary immune response to tetanus toxoid with a review of war reports and observations on simultaneous injection of toxoid and antitoxin. Pediatrics 3, 64—74 (1949).

MIR CHAMSY, H.: Mise en application des théories sur l'extraction de la toxine endocellulaire. III. La cryoextraction de la toxine tétanique endobacillaire. Ann. Inst. Pasteur 94, 402—403 (1958 c).

— M. LAFITY et Mlle A. MOHEBZADEH: Titrage approximatif de l'antitoxine tétanique chez les chevaux vaccinés contre le tétanos par diffusion dans le milieu gélifié. Rev. Immunol. (Paris) 21, 174—180 (1957).

—, et F. NAZARI: Mise en application des théories sur l'extraction de la toxine tétanique endocellulaire. II. Action de la trypsine. Ann. Inst. Pasteur 94, 399—401 (1958 b).

—, et A. SADEGH: Mise en application des théories sur l'extraction de la toxine tétanique endocellulaire. I. Action de la pénicilline. Ann. Inst. Pasteur 94, 396—399 (1958 a).

MODERN, F., et G. RUFF: Concentration et purification des toxines et des toxoïdes par ultrafiltration. (Soc. Biol. Buenos Aires.) C. R. Soc. Biol. (Paris) 123, 69—70 (1936).

— — y A. GATTI: Purificatión y concentratión del toxoido tetánico. Rev. Soc. argent. Biol. 24, 181 (1948). Réf. Bull. Inst. Pasteur 48, 370 (1950).

— — — Floculación de toxoides tetánicos y su aplicación en la medición de sangrias exploradoras. Rev. Soc. argent. Biol. 25, 263—269 (1949).

MODERN, F., G. RUFF y A. GATTI: Die Reinigung und Konzentration des Tetanustoxoids. Mh. Chem. **81**, 165—173 (1950).
— — — Floculación de toxoides tetánicos y su applicación en la medición di sangrias exploradoras. Rev. Inst. bact. Malbrán **15**, 66—72 (1950—1953).
— — — Concentration et titration des toxoïdes tétaniques. C. R. Soc. Biol. (Paris) **150**, 2016 (1956).
MÖRCH, J. R., et S. SCHMIDT: Sur la relation entre le pouvoir antigène intrinsèque de la toxine diphthérique et la vitesse de floculation entre toxine et antitoxine. C. R. Soc. Biol. (Paris) **103**, 1293—1296 (1930).
MÖRL, F.: Experimentelles zur Prophylaxe des Tetanus (Discussion). Zbl. Chir. **80**, 322 (1955).
— Weitere Untersuchungen und Erkenntnisse zur Prophylaxe des Wundstarrkrampfes. Dtsch. Z. Chir. **284**, 125—130 (1956).
MÖSE, J. R.: Zur Beeinflussung von Tetanus- und Diphtherie-Toxin durch Antibiotica, insbesondere Usninsäure. Arzneimittel-Forsch. **7**, 65—69 (1957).
MOLLO, L.: I portatori di germi tetanici sulla cute. G. Batt. Immun. **15**, 417—431 (1935).
MOLONEY, P. J., and J. N. HENNESSY: Purification of tetanus toxoid. Biochem. J. **36**, 544—547 (1942).
— — Titration of tetanal toxin and toxoids by flocculation. J. Immunol. **48**, 345—354 (1944).
MONTANT, R., et G. MOTTIRONI: Réflexions à propos de quelques cas de tétanos. Schweiz. med. Wschr. **1955**, 108—111.
MORENO DE VEGA, F.: Inmunización activa contra el tétanos. An. Inst. Llorente **14**, 43—46 (1957).
MORGUNOV, I. N., and V. V. KHATUNTSEV: The singnificance of the dose and of the intervals in the total stimulation by tetanus toxin. J. Microbiol. Epidemiol. Immunobiol. **5**, 34—38 (1955).
MORRELL, C. A., and L. GREENBERG: The practical value of sound methods of biological assay. Fed. Proc. **13**, 808—814 (1954).
MRAVUNAC, B.: Accidents and incidents in the most common vaccinations. Higijena **8**, 199—211 (1956).
M.R.C.: Poliomyelitis and prophylactic inoculation against diphtheria, whopping-cough, and smallpox. Report of the Medical Research Council Committee on inoculation, procedures and neurological lesions. Lancet **1956** II, 1223—1231.
MUELLER, J. H., and P. A. MILLER: Tetanus toxin production on a simplified medium. Proc. Soc. exp. Biol. (N.Y.) **43**, 389—390 (1940).
— — Large-scale production of tetanal toxin on a peptone-free medium. J. Immunol. **47**, 15—22 (1943).
— — Production of tetanal toxin. J. Immunol. **50**, 377—384 (1945).
— — Factors influencing the production of tetanal toxin. J. Immunol. **56**, 143—147 (1947).
— — Factors affecting the production of tetanus toxin: temperature. J. Bact. **55**, 421—423 (1948a).
— — Unidentified nutrients in tetanus toxin production. J. Bact. **56**, 219—233 (1948c).
— — Inhibition of tetanus toxin formation by D-serine. J. Amer. chem. Soc. **71**, 1865—1866 (1949a).
— — Glutamine in the production of tetanus toxin. J. biol. Chem. **181**, 39—41 (1949b).
— — Variable factors influencing the production of tetanus toxin. J. Bact. **67**, 271—277 (1954).
— — Separation from tryptic digests of casein of some acide-labile components essential in tetanus toxin formation. J. Bact. **69**, 634—642 (1955).
— — Essential role of histidine peptides in tetanus toxin production. J. biol. Chem. **223**, 185—194 (1956).
— — and E. M. LERNER: Factors influencing the production of tetanus toxin: gaseous products of growth. J. Bact. **56**, 97—98 (1948b).
MURRAY, R.: Simultaneous immunization against diphtheria, tetanus, and pertussis. Amer. J. publ. Hlth **40**, 686—690 (1950).

MUTERMILCH, S., M. BELIN et E. SALAMON: Technique permettant de déterminer la toxicité réelle de la toxine tétanique. C. R. Soc. Biol. (Paris) 114, 1005—1008 (1933).

— — — Etude de l'absorption de la toxine tétanique par le verre. C. R. Soc. Biol. (Paris) 116, 1245—1247 (1934).

—, et E. SALAMON: Sur la vaccination du lapin et du cobaye contre le tétanos cérébral. Ann. Inst. Pasteur 45, 85—101 (1930).

NASH, T.: Colorimetric estimation of formaldehyd by means of the Hantzsch reaction. Biochem. J. 55, 416—421 (1953).

NATTAN-LARRIER, L., G. RAMON et E. GRASSET: De l'immunité antitétanique chez le nouveau-né. C. R. Acad. Sci. (Paris) 183, 458—460 (1926).

— — — Recherches sur le passage des toxines et des antitoxines à travers le placenta. C. R. Soc. Biol. (Paris) 96, 241—243 (1927).

NÉLIS, P.: Recherches sur la vaccination antidiphtérique. C. R. Soc. Biol. (Paris) 92, 1114—1116 (1925).

— Nouvelles recherches sur la vaccination antidiphtérique. C. R. Soc. Biol. (Paris) 94, 142—144 (1926).

— Sur les propriétés de l'anatoxine diphthérique. C. R. Soc. Biol. (Paris) 104, 1060—1062 (1930).

— Contribution à l'étude de la présence et de la formation des anticorps dans le liquide céphalo-rachidien. IV. Formation locale des anticorps au niveau des méninges. Rev. belge Sci. méd. 12, 191—245 (1940).

NETO, J. C. F.: Evaluation of active immunity against tetanus by means of the mouse protection test. Arch. Inst. Biol. Exérc. (Rio de J.) 1947, 107. Réf. Bull. Inst. Pasteur 47, 293 (1949).

NEYROUD, M.: Sur la concentration et la purification de la toxine et du toxoïde formolé tétaniques par le sulfate de soude et la chromatographie. Thèse de doctorat, Berne 1944.

NICOL, L.: De l'emploi des substances dites „adjuvantes et stimulantes" dans l'immunisation et l'hyperimmunisation actives. Bull. Acad. vét. Fr. 23, 47—57 (1950).

— Le tétanos. Etiologie, pathogénie et prophylaxie. Cinquième congrès national de médecine rurale. Reims 12 juin 1955.

— O. GIRARD, R. CORVAZIER, M. CHEYROUX et P. RECULARD: Sur l'étude comparée de la valeur antigénique des anatoxines par la méthode de floculation et par la méthode du pouvoir de combinaison. Troisième rencontre internationale de standardisation biologique. Opatijà 2—6 septembre 1957.

NICOLLE, M., E. DEBAINS et E. CÉSARI: Etudes sur la précipitation des anticorps et des antigènes. Sérums « anti-sérums». Ann. Inst. Pasteur 34, 149—152 (1920).

N.I.H.: Minimum requirements: Tetanus toxoid. U.S. Department of Health, Education and Welfare. Public Health Service. National Institutes of Health. Bethesda 14, Maryland. 4th Revision, December 15, 1952.

NISHIURA, Y.: Über die Wirkung des Tetanusserums bei der spezifischen Intoxikation und Infektion. Zbl. Bakt., I. Abt. Orig. 113, 454—468 (1929).

NOEGGERATH, C., u. E. SCHOTTELIUS: Serologische Untersuchungen bei Tetanuskranken. Münch. med. Wschr. 1915, 1293—1295.

NORD, F.: Über die Antikörperbildung bei Immunisierung mit Tetanusbazillen. Z. Immun.-Forsch. 43, 399—408 (1925).

NORMAN, H. B.: Tetanus in the immunized subject. Lancet 1943 I, 557—558.

NOUREDDINE, O.: Essais sur le dosage du pouvoir toxique des toxines diphtérique et tétanique à l'aide d'une réaction colorimétrique. C. R. Soc. Biol. (Paris) 98, 498—499 (1928).

NOVAK, M., M. GOLDIN and W. I. TAYLOR: Tetanus prophylaxis with penicillin-procaine G. Proc. Soc. exp. Biol. (N.Y.) 70, 573—576 (1949).

OAKLEY, C. L.: Bacterial toxins. Ann. Rev. Microbiol. 8, 411—428 (1954).

OGLOBLINA, L., and N. PONOMAREVA: The action of low temperatures on the immunogenous properties of tetanus anatoxin. J. Microbiol. Epidemiol. Immunolbiol. 9, 51 (1944). Réf. Bull. Inst. Pasteur 47, 288 (1949).

OKUDA, K.: On the prophylactic value of iodized tetanus toxin. J. Japan. Soc. vet. Sci. 1923. Réf. Bull. Inst. Pasteur 21, 694 (1923).

O'MEARA, R. A. Q.: Some factors which influence reduction in broth and their bearing on the growth of «Cl. tetani». J. Path. Bact. 45, 541—553 (1937).

Ommyoji, E., and Chü Ping Shan: On the distribution of tetanus bacilli, particularly in the faeces of animals and in soil. J. orient. Med. 11, 156 (1929).

O.M.S.: Second memorandum on tetanus toxoid. Unpublished Working Documents WHO/BS 125 du 25 novembre 1951.

— Comité d'experts de la standardisation biologique. Dixième rapport. Org. mond. Santé, série des rapports techniques no 127, p. 24 et 25, Genève 1957.

Otten, L., and J. P. Hennemann: Combined (simultaneous) immunisation against tetanus. J. Path. Bact. 49, 213—230 (1939).

— — De gecombineerde (simultane) immunisatie tegen tetanus. Geneesk. T. Ned.-Ind. 80, 194—240 (1940). Réf. Bull. Inst. Pasteur 39, 448 (1941).

Ottensooser, F.: Über die Gruppensubstanz A des Peptons und des Diphtherietoxins. Klin. Wschr. 1932, 1716.

— Über Alauntoxin. Der Mechanismus der Entgiftung durch Alaun. Schweiz. Z. allg. Path. 1, 315—326 (1938).

Ottino, C.: La presenza del bacillo del tetano nelle feci dell'uomo e del cavallo. G. Batt. Immun. 2, 193—197 (1927).

Otto, R.: Vermeidung von Serumzufällen bei Verwendung von Tetanus-Antitoxin. Z. ärztl. Fortbild. 31, 310—312 (1934).

Paoletti, A.: Ricerca del bacillo del tetano nelle ragnatele. Considerazioni sulla consuetudine popolare di usare le ragnatele come coadiuvanti e favorenti la epitelizzazione di piaghe estese. Nuovi Ann. Ig. 2, 330—335 (1951).

Pappenheimer jr., A. M.: Proteins of pathogenic bacteria. Advanc. Protein Chem. 4, 123—153 (1948).

—, and E. S. Robinson: A quantitative study of the Ramon diphtheria flocculation reaction. J. Immunol. 32, 291—300 (1937).

Parish, H. J.: Discussion on active immunization. Some present — day problems. Proc. roy. Soc. Med. 34, 247—256 (1941).

— Some aspects of immunity following injections of toxoids. J. roy. Army med. Cps 98, 95—98 (1952).

—, and C. L. Oakley: Anaphylaxis after injection of tetanus toxoid. Report of a case. Brit. med. J. 1, 294—295 (1940).

Pedrosa, E.: Simplificaçao na prepareçao do caldo obtençao da toxina tetanica. Arch. Inst. Biol. Excérc. (Rio d. J.) 9, 73—76 (1948). Réf. Bull. Inst. Pasteur 48, 363 (1950).

Pelloja, M.: Sulla attivazione della tossina tetanica da parte del muscolo e degli estratti muscolari. Med. sper. 21, 377—384 (1949a).

— Effetto delle dosi subletali ripetute di tossina tetanica. Boll. Ist. sieroter. milan. 28, 231—233 (1949b).

— Velocità di insorgenza e di evoluzione del tetano sperimentale in rapporto alla via di somministrazione e alla quantità di tossina iniettata. Boll. Ist. sieroter. milan. 28, 227—230 (1949c).

— Mesomucinasi (jaluronidasi) e intossicazione tetanica sperimentale. Atti Accad. Fisiocr. Siena 18, 51—52 (1950a).

— Toxine tétanique et période d'incubation du tétanos expérimental. Rev. Immunol. (Paris) 14, 123—129 (1950b).

*— Le tétanos expérimental par la toxine tétanique. Paris: Masson & Cie. 1951.

Penso, G., e G. Vicari: Studio dei fenomeni immunitari per mezzo delle colture di tessuto. R. C. Ist. sup. Sanità 20, 655—658 (1957a).

— — Immunological phenomena studied by the tissue culture method. Troisième rencontre internationale de standardisation biologique. Opatijà 2—6 septembre 1957b.

Perry, H. M.: Prevention of tetanus. Brit. med. J. 2, 364—365 (1940).

— W. L. M., Introductory report on the importance of control during production as compared with control of finished products. Quatrième congrès de standardisation biologique, Bruxelles 24—30 juillet 1958a.

— Communication personnelle 1958b.

Peshkin, M. M.: Immunity to tetanus induced by combined alum-precipitated diphtheria and tetanus toxoids. Based on a study of one hundred and eighty-six allergic children. Amer. J. Dis. Child. 62, 9—25 (1941a).

Peshkin, M. M.: Immunity of tetanus induced by combined alum-precipitated diphtheria and tetanus toxoids. Based on a study of sixty-five allergic children given a third, or «repeat» dose. Amer. J. Dis. Child. 62, 309—319 (1941 b).
— Immunity to tetanus induced by a third dose of toxoid two years after basic immunization based on a study of thirty-one allergic children. Amer. J. Dis. Child. 65, 873—881 (1943).
— Immunity to tetanus induced by a third dose of toxoid three years after basic immunization based on a study of thirty eight allergic children. Amer. J. Dis. Child. 67, 22—29 (1944).
— Immunity to tetanus induced by a third dose of toxoid four years after basic immunization. Amer. J. Dis. Child. 69, 83—88 (1945).
Peterson, J. C., and A. Christie: Immunization in the young infant: Response to combined vaccines. Amer. J. Dis. Child. 81, 483—529 (1951).
— — and W. C. Williams: Tetanus immunization. XI. Study of the duration of primary immunity and the response to late stimulating doses of tetanus toxoid. Amer. J. Dis. Child. 89, 295—303 (1955).
Pettenella, G., e A. Sella: Produzione industriale di tossina tetanica ad alto titolo. Boll. Ist. sieroter. milan. 37, 401—413 (1958).
Petrović, D.: Recherches comparées sur la valeur du tétalpan et du vaccin combiné contre le tétanos, la typhoïde et la paratyphoïde A et B, aussi bien que sur l'effet des injections de rappel des mêmes vaccins. Rev. Immunol. (Paris) 20, 231—244 (1956).
Pickett, M. J., K. P. Hoeprich and R. O. Germain: Purification of high titer tetanus toxin. J. Bact. 49, 515—516 (1945).
Pierret, H.: Un cas de trismus après injections préventives chez un enfant blessé non vacciné de sérum antitétanique et d'anatoxine. Maroc méd. 33, 268—269 (1954).
Pike, C.: Anaphylaxis after tetanus toxoid. Brit. med. J. 1, 546 (1940).
Pillemer, L.: The immunochemistry of toxins and toxoids. I. The solubility and precipitation of tetanal toxin and toxoid in methanol-water mixtures under controlled conditions of pH, ionic strength and temperature. J. Immunol. 53, 237—250 (1946 a).
— The preparation and properties of purified toxins and toxoids. Bull. N.Y. Acad. Med. 24, 329—330 (1948 a).
—, and J. Bentoff: The immunochemistry of toxins and toxoids. VIII. Further studies on the solubility of tetanal toxoid in methanol-water mixtures of controlled pH, ionic strength and temperature. J. Immunol. 65, 599—603 (1950).
— D. B. Grossberg and R. G. Wittler: The immunochemistry of toxins and toxoids. II. The preparation and immunologic evaluation of purified tetanal toxoid. J. Immunol. 54, 213—224 (1946 b).
—, and D. H. Moore: The spontaneous conversion of crystalline tetanal toxin to a flocculating atoxic dimer. J. biol. Chem. 173, 427—428 (1948 b).
—, and K. C. Robbins: Chemistry of toxins. Ann. Rev. Microbiol. 3, 265—288 (1949).
—, and W. B. Wartman: The clinical behavior, incubation period, and pathology of tetanus induced in white swiss mice by injection of crystalline tetanal toxin. J. Immunol. 55, 277—281 (1947).
— R. G. Wittler, J. I. Burrell and D. B. Grossberg: The immunochemistry of toxins and toxoids. VI. The crsytallization and characterization of tetanal toxin. J. exp. Med. 88, 205—221 (1948 c).
— — and D. B. Grossberg: The isolation and crystallization of tetanal toxin. Science 103, 615—616 (1946 c).
Pilod: L'application de la vaccination associée triple antityphoïdique, antidiphtérique et antitétanique dans l'armée. Rev. méd. franç. 19, 377—386 (1938). Réf. Bull. Inst. Pasteur 37, 979 (1939).
Pinheiro, D.: Tetanus. General considerations on 1047 cases admitted to the Hospital Das Clinicas de São Paulo. J. Pediat. 51, 171—180 (1957).
Piringer, W.: Über den Nachweis von Tetanusbazillen im Herzblut und in der Milz. Zbl. Bakt., I. Abt. Orig. 141, 375—379 (1938).
Pletsityii, D. F., A. S. Labinskaia et A. S. Aksenova: Rapidité de l'accumulation d'anticorps après la revaccination. J. Microbiol. Epidemiol. Immunobiol. 1, 32—36 (1956). Cité par P. Grabar 1957, p. 61.

Pletsityn, D. F., E. M. Shever, A. M. Monaenkov and co-work.: Comparative effect of subcutaneous and intramuscular injection of tetanus anatoxin in vaccinating people against tetanus. J. Microbiol. Epidemiol. Immunol. 4, 3—10 (1957).

Ployé, M.: Le tétanos et les injections de quinine. Rev. Palud. Méd. trop. 1948, 267—271.

— Vaccination antitétanique en pays tropical. Bull. Acad. nat. Méd. (Paris) 133, 584—586 (1949).

Pochon, J.: Recherches sur l'origine de l'immunité antitétanique naturellement acquise des ruminants. Ann. Inst. Pasteur 57, 82—90 (1936a).

— Rôle de la panse dans l'immunisation antitétanique, progressive et naturelle, des ruminants. C. R. Soc. Biol. (Paris) 121, 300—302 (1936b).

—, et G. Amoureux: Application de la mesure du degré de dégradation protéique à l'étude des milieux de culture pour le bacille tétanique. C. R. Soc. Biol. (Paris) 135, 1063—1065 (1941).

Polson, A., et M. Sterne: Production of potent botulinus toxins and formol-toxoids. Nature (Lond.) 158, 238—239 (1946).

Pons, R.: Affinités comparées de l'antitoxine diphtérique pour la toxine et l'anatoxine diphtériques. Rev. Immunol. (Paris) 5, 557—562 (1939).

Pontano, T.: Si può associare l'immunità passiva serica con l'immunità attiva anatossica? Minerva med. (Torino) 26(I), 801—803 (1935a).

— Sulle associazioni dei sieri antitossici con le rispettive anatossine. Minerva med. (Torino) 26(II), 399—400 (1935b).

— L'associazione siero-anatossina tetanica. Ann. Igiene 45, 678—694 (1935c).

Pontecorvo, M., e D. Soprano: La vaccinazione combinata contro il tifo, i paratifi ed il tetano. Riv. Ist. sieroter. ital. 32, 66—78 (1957).

Ponzoni, R., e A. Giberti: Sulla azione esercitata dalla anatossina specifica nella intossicazione tetanica sperimentale della cavia. Riv. Ist. sieroter. ital. 26, 449—460 (1951).

Pope, C. G., F. M. Stevens, E. A. Caspary and E. L. Fenton: Some new observations on diphtheria toxin and antitoxin. Brit. J. exp. Path. 32, 246—258 (1951).

Potter, F. de: La vaccination antidiphtérique à l'aide de la toxine chauffée. C. R. Soc. Biol. (Paris) 91, 895—898 (1924).

Poulain, P.: Cinq années de vaccination antidiphtérique-antitétanique obligatoire dans une grande ville. Bull. Acad. Méd. (Paris) 132, 142—145 (1948).

Prévot, A. R.: Potentiel d'oxydo-réduction et toxinogenèse du bacille tétanique. C. R. Soc. Biol. (Paris) 127, 685—687 (1938a).

— Discrimination graphique entre floculation vraie et paradoxale dans le titrage de la toxine tétanique par la méthode de Ramon. C. R. Soc. Biol. (Paris) 127, 1166—1168 (1938b).

— Evolution des floculations tétaniques vraie et flagellaire en fonction de l'âge des cultures. C. R. Soc. Biol. (Paris) 128, 307—311 (1938d).

— Hémolysines et antihémolysines bactériennes. Sang 21, 565—587 (1950).

*— Biologie des maladies dues aux anaérobies. Collection de l'Inst. Pasteur. Ed. médicales. Paris: Flammarion 1955.

—, et H. J. Boorsma: Action stimulante des extraits aqueux de cervelle sur la toxinogenèse tétanique. C. R. Soc. Biol. (Paris) 130, 1254—1256 (1939a).

— — Parallélisme entre le rapport COOH polypeptidique/COOH aminé + diaminé + polypeptidique des milieux de culture et la toxinogenèse tétanique. C. R. Soc. Biol. (Paris) 131, 1134—1137 (1939c).

— — Répartition de l'azote et métabolisme azoté dans la toxinogenèse tétanique. Ann. Inst. Pasteur 63, 600—610 (1939d).

—, et E. Kirchheiner: Bilan d'utilisation du glucose par le bacille tétanique au cours de la toxinogenèse. C. R. Soc. Biol. (Paris) 128, 840—842 (1938e).

— — Etude quantitative de l'utilisation des glucides par le bacille tétanique. C. R. Soc. Biol. (Paris) 129, 158—159 (1938f).

— — Action stimulante des extraits de foie et de l'acide nicotinique sur la toxinogenèse tétanique. C. R. Soc. Biol. (Paris) 130, 1429—1431 (1939b).

—, et J. Pochon: Cause et signification de la floculation dite paradoxale dans le titrage de la toxine tétanique par la méthode de Ramon. C. R. Soc. Biol. (Paris) 128, 152—153 (1938c).

—, et J. Taffanel: Action des vitamines hydrosolubles sur la fermentation du glucose de « Plectridium tetani ». C. R. Soc. Biol. (Paris) 136, 384—385 (1942).

PRIGGE, R.: Diphtherie-Schutzimpfung mit hochaktiven Impfstoffen. Ergebn. Hyg. Bakt. **22**, 1—68 (1939).

— Experimentelle Untersuchungen über aktive Tetanus-Immunität. IV. Mitt. Die Aktivierung der Tetanus-Impfstoffe durch Aluminiumverbindungen. Zbl. Bakt., I. Abt. Orig. **145**, 241—248 (1940).

— Standardization of diphtheria and tetanus toxoids. Bull. Org. mond. Santé **9**, 843—849 (1953).

— Die Beziehung zwischen dem Antigengehalt und der Wirksamkeit von Diphtherie- und Tetanus-Impfstoffen. Untersuchungen über eine biologische Konstante. Arb. Staatsinst. exp. Ther. Frankfurt **51**, 108—123 (1954).

RAFYI, A., H. MIR CHAMSY et J. L. DELSAL: Application de la méthode par diffusion d'Oudin-Oakley pour le dénombrement des antigènes tétaniques. Rev. Immunol. (Paris) **18**, 391—398 (1954).

RAINSFORD, S.G.: The preservation of Vi antigen in T.A.B.C. vaccine with a note on combined active immunization with T.A.B.C. vaccine in tetanus formol toxoid. J. Hyg. (Lond.) **42**, 297—322 (1942).

RALSTON, G.: Tetanus prophylactic toxoid: a personal experience. Brit. med. J. **1**, 832—833 (1940).

RAMON, G.: Floculation dans un mélange neutre de toxine-antitoxine diphtériques. C. R. Soc. Biol. (Paris) **86**, 661—663 (1922a).

— Sur une technique de titrage in vitro du sérum antidiphtérique. C. R. Soc. Biol. (Paris) **86**, 711—712 (1922b).

— Sur le pouvoir floculant et sur les propriétés immunisantes d'une toxine diphtérique rendue anatoxique (anatoxine). C. R. Acad. Sci. (Paris) **177**, 1338—1340 (1923).

— Sur l'anatoxine diphtérique. A propos d'une note de P. NÉLIS, sur la vaccination antidiphtérique. C. R. Soc. Biol. (Paris) **92**, 1432—1434 (1925a).

— Sur la production des antitoxines. C. R. Acad. Sci. (Paris) **181**, 157—159 (1925e).

— Procédés pour accroître la production des antitoxines. Ann. Inst. Pasteur **40**, 1—10 (1926e).

— Essais sur l'immunité antitétanique. Sur la chute du pouvoir antitoxique après l'injection d'antigène spécifique et sur la signification de la « phase négative » au cours des vaccinations. C. R. Soc. Biol. (Paris) **100**, 783—785 (1929b).

— A propos de la réaction de floculation dans les mélanges de toxines avec les sérums-anti correspondants. C. R. Soc. Biol. (Paris) **101**, 1033—1035 (1929c).

— Sur la production de l'antitoxine tétanique. C. R. Acad. Sci. (Paris) **191**, 1393—1394 (1930).

— Sur l'irréversibilité du processus de transformation des toxines diphtérique et tétanique en anatoxines. C. R. Soc. Biol. (Paris) **113**, 373—377 (1933d).

— La vaccination antidiphtérique et la vaccination antitétanique au moyen des anatoxines spécifiques et les vaccinations associées dans la pratique. Rev. Immunol. (Paris) **1**, 37—73 (1935g).

— Les problèmes de l'immunité. Sur l'immunité antitoxique naturellement acquise. Son existence, son mécanisme. Rev. Immunol. (Paris) **2**, 305—344 (1936b).

— L'immunité et l'influence des « substances adjuvantes et stimulantes » injectées en mélange avec l'antigène. Introduction à une étude d'ensemble. Rev. Immunol. (Paris) **3**, 193—201 (1937d).

— L'utilisation des anatoxines dans le traitement des toxi-infections en évolution. La séro-anatoxithérapie. C. R. Acad. Sci. (Paris) **205**, 469—471 (1937e).

— L'anatoxine tétanique et la vaccination contre le tétanos. Ann. Méd. **42**, 358—380 (1937f).

— Usage courant de la floculation pour l'évaluation de l'activité antigénique de la toxine et de l'anatoxine tétanique. C. R. Soc. Biol. (Paris) **127**, 1163—1166 (1938b).

— La floculation dans les mélanges de bouillon tétanique et de sérum spécifique et la mesure de la valeur antigène de la toxine et de l'anatoxine tétanique. C. R. Soc. Biol. (Paris) **129**, 539—541 (1938c).

— Sur les méthodes de production rapide et intensive des antitoxines diphtérique et tétanique chez le cheval. C. R. Acad. Sci. (Paris) **207**, 1260—1262 (1938e).

RAMON, G.: L'immunité conférée par l'anatoxine tétanique chez l'homme et chez le cheval. Précisions d'ordre immunologique et épidémiologique. Conséquences. Rev. Immunol. (Paris) **5**, 477—490 (1939a).

— L'anatoxine tétanique et les nouvelles méthodes de lutte contre le tétanos. Bull. Acad. Méd. (Paris) **121**, 609—636 (1939b).

— Sur l'obtention de sérums antidiphtérique et antitétanique de valeur antitoxique élevée. Données expérimentales et techniques. Conséquences pratiques. Bull. Soc. méd. Hôp. Paris **55**, 617—624 (1939c).

— La méthode de floculation et le dosage des toxines et anatoxines des bacilles diphtérique et tétanique et de certains germes anaérobies de la gangrène gazeuse (Perfringens, Vibrion septique, Oedematiens, Histolytique). Rev. Immunol. (Paris) **6**, 65—85 (1940f).

— Les nouvelles méthodes spécifiques de lutte contre le tétanos, vaccination, séro-vaccination préventives, séro-anatoxithérapie du tétanos déclaré. Le mouvement sanitaire **18**, 3—30 (1940k).

— La vaccination contre le tétanos au moyen de l'anatoxine tétanique et les vaccinations associées avant et pendant la deuxième guerre mondiale; résultats. Presse méd. **1951**, 1257—1260.

— A propos des « vaccinations associées » contre la diphtérie, le tétanos, les infections typho-paratyphoïdes. Rev. Path. comp. **52**, 73—82 (1952b).

*— *Quarante années de recherches et de travaux.* Imprimerie régionale, Toulouse 1957a.

— Ferments, anaferments, antiferments. Etude immunologique. Rev. Immunol. (Paris) **21**, 221—236 (1957b).

— Les substances adjuvantes et stimulantes de l'immunité. Bases. Etude expérimentale. Applications. Rev. Path. gén. **1957**c, 2—62.

—, et G. AMOUREUX: Sur une production économique des toxines microbiennes au moyen de bouillon à base de viande de cheval impropre à la consommation. C. R. Acad. Sci. (Paris) **211**, 304—306 (1940i).

— — et J. POCHON: De la production dans la période actuelle des toxines microbiennes destinées à l'obtention, en grandes quantités, des anatoxines, et, en particulier de la production des toxines diphtérique, staphylococcique, tétanique, au moyen d'un nouveau milieu de culture à base de digestion papaïnique et de viande de cheval impropre à l'alimentation de l'homme. Rev. Immunol. (Paris) **7**, 1—15 (1942b).

— — — De la production de la toxine tétanique à l'aide d'un milieu de culture à base de digestion pepsique et de digestion papaïnique de viande et de foie de cheval. Utilisation des « extraits de malt » comme source glucidique. C. R. Soc. Biol. (Paris) **137**, 8—9 (1943a).

— — — Préparation de la toxine tétanique à l'aide de digestion papaïnique de viande et de foie de cheval. C. R. Soc. Biol. (Paris) **137**, 350—351 (1943b).

— — — Production de la toxine tétanique destinée à l'obtention de l'anatoxine spécifique au moyen de milieux à base de digestion papaïnique de viande et de foie de cheval. Rev. Immunol. (Paris) **8**, 115—118 (1943c).

— A. BERTHELOT, E. GRASSET et Mlle AMOUREUX: Production d'anatoxine tétanique par culture du bacille tétanique en bouillon bilié. C. R. Soc. Biol. (Paris) **96**, 30—32 (1927b).

—, et A. BOIVIN: Sur la préparation du vaccin triple associé «antidiphtérique, antitétanique, antityphoparatyphoïdique ». C. R. Soc. Biol. (Paris) **135**, 12—16 (1941a).

— — A. LAFFAILLE et E. LEMÉTAYER: Résultats immunologiques comparatifs obtenus chez l'homme au moyen du vaccin triple associé « antidiphtérique, antitétanique, antityphoparatyphoïdique » préparé selon deux formules différentes. C. R. Soc. Biol. (Paris) **135**, 784—789 (1941c).

— — et R. RICHOU: Sur les propriétés floculantes et immunisantes des anatoxines purifiées par précipitation à l'acide trichloracétique. C. R. Acad. Sci. (Paris) **203**, 634—636 (1936c).

— — — L'anatoxine tétanique purifiée par l'acide trichloracétique et son pouvoir antigène « in vitro » et « in vivo ». C. R. Soc. Biol. (Paris) **124**, 32—35 (1937a).

— — — M. DJOURICHITCH et R. MACCOLINI: La séro-anatoxithérapie des toxi-infections en évolution. Ses bases expérimentales. Rev. Immunol. (Paris) **4**, 24—39 (1938d).

— G. CHASSIGNEUX, R. RICHOU et C. GERBEAUX: Le nucléinate de soude, substance adjuvante et stimulante de l'immunité. C. R. Acad. Sci. (Paris) **235**, 111—114 (1952a).

RAMON, G., et P. DESCOMBEY: Sur l'immunisation antitétanique et sur la production de l'antitoxine tétanique. C. R. Soc. Biol. (Paris) **93**, 508—509 (1925b).
— — Sur l'immunisation antitétanique et sur la production de l'antitoxine tétanique. C. R. Soc. Biol. (Paris) **93**, 898—899 (1925d).
— — Sur l'appréciation de la valeur antigène de la toxine et de l'anatoxine tétaniques par la méthode de floculation. C. R. Soc. Biol. (Paris) **95**, 434—436 (1926b).
— — L'anatoxine tétanique et la prophylaxie du tétanos chez le cheval et les animaux domestiques. Ann. Inst. Pasteur **41**, 834—847 (1927g).
— M. DUCOSTE, R. RICHOU et M. BUISSON: Sur la production des antitoxines diphtérique et tétanique chez les sujets immunisés, par voie cérébrale, au moyen des anatoxines spécifiques. C. R. Soc. Biol. (Paris) **134**, 12—19 (1940d).
— — — — Le développement des antitoxines diphtérique et tétanique chez les sujets immunisés, par voie cérébrale, avec chacune des anatoxines spécifiques ou avec le mélange des deux anatoxines. C.R. Soc. Biol. (Paris) **134**, 72—76 (1940e).
— — — — Développement et localisation des antitoxines diphtérique et tétanique chez l'homme soumis aux injections intracérébrales d'anatoxine spécifique. Considérations sur la prétendue « formation locale » des antitoxines. Rev. Immunol. (Paris) **6**, 145—158 (1940g).
—, et E. FALCHETTI: Recherches expérimentales sur l'immunisation antitétanique passive. Elimination rapide de l'antitoxine lors des réinjections de sérum antitétanique. C. R. Soc. Biol. (Paris) **119**, 6—10 (1935c).
—, et E. GRASSET: La réaction de floculation et le dosage du pouvoir antitoxique du sérum antidiphtérique purifié. C. R. Soc. Biol. (Paris) **95**, 436—438 (1926c).
— — Sur l'immunité antitoxique active, par voie buccale, chez l'animal d'expériences. C. R. Soc. Biol. (Paris) **95**, 1405—1407 (1926d).
— P. JOANNON, R. RICHOU et L. CORRE: Une nouvelle substance adjuvante et stimulante de l'immunité: le tannin. C. R. Soc. Biol. (Paris) **135**, 45—48 (1941b).
— R. KOURILSKY, R. RICHOU et S. KOURILSKY: Essais de séro-vaccination antitétanique. Etude immunologique. Application éventuelle à la séro-anatoxithérapie spécifique du tétanos. Bull. Soc. Méd. Hôp. Paris **54**, 1287—1296 (1938f).
— — — — A propos du procès-verbal: sur la séro-vaccination anatoxique et sur la séro-anatoxithérapie tétanique. Bases. Résultats immunologiques. Bull. Soc. Méd. Hôp. Paris **54**, 1442—1445 (1938g).
— — — — Recherches immunologiques sur la séro-anatoxithérapie tétanique. Rev. Immunol. (Paris) **5**, 132—150 (1939d).
—, et A. LAFFAILLE: Sur l'immunisation antitétanique. C. R. Soc. Biol. (Paris) **93**, 582—584 (1925c).
—, et E. LEMÉTAYER: Sur une méthode de production intensive de l'antitoxine tétanique. C. R. Soc. Biol. (Paris) **106**, 23—25 (1931).
— — Sur la spécificité rigoureuse des interactions « in vitro » et «in vivo » des toxines, anatoxines et antitoxines. C. R. Soc. Biol. (Paris) **112**, 136—139 (1933a).
— — Immunité antitétanique naturellement acquise chez les animaux des espèces bovine, ovine et caprine. C. R. Soc. Biol. (Paris) **112**, 1157—1160 (1933c).
— — Sur l'immunité antitétanique naturellment acquise chez les animaux d'espèce bovine et caprine. Son existence. Son mécanisme. Essai d'immunologie comparée. Bull. Acad. vét. Fr. **6**, 183—192 (1933e).
— — Sur l'immunité antitétanique naturellement acquise chez quelques espèces de ruminants. C. R. Soc. Biol. (Paris) **116**, 275—277 (1934).
— — Sur l'immunité produite, chez le lapin, par injection de toxine tétanique enrobée dans la lanoline, et sur son mécanisme. C. R. Soc. Biol. (Paris) **118**, 935—938 (1935b).
— — Recherches comparatives sur le sort de la toxine tétanique injectée à l'animal soit seule, soit enrobée dans la lanoline. C. R. Soc. Biol. (Paris) **119**, 906—909 (1935d).
— — Sur les suites des injections répétées, chez le lapin et dans différentes conditions, de très petites doses de toxine tétanique non atténuée. C. R. Soc. Biol. (Paris) **119**, 909—912 (1935e).
— — Recherches sur l'immunité antitétanique naturellement acquise chez l'homme et chez différentes espèces animales, en particulier chez les ruminants. Rev. Immunol. (Paris) **1**, 209—227 (1935i).

RAMON, G., et E. LEMÉTAYER: Sur l'action immunisante de la toxine tétanique, enrobée dans la lanoline, chez l'animal d'expérience. C. R. Acad. Sci. (Paris) **200**, 592—594 (1935k).

— — et J. BORCILA: Sur la valeur de l'immunité spécifique provoquée chez le lapin par des injections répétées de petites doses d'anatoxine tétanique effectuées dans diverses conditions. C. R. Soc. Biol. (Paris) **119**, 913—915 (1935f).

— — et A. GUHATHAKURTA: Nouvelles recherches sur le sort de la toxine tétanique injectée au cobaye après incorporation dans la lanoline. C. R. Soc. Biol. (Paris) **122**, 866—868 (1936a).

— — et L. NICOL: De l'instabilité des dilutions de toxine tétanique. Influence de l'agitation sur les propriétés du poison tétanique en solution dans l'eau physiologique. C. R. Soc. Biol. (Paris) **127**, 1160—1163 (1938a).

— — et R. RICHOU: Sur les résultats de l'injection à l'animal d'expérience de la toxine tétanique mélangée à diverses substances et en particulier à la lanoline. C. R. Soc. Biol. (Paris) **118**, 112—114 (1935a).

— — — De l'influence de diverses substances ajoutées à l'antigène anatoxique dans la production de l'immunité antitoxique. Rev. Immunol. (Paris) **1**, 199—208 (1935h).

— — — Sur l'évaluation du pouvoir antigène intrinsèque de la toxine et de l'anatoxine tétaniques par la floculation. C. R. Soc. Biol. (Paris) **124**, 416—420 (1937b).

— — — et L. NICOL: Des causes d'erreur dans la détermination, chez l'animal d'expérience, de l'activité de la toxine tétanique. C. R. Soc. Biol. (Paris) **127**, 1157—1160 (1937c).

— J. POCHON, G. AMOUREUX et R. RICHOU: De l'emploi des « extraits de malt » dans la production des toxines microbiennes et spécialement de la toxine diphtérique. C. R. Soc. Biol. (Paris) **136**, 473—474 (1942a).

—, et R. RICHOU: Recherches sérologiques sur les suites des injections, à l'animal d'expérience, de sérum antitétanique et de « solution d'antitoxine tétanique ». C. R. Soc. Biol. (Paris) **133**, 18—21 (1940a).

— — Solution d'antitoxine tétanique et prophylaxie des accidents sériques dans la prévention du tétanos. Recherches expérimentales et résultats cliniques. Rev. Immunol. (Paris) **6**, 209—223 (1940/41h).

— — Sur l'antitoxine tétanique et les variations de ses qualités au cours de l'immunisation. C. R. Soc. Biol. (Paris) **135**, 1281—1285 (1941d).

— — Sur le titrage des toxines, des anatoxines et des antitoxines diphtérique, tétanique et staphylococcique. Rev. Immunol. (Paris) **14**, 161—194 (1950a).

— — et G. MONOURY: Les manifestations anaphylactiques, chez l'animal d'expérience, à la suite d'injections de sérum antitétanique et de « solution d'antitoxine ». Recherches comparatives. C. R. Soc. Biol. (Paris) **133**, 21—24 (1940b).

— — et J.-P. THIÉRY: De l'immunité provoquée chez les animaux par un mélange d'anatoxine tétanique et de latex d' « Hevea brasiliensis ». Conséquences théoriques et pratiques. C. R. Acad. Sci. (Paris) **227**, 575—577 (1948).

— — — C. GERBAUX et J. LEPLATRE: Sur les substances adjuvantes et stimulantes de l'immunité. Nouvelles recherches. Rev. Immunol. (Paris) **14**, 205—214 (1950b).

— A. SAENZ et R. RICHOU: Le développement et les fluctuations du taux des antitoxines diphtérique, tétanique et staphylococcique chez les animaux immunisés soit avec chacune des anatoxines spécifiques, soit avec le mélange des trois anatoxines. Déductions d'ordre théorique. C. R. Soc. Biol. (Paris) **133**, 184—188 (1940c).

—, et CH. ZOELLER: Les « vaccins associés » par union d'une anatoxine et d'un vaccin microbien (TAB) ou par mélange d'anatoxines. C. R. Soc. Biol. (Paris) **94**, 106—109 (1926a).

— — De la valeur antigène de l'anatoxine tétanique chez l'homme. C. R. Acad. Sci. (Paris) **182**, 245—247 (1926f).

— — Essai d'immunisation antitoxique, active et passive, par voie buccale chez l'homme. C. R. Soc. Biol. (Paris) **95**, 1409—1411 (1926g).

— — Nouveaux résultats concernant les rhino-vaccinations antitoxiques. C. R. Soc. Biol (Paris) **97**, 701—703 (1927c).

— — De l'immunisation antitoxique par voie nasale chez l'homme et du mécanisme de l'immunisation occulte. C. R. Soc. Biol. (Paris) **96**, 757—759 (1927d).

— — Existe-t-il une immunité occulte à l'égard de l'infection tétanique? C. R. Soc. Biol. (Paris) **96**, 762—763 (1927e).

Ramon, G., et Ch. Zoeller: L'anatoxine tétanique et l'immunisation active de l'homme vis-à-vis du tétanos. Ann. Inst. Pasteur 41, 803—833 (1927f).

— — Nouveaux résultats concernant la vaccination de l'homme contre le tétanos. C. R. Soc. Biol. (Paris) 100, 92—93 (1929a).

— — Sur la valeur et la durée de l'immunité conférée par l'anatoxine tétanique dans la vaccination de l'homme contre le tétanos. C. R. Soc. Biol. (Paris) 112, 347—350 (1933b).

Ramshorst, J. D.: De bereiding van diphtherie-entstoffen. Thèse de doctorat, Utrecht 1951.

— Onderzoek naar de houdbaarheid van aan aluminiumphosphaat geatsorbeerde entstoffen. Ber. Rijksinst. Volksgezondheid, Utrecht 1954, S. 143—144.

— Potency and purification of tetanus toxoid obtained by means of the «Mueller medium». Antonie v. Leeuwenhoek 23, 97—108 (1957).

Raubitschek, H., u. V. K. Russ: Über entgiftende Eigenschaften der Seife. Z. Immun.-Forsch. 1, 395—407 (1909).

Rauss, K., I. Kétyi, L. Réthy et I. Joó: Le rôle de l'intervalle entre les vaccinations par les vaccins polyvalents antidysentériques et par des vaccins combinés antityphoïdiques-antidysentériques-antitétaniques. Quatrième congrès international de standardisation biologique. Bruxelles 24—30 juillet 1958a.

— — — et J. Maroczi: Recherches sur l'effet immunisant des antigènes tétaniques et dysentériques combinés à des antigènes « à l'action de rappel ». Recherches sur terrain. Quatrième congrès international de standardisation biologique. Bruxelles 24—30 juillet 1958b.

Ray, N. N., and G. C. Das: The immunisization of horses for the production of high-titre tetanus antitoxin. Indian J. med. Res. 25, 617—621 (1938a).

— — The immunization of horses for the production of high-titre tetanus antitoxin. Indian J. med. Res. 26, 317—320 (1938b).

Raynaud, M.: Extraction de la toxine tétanique et de la toxine de « Clostridium sordelli » à partir des corps bactériens. C. R. Acad. Sci. (Paris) 225, 543—544 (1947).

— Extraction de la toxine tétanique à partir des corps microbiens. Ann. Inst. Pasteur 80, 356—377 (1951a).

— J. Blass et A. Turpin: Mécanisme de la détoxification des toxines par le formol. Etude de deux nouveaux dérivés atoxiques antigéniques: 2.4-dinitrofluorobenzène toxoïde et β-propiolactone toxoïde. C. R. Acad. Sci. (Paris) 245, 862—863 (1957).

— E. Lemétayer, A. Turpin, L. Nicol et M. Rouyer: Installation rapide d'un état de résistance à l'intoxication tétanique expérimentale par l'influence de doses massives d'anatoxine tétanique. C. R. Acad. Sci. (Paris) 233, 586—588 (1951b).

— B. Nisman et R. O. Prudhomme: Extraction de la toxine tétanique à partir des corps microbiens par les ultrasons. C. R. Acad. Sci. (Paris) 230, 1370—1372 (1950).

— R. Saissac, A. Turpin et M. Rouyer: Formation de toxine tétanique par des suspensions de corps microbiens. Ann. Inst. Pasteur 83, 693—704 (1952).

—, et L. Second: Extraction des toxines botuliniques à partir des corps microbiens. Ann. Inst. Pasteur 77, 316—319 (1949).

— A. Turpin et E. Lemétayer: Etude des anatoxines tétaniques concentrées. Ann. Inst. Pasteur 85, 376—379 (1953a).

—, and E. A. Wright: Rapid specific preventive action of tetanus toxoid. Nature (Lond.) 171, 797 (1953b).

Redi, R.: Contributo allo studio della durata e del valore dell'immunità antitossica in soggetti iniettati con siero ed anatossina tetanica. Pathologica 27, No 522 (1935).

Reed, L. J., and H. Muench: A simple method of estimating fifty per cent endpoints. Amer. J. Hyg. 27, 493—497 (1938).

Regamey, R. H.: Etude in vivo et in vitro sur le sort de la toxine tétanique dans le tube digestif. Ann. Inst. Pasteur 56, 87—100 (1936a).

— Le sort de la toxine tétanique dans le tube digestif. Concours Marc Dufour de l'Université de Lausanne, 1936b, p. 1—116.

— Les gangrènes gazeuses après injections médicamenteuses. Schweiz. med. Wschr. 1939, 874—876.

— Résultats sérologiques chez les soldats suisses vaccinés contre le Typhus-Paratyphus-Tétanos (T.P.T.). Schweiz. Z. allg. Path. 3, 304—317 (1941a).

Regamey, R. H.: Studien über die Wirksamkeit der antitetanischen «injection de rappel» bei T.P.T.-Geimpften. Schweiz. Z. allg. Path. **4**, 177—192 (1941b).

— Etude immunologique sur les tests de Schick et de Reh. Schweiz. Z. allg. Path. **6**, 409—412 (1943c).

— Essais de vaccination antitypho-paratypho-tétanique avec l'Anendo T.P.T. Rev. méd. Suisse rom. **63**, 841—859 (1943d).

— Etude expérimentale sur la résorption et l'élimination du sérum antitétanique chez le lapin. Schweiz. Z. allg. Path. **7**, 500—504 (1944a).

— L'immunité antitétanique conférée par le vaccin T.P.T. Schweiz. med. Wschr. **1944b**, 230—234.

— Etude sur l'immunité antitétanique conférée par le toxoïde formolé (anatoxine) tétanique. Schweiz. med. Wschr. **1945**, 641—644.

— Expériences de désensibilisation avec un antihistaminique de synthèse, le 2-phénylbenzyl-aminométhyl-imidazoline (Antistine Ciba). Schweiz. Z. allg. Path. **10**, 429—432 (1947a).

— Etude sur la relation entre le titrage direct et le titrage par floculation de la toxine tétanique. Schweiz. Z. allg. Path. **10**, 492—496 (1947b).

— Exposé général des procédés modernes de prophylaxie des maladies épidémiques aux armées. 11ème congrès intern. de Médecine et de Pharmacie militaires. Bâle, juin 1947c.

— Bases expérimentales d'un vaccin antitétanique de type nouveau en Suisse. 11ème congrès intern. de Médecine et de Pharmacie militaires. Bâle, juin 1947. In: J. trim. off. suisses Serv. santé **24**, 62—69 (1947e).

— Essai de vaccination associée contre la diphtérie et le tétanos. Schweiz. med. Wschr. **1948**, 303—307.

— Effets, chez le lapin, d'injections répétées de sérum antitétanique issu d'animaux apparte-nant à des espèces différentes. Schweiz. Z. allg. Path. **16**, 873—882 (1953).

*— *Tétanos. Incidences immunologiques.* Praxis **44**, 268—272, 288—294 (1955a).

— Beitrag zur Tetanustoxin-Antitoxin-Flockung nach Ramon. Schweiz. Z. allg. Path. **9**, 568—572 (1956).

— Détermination rapide de la fraction tétanique dans les vaccins associés. Troisième ren-contre internatinale de standardisation biologique. Opatijà 2—6 septembre 1957a.

— Méthode accélérée pour déterminer l'antigénicité des vaccins tétaniques. Schweiz. Z. allg. Path. **20**, 628—633 (1957b).

— Introduction aux contrôles en cours de production des substances immunobiologiques. Quatrième congrès international de standardisation biologique. Bruxelles 24—30 juillet 1958.

—, et W. Aegerter: La séroanatoxithérapie expérimentale du tétanos. Schweiz. Z. allg. Path. **14**, 554—559 (1951b).

—, et M. Bertschmann: Etude de 10 anatoxines tétaniques mises en présence de 10 sérums antitétaniques. 1959. (En préparation.)

— L. Calpini, R. Ballinari et M. Neyroud: L'application de la chromatographie à la purification des toxines et anatoxines diphtériques et tétaniques. Schweiz. Z. allg. Path. **6**, 405—406 (1943b).

— M. Neyroud et L. Calpini: Recherches sur la purification et la concentration des anatoxines diphtérique et tétanique par la précipitation au sulfate de soude. Schweiz. Z. allg. Path. **6**, 402—405 (1943a).

—, u. H. J. Schlegel: Betrachtungen zur Sero- bzw. Anatoxiprophylaxe des Tetanus. Schweiz. med. Wschr. **1950**, 919—920.

— — L'immunité antitétanique dans les 10 ans qui suivent l'immunisation de base. Schweiz. Z. Path. **14**, 550—554 (1951a).

— K. Simon et M. Wantz: Prophylaxie du tétanos: injection associée de sérum et d'ana-toxine. Schweiz. Z. allg. Path. **18**, 1157—1163 (1955b).

—, et M. Wantz: Sur l'inexactitude des titrages par floculation de la toxine tétanique. Schweiz. Z. allg. Path. **10**, 488—491 (1947d).

Reitani, U.: Spore di anaerobi nella cavità buccale di soldati. G. Med. milit. **77**, 279—283 (1929).

Renaux, E.: Sur la floculation de la toxine diphtérique par le sérum antidiphtérique. C. R. Soc. Biol. (Paris) **90**, 964—966 (1924).

Réthy, L., I. Joó and I. Kétyi: Preparation and potency test of differently composed combined typhoid — tetanus vaccines. Troisième rencontre internationale de standardisation biologique. Opatijà 2—6 septembre 1957.

Reymann, G. C.: Untersuchungen über Tetanospasmin und Tetanolysin. I, II et III. Z. Immun.-Forsch. I: **50**, 31—70 (1927); II: **50**, 405—461 (1927); III: **51**, 1—26 (1927).

Richou, R.: La séro-anatoxithérapie en médecine vétérinaire. Résultats d'ensemble. Rec. Méd. vét. **123**, 112—118 (1947).

—, et P. Thibault: Augmentation de la production de l'antitoxine tétanique, chez le cobaye, par addition à l'anatoxine spécifique de quantités variables de saponine. C. R. Soc. Biol. (Paris) **121**, 748—749 (1936).

—, et G. Torrisi: Absence d'antitoxine tétanique d'origine naturelle chez les animaux d'espèces porcine et canine. C. R. Soc. Biol. (Paris) **114**, 595—596 (1933).

Richter, S.: Die Simultanprophylaxe des Tetanus. Zbl. Chir. **80**, 289—293 (1955).

Ritossa, P.: Sulla immunizzazione combinata attiva e passiva nelle infezione tetanica e difterica. Pediatria (Napoli) **42**, 1411—1420 (1934).

Riva, G., S. Barandun, H. Cottier u. A. Hässig: Über Defektdysproteinämien und andere Anomalien der Plasmaeiweiße. Schweiz. med. Mschr. **1958**, 1025—1034.

Robert, H.: Etude sur la production de toxines hémolytiques par les microbes anaérobies et sur les propriétés antihémolytiques des sérums naturels. Description d'un phénomène de zone. Thèse de doctorat, Berne 1958.

Robertson, P. W., and B. J. Leonard: Some delayed complications of inoculation. Brit. med. J. **2**, 1029—1032 (1956).

Rodriguez, J. F., and J. C. F. Neto: Plan for immunization against tetanus and typhoid infections in the Brazilian Army. Arch. Inst. Biol. Excérc. (Rio de J.) **1947**, 95. Réf. Bull. Inst. Pasteur **47**, 293 (1949).

Römer, P. H.: Über das Vorkommen von Tetanusantitoxin im Blute normaler Rinder. Z. Immun.-Forsch. **1**, 363—386 (1909).

Ross, V., F. L. Clapp and C. H. Parsons: Protamine tetanus toxoid. Amer. J. Dis. Child. **79**, 10—16 (1950).

*Rostock, P.: *Tetanus.* Berlin: W. de Gruyter & Co. 1950.

Roux, E., et A. Yersin: Contribution à l'étude de la diphtérie. Ann. Inst. Pasteur **2**, 629—661 (1888).

— — Contribution à l'étude de la diphtérie. Ann. Inst. Pasteur **3**, 273—288 (1889).

Ruggerini, G.: L'azione delle sostanze aspecifiche sopra organismi precedentemente immunizzati contro il tetano mediante anatossina tetanica, a immunità antitetanica scomparsa o fortemente diminuita. Boll. Ist. sieroter. milan. **8**, 765—778 (1929a).

— Se il potere vaccinante dell'anatossina tetanica si conserva inalterato dopo un suo contatto più o meno lungo con estratti batterici (tuberculina, malleina), e se detto potere si altera o si modifica colla transformazione dell'anatossina liquida in polvere secca. Clin. vet. (Milano) **1929b**. Rif. Bull. Inst. Pasteur **28**, 140 (1930).

— Sulla elevazione dell'antitossina tetanica in immediata vicinanza di iniezioni aspecifiche. G. Batt. Immun. **5**, 1745—1758 (1930).

— Influenza dell'aggiunta di aminoacidi al terreno di coltura sopra 10 sviluppo e la produzione di tossina del bacillo del tetano e del perfringens. Biochem. Terap. sper. **20** (1933). Rif. Bull. Inst. Pasteur **32**, 475 (1934).

Sachs, H.: Anaphylaxis after injection of tetanus toxoid. Brit. med. J. **1**, 588 (1940).

— A.: Modern views on the prevention of tetanus in the wounded. Proc. roy. Soc. Med. **45**, 641—652 (1952).

Sacquépée, E.: Immunisation contre le tétanos par l'emploi simultané de sérum et d'anatoxine antitétanique. Paris méd. **1933**, No 22 du 3 juin.

—, et A. Jude: Sur la valeur et la durée de l'immunité antitétanique, après injection de rappel, chez l'homme immunisé contre le tétanos par l'emploi simultané du sérum et de l'anatoxine tétanique. C. R. Soc. Biol. (Paris) **125**, 711—712 (1937).

— M. Pilod et A. Jude: L'immunité antidiphtérique et antitétanique chez l'adulte soumis à la revaccination associée triple antityphoparatyphoïdique-antidiphtérique-antitétanique. Rev. Immunol. (Paris) **4**, 389—404 (1938).

Saegesser, M.: Der heutige Stand der Tetanusbehandlung unter besonderer Berücksichtigung der Magnesiumsulfattherapie. Ergebn. Chir. Orthop. **26**, 1—62 (1933).

SAERSOY et S. GÖREN: Tetanoz toksini antitoksini ve anatoksini. Askeri tibbi baytari mecmuasi **13**, 126 (1936). Réf. Bull. Inst. Pasteur **34**, 774 (1936).

SAID BILÂL: Quelques notes sur l'emploi des anatoxines en Turquie. Rev. Immunol. (Paris) **12**, 261—265 (1948).

SANT'AGNESE, P. A. DI: Combined immunization against diphtheria, tetanus and pertussis in newborn infants. II. Duration of antibody levels. Antibody titers after booster dose. Effect of passive immunity to diphtheria on active immunization with diphtheria toxoid. Pediatrics **3**, 181—194 (1949).

— Simultaneous immunization of newborn infants against diphtheria, tetanus, and pertussis. Production of antibodies and duration of antibody levels in an eastern metropolitan area. Amer. J. publ. Hlth **40**, 674—680 (1950).

SASKI, ST., et ST. STETKIEWICZ: Immunisation au moyen de l'anatoxine tétanique chez l'homme. Med. Loswiadz. ispol. **17**, 355 (1933). Réf. Bull. Inst. Pasteur **32**, 1143 (1934).

SAUER, L. W., and W. H. TUCKER: Simultaneous immunization of young children against diphtheria, tetanus, and pertussis. Amer. J. publ. Hlth **40**, 681—685 (1950).

SAVOLAINEN, T.: Über das Vorkommen von Tetanus nach kriminellem Abort. Ann. Med. exp. Fenn. **27**, 178—184 (1949).

— Catgut Tetanus. Arb. Sero-Bakt. Inst. Univ. Helsinki 1950a. Réf. Bull. Inst. Pasteur **50**, 1164 (1952).

— Catgut Tetanus. Ann. Med. exp. Fenn. **28**, 55—71 (1950b).

— On sterility of catgut and catgut tetanus. Acta path. microbiol. scand. **91**, 177—180 (1951).

SBARSKY, B., et Z. IERMOLJEIVA: De l'influence de certains aminoacides sur la toxine tétanique. J. Biol. Méd. exp. (russe) **5**, 58—66 (1927). Réf. Bull. Inst. Pasteur **25**, 853 (1927).

SCARTOZZI, C.: L'immunità antitetanica naturale in rapporto con la questione dei portatori. G. Batt. Immun. **21**, 261—280 (1938).

SCHAAFSMA, A. W.: Tetanus toxin in cellophanebags. A.R.S. Afr. Inst. med. Res. **1954**, 34.

— Production of tetanus toxin and antitoxin. A.R.S. Afr. Inst. med. Res. **1957**, 39.

SCHEER, J. VAN DER, et R. G. WYCKOFF: An electrophoretic study of tetanus antitoxin sera. Proc. Soc. exp. Biol. (N.Y.) **43**, 427—428 (1940).

SCHEIBEL, I.: The uses and results of active tetanus immunization. Bull. Org. mond. Santé **13**, 381—394 (1955).

— Die aktive Tetanusschutzimpfung. Zbl. Bakt., I. Abt. Ref. **163**, 251—258 (1957a).

— Control of the Danish combined diphtheria-tetanus vaccine with special reference to the assaying of its antigenic potency by international units. Troisième rencontre internationale de standardisation biologique. Opatijà 2—6 septembre 1957b.

—, and M. WEIS BENTZON: Sterility control of sera and vaccines. Troisième rencontre internationale de standardisation biologique. Opatijà 2—6 septembre 1957c.

SCHERRER, F.: Zur Tetanusprophylaxe. Dtsch. Z. Chir. **284**, 133—136 (1956).

SCHLAFER, J.: Le problème des vaccinations chez les allergiques. Réf. Méd. et Hyg. (Genève) **1957**, 495.

SCHLEGEL, J. J.: Klinische Ergebnisse der Tetanusimmunisierung. Helv. chir. Acta **18**, 378—388 (1951).

*— *Tetanustherapie und Prophylaxe*. Dtsch. Z. Chir. **284**, 80—102 (1956).

SCHMIDT, H.: Die Schutzimpfung gegen Diphtherie mit einem neuen Impfstoff T.A.F. Zbl. Bakt., I. Abt. Orig. **97**, 63*—68* (1925).

*— *Die Praxis der Auswertung von Toxinen und Antitoxinen*. Jena: Gustav Fischer 1931.

— Die Resorption und Ausscheidung antitoxischer Sera in Bezug auf die Serumtherapie, die Dauer der passiven Immunität und die Serumkrankheit. Behringwerk-Mitt. **6**, (1934).

— Zur Serumprophylaxie des Tetanus. Med. Welt. **1937**, 604—606.

*— *Pathogenese, Therapie und Prophylaxe des Tetanus*. Marburg a. d. Lahn: N. G. Elwert Universitäts- und Verlagsbuchhandlung 1952a.

*— *Die aktive Immunisierung gegen Tetanus*. Behringwerk-Mitt. **25**, 9—60 (1952b).

— Fortschritte der Serologie. Darmstadt: Dr. Dietrich Steinkopff 1955.

SCHMIDT, S.: Vitesse de floculation et vitesse de neutralisation du sérum antitétanique vis-à-vis de la toxine tétanique. C. R. Acad. Sci. (Paris) **184**, 1138—1140 (1927a).

Schmidt, S.: Contribution à l'étude du processus de neutralisation entre toxines et antitoxines (Diphtérie et Tétanos). C. R. Acad. Sci. (Paris) **185**, 1080—1082 (1927b).
— Sur l'adsorption de la toxine diphtérique par l'hydroxyde d'aluminium. C. R. Soc. Biol. (Paris) **103**, 1296—1298 (1930).
— Sur la toxicité des toxines additionnées d'hydrate d'aluminium ou de tapioca. C. R. Soc. Biol. (Paris) **107**, 330—332 (1931b).
— Influence de certains composés chimiques sur la toxine tétanique. C. R. Soc. Biol. (Paris) **113**, 882—883 (1933a).
— Über die Einwirkung von Formaldehyd auf Diphtherietoxin. I. Die Bedeutung der Formaldehydkonzentration, der Temperatur und der Wasserstoffionenkonzentration für die Anatoxinbildung. Z. Immun.-Forsch. **78**, 27—45 (1933b).
— Über den Mechanismus der Anatoxinbildung. Z. Immun.-Forsch. **78**, 339—354 (1933c).
—, et A. Hansen: Sur l'adsorption de la toxine et de l'anatoxine tétaniques au moyen de l'hydrate d'aluminium. C. R. Soc. Biol. (Paris) **106**, 314—316 (1931a).
—, et E. Steenberg: Comparaison entre l'effet stimulant du tapioca et de l'hydroxyde d'aluminium additionnés à l'anatoxine tétanique destinée à l'immunisation du cheval. Acta path. microbiol. scand. **13**, 401—403 (1936).
Schmidt-Burbach, A., u. H. Dehmel: Über Simultan-Schutzimpfung gegen Diphtherie. Zbl. Bakt. I. Abt. Orig. **140**, 237*—240* (1937).
Schober, K. L.: Klinische und serologische Untersuchungen zur simultanen Tetanusprophylaxe. Dtsch. Z. Chir. **284**, 131—133 (1956).
Schubert, R.: Neue Wege der Entgiftung durch Infusion niedermolekularer Kollidonfraktionen. Dtsch. med. Wschr. **1948**, 551—553.
— Einfluß von Kollidon auf Tetanustoxin. Ärztl. Forsch. **1949**, 425—428.
Schürmann, J.: Zur aktiven Immunisierung gegen Tetanus (Gefahr falscher Orientierung des Arztes durch den Patienten). Praxis **1954**, 159—160.
Schultze, H. E.: Über Proteasen des Tetanustoxins und ihre Beziehungen zum Tetanusspasmin. Hoppe-Seylers Z. physiol. Chem. **274**, 157—174 (1942).
Schulz, W.: Zur Problematik statistischer Methoden bei der Auswertung biologischer Befunde. Arch. int. Pharmacodyn. **88**, 94—97 (1951).
Sédallian, P., et Ch. Clavel: Sur l'évolution des taux antitoxiques des animaux immunisés continuellement. Rev. Immunol. (Paris) **9**, 89—92 (1944/45).
Seemüller, H.: Über Tetanus-Normal-Antitoxin bei Schweizer Rindern. Schweiz. Z. allg. Path. **6**, 120—124 (1943).
Seibold, M., u. H. Bachmann: Die Wirkungsdauer der aktiven Tetanusschutzimpfung. Dtsch. med. Wschr. **1955**, 1188—1189.
Seidl, G., u. E. Vogler: Zur Aetiologie des postoperativen Tetanus. Wien. med. Wschr. **1949**, 86—88.
Sellers, A. H., G. D. Caldbick, D. T. Fraser, M. H. Brown and P. J. Moloney: The use of a combined antigen TABTD. Canad. J. publ. Hlth **41**, 141—145 (1950).
Sevitt, S.: Source of two hospital-infected cases of tetanus. Lancet **1949**II, 1075—1078.
— Gas-gangrene infection in an operating-theatre. Lancet **1953**II, 1121—1123.
Shewell, J., and D. A. Long: A species difference with regard to the effect of cortisone acetate on body weight, γ-globulin and circulating antitoxin levels. J. Hyg. (Lond.) **54**, 452—460 (1956).
Siegel, M.: Susceptibility of mongoloids to infections. II. Antibody response to tetanus toxoid and typhoid vaccine. Amer. J. Hyg. **48**, 63—73 (1948).
Silberschmidt, W.: Essais d'immunisation par inhalation. I. Diphtérie et tétanos. Ann. Inst. Pasteur **52**, 690—708 (1934).
Silverman, M. S., and P. H. Chin: The effect of whole body-X-irradiation of mice on immunity to tetanus toxoid. I. The effectiveness of pre- and postirradiation injections of tetanus toxoid with respect to the development of immunity. J. Immunol. **75**, 321—325 (1955).
— — The effect of whole body X-irradiation of mice on immunity to tetanus toxoid. II. The delayed immune response to injections of tetanus toxoid. J. Immunol. **77**, 266—270 (1956).
Siméon, A., L. Déjou et J. Léniaud: Sur le traitement du tétanus déclaré par la technique de G. Ramon: 296 cas personels. Mém. Acad. Chir. **76**, 846—848 (1950).

Simon, R., et G. A. Patey: Le tétanos de guerre (A propos des 14 cas observés au centre sanitaire français de Besançon), action des infiltrations anesthésiques du sympathique. Presse méd. **1940**, 935—937.

Smith, M. L.: Note sur le caractère complexe de la toxine tétanique. Bull. Org. mond. Santé **10**, 138—150 (1942/43).

Sneath, P. A. T.: Development of tetanus antitoxin following administration of tetanus toxoid. J. Amer. med. Ass. **102**, 1288—1289 (1934).

—, and E. G. Kerslake: Observations following administration of tetanus toxoid. Canad. med. Ass. J. **32**, 132—134 (1935).

— — and F. Scruby: Tetanus immunity: resistance of guinea pigs to lethal spores doses induced by active and passive immunization. Amer. J. Hyg. **25**, 464—476 (1937).

Soc. Path. comp.: Le tétanos. Compte-rendu de la Société de Pathologie comparée. Réf. Presse méd. **1957**, 1331—1332.

Sohier, R.: A propos d'un nouveau sérum antitétanique utilisé pour la prévention du tétanos et de la rareté des accidents sériques observés après son emploi. Bull. Soc. méd. Hôp. Paris **55**, 642—644 (1939).

— Vaccination associée antityphoparatyphoïdique, antidiphtérique, antitétanique, chez l'enfant. Paris méd. **1944a**, 121—124.

—, et M. Alaize: Sur la pathogénie des manifestations pathologiques observées après la vaccination antityphoparatyphoïdique. Rev. Immunol. (Paris) **8**, 173—188 (1943).

— C. Marchetti et R. Poulin: Sur l'immunisation des enfants par le vaccin triple associé antityphoparatyphoïdique, antidiphtérique, antitétanique. Bull. Acad. Méd. (Paris) **128**, 444—446 (1944b).

Sommer, H.: Purification of botulinus toxin. Proc. Soc. exp. Biol. (N.Y.) **35**, 520—521 (1937).

Spath, F., u. W. Köle: Zur Prophylaxe der Lyssa und des Tetanus. Wien. med. Wschr. **1952**, 709—711.

Stafford, E. S., T. B. Turner and L. Goldman: On the permanence of anti-tetanus immunization. Ann. Surg. **140**, 563—568 (1954).

Stern, P., u. S. Hukoviš: Substanz P und Tetanustoxin. Naturwiss. **43**, 538 (1956).

Stone, J. L.: The effect of lysozyme on the production of tetanus toxin. I. Studies with flocculation. J. Bact. **64**, 299—303 (1952).

— Increasing color density with decreasing toxicity of tetanus toxin in the Mueller medium. Appl. Microbiol. **1**, 160—161 (1953a).

— A modified Mueller medium without native protein for tetanus toxin production. Appl. Microbiol. **1**, 166—168 (1953b).

— The effect of lysozyme on the production of tetanus toxin. II. Mouse M.L.D. Yale J. Biol. Med. **25**, 239—244 (1953c).

— On the mode of release of tetanus toxin from the bacterial cell. J. Bact. **67**, 110—116 (1954c).

—, and T. Berman: Comperative antigenicity of tetanus toxoids prepared in the presence and absence of veal infusion. J. Immunol. **73**, 259—263 (1954a).

—, and L. Levine: Effect of the elimination of native proteins on the yield and purifiction capacity of tetanus toxoid. Appl. Microbiol. **2**, 262—263 (1954b).

Surján, M.: The large scale production of tetanus toxin. Schweiz. Z. allg. Path. **19**, 311—317 (1956).

—, u. B. Gorzó: Vergleichende Wertbemessung des Tetanustoxins und Anatoxins. Z. Immun.-Forsch. **112**, 325—332 (1955).

Suter, R.: Etude sur la flore anaérobie pathogène du sol en Suisse. (1943, non publié.)

Swift, S.: Anaphylactic shock due to anti-tetanus serum. S. Afr. med. J. **26**, 714—716 (1952).

Szélyes, L.: Experimentelle und praktische Erfahrungen über die Wirksamkeit des Tetanusantitoxins. Allatorvosi Lapok **1930**. Réf. Bull. Inst. Pasteur **29**, 314 (1931).

— P. Elek u. J. Köves: Schutzimpfung mit kombinierten Impfstoffen gegen Tetanus + Milzbrand und Gasbrand + Milzbrand. Acta vet. Acad. Sci. hung. **2**, 97—106 (1952). Réf. Bull. Inst. Pasteur **52**, 160 (1954).

Tasman, A.: Welke tetanus-entstof moet men gebruiken als «injection de rappel» bij patiënten, di vooraf tegen tetanus zijn geïmmuniseerd. Ber. Rijksinst. Volksgezondheid, Utrecht **1955**, S. 131—141.

TASMAN, A., et A. PONDMAN: Sur la fermentation du glucose par « Cl. tetani ». Antonie v. Leeuwenhoek 7, 169—242 (1941).

—, and J. D. VAN RAMSHORST: On the preparation and properties of a new tetanus-prophylactic, the «tetanus P T». Antonie van Leeuwenhoek 18, 357—363 (1952).

— Experimentelle Untersuchungen über die Möglichkeit der Tetanus-Simultanprophylaxe. Med. Wschr. 1958, 826—831.

TAYLOR, E. M.: Production of tetanal toxin. J. Immunol. 50, 385—389 (1945).

TEALE, F. H., and D. EMBLETON: Studies in infection. II. The paths of spread of bacterial exotoxins with special reference to tetanus toxin. J. Path. Bact. 23, 50—68 (1919).

TEICHMANN, J.: Die aktive Schutzimpfung mit kombinierten Impfstoffen. Wien. klin. Wschr. 1952, 425—429.

TENBROECK, C., and J. H. BAUER: The tetanus bacillus as an intestinal saprophyte in man. J. exp. Med. 36, 261—271 (1922).

— — Studies on the relation of tetanus bacilli in the digestive tract to tetanus antitoxin in the blood. J. exp. Med. 37, 479—489 (1923).

— — The immunity produced by the growth of tetanus bacilli in the digestive tract. J. exp. Med. 43, 361—377 (1926).

THIBAULT, P., et R. RICHOU: La réaction locale après injection d'anatoxine tétanique seule ou enrobée dans la lanoline. C. R. Soc. Biol. (Paris) 121, 810—813 (1936).

THIÉRY, J. P., et R. RICHOU: Sur le contrôle de l'innocuité et de la valeur immunisante de l'anatoxine tétanique. Bull. Off. int. Epizoot. 33, 286—292 (1950).

TOKGÖZ, S. K., et S. BILÂL: Beygirlede difteri ve tetanos a karsi musterek hiperimmunisasyon tecrubeleri. Turkiye Cumhyriyeti 1937, No 7. Réf. Bull. Inst. Pasteur 36, 1133 (1938).

— SEVER KÂMIL, u. SAID BILÂL GOLEM: Versuche über Immunisierung von Meerschweinchen auf cutanem und nasalem Wege mit Lanolinemulsion von Tetanusanatoxin. Türk. Hifzissihha Tecrübî Biyol. Mecmuasi 1, 30—34, dtsch. Zus.fass. 35 (1939). Réf. Zbl. ges. Hyg. 47, 159 (1941).

TRÉFOUËL, J.: Réponse à M. RAMON au sujet de la lecture de M. M. LEMÉTAYER, LAFFAILLE, NICOL, LAMY, VALLÉE et GIRARD. Bull. Acad. Méd. (Paris) 129, 514—521 (1945).

— In P. CHAVANON: Nous, les … cobayes, p. 69. Paris: Médicis 1946.

TULLOCH, W. J.: The distribution of the serological types of «B. tetani» in wounds of men who received prophylactic inoculation, and a study of the mecanism of infection in, and immunity from, tetanus. Proc. roy. Soc. B 90, 529—541 (1919).

TURNER, T. B., E. S. STAFFORD and L. GOLDMAN: Studies on the duration of protection afforded by active immunization against tetanus. Bull. Johns Hopk. Hosp. 94, 204—217 (1954).

TURPIN, A., M. RAYNAUD et M. ROUYER: Purification de la toxine et de l'anatoxine tétanique. Ann. Inst. Pasteur 82, 299—310 (1952).

TZAMALOUKAS, G.: Einige Erfahrungen über Prophylaxe des Tetanus bei gleichzeitig angewandter aktiver und passiver Immunisierung. Zbl. Chir. 75, 317—327 (1950).

UNGAR, J.: The effect of added vaccines on diphtheria antitoxin production. Proc. roy. Soc. Med. 45, 674—676 (1952).

— Discussion on combined immunization. Proc. roy. Soc. Med. 47, 351—360 (1954).

— Effect of added toxoids on the antigenicity of «H. pertussis» vaccines. Brit. med. J. 1, 841—842 (1956).

— Preparation, composition and control of combined vaccine: B. pertussis vaccine, diphtheria toxoid and tetanus toxoid. Troisième rencontre internationale de standardisation biologique. Opatijà 2—6 septembre 1957.

VAHLQUIST, B., L. HESSELVIK et H. ERICSSON: Active basal immunity and its application to epidemiology. III. Combined immunization against diphtheria, tetanus and pertussis in infancy. Acta paediat. (Uppsala) 43, 15—21 (1954).

VALCARENGHI, E., et R. RICHOU: Sur la présence d'antitoxine tétanique d'origine occulte dans le sérum des animaux d'espèces bovine et ovine. C. R. Soc. Biol. (Paris) 114, 597—598 (1933a).

— — De la présence de l'antitoxine tétanique d'origine occulte dans le sérum de certains animaux domestiques et des bovidés en particulier. Bull. Acad. vét. Fr. 6, 386—389 (1933b).

VALLÉE, H., et L. BAZY: Essai, chez l'homme, de vaccination active contre le tétanos. Bull. Soc. nat. Chir. 43, 1445—1451 (1917).

VELLUZ, L.: Recherches sur le phénomène de neutralisation des toxines. C. R. Soc. Biol. (Paris) **102**, 745—747 (1929a).
— Sur le mécanisme de l'inactivation des toxines par les aldéhydes. C. R. Soc. Biol. (Paris) **102**, 819—821 (1929b).
— Sur l'adsorption de la toxine tétanique par les hydroxydes d'aluminium. C. R. Soc. Biol. (Paris) **106**, 887—889 (1931a).
— Sur les propriétés des hydrates d'aluminium vis-à-vis de la toxine tétanique. C. R. Soc. Biol. (Paris) **107**, 674—676 (1931b).
— Action du pyrophosphate de sodium sur la toxine tétanique. C. R. Soc. Biol. (Paris) **112**, 556 (1933).
— Action comparée du sulfure de carbone sur les toxines tétanique et diphtérique. Essai d'interprétation. C. R. Soc. Biol. (Paris) **128**, 11—13 (1938).
VERONESI, R.: Clinical observations on 712 cases of tetanus subject to four different methods of treatment: 18,2% mortality rate under a new method of treatment. Amer. J. med. Sci. **232**, 629—647 (1956).
VIGODTCHIKOV, G. V., S. K. SOKOLOV, M. K. KOLESNIKOVA et coll.: Etude comparée de différentes méthodes prophylactiques contre le tétanos chez les non-vaccinés. La prophylaxie active et passive. J. Microbiol. Epidemiol. Immunbiol. **27**, No 12, 77—83 (1956). Réf. Bull. Inst. Pasteur **55**, 2469 (1957).
VINCENT, H.: Nouvelles recherches sur les cryptotoxines microbiennes. C. R. Acad. Sci. (Paris) **184**, 921—923 (1927).
VINET, G., et V. FREDETTE: Acides aminés libres du bouillon Vf et toxinogenèse chez Plectridium tetani. Ann. Inst. Pasteur **94**, 530—533 (1958).
VOLK, V. K.: Observations on the safety of multiple antigen preparations. Amer. J. Hyg. **48**, 53—63 (1948).
— Safety and effectiveness of multiple antigen preparations in a group of free-living children. Amer. J. publ. Hlth **39**, 1299—1313 (1949).
— F. H. TOP and W. E. BUNNEY: Reinoculation with multiple antigen preparations of free-living children previously inoculated with multiple antigen preparations. Amer. J. publ. Hlth **43**, 821—832 (1953).
VOROBIEV, A. A.: The changes of immunogenesis of the purified adsorbed tetanus anatoxin in storage. J. Microbiol. Epidemiol. Immunobiol. **28**, No 4, 31—34 (1957a). Réf. Bull. Inst. Pasteur **56**, 1477 (1958).
— Immunological mechanisms in the two-stage immunization with deposited tetanus anatoxin. Réf. Biol. Abstr. Sect. C, **32**, No 5658, 494 (1958).
— B. S. GONCHAROV, L. P. LUBIJANSKY and A. V. MARKOVITCH: Interrelations between the level of specific antitoxine in blood and immunity intensity against tetanotoxine in the case of withe mice immunization with sorbed tetano-anatoxine. Byoull. eksp. Biol. Med. (russe) **43**, 63—66 (1957b). Réf. Bull. Inst. Pasteur **55**, 2467 (1957).
WADSWORTH, A., and M. C. PANGBORN: The reaction of formaldehyde with amino acids. J. biol. Chem. **116**, 423—436 (1936).
— J. J. QUIGLEY and G. R. SICKLES: Preparation of diphtheria toxoid. The action of formaldehyde: precipitation by calcium. J. infect. Dis. **61**, 237—250 (1937).
WARD, H. K.: The persistence of antibodies in the absence of antigenic stimulus. Aust. J. exp. Biol. med. Sci. **35**, 499—506 (1957).
WARTMAN, W. B., et L. PILLEMER: Effect of injecting crystalline tetanal toxin and tetanal antitoxin into mice. Proc. Soc. exp. Biol. (N.Y.) **70**, 65—66 (1949).
WATTS, P. S.: Note on the isolation of «Cl. tetani» from the intestines of normal sheep in Combridgeshire. Brit. J. exp. Path. **19**, 422—424 (1938).
WEINBERG, M., et B. GINSBOURG: Données récentes sur les microbes anaérobies et leur rôle en pathologie. Bull. Inst. Pasteur **23**, 689—698, 737—748 (1925).
WELCH, H., G. G. SLOCUM and J. J. DURRETT: Production of tetanus in guinea pigs by subcutaneous implantation of pellets contaminated with tetanus spores. J. Amer. med. Ass. **119**, 1396—1401 (1942a).
— — and R. P. HERWICK: Tetanus from sulfonamide dusting powders. Development in guinea pigs treated with sulfanilamid powder contaminated with washed tetanus spores. J. Amer. med. Ass. **120**, 361—364 (1942b).
WHITTINGHAM, H. E.: Anaphylaxis following administration of tetanus toxoid. Brit. med. J. **1**, 292—293 (1940).

WILCOX, H. L.: A note on the comparative value of fresh and aged tetanus toxoid as an immunizing agent. J. Immunol. 27, 195—198 (1934).

WILDFÜHR, G.: Über die Toxinbildung der Tetanusbazillen unter der Einwirkung von Vitamin-H' (p-Aminobenzoesäure). Zbl. Bakt., I. Abt. Orig. 152, 96—98 (1947).

WILSON, G. S.: Brit. med. Assoc.-Clinical and scientific proceedings. Recent developments in immunization. Brit. med. J. 1, 593—594 (1941).

WINKELBAUER, A.: Neue Wege in der Therapie des Tetanus. Dtsch. med. Rdsch. 2, 41—46 (1948).

WISHART, F. O., and L. K. JACKSON: Recall dose of tetanus toxoid by the intranasal or intraocular route. Canad. J. publ. Hlth 41, 43 (1950).

— — Tetanusantitoxin titres following a recall dose of tetanus toxoid. Canad. J. publ. Hlth 42, 384—389 (1951).

WOLFF-EISNER, A.: Versuche über die Serumwirkung bei der Tetanus-Infektion und Tetanus-Toxinvergiftung. Zbl. Bakt., I. Abt. Orig. 106, 348—388 (1928).

—, et I. IAHR: Inefficacité du sérum antitétanique dans l'infection tétanique expérimentale. C. R. Soc. Biol. (Paris) 96, 690 (1927).

WOLTERS, K. L., u. H. DEHMEL: Versuche zur aktiven Immunisierung gegen Tetanus mit nativem und alaunpräzipitiertem Tetanustoxoid. Zbl. Bakt., I. Abt. Orig. 138, 221—226 (1936).

— — Über die aktive Immunisierung gegen Tetanus. 17. Tagg der Dtsch. Ver.igg. für Mikrobiol. Zbl. Bakt., I. Abt. Orig. 140, 249*—258* (1937).

— — Über den Verlauf der aktiven und aktiv-passiven Immunität bei Tetanus. Z. Infekt.-Kr. Haustiere 53, 140—147 (1938a).

— — Experimentelle Versuche zur aktiven Immunisierung mit Alaunpräzipitat-Misch-impfstoffen. Dtsch. tierärztl. Wschr. 1938b, 342—345.

— — Die Tetanusprophylaxe durch aktive Immunisierung. Z. Hyg. Infekt.-Kr. 122, 603—612 (1940).

— — Abschließende Untersuchungen über die Tetanusprophylaxe durch aktive Immunisierung. Z. Hyg. Infekt.-Kr. 124, 326—332 (1942).

— — Experimentelle Untersuchungen zur aktiv-passiven Immunisierung und zur Serum-Toxoid-Therapie bei Tetanus. Z. Hyg. Infekt.-Kr. 132, 582—594 (1951).

—, u. E. FISCHOEDER: Über die Bindung von Tetanustoxin an Hirnsubstanz ohne und nach Vorbehandlung mit Tetanustoxoid. Z. Hyg. Infekt.-Kr. 139, 541—544 (1954).

WORMS, R., et B. DREYFUS: Plasmocytose sanguine des accidents sériques. Presse méd. 1953, 21—23.

WRIGHT, E. A.: The fixation of tetanus toxin to the brain stem of guinea pigs. J. Path. Bact. 68, 131—135 (1954).

*— G. P.: Botulinus and tetanus toxins. In: Mechanisms of microbial pathogenicity. Cambridge: Cambridge University Press 1955.

YEAZELL, L., and W. DEAMER: Response to a stimulating injection of tetanus toxoid. Report of a study of children previously immunized with combined diphtheria and tetanus toxoid. Amer. J. Dis. Child. 66, 132—142 (1943).

YOMTOV, M., et K. SOLOMONOVA: Concentration et purification de l'anatoxine tétanique. Sborn. Troudove 3, 175—179 (1956). Réf. Bull. Inst. Pasteur 55, 2466 (1957).

ZANNIOL, A.: Anaphylaktischer Schock nach erster Seruminjektion. Rif. med. 1937, 803.

ZDRODOWSKI, P.: Essai d'analyse physiologique de quelques processus infectieux et immunologiques. Rev. Immunol. (Paris) 21, 341—355 (1957).

ZOELLER, C., et G. RAMON: L'immunité antitétanique par l'anatoxine chez l'homme. Presse méd. 1926, 485—486.

ZUGER, B., and U. FRIEDEMANN: Investigations on the pathogenesis of tetanus. Proc. Soc. exp. Biol. 38, 283—285 (1938).

— C. K. GREENWALD and H. GERBER: The antitoxin response of partially immunized guinea pigs to infection with tetanus spores. J. Immunol. 38, 431—447 (1940).

— — — Tetanus immunization: effectiveness of stimulating dose of toxoid under conditions of infection. J. Immunol. 44, 309—317 (1942).

— A. HOLLANDER and U. FRIEDEMANN: Potentiation of tetanus toxin by tissues and chemicals. J. infect. Dis. 65, 86—91 (1939).

ZYLKA, N.: Tétanos après extraction dentaire. Neue Zahnheilk. 3, 96 (1957). Réf. Méd. et Hyg. (Genève) 1957, 600.

Namenverzeichnis

Die *kursiv* gedruckten Seitenzahlen beziehen sich auf die Literatur

Evans, E. E., L. J. Sorensen
u. K. W. Walls *67*
— u. R. J. Theriault *67*
— s. Bocobo, F. C. *65*
Eveland, W. C. 59
— J. D. Marshall u. A. M. Silverstein *72*
Everitt, P. N. 104
— Bhatt u. J. P. Fox *143*
— M. G. s. Fox, J. P. *143*
Eyer, H. s. Aschenbrenner, R. *138*

Fabre, M. 280
— s. Guillaumie, M. *355*
Faia, M. de 110
— s. Sampaio, A. *154*
Faillard, H. 6, 7, 8, *19*
— s. Klenk, E. *21*
Faivre, G. s. Lavergne, V. de *360*
Falchetti, E. 277, *352*
— s. Ramon, G. *371*
Faragó, F. 301, 302, 333, *352*
— u. S. Pusztai *352*
— u. K. Ujhelyi *352*
Farbstein, M. E. 248, *261*
Fasal, P. 276, *352*
Fastier, L. B. 114, *143*
Fauconnier, J. 210, 232, *261*
Favarissova, B. I. 201, *261*
Fazekas de St. Groth, S. 4, 5, 12, 14, 15, 16, *19*
— u. D. M. Graham *19*
— s. Anderson, S. G. *18*
Fedorowa, N. I. 113
— T. A. Bektemirow, I. V. Tarasewitsch, E. B. Kerbabaew u. L. L. Semaschko *143*
Feduchy 27
Feeney, R. E. 293
— J. H. Mueller u. P. A. Miller *352*
Feierabend, B. 318, *352*
Feldman, W. H. 200, 225, 226, 227, 230, 235, 239, 243, *261*
Felix, A. 279
— u. M. Robertson *352*
Felsenfeld 164
— Young u. J. Yoshimura *188*
Felton 335
Fenner, F. 112, *143*, 315
— s. Burnet, F. M. *349*
Fenton, E. L. 279
— s. Pope, C. G. *368*
Fernan-Nunez, M. 309, *352*
Ferreira, J. M. 115

Ferreira, J. M. s. Ribeiro do Valle, L. A. *154*
Fevurly, J. 88
— s. Downs, C. M. *142*
Fey, H. 116, 203, *261*
— s. Wiesmann, E. *159*
Fiala, S. 296, *352*
Fiedler, C. *188*
Fildes, P. 273, *352*
Finter, N. B. 16
— s. Hoyle, L. *20*
Firor, W. M. 288
— s. Lamont, A. *359*
Firth, M. E. 8
— s. Cornforth, J. W. *19*
Fischer, J. B. 55
— u. N. A. Labzoffsky *67*
— R. 238
— s. Weltmann, O. *269*
Fischoeder, E. 342
— s. Wolters, K. L. *381*
Fisek, N. H. 293, 305, 307, 317, 327, *352*
— J. H. Mueller u. P. A. Miller *352*
Fiset, P. 84
— s. Stocker, M. G. P. *157*
Fisher, A. M. 39, *67*
Flamm, H. 220, 248, 249, 250, 251, 252, 254, *262*
— u. W. Kovac *262*
Fleming, D. S. 307, 309, 333
— u. L. Greenberg *352*
— s. Greenberg, L. *354*
— W. L. 280, *352*
Flick, J. A. 11
— B. Sanford u. S. Mudd *19*
Flickinger, M. H. 41
— s. Wickerham, L. J. *72*
Floch, H. 276, *353*
— u. R. Barrat *353*
Florestano, H. J. 40
— u. M. E. Bahler *67*
Florey, H. 3, *19*
Floyd, Th. M. 170, 181
— u. G. B. Jones *189*
Foder, I. 249, 250
— s. Pedhragyai, L. *267*
Foley, G. E. 30, 33
— u. L. L. Uzman *67*
— s. Uzman, L. L. *72*
Forster, L. E. 258
— s. Quan, S. F. *267*
Foryś, S., R. Lutyński u. Z. Raginis *143*
Foster, W. D. *352*
Foshay, L. 231
— s. Hesselbrock, W. *263*

Fournier, R. 137
— s. Varela, G. *158*
Fowler, R. 39
— s. Mezey, C. M. *69*
Fox, J. P. 83, 85, 87, 94, 101, 104, *143*
— M. G. Everritt, T. A. Robinson u. D. P. Conwell *143*
— M. E. Jordan, E. P. Conwell u. T. A. Robinson *143*
— M. E. Jordan u. H. M. Gelfand *143*
— J. A. Montoya, M. E. Jordan u. M. Espinosa *143*
— s. Clarke, D. H. *141*
— s. Everitt, P. N. *143*
— s. Pérez Gallardo, F. *152*
— s. Ris, H. *154*
Fraenkel, E. 237, 254, *262*
Franklin, J. 120
— E. Wassermann u. H. S. Fuller *143*
Fraser, D. T. 332
— s. Sellers, A. H. *377*
Fredette, V. 293, 294
— u. G. Vinet *352*
— s. Vinet, G. *380*
Fredricson, M. 235
— s. Wramby, G. *269*
Freeman, G. 129
— G. Varela, H. Plotz u. C. Ortiz Mariotte *143*
— M. 74
— s. Burnet, F. M. *140*
— W. 26, 29, *67*
Freitag, R. 165, 179, *189*
French, D. 295, 296
— u. J. T. Edsall *353*
Freni, S. 331
— s. Mazetti, G. *362*
Freudenberg, K. 272
— s. Hübner, A. *356*
Freund, J. 308, 311, 324, *353*
— u. M. V. Bonanto *353*
— E. M. Schryver, M. B. McGuiness u. M. B. Geitner *353*
Freyche, M. J. 114, 131
— u. Z. Deutschman *143*
Frick, L. P. 124
— s. Traub, R. *158*
Friedemann, U. 283, 284, 285, 286, 317, *353*
— u. A. Hollander *353*
— B. Zuger u. A. Hollander *353*
— s. Zuger, B. *381*
Friedman, M. 133

Sachverzeichnis